Advanced PHYSICS

Steve Adams
Jonathan Allday

OXFORD
UNIVERSITY PRESS

OXFORD
UNIVERSITY PRESS

Great Clarendon Street, Oxford OX2 6DP

Oxford University Press is a department of the University of Oxford.
It furthers the University's objective of excellence in research, scholarship,
and education by publishing worldwide in

Oxford   New York

Auckland   Cape Town   Dar es Salaam   Hong Kong   Karachi
Kuala Lumpur   Madrid   Melbourne   Mexico City   Nairobi
New Delhi   Shanghai   Taipei   Toronto

With offices in

Argentina   Austria   Brazil   Chile   Czech Republic   France   Greece
Guatemala   Hungary   Italy   Japan   Poland   Portugal   Singapore
South Korea   Switzerland   Thailand   Turkey   Ukraine   Vietnam

Oxford is a registered trade mark of Oxford University Press
in the UK and in certain other countries

First published 2000
Reprinted with revisions 2013

ISBN: 978-0-19-839292-7

10 9 8 7 6 5 4 3 2 1

Typeset in New Aster

by Mark Walker Design, Plymouth, Devon

Artwork by Mark Walker

Printed by Vivar Printing Sdn Bhd, Malaysia

Paper used in the production of this book is a natural, recyclable product
made from wood grown in sustainable forests. The manufacturing process
conforms to the environmental regulations of the country of origin.

Introduction

With so many excellent physics textbooks on the market, there has to be a good reason for introducing a new one. The only valid reason can be that it is in some way distinctive from others that have gone before. In the case of *Advanced Physics*, its primary distinctive quality is that it has been written in spreads – double-page 'quanta' of text covering a particular piece of the material in a largely self-contained unit.

The discipline of writing in this manner has forced us to re-think the conventional way of explaining certain topics and to look for ways in which the material can be broken down into parts that will fit on spreads. It has also allowed us a certain freedom in writing. We wanted this book to contain material that would stretch the most able A-level candidates but did not at the same time wish to force others to sift through the text looking for the information relevant to them. The sub-division into spreads has allowed us to cater for the whole ability range in a very natural way.

A challenge facing many students of A-level physics is the mathematical nature of the subject. A separate appendix (the 'Mathematics toolbox') takes the commonly required mathematical techniques and explains them in a manner appropriate for physicists.

One of the great pleasures of working on this project has been the discussion and debate that has gone on between us. Although the different sections were divided between us, each section bears the imprint of the other author. Physics is an open-ended subject, and despite the ambitious talk about 'theories of everything', there is still a great deal to do. We hope that some of the excitement and stimulation we felt when writing this book has carried through to the page and that some of our readers may go on to participate in the debate.

Steve Adams, Jonathan Allday

Introduction to update

Welcome to this update of *Advanced Physics*.

We are delighted to be have been given the chance to revisit this text. As a result we have updated the material to include information for new syllabuses (especially the Pre-U and international syllabuses). Some of the questions have been revised, and we have extended the range available to include examination-style questions for the Pre-U (marked*). The new chapter 14 includes syllabus material, but also gives us the chance to consider advances and discoveries made since the text was first published. Each spread has been looked at again, and revised where necessary.

The original printing of this text was very well received and we obviously hope that new generations of physics students will enjoy this version. One aspect of being involved with this project, aside from the joy of working together on re-thinking conventional ways of explaining things, has been the positive comments received from readers who have contacted us directly or posted reviews on various websites. It was always at the back of our minds when revising the text that as much as possible of the original 'feel' of each spread should be retained.

Once again we are delighted to acknowledge the help and support received from a variety of groups:

- The original editorial team, who have moved on to bigger and better things, and their more than able replacements at OUP

- Ken Price for revising and extending the questions (and checking them for accuracy)

- Our families who have had to go through it all again!

Jonathan Allday, Steve Adams
2013

Contents

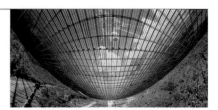

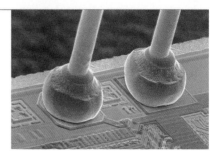

CHAPTER 9   THE PHYSICS OF PARTICLES

CHAPTER 10   THE PHYSICS OF MATERIALS

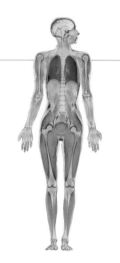

1

The scope of **PHYSICS**

The most beautiful and deepest experience a person can have is the sense of the mysterious. It is the underlying principle of religion as well as of all serious endeavour in art and in science…He who never had this experience seems to me, if not dead, then at least blind. The sense that behind anything that can be experienced there is a something that our mind cannot grasp and whose beauty and sublimity reaches us only indirectly and as feeble reflection, this is religiousness. In this sense I am religious. To me it suffices to wonder at these secrets and to attempt humbly to grasp with my mind a mere image of the lofty structure of all there is.

Albert Einstein (1879–1955)

For I can end as I began. From our home on Earth we look out into the distances and strive to imagine the sort of world into which we were born. Today we have reached far out into space. Our immediate neighbourhood we know intimately. But with increasing distance our knowledge fades…until at the last dim horizon we search among ghostly errors of observations for landmarks that are scarcely more substantial. The search will continue. The urge is older than history. It is not satisfied and it will not be suppressed.

Edwin Hubble (1889–1953)

A view from underneath the dish antenna of the world's largest radio telescope at Arecibo, Puerto Rico. Measuring 350 m across with an area of about 70 000 m², the dish is made of 38 800 aluminium panels built into a volcanic crater. The panels reflect radio waves up to a receiving antenna suspended above them, which can just be seen at upper centre.

The startling variety of the materials that surround us derives from just three fundamental particles: the electron and two different types of quark (up quarks and down quarks).

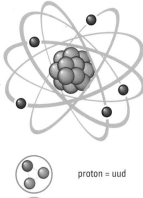

Helios personified the Sun in Greek mythology. Each morning he rose from a swamp in the east and climbed into his golden chariot. This was pulled across the sky by dazzling white winged horses whose nostrils breathed fire. In the evening he arrived in the west at the land of the Hesperides, where a ship was waiting for him. During the night he would sail back to the original swamp, to begin his journey again the following morning.

proton = uud

neutron = ddu

e ● electron

u ● up quark

d ● down quark

A representation of the atom and its constituents. Electrons orbit in a cloud about the nucleus, which is made up of protons and neutrons. In turn the neutrons and protons are made of combinations of quarks. Electrons and quarks are fundamental.

EXPLAINING EVERYTHING

Why?

If you asked a medieval pilgrim or a modern physicist why the Sun shines, you would probably get rather different responses. The pilgrim would say that the Sun shines for some *purpose*, perhaps because God placed it there to keep us warm.

The physicist knows a great deal more about the radiation from the Sun, its temperature, its structure, and the elements from which it is made. These, however, say nothing about the *purpose* of the Sun. Physics tells us *what* the Sun is and *how* it comes to shine, and stops at that.

This is worth remembering. You are living at a time when orbiting telescopes send us images of the universe as it appeared soon after the Big Bang and theoretical physicists regularly claim to be working on a 'theory of everything'. Knowing everything may well rule out some possibilities, but it doesn't provide a moral framework for human life or a purpose for existence. Beware of those who claim otherwise!

On the other hand, the physicist might say that the Sun shines because of **nuclear fusion** and that the energy released is radiated into space as **electromagnetic waves**. What she has really done is explain *how* the Sun works, not why.

What?

If you ask *what* something is, a physicist will explain it in terms of something else. For example:

• If you ask what water is, she will reply that it is matter in a liquid state.

• If you press her further, she will tell you this means it is made of **molecules** packed together closely but irregularly (the molecules can move past each other – this is why liquids flow to fit their container but have a constant total volume).

• She might then tell you that molecules are made of regular groups of **atoms**, and that the atoms are bound together because they share **electrons**.

• These electrons are subatomic particles, like **neutrons** and **protons**: neutrons and protons cluster in a **nucleus** less than a million millionth of the volume of the atom.

• If you then ask what neutrons and protons are, she will say that say that they are triplets of **quarks**.

• 'What are quarks?', you ask, feeling a little dizzy. She looks slightly sheepish and says that they are probably **fundamental particles**.

• 'What are electrons then?', you try again. They too, she admits, are probably fundamental. You have reached the end of explanation.

Any attempt to explain what A is must lead to B; the question of what B is leads to C; and so on. This chain of question and answer will stop at something 'fundamental', lead you in a circle (which is not very helpful), or never end. Physics tries to reduce everything in the universe to fundamentals, whether it is a rock, a star, or a human being. (Explaining complex things in terms of simple parts is called **reductionism**.)

At the moment, everything can be reduced to three types of fundamental particle:

- **leptons**, which include electrons;
- quarks, which make up protons and neutrons;
- force-carriers, such as the **photon**.

There is nothing else, at least not yet. This is a fairly grand idea, if somewhat baffling. Why did physics settle on such an unlikely group of fundamental particles? How do we know if these are truly fundamental or just a link in the chain of explanation?

Both questions are difficult to answer. However, the guiding principle seems to be that nature has a beautiful mathematical simplicity, which reveals itself in patterns of symmetry. Leptons, quarks, and force-carriers seem to fit such a pattern. Combinations of them generate every object we will ever experience (and many others besides).

How?

How does a steam engine work? How do stars form? These questions need some mechanism or process, some chain of cause and effect that stretches from the starting condition to the final outcome.

Stars, for example, form when gas clouds collapse under the influence of their own gravitational fields. The temperature and pressure at their centres rise to such an extent that nuclei collide violently and join together (nuclear fusion). The energy released 'ignites' the star, and it 'burns' for millions of years, radiating energy in electromagnetic waves like light. (The question can of course be answered in much greater detail, if we deal with the complex nuclear and quantum physics involved in fusion.)

Often, physical laws describe transformations, giving recipes for converting one state of affairs into another, whilst conserving certain quantities like **mass-energy** or momentum. **Conservation laws** give us something to hold on to, an unchanging value in a changing world.

A complete answer to any 'how' question will reduce it to an explanation in terms of fundamental particles and their interactions. However, it is usually unnecessary to go so far: physics operates on many levels. A useful answer to how a steam engine works, for example, can be given without even knowing about the existence or behaviour of molecules.

All answers to 'how' questions in physics use some physical principle (such as Boyle's law or Newton's laws of motion) to relate the specific example under consideration to more general patterns of behaviour.

Limits to explanation

Some systems are very sensitive to their initial conditions. For example, very similar weather conditions on a Monday can lead to completely different weather by Friday. All measurements are inaccurate to some extent. This means that reality diverges from our predictions. Systems that diverge rapidly are called **chaotic** and are commonplace in the world around us; even the rhythms of a beating heart are chaotic.

Chaos limits the ability to predict. But it does not rule out the possibility that the universe is **deterministic** (meaning that its future is entirely determined by its present state).

Quantum theory, on the other hand, *may* rule this out. On a fundamental level, what happens depends entirely on probabilities. Similar initial conditions can lead to quite different outcomes.

For example, two radioactive nuclei, identical when we begin to observe them, are likely to decay at different times. This means that any prediction about the behaviour of a collection of nuclei is at best statistical.

When large numbers of nuclei are involved, there is often only one highly likely outcome. To say that the half-life of uranium-238 is 4.5 billion years means that close to half the nuclei in a large sample will decay in that time. However, we cannot say which nuclei will decay and which will survive.

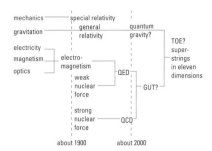

The unification of physics.
Key:
QED: quantum electrodynamics
QCD: quantum chromodynamics
GUT: grand unified theory
TOE: theory of everything

The Great Red Spot was discovered over 300 years ago by Giovanni Domenico Cassini. It is a long-lived cyclonic disturbance of Jupiter's atmosphere and shows the complexity and also the stability of structures that arise in turbulent fluids.

A computer-generated spiral fractal derived from the Julia set, which later led to the more famous Mandelbrot set.

- the different types of field
- how fields work

FIELDS

The Starry Night *(1889)*
by Vincent Van Gogh.

Electric fields

Electric fields are made by *stationary* charges, and magnetic fields are made by moving charges. Imagine an electric current carried by a stream of electrons moving past you at a constant velocity. For anyone who happens to be moving *with* the electrons, they are relatively at rest and so create *no* magnetic effect. For you the electrons create *both* electric *and* magnetic fields, but for the moving observer they create only an electric field. This emphasizes the intimate connection between electricity and magnetism; the *electromagnetic field* can reveal itself differently to different observers and is more fundamental than the electric or magnetic fields alone. Einstein's **theory of relativity** reconciles the apparent contradiction between the two observations.

Michael Faraday and Vincent Van Gogh

The Starry Night conveys the idea of a **field** brilliantly. Van Gogh's sky is alive with an ebb and flow of energy; the stars betray themselves by the way they disturb it, and everything is connected to everything else.

Michael Faraday died when Van Gogh was 14, but his idea of describing electrical and magnetic interactions as fields revolutionized physics and is still central to its development.

The crucial idea of a **field theory** is that certain particles disturb the space around them in such a way that another particle placed in that space will experience a force. Before Faraday, electric, magnetic, and gravitational forces were thought to act 'at a distance', even though there was no idea how they achieved this. Fields act *where the particle is* (they are *local*).

Fields are distinguished from one another by the class of particle on which they act. The fields included in the following table are those that have the most obvious direct effect upon us.

Properties of various fields.

Field theory	Source of field	What is affected
electrostatics	charges	charges
magnetism	moving charges	moving charges
gravity	masses	masses (i.e. everything)

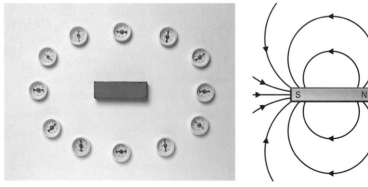

The compass needles indicate magnetic field vectors at various points around the bar magnet. The same field can be represented by Faraday's field lines.

There are also fields associated with the strong and weak nuclear forces. These are only important on the scale of the nucleus.

The invisible influence

Nobody has ever seen a field. Field theory was invented to explain the influence of one particle on another, so fields only reveal themselves in the motions of these particles. The essential property of a field is that it has *magnitude* and *direction* at each point in space. The gravitational field is the one most familiar to us. It reveals itself every time we drop something. Locally this defines what we mean by up and down. Globally, however, the field has nearly spherical symmetry.

The strength of a field is measured by the force it exerts on things that are placed in it. For example, the gravitational field affects masses, so its strength at a point is equal to the force exerted per kilogram placed at that point. The electric and magnetic field strengths are defined in a similar way.

Disturbing the field

Picture the surface of a smooth pond as one drop of water hits its centre. There will be a temporary disturbance at the point of impact: the surface moves up and down several times, dissipating (scattering) the energy it has received.

The surface forms a connected sheet over the water, so what happens at one point affects adjacent points a little later. If the centre moves down, all points some distance away will also move down later, and then up again, and so on. This delayed response is directly proportional to the distance from the impact: points the same distance from the impact site will always move in the same way. This produces circular ripples travelling out from the centre, carrying energy with them.

A field is connected to itself in a very similar way to the surface of the pond.

- If the field is disturbed at some point, waves transmit energy through the field, away from the source.

- The speed of the waves is determined by the rate at which the field passes on the disturbance from one point to another. The speed can be calculated from the field equations.

In the 1860s, James Clerk Maxwell developed a mathematical description of Faraday's field lines. **Maxwell's equations** describe:

- how the field depends on the distribution and movement of charges in space;

- how the field connects to itself.

From these equations, Maxwell derived an equation for electromagnetic waves in a vacuum and showed that these would travel at a velocity equal to the speed of light. This confirmed that light is an electromagnetic wave and that other invisible forms of light are possible. All electromagnetic waves are produced by accelerating charges.

In 1916 Albert Einstein published his **general theory of relativity**. This included equations for the gravitational field. He too predicted that disturbances of the gravitational field should generate gravitational waves that would travel at the speed of light and transmit energy from the source to objects disturbed by the waves.

Gravitation is very much weaker than electromagnetism, and gravitational waves have not yet been detected. However, indirect evidence does exist. Russell Hulse and Joseph Taylor, Jr received the 1993 Nobel Prize for physics for their studies of the binary pulsar PSR B1913+16. This is a system of two stars bound by gravity and rotating rapidly about their common centre of mass. Since 1974, the period of this rotation (i.e. the time for one rotation) has decreased by about 1 μs. This value agrees with Einstein's theory, assuming that the decrease is due to energy lost to gravitational radiation.

Definitions
Gravitational field strength (g): the force per unit mass ($N\,kg^{-1}$) at a point in space
Electric field strength (E): the force per unit charge ($N\,C^{-1}$) at a point in space
Magnetic field strength (B): the force per unit 'current-length' ($N\,A^{-1}\,m^{-1}$, or T, tesla)

Magnetic fields
The magnetic field is a little awkward. If isolated north poles existed, the field strength would be defined as force per unit magnetic pole, but so far, such magnetic **monopoles** have never been detected (although some **grand unified theories** require them). Since all magnetic fields are created by, and act upon, *moving* charges (currents), the force on moving charges is used to define magnetic field strength. In fact all magnetic effects, including **ferromagnetism**, result from moving charges. Ferromagnetic materials are those strongly affected by magnetism.

Gravity waves should make this 3 m aluminium bar vibrate with a resonant amplitude of about 10^{-18} m.

OBJECTIVES

- quantum mechanics and common sense
- the interference experiment

I think that I can safely say that nobody understands quantum mechanics.

Richard Feynman

Richard Feynman.

QUANTUM MYSTERIES

Richard Feynman was one of the greatest physicists of modern times. He was also renowned as a teacher, but his often-quoted statement (left) is not guaranteed to encourage students.

Quantum mechanics is the most successful theory in the history of physics. Its predictions have been checked to 15 decimal places – an extraordinarily precise agreement between theory and experiment. It is the only theory that successfully explains the behaviour of atoms and subatomic particles.

Yet it is not simply of interest to the professional physicist. Without quantum mechanics, the way in which the silicon chips inside a computer work would not be understood.

Given the importance and success of quantum mechanics, Feynman's statement seems bizarre. He did not mean that nobody could understand how to use the theory to make useful calculations and predictions. Nor did he mean that the theory was so difficult that only a few people had the technical skill to apply it.

He was referring to the underlying mystery of quantum behaviour, which seems to go against all the common-sense ideas that we have about how objects should behave. If understanding means being comfortable with the properties of quantum objects, having a clear picture in one's mind of exactly what is happening, or being able to describe in words the nature of the particles of matter, then Feynman was right.

Can an object be in two places at the same time?

Most of the puzzling aspects of quantum behaviour are beautifully illustrated in a simple experiment in which a beam of electrons is directed at a thin metal sheet with two slits cut into it, behind which is a photographic film.

The purpose of the experiment is to find the pattern of hits on the photographic film, which records where the electrons arrive.

This is a version of a standard experiment used to demonstrate that light is a wave. From a wave point of view, it is easily understood. The pattern produced is a series of light and dark bands. This shows the **interference** of light waves. Interference can only arise if the wavefronts from the light source arrive at the two slits and split. This is not a problem for a wave. We can imagine that the wave can pass through both slits at the same time.

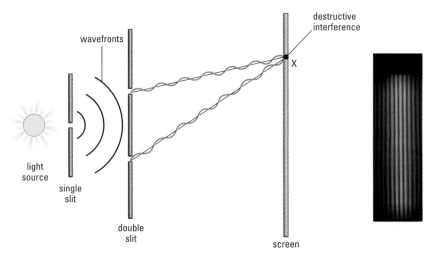

The double-slit experiment with a light source, demonstrating the interference of light. To the right is an actual optical interference pattern produced in this way.

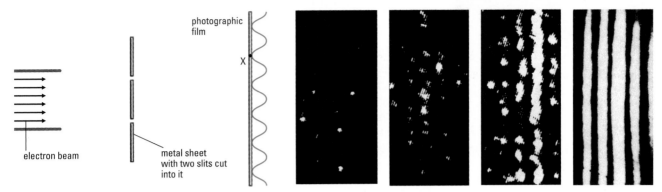

The double-slit experiment with electrons. To the right are images showing the successive build-up of an actual interference pattern produced by electrons.

This experiment can also be carried out with electrons. The result is the same. When the film is developed, it shows a series of light and dark bands. Somehow interference is arising with the electrons, which we thought were particles.

- If the intensity of the electron beam is turned down so low that only one electron at a time is involved in the experiment, an individual dot is recorded on the film each time an electron arrives (so the electron has not been split up by the screen).

- However, when negatives are stacked on top of one another to build up the pattern produced by several electrons, the bands of the interference pattern are produced. Each individual electron has some way of 'knowing' that it should never arrive at a dark band and must always arrive at a light band – even though the pattern is not complete. Like the wave, this can only happen if each electron somehow passes through both holes at the same time!

The mystery deepens

The experiment can also be done with one of the holes blocked. This produces a distribution of electron hits opposite the unblocked hole, with no sign of any interference pattern – exactly what one would expect for a particle. However, for an electron that passes through the unblocked hole, how is the experiment any different from one in which *both* holes are unblocked? How does it 'know' that the other hole is blocked? Why is this electron not following the interference pattern?

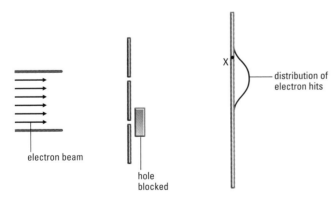

Blocking one of the holes in the double-slit experiment with electrons. Note that the number of electrons arriving at point X has increased.

It seems that the only way to 'explain' this is for each electron to be somehow aware of what is happening at both of the holes. Again we are forced to conclude that an electron arrives at both holes at the same time.

What is the world made of?

Similar experiments have now been carried out with particles as large as substantial molecules, and they show exactly the same results. Quantum mechanics forces us to think of supposedly solid matter as being more ghost-like, and its behaviour sometimes goes against common sense and logic.

Like it or not, we too are made of this strange stuff.

- size scales in the universe

- the importance of the very small and the very big

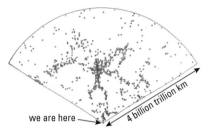

A map of a slice of the universe. The slice extends over 4 billion trillion km into space and is 120° across. Each point is a galaxy roughly equivalent to the Milky Way.

The very large: the Andromeda galaxy, a neighbour of our own Milky Way, is 1 million billion km across.

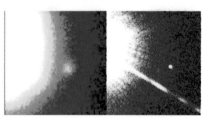

The brown dwarf Gliese 229B. Left: the very first image, taken at Palomar Observatory, California, USA on 27 October 1994. Right: Hubble Space Telescope image, 17 November 1995.

The very small: a scanning tunnelling micrograph of a circle of iron atoms, just 7 billionths of a metre across.

Nuclear forces

The strong nuclear force holds the protons and neutrons together inside the nucleus of an atom, against the electrostatic repulsion of the protons. The weak nuclear force converts protons into neutrons and vice versa, allowing radioactive decay. It is also responsible for some of the reactions that take place inside an exploding star, or **supernova**.

THE SCALE OF PHYSICS

The picture opposite is known as Glashow's snake, named after the theoretical physicist Sheldon Glashow (Nobel Prize winner in 1979). It shows the various structures in the universe that physics holds to be important and the scales at which they exist. The snake is shown swallowing its own tail because we believe that the forces and particles that exist on a very small scale ($<10^{-24}$ m) influenced the development of the universe in its very early stages: hence they may also have set the scale of the largest structures in the universe.

Descending through the universe

In January 1986 the results of a detailed survey of a small region of space were first published. On the scale of 10^{18} m, **stars** are held together by mutual gravitational attraction to form **galaxies**. To everyone's surprise, the survey showed that galaxies were not distributed uniformly throughout space. Instead they are clustered along thin **filaments** between which there are great **voids**. These are the largest structures known to humankind. A typical void can be up to 10^{24} m across.

The structure of galaxies is something of a mystery.

- Some seem to have incredible energy sources at their centres, perhaps powered by giant **black holes**, billions of times heavier than our Sun.

- The stars on the fringes of galaxies rotate round the centre faster than expected. This suggests that there is some unseen 'dark matter' in a halo round the galaxy, exerting a gravitational pull that keeps the stars in place.

At 10^{12} m, the gravity of an individual star holds **planets** in orbit. The planets, the Sun, and some other bodies form the **Solar System**. On the left is the first photograph of a planet in orbit about another star. This is a very large planet, some 6000 times more massive than Earth. It is nearly a star in its own right and is composed mostly of gas, like the planet Jupiter.

The typical sizes of stars and planets range from 10^8 to 10^6 m. Gravity also holds these structures together. In the case of a star, the gravity is resisted by the energy generated by internal nuclear reactions. If the nuclear reactions stop, the star will collapse – perhaps into a **black hole**.

On the scale of mountains and people (10^3–10^0 m), gravity is no longer strong enough to govern the formation of structures, and electromagnetic forces give objects rigidity and solidity. Within the atom (10^{-10} m) electrons are held in a complex dance round the charged nucleus by electrostatic forces.

At the next level down (10^{-15} m), the **strong** and **weak nuclear forces** start to have an influence.

Inside protons and neutrons, the strong force holds quarks together so tightly that they can never be extracted. Nobody knows how big quarks are, but they are certainly less than 10^{-17} m in size. They may be fundamental particles, in which case they would be very much smaller than that.

On an even smaller scale of 10^{-33} m, mysterious objects called **strings** may exist. If so, they had a fundamental impact on the evolution of the early universe. The universe was created in the **Big Bang**, and during the first 10^{-43} s after that event, the structure of space and time was influenced by strings. It is possible that during these fleeting instants the expansion of the universe fluctuated in different regions. These fluctuations may have led to the construction of the voids and filaments that we now see at scales of 10^{24} m.

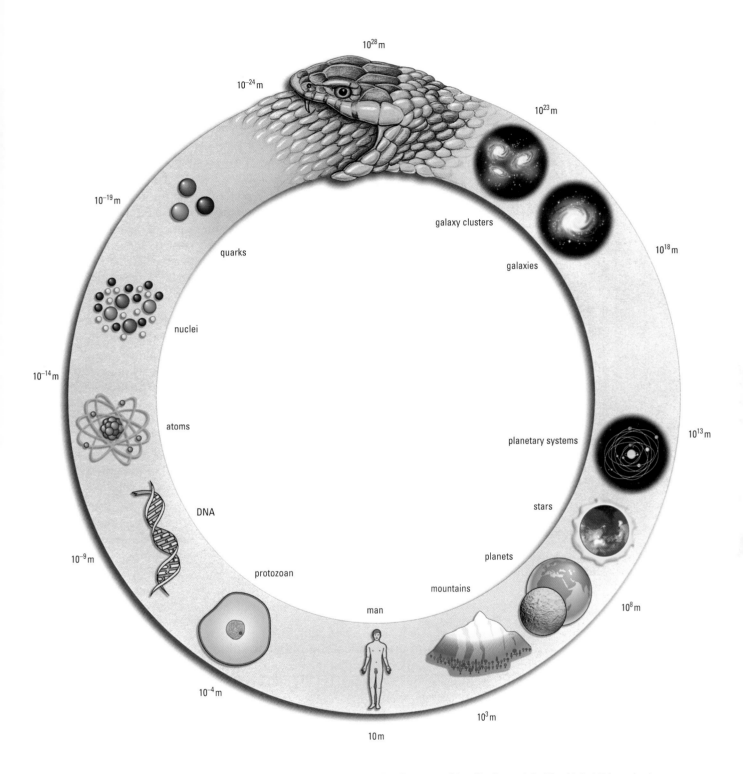

10^{28} m

10^{-24} m

10^{23} m

10^{-19} m

galaxy clusters

quarks

galaxies

10^{18} m

nuclei

10^{-14} m

atoms

planetary systems

10^{13} m

DNA

stars

10^{-9} m

planets

protozoan

mountains

10^8 m

man

10^{-4} m

10^3 m

10 m

Glashow's snake: the cosmic puzzle. The concept of a snake swallowing its own tail is of Indian origin. The Nobel Prize-winning physicist Sheldon Lee Glashow drew on the idea to depict our view of the universe, from the very large to the very small. Starting at the snake's head, we are beyond the scale of clusters of galaxies. Humanity exists on a more comfortable scale, some 28 powers of ten smaller. Descending another 25 powers of ten takes us to the world of the quark and possibly smaller fundamental particles. (Below the size of the atom, visual depictions do not correspond to reality in the same way that they do above the size of the atom, because we are approaching the wavelength of the very light that mediates those images.) Today, the snake could be drawn even longer – the effects of quantum gravity are believed to operate at 10^{-45} m, while we are just beginning to map the very large-scale structure of the cosmos at scales of 10^{30} m or more. At these extremes, the serpent swallows its own tail, completing the picture. The suggestion is that the largest-scale structures cannot be understood without reference to those on the smallest scale, and vice versa. The possible intimate link between the microstructure at the time of the Big Bang and the macrostructure of the universe today is central to the work of some cosmologists and particle physicists, who are searching for the holy grail of physics: the unification of all known physical forces in a single, all-encompassing theory.

- predictability in classical mechanics
- the uncertainty principle
- chaotic systems

Could the flapping of this White Nymph's wings lead to a hurricane?

The complex movement of air is probably a result of the physics of complex systems that obey simple rules.

Initial conditions

If by some means I am able to measure the exact position and velocity of all the particles in a system at any moment, then by using Newtonian mechanics (and a very fast computer), I should be able to calculate the position and velocity of all the particles at any time in the future. However, without the starting information (the initial conditions), the Newtonian mechanic has nothing to work on.

The Great Red Spot on Jupiter (lower left on the planet's disc) is an example of a complex self-sustained weather system.

CHAOS

In one of his Discworld books, Terry Pratchett comments that much of the money used to fund research into chaos would be better spent sending an expedition to the Amazon to find the butterfly that, by flapping its wings, is able to generate hurricanes over Europe *and then getting it to stop.*

This mythical butterfly has become one of the central symbols of chaos. The idea that a small disturbance (wings moving the air) in one part of a system (the atmosphere) can have enormous consequences elsewhere (a hurricane) is central to **chaos theory**. Any system that behaves in such a manner is chaotic. Its behaviour is unpredictable even in principle.

What has delighted and stimulated researchers in this field is the discovery that very simple systems, governed by quite straightforward laws of nature, can be chaotic. This implies that some of the highly complex behaviours that we see in nature (the weather, the turbulent flow of liquid, the motion of the small objects that make up the rings of Saturn, etc.) may not be random but chaotic and so be governed by simple laws.

Classical physics

The cornerstone of classical physics is **Newtonian mechanics**, and the cornerstone of Newtonian mechanics is the idea of **predictability**.

Isaac Newton's laws of nature show how objects respond to being acted upon by forces. The various force laws (Newton's law of gravity, Coloumb's law of electrostatics, etc.) show how forces arise. If the two are put together, we get a system for calculating how objects move in the presence of other objects that generate forces.

This is such a general idea that it should apply to all branches of physics and to all natural phenomena. However, a third piece needs to be added to the jigsaw to make calculation possible – the **initial conditions**.

The great mathematician Pierre Simon de Laplace was so impressed with this idea that he boasted that, given the position and momentum of all the particles in the universe, he could, in principle, predict the future with absolute accuracy. This is a rather bleak philosophy; it destroys any idea that human beings are free to make decisions about their lives.

Modern physics has eroded this point of view. Even in principle, the sort of calculation that Laplace imagined is not possible. There are two reasons for this – the implications of quantum mechanics and the discovery of chaos.

Quantum mechanics and initial conditions

The primary impact that quantum mechanics has on Laplace's programme is to show that the initial conditions that he required can never be provided.

Werner Heisenberg discovered that there is a fundamental limit to our ability to specify the exact position and exact velocity of a particle at the same moment. Heisenberg's famous **uncertainty principle** is represented by the relationship

$$\Delta x \Delta p \geq \hbar$$

where Δx is the uncertainty in the particle's position, Δp is the uncertainty in its momentum at the same moment, and $\hbar$ stands for $h/2\pi$ (where h is the **Planck constant**).

In a way, the uncertainty principle is misnamed. Saying that there is an uncertainty in the particle's position, for example, implies that the particle *has* a definite position but that our measurement equipment is not good enough to detect exactly where it is. The uncertainty principle is actually much deeper than this.

- It implies that if you could produce a particle moving with a unique velocity, then the particle would not *have* a precisely defined position.

- Conversely, if you could localize a particle to a unique point in space, then it would be impossible for you to define what you meant by its momentum at that moment.

Laplace's programme requires perfectly specified positions and momenta of all particles at the same moment. In principle, quantum mechanics does not allow such information to exist. In practice, the quantum uncertainties for large objects are so small that Newtonian mechanics can ignore them.

Chaos in celestial mechanics

When Laplace made his famous boast, he was undoubtedly motivated by the success of celestial mechanics. The Sun and the planets are apparently perfectly behaved objects that move like clockwork. The system is so regular that it is possible to predict the date and time of solar eclipses hundreds of years in advance. Yet even in such a well-behaved system as this, chaos exists.

Johannes Kepler's **laws of planetary motion** laid down the rules to which planetary orbits must conform. One of Newton's great successes was to prove Kepler's laws from his own law of gravity and laws of motion.

However, to do this, Newton had to make an assumption. His calculations assume that the only two objects in the Solar System are the Sun and the planet whose orbit is being calculated (the **two-body problem**). Now, the Sun is so much more massive than all the planets put together that its gravity dominates the system, but the planets *do pull on each other* by gravity as well. As a result, the planetary orbits deviate from the true Keplerian rules.

Newton tried to take this into account by solving the **three-body problem** (the Sun and two planets) but failed. We now know that this problem does not have a unique solution. The difficulty lies in the initial conditions.

Consider the **asteroids**. These are small rocky bodies that drift about the Solar System. Many of them lie in a band of orbits between Mars and Jupiter. If you consider two asteroids next to one other and moving along in slightly different orbits, it is natural to assume that they will always be following each other round. Yet this need not be the case. A small difference in the orbits now can increase very rapidly with time. The asteroids may end up nowhere near each other in the future. This is chaos at work.

Newton and Laplace knew that it would never be possible to specify initial conditions exactly. They assumed that if your information was slightly wrong, then the calculation would give a slightly wrong result but that this error would not be too bad over a reasonable time scale. This is true for a two-body problem. It is not true for a three-body problem.

Chaotic systems are extremely sensitive to initial conditions. With them the error does not stay manageable: it grows exponentially with time.

- An asteroid in a non-chaotic orbit can stay monotonously orbiting the Sun for billions of years.
- However, an identical asteroid in an orbit only centimetres away may quickly fly off into some new part of space.
- The second asteroid's orbit is still determined by mechanics, but it is totally different in shape from the orbit just next to it.

A comprehensive chart of the asteroids' orbits shows that there are bands in which no asteroids are orbiting. Observing the rings of Saturn also shows orbits where there are very few objects. In both cases, these bands correspond to the positions where one would expect the orbits to be chaotic. At some time in the past both the asteroids and the objects that make up the rings of Saturn were uniformly distributed. Some of them lay in chaotic orbits. These chaotic orbits have since taken them away from that part of space, leaving their more regular partners behind.

| The chaotic pendulum |

A pendulum with a free-swinging metal bob hangs above a table upon which are two magnets ('blue' and 'white'). The pendulum is pulled away from the vertical and released. After swinging under the influence of gravity and the magnets, it becomes trapped near one of them; which one depends on the bob's release point. That point is marked on the table blue or white accordingly. Computer simulation of the experiment repeated for many different release points gives the plot below. The centre of the system is an unstable rest point, while the points near the magnets are stable ones. Such unpredictability and sensitivity to initial conditions are characteristic of chaotic systems. The pattern is actually a fractal: zooming into the borders between blue and white regions, we would see complex patterns that repeat the pattern of the overall figure.

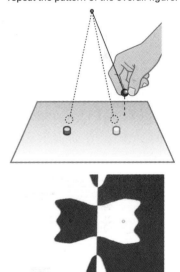

This is the asteroid Ida (as photographed by the Galileo *space probe). It is the only asteroid that we know has its own 'moon' – which is the small object on the right.*

The rings of Saturn.

2

Handling **DATA**

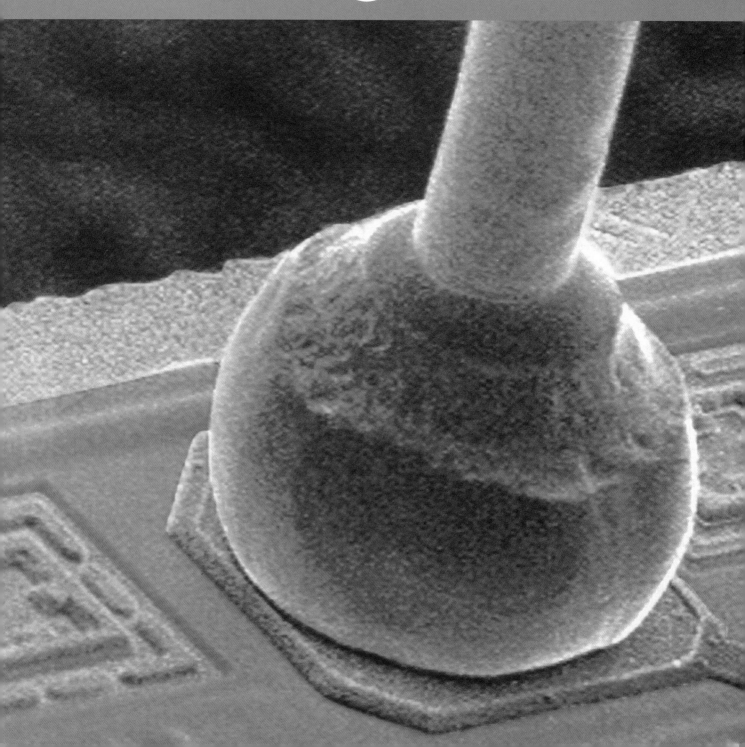

Work of the precision needed to position the gold-wire connections on an integrated circuit (IC) is only possible as the result of a meticulous approach to measurement and the elimination of error in the instruments used to assemble the IC.

In this chapter you will find out how to use units and dimensions correctly and how to work with uncertainties. Physics relies on experiment and a mathematical framework, so you will also learn how to carry out investigations and analyse data as well as how to use scientific calculators and datalogging and computer interfacing equipment. Any other mathematics required is covered in the Mathematics toolbox appendix at the end of the book.

A coloured scanning electron micrograph of the bonded ends of two microwires on a silicon chip. These wires link the integrated circuit on the chip to the pins that connect the chip to the circuit board on which it sits, for example, in a computer. (Magnification 1400×.)

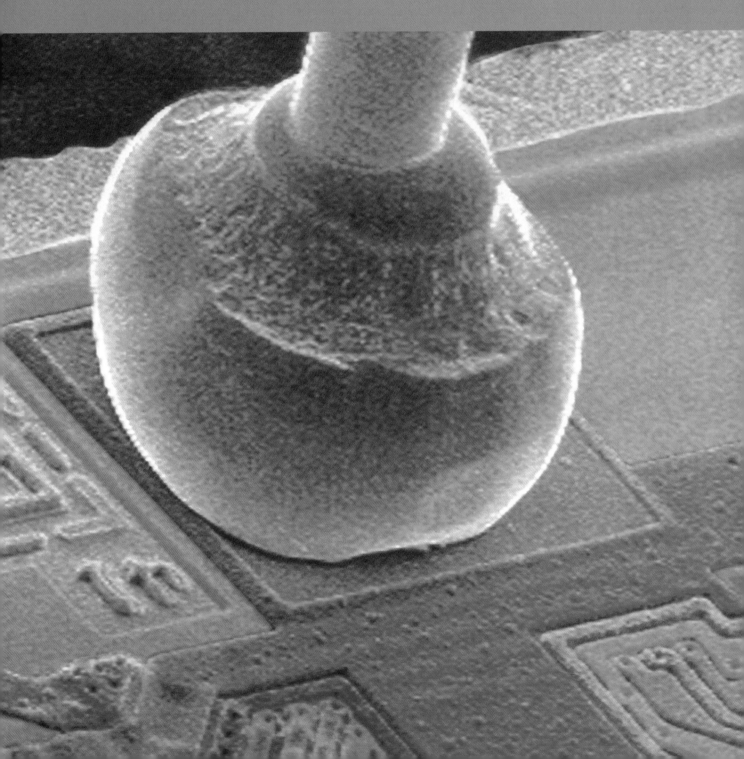

BASE UNITS AND DERIVED UNITS

Imagine walking into a bookshop to enquire the price of a new physics textbook. The owner tells you that the best on the market costs 3500. You would stop to catch your breath. However, you soon realize that the price was in pence, not pounds.

This illustrates an important point. Quoting the result of a calculation or measurement without a unit is useless. This is a very common reason for students losing marks in examinations.

By convention, scientists and engineers use the International System of Units (SI – for Système Internationale d'Unités). This set of rules defines seven **base units** and various **derived units** obtained from them.

Base units.

Quantity	Unit	Definition
time	second (s)	The time taken for 9 192 631 770 periods of the radiation emitted when an electron makes a transition between two specified energy levels of the ground state of a ^{133}Cs atom.
mass	kilogram (kg)	The mass of a platinum–iridium cylinder held at the International Bureau of Weights and Measures in Sèvres, France.
length	metre (m)	The distance travelled by light in a vacuum during a time interval of 1/299 792 458 s.
current	ampere (A)	The constant current which, if maintained in two straight parallel conductors of infinite length and negligible cross-section, placed 1 m apart in a vacuum, would produce a magnetic force between the conductors of 2×10^{-7} N for every metre length of conductor.
temperature	kelvin (K)	The thermodynamic temperature of the triple point of water (the conditions in which water, ice, and steam are in equilibrium) is defined as 273.16 K above absolute zero.
amount of substance	mole (mol)	The number of atoms in 0.012 kg of ^{12}C.
intensity	candela (cd) (The candela is a fairly specialist unit used in the study of luminous objects. It will not be used in this book.)	The luminous intensity in a given direction of a source that emits monochromatic radiation (i.e. radiation of one colour or wavelength) of frequency 540×10^{12} Hz having a radiant intensity in that direction of 1/683 W per **steradian** (an SI unit for solid angles).

Base units

Scientists have worked out that they can measure any quantity in nature in terms of a small number of base units. The challenge is to reduce the number of base units to a minimum and still be able to measure anything that crops up. So far it has been possible to keep to combinations of the seven base units that are listed above.

The main advantage of using a set of agreed units is that scientists from all over the world can exchange ideas and designs for experiments without having to translate specifications into different units. It is rather like having a common language.

If this is to work, there must be an agreed set of standards against which people can check their measuring devices. The SI base units rely on simple physical effects. Anyone with sufficient patience and equipment can set up a comparison between their measuring device and the agreed standard.

Derived units

The derived units enable us to measure more than the basic quantities of length, time, mass, etc. For example, there is no unit for speed among the base units. However, a suitable unit can be derived from the equation for speed:

$$\text{average speed} = \frac{\text{distance (m)}}{\text{time (s)}}$$

This suggests that the unit of speed is metres divided by seconds; unfortunately, such a division is impossible. Division can only happen when the quantities are of the same type. We say instead that the unit of speed is metres *per* second, and write it as $m\,s^{-1}$. When a car is moving at a constant speed of 100 m in 10 s, what we are really saying is that it must cover a distance of 10 m in 1 s, or $10\,m\,s^{-1}$.

Acceleration is another important quantity. Acceleration is the rate at which speed is changing:

$$\text{average acceleration} = \frac{\text{final speed (m s}^{-1}) - \text{initial speed (m s}^{-1})}{\text{time (s)}}$$

This is a little trickier to handle. The top line is the difference between two quantities measured in $m\,s^{-1}$. The bottom line is in seconds. So the units of acceleration are 'metres per second per second', or $m\,s^{-2}$.

This causes some people problems: they cannot visualize a 'square second'. (It is not at all like a square metre!) We are not dividing by a 'square second'. This is just as impossible as dividing metres by seconds. We are using an abbreviated way of speaking. An acceleration of $10\,m\,s^{-2}$ really means that in every second, the speed increases by 10 metres per second.

The following table shows some of the derived units that are used in this book.

Derived units.

Quantity	Derived unit	Name
speed	$m\,s^{-1}$	–
acceleration	$m\,s^{-2}$	–
force	$kg\,m\,s^{-2}$	newton (N)
pressure	$kg\,m^{-1}\,s^{-2}$	pascal (Pa)
energy	$kg\,m^2\,s^{-2}$	joule (J)
charge	$A\,s$	coulomb (C)
potential difference	$kg\,m^2\,A^{-1}\,s^{-3}$	volt (V)
resistance	$kg\,m^2\,A^{-2}\,s^{-3}$	ohm (Ω)

Secondary standards

To make it easier for everyone to compare their instruments to the standard units, there are a set of **secondary standards** throughout the world that are checked against the primary standards. These are located at universities and scientific institutions.

NASA's expensive mistake

Even professionals make mistakes. In 1999 the *Mars Climate Orbiter* spacecraft was lost because one of the teams involved used imperial units while another used metric units for a key spacecraft operation. This information was critical to the manoeuvres required to place the spacecraft in the proper Mars orbit. The sister mission, the *Mars Polar Lander*, was later lost for additional reasons. The total cost of the project was $327.6 million.

- all equations balance dimensionally

- using dimensions to check and predict equations and to determine units

DIMENSIONS

Balancing equations

A simple arithmetic equation such as

$$3 + 5 = 8$$

balances because the *value* of the numbers on the left-hand side is equal to the value of the numbers on the right-hand side. In physics, equations usually equate quantities that have magnitudes (or values), dimensions, and units. All three must balance for the equation to make sense.

Dimension is the *type* of quantity we are dealing with independent of its units or value. For example, 100 cm, 1 m, 2 miles, and 3 light-years all have the *dimension* of length but are expressed in different *units*. Simple numbers, as in the above arithmetic equation, are dimensionless, whereas dimensions are valueless.

The examples below show situations in which values balance but dimensions and/or units do not:

$$3\,\text{kg} + 5\,\text{kg} = 8\,\text{m} \tag{1}$$

$$3\,\text{m} + 5\,\text{m} = 8\,\text{cm} \tag{2}$$

- Equation (1) is meaningless because the left-hand side has the dimension of mass while the right-hand side has the dimension of length.

- Equation (2) is also incorrect. Both sides of the equation have the same dimension (length), but the units are different.

Equations whose dimensions and units balance are said to be **homogeneous**. (Being homogeneous does not guarantee that an equation is correct. An example of a homogeneous but incorrect equation is $F = 2ma$ with all quantities measured in SI units.)

Dimensional analysis

All the quantities used to describe mass and motion are combinations of the three fundamental dimensions of mass (M), length (L), and time (T). To find the dimensions of any particular quantity, you must find an equation that relates it to quantities you do know and then balance the dimensions.

It helps to introduce square brackets to mean 'the dimensions of …'. For example: $[v]$ means 'the dimensions of velocity'. Then $[v] = \text{L T}^{-1}$ means 'the dimensions of velocity are length per unit time'. This makes sense, because velocity is calculated from displacement over time.

The units of a quantity follow from its dimensions. In principle, there is a free choice of all possible units for each dimension. That is, velocity could be measured in metres per second, or centimetres per century, or feet per hour. In practice, SI is usually adopted, so that all lengths are in metres, masses in kilograms, and times in seconds.

Balancing vector quantities

Equations must balance in *all* respects. If two people push you with forces of 100 N each, the resultant force is anywhere from 0 N (if they oppose one another) to 200 N (if they both push in the same direction). Forces have *direction* as well as *magnitude* (they are **vectors**). If an equation includes vector quantities, then the directions of the things equated must be the same, as well as their magnitudes, dimensions, and units. For example:

100 N to the left + 100 N to the right

= 0 N

100 N to the right + 100 N to the right

= 200 N to the right

Beyond mechanics

All electromagnetic quantities can be included in dimensional analysis by introducing one more dimension, current I. Charge would be an alternative, but electrical measurements are based on the ampere rather than on the coulomb, so current is preferred.

Worked example 1

Find the dimensions and units for acceleration.

$$a = \frac{v - u}{t}$$

so

$$[a] = \frac{[v - u]}{[t]} = \frac{\text{L T}^{-1}}{\text{T}} = \text{L T}^{-2}$$

Suitable SI units are m s^{-2}.

Worked example 2

Find the dimensions and units for energy.

Any equation for energy would do, for example

$$KE = \tfrac{1}{2}mv^2$$

Hence

$$[\text{energy}] = [\tfrac{1}{2}][m][v]^2 = \text{M L}^2\,\text{T}^{-2}$$

Note that the number $\tfrac{1}{2}$ is simply replaced by unity (1) because it is dimensionless. Suitable SI units for energy are therefore $\text{kg}\,\text{m}^2\,\text{s}^{-2}$. This is cumbersome and so is renamed the joule: $1\,\text{J} = 1\,\text{kg}\,\text{m}^2\,\text{s}^{-2}$.

Using dimensions

Although dimensions are more fundamental than units, either can be used in a similar way to check equations and to predict the form of new relationships. In complex problems involving a great deal of algebra, it is very easy to lose a term and end up with a result that looks convincing but is actually wrong. A quick dimensional check will often show up this kind of error but will not find any arithmetic errors, because dimensions have no numerical value.

Worked example 3

A student derives an expression for frictional drag on a car body and comes up with

$$F = \tfrac{1}{2}C\rho Av$$

where ρ is the density of the air through which the vehicle moves at a velocity v, A is the cross-sectional area of the vehicle, and C is a dimensionless constant of proportionality.

At first glance this seems reasonable, but a quick dimensional check shows that it must be incorrect:

LHS: $[F] = MLT^{-2}$ (taken from the dimensions of $F = ma$)

RHS: $[\tfrac{1}{2}][C][\rho][A][v] = 1 \times 1 \times \text{M L}^{-3} \times \text{L}^2 \times \text{L T}^{-1} = \text{M T}^{-1}$

The two sides have different dimensions (by L T^{-1}). The difference corresponds to the dimensions of velocity. The correct relation is

$$F = \tfrac{1}{2}C\rho Av^2$$

Worked example 4

The period T of a mass–spring oscillator is likely to depend on its mass m and spring constant k, so a possible equation might be

$$T = Ck^x m^y$$

where x and y are powers to be determined and C is a dimensionless constant. (The spring constant k is the tension required to produce unit extension.) To find x and y, compare the dimensions of each side of the equation:

$[\text{LHS}] = \text{T}$

$[\text{RHS}] = [C][k]^x[m]^y = 1 \times (\text{M T}^{-2})^x (\text{M})^y = \text{M}^{x+y}\,\text{T}^{-2x}$

Now balance powers of each dimension separately:

$1 = -2x \quad \therefore x = -\tfrac{1}{2}$ (time)

$0 = x + y \quad \therefore y = -x = \tfrac{1}{2}$ (mass)

and hence

$$T = Ck^{-\frac{1}{2}}m^{\frac{1}{2}} = C\sqrt{\frac{m}{k}}$$

This is dimensionally correct and so could be the required equation. In fact the correct equation is

$$T = 2\pi\sqrt{\frac{m}{k}}$$

The factor of 2π cannot be determined from dimensional analysis.

The Nairn Falls, British Columbia, Canada. Sometimes the only way to deal with the complexity of fluid flow is to use arguments derived from dimensional analysis.

PRACTICE

1 Express the newton in terms of the base SI units kg, m, and s.

2 Find the dimensions of each of the following quantities:

a pressure

b power

c linear momentum

d current

e potential difference

f resistance.

3 Suggest suitable units for viscosity η in Stokes's law for the force of viscous drag on a sphere of radius r moving through a fluid:

$$F = 6\pi\eta rv$$

η is the viscosity of the fluid and v is the velocity of the sphere.

4 The frequency of a simple pendulum depends only on its length and the gravitational field strength. Use dimensions to derive a possible form of the equation for this frequency.

- standard prefixes for SI units
- conversions
- common non-SI numbers and units

Choosing the right measuring units can be all important.

CUSTOMIZED UNITS

Prefixes

It would make little sense to measure your height in kilometres or light-years, or your mass in tonnes. It is much more convenient to choose a unit comparable in size to the quantity that is being measured.

Physical quantities cover an incredibly wide range of values, so prefixes are used to give multiples and submultiples of basic units. For example, the thickness of a hair might be measured in micrometres (millionths of a metre), while the voltage on a Van de Graaff generator is measured in kilovolts (thousands of volts).

The standard multiples and submultiples are in steps of 10^3. For example, a millimetre is one thousandth of a metre and a kilometre is one thousand metres. However, there are prefixes that are sometimes used but that do not fit this pattern. The centimetre is the most common of these (one hundredth of a metre).

Standard prefixes.

Prefix	Symbol	Multiplies unit by	Example
Multiples			
kilo-	k	10^3	$1\,kV = 1000\,V$
mega-	M	10^6	$1\,MW = 10^6\,W$
giga-	G	10^9	$1\,GW = 10^9\,W$
tera-	T	10^{12}	$1\,TW = 10^{12}\,W$
peta-	P	10^{15}	$1\,Pm = 10^{15}\,m$
exa-	E	10^{18}	$1\,Em = 10^{18}\,m$
Submultiples			
milli-	m	10^{-3}	$1\,mA = 10^{-3}\,A$
micro-	μ	10^{-6}	$1\,\mu V = 10^{-6}\,V$
nano-	n	10^{-9}	$1\,nm = 10^{-9}\,m$*
pico-	p	10^{-12}	$1\,pJ = 10^{-12}\,J$
femto-	f	10^{-15}	$1\,fm = 10^{-15}\,m$**
atto-	a	10^{-18}	$1\,am = 10^{-18}\,m$
Other common prefixes			
deci-	d	10^{-1}	$1\,dm = 0.1\,m$
centi-	c	10^{-2}	$1\,cm = 0.01\,m$

* An older unit, the **angstrom**, is $10^{-10}\,m$ and sometimes appears in textbooks. It is still often used to measure wavelengths in astronomy.

** 1 fm is sometimes called a **fermi**.

Converting units

To carry out a calculation, it is often necessary to convert measurements using units in different forms to a common form. For example, the length of a wire is likely to be measured in metres or centimetres using rulers, whereas its diameter will be measured in millimetres, using a micrometer screw gauge. To find its volume, all dimensions must be in terms of the same unit of length. If this is millimetres, then the calculated volume will be in mm^3; if it is metres, the volume will be in m^3. It is surprising how often people forget to use consistent (coherent) units in a calculation.

Worked example 1

Calculate the volume in mm^3 of a wire of length 1 m 25 cm and diameter 0.65 mm.

length $\quad h = 1.25\,m = 1.25 \times 10^3\,mm$
diameter $\quad d = 0.65\,mm$
volume of a cylinder $= \pi r^2 h$

$$= \pi \times \left(\frac{0.65\,mm}{2}\right)^2 \times 1.25 \times 10^3\,mm$$

$$= 410\,mm^3 \text{ to two significant figures (2 sig. figs)}$$

Worked example 2

Convert a speed of $85\,km\,h^{-1}$ to a speed in $m\,s^{-1}$.

$$1\,km = 10^3\,m \quad 1\,h = 3600\,s$$

$$80\,km\,h^{-1} = \frac{80\,km}{1\,h} = \frac{80 \times 10^3\,m}{3600\,s} = 22\,m\,s^{-1}\ (2\text{ sig. figs})$$

Worked example 3

How many cubic millimetres are there in a cubic metre?

$$1\,m = 10^3\,mm$$

so

$$(1\,m)^3 = (10^3\,mm)^3 = 10^9\,mm^3$$

> **Don't confuse**
>
> Beware the difference between expressions like $100\,cm^2$ and a $100\,cm$ square. The former is an area equal to 100 individual 1 cm by 1 cm squares, whereas the latter is a square 100 cm long by 100 cm across and so has an area equal to $10\,000\,cm^2$.

Scale models

A calculation very similar to that in worked example 3 is involved whenever we scale up a model. If a model dinosaur has height 12 cm and is to a scale of 1:50, then the real dinosaur has a height of $12 \times 50\,cm = 6\,m$. However, its volume depends on the cube of its linear dimensions and so is $50^3 = 12\,5000$ times greater than the volume of the model. Its mass will also be increased by a similar factor, assuming that the density of the modelling material is close to that of a real dinosaur. Great care must be taken when results achieved using a scale model (a model aircraft in a wind tunnel, for example) are to be applied to the real thing.

A critical assembly

It is not possible to build a tiny nuclear reactor. There is a minimum size or **critical assembly** for any particular type of reactor. A scaled-down model will not produce a self-sustaining **chain reaction**. Why not?

It is all a question of scale. For the reaction to go critical, at least one of the neutrons released when a nucleus splits must go on to split another nucleus (this causes a chain reaction). However, many neutrons simply escape from the surface of the reactor core.

Now, if the linear dimensions of the core are increased by a factor of 10, the surface area increases 10^2 times. This sounds like a bad thing. Surely more neutrons will be lost. This is true, but the volume of the core has increased 10^3 times, and the number of fissions taking place depends on the volume. The net effect is that as the reactor core is made larger, the ratio of neutrons lost from the surface to neutrons produced falls, and at some point a chain reaction becomes possible. The actual size will depend on the type and arrangement of fuel, the design of the reactor, and its geometry.

PRACTICE

1 Estimate the volume of your body in m^3, mm^3, and cm^3.

2 Convert the density of mercury, $1.36 \times 10^4\,kg\,m^{-3}$, to $g\,cm^{-3}$.

3 How many minutes are there in a microcentury?

4 How many 250 ml (millilitre) beakers of water are needed to completely fill a cylindrical bucket of radius 15 cm and depth 0.35 m?

5 Convert the following:

  a $10\,m\,s^{-1}$ to $km\,h^{-1}$;

  b $120\,km\,h^{-1}$ to $m\,s^{-1}$.

6 a What is the volume of the Earth in km^3? (Take the Earth to be a sphere of radius $6.4 \times 10^6\,m$.)

  b The mass of the Earth is $6.0 \times 10^{24}\,kg$. Calculate its density in $kg\,m^{-3}$ and $g\,cm^{-3}$.

  c What is its density if it is compressed to a diameter of 1 cm (to make it a **black hole**)?

7 A model showhouse constructed to a scale of 1:120 has a swimming pool 20 cm by 10 cm that is filled with water to a depth of 2.5 cm.

  a What is the volume of water in the model swimming pool?

  b What is the volume of water in the real swimming pool?

  c If the model swimming pool takes 1 minute to fill through a model hosepipe, suggest how long it would take to fill the real swimming pool, stating any assumptions you make.

8 Could mammals ten times the size of dinosaurs survive?

- the range of values encountered in physics

- using powers of 10 to write very large and very small numbers concisely

The Hubble deep-field view is a window on creation, revealing an amazing variety of galaxies 'as they were' when light left them, up to 9 billion years ago.

SCIENTIFIC NOTATION

From deep field to strings

In January 1996, Bob Williams, Director of the Space Telescope Science Institute, USA, trained the **Hubble Space Telescope (HST)** on an area of apparently empty space about 1/10 the size of the full Moon. Over 48 orbits in 10 days the telescope built up the long-exposure photographs now known as the **'Hubble deep-field'** view. It was far from empty. Many familiar and not so familiar structures were found, some so far away that the photons entering the HST left their sources over nine billion years ago, when the universe was less than half its present age.

Whilst the deep-field images reveal structures influenced by long-range gravitational forces, the nature of these forces is still mysterious. Unlike the strong, weak, and electromagnetic forces, there is still no satisfactory quantum theory of gravity. The best description is given by Einstein's general theory of relativity, but the smooth spacetime of Einstein's theory is at odds with the intrinsic uncertainties of quantum mechanics. This has led some physicists (e.g. John Wheeler) to suggest that spacetime itself becomes a kind of 'foam' at the smallest or 'Planck' scales so that it is impossible to distinguish different locations or times to greater precision than this. This is also the scale at which the 'strings' in string theory have structure. Whilst this is one hundred billion billion times smaller than a proton it might not be entirely beyond the reach of physics. In fact, in 2012 Jacob Bekenstein, a physicist at the Hebrew University of Jerusalem, suggested an experiment that could reveal effects on this scale!

These examples make a very important point. Physics deals with phenomena that occur over a mind-boggling range of magnitudes. The light reaching us from distant galaxies in the Hubble deep-field view has travelled for nine billion years at the speed of light and covered about 85 000 000 000 000 000 000 000 000 m. On the other hand the proton is about 0.000 000 000 000 001 m across, and quarks are much smaller than this.

Our standards of measurement are based on human-sized distances (the metre), so it is handy to have a shorthand way to write down very large and very small numbers. This is usually done by using powers of ten.

Scientific notation reduces all numbers to a value between 0 and 10 multiplied by ten raised to some power. The power can be determined simply by counting how many places the decimal point must move to produce a value between 0 and 10. For example:

$$52\,000 = 5.2 \times 10^4 \quad \text{and} \quad 0.000\,52 = 5.2 \times 10^{-4}$$

(In both the above cases the final value has been quoted to just two significant figures.)

Manipulating powers

There are some simple rules that make working with scientific notation much easier:

- When two numbers in scientific notation are multiplied together, the powers of ten are added. For example:

$$(3.0 \times 10^8) \times (2.5 \times 10^4) = 7.5 \times 10^{12}$$
$$(3.0 \times 10^8) \times (2.5 \times 10^{-4}) = 7.5 \times 10^4$$
$$(3.0 \times 10^8) \times (2.5 \times 10^{-10}) = 7.5 \times 10^{-2}$$
$$(3.0 \times 10^8) \times (6.2 \times 10^5) = 18.6 \times 10^{13} = 1.9 \times 10^{14}$$

Notice in the last example that the final result is given to two significant figures to be consistent with the data used in the calculation. An extra step has been necessary to reduce it to the standard form.

- When two numbers in scientific notation are in a quotient, their powers of ten are subtracted from one another. For example:

$$\frac{5.6 \times 10^{18}}{2.8 \times 10^{12}} = 2.0 \times 10^{(18-12)} = 2.0 \times 10^6$$

$$\frac{5.6 \times 10^{18}}{2.8 \times 10^{-12}} = 2.0 \times 10^{(18-(-12))} = 2.0 \times 10^{30}$$

- Raising a power of ten to some power multiplies the power of ten by the new power. For example:

$$(10^4)^2 = 10^{(4 \times 2)} = 10^8$$
$$(10^4)^3 = 10^{(4 \times 3)} = 10^{12}$$
$$(10^4)^{0.5} = 10^{(4 \times 0.5)} = 10^2$$

(Raising something to the power 0.5 is the same as taking its square root.)

Significant figures

The difference in quoting the magnitude of gravitational field strength as $9.8\,N\,kg^{-1}$ or $9.81\,N\,kg^{-1}$ is important.

- In the first case, you are saying that the gravitational field strength is closer to $9.8\,N\,kg^{-1}$ than to $9.7\,N\,kg^{-1}$ or $9.9\,N\,kg^{-1}$ (i.e. it is within $0.05\,N\,kg^{-1}$ of $9.8\,N\,kg^{-1}$).
- In the second, your claim is much tighter – within $0.005\,N\,kg^{-1}$ of $9.81\,N\,kg^{-1}$ (that is, closer to $9.81\,N\,kg^{-1}$ than to $9.80\,N\,kg^{-1}$ or $9.82\,N\,kg^{-1}$).

The first value would include the field strength at all points on the Earth's surface, but there are places where the field strength varies above and below the range allowed by the second value. In a similar way, in calculations, it would be wrong to go from data quoted to two significant figures to a result (calculated from that data) given to three or more significant figures.

For example, if I walk 10 m in 6.0 s, my calculator tells me that

$$10 \div 6.0 = 1.666\,666\,7$$

(It only stops there because it runs out of display!) My average speed should not be quoted as $1.666\,666\,7\,m\,s^{-1}$, because the data used to calculate it is not this precise. A reasonable result would be $1.7\,m\,s^{-1}$.

It is a good general rule to quote calculated values to the same number of significant figures as the data used in the calculation. Of course, some whole numbers are given to just one figure since it is obvious they are exact. (For example, if eight springs are connected in series, we do not have to limit later results to one significant figure on account of the eight.)

If the result of a calculation involves a decimal less than one, then none of the zeroes before the first non-zero digit are significant. For example, $0.000\,000\,007\,35$ is given to three significant figures. On the other hand, a number like $735\,000\,000$ might have been rounded to three significant figures or might be accurate to nine significant figures. You can't tell just by looking at it. However, if the number is given in scientific notation, then all the figures used are significant. For example:

7.35×10^{-9}	(3 sig. figs)
7.355×10^8	(4 sig. figs)
$7.350\,05 \times 10$	(6 sig. figs)

Using calculators

A lot of people make mistakes when keying powers of ten into their calculator. What you need to do depends on your calculator, but to see if you have a problem try this:

$(5.7 \times 10^7) \times (1.6 \times 10^{-19}) \times (2.6 \times 10^3)$

You should get 2.4×10^{-8}. One of the commonest mistakes would lead you to 2.4×10^{-5} instead. (See spread 2.5.)

Frighteningly big numbers

The number of electrons passing each point in a circuit when 1.0 A flows is about

$6\,300\,000\,000\,000\,000\,000$

In scientific notation this becomes 6.3×10^{18}, which is much more concise. Although this is a big number, it is tiny compared with the numbers we come across when considering the distribution of particles in the universe as a whole. It is estimated that there are about 1080 protons and neutrons in the universe, but the really overwhelming number is the number of ways all the particles present at the Big Bang might have been arranged. Professor Sir Roger Penrose of the University of Oxford estimates that the initial state of our universe is one arrangement in 10 to the power 10^{123}. This number is unimaginably large. If we tried to write it down and used a single proton or neutron for each zero, there would not be anywhere near enough particles in the entire universe to complete the number! And yet we can write it in a few characters as $10^{10^{123}}$.

PRACTICE

1 Estimate the number of cubic millimetres in your physics laboratory. Write down your answer in full and in scientific notation. How many significant figures are appropriate?

2 A current of 0.167 mA flows for 120 ms. How much charge is transferred?

3 How far is a light-year in metres? (The speed of light is $3.00 \times 10^8\,m\,s^{-1}$ and one year is $3.16 \times 10^7\,s$.)

4 How many seconds have passed since you were born? Write down your answer in full and in scientific notation. What is a sensible number of significant figures for your answer?

5 A block of wood is 3.55 cm by 12.2 cm by 1.05 m in size. Its mass is 5.2 kg. What is its volume in cm^3

and its density in $kg\,m^{-3}$? Make sure that you quote your answer to an appropriate number of significant figures.

6 The second is defined as 9 192 631 770 periods of the radiation corresponding to the transition between the two hyperfine levels of the ground state of the caesium-133 atom.

a Put this number into standard form.

b Round it off to four significant figures.

c The metre is defined as the distance travelled by light in 1/299 792 458 of a second. How many wavelengths of caesium-133 radiation, used to define the second, is this? Give your answer as accurately as you can.

USING SCIENTIFIC CALCULATORS

The scientific calculator is a basic tool for a physics student. Many people rely on such calculators totally for producing the answers to numerical problems, but they must be used properly or they will give incorrect results. It is well worth taking some time to learn how to use your calculator correctly.

There is such a wide variety of calculators on the market that it would be impossible to explain how to use all of them. It is therefore important that you also work through your calculator's instruction manual. This section will point out the features and problems that crop up in using calculators to solve numerical problems in physics.

Entering numbers

You will often have to enter very large or very small numbers. It is best to write these down in scientific notation, that is, as a number between 1 and 10 multiplied by a power of ten. For example:

$$300\,000\,000 = 3 \times 10^8$$
$$0.000\,678 = 6.78 \times 10^{-4}$$

Entering such numbers into a calculator is quite simple. The calculator will have a key marked EXP, EE, or 10^x, which tells it that you are working in scientific notation.

To enter a number such as 3×10^8, type

$$\boxed{3}\ \boxed{\text{EXP}}\ \boxed{0}\ \boxed{8}$$

Using the +/– key allows you to enter numbers such as 6.78×10^{-8}:

$$\boxed{6}\ \boxed{.}\ \boxed{7}\ \boxed{8}\ \boxed{\text{EXP}}\ \boxed{+/-}\ \boxed{0}\ \boxed{8}$$

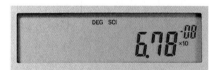

A typical scientific calculator and two ways of displaying an exponent.

Calculator modes

Some calculators have different modes of operation. These affect the way in which calculations are carried out and the results that are displayed on the screen.

• FIX displays answers to a specified number of decimal places.

• SCI displays a number in scientific notation.

• ENG or engineering notation is like scientific notation, except that the powers of ten are restricted to multiples of three, so 3×10^8 would be displayed as 300×10^6. This makes it easier to work with the standard prefixes such as kilo-, mega-, and milli-.

Some sophisticated calculators have graphical capabilities.

- NORM sets the display into normal mode. This displays numbers in ordinary notation, unless the number requires more digits than the calculator screen has available, in which case it will switch to scientific notation. It will also switch to scientific notation if the number is smaller than a set limit, say $<10^{-4}$.
- DEG sets the trigonometric functions (sin, cos, tan) to interpret numbers as angles in degrees.
- RAD sets the trigonometric functions to interpret numbers as angles in radians.
- GRAD is not generally used by scientists and is another way of measuring angles.
- SD (standard deviation) mode, on some calculators, provides access to the pre-programmed functions that calculate averages, standard deviations, etc.

If the numbers given in a question are presented using prefixes, for example 30 kN, or in scientific notation, such as 3×10^4 N, it is better to use either SCI or ENG mode.

Helpful hint

Examination questions that involve forces or motion generally quote angles in degrees. Radians are used in calculations to do with simple harmonic motion, oscillations, and circular motion.

Calculator functions

The calculator functions most often used in physics are as follows:

- SIN, COS, and TAN calculate the sine, cosine, and tangent of a number taken to be an angle in DEG or RAD, depending on the mode that is set.
- ASIN, ACOS, and ATAN (that is arcsine or $\sin^{-1}$, etc.) give the angle that corresponds to a value of sine, cosine, and tangent. (NB These functions are sometimes written on the keyboards as $\sin^{-1}$, $\cos^{-1}$, and $\tan^{-1}$. They should not be confused with 1/sin, etc.) For example, SIN 30° = 0.5, ASIN 0.5 = 30°. The calculator will display the angle in degrees or radians depending on the mode that is set. For trigonometric functions, more than one angle can have the same value:

    sin 30° = 0.5   sin 150° = 0.5   sin –210° = 0.5

    The calculator is programmed to give the answer within a set angle range:

    ASIN is between –90° and +90°;

    ACOS is between 0° and 180°;

    ATAN is between –90° and +90°.

    This may not give you the answer that you need, so you will have to convert to the angle range that you want. You should have a good idea of the angle that is correct, so that you can check it against the calculator's answer.

Helpful hint

See the Mathematics toolbox for information on radians, negative angles, and calculating trigonometric functions of angles >90°.

- LN calculates the logarithm to base e (e = 2.718 281 8...) of a number. Base e is so often used that physicists tend to refer to it as 'log' without mentioning the base. This confuses students at first as there is also a LOG function on a calculator, which gives the logarithm to base 10. You should assume that LN is always meant unless otherwise specified.
- e^x raises e to any power. This function is useful in questions that cover radioactive decay or the discharging of capacitors, where the equations $N = N_0 e^{-\lambda t}$ (radioactivity) and $I = I_0 e^{-t/RC}$ are used.
- x^y raises any number (x) to any power (y). This function is far more useful than people realize. For example, using a power of 1/3 calculates the cube root of a number.
- 10^x raises 10 to any power. It is wrong to use this to enter numbers in scientific notation. This function is rarely used in physics.

UNCERTAINTIES

When scientists say that there is an **uncertainty** in a measurement, they do not mean that they are unsure of its value. Uncertainties, which are sometimes less helpfully called errors, are not mistakes. They are natural variations in measurements that come about for a variety of reasons:

• No instrument is exactly precise.

• Different people may be using different types of instrument.

• No two people read an instrument in exactly the same way.

• Sometimes instruments are read wrongly.

• The instrument's adjustment may have changed.

No matter how carefully we set up experiments, problems like these always arise. Scientists try to estimate how big an effect such problems may have on an experiment, and they quote an uncertainty in the measurement.

Quoting uncertainties

If you read a scientific paper, you will find that the numerical results include a second number and a '±' symbol. For example,

the lifetime of the D^0 particle = $4.6 \pm 0.6 \times 10^{-13}$ s

(The D^0 is a subatomic particle containing a **charm quark**.) The paper's author believes that the measurement result is 4.6×10^{-13} s but suspects that it might be as large as 5.2×10^{-13} s $(4.6 + 0.6)$ or as small as 4.0×10^{-13} s $(4.6 - 0.6)$. This defines the range of uncertainty that is unavoidably part of the experiment.

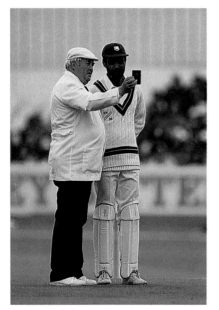

Using a light meter.

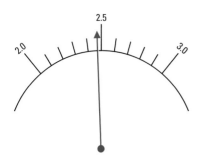

The reading on the meter.

Worked example 1

A class of pupils measures the length of a block of wood using a 30 cm ruler calibrated in millimetres. They obtain the following results:

4.90 cm, 4.95 cm, 4.92 cm, 4.96 cm, 4.93 cm, 4.91 cm

What is the length of the block?

The average value of these results is 4.93 (3 sig. figs), and this is the result to quote as the length of the block. However, only one result agrees with this average, so are the others wrong?

The different readings have arisen for the reasons listed above and show that there is an uncertainty associated with this measurement. A professional scientist would use a mathematical technique to estimate the uncertainty in an important measurement. In this case, we can use common sense.

The range of the data is between 4.90 cm (smallest) to 4.96 cm (largest), which is 0.06 cm. The uncertainty is taken as half the range, so the length of the block is quoted as 4.93 ± 0.03 cm.

Worked example 2

An umpire at a cricket match has to worry about the level of light. He measures the light using a light meter, and the reading is 2.48. What uncertainty should he quote in the value (being a good scientist as well as an umpire)?

Single measurements are always a problem because there are no others with which to compare them. In this case, the umpire has to look carefully at the scale on the instrument.

In the diagram, the pointer lies between 2.4 and 2.5. We can be sure that the light is between these values. However, this would be a cautious measurement that does not use the scale to its full potential. The umpire has tried to estimate how near the pointer is to 2.5 and has come up with an answer of 2.48. We can probably agree that it is nearer to 2.5 than 2.4, but other people may disagree about the exact value to quote.

As a rule, people always agree on estimations to within half the smallest scale division. In this case the scale divides into intervals of 0.1, so half the smallest scale division is 0.05. The umpire should quote the light as being 2.48 ± 0.05.

Estimating uncertainties

Repeating a measurement several times and charting the results produces a **distribution** from which the uncertainty can be obtained. This is the technique used by professional scientists. In school physics, measurements are often not repeated. In this case some rules of thumb help to estimate the uncertainty, as shown in the following table.

Estimating uncertainties.

Measurement	Typical equipment	Rule of thumb
reading scales	rulers, Verniers, dials, etc.	halve the smallest division
timing	clocks, stopclocks, etc.	halve the smallest scale division or reaction time (typically ~0.5 s), whichever is the larger
counting	Geiger counter	if the number of counts is N, then the uncertainty is always $\sqrt{N}$

> ### Charting counts
>
> If you make a chart of the number of counts from a Geiger counter measuring identical samples, they will not follow a normal distribution. The shape produced is called a **Poisson distribution**, and it can be shown that the uncertainty is $\sqrt{N}$.

Precision instruments

Several instruments have features that enable them to make measurements to a high degree of precision. The most common feature is the **Vernier scale**.

The diagram on the right shows a Vernier scale on a travelling microscope. There are two scales. One is fixed, while the other moves with the instrument.

- The position of the arrow against scale A (that is, between 4.9 and 5.0) gives the first part of the measurement.

- The scale line on B that exactly lines up with one of the lines on A (it does not matter which one) gives the next decimal place. In this case it is the second line on scale B, so the measurement is 4.92.

The intervals on scale A are 1 mm, and those on scale B are 0.9 mm. The arrow points to a position 0.2 mm from 4.9, and each line on scale B will be 0.1 mm nearer to the scale A line. By the second line they are opposite each other.

There are a variety of instruments that provide precise measurements in different circumstances. These are shown in the table below.

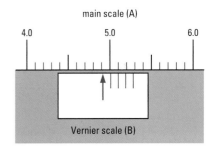

A Vernier scale is used to make high-precision measurements.

Instruments for precise measurements.

Instrument	Typical use
Vernier callipers	lengths and diameters
micrometer	small diameters and thicknesses
travelling microscope	distances, depths, etc.
spectrometer	wavelengths of light

PRACTICE

1 How would you quote the following results from a repeated experiment?

2.00 cm, 1.98 cm, 1.99 cm, 2.02 cm, 1.97 cm, 2.01 cm.

2 The length of a block is found by placing it against a 30 cm ruler (calibrated in mm). One end is judged to be next to 6.2 cm on the scale. The other end is next to 1.5 cm on the scale. What is the length of the block? What is the uncertainty in the length?

3 A production line manufactures blocks of metal of mass 4.0 kg. As the blocks leave the machinery, their mass is measured. In a production run the following masses were measured:

3.8 kg, 4.3 kg, 4.3 kg, 5.0 kg, 4.7 kg, 4.0 kg, 4.3 kg, 4.7 kg, 4.7 kg, 5.2 kg, 4.3 kg, 4.0 kg, 4.7 kg, 5.0 kg, 4.7 kg.

Make a chart of the results. What is the uncertainty in the manufacturing process?
Does the machinery need re-calibrating?

OBJECTIVES

- accuracy and precision
- systematic and random uncertainties
- combining uncertainties

WORKING WITH UNCERTAINTIES

I read my horoscope in the paper yesterday. It said that I would meet a tall stranger. Later in the day I went to see a fortune teller. She said that I would meet a tall, dark-haired stranger with a mole on his left arm, and that he would tell me something important.

The purpose of this story is not to suggest that physics is like telling fortunes, but to illustrate the difference between **accuracy** and **precision**.

- Precision is the degree of uncertainty in a measurement.
- Accuracy is how close a measurement is to the true value.

Both predictions turned out to be quite *accurate*. (Steve Adams is quite tall and he is dark haired, but he is not strange.) They contained some *correct information*, but the fortune teller's statement was more *precise* than the horoscope. It contained *more information*.

In scientific terms, accuracy and precision refer to the uncertainties that exist in the data. These uncertainties can be either systematic or random.

Random uncertainties

Random uncertainties show no pattern from one measurement to another. Charting a set of results produces a cluster round an average value, and in most cases the chart will be a **normal distribution**. The width of the distribution is a measure of the random uncertainty in the data.

For a normal distribution the width is one **standard deviation** and covers about two-thirds of the data. It is not possible to define a width that would cover all the data, because there is no limit to how far away from the average a measurement can be.

Random uncertainties come about for many different reasons. Different people may read the dials slightly differently, and the measuring equipment may change slightly from one measurement to the next. Also, different experiments may use slightly different equipment.

A precise measurement has a small random uncertainty. For example, 5.2 ± 1.5 is not very precise, whereas 5.21 ± 0.01 is very precise.

In the first result there is no point in quoting more than two significant figures, because the uncertainty is so large. The second result uses more significant figures because the uncertainty is smaller. Random uncertainties limit the precision of a measurement. One or two readings a long way from the average can have a big influence on the measurement if only a few readings are taken. However, if there are hundreds of readings to average, the few that are a long way off will have little effect. By taking more readings it is always possible to reduce the size of a random uncertainty.

Systematic uncertainties

A **systematic uncertainty** shifts *all* the measurements away from their true value by the same amount, and it can significantly affect the accuracy of a measurement. This can happen if a measuring device goes out of alignment or if it is not calibrated properly. If a metre ruler was actually 99.8 cm long, it would introduce a systematic uncertainty.

Closing the jaws of a micrometer should make the dial read zero. If it does not this will produce a **zero error** and cause a systematic uncertainty. Always check the zero error of such devices before taking readings. Sometimes it is possible to adjust the device to remove the error, but if this is not possible you must *subtract the zero error from every measurement*.

Systematic uncertainties are hard to avoid. Taking more readings will not affect the systematic uncertainty, because it is present in all of the readings. The best way of seeing if there is a systematic uncertainty in an experiment is to use the equipment to measure a known quantity first. This is a very important part of setting up an experiment and is known as **calibrating** the equipment.

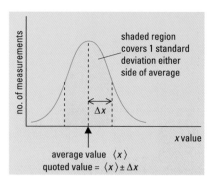

Random uncertainty in measurements.

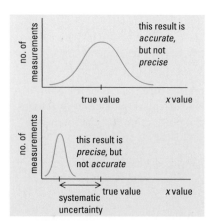

The difference between accuracy and precision.

Percentage uncertainty

Sometimes it is useful to know how large an uncertainty is as a percentage of the measurement's value:

$$\% \text{ uncertainty} = \frac{\text{size of uncertainty}}{\text{size of measurement}} \times 100\%$$

Worked example 1

The Hubble constant, H_0, measures the rate at which the universe is expanding. If $H_0 = 23 \pm 7 \text{ km s}^{-1}$ million light-years^{-1}, what is the percentage uncertainty in H_0?

$$\% \text{ uncertainty} = \frac{7}{23} \times 100\% = 30\%$$

Combining uncertainties

Scientists must often combine measurements to obtain a final result. The uncertainty of this result will depend on the uncertainties of the individual measurements. In the simplest cases results are added or subtracted. In either case the uncertainties must be added together.

Worked example 2

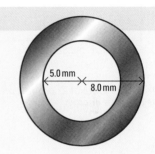

The inner radius of a washer is 5.0 ± 0.2 mm, and the outer radius is 8.0 ± 0.2 mm. What is the width of the washer?

width = 8.0 – 5.0 mm = 3.0 mm

However, it could be as big as $(8 + 0.2) - (5 - 0.2)$ mm = 3.4 mm or as small as $(8 - 0.2) - (5 + 0.2)$ mm = 2.6 mm.

The range of possible values is 2.6 mm – 3.4 mm, that is, 3.0 ± 0.4 mm.

(NB This is the uncertainty one would get by adding the separate uncertainties.)

In more complicated examples, measurements may have to be multiplied, divided, or operated on by some function, for example sin or log. It is possible to derive these rules for calculating uncertainties using calculus, but in most cases simple rules of thumb provide answers that are good enough.

Combining uncertainties.

Situation	Uncertainty
adding/subtracting measurements	add uncertainties
multiplying/dividing measurements	add % uncertainty
functions of measurements (log, sin, etc.)	same % uncertainty

PRACTICE

1 A set of students independently measure the period of a pendulum and obtain the following results:

1.00 ± 0.01 s, 1.03 ± 0.02 s, 1.01 ± 0.01 s, 0.99 ± 0.02 s, 1.02 ± 0.03 s.

What would be quoted as the period and the uncertainty?

2 A cylindrical block of aluminium has a radius of 3.0 ± 0.1 cm, a height of 6.0 ± 0.2 cm, and a mass of 460 ± 5 g. Calculate the density of aluminium and the uncertainty in the density.

3 An angle is measured as $30.0 \pm 0.1°$. What is the sine of the angle and the uncertainty in the sine?

4 The equation for the period of a pendulum is

$$T = 2\pi \sqrt{\frac{l}{g}}$$

where l is the length of the string and g is the acceleration due to gravity.

If $l = 54.1 \pm 0.05$ cm and $g = 9.8 \pm 0.1 \text{ m s}^{-2}$, what is the period and the uncertainty in the period?

O B J E C T I V E S

- good practice

- use of graphs

- testing formulae

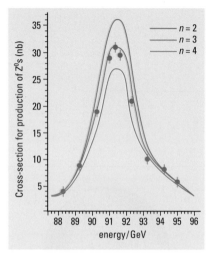

Graph showing rate of Z^0 production versus collision energy.

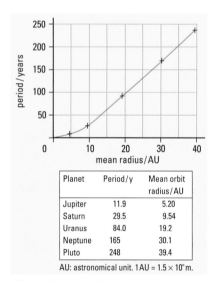

Planet	Period/y	Mean orbit radius/AU
Jupiter	11.9	5.20
Saturn	29.5	9.54
Uranus	84.0	19.2
Neptune	165	30.1
Pluto	248	39.4

AU: astronomical unit. 1 AU = 1.5×10^{11} m.

Graph showing orbital period versus mean orbital radius for the outer planets.

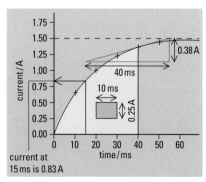

current at 15 ms is 0.83 A

Graph of current versus time for an electromagnet. At 40 ms, rate of current increase = 0.38 A/40 ms = 0.95 A s^{-1}. The area under the graph line represents the amount of charge passed through the coil. The dark blue rectangle represents 0.25 A × 10^{-2} s = 0.0025 C.

GRAPHICAL ANALYSIS

Deciding between competing theories

Until a few years ago there was a very important unanswered question in particle physics: how many generations of fundamental particles are there? Particle physicists had discovered that the electrons and quarks that make up all the matter around us have similar but more massive relations that are produced in high-energy accelerators or cosmic rays.

In fact, there was evidence for three generations of electron-like particles (**leptons**) mirrored by three generations of quarks. Could there be a fourth or higher generation? Do the laws of physics allow for an infinite hierarchy of particles at ever higher energy?

Theorists predicted that the rates of production and decay of a particular type of particle – the Z^0 – would depend on the number of generations, so an experiment was set up to test their predictions. The results of this experiment (in red) are shown in the first graph on the left alongside theoretical predictions of the rates for 2, 3, or 4 generations.

At the low- or high-energy ends of the graph, the experimental results cannot distinguish between predictions. However, close to the peak the experimental data is only consistent with $n = 3$; the theoretical lines for $n = 2$ or $n = 4$ are well beyond the limit of the error bars and so can be ruled out. These results agree with astronomical evidence, and now it is accepted that there are just three generations of fundamental particles. (Note that without error bars, the data would suggest $n = 3$ but would be less certain.)

Plotting a graph

The second graph on the left illustrates some important points to bear in mind when plotting a graph.

- Period is plotted on the y axis to show how it *depends on* radius. In general, the **dependent variable** goes on the y axis and the **independent variable** on the x axis.

- The graph has a title, here set as a caption.

- The scale makes good use of the space available on the graph paper.

- Axes are labelled with quantity and unit: *period/years* means the numbers represent time period divided by years, so that 50 years becomes simply 50 on the axis.

- Data points are distinct vertical–horizontal crosses centred on the value. An alternative is a dot at the value with a small surrounding circle. In either case the points remain visible when a line passes through them.

- The points represent a non-linear relationship. A *smooth* curve is drawn to show this.

- Error bars are not shown because they would be too small to be significant.

Values from graphs

The third graph on the left shows how current changes with time when an electromagnet is switched on. Graphs contain a great deal of information.

- Values of current at any time (not just measured times) can be read off.

- The gradient at any time is equal to the rate at which current is changing at that time.

- The area up to a certain time is the charge that has passed through the coil. (Area is related to the product of quantities on the axes; here this is current multiplied by time.)

- The shape of the graph suggests that the current will reach a limiting value of about 1.5 A. Projecting beyond the limit of measured data is called **extrapolation**.

Graphs and formulae

A smooth graph usually means there is a nice mathematical relation between the things that are plotted. The most useful type is a straight-line graph. If this passes through the origin it means that the variables are **directly proportional**, and the gradient gives the constant of proportionality. See, for example, the graph on the right. This is a straight line through the origin, so F is proportional to e. We can then say that $F = ke$, where k is a constant equal to the gradient:

$$k = \text{gradient} = \frac{2.0\,\text{N}}{3.6\,\text{cm}} = 0.56\,\text{N cm}^{-1}$$

If the line does not pass through the origin, and the relation between variables is *linear*, the gradient and intercepts (where the line hits the axes) both give information.

Straight-line graphs are used to test for all kinds of mathematical relationships, as the following table shows.

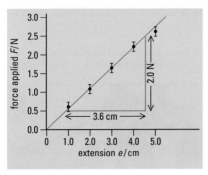

Graph of applied force versus extension for a spring obeying Hooke's law.

Using straight-line graphs.

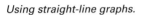

To test (relationship)	Plot	Successful result	Equation	Comments
direct proportion ($y \propto x$)	y vs x	straight line through origin	$y = mx$	m is gradient
linear	y vs x	straight line	$y = mx + c$	m is gradient, c is y intercept
inverse proportion $\left(y \propto \dfrac{1}{x}\right)$	y vs $\dfrac{1}{x}$	straight line through origin	$y = \dfrac{m}{x}$	m is gradient
power law ($y \propto x^n$)	log y vs log x	straight line	$y = kx^n$	n is gradient, log k is log y intercept
exponential ($y \propto e^{kx}$)	ln y vs x	straight line	$y = ce^{kx}$	k is gradient, ln c is ln y intercept

Uncertainties and graphs

One way in which a systematic uncertainty may show up is to prevent a best-fit line passing through the origin when the theory suggests it should. This will not harm the value of the gradient, provided the best-fit line is not forced through the origin.

Uncertainties in gradients and intercepts can be calculated automatically by computers or graphical calculators, but they can also be estimated 'by hand'. You have to draw two lines through the data – the steepest and shallowest that could both be accepted as lines of best fit. The two lines will give two values each for the gradient, G_1 and G_2, and the intercept, I_1 and I_2. The average gradient or intercept is the value to quote, and half the difference between the extreme values is the associated uncertainty. For the gradient,

$$G = \frac{G_1 + G_2}{2} \quad \text{and} \quad \Delta G = \frac{G_1 - G_2}{2}$$

so the quoted result is $G \pm \Delta G$.

Helpful hint

Error bars (see the graph above) help to show where the steepest and shallowest lines that could both be accepted as lines of best fit should be drawn. Both lines should be consistent with the error bars.

PRACTICE

1 Explain why a graph plotted from two points is pretty useless.

2 What conclusion could be drawn if all three lines in the first graph on this spread had been covered by the error bars at every point?

3 What (roughly) is the initial rate of increase of current in the third graph on this spread?

4 Boyle's law says that the pressure of an ideal gas is inversely proportional to its volume as long as the temperature is constant. Suggest one graphical and one non-graphical way of testing the law using measured values of pressure and volume.

5 Kepler's third law of planetary motion says that r^3/T^2 = constant, for all satellites orbiting a common central body (r is the mean radius and T is the period). Test this using the data for the second graph on this spread:

a by calculating the ratio r^3/T^2 for each planet;

b by plotting r^3 against T^2;

c by plotting log r against log T.
   In **b** and **c** explain how you use the graphs to confirm the relation.

d Which of **a**, **b**, and **c** would be most use in finding Kepler's law, if you didn't already know it?

Sir Karl Popper (1902–1994). Popper was probably the most influential and accessible philosopher of science in the twentieth century. He saw the essence of science in making bold conjectures and testing them by experiment. If a theory is not open to testing, it is not science. There must be experiments that could in principle refute the theory. Scientific theories can never be proven but can be falsified, so science progresses by making bolder and bolder conjectures and attempting to refute them.

Why experiment?
• To investigate physical phenomena and test predictions.
• To make an accurate measurement of some property or physical constant.
• To reproduce results and relationships that are met in theory.
• To practise and become familiar with techniques and apparatus.

The written report
The point of the written report is to give a clear, accurate, and concise account of your investigation.
• If your handwriting is dreadful, use a computer.
• Structure your work using subheadings.
• Include well-labelled diagrams.
• If you quote from other students or textbooks, give references and credit.

INVESTIGATIONS

The essential characteristic of science is that what it says about the world can be tested. We are free to concoct all kinds of theories about how the world works and what might happen in the future, but the ultimate test of any theory is whether its predictions correspond to actual events. Precise measurements, accurate observations, and intelligent analysis enable us to discriminate between alternative theories. Experiments, therefore, are essential to progress in science.

Sometimes experiments produce unexpected results that lead to new theories – radioactivity and X-rays were discovered in this way – but you are only likely to notice something unexpected if you know what you are expecting. Whenever you carry out an experiment, keep asking yourself whether what you are observing makes sense and agrees with reasonable expectations based on good physics.

A plan of action

There are many equally good ways to organize and carry out an investigation, but they should all include the steps below.

Choosing a project In many courses you have a free choice of experiments. Choose something that will produce plenty of **quantitative** data. (There is nothing wrong with **qualitative** observations, but they are difficult to analyse, and your conclusions are likely to end up rather weak and woolly.) Good data allows you to show what you can do graphically and mathematically and can be analysed for errors and accuracy.

Before you start Do some preliminary research. Read up on the physics involved in your project, even if the topics are not in your particular course! Don't be afraid to ask your teacher for suggestions. The earlier you begin to plan your investigation the better. Some major investigations may need special pieces of apparatus to be ordered or constructed. It is essential that everything is ready and working when you start – if it is not you will soon lose a great deal of valuable time.

Planning and predicting You cannot start an experiment unless you have a pretty good idea of what you are trying to do and what you think will happen. This means that you need some kind of *physical model* in your head. Once you start the experiments, it is very likely that some unexpected things will happen. These may deflect you along a new path leading to new ideas and experiments, but all the time you should be making predictions (forming hypotheses) and planning how to test them. In many cases, it will be possible to predict simple relationships between variables directly from a physical model. In more complex situations, the prediction may be little more than an *informed* guess.

Controlling variables Variables are things that can change in an experiment. Many experiments are designed to see how changing one variable (the *independent* variable) affects another (the *dependent* variable). For example, changing the mass suspended from a parachute (independent variable) will affect its terminal velocity (dependent variable). In most experiments there are many possible variables (the parachute area and shape are two more), and so it is important to ensure that everything other than the pair of variables being tested is held constant. This is sometimes difficult to do. For example, if the number of coils on an electromagnet is increased by winding them in series, the length of the coil is also increased, so this is not a valid way to test the dependence of field strength on number of turns. (One way around this is to use a different variable such as turns per unit length.)

Risk assessment You must try to identify any hazards in your proposed experiments. This should also be discussed with your teacher. If there are significant hazards, appropriate precautions should be taken, or an alternative strategy or experiment must be designed. There are strict guidelines concerning the use of radioactive sources, **evacuated** containers (i.e. those containing a vacuum), and mains electricity. It is easy, however, to overlook hazards from more innocent things such as masses on springs that can drop on your foot, glassware that can break and cut, hot water that scalds, sharp knives, bright lights, stroboscopes, and loud sounds. Always err on the side of caution!

Measurements Having decided what you are going to measure, you must now decide **1** what instruments are appropriate, **2** what range of values to use, and **3** how many readings to take.

1 An appropriate instrument is one that measures the variable to acceptable precision and over a wide enough range. For example, an ammeter with full-scale deflection of 1.0 A would be inappropriate for measuring currents that never exceed 50 mA. A metre ruler may be fine for measuring the length of a steel wire, but it would be inappropriate for its diameter, and a micrometer screw gauge should be used. You should also consider whether to use automatic data recording equipment. This makes it possible to take large numbers of readings over a long period of time and then to process the data using a computer.

2 The range of values will depend on the experiment, but remember that all non-linear graphs will look linear if you only look at a small part of them! It is usually a good idea to test as wide a range of values as possible. Extreme values often also give additional information.

3 Aim for an absolute minimum of five well-spaced readings across the entire range of values. Usually, the more readings you take, the better. Use your judgement as you carry out the experiments. If you are measuring temperature as something cools (for example, investigating rates of heat loss through various layers of clothing), there is no point in taking readings every ten seconds if the temperature is only dropping one degree per minute.

In most cases you should repeat measurements and calculate an average. This also gives a check on **gross errors** (for example, miscounting oscillations when measuring time periods) and allows you to spot anomalous or unusual results that may need checking or even discarding.

Recording data All useful data gathered during an investigation should be recorded and included in the write-up, even if raw data is simply tabulated in an appendix. This enables calculated values to be checked later and is very useful if the experiment seems to be saying something that disagrees with expectations. It is therefore very important that all tables of data are clearly labelled, not just with the quantities that are measured but also with an indication of the experiment to which they belong, and that any particular features or values that might be useful are also recorded. The table shown on the right is not the only way to construct a table, but it serves as a model of good practice.

Analysing data The main things that you will do with data are:

1 calculate rates of change;

2 construct equations to represent these relationships;

3 calculate values based on the data; for example, a value for the wavelength of a spectral line;

4 analyse the errors in the data to determine their bearing on your conclusions.

Conclusions Your conclusions should give a concise summary of what you achieved. They should refer back to any predictions you made at the start and say to what extent these have been borne out. If you discovered mathematical relationships, they should be stated here. So should calculated values, along with estimated final errors or at least a comment on how strongly the data that you collected supports these values. It is also important to interpret your conclusions in terms of physical theories or models.

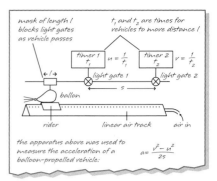

A clearly labelled diagram is a good way to convey complex information. It saves you a lot of writing and gives the reader a clear idea of what your experiment involved.

Taking readings automatically

Consider the use of computer interfaces, chart recorders, or data loggers to take readings automatically, but always think about how the apparatus you use is likely to affect the quantity you are trying to measure. For example, ammeters have resistance and so reduce current.

---

heading essential
TABLE 1: period of oscillation for masses suspended from a steel spring

specific information relevant to this experiment
Data: measured spring constant = 20.0 N m⁻¹ (See expt 3(a))

extra information

Mass	TRIAL 1	TRIAL 2	TRIAL 3	AVERAGE	physical quantities
kg	$\frac{10T}{s}$	$\frac{10T}{s}$	$\frac{10T}{s}$	$\frac{T}{s}$	unit
0.10	4.54	4.48	4.50	0.451	all data values
0.20	6.30	6.32	6.32	0.631	recorded as
0.30	7.71	7.68	7.73	0.771	pure numbers

no. of significant figures consistent with measurements
NB Use of averages

Tables should be as clear and informative as possible.

Investigation plan or checklist

Heading: for example, 'An investigation into crater formation'.

Aim: for example, 'To investigate crater formation by dropping marbles into a sand tray'.

Predictions: ideally these should be quantitative, based on good physics.

Proposed/actual experiments: include informative diagrams and/or photographs.

Results: organize numerical data into tables; include qualitative observations.

Analysis: use graphs and calculations.

Conclusions: try to make these as clear-cut as possible, and include errors.

Evaluation: was it a fair test; how confident are you about the conclusions?

Appendix: include references, extra raw data, and copies of relevant articles.

INTERFACING AND DATALOGGING

Technology in experimental work

The ongoing revolution in information technology has made it possible to use electronic sensors to measure variables (such as temperature, pressure, or position), convert the measurements to digitally encoded voltages, and store them in a computer. Software packages can then be used to:

• carry out calculations on the data;

• plot graphs;

• search for mathematical relationships between variables.

This takes much of the drudgery out of experimental work, but it does more than that – readings can be taken automatically over extremely short periods (fast timing) or at long intervals over an extended period of time (you could set up the apparatus on a Friday evening and collect your results on Monday morning). Also, the speed and power of the processing means that virtually no time is spent plotting graphs or recording data, so much more time is available to extend the investigation or consider alternative interpretations.

The description above has assumed you are using an interface to convert the outputs of your sensors into digital information for a computer. However, you do not even need a computer. Many datalogging packages will run on batteries and contain their own processor, so they can be used away from the laboratory. The datalogger itself will store, analyse, and display the data (which can also be downloaded to a computer if required).

Some systems are based on graphical calculators – the sensors plug straight into the calculator, and it stores, displays, and analyses the data.

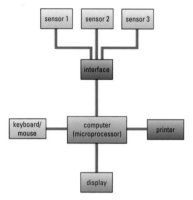

Combinations of equipment like this can improve your experimental work and make it easier.

Essential components

The following items of equipment are available.

• **Sensors** A wide range of sensors is available, including: voltage, current, temperature, position, motion, rate (for radioactivity experiments), magnetic field strength, infrared intensity, light intensity, ultraviolet intensity, pressure, and sound level. In each case, a change in the variable being measured creates a change in voltage, which is transferred to the output. Sensors may have alternative ranges and must be calibrated so that the datalogger or interface to which they are connected stores the appropriate data. In some cases the datalogger or interface automatically detects the type and range of the sensor. In others you would have to key this in or select it on screen by clicking and pointing with a mouse.

• **Interface** This accepts the analogue voltage output from the sensor and converts it to a digitally encoded signal which 'makes sense' to the computer. To do this it must contain an analogue-to-digital converter and an amplifier. Most interfaces have several channels, so that data from more than one sensor can be recorded simultaneously. Each sensor feeds into a separate channel.

• **Datalogger** This is an interface and processor combined. It allows more flexibility, because you are not tied to a particular computer. The processing power may be rather more limited, however, so it is sometimes worthwhile to use a datalogger to record and store the data, and then transfer it to a computer on which more powerful software is available to analyse it.

• **Computer** This is used to store data and analyse it. To do this you need a software package that can accept and manipulate the input data. You may also wish to transfer the data to a spreadsheet program such as Microsoft® Excel. Most manufacturers offer their own packages, which are compatible with their sensors and interface.

Example 1: Investigating how cells run down

This is a simple investigation in which the terminal voltage of four cells is monitored continuously over a period of three days, as they drain through resistors of different value.

You would have to:

- set the range and type of sensor;
- select the timing interval at which measurements should be taken (e.g. 30 minutes).

Once the data is collected, it is displayed in a table, and new data can be generated from it. In this experiment, voltage V is recorded against time for each of the four cells. Current can be calculated (from $I = V/R$) simply by dividing each voltage column by the resistance used to generate it. These new values can also be plotted by the computer, as can functions (e.g. logarithms) of the original data.

Example 2: Electromotive force (e.m.f.) and internal resistance

So far it has been assumed that the variable concerned will be measured at different times, and eventually something will be plotted against time. However, you may want to investigate the variation of current with voltage across a component, or the variation of magnetic field strength with current through a solenoid, or temperature with position along a rod. All of these are possible using interfaces and dataloggers, and in many cases the experiments can be carried out incredibly quickly.

Here is an example of a very standard experiment to determine the e.m.f. and internal resistance of a cell.

If the e.m.f. of the cell is E and the internal resistance r, then

$E = V + Ir$

$V = E - Ir$

So if V is plotted against I, the gradient is $-r$ and the intercept on the V axis is E.

The usual way to carry out this experiment is to vary the resistance R and take measurements of V and I, record these in a table, and then plot the graph of V versus I. Once it is interfaced, the experiment can be carried out in a couple of seconds. All you need to do is sweep the variable resistor from one extreme to the other, and readings will be taken automatically as V changes. All the data for the entire experiment is recorded in a single pass along the resistor!

What is more, the results can be displayed as you carry out the experiment, so you will see a line 'drawn' on the graph as you adjust the resistor. A couple of seconds later the graph is plotted, the gradient and intercept are calculated and displayed, and you can print the whole thing out to put in your notes.

Example 3: Investigating the variation of magnetic field strength along the axis of a solenoid

This experiment can be carried out using a **Hall probe**. The magnetic field sensor is based on a Hall probe and can be placed at various positions along the axis of the solenoid.

The position can be obtained in one of two ways:

- It can be measured using a ruler. The value can then be keyed into the computer to store alongside the value for magnetic field strength.
- Alternatively, a distance sensor can be used to input to the second channel. This type of sensor works by timing the return of reflected ultrasonic pulses, so some kind of reflector must be fixed to the Hall probe.

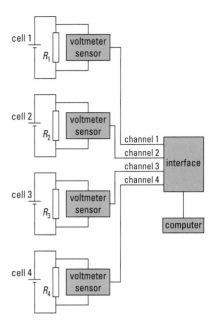

Investigating cells.

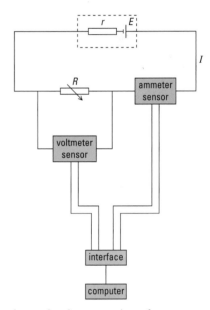

Internal resistance and e.m.f.

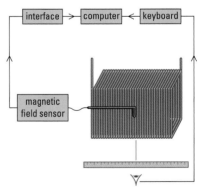

Investigating variations of magnetic field strength on the axis of a solenoid. The method shown combines human and electronic measurements.

3

MECHANICS

At one time, physicists believed that the concept of a mechanical universe held the key to explaining everything. In this amazingly simple model, everything is made of particles. Each particle has mass, position, and motion and responds to applied forces. The central ideas were developed by Galileo Galilei (1564–1642) and Isaac Newton (1642–1727) over 300 years ago and have been remarkably successful in explaining the properties of such diverse phenomena as gases and galaxies. Indeed, despite the discoveries and theories of modern physics, space scientists and engineers still turn to the classical laws of mechanics and gravitation when designing buildings, spacecraft, or amusement park rides. Even in the quantum world, mechanics provides a foundation for a deeper mathematical understanding of the universe around us.

In this chapter you will learn how to analyse linear, circular, and oscillatory motion and predict the effects of resultant forces. Along the way, many key ideas and models are described, including vectors, linear momentum, and work and energy.

A good old-fashioned fairground ride. How fast must it be rotating for the chairs to swing out as high as they have done?

Rest or motion?

GATSO speed cameras measure the instantaneous speed of a vehicle.

Maths box: rates of change

Δ means 'change in ...'.

δ means 'small change in ...'.

$\dfrac{\Delta s}{\Delta t}$ is the average speed or velocity during a significant time interval.

$\dfrac{\delta s}{\delta t}$ is the average speed or velocity during a very small time interval.

$\dfrac{\mathrm{d}s}{\mathrm{d}t}$ is the limiting value this ratio approaches as δt gets closer to zero.

This is written more formally as

$\dfrac{\mathrm{d}s}{\mathrm{d}t} = \lim_{\delta t \to 0} \dfrac{\delta s}{\delta t}$

and represents *instantaneous* speed or velocity.

Geometrically, this is the *gradient* of the graph of displacement versus time.

Instantaneous acceleration is the rate of change of velocity: $a = \dfrac{\mathrm{d}v}{\mathrm{d}t}$

CAPTURING AND DISPLAYING MOTION

The fleeting instant

About two thousand years ago, the Greek philosopher Zeno argued that motion is impossible because a moving object, for example an arrow, is at rest at each instant of its motion. Film and video do capture motion in still images, but they reproduce it by projecting them one after another. The essential quality of motion is that something is in a different place at a different time, and Zeno's argument fails because it removes time altogether.

Nowadays we have the technology to capture sequences of fleeting instants and analyse the motion they reveal. High-speed flash photography can 'freeze' a moving object, and multi-flash or **stroboscopic** photography can show us multiple images as the object passes. Other devices, such as ultrasonic or infrared rangefinders, can be interfaced with a computer to measure an object's position at different times, and so calculate its speed and acceleration.

Definitions: speed, displacement, velocity, and acceleration

Displacement is distance moved in a particular direction, for example, 120 m east.

Speed is the rate at which distance is covered, for example, 150 m s⁻¹.

$$\text{average speed (m s}^{-1}) = \frac{\text{total distance covered (m)}}{\text{total time taken (s)}} = \frac{s}{t} \ \text{ or } \ \frac{\Delta s}{\Delta t}$$

Velocity is speed in a specified direction and is defined as the rate of change of displacement. It is also measured in m s⁻¹; for example, 150 m s⁻¹ vertically upwards.

Acceleration is the rate of change of speed or velocity.

$$\text{average acceleration (m s}^{-2}) = \frac{\text{change in velocity (m s}^{-1})}{\text{time taken (s)}} = \frac{\Delta v}{\Delta t}$$

A sprinter who reaches 10 m s⁻¹ from rest in 2 s has an average acceleration of 5 m s⁻².

Average and instant values

There are two main types of speed camera in use on roads in the UK.

A GATSO speed camera measures the instantaneous speed of the car as it passes a single camera. It does this by capturing two images a short time apart (typically $\delta t = 0.50\,\text{s}$). By comparing the position of the car against road markings the distance moved in this time (δs) can be determined. The ratio of these values gives the car's speed:

$$v = \frac{\delta s}{\delta t}$$

Whilst we have called this the car's 'instantaneous' speed it is in fact still an average speed over a short time interval. To get a truly instantaneous speed we would have to make that time interval vanishingly small. The instantaneous speed is the value of v in this limit and is written as:

$$v = \frac{\mathrm{d}s}{\mathrm{d}t}$$

where d/dt (pronounced 'dee-by-dee-tee') can be read as the 'rate of change of....'

This will be familiar to you if you have studied calculus. Speed can be calculated by differentiating the distance travelled with respect to time. It is equal to the gradient of a distance–time graph.

Average speed cameras use two cameras placed some distance Δs apart (e.g. at the start and end of a section of road works). Each camera captures an image of the car along with a time stamp and its registration number. The time Δt between the two images is then used to calculate the average speed:

$$v = \frac{\Delta s}{\Delta t}$$

Graphs of motion

Motion data is usually displayed graphically. The most useful graph is one of speed or velocity against time. This can also be used to work out displacement and acceleration.

- Displacement is equal to the area under a velocity–time graph.
- Acceleration is equal to the gradient of a velocity–time graph.

Example 1. The three graphs on the right all show the motion of the same object – a car accelerating uniformly from rest and then travelling at a constant velocity. In reality the acceleration is unlikely to be uniform; at higher speeds there is more drag and the acceleration is smaller. More realistic velocity–time and acceleration–time graphs showing non-uniform acceleration of the car are shown below.

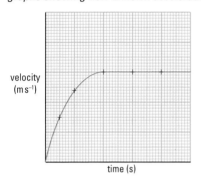

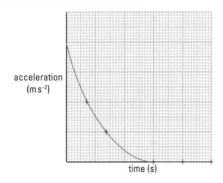

Example 2. The two graphs below show displacement–time and velocity–time graphs for a ball thrown vertically upwards from the Earth's surface. The timing continues until it returns to the same height from which it was thrown. Throughout the analysis the upward direction is taken as positive.

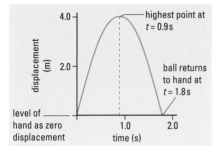

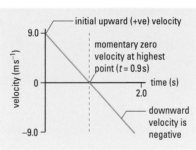

(a)

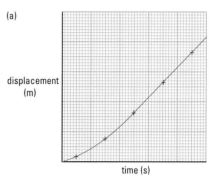

(b)

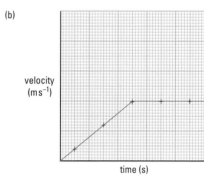

(c)

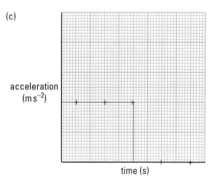

PRACTICE

1 Marita Koch's 1985 world record for the women's 400 m is 47.6 s. What was her average speed and velocity as she set this record?

2 A cyclist travels 1 km at a steady $14\,\mathrm{m\,s^{-1}}$, turns straight around and cycles back at a steady $8\,\mathrm{m\,s^{-1}}$. What was his average speed?

3 Sketch graphs of displacement, velocity, and acceleration versus time for a high-diver who dives vertically down into a deep pool. Take downwards as positive and include the effects of friction. Stop when the diver surfaces and assume all motion is vertical.

4 What is the acceleration of the ball in example 2 above at the top of its motion? Sketch a graph of acceleration versus time for the ball.

5 The Earth's radius is 6400 km. How fast is a point on the equator moving? What is the speed of Beirut at latitude 34°?

6 If a spacecraft in deep space had a constant acceleration of 1 g (about $10\,\mathrm{m\,s^{-2}}$), its passengers would experience comfortable artificial gravity throughout their journey. How long would it take this spacecraft to accelerate from rest to half the speed of light ($1.5 \times 10^8\,\mathrm{m\,s^{-1}}$)? (According to Einstein's theory of relativity, it is not possible for any material body to reach the speed of light.)

- derivation of equations for constant acceleration

- use of equations for constant acceleration

Definitions

s = displacement

u = initial velocity (m s⁻¹), at $t = 0$ s

v = final velocity (m s⁻¹)

a = acceleration (m s⁻²)

t = time (s)

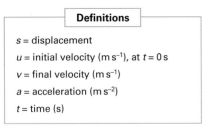

Walking in a straight line at 2 m s⁻¹.

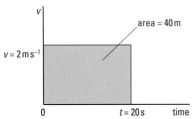

If δt is small, velocity v at time t is almost constant during δt.

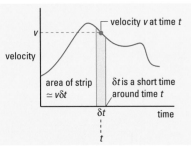

The gradient of this line gives the acceleration.

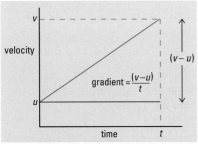

Graph 1.

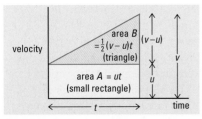

Graph 2.

EQUATIONS OF MOTION

Imagine that you are walking in a straight line, taking two steps every second and each step being 1 m long. Your velocity would be 2 m s⁻¹, and a graph of your velocity versus time would be a straight, horizontal line at this value. After 20 s your displacement is 40 m. This is exactly equal to the shaded area under the velocity–time graph between $t = 0$ s and $t = 20$ s, as shown in the diagram. This isn't really surprising. The area is equal to the product of velocity and time, or the y value times the x value. This is exactly how we would calculate displacement.

The area under any velocity–time graph is equal to displacement, even if the velocity is changing. To understand this, imagine dividing the graph into vertical strips which are so narrow that the velocity is effectively constant during each time interval. Each strip is now a tall thin rectangle whose area equals the displacement during that interval. The total area is the sum of all the strip areas and so must equal the sum of all the individual displacements. It is as if you walked with varying step-length; the total distance covered is still the sum of the lengths of each step taken.

Equations for constant acceleration

Velocity–time graphs are particularly useful because their gradient is the acceleration and their area is the displacement. Here they are used to derive equations for objects moving with constant acceleration.

Think of something accelerating past you, a racehorse or a sports car perhaps. It is already moving with velocity u when you first look (time zero) and it has changed its velocity to v when you stop looking (t seconds later). Its velocity–time graph has been drawn in the margin.

The gradient of this line is

$$a = \frac{(v - u)}{t}$$

giving

$$v = u + at \tag{1}$$

This is the first **equation of motion**.

The area under this graph is the displacement. There are three different ways to calculate the area and each one gives a different equation.

(i)   Using the trapezium rule, the area is equal to the average of the parallel sides times their separation:

$$s = \frac{(v + u)}{2} t \tag{2}$$

This equation has a simple interpretation. It says that total displacement equals average velocity times time.

(ii)   Adding the areas of the small rectangle and the triangle (graph 1):

$$s = ut + \frac{1}{2}(v - u)t$$

but $(v - u) = at$ from equation (1), so

$$s = ut + \frac{1}{2}at^2 \tag{3}$$

(iii)   Subtracting the area of the 'missing triangle' from the large rectangle (graph 2) gives, by similar algebra to (ii)

$$s = vt - \frac{1}{2}at^2 \tag{4}$$

There is a fifth equation. All of those derived so far include t. You can derive the final equation by using equation (1) to get an expression for t and then substituting it into any of the others (and then simplifying). This will give you

$$v^2 = u^2 + 2as \tag{5}$$

Why are there *five* equations of motion? Because there are *five* variables; each equation eliminates one of them. This means that if we know the values of any three of them, we can calculate the other two. (These equations are sometimes called the **suvat equations**.)

48

Worked example 1

In an experiment to measure the acceleration of free fall a ball bearing is released from an electromagnet and falls through a light gate placed 2.00 m below. An electronic timer starts when the ball bearing is released and is stopped when the ball bearing passes through the lower light gate. Three successive measurements are: 0.642 s; 0.636 s; 0.641 s.

What value does this data give for the acceleration of free fall?

s	u	v	a	t	
2.00 m	0.000 m s^{-1}	?	?	0.640 s	(average)

Use equation (3):

$$s = ut + \frac{1}{2}at^2$$

But $u = 0$, so:

$$s = \frac{1}{2}at^2$$

Rearranging this gives:

$$a = \frac{2s}{t^2} = \frac{4.00}{0.4096} = 9.77 \text{ m s}^{-2} \text{ (3 sig. fig.)}$$

How to solve problems

Step 1: Write down, s, u, v, a, t.

Step 2: Identify those values you know (beware of starting or finishing at rest – this merely means $u = 0$ or $v = 0$), and convert to consistent units, usually SI units.

Step 3: Identify the equation containing three known quantities and the unknown quantity you wish to calculate.

Step 4: Rearrange the equation to make the unknown its subject.

Step 5: Now put in known values and solve the problem.

Worked example 2

A stone falls vertically for 2 s and embeds itself 2 cm in soft mud. What is its average deceleration in the mud?

This problem must be split into two parts to solve it.

(i) For the free fall:

s	u	v	a	t
?	0.00 m s^{-1}	?	9.8 m s^{-2}	2.0 s

Use equation (1):

$$v = u + at$$

With $u = 0$

$$v = at = 9.8 \text{ m s}^{-2} \times 2.0 \text{ s} = 19.6 \text{ m s}^{-1}$$

(ii) For the deceleration in mud (note that v for (i) becomes u for (ii)):

s	u	v	a	t
0.02 m	19.6 m s^{-1}	0.00 m s^{-1}	?	?

Use equation (5):

$$v^2 = u^2 + 2as$$

Rearrange to give

$$a = \frac{v^2 - u^2}{2s} = \frac{0 - 19.6^2 \, (\text{m s}^{-1})^2}{2 \times 0.02 \text{ m}} = -9600 \text{ m s}^{-2}$$

This is a negative downward acceleration, that is, a downward deceleration of 9600 m s^{-2}.

Sign convention

If all displacements, velocities, and accelerations are in the same direction, you can use positive values for all of them. However, if they are not, you need a **sign convention**. One direction is chosen as positive, and quantities in the opposite direction are negative. For example, if 'right is positive', then an initial velocity of 3 m s^{-1} to the left is given by $u = -3$ m s^{-1}. A deceleration in one direction is a negative acceleration in that direction, so $a = -4$ m s^{-2} to the right is a deceleration of 4 m s^{-2} to the right or an acceleration of 4 m s^{-2} to the left.

PRACTICE

1 A passenger jet carrying 150 passengers needs to reach a take-off speed of 80 m s^{-1} along a 1500 m runway. What is the minimum average acceleration necessary to achieve this?

2 On the Moon free-fall acceleration is about 1.7 m s^{-2}. How far does a stone fall from rest in 1 s?

3 A train moving at 20 m s^{-1} takes 10 s to accelerate to 25 m s^{-1}. Calculate its average acceleration and the distance it travels while it accelerates.

4 A free-fall parachutist is falling at 32 m s^{-1} when her parachute opens. The parachute then slows her descent to 5 m s^{-1} in a further 3 s. What was the magnitude and direction of her acceleration when the parachute opened?

5 A tennis racquet imparts an average acceleration of 3000 m s^{-2} to a tennis ball approaching at 25 m s^{-1}, and sends it back in the opposite direction at 20 m s^{-1}. How long was the racquet in contact with the ball? Sketch a graph to show how the velocity of the ball changed from just before contact to just after contact with the racquet strings. (Hint: be careful about signs and directions.)

6 Electrons in a cathode ray tube have an acceleration of 10^{17} m s^{-2}. Calculate their final velocity if they accelerate over a distance of 1.5 cm.

3.3

OBJECTIVES

- distinguishing scalars from vectors

- adding and resolving vectors

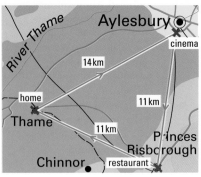

Vectors have both magnitude and direction.

Vector notation

To make it clear that we are dealing with a vector quantity we underline it in handwriting or set it in bold in print. For example, $\underline{F}$ or **F** is a vector force of magnitude F. Sometimes it is unnecessary to specify that something is a vector, for example when using $F = ma$ in a problem of straight-line motion.

Drawing resultant vectors

Draw all the vectors end to end. The resultant is then the vector that joins the start of the first vector to the end of the last vector. Vectors are **commutative** (the order in which they are drawn makes no difference).

If you are combining just two vectors, then the 'parallelogram rule' is useful. Here, the two vectors start at the same point and the resultant is the diagonal of the completed parallelogram.

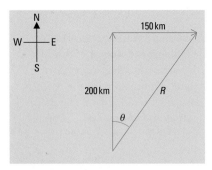

The resultant displacement has direction as well as magnitude.

VECTORS

Funny sums

I had a lucky week. I won £10 on the lottery and £5 on the football pools. My total winnings for the week are £15. This is an example of familiar arithmetic addition.

To celebrate my good fortune I went first to the cinema 14 km away (as the crow flies) and then to a restaurant, a straight 11 km drive from the cinema. In the restaurant it suddenly occurred to me that I had just 11 km to drive home even though I had driven a total of 25 km to get there. How could this be? Why don't displacements add like ordinary numbers?

The answer is obvious if you look at the map of my trip. Each displacement has direction as well as magnitude, and combining displacements of the same magnitude will have different results if their relative directions change. For example, if you take two 1 m steps one after the other, you could end up 2 m from your starting position or even return to it. The latter is an example where 1+1 certainly does not equal 2.

Scalars and vectors

Quantities which have magnitude only are called **scalars**. Quantities which have both magnitude and direction are called **vectors**. Scalars and vectors are both very important in physics. The following are examples of scalars and vectors.

- **Scalars:** mass, energy (including kinetic energy), distance, speed, temperature, potential.

- **Vectors:** displacement, velocity, acceleration, force, momentum, field strength.

Dealing with scalars is easy. They obey the rules of simple arithmetic. Vectors combine according to different, but related rules. The simplest way to tackle vector problems is to start with a vector diagram like the map used in the example above. The beauty of vectors is that they all behave in the same way, so anything we say about combining displacements will also hold for forces, momenta, or field strengths. For this reason displacements will be used to introduce vector techniques.

Combining vectors

The combined effect (or sum) of two or more vectors is called the **resultant** vector.

Worked example 1

What is the resultant displacement when a ship sails 200 km due north and then 150 km due east?

Using Pythagoras's theorem we have

$$R^2 = 200^2 + 150^2 = 62\,500$$

$$R = 250 \text{ km}$$

Since the resultant displacement is itself a vector, we must state its direction as well as its magnitude. This can be done by giving a bearing (angle to the east of north) in this case. From simple trigonometry

$$\tan \theta = \frac{150}{200} = 0.75, \qquad \theta = 37°$$

In this example (which is actually a 3:4:5 triangle), we used Pythagoras's theorem because the two vectors were perpendicular. In general, we might have to use the sine rule or cosine rule (see the Mathematics toolbox). If you find this difficult you can always use a scale drawing.

---

advanced **PHYSICS**

Worked example 2

What is the tension in the rope supporting the tightrope walker (of weight 750 N) in the photograph? The rope on either side of his foot makes an angle of 20° to the horizontal.

Focus on the point where the rope and foot make contact. At this point the resultant effect of the tension on either side of the foot is to produce a vertically upward force equal but opposite to the force exerted by the foot on the rope (750 N downwards).

Upward component of T is $T \sin 20°$ so the resultant upward force is

$$R = 2T \sin 20°$$

This must balance the weight, so

$$2T \sin 20° = 750\,\text{N}$$

and

$$T = \frac{750\,\text{N}}{2 \sin 20°} = 1100\,\text{N} \quad \text{(to 2 sig. figs)}$$

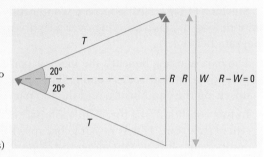

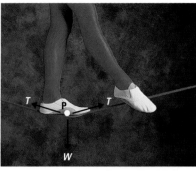

The forces acting on the rope at point P are in equilibrium.

Resolving vectors into components

In the example above, the two tension forces act partly vertically and partly horizontally. These 'parts' are called **components** of the vectors. In this case the two vertical components act upwards and so add together to support the tightrope walker. The horizontal components act in opposite directions and so cancel out.

Any vector quantity can be **resolved** (split) into components along any axes. This helps because components along any given axis add like ordinary numbers. To find a resultant of several vectors, it is simplest to resolve them all along a common set of axes, find the resultant components, and then reconstruct the resultant vector. A sensible choice of axes will usually simplify the problem. For example, it is useful to choose axes parallel and perpendicular to some of the vectors involved in the problem:

$$\cos \theta = \frac{v_x}{v}, \quad \text{so} \quad v_x = v \cos \theta$$

This is the x component of v.

$$\sin \theta = \frac{v_x}{v}, \quad \text{so} \quad v_y = v \sin \theta$$

This is the y component of v.

Reconstructing the vector v we have

$$v^2 = v_x^2 + v_y^2 \quad \text{and} \quad \tan \theta = \frac{v_y}{v_x}$$

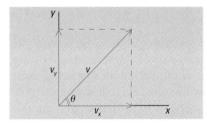

Resolving a vector.

How to combine vectors

Step 1: Choose convenient perpendicular axes.

Step 2: Resolve all vectors along each axis.

Step 3: Find the resultant components along each axis.

Step 4: Find the magnitude of the resultant vector using Pythagoras.

Step 5: Find the direction of the resultant using trigonometry.

Scale drawings

Vectors can also be added (combined) using scale drawings.

Step 1: Choose a suitable scale (e.g. 1 cm for 1 km, or 1 cm for 50 m s⁻¹).

Step 2: Draw all vectors end to end in their correct orientations.

Step 3: Construct and draw the resultant vector.

Step 4: Measure the length of the resultant and use the scale to work out its magnitude.

Step 5: Measure the direction of the resultant using a protractor.

A similar method can be used to resolve a vector into its components.

PRACTICE

1 A bird flies 5 km on a bearing of 35°. How far north and east of its starting point is its final position?

2 A stone is thrown horizontally with an initial velocity of 5 m s⁻¹. What is the magnitude and direction of its velocity 0.2 s later? Take the acceleration of free fall to be 9.8 m s⁻² and ignore friction.

3 A football is kicked at 15 m s⁻¹ at 20° to the ground. What are its initial horizontal and vertical components of velocity?

4 A river flows due east at 2 m s⁻¹. A fish has a velocity relative to the water of 4 m s⁻¹ and swims due north from one bank to the other. Assuming the fish swims in the same direction as its body points, calculate the angle of the fish's body to the bank.

5 One ship sails due south at 7 m s⁻¹ while another sails due east at 4 m s⁻¹. What is the velocity of the second ship relative to the first? (Hint: a vector diagram will help.)

6 In the UK the angle of dip of the Earth's magnetic field is about 20° to the vertical. What is the ratio of the vertical to horizontal components of the field here?

7 A man drags a heavy fishing net along the ground by pulling on a rope that makes an angle of 15° to the ground. If he pulls with a force of 400 N, what horizontal force is applied to the net? Does the vertical component of the applied force help in any way?

- resolving forces
- free-body diagrams
- equilibrium

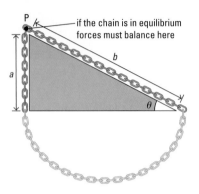

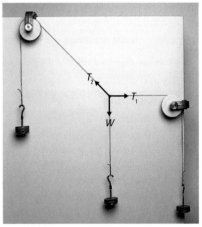

Stevinus used this idea to show that forces combine as vectors. All the forces acting on the chain must balance, otherwise it would begin to slide one way or the other.

This apparatus can be used to check that forces in equilibrium combine as vectors. In this case the horizontal component of T_2 balances T_1 and the vertical component of T_2 balances W.

Forces along a slope

For forces in the chain to balance at P the effect of the part lying on the slope must balance the weight of the vertical part. Since the lengths are in a ratio b/a the sloping part must exert a fraction a/b of its weight down the slope.

But

$$\frac{a}{b} = \sin \theta$$

So the component of its weight W parallel to the slope is

$$W \sin \theta$$

... exactly what we get by resolving vectors!

FORCE AS A VECTOR

Stevinus's wonderful proof

You may well ask why forces should combine in the same way as displacements. It is not difficult to test this in the laboratory, as shown in the photograph, but it would be more satisfying to prove it from some basic ideas.

Stevinus (1548–1620) offered a wonderfully simple proof. Imagine a smooth loop of chain hanging over a right-angled block, as shown in the diagram below left.

- The part hanging beneath the block is perfectly symmetrical, so it must pull down on both sides equally. This means it can be removed without upsetting the equilibrium and the top part will stay where it is.

- In this condition, the forces in each part of the chain must balance out where they join. This is possible because the inclined part exerts only a component of its weight along the slope. A simple calculation confirms that this component must be proportional to the sine of the angle of incline, and so is equal to the weight of the vertical section. This is exactly the result you obtain by treating force as a vector and resolving the weight of the long section of chain parallel to the slope.

Tension

A lot of people are confused by the idea of tension. Test yourself: write down what you think the tension is at point A in the string in the left-hand diagram below. Then read on.

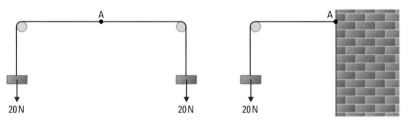

Calculating tension.

If you thought the tension was 40 N, answer these questions. The right-hand part of the system is replaced and the string is attached to a wall at A. What force does the string exert on the wall? What is the tension in the string now? Has attaching the string to the wall affected the tension at any point to the left of A?

The tension in the string – at all points in the string – is 20 N *in both examples*. To understand tension, it is useful to think of the force that one end of the string would exert if it were cut at the point concerned. It is wrong to add the tensions from both sides. These are produced by different parts of the string, and would cancel if applied to a single object.

Forces in equilibrium

When several forces act they may create a resultant force that accelerates the body or they may cancel out so that the body remains at rest or moves at a constant velocity. If they do cancel out there is no resultant force, and the body is **in equilibrium** (assuming the forces do not create any resultant turning effect).

The condition for forces to be in equilibrium is that their **vector sum** is equal to zero. This can be written

$$\Sigma \boldsymbol{F} = 0$$

where 'Σ' means the 'sum of ...'.

The idea of equilibrium is shown using a bungee jump. If friction is ignored, the only forces acting on the jumper are his weight and the tension in the rope. This means that he is in equilibrium when these two forces are equal in magnitude but opposite in direction, so that their vector sum is zero. This is the position the jumper ends in when he stops moving up and down. Above this point *weight* is greater than *tension*, and below it *tension* is greater than *weight*.

Being in equilibrium is not the same as being at rest. The forces on the bungee jumper are momentarily in equilibrium as he falls through the position in which he will finally come to rest. Conversely, the forces are *not* in equilibrium when he momentarily stops at the bottom, because the stretched rope applies a larger upward force than his weight, causing him to accelerate upwards.

At the lowest point of his motion, this bungee jumper is momentarily at rest but is not in equilibrium.

Using free-body diagrams

The way motion changes is controlled by the resultant force. It is a great help in analysing motion to draw a **free-body diagram**, which isolates all the forces acting on a particular object. These forces are then combined to calculate the resultant force. Free-body diagrams are shown alongside each photograph of the bungee jumper to the right.

- If forces in the same plane are in equilibrium, then adding them by placing them end-to-end will always produce a closed figure. It is quite common to come across situations involving three forces. These produce a **triangle of forces** if they are in equilibrium. The photograph on the right shows a rucksack hanging from a rope. The triangle of forces is particularly useful because it is easy to solve a triangle for unknown angles and sides.

- If there are more than three forces, it is easier to solve problems by resolving.

When the rucksack hangs in equilibrium forces T_1, T_2, and W add to zero. The free-body diagram is at upper right.

Worked example

A car of weight 9000 N climbs a 10° hill at a steady speed of $20\,\mathrm{m\,s^{-1}}$ against a constant drag of 400 N. What is the driving force of the car?

Choose axes parallel and perpendicular to the slope. The car is travelling at constant velocity, so the forces acting on it are in equilibrium.

Resolving perpendicular to the slope:

$C - W\cos 10° = 0$ (1)

Resolving parallel to the slope:

$T - (D + W\sin 10°) = 0$ (2)

Equation (1) could be used to calculate the contact force C but is not needed here. Equation (2) gives:

$T = D + W\sin 10° = 400\,\mathrm{N} + 9000\,\mathrm{N} \times 0.174 = 1960\,\mathrm{N}$

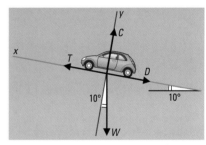

A car climbing a hill.

PRACTICE

1 A wooden block weighing 30 N floats in water. What is the buoyancy force from the water?

2 A lift weighs 4000 N and has two passengers each of weight 700 N. What is the tension in the lift cable when the lift is **a** at rest, **b** rising at a steady speed, and **c** moving downwards at a steady speed?

3 A picture of weight 25 N hangs from a hook by two strings which each make an angle of 35° with the vertical. What is the tension in each string?

4 A child drags a sledge along horizontal ground at constant velocity by pulling on a rope inclined at 30° to the ground. The drag on the sledge is 50 N. What is the tension in the rope?

5 Calculate the magnitude and direction of the frictional force between the tyres and the road if the car in the worked example above were to park on the hill described.

6 Why is it impossible for a tightrope to be horizontal (even before the tightrope walker steps on it)?

7 A parent holds a small child in a swing so that the angle of the swing support is at 20° to the vertical. The child's weight is 200 N and the swing seat weighs 50 N. What horizontal force must the parent apply?

8 A 1 kg camera is supported on a light symmetrical tripod whose legs each make an angle of 25° with the vertical. What is the compressive force in each leg and what horizontal force does each leg exert on the ground?

OBJECTIVES

- terminology
- the principle of moments
- equilibrium of coplanar forces

Hitting the cue ball near the top will create 'topspin'; hitting it near the bottom will create 'backspin'.

TURNING EFFECTS

Making things spin

Any snooker player knows that putting spin on the cue ball controls what it does after hitting another ball. But how do you put spin on it in the first place? The photographs on the left show that it all depends on the line of action of the cue. If the same force is exerted along different lines, a different turning effect is imparted to the ball. The same principle is used to impart spin in table tennis, tennis, and many other sports.

Definition: the moment of a force

The turning effect of a force F about some axis is called its **moment**. This can be thought of as the 'leverage' of the force, and is increased if the force is made larger or if its line of action is farther from the point considered. If the force acts in a plane perpendicular to the rotation axis, the moment is defined by

moment = magnitude of force × perpendicular distance of the line of action of the force from the axis of rotation

moment = Fr (units are N m)

Two examples are shown below. The first shows the general case, the second shows how this applies in a simple balance.

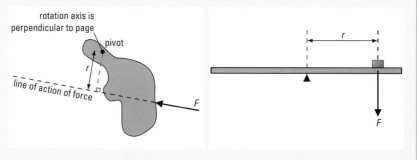

Couples and torques

If a single force is applied off-centre to something, it will tend to make it rotate and move off in the direction of the force (**translate**). The object changes its translational and rotational motion simultaneously – as a pool ball does while being struck by a cue, or a table tennis ball does while being hit by a bat.

If pure rotation is desired, it is best to apply a pair of equal but opposite forces acting along parallel but different lines. This is called a **couple**. Sometimes a single force is applied, but a couple arises in combination with a reaction force at the hinge or pivot. For example, when you push a roundabout it doesn't move away from you because the axle resists translation. The combined effect of the two forces is a couple, and the roundabout rotates.

The moment of a couple is called a **torque**.

$$\text{moment of couple} = F\frac{d}{2} + F\frac{d}{2} = Fd$$

where d = perpendicular distance between the forces.

Torques can also arise from continuous contact forces, as when we turn a screwdriver or in the frictional resistance of a stiff hinge.

Equilibrium of coplanar forces

If something is in equilibrium there must be no resultant force on it, and no resultant moment about any point in it. Conditions for equilibrium of coplanar forces (i.e. forces acting only in one plane) are that

$$\Sigma \boldsymbol{F} = 0 \quad \text{and} \quad \Sigma \text{ moments} = 0$$

To analyse situations with forces and moments you must resolve the forces and then take moments about some point. The point about which moments are taken should be chosen carefully – if a force passes through

---

Don't confuse

Work done is also calculated from force times distance, but then the force and distance are parallel. The N m unit for moment cannot therefore be reduced to the joule.

A couple is used to apply a turning effect but no resultant force. If you steer one-handed there must be an equal and opposite force from the steering wheel support acting at its centre.

advanced **PHYSICS**

that point then that force will have no moment about the point, since $d = 0$. For coplanar forces this will give three independent equations, which can be solved simultaneously for up to three independent unknowns.

Worked example

A simple beam bridge across a stream is supported at its ends by rigid blocks 4 m apart. The beam itself weighs 800 N. What are the reaction forces at each support if a man of weight 700 N stands on the beam 1 m from the left-hand support?

There are no horizontal forces.

Resolving vertically:

$L + R = W_m + W_b = 1500\,\text{N}$ (1)

Moments about X:

$(W_m \times 1\,\text{m}) + (W_b \times 2\,\text{m}) = R \times 4\,\text{m}$

$700\,\text{N m} + 1600\,\text{N m} = R \times 4\,\text{m}$ so $R = 575\,\text{N}$

From equation (1):

$L = 1500\,\text{N} - 575\,\text{N} = 925\,\text{N}$

It would be equally good to take moments about X or Y, and certainly better than using the centre of gravity.

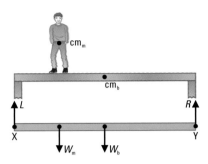

Calculating reaction forces.

Archimedes and the principle of the lever

Symmetry is a powerful guiding principle in physics. More than 2000 years ago Archimedes of Syracuse (287–212 BC) used symmetry to derive the principle of moments. He showed that balance conditions are equivalent to a perfectly symmetrical set-up, and if a balance is set up symmetrically it will have no reason to tip one way or the other, so it must be in equilibrium.

The argument is best shown by a sequence of diagrams for a particular set-up. It is included here not because it tells us any new results, but because of the compelling simplicity and beauty of the method.

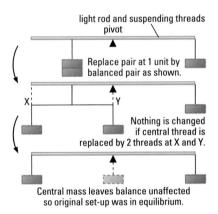

Archimedes's argument.

PRACTICE

1 A spanner 15 cm long is used to undo a nut against a maximum frictional torque of 10 N m. What is the minimum force that must be applied to the spanner to undo the nut? In practice, the minimum force will be larger than this. Why?

2 Explain why, when lifting a heavy object, it is advisable to keep a straight back and bend your legs rather than keeping your legs straight and bending at the waist.

3 Tall vehicles should have as low a centre of mass as possible. Why?

4 A crane is lifting a load as shown in the diagram.

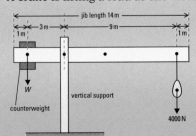

a Ignoring the weight of the jib, what is the ideal weight of the counterbalance?

b A bird of mass 2 kg lands on the counterbalance. Does the crane collapse? Explain.

c What would be the ideal mass of the counterbalance in this example if the jib has a weight of 7000 N and its centre of gravity is halfway along its total length?

d If the vertical structure of the crane has a weight of 15 000 N, calculate the total vertical support force from the ground (do not ignore the weight of the jib).

5 A wardrobe is 2 m high and 1.6 m wide. When empty it has a mass of 110 kg and its centre of gravity is 0.8 m above the centre of its base. What is the minimum angle through which it must be tipped before it will continue to fall by itself?

6 Show that the moment of a couple does not depend on the position of the point about which it is calculated.

7 Derive formulae for L and R in the worked example above when the man stands a distance x from the left-hand support.

8 Use Archimedes's symmetry approach to prove that 3 kg at 1 m from the pivot will balance 1 kg at 3 m from it. (Hint: split the 3 kg so that one mass remains where it is placed.)

- definitions of mass, weight, and density

Standard kilogram

The international standard kilogram is a particular cylinder of platinum–iridium alloy kept at the International Bureau of Weights and Measures at Sèvres, near Paris, France.

Equal forces applied for equal times change the velocities so that:

$$\frac{v_1}{v_2} = \frac{a_1}{a_2} = \frac{m_2}{m_1}$$

Using acceleration to compare masses.

Measuring the mass of ions

Inertia is used to measure the mass of ions. This is done in a mass spectrometer where ion beams are made to move in curved paths by a magnetic field. The more inertia that the ion has, the harder it is to deflect it, and the larger is the radius of its path (compared with other ions of the same speed and charge).

Comparing densities

'Intergalactic space' has a density of about 10^{-20} kg m^{-3} (mainly due to hydrogen atoms). Air density is about 1.2 kg m^{-3} near the Earth's surface, so a typical school laboratory contains several hundred kilograms. Body density is similar to the density of water, about 1000 kg m^{-3}, whilst the element osmium has a density of 22 500 kg m^{-3}. The density of a neutron star can exceed 10^{18} kg m^{-3}.

Other units used for density

The alternative (and non-SI) unit for density is g cm^{-3}. It is easy to convert to or from kg m^{-3}:

$$1\,\text{kg m}^{-3} = \frac{1\,\text{kg}}{1\,\text{m}^3} = \frac{10^3\,\text{g}}{10^6\,\text{cm}^3} = 10^{-3}\,\text{g cm}^{-3}$$

$$1\,\text{g cm}^{-3} = \frac{1\,\text{g}}{1\,\text{cm}^3} = \frac{10^{-3}\,\text{kg}}{10^{-6}\,\text{m}^3} = 10^3\,\text{kg m}^{-3}$$

There is a conversion factor of 1000 in either direction.

MASS, WEIGHT, AND DENSITY

What is **mass**? Newton found this a very difficult question to answer. He said that mass was the 'quantity of matter' in a body. However, mass plays a central role in dynamics and gravitation. Larger masses have greater weights and produce stronger gravitational fields than smaller masses, and something with a lot of mass is difficult to accelerate or decelerate. This built-in 'reluctance' to change state of motion is called **inertia**. Inertia can therefore be used to measure mass.

Measuring mass

An unknown mass could be measured by comparing its acceleration with that of a standard kilogram mass when the same resultant force is applied to both. If the unknown mass has an acceleration twice that of the standard kilogram, it must have half the inertia (because it has half the 'reluctance' to accelerate) and so half the mass. In general:

$$\frac{m_1}{m_2} = \frac{a_1}{a_2}$$

In practice this is rarely done with large masses. It is easier to weigh them. However, weighing them uses the force of gravity as a measure of their mass. This will only be accurate if they are both in the same gravitational field. On Earth this is a very good assumption but the method would not work in deep space or in a satellite.

Mass, weight, and gravity

Weight is the *force* that gravity exerts on a mass, so it is measured using the newton. In common language 'mass' and 'weight' are often confused. For example, an overweight person is really 'overmass'. It would be possible for them to lose weight by going to the Moon, because the Moon's gravitational field strength is about one sixth of the Earth's at the surface. On the Moon they would still have exactly the same amount of mass and still look 'fat'. Dieting is really an attempt to lose mass.

Gravitational field strength is defined as the gravitational force exerted per unit mass at a point in space:

$$g = \frac{F}{m}$$

The units of g are N kg^{-1} (close to the Earth's surface, the magnitude of g is about 9.8 N kg^{-1}).

Weight is determined by two things: the mass of the body (number of kilograms) and the gravitational field strength:

$$W = mg$$

The units of W are N. A 5 kg mass on Earth will weigh about 5 kg × 9.8 N kg^{-1} = 49 N.

Weight will depend on where you are in the universe (even on Earth it varies from about 9.81 N kg^{-1} at the poles to around 9.78 N kg^{-1} at the equator). On the other hand, mass is a property of the object you are dealing with, and so stays the same everywhere. A 5.00 kg mass is 5.00 kg at the poles and at the equator, even though its weight is 49.1 N at the poles and 48.9 N at the equator.

Definition: density

Density is mass per unit volume. It is defined by the equation

$$\text{density (kg m}^{-3}) = \frac{\text{mass (kg)}}{\text{volume (m}^3)} \quad \text{or} \quad \rho = \frac{m}{V}$$

Centre of mass

Imagine pushing a freely floating astronaut. The astronaut will begin to drift away from you, and may rotate as she does so. However, if the force is applied along a particular line it is possible to make her drift in a straight line without rotating. You don't have to go into space to try this. Place a ruler on a table and try pushing it along the table. There is only one line of action of the applied force that causes pure translation and no rotation.

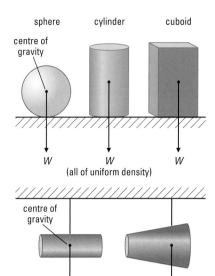

> **Definition: centre of mass**
>
> The **centre of mass** of a body is the point through which *any* applied force produces translation but no rotation.

Something like a cubic block of stone is uniform and symmetrical, so its centre of mass is at its geometric centre. It is possible, however, for the centre of mass to lie *outside* the material of the body itself, for example at the centre of a hollow sphere or an empty box. Often in mechanics a problem is drastically simplified by treating large objects like the Earth as if they are point masses concentrated at their centre of mass. If this simplification was not justified, most of mechanics would be impossibly complicated.

Centres of gravity

Centre of gravity

The **centre of gravity** is the point at which the resultant force of gravity acts – that is, it is the place where a single force equal to the weight of the object can replace the billions upon billions of individual gravitational forces acting on each particle that makes up the object.

In almost all problems, the centre of gravity and the centre of mass are in the same place. However, they are not the same thing. Using the centre of gravity makes life very much easier than having to deal with all parts of a large body separately.

> **Centre of gravity and centre of mass**
>
> If a uniform beam balance is in a non-uniform gravitational field, one side will be pulled down more strongly than the other. Its centre of mass is still above the pivot but its centre of gravity is not.

PRACTICE

1 Why can't you weigh something in a satellite or in deep space?

2 An astronaut in an orbiting satellite is not really weightless. Why not?

3 Air at room temperature and pressure has a density of about $1.2 \, \text{kg} \, \text{m}^{-3}$.

   a A room measures 4.0 m by 4.0 m by 2.5 m. What is the mass and weight of air in the room?

   b If all this air is compressed so that it fits into a cylinder of height 1.0 m and diameter 0.75 m, what is the mass, weight, and density of air in the cylinder?

   c How would your answers to **b** change if the sealed cylinder of air was transported to the Moon?

   (Take the magnitude of g as $9.8 \, \text{N} \, \text{kg}^{-1}$ on Earth and $1.7 \, \text{N} \, \text{kg}^{-1}$ on the Moon.)

4 A neutron star has radius 15 km and mass $2 \times 10^{24} \, \text{kg}$. What is its density? What volume would a litre of water occupy if crushed to the density of a neutron star?

5 The table below gives the radii and masses of some nuclei. Calculate the density of each nucleus. What does this tell you about the way nucleons (protons and neutrons) pack together in the nucleus? ($1 \, \text{fm} = 10^{-15} \, \text{m}$, $1 \, \text{u} = 1.66 \times 10^{-27} \, \text{kg}$; assume that the nuclear mass is approximately equal to A times $1 \, \text{u}$.)

nucleus	Si	Ca	Co	Sr	Au
radius/fm	3.92	4.54	4.94	5.34	6.87
mass no. A	28	40	59	88	197

6 A brass alloy is made by mixing 85% copper with 15% zinc (by mass). What is its density? What assumptions have you made? ($\rho_{\text{Cu}} = 8920 \, \text{kg} \, \text{m}^{-3}$, $\rho_{\text{Zn}} = 7140 \, \text{kg} \, \text{m}^{-3}$.)

7 Before a London double-decker bus goes into service its stability is tested by loading the top deck with sandbags and then tilting it from the vertical. It must keep all four wheels in contact with the ground up to an angle of at least 280° to the vertical. The height of the bus is 4.4 m and its width is 2.4 m.

   a Calculate the height of the centre of mass of the loaded bus by assuming it would topple over at larger angles.

   b Explain why the top deck of the bus is filled with sandbags but the lower deck is not when this test is carried out.

FORCES IN BUILDINGS

Buildings are designed so that the forces exerted upon them by weight, added loads, and external influences (like wind, rain, snow, and earth tremors) are supported in equilibrium.

A roof over your head

A flat roof placed on vertical supports will tend to sag in the middle. This puts its top surface into compression and its bottom surface into tension. Materials like stone and concrete are dense, weak in tension, and brittle (that is, they tend to crack). The span of a flat roof using these materials is limited to a maximum of about 2.5 m.

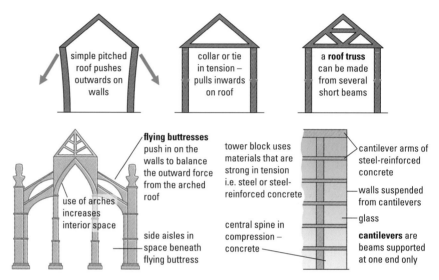

Supporting buildings.

The ancient Greeks often made temple roofs by laying wooden beams horizontally across the walls and then boarding over them to cover the space beneath. The top of the flat roof would then be covered with a composite of clay, soil, and straw, and then tiled. To support the weight of such a roof it was often necessary to add many supporting columns.

A pitched roof solves the problem of shedding water and increases the space inside the building, but pitched roofs push outwards on the walls that support them, so ties or collars are added. These are in tension and pull inwards on the roof beams. The shortage of long wooden beams may have led to the invention of the **roof truss**, which can be constructed from several shorter beams. The simple A-frame design was developed in the Middle Ages. The diagrams above show some ways in which a pitched roof can be supported.

The problem with Pisa

Construction of the Tower of Pisa began in 1173. Soon after, the Tower began to lean to the north. Work stopped for nearly 100 years, during which time the lean shifted to the south. Construction began again in 1272 and the eighth floor and bell chamber were completed in 1370. By 1990 the top of the Tower was more than 5 m off centre and there were fears that its collapse was imminent. It was closed to visitors.

The Tower is leaning because of movement in its foundations. In fact, the builders tried to allow for this as they constructed it, and their corrections can be seen in the shape of the building today. The Tower is not straight but banana-shaped.

The main danger is not that the centre of gravity will fall outside the base. It is that the southern walls are bearing a major proportion of the building's weight and, even though stone is strong in compression, any faults in the masonry could fail explosively, leading to sudden collapse. To protect against this the first level of the building was given a corset of steel straps in 1992.

The Temple of Zeus at Olympus, Greece. Cracks form where the stone is in tension. This may lead to collapse or the stone might move to adopt another stable configuration. The weakness of stone in tension limits the span of stone lintels.

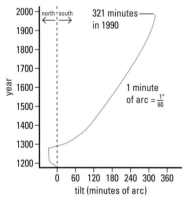

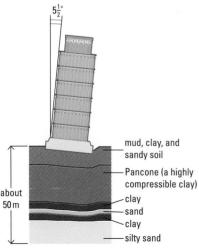

The tilt of the Tower of Pisa.

The next problem was how to arrest the tilt. The first step was to assemble over 750 tonnes of lead blocks on the northern side to compress the foundations there (the total mass of the building is about 15 000 tonnes). This resulted in a northward correction of about 2.5 cm between 1993 and 1994.

The next step is to surround the base with a massive concrete collar, and anchor this to a level of sand about 50 m below the ground. Once the building is stabilized a little, the main problem of the foundations can then be attacked. This means carefully removing some of the soil beneath the northern side of the building, allowing the Tower to settle back into a vertical and more stable position.

Shaking and breaking

It is said that earthquakes don't kill people, but buildings do. Some indication of this can be seen by the following comparison: the 1989 earthquake in San Francisco reached a magnitude of 7.1 on the Richter scale and killed 62 people, while the 1988 earthquake in Armenia, with a magnitude of 6.7, killed 25 000 people. However, whilst building construction undoubtedly saved lives in San Francisco, this is not the only factor that affects the death toll.

In the Mexico earthquake of 1985 about 75% of the 20 000 killed were crushed by masonry, but not all buildings collapsed. Those between 8 and 15 storeys were most vulnerable.

- A typical ten-storey building sways with a natural frequency of about 1 Hz if it is shaken. At this frequency its swaying motion grows just like the motion of a child on a swing being pushed once per swing. Tragically, the frequency of the seismic waves (which were actually quite weak when they reached Mexico City from the epicentre 400 km away) matched both the natural frequency of these buildings and the natural frequency of the soil on which the city is built.

- Mexico City is built on a lake bed with mountains surrounding it. The seismic waves made the soil in the lake bed resonate, and this produced intense oscillations of the surface over an area of 20 blocks by 20 blocks near the centre of the city. Buildings in tune with this vibration were most vulnerable.

Most seismic waves have frequencies between 0.5 Hz and 5 Hz, so earthquake building codes are designed to reduce the response of buildings at these frequencies. One of the modern approaches to the problem is to build **base-isolated buildings**, that is, buildings connected to the Earth's surface by rubber bearings that allow the building to move as a whole rather than suffer shear forces when it is attached directly to the ground. Research is being carried out into active control in which buildings respond to an earthquake by shifting large masses across their roofs to counteract the swaying of the building.

The success of these precautions can be gauged by comparing the acceleration of parts of the building with the acceleration of the ground during the earthquake. Acceleration is proportional to force, so reducing the acceleration of a building reduces the force to which it is subjected, and reduces the danger of damage or collapse. When resonance occurs the acceleration at the top of a building might be several times greater than at ground level.

The use of **smart materials** that can respond to changes in strain or temperature by producing voltages or changing shape is becoming important in monitoring the behaviour of buildings. In principle, these materials could be used to change the stiffness of a building and so alter the way it responds to applied forces.

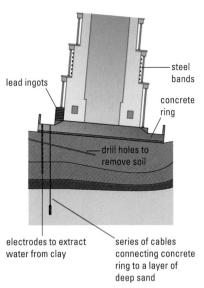

Some of the techniques used or being considered to stabilize the Tower of Pisa.

An earthquake measuring 9.0 on the Richter scale hit Tōhoku, Japan in 2011 triggering tsunami waves up to 40 m high. Over 15 000 people died, millions of buildings collapsed, and nuclear accidents occurred at the Fukushima Daiichi nuclear power plant.

Base-isolated buildings

The Foothills Communities Law and Justice Centre, about 100 km east of Los Angeles, was opened in 1986 and is mounted on 98 rubber bearings. It is designed to survive earthquakes up to 8.3 on the Richter scale (as big as any recorded earthquake to hit California). Japan is also investing in base-isolated buildings.

- force is proportional to extension
- the spring constant
- combinations of springs

Buildings are like very stiff springs. They compress when loaded until the forces resisting compression balance the weight of the load.

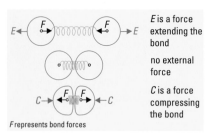

The action of bond forces.

Extension, tension, and compression

Extension is the difference between the stretched length of something and its *original* unstretched length. Be careful not to confuse total extension with the increase in extension produced when an extra load is added to an already stretched object.

Tension is a force that acts along the axis of an object and makes it longer.

Compression is a force that acts along the axis of an object and makes it shorter.

For many objects Hooke's law applies for both extension and compression.

HOOKE'S LAW

What holds things together?

Everything must give a little when we push it. This is more obvious for rubber or springs than for apparently rigid substances like stone or concrete. Nonetheless it is true for all materials.

Think of an empty building gradually filling with people. Soon its floors, walls, and foundations support not just the weight of the building but all of the people in it as well. More than that, the floor beneath each person must push up on the soles of their shoes with exactly the right force to balance their weight. If this were not so they would either sink into the floor or accelerate upwards. How does the floor 'know' exactly how hard to push up?

Take a ruler and try to break it by pulling along its length. It will resist you. Pull harder and it pulls harder. Unless it does break, it will always pull back exactly as hard as you pull on it. How does it 'know' how hard you are pulling?

The force exerted on a structure distorts it slightly, and materials in tension or compression exert forces which depend on the amount they are distorted. A building is like a very stiff spring. It balances the compressing force with a force generated by its own particles pushing one another apart. The ruler generates tension from the attractive forces between its particles as they are moved slightly further apart.

How do these forces arise? Solids are made of atoms or molecules bound together by electromagnetic bonds. Pushing them closer together produces a repulsion as the electron orbits get distorted. Pulling them apart increases the attractive force of the bonds (up to a point). These ideas are discussed in more detail in chapter 10.

Hooke's law

For many objects the force F required to maintain an extension e is directly proportional to the extension:

$$F \propto e \qquad \text{leading to} \qquad F = ke$$

where k is the **spring constant** and is a measure of 'stiffness' in N m^{-1}.

Hooke's law is usually obeyed up to some maximum value of the applied force, and beyond that the relation between force and extension is *non-linear*.

Worked example

A car sinks 5.0 cm when a 20 kg suitcase is put into its boot directly above the rear suspension spring. What is the spring constant of the rear suspension?

Assuming the spring obeys Hooke's law:

$$k = \frac{F}{e} = \frac{20\,\text{kg} \times 9.8\,\text{N kg}^{-1}}{0.05\,\text{m}} = 3900\,\text{N m}^{-1}$$

Stretching a helical steel spring

The graph shows how the force needed to balance the tension in a spring depends on the spring's extension. Key ideas are listed below:

Spring constant: $k = F/e$ equals the gradient of the graph in the linear region (OA).

Limit of proportionality: beyond this point (A) the graph is non-linear.

Elastic behaviour: means that a spring returns to zero extension when load is removed (region OB).

Elastic limit: beyond this point (B) the spring suffers permanent deformation, and does not return to zero extension when the load is removed.

Plastic deformation: permanent structural change, after which the spring will not return to zero extension when the load is removed.

In the elastic region the applied forces extend the bonds between particles in the material. Permanent or plastic deformation occurs when bonds break and particles move or flow past one another to new positions.

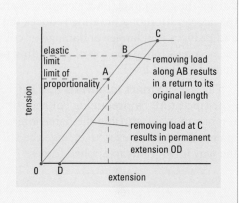

The Pont de Brotonne over the River Seine is a cable-stayed bridge with a central span of 456 m. The composite steel and concrete deck is supported by the tension in taut steel cables that transfer their loads to vertical supporting columns. Steel is ideal because it is strong in tension and reasonably stiff. In the UK Government regulations mean that the cables used on British bridges are often thicker than those on comparable foreign bridges, and are also more expensive to build.

Systems of support

It is hard for one person to move a piano, but four people might manage it. If each lifts one corner they will each have to apply an upward force equal to about 1/4 of the piano's weight. They are pushed down by a similar force. If something hangs from a number of symmetric supports, these supports 'share the load'. This means that the extension or compression of the supports is reduced. The system of support is *stiffer* than a single support. This is modelled in the diagram by several springs in parallel.

Assuming the springs obey Hooke's law, the system of n springs will extend/compress $1/n$ times the extension/compression of a single spring. The system is equivalent to a single support that is n times stiffer than the original spring:

The effective spring constant of n springs in parallel is equal to nk.

On the other hand, if something is suspended from a long support rather than a short support, each part of the support has the same load to support (ignoring the weight of the support itself), so the total extension will be in proportion to the length. This is modelled by three springs connected in series. When in equilibrium points A and B are at rest. Spring 2 must support spring 1 plus the load. Ignoring the weight of spring 1, it must stretch just as much as if it were the only spring supporting the load. The same is true for springs 1 and 3, so the system stretches three times as much as a single spring supporting the same load. It is 1/3 as stiff as the single spring:

The spring constant of n springs in series is equal to k/n.

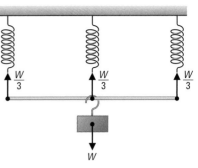

This system is stiffer than a single spring.

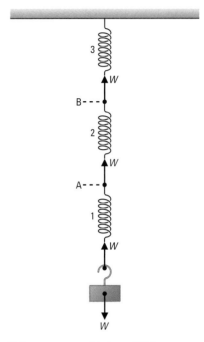

This arrangement is less stiff than a single spring.

PRACTICE

1 A spring is extended 4 mm by an applied force of 2.0 N.

  a What is its spring constant?

  b How much would it extend if a load of 150 g was suspended from it?

  c Predict its extension for an applied force of 8.0 N, stating any assumptions.

2 The human adult tibia (shin bone) compresses by about 1 mm per 1000 N applied force.

  a What is the compressive spring constant for the tibia?

  b Estimate the compression of a 75 kg man's tibia when he stands on one leg.

  c Why is your answer to b an estimate?

3 A spring has unstretched length 10 cm and spring constant 40 N m^{-1}, and a limit of proportionality at 5.0 N.

  a What is its extension at the limit of proportionality?

  b What is its extension for a load of 350 g?

c What is its extension for a stretching force of 2.0 N if the spring support is held stationary?

4 A particular spring has a spring constant of 30 N m^{-1}. Its limit of proportionality and its elastic limit are both at 5 N. Two such springs are suspended in series from two more in parallel. What is the extension of the system for a load of 4 N? What is the spring constant and elastic limit for the whole system?

5 If the extension of a metal wire is proportional to its tension, what conclusions can be drawn about the bonds between atoms in the wire?

6 How will the spring constant for a cylindrical wire obeying Hooke's law depend on the length and cross-sectional area of the wire?

7 A climber who slips and falls is saved by his rope. The rope has unstretched length l and gives him a maximum upward acceleration of 9g (that is 9 times the free-fall acceleration) as it extends by 10% to stop him falling further. If the rope obeys Hooke's law and the climber has a mass m, find an expression for the spring constant of the rope.

3.9

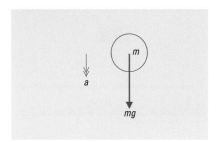

Free-fall acceleration.

FALLING

Free fall

Apollo astronauts dropped a hammer and a feather on the Moon to show that all objects fall with the same acceleration in the same gravitational field. To repeat the experiment on Earth you would need to remove resistive forces by dropping things inside an evacuated container. During free fall the resultant force on a body is its own weight.

$$\text{resultant force} \qquad F = mg$$

Newton's second law: $\quad F = ma$ (see spread 3.21)

$$mg = ma$$

$$a = g$$

This shows that everything falls freely with the same acceleration in the same gravitational field.

If the distance through which something falls is small enough that we can neglect changes in g, then everything falls with the same constant free-fall acceleration. Near the Earth's surface this is about $9.8\ \text{m s}^{-2}$; on the surface of the Moon it is closer to $1.7\ \text{m s}^{-2}$. Free fall can be analysed either by using the 'suvat equations' or by considering energy changes. For this we need to consider **kinetic energy** and **gravitational potential energy**, which are defined in spread 3.13.

Worked example

A rock falls from a ledge 10 m above the ground. How fast is it moving just before it hits the ground?

Method 1:

We know

$$u = 0\ \text{m s}^{-1};\ s = 10\ \text{m s}^{-1};\ a = 9.8\ \text{m s}^{-2}.$$

We want v. Use

$$v^2 = u^2 + 2as$$
$$= 0 + 2 \times 9.8\ \text{m s}^{-2} \times 10\ \text{m}$$
$$= 196\ \text{m}^2\,\text{s}^{-2}$$

So

$$v = 14\ \text{m s}^{-1}$$

Method 2:

Initial gravitational potential energy (GPE) with respect to the ground = mgh; initial kinetic energy (KE) = 0.

$$\text{Final GPE} = 0 \quad \text{Final KE} = \tfrac{1}{2}mv^2$$

Total energy is conserved so $GPE + KE = $ constant.

$$\tfrac{1}{2}mv^2 = mgh$$

(m cancels – final velocity doesn't depend on rock mass.) Hence

$$v = \sqrt{2gh} = 14\ \text{m s}^{-1}$$

Aristotle and Galileo on free fall

Aristotle was a highly influential ancient Greek philosopher, but his ideas about falling bodies were wrong. He thought that bodies fall with speeds that are directly proportional to their weight. This implies that heavier bodies fall faster than lighter bodies. Galileo, in his great book *Dialogues concerning two new sciences* (1638), used a brilliant argument to refute Aristotle's claim.

Galileo imagined dropping a heavy body and a light body and then joining them together and dropping the combined body. According to Aristotle the combined body would be heavier and so should fall faster than either of the smaller bodies from which it is made. However, in joining the two smaller bodies together surely the lighter one ought to retard the heavier one so that the combined body falls *more slowly* than the heavier of the two smaller bodies? This leads to a contradiction, so Aristotle's original assumption must be wrong. Galileo argued that all objects must fall at the same rate.

It has been claimed that Galileo provided an experimental demonstration of this by dropping cannonballs of different mass from the Tower of Pisa and observing that they landed at the same time. The *Apollo* experiment described above is a modern version of Galileo's experiment.

cannonballs dropped at same time

cannonballs hit the ground at same time

Galileo may or may not have actually carried out this experiment. If air resistance can be ignored, all objects fall at the same rate in the same gravitational field.

Terminal velocity

When things move through a fluid, drag increases with speed. This is because they encounter more fluid per second and hit it harder. For low velocities, drag is proportional to speed, and for higher velocities it increases with velocity squared (see spread 3.25). In either case the motion will differ from free fall.

The resultant downward force will now be the difference between weight W (= mg) and drag D (which increases with velocity) so the equation of motion is:

resultant $\ F = W - D = mg - D = ma$

so the acceleration is

$$a = \frac{mg - D}{m} = g - \frac{D}{m}$$

When $v = 0$, $D = 0$ so $a = g$. (This is the initial acceleration.) As v increases, D increases, so a falls. Eventually a velocity is reached when $D = mg$, and then $a = 0$. The object continues to fall at a constant **terminal velocity**.

There is another force acting on the falling body that we have neglected so far. This is a buoyancy force which is equal to the weight of fluid displaced. For a solid object falling through air this would be negligible, but in water or some other dense liquid it must be included. One way to do this is to replace the actual weight of the falling body with its effective weight in the fluid.

effective weight = actual weight – weight of the same volume of fluid

$$W' = \rho V g - \rho' V g = (\rho - \rho')Vg$$

where ρ and ρ' are the density of the solid and fluid respectively, and V is the volume of the falling body.

The equation of motion is now:

$$W' - D = ma$$

from which equations for a can be derived.

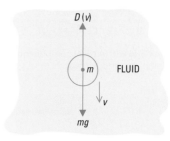

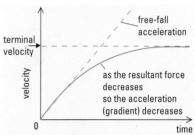

Terminal velocity.

A free-fall parachutist can change his speed by altering his body shape to increase or decrease drag.

PRACTICE

1 A stone is dropped into a deep well and takes 4 s to reach the bottom.

   a How deep is the well?

   b How fast is the stone moving just before it hits the bottom?

2 A simple way to measure someone's reaction time is to hold a ruler vertically and ask them to put their finger and thumb level with the lower end of the ruler, and prepare to catch it when it falls. You then drop it and measure how far up the ruler they catch it. The further up they catch it, the further it has fallen and the slower their reaction.

   a What is their reaction time if they catch it after it has fallen 10 cm?

   b How would you calibrate a 50 cm ruler so that reaction times can be read off it?

3 A ball is thrown vertically upwards with an initial velocity of $5.0 \, \text{m s}^{-1}$.

   a How long is it before it falls back to its original level?

   b What is its maximum height?

4 A parachutist leaps from a hovering helicopter and reaches terminal velocity before opening his parachute. What energy transfers take place as he falls?

5 If drag is given by $D = kv$, where k is a constant, what is the terminal velocity of a stone of mass m falling in a gravitational field of strength g?

6 A tennis ball is thrown horizontally at $10 \, \text{m s}^{-1}$ from a height of 2 m above flat horizontal ground.

   a How long does it take to hit the ground?

   b How far away from the thrower does it first hit the ground?

7 A fountain pumps $0.1 \, \text{m}^3$ of water per second and ejects it vertically at $12 \, \text{m s}^{-1}$.

   a Calculate the minimum power of the pump.

   b Calculate the maximum height reached by the water.

   c State any assumptions you have made.

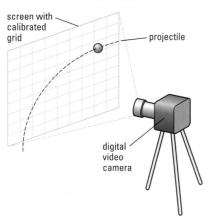

- horizontal and vertical components of motion are independent

Projectile motion can be captured using a digital camera. The video file can be analysed frame by frame using computer software.

PROJECTILE MOTION

Vertical and horizontal components of motion

When a stone is dropped from the top of a ship's mast it falls parallel to the mast (ignoring effects of wind, etc.) until it hits the deck. This is true whether the boat is at rest or moving with a constant speed on a smooth sea.

If this does not seem obvious to you think about your experiences inside a moving train or, better still, a jet aircraft. If you drop something in either situation it will fall straight down to the floor, even though an outside observer would see it follow a very extended curved path. It certainly doesn't fly out of your hand at the speed of the vehicle! This is because of the vector nature of velocity. Gravity acts vertically and so only affects the vertical component. The horizontal component continues at constant velocity in a straight line.

- The horizontal and vertical components of motion are completely independent of one another.

The best way to approach projectile problems is to resolve the motion into vertical and horizontal components, and deal with each component as a one-dimensional problem.

Worked example 1

A girl drops a bag inside a moving train. Her friend sees this happen from the platform. The bag drops 1 m from rest when the train is moving steadily along the platform at $2\,\mathrm{m\,s^{-1}}$. How will the path of the falling bag appear to each of the two girls?

According to the girl in the train:

For the purely vertical motion $s = 1$ m; $u = 0\,\mathrm{m\,s^{-1}}$; $a = 9.8\,\mathrm{m\,s^{-2}}$.

We need v and t. We can get these from the 'suvat equations':

$$v^2 = u^2 + 2as \qquad v = \sqrt{2as} = 4.4\,\mathrm{m\,s^{-1}}$$
$$s = ut + \tfrac{1}{2}at^2 \qquad t = \sqrt{\frac{2s}{a}} = 0.45\,\mathrm{s}$$

The bag falls vertically and takes 0.45 s to reach the ground.

According to the girl on the platform:

The same vertical motion occurs as for the girl on the train, but a horizontal velocity of $2\,\mathrm{m\,s^{-1}}$ is superimposed on it.

For this horizontal motion:

$$s = ut + \tfrac{1}{2}at^2 = ut = 0.9\,\mathrm{m} \quad \text{(no horizontal acceleration)}$$

So the bag moves 0.9 m horizontally as it falls 1 m in a time of 0.45 s.

Viewed from the platform, the bag ends up with an inclined velocity having two components:

horizontal velocity = $2.0\,\mathrm{m\,s^{-1}}$
vertical velocity = $4.4\,\mathrm{m\,s^{-1}}$

These can be combined to find the resultant velocity seen from the platform.

$$v^2 = 2.0^2 + 4.4^2 \qquad v = 4.8\,\mathrm{m\,s^{-1}}$$

This velocity is at angle θ to the vertical:

$$\tan\theta = \frac{2.0}{4.4} \qquad \theta = 24°$$

The bag follows a curved (parabolic) path before striking the ground at $4.8\,\mathrm{m\,s^{-1}}$ at an angle of 24° to the vertical.

advanced **PHYSICS**

Maths box: a parabolic path

Taking the point at which the bag is dropped as the origin, we can draw horizontal and vertical axes x and y to determine the equation of the bag's path as seen from the platform. From the previous analysis:

$$x = ut \qquad y = \tfrac{1}{2}at^2$$

Separate equations like these are called **parametric** equations. They can be combined to give y in terms of x:

$$t = \frac{x}{u} \quad \text{therefore} \quad y = \frac{ax^2}{2u^2} \quad \text{or} \quad y = mx^2 \quad \text{(where } m \text{ is a constant)}$$

This is the equation of a parabola.

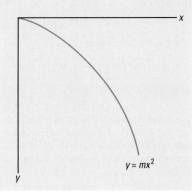

A parabolic path.

Falling bullets

If a rifle is fired horizontally the bullet leaves the barrel with a large horizontal velocity. Its initial vertical velocity is zero, but it will immediately begin to fall because of gravity.

- Since this motion is independent of the horizontal motion, it will hit the ground at exactly the same time as a bullet dropped vertically beside the barrel at the instant of firing (assuming the ground is horizontal).

- The range of the gun is determined by the product of how long it takes the bullet to fall to the ground and the initial horizontal velocity. (Actually, it will be less than this because of air resistance.)

If this seems strange, think of the example on the previous page. If the train was travelling at the speed of a bullet then dropping a bullet inside the train would be equivalent to firing the gun outside; both projectiles would follow the same parabolic trajectory.

The independence of vertical and horizontal velocities is illustrated by trying to shoot a falling target. Imagine William Tell firing his crossbow at an apple held by his son. If his bow is on the same level as the apple and pointed directly at it (not aimed above it as it should be to hit a stationary apple), he will miss the apple because the bolt falls. However, if his son drops the apple at exactly the same time that the bolt leaves the bow, they will both fall the same vertical distance in the time that it takes for the bolt to cross the horizontal distance to the apple. It will hit the apple.

In fact, the same result occurs even if the apple and bow are initially at different heights. All that is essential is that the bolt is aimed directly at the apple, and that the apple is dropped at the instant the bolt is fired.

Maths box: the range of a projectile

If a projectile is fired at velocity u at an angle θ to the horizontal it has initial horizontal and vertical velocity components:

$$u_{\mathrm{h}} = u \cos \theta \qquad u_{\mathrm{v}} = u \sin \theta$$

(Up is positive.)

Vertical motion:

It will reach its highest point when the vertical velocity $v = 0$.

This will be after a time t where

$$t = \frac{v_{\mathrm{v}} - u_{\mathrm{v}}}{a_{\mathrm{v}}} = \frac{0 - u \sin \theta}{-g} = \frac{u \sin \theta}{g}$$

Horizontal motion:

Ignoring air resistance, this is half the time of flight so its range is:

$$R = 2tu_{\mathrm{h}} = \frac{2u^2 \sin \theta \cos \theta}{g} = \frac{u^2 \sin 2\theta}{g}$$

This has a maximum value when

$2\theta = 90°$ or $\theta = 45°$.

PRACTICE

1 An air rifle fires pellets at $100\,\mathrm{m\,s^{-1}}$ at a target $25\,\mathrm{m}$ away. If no allowance is made for gravity, what error will be introduced? (Ignore air resistance.)

2 A car manufacturer tests crash resistance by driving test vehicles off a horizontal platform so that they fall to a concrete surface below. If the car is driven off at $20\,\mathrm{m\,s^{-1}}$ and the platform is $15\,\mathrm{m}$ above the ground calculate the impact angle and speed, and the horizontal distance of the impact from the edge of the platform.

3 A grass-hopper jumps at an angle of $30°$ to the horizontal with a take-off speed of $3\,\mathrm{m\,s^{-1}}$.

  a What is the height of its jump?

  b How long is it above the ground?

  c What is the range of its jump?

  d How would your answers be affected if it jumped

in the same way on the Moon, where the surface gravity is 1/6 as strong as on Earth?

4 It is raining steadily and vertically, and you have to cross an open courtyard. Will you get wetter if you walk or if you run? Explain your answer.

5 The value of g can be measured very accurately by timing the vertical motion of a ball as it rises past a pair of light gates separated by a known vertical distance h.

  a If the ball spent a total time T_1 above the lower light gate and a total time T_2 above the upper one, show that

$$g = \frac{8h}{(T_1^2 - T_2^2)}$$

  b What precautions must be taken to ensure that this gives as accurate a result as possible?

O B J E C T I V E S

- energy conservation
- energy transfer by work and by heating
- forms of energy

The physicist uses ordinary words in a peculiar manner ... a conservation law means that there is a number which you can calculate at one moment, then as nature undergoes its multitude of changes, if you calculate this quantity again at a later time it will be the same as it was before, the number does not change. An example is the conservation of energy. There is a quantity that you can calculate according to a certain rule, and it comes out the same answer always, no matter what happens.

Richard Feynman, *The Character of Physical Law* – an excellent and very readable book!

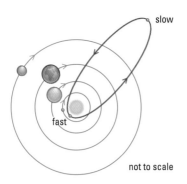

slow

fast

not to scale

The orbit of a comet around the Sun.

Visible comets

The fact that a comet moves rapidly near the Sun (when it is visible from the Earth) and slowly when it is far away from the Sun explains why we can see comets for only a small fraction of their orbital periods.

WHAT IS ENERGY?

The law of conservation of energy

A comet orbiting the Sun illustrates this very well. As the comet moves through space the only significant forms of energy to consider are its kinetic energy and its gravitational potential energy. The comet and Sun together are a closed system, so we can write down a very simple equation:

$$(\text{comet's KE}) + \left(\begin{array}{c}\text{comet's}\\\text{gravitational PE}\end{array}\right) = \left(\begin{array}{c}\text{comet's}\\\text{total energy}\end{array}\right) = (\text{a constant})$$

The fact that they form a closed system is important. If they didn't then energy could enter or leave from outside, and there would be no reason for the total energy of the system to stay constant.

The comet's orbit is an eccentric ellipse (see diagram) so its distance from the Sun changes. When it is far from the Sun it has a large amount of gravitational potential energy, so its kinetic energy must be low and it moves slowly. Close to the Sun it has less gravitational potential energy and more kinetic energy, so it moves rapidly in this region of its orbit. In moving away from the Sun it is transferring kinetic energy to gravitational potential energy, and as it approaches the Sun it does the opposite. In the first case it is doing work to increase its energy in the Sun's gravitational field, and in the second case the Sun's gravitational field is doing work on the comet as it accelerates it.

Conservation of energy

The total energy of a closed system is constant.

This is a deceptively simple statement; it took a long time before energy conservation was discovered. This is because it was not obvious that the energy of motion, which seems to disappear when something stops, might have been transferred to internal energy in the objects that rubbed against it as it stopped, or that the mechanical work we do might derive from energy released in chemical reactions as we digest our food. Some forms of energy are 'hidden'.

Once physicists realized that conserving energy was a good 'book-keeping technique' for some kinds of change, they extended the idea to cover all kinds of change and found that it worked. When we define how different forms of energy are measured we are simply 'book-keeping'. Another way to define the law of conservation of energy is:

- Energy is never created or destroyed but it can be transferred from one form to another.

(Some people think that Einstein's famous $E = mc^2$ changes all this and that energy can be *converted* to mass and vice versa. This is not true. The equation tells us that mass has energy and energy has mass. Mass and energy are still *both* conserved in all physical changes.)

Joule's paddle-wheel (water friction) apparatus. The essential clue to energy conservation was found by James Joule (1818–1889). He showed that the mechanical work done turning submerged paddles is directly proportional to the temperature rise in the water. This led to the idea that energy is never lost, simply transferred from one thing to another.

Types of energy.

Energy type	Description	Example
kinetic energy	Energy because of motion.	A served tennis ball.
radiant energy	Energy transferred by electromagnetic waves.	Solar heating of the Earth.
internal energy	Sum of internal kinetic and potential energies of particles in something.	Compressed steam in a reactor.
potential energy (many forms – several below)	'Hidden energy' stored in a field by virtue of position.	(see below)
gravitational potential energy	Energy of a mass because of its position in a gravitational field.	Water in a mountain lake.
electrostatic potential energy	Energy of charges because of their position in an electric field.	Charges on a charged capacitor plate.
chemical potential energy	Energy able to be released by atomic bond formation or rearrangement.	Petrol as a fuel; electric cells.
nuclear potential energy	Energy able to be released by nuclear bond formation or rearrangement.	Binding energy.
elastic potential energy	Energy stored in something as a result of reversible deformation.	A stretched spring.

This explosion was created with 23 kg of gunpowder during research into explosives. Instruments around the fireball measure its thermal output. Each red and white marking on the poles around the explosion is 1 m long.

Energy transfers.

Means of energy transfer	Description	Example
working	Energy transferred when an applied force moves.	Pulling a plough.
heating	Energy transferred by a temperature difference.	Using a Bunsen burner to warm a beaker of water.

Work and heat are not forms of energy but the means by which energy is transferred. They differ in a very important way – working involves organized energy (for example, making all particles in something move the same way at the same speed), whereas heating involves disordered energy (for example, making all the particles in something move faster but in random directions).

'Heat' by another name

Some physicists get rather steamed up about 'heat'. They say it is as wrong to say something possesses heat as it is to say that it contains work. The correct term for this random motion energy of the particles in a body is **internal energy**. However, there are probably even more physicists who are perfectly happy with heat both as a form of energy and as a means of energy transfer. You have been warned!

PRACTICE

1 A cyclist claims that his brake blocks wear out because 'they do all the work to stop the bike'. Is this true? Explain. (Hint: think about the energy transfer that takes place when the bike stops.)

2 Describe the energy transfers that occur in a pocket torch that runs from a dry cell.

3 Water vapour in the air stores solar energy that can be released when the vapour condenses. How does this explain the fact that all cyclones begin over water?

4 If energy is conserved why do we worry about world energy resources?

5 Does the Earth's gravity do any work on the Moon? Assume the Moon's orbit is circular.

6 Tidal energy can be harnessed to generate electricity. Where does the energy of the tides come from? Is this an inexhaustible supply?

7 A perpetual motion machine could be used to drive a dynamo and generate unlimited electricity. Why is this impossible?

8 Describe the energy changes that take place when an ice cube melts in a thermally isolated beaker of water. Why does this result in a lower final temperature than if the same quantity of water at 0 °C had been added to the same beaker of water?

- calculating work done by constant and variable forces

- calculating strain energy

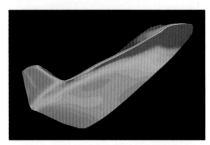

A computer-generated temperature map of Hermes, the European Space Agency's reusable space plane design concept, doing work against frictional forces as it re-enters the atmosphere. The work done transfers some of its kinetic energy to heat.

Changing non-parallel forces

In general the applied force may vary in magnitude and the displacement of the object may not be parallel to the force. You then have to use the component of displacement parallel to the force, and add up the contributions to total work from short sections in which the force stays approximately constant.

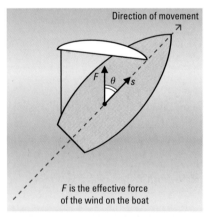

Direction of movement

F is the effective force of the wind on the boat

If the boat is sailing at constant speed its kinetic energy is constant. What happens to the work done on the sail by the wind? (The force from the keel is not shown; it enables the boat to sail away from the wind but does no work on it. Why not?)

WORK

Transfer of energy

When the space shuttle re-enters the Earth's atmosphere, a layer of heat-resistant tiles on the hull is vaporized. The same effect could be produced with stationary tiles by putting them in contact with a flame which is at a very high temperature. In both cases energy is transferred to the tiles, but the mechanism is very different. In the first case the tiles move against a frictional force from the atmosphere; in the second case energy flows from hot to cold. This illustrates two distinct ways in which energy is transferred.

- **Working** is an energy transfer when an applied force moves.

- **Heating** is an energy transfer resulting from a temperature difference.

Notice that the effect on the tiles is the same in both cases: to increase their internal energy, raise their temperature, and eventually vaporize them. On re-entry this is a by-product of the work done by the space shuttle as it pushes through the atmosphere.

Definition: work done

Work is defined as the product of force F applied and displacement s moved in the direction of the force:

Work done (J) = force applied (N) × distance moved parallel to force (m)

$$W = Fs$$

The equation above applies for a constant force and a displacement parallel to the force. The unit of energy is therefore the newton metre (N m) which is renamed as the joule (J).

Worked example 1

How much work is done if you push a shopping trolley with a constant force of 60 N as it moves 5 m parallel to the applied force?

$$W = Fs = 60\,\text{N} \times 5\,\text{m} = 300\,\text{N m} \quad \text{or} \quad 300\,\text{J}$$

This means that 300 J of energy have been transferred from you to the trolley. In this case the trolley is also continually doing work against frictional forces and so 'passing on' this energy to other objects.

Worked example 2

The diagram on the left shows a bird's-eye view of a sailing boat moving with the wind. However, the directions of the motion of the boat and the force from the wind are not parallel. How much work is done by the wind on the boat if it applies a steady force F as the boat moves a distance s?

$$W = Fs \cos \theta$$

$F \cos \theta$ can be thought of as the component of force in the direction of motion, or $s \cos \theta$ can be thought of as the displacement in the direction of the applied force.

Working to deform things

When something is stretched its tension increases. This means the stretching force must also increase. The work done to increase extension from 0 to a maximum value e is equal to the average force multiplied by e. If an object obeys Hooke's law the required force F is directly proportional to the extension and is parallel to it. In this case the average force is simply half the maximum force:

$$F_{\text{max}} = ke$$

$$F_{\text{av}} = \frac{1}{2}ke$$

$$W = F_{\text{av}}e = \frac{1}{2}ke^2$$

Another way to look at this is to realize that the work done on the spring is equal to the area under a graph of applied force versus extension. This approach is useful if the force varies in an irregular way – we can then use the graph to estimate work done.

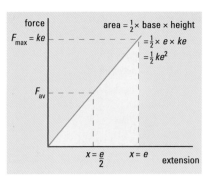

The shaded area equals work done to produce an extension e.

Strain energy

The analysis above gives the elastic energy stored in a *particular object*. It is often useful to calculate the elastic or strain energy stored in a particular *material*. This is done by recasting the equations in terms of **stress** and **strain** (see chapter 10) instead of force and extension. Take a wire of length l and cross-sectional area A, and apply a force F to it. It will extend an amount e and (if it obeys Hooke's law) the energy stored will be given by $EPE = \frac{1}{2}ke^2$ (EPE stands for elastic potential energy).

$$W = \frac{1}{2}Fe \qquad (F = ke)$$

Now use the definitions of stress and strain to replace F and e:

$$F = A\sigma \qquad (\sigma \text{ is stress}) \qquad e = l\varepsilon \qquad (\varepsilon \text{ is strain})$$

$$EPE = W = \frac{1}{2}\sigma\varepsilon Al$$

but Al is the volume, V, of the sample so the strain energy per unit volume is

$$\frac{EPE}{V} = \frac{1}{2}\sigma\varepsilon = \frac{1}{2}E\varepsilon^2 = \frac{1}{2}\frac{\sigma^2}{E}$$

The alternative forms for this equation come by substituting for σ or ε using $\sigma = E\varepsilon$ (where E is the Young modulus, specific to each material – see chapter 10), but the simplest way to remember the equation is:

$$\text{strain energy} = \frac{1}{2}\text{stress} \times \text{strain}$$

Worked example 3

Compare the strain energy per unit volume of mild steel and copper
(a) when equal stresses are applied, and (b) when they have equal strains.

$(E_{steel} = 2.1 \times 10^{11}\,\text{Pa}, E_{copper} = 1.1 \times 10^{11}\,\text{Pa})$

Let S represent the strain energy per unit volume:

(a) $\dfrac{S_s}{S_c} = \dfrac{\sigma^2}{2E_s}\dfrac{2E_c}{\sigma^2} = \dfrac{E_c}{E_s} = 0.52$ (same stresses)

(b) $\dfrac{S_s}{S_c} = \dfrac{E_c\varepsilon^2}{2}\dfrac{2}{E_c\varepsilon^2} = \dfrac{E_s}{E_c} = 1.9$ (same strains)

These results are easy to interpret. If steel and copper wires of similar dimensions are compared then the steel will extend less for the same applied force and so less work will be done in stretching it. This is why the answer to (a) is less than 1. On the other hand, to extend the steel wire as far as the copper wire requires a greater force so (b) is greater than 1.

Maths box

Calculating the area under a graph of force versus displacement is the same as integrating force with respect to displacement, so the general expression for calculating work done is:

$$W = \int_A^B F\,ds$$

A and B are the limits of the integral, that is, the start and end points for the sum. F varies with displacement and the infinitesimal displacements ds are parallel to F.

In this case limit A is $x = 0$ and limit B is $x = e$, and F varies according to Hooke's law, $F = kx$, so:

$$W = \int_0^e kx\,dx = \left[\frac{1}{2}kx^2\right]_0^e = \frac{1}{2}ke^2$$

PRACTICE

1 How much work must be done:

 a To lift a 5 kg bag 1.2 m and place it on a table?

 b To extend a spring of spring constant 25 N m⁻¹ by 3 cm?
 (Assume Hooke's law is obeyed.)

 c To drag a bag of rubbish 10 m across a field by pulling on a string at 40° to the horizontal with a force of 250 N?

2 The total stopping distance for a car is calculated on the assumptions (i) that there will be a constant thinking time once the driver sees a problem ahead and (ii) that he will then apply a constant maximum braking force to stop the car. Explain why the 'thinking distance' is proportional to initial speed whilst the 'braking distance' is proportional to the square of this speed.

3 An 80 kg baseball player slides to a halt from a speed of 8 m s⁻¹ in a distance of 4 m. What is the average horizontal stopping force exerted on him by the ground?

4 A 50 g ball bearing is dropped from a height of 0.50 m into a tray of fine sand and embeds itself to a depth of 1.5 cm. What was the average vertical force exerted on the ball bearing by the sand?

- work done used to find formulae for kinetic and potential energies

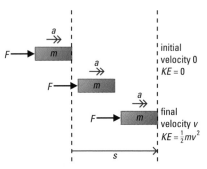

Transferring work to kinetic energy.

KINETIC AND POTENTIAL ENERGY

Kinetic energy

Anything that moves has the potential to do work. We can find an equation for this motion or **kinetic** energy from the work that must be done to accelerate something to its final velocity. The simplest way to do this is to assume it is accelerated by a constant force from rest. The work done is then just the force F times the distance s moved whilst accelerating. This distance is calculated from the equations for constant acceleration:

$$s = \frac{v^2 - u^2}{2a} = \frac{v^2}{2a} \quad \text{since } u = 0$$

$$W = Fs = \frac{Fv^2}{2a}$$

but $F = ma$ (see spread 3.21), so

$$W = \frac{1}{2}mv^2$$

The work done is transferred entirely to kinetic energy so

$$KE = \frac{1}{2}mv^2$$

The work done accelerating from u to v when u is not zero is:

$$W = \frac{1}{2}mv^2 - \frac{1}{2}mu^2 = \text{final KE} - \text{initial KE}$$

Often it is possible to solve a mechanics problem either from the equations of motion or by considering work and energy. In many cases it is quicker to use the energy approach. Here is a problem solved by both methods for comparison.

Worked example 1

Small dust particles of mass 10 mg collide with a spacecraft with a relative velocity of 40 km s⁻¹ and embed themselves in an aluminium shield of thickness 50 cm. What is the average stopping force exerted on the particles if they just fail to penetrate the shield?

Method 1 – 'suvat equations':

First calculate the acceleration. We know u, v and s so use $v^2 = u^2 + 2as$:

$$a = \frac{v^2 - u^2}{2s} = \frac{-u^2}{2s}$$

$$F = ma = \frac{-mu^2}{2s} = \frac{-10^{-5}\,\text{kg} \times (40 \times 10^3\,\text{m s}^{-1})^2}{2 \times 0.50\,\text{m}} = -1600\,\text{N}$$

Method 2 – energy:

The work done on the dust particle in stopping equals its initial kinetic energy:

$$\frac{1}{2}mu^2 = Fs \quad \text{therefore} \quad F = \frac{mu^2}{2s} = \frac{10^{-5}\,\text{kg} \times (40 \times 10^3\,\text{m s}^{-1})^2}{2 \times 0.50\,\text{m}} = 1600\,\text{N}$$

Gravitational potential energy

To lift something you have to apply a constant upward force at least equal to the object's weight. This force moves through the distance you lift the object, and the work you do is transferred to **gravitational potential energy** (**GPE**). (It is called *potential* because the gravitational field acts as a store for 'hidden' but retrievable energy.)

$$F = mg \qquad \text{(magnitude of applied force equals object's weight)}$$

$$W = Fh = mgh \qquad \text{(work done in lifting through height } h\text{)}$$

Since this work transfers energy entirely to gravitational potential energy:

$$\Delta GPE = mgh \qquad \text{(increase in gravitational potential energy)}$$

This formula is useful close to the Earth's surface where g is roughly constant, but would not be much use for calculating the work done in sending a spacecraft out into deep space, since g gets weaker with distance. We shall return to this problem when we look at gravitation a little later (see spread 5.5).

Worked example 2

A fountain shoots water to a height of 30 m. What is the minimum speed of water at the bottom of the fountain?

To rise 30 m the water must do work against gravity and store energy in the gravitational field. The amount of work is:

$$W = mgh$$

This work transfers energy from the initial kinetic energy of the water to gravitational potential energy, so the minimum initial kinetic energy of a mass m of water is:

$$KE = \frac{1}{2}mu^2 = mgh$$

so

$$v^2 = 2gh$$

$$v = \sqrt{2 \times 9.8\,\text{N kg}^{-1} \times 30\,\text{m}} = 24\,\text{m s}^{-1}$$

Hard work or no work?

Holding a heavy load above your head is very tiring, but actually does no work on it. The physical reason for this is that the support force does not move, but people often find this unconvincing. Another way of looking at it is to think of other ways of supporting the load, for example, by placing it on a shelf. The shelf has no source of energy and so obviously is unable to do work on it. So why do we feel tired? Why can a weight-lifter only support a raised load for a few seconds?

Some work is being done, but it is not transferred to the load. We are transferring chemical energy from our food to heat in metabolic reactions and the mechanical changes that keep our muscles tense – a continual tensing and relaxation of muscle fibres does work and converts some of our chemical energy to heat, which makes us tired.

In ordinary language this is 'hard work', but it is obvious that no work (force times displacement) is being done on the raised masses; their kinetic energy and gravitational potential energy are constant.

PRACTICE

1 Estimate the kinetic energy of **i** a Grand Prix car at full speed, **ii** an Olympic sprinter, **iii** an ant, and **iv** the Earth in its orbit.

2 Is a collision at 40 km h^{-1} twice as bad as a collision at 20 km h^{-1}?

3 A spacecraft is placed in a low Earth orbit about 300 km above the surface. Approximately how much gravitational potential energy (GPE) does it have with respect to the surface? (Assume g remains close to 9.8 N kg^{-1}.)

4 Which gains the greatest velocity, a block dropped vertically through 3 m or a similar block allowed to slide down a smooth slope with a vertical drop of 3 m? Ignore friction.

5 A swimming pool is 20 m by 30 m and contains water to a depth of 2 m. What is the change in GPE of the water when the pool is drained? (Ignore any further changes after the water leaves the pool, and assume it is drained by pulling a plug from the bottom of the pool. Density of water = 1000 kg m^{-3}.)

6 A tennis ball has a mass of 0.058 kg and is thrown vertically up as part of the service action. If its initial vertical velocity is 5.0 m s^{-1} what is the maximum height through which it rises?

7 Show that the total kinetic energy of the air passing through a wind turbine per second is proportional to the cube of the wind speed. What implications does this have for the operation of wind turbines?

8 The largest pumped storage power station in Europe is Dinorwig in Wales. Energy is stored by pumping water into a high reservoir during times of low electricity demand. When the demand increases Dinorwig operates as a hydroelectric power station, allowing the water to fall to a lower reservoir through a drop of 500 m, passing through six turbines on the way which each generate 310 MW of power.

a Water has a density of 1000 kg m^{-3}; what is the minimum volume of water that must flow through each turbine per second?

b Dinorwig can generate power continuously for 5 h. What is the minimum work that must be done to replenish the upper reservoir?

c In practice, when the turbines are used to pump water uphill they need an input of 280 MW for 6 h to refill the reservoir. What percentage of the energy stored at Dinorwig is returned to consumers?

d What happens to the rest of the energy?

Drax Power Station, Selby. Maximum power ouput is 4000 megawatts. Efficiency is 39%. It would be 40% except that the flue gas desulphurization plant drops the efficiency by 1%.

Running up stairs

Running up stairs increases your gravitational potential energy by the same amount as walking up them, but it transfers energy at a greater rate. When you run up stairs your power output is greater than when you walk up.

It takes more power to wade quickly than slowly.

POWER AND EFFICIENCY

Power

Sometimes we are more concerned about the *rate* at which energy is transferred than with the amount that is transferred. For example, a 40 W bulb may leave a room looking dark and dingy, but a 120 W bulb will make it seem bright. The 40 W lamp will emit as much energy as the 120 W lamp, but only if it is left on for a much longer time.

This is of no use to us. It is the amount of light per second that determines how bright the room will be. The 120 W lamp transfers more electrical energy to heat and light per second than the 40 W lamp. It is more powerful.

Definition: power

Power is the rate of transfer of energy.

If the energy transfer is by working then:

$$\text{average power (watts)} = \frac{\text{work done (joules)}}{\text{time taken (seconds)}}$$

$$P = \frac{W}{t}$$

The *instantaneous* power is the rate of transfer of energy:

$$P = \frac{dW}{dt}$$

A power of 1 watt transfers 1 joule of energy per second.

Power and motion

If something moves a distance s against a constant retarding force F (a swimmer moving through water, for example), then work is being done at a constant rate:

$$P = \frac{W}{t} = \frac{Fs}{t}$$

but

$$\frac{s}{t} = v$$

so

$$P = Fv \quad \text{or} \quad \text{power (W) = force (N)} \times \text{velocity (m s}^{-1})$$

Efficiency

Machines that transfer energy invariably transfer some of the input energy into unwanted forms. For example, a filament lamp gets hot as well as glowing. An electric motor vibrates, gets hot, and emits sound. There is a great deal of waste heat from a power station because more of the fuel's input energy ends up as heat than as electricity. A Sankey diagram can be used to follow the energy changes that take place in such systems. The diagram below is for an electric motor used to winch up a load. The width at each point represents the amount of energy in that form.

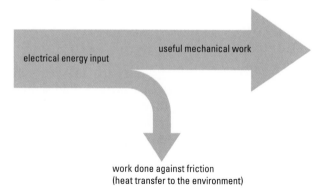

electrical energy input

useful mechanical work

work done against friction
(heat transfer to the environment)

A Sankey diagram for an electric motor.

Definition: efficiency

$$\text{efficiency} = \frac{\text{useful energy output}}{\text{total energy input}} \times 100\% = \frac{\text{useful energy output}}{\text{total power input}} \times 100\%$$

Limits to efficiency

Power stations, steam engines, and internal combustion engines are all **heat engines**. They convert some of the internal energy released by burning fuel into electrical energy that can be used to do work. All heat engines depend upon a temperature difference between the source of energy and the sink into which the heat flows.

For example, the temperature in the furnace of a power station is typically 1400 K and operates a steam cycle at 800 K. The sink is effectively the atmosphere at around 300 K. There is a theoretical limit (set by the **second law of thermodynamics**; see spread 7.13) to the efficiency of a heat engine and it depends on the ratio of the sink temperature T_2 and source temperature T_1.

efficiency is less than or equal to $(1 - T_2/T_1) \times 100\%$

For the steam cycle temperatures quoted above, which are typical of a fossil-fuelled power station or advanced gas-cooled nuclear reactor:

efficiency is less than or equal to $(1 - 300/800) \times 100\% = 62.5\%$

In practice, efficiencies closer to 40% are achieved. Energy transfers which do not involve heat are not limited in this way, so turbogenerators can be highly efficient at transferring kinetic energy to electrical energy.

PRACTICE

1 A 100 W filament lamp is 20% efficient at converting electrical energy to light.

  a What is the power of the light it radiates?

  b What is the power transferred to forms other than light?

  c How much electrical energy is transferred in one minute?

2 A 700 kg car can accelerate from rest to 30 m s^{-1} in 12 s.

  a What is its power output?

  b The same car is race-tuned and can now accelerate to the same speed in half the time. What is its new power output?

3 The intensity of solar radiation reaching the edge of the Earth's atmosphere is about 1400 W m^{-2}. Use this to estimate the total power output of the Sun. (Hint: assume no energy is absorbed between the Sun and the Earth, and that the radiation from the Sun spreads evenly over the surface of a sphere whose radius is equal to the distance from the Earth to the Sun, about 1.5×10^{11} m.)

4 A 50 tonne train can move at a steady 20 m s^{-1} on a horizontal track against a total resistance of 50 000 N. What is the train's power output?

5 An electric motor is run from the mains supply and is used to lift a 150 kg mass through 3 m vertically. The motor is only 50% efficient and the electrical supply system is 95% efficient. If the electricity is generated by a gas-powered power station with an efficiency of 40%, calculate the energy input from primary fuel required to lift this load. What is the overall efficiency of the process?

6 A thunderstorm deposits 1.5 cm of rain over an area of 25 km^2. The rain falls from an average height of 2 km.

  a What is the total change of gravitational potential energy (GPE) of the rain?

  b If the storm lasts 30 min what is its average power? (Ignore other energy transfers.)

  c How did this amount of energy get stored in the storm clouds?

7 A 500 MW hydroelectric power station operates with water falling 100 m and passing through turbogenerators. Assuming the transfer of GPE to electricity is 100% efficient, calculate the mass flow-rate of water through the turbines.

8 The power of a car is often expressed in horsepower (HP). A typical saloon might have a maximum power of 80 HP. Use this to make an estimate of 1 HP in watts.

9 Estimate your mechanical power when you make a sprint start. Do you think this is equal to the power consumption inside your body as you accelerate?

- the definition of pressure
- the variation of pressure with depth
- flotation

Definition: pressure

Pressure is the normal force per unit area.

$$\text{pressure (Pa)} = \frac{\text{force (N)}}{\text{area (m}^2)}$$

$$p = \frac{F}{A} \qquad 1\,\text{Pa} = 1\,\text{N m}^{-2}$$

Units of pressure

The pascal (Pa) is the SI unit of pressure, but two other units are also used – millimetres of mercury (mm Hg) and bars (or millibars). The first is the height of a mercury column that could be supported by the measured pressure, and the second is a unit related to atmospheric pressure and used in meteorology.

1 bar = 750 mm Hg = 1.00×10^5 Pa
(approximately 1 atmosphere)

1 mm Hg = 133 Pa

Standard atmospheric pressure =
101 325 Pa = 760 mm Hg

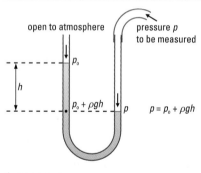

A manometer measures pressure difference. The pressure difference is measured by the difference in height between the two columns.

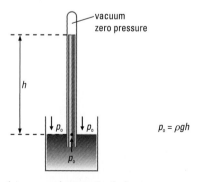

A mercury barometer balances a column of mercury against the atmosphere. The space at the top of the tube is a near vacuum (zero pressure), so the column rises or falls to a height at which the pressure at its base, arising from its own weight, equals the pressure in an open mercury reservoir.

FLUID PRESSURE

We live in a pressurized environment, but only really respond when the pressure changes. On a large scale, pressure differences cause winds and affect the weather. On a smaller scale, a sudden change of pressure when a train enters a tunnel can make our ears 'pop'.

At extremes of pressure the physical properties of substances around us can change dramatically. Sound cannot pass through a vacuum, and water can be made to boil at low temperatures. At 2.5 million atmospheres hydrogen becomes a metal, and under the influence of gravity the enormous pressure at the core of collapsing stars can turn hydrogen atoms into neutrons and even crush these into black holes. Closer to home, coffee beans can be decaffeinated by exposure to hot high-pressure carbon dioxide which acts as a solvent for the caffeine.

Variation of pressure with depth

If you swim down to the bottom of a swimming pool the pressure difference across your eardrum can be very painful. This is because the outside water pressure increases with depth and pressure in a fluid acts in all directions.

Scuba divers periodically equalize pressure by making their ears 'pop'. This opens the Eustachian tube which connects the middle ear to the throat, which has already equalized with the external pressure. If this tube is blocked, pressure on the eardrum causes it to bulge into the middle ear and can cause bleeding or even rupture of the eardrum.

The pressure difference is much greater on submarines and diving bells because they go much deeper, so they must have very strong hulls. For an aircraft, the pressurized air in the cabin exerts an outward force on the fuselage (balanced by elastic forces in the fuselage), and any puncture would cause air to rush out.

The increase of pressure with depth is due to the weight of fluid above. If the fluid density and gravitational field strength are constant, the **excess pressure** is easily calculated.

Imagine a vertical column of fluid of cross-sectional area A.

Volume of column = Ah

Mass of column = ρAh (ρ is the fluid density)

Weight of column = ρAhg

Pressure on area A: $p = \dfrac{\rho Ahg}{A} = \rho hg$

This is described as **excess** pressure because it adds to any pressure acting on the surface of the fluid. For example, the total pressure at the bottom of a swimming pool is the pressure of the water plus the atmospheric pressure above it. The excess pressure resulting from a column of liquid can be used to measure pressure differences.

- A **manometer** can be used to compare pressures. It consists of a U-tube of fluid that is displaced when the pressures at each end of the tube are different.

- In the **mercury barometer**, atmospheric pressure supports a column of mercury in a tall, sealed tube.

Flotation

Push a ping-pong ball down in a cup of water and release it. It bobs back up. Do the same thing with a stone of similar volume and it sinks. Why?

When something is fully or partly submerged there is a buoyancy force acting on it. This is the resultant of all the forces exerted by fluid pressure over the submerged surface of the object. It will float if the buoyancy force is at least equal to its weight.

For a rectangular block of density d, height x, and cross-sectional area A fully submerged at a depth h in a liquid of density ρ:

Force on bottom surface = $\rho g h A$

Force on top surface = $\rho g(h - x)A$

Resultant force = $\rho g h A - \rho g(h - x)A = \rho g A x$

The resultant force is called buoyancy or **upthrust**.

Volume of block = Ax, mass of block = Ax, weight of block = gAx. Block floats if:

buoyancy > weight

$\rho g A x > d g A x$

$\rho > d$

Things float in liquids which are denser than they are. (This is proved here for a rectangular block, but it applies to all shapes.)

Maths box: atmospheric pressure

The equation $p = \rho g h$ cannot be applied to the atmosphere, because density changes with height (g also changes, but insignificantly). To model how atmospheric pressure changes, some assumptions must be made. The simplest is to assume that temperature remains constant and use the ideal gas equation (see chapter 7) to find how density depends on pressure. The final step is to integrate the pressure differences across each small section of a column of air, from the surface out to some height. (In reality the pressure is negligible after about 100 km.)

The pressure difference across a thin horizontal slice of air at altitude x is:

$\delta p = -\rho(x)g\delta x$ (this is negative as p decreases with height x)

From the gas equation, $p = \frac{1}{3}\rho\overline{c^2}$, so

$\rho(x) = \frac{3p(x)}{\overline{c^2}}$ and $\delta p = \frac{-3p(x)g\delta x}{\overline{c^2}}$ ($\overline{c^2}$ is constant at a constant temperature)

Separating the variables, we can integrate this:

$\int_{p_0}^{p(h)} \frac{dp}{p(x)} = -\int_{x=0}^{x=h} k\,dx$ $\left(k = \frac{3g}{\overline{c^2}}$ and p_0 is the pressure at the Earth's surface$\right)$

$\therefore [\ln p]_p^{p(h)} = [kx]_0^h$ and $\ln\frac{p(h)}{p_0} = -kh$

Hence

$p(h) = e^{-kh}$

Atmosphere pressure decreases exponentially with height. The value of $\overline{c^2}$ can be calculated from the gas equation using sea-level values for p and ρ.

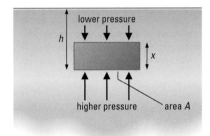

Upthrust is created by the increase of pressure with depth.

Heavy objects can still float.

PRACTICE

1 A rectangular block is 10 cm by 20 cm by 50 cm and has density 2 g cm^{-3}. Calculate the different pressures it could exert on a horizontal surface in each possible orientation.

2 When a doctor measures your blood pressure as '120 over 80' she is referring to the ratio of the maximum (systolic) to minimum (diastolic) pressure in mm Hg above atmospheric pressure.

   a Convert '120 over 80' to excess pressures in pascals. (Density of mercury = 13 600 kg m^{-3}.)

   b Blood has a similar density to water (1000 kg m^{-3}). Does the vertical distance between your head and feet make a significant difference to pressure in your body?

   c Astronauts in orbiting spacecraft initially experience an increased pressure in their heads. Why?

   d On returning to Earth there is a danger that astronauts will faint. Why?

3 Hydraulic braking systems transmit force through a liquid.

   a Give two reasons why liquids are suitable for this.

   b When the brakes are pressed a force of 150 N is applied to a piston in the master cylinder which has cross-sectional area of 4 cm^2. The force is transmitted to the slave cylinder and moves a piston there which operates the brakes. The cross-sectional area of the slave cylinder is 60 cm^2. What is the force on the slave cylinder piston?

   c How is energy conserved in the system that operates the brakes?

4 Why are gravity dams – that is, dams that stay in place because of their own weight – wider at the bottom than at the top? Does the size of the dam depend on the area of the lake it dams? Explain.

Incompressible fluids

Something is regarded as incompressible if its volume does not change significantly when it is pressurized. Liquids are usually incompressible. Even though gases are obviously compressible, we can still use the Bernoulli equation provided that the velocity of the object moving through the gas, or the velocity of the gas as it flows past an object, is small compared with the velocity of sound through the gas.

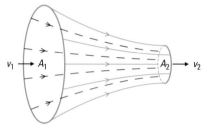

If the fluid is assumed to be incompressible, as much fluid leaves the pipe each second as enters it.

THE BERNOULLI EQUATION

The mechanics of fluid flow are complex and produce many unexpected effects, some of which are still not fully understood. The onset of turbulence when smooth flow is suddenly disrupted by complex motions like vortices and eddies is very difficult to predict or model effectively. In some situations turbulence is desirable; for example, to provide rapid mixing of fuel and oxygen inside a jet engine. In other situations it is disastrous. Turbulent airflow over a wing destroys lift.

However, some useful results can be found if the problem is simplified in the following ways.

- Assume that the fluid is *incompressible* (see left).
- Ignore fluid friction, that is, **viscosity**.
- Assume flow is smooth (laminar) and not turbulent.

Fluid flow is described using **streamlines**. These are arrows that represent the velocity of the fluid at each point. In steady flow they correspond to **lines of motion** which follow the paths of the particles.

The equation of continuity

Imagine an ideal fluid flowing through a frictionless pipe which becomes narrower and narrower. The fluid is incompressible, so as much fluid enters the pipe each second as leaves it. If the cross-sectional areas and velocities at entry and exit are A_1 and A_2 and v_1 and v_2 respectively then:

volume entering per second = $A_1 v_1$ = volume leaving per second = $A_2 v_2$

This is the **equation of continuity** and leads to:

$$\frac{v_2}{v_1} = \frac{A_1}{A_2}$$

The flow speed is greater in the thinner part of the pipe.

The Bernoulli equation

If a fluid accelerates along a constricting tube then there must be some force which increases its momentum and energy. This force arises from a pressure difference along the tube – high pressure where the flow is slow and low pressure where it is fast. Many surprising things are linked to this reduction in pressure for fast flow. It is known as the **Bernoulli effect**.

When fluid flows through a horizontal constricting pipe, more work is done on it by the pressure p_1 at entry than the fluid does against the pressure p_2 on exit. Neglecting friction, this increases the kinetic energy of the fluid.

In time δt a volume $A_1 v_1 \delta t = A_2 v_2 \delta t$ enters and leaves the tube.

Force on fluid at entry: $F_1 = p_1 A_1$

Force exerted by fluid at exit: $F_2 = p_2 A_2$

Since the fluid is moving, F_1 does work on the fluid and F_2 is responsible for work done by it.

Work done on fluid at entry: $W_1 = p_1 A_1 v_1 \delta t$

Work done by fluid on exit: $W_2 = p_2 A_2 v_2 \delta t = p_2 A_1 v_1 \delta t$

(using the equation of continuity)

Kinetic energy on entry: $KE = \frac{1}{2}(\rho A_1 v_1 \delta t) v_1^2$

Kinetic energy on exit: $KE_2 = \frac{1}{2}(\rho A_2 v_2 \delta t) v_2^2 = (\rho A_1 v_1 \delta t) v_2^2$

Conservation of energy: $W_1 - W_2 = KE_2 - KE_1$

$$(p_1 - p_2) A_1 v_1 \delta t = \frac{1}{2}(\rho A_1 v_1 \delta t)(v_2^2 - v_1^2)$$

or

$$(p_1 - p_2) = \frac{1}{2}\rho(v_2^2 - v_1^2)$$

$$p_1 + \frac{1}{2}\rho v_1^2 = p_2 + \frac{1}{2}\rho v_2^2$$

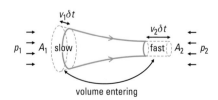

in δt the volume leaving = $A_2 v_2 \delta t$
in δt the volume entering = $A_1 v_1 \delta t$

Deriving the Bernoulli equation.

Along any horizontal streamline:

$$p + \frac{1}{2}\rho v^2 = \text{constant.}$$

If the streamline is not horizontal the fluid will be accelerated or decelerated by gravity. There is another term to this equation:

$$p + \rho gh + \frac{1}{2}\rho v^2 = \text{constant}$$

This is the **Bernoulli equation**. The height h is measured from some convenient reference level, for example at the surface of the liquid. The equation is really a statement of conservation of energy per unit volume in the fluid.

- $\rho gh + \frac{1}{2}\rho v^2$ is called **total pressure**.
- ρgh is called **static pressure**.
- $\frac{1}{2}\rho v^2$ is called **dynamic pressure**.

Total (or 'stagnation') pressure can be measured using a **Pitot tube**. This consists of a small open tube facing into the oncoming fluid. The fluid is brought to rest in the tube and the pressure inside the tube can be measured by connecting it to a pressure transducer. A **Pitot-static** tube has secondary tubes pointed at 90° to the flow. These measure static pressure. The pressure difference between the Pitot and static tubes is therefore equal to the dynamic pressure in the fluid, from which the fluid velocity can be derived. This is the basis of air-speed meters connected to jet aircraft. A **Venturi meter** measures the difference in pressure between two points in a fluid. This can also be used to find the fluid flow velocity.

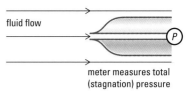

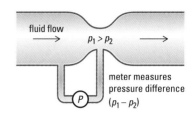

The Pitot tube.

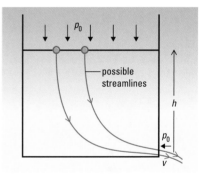

The Venturi meter.

Flow from a reservoir

Imagine a large container with a small hole near the base from which liquid flows. The external pressure at the hole and the pressure at the top surface of the liquid will both be about one atmosphere, so the change in dynamic pressure along a streamline going from the surface to the hole arises entirely from the gravitational term in the Bernoulli equation:

$$p_0 + \rho gh + 0 = p_0 + 0 + \frac{1}{2}\rho v^2 \text{ (h is measured from the base up)}$$
$$\frac{1}{2}\rho v^2 = \rho gh$$
$$v = \sqrt{2gh}$$

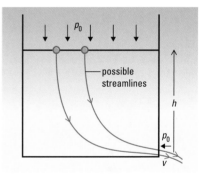

Flow from a reservoir.

PRACTICE

1 The destructive power of tornadoes in urban areas is mainly due to the sudden decrease in pressure they cause as they pass. This can be 10% of atmospheric pressure. Atmospheric pressure is about 10^5 Pa.

  a Explain why the pressure drops.

  b A 10% drop in pressure doesn't sound too bad; why is it so destructive?

2 A crowd of people is queuing to pass through a narrow tunnel leading into a football ground. Describe how it would feel to be one of these people as you approach the entrance and then pass along the tunnel (the experience is very closely analogous to water entering a constricted tube). Say where you are likely to feel crushed and how your rate of progress changes at each stage.

3 The speed of an aircraft can be measured by comparing the pressure of air moving past the plane with static atmospheric pressure.

  a How does this work?

b If the aircraft is flying at an altitude where the density of the air is $0.45\,\text{kg m}^{-3}$ and the pressure difference is 9000 Pa, what is the air speed of the plane?

4 A measuring cylinder contains $1000\,\text{cm}^3$ of water which fills it to a depth of 20 cm, and it has a small hole near its base through which water leaks out. The hole has a diameter of 0.8 mm.

  a What is the initial velocity of the escaping water? State any assumptions.

  b What is the initial rate of loss of water ($\text{cm}^3\,\text{s}^{-1}$)?

  c Repeat the calculations for **a** and **b** when the depth of water in the cylinder is x cm.

  d Sketch a graph showing how the level of water changes with time as the cylinder empties.

  e How would you modify this to make a practical water clock? What are its limitations as a timekeeper?

Anyone for tennis? Bernoulli in action.

THE BERNOULLI EFFECT

Bernoulli in action

1 Spin Table tennis players, golfers, tennis players, and footballers all know how spin can be used to control the flight of a ball. (The dimples on the surface of a golf ball are there to increase this effect.) Backspin causes the ball to rise, and topspin makes it dip sharply. Sidespin causes it to swerve, and is very useful for bending a ball round a defensive wall at free kicks! All of these are explained by the Bernoulli effect.

If a smooth non-spinning ball moves through the air then the streamlines split symmetrically either side of the ball, and there is no tendency to deviate from its straight-line path. If the ball is spinning and has a rough surface, the result is quite different. Air close to the ball is dragged round in the direction of the ball's spin, producing a larger velocity relative to the rest of the air on one side of the ball than on the other side. On the side on which the dragged air is pushed in the same direction as the outside air that is moving past the ball there is a lower pressure, and so a resultant force acts to deflect the ball from its straight path.

2 The 'Flettner ship' The sideways force generated by wind blowing past a tall rotating vertical cylinder (called the **Magnus effect**) can be used to propel a ship. The force arises in the same way as the force on a spinning ball. Before his death in 1997, Jacques Cousteau, the well-known French marine biologist, was planning to use this method to propel his latest craft.

3 The levitating ball If air is blown through a tube, a light ball (for example, a ping-pong ball) can be suspended in the airflow and remains there even when the tube is tipped sideways.

4 The lifting page Place a sheet of paper over the edge of a table so it sags downwards and then blow gently over the top of the paper parallel to the table. The paper lifts into the air flow.

5 Lifting by blowing Take a piece of card and push a drawing pin through its centre. Place the card on a table so the pin sticks up through its centre. Now blow down through a tube placed lightly over the pin so that the air spills out sideways around it. It should be possible to lift the card!

The last three examples above can be explained in much the same way as the previous ones. Try them.

The first Flettner ship, built in 1925.

Viscosity and turbulence

Fluids are characterized by their ability to shear easily. The **viscosity** of a liquid is a measure of its resistance to shear, and has a dramatic effect on its properties.

- A liquid with *low* viscosity will move from smooth laminar flow to turbulent (or eddying) flow, in which the streamlines behave unpredictably, at fairly low velocities.

- On the other hand, a liquid of higher viscosity will remain laminar to higher flow velocities.

The onset of turbulence is not well understood, but Reynolds carried out a series of experiments that allow us to predict roughly when turbulence will begin. The **Reynolds number** R is the critical parameter that determines when the transition to turbulence occurs:

$$R = \frac{vr\rho}{\eta}$$

where v is flow velocity, ρ is density, η is viscosity, and r is a characteristic dimension, for example, the radius of a tube. Air has a higher ratio of viscosity to density than water, and will remain laminar to higher flow velocities. The ratio

$$\frac{\eta}{\rho}$$

is called **kinematic viscosity**. Turbulence does not occur if the Reynolds' number is below 1000, but laminar flow can be maintained to much higher values if care is taken to avoid disturbances. (For more on viscosity see spread 3.25.)

The transition from laminar to turbulent flow can have great significance. For example, in normal flight the flow across a wing should be near laminar. Stalling occurs when it becomes turbulent above the wing. Aerofoils are shaped to increase the air speed and angle of attack at which this occurs, and this increases lift at a particular airspeed. Stalling can set in suddenly as the boundary layer of laminar flow breaks away from the wing's upper surface.

Windsurfers make use of the Bernoulli effect all the time – how?

streamline in laminar flow

streamline in turbulent flow

Laminar and turbulent flow.

normal flight

turbulent wake

stalling

boundary layer has broken away from wing

What happens when a plane stalls.

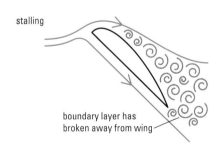

PRACTICE

1 Explain examples 3, 4, and 5 of the Bernoulli effect on the previous page.

2 Make estimates of relevant quantities and say whether you would expect laminar or turbulent flow in each of the following situations:

  a water flow in a pipe leading to a tap as it fills a bath (viscosity of water = 0.001 N s m⁻²,
    water density = 1000 kg m⁻³);

  b a major oil pipeline 1 m in diameter carrying 60 000 m³ of oil per day (viscosity of oil = 0.01 N s m⁻², oil density = 850 kg m⁻³);

  c a breeze blowing between tall buildings in a city
    (viscosity of air = 2 × 10⁻⁵ N s m⁻², density of air = 1 kg m⁻³);

  d blood flow in an artery;

  e water flow in a wide river;

  f water flow in a fast mountain stream;

  g air flow over a car;

  h water flow around a diver.

A McDonnell–Douglas C-17A. Even this can fly!

THE PHYSICS OF FLIGHT

Here is a free-body diagram showing the forces on an aircraft in uniform horizontal flight.

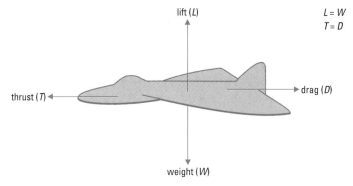

$L = W$
$T = D$

For uniform horizontal flight there is no resultant force.

Gravity creates weight; the other three forces arise from interactions between the aircraft and the air that surrounds it.

* **Thrust** Jets or propellors push back on the air and the air exerts an equal but opposite force back on them.

* **Lift and drag** As the aircraft moves forwards it collides with the air in front of it, pushing the air forwards and downwards. The air exerts a reaction force of equal magnitude backwards and upwards on the aircraft.

Lift and drag

A stationary wing experiences no lift or drag. Both these forces arise when the wing moves forwards through the air. The resulting upward and backward force can be resolved in two parts, lift and drag.

Another way to think about this is in terms of pressure. The stationary wing is surrounded by air at atmospheric pressure, so there is no resultant force from the air on the wing. When the wing moves, the pressure in front of and below it is increased whilst the pressure behind and above it is reduced. The net effect of these varying pressures over the entire surface of the wing is to create the resultant force described above. The distribution of pressure over the wing is quite complicated, and the pressure differences are greater near the leading edge of the wing, so the resultant force acts at a point closer to the front of the wing called the **centre of pressure**.

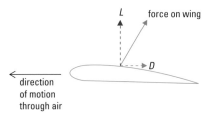

Lift and drag.

Miss America, a P-51D. The classic wing cross-section can just be seen at the wing tip.

advanced **PHYSICS**

The origin of lift

Lift is generated in two ways:

- By mounting the wing at an angle to the direction of motion so it deflects the oncoming air downwards. This angle is called the '**angle of attack**'.
- By using a curved surface to allow the air to flow smoothly over the wing and reduce **turbulence**. Turbulence (vortices created in the air behind the wing) carries away energy and so increases drag. A curved wing is called an **aerofoil** and its use greatly increases the ratio of lift to drag. Air leaving the wing flows downwards, so its reaction on the wing is an upward force.

Surprisingly, for most wings, the decrease in pressure above the wing is considerably greater than the increase of pressure below it.

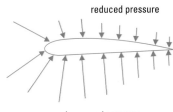

Pressure on a moving wing.

Aerofoils and the Bernoulli effect

The air flow past an aerofoil can also be regarded as an example of the Bernoulli effect. The air above the aerofoil moves faster than the air below it, so there is a lower pressure above the wing, and the pressure difference creates a resultant upward force on it. Sometimes this is explained by saying simply that the air above has to travel farther than the air below and so must move faster, but this is not really satisfactory by itself.

Imagine you are a tiny observer riding on a horizontal aerofoil. In your reference frame the air is flowing past a stationary aerofoil. Now visualize a large imaginary 'tube' through which the air flows with the aerofoil on its lower surface. The aerofoil acts like a constriction in this 'tube'. However, all the air entering the tube each second must leave it, so it must flow faster through the constricted region (i.e. above the aerofoil) than in the unblocked part of the 'tube'. This increase in speed is brought about by forces from the surrounding air created by differences in pressure. The result is that the fast-moving air over the wing exerts a smaller pressure on the surface of the aerofoil than the slower-moving air below it – the Bernoulli effect.

Incompressible air?

It is natural to expect the compressibility of air to affect the behaviour of an aerofoil. However, for subsonic flight, air acts as an almost incompressible fluid, so this is not a problem.

Streamlining

Imagine a circular plate of area A being pushed through the air at speed *v*. Relative to the plate the air is arriving at speed *v* and giving up all its kinetic energy in the collision. The dynamic pressure on the plate (from Bernoulli's equation) is

$\frac{1}{2}\rho v^2$ and the drag force is: $\frac{1}{2}A\rho v^2$

Replacing the plate with a streamlined object of the same cross-sectional area can reduce the drag by a factor of more than 20. The streamlined shape allows the air to have laminar flow around the object (so that it does not give up all its kinetic energy) and reduces the turbulence in the air behind the object. The streamlined shape of an aerofoil dramatically reduces drag and increases the lift-to-drag ratio.

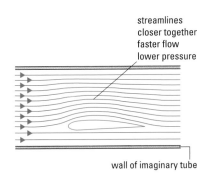

Visualizing the Bernoulli effect.

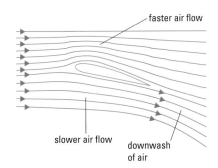

The air is pushed down so the wing is pushed up.

Stalling.

Stalling

At the angle of attack increases, so does the lift – up to a point. At about 15° (this varies slightly for different aerofoils) the lift drops dramatically and the aircraft **stalls**. At this angle the laminar flow over the aerofoil is suddenly lost and the air breaks away from the surface of the wing with a sudden onset of turbulence.

If I have seen farther [than others] it is by standing upon the shoulders of giants.

Isaac Newton in a letter to Robert Hooke, 1675

Isaac Newton (1642–1727).

Aristotle (384–322 BC).
Galileo Galilei (1564–1642).
Nicolaus Copernicus (1473–1543).
Johannes Kepler (1571–1630).

THE NEWTONIAN REVOLUTION

Newton revolutionized physics by unifying all of mechanics in three laws of motion and inventing the mathematics – calculus – needed to solve the equations they produced. He also formulated the law of universal gravitation which, when combined with the laws of motion, explains the orbits of planets and comets. The successful extension of physics to the cosmos led to the dominance of Newtonian mechanics through to the turn of the twentieth century. Newtonian physics will continue to dominate practical engineering, since the new theories of relativity and quantum mechanics lead to essentially identical results on a human scale.

Giants

Physical theories do not emerge completely unannounced. Sometimes history records the names of those who complete theories at the expense of those who contributed to them.

- Newton was probably the greatest physicist who ever lived, but he lived at the right time to benefit from the research of Simon Stevinus (vectors and equilibrium problems), René Descartes (Cartesian coordinates and analytic geometry), and especially Galileo Galilei (analysis of motion and force).

- He was also able to criticize classical ideas about time, space, and motion, particularly those of the Greek philosopher Aristotle, who in turn responded to ideas from Plato and Zeno.

- Newton's theory of gravitation assumed that planets orbit the Sun (Copernicus's heliocentric or Sun-centred 'universe'), and would have been far less obvious if applied to the earlier geocentric (Earth-centred) 'universe' of Ptolemy.

- The idea of elliptical orbits had already been proposed by Johannes Kepler as the best way to account for the accurate measurements of planetary motions made by Tycho Brahe. Kepler's three laws of planetary motion were later derived from Newton's theory.

The development of calculus and the concept of vectors were crucial to the new mechanics. Without them the idea of an *instantaneous* velocity made little sense, and projectile motion could not be analysed. Calculus is the mathematics of change, and Newton's laws of motion are recipes for change in Nature. The fact that the German mathematician, philosopher, and physicist Gottfried Leibniz developed the calculus independently of Newton suggests that science and mathematics had reached the point where this was a logical next step. Someone had to invent it to allow further progress.

Newton's other life

Newton was born on Christmas Day 1642, and spent his early years with his grandmother. He went to school in Grantham. He went up to Trinity College Cambridge in 1661 but was forced to return to his home village of Woolsthorpe when plague swept through the city in 1665. It was during the following year that he began to develop his ideas on calculus (or 'fluxions' as he called it), the laws of motion, and optics. It was in 1665 that the famous apple is supposed to have fallen from a tree and made Newton think of gravity.

He was elected a fellow of Trinity College in 1667 and became Lucasian Professor in 1669. Throughout his life he was reluctant to publish his own work and this often led to priority disputes (as with Leibniz). His great work, the *Philosophiae Naturalis Principia Mathematica* (Mathematical Principles of Natural Philosophy, usually called simply the *Principia*) was published in 1687, *Opticks* was published in 1704, and *Methodis Fluxionum* (calculus) was only published posthumously in 1736.

In 1696 Newton became warden of the Royal Mint and was made Master of the Royal Mint in 1699. He resigned his Cambridge post in 1701 and was knighted in 1705. He became president of the Royal Society in 1703 but did virtually no new science from then on.

Newton is often held responsible for the development of a mechanistic philosophy that regards everything as deterministic and predictable, with living things as complex automata. However, Newton was also fascinated by religious and mystical questions. He left over a thousand manuscript pages containing his own research into biblical chronologies and over 600 000 words on alchemy. His library contained 138 books on alchemy.

As the world's greatest physicist Newton claimed that he simply wrote down laws and derived results from them, but that he would not speculate idly about the underlying causes of gravity or the reasons for the laws of motion. However, in private he indulged in wild speculations, looked for signs in the Book of Revelations, and believed that many of his discoveries were already known to the ancients.

After Newton

Newton's theories were applied to astronomical and terrestrial physics with unprecedented success in the two centuries following his death. They were also developed mathematically, particularly by Lagrange and Hamilton, who produced generalized equations of great beauty and mathematical elegance. There had never been a theory like it, and the mechanical model was copied in the other sciences and even adapted for disciplines like economics and politics. However, there is physics beyond Newton.

Towards the end of the nineteenth century it seemed as if physics would present a final coherent and complete theory of everything. This theory would involve atoms as fundamental mechanical particles interacting by gravitational and electromagnetic forces.

Unfortunately:

- There were problems accounting for the spectrum of heat radiation given off by a hot body (**black-body radiation**). According to classical physics this should depend on mechanical vibrations of atoms causing electromagnetic waves of the same frequency, but careful calculations based on these ideas led to the prediction of infinite energy radiated at very high frequencies (the **ultraviolet catastrophe**).

- J.J. Thomson's discovery of the electron showed that atoms are not fundamental, and subsequent developments in particle physics generated a host of new and previously unexpected particles.

- The newly discovered **photoelectric effect** in which electrons are knocked off the surface of a metal by light was also proving impossible to explain using the wave theory of light.

- Attempts to explain light as a mechanical wave made of vibrations in some all-pervasive medium (the 'lumiferous aether') failed when the medium proved undetectable (the Michelson–Morley experiment).

New physics was needed, and several revolutionary theories were created in the early part of the twentieth century:

- Einstein's **special theory of relativity** (1905), dealing with the mechanics of rapidly moving objects;

- Einstein's **general theory of relativity** (1915), dealing with the geometry of space and time and the nature of gravity;

- **quantum theory**, developed by many physicists including Planck, Einstein, Bohr, Schrödinger, and Heisenberg (between 1900 and 1925), dealing with the nature of matter on a subatomic scale.

To explain all nature is too difficult a task for any one man or even for any one age. 'Tis much better to do a little with certainty, and leave the rest for others that come after you, than to explain all things.

Isaac Newton

Albert Einstein (1879–1955). 'Do not worry about your problems with mathematics, I assure you mine are far greater.'

Quantum theory cannot be attributed to one physicist alone. Planck (1858–1947) (top left) was the first to introduce quantization (1900). Einstein (above) proposed the photon theory of light (1905). Bohr (1885–1962) (top right) quantized the atom (1913). Schrödinger (1887–1961) (bottom left) and Heisenberg (1901–1976) (bottom right) proposed alternative but equivalent mathematical theories (1925). Many other twentieth-century physicists made major contributions.

Newton for everyday life

Special relativity diverges from Newtonian physics when things move with velocities comparable to the speed of light, for example electrons in a particle accelerator. **General relativity** disagrees with Euclidean geometry and Newtonian gravitation over extremely large distances, or in extremely strong gravitational fields. **Quantum theory** diverges from Newtonian physics on a subatomic scale. However, ordinary experience is in complete agreement with Newton's picture of the world.

NEWTON'S FIRST LAW OF MOTION

Who is moving?

A natural state of motion

Imagine yourself lost in deep space. All you can see are the distant stars, but they are too far away to tell if they are moving. Then something comes into view. It is small and far away but it is getting closer. As it drifts past, you recognize your twin, also lost in space, and you wave sadly to each other. Some time later a thought occurs to you: is it you who is moving or is it your twin, or are you both moving? Is there any way of telling?

A similar thought occurred to Galileo. He imagined being locked below decks in a windowless room on a ship which may or may not be sailing smoothly across the sea. Is there any experiment that could be carried out inside the room to determine whether the ship is moving or not? He considered jumping up and down, pouring water into a jug and throwing a ball, but a moment's reflection shows that none of these would show the slightest sign of the ship's motion. (If you doubt this and think that jumping vertically in a moving boat would cause you to land behind your take-off point, transfer the argument to a jet aircraft travelling at hundreds of metres per second, or try it on a train!)

Galileo concluded that uniform motion in a straight line is just as natural as being at rest, and so rest itself is not special. All it means is *at rest with respect to your surroundings*. We are used to using the Earth as our frame of reference, but this is simply a small planet orbiting a minor star some way from the centre of a rotating spiral galaxy which is itself moving in the gravitational fields of other galaxies. In other words, our Earth is not so special after all or, as Einstein might have said:

There is no such thing as absolute rest or absolute motion; all motion is relative. The laws of mechanics are the same in a laboratory at rest or in uniform motion.

This conclusion was revolutionary. It meant that *all* motion is relative. For two thousand years people had argued that all bodies tend to come to rest naturally. Galileo disagreed.

(a) real world

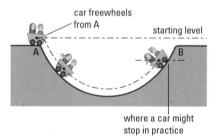

(b) ideal world – no friction

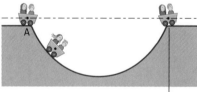

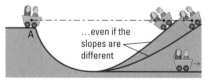

if the track *never* rises back up, the car continues for ever

Galileo's thought experiment.

A roller-coaster ride to infinity!

Galileo described a simple experiment in which a ball rolls down a U-shaped track and then rises to very nearly the same height on the far side of the track. Galileo assumed that the only reason it failed to reach the original height was because of frictional forces from the track and air. If these could be removed it really would reach the same height. He then reduced the slope on the far side so that the ball had to travel *further* to rise through the same vertical height. It did. What if the far side *never* rises up again? The ball will continue rolling forever at a constant speed in a natural state of motion every bit as significant as being at rest on Earth!

Most of the movement we see around us is affected by frictional forces that stop things moving past their surroundings. However, it is possible to simulate an ideal friction-free environment on an air track or air table, and then the riders do continue with almost constant velocity between collisions. Images from inside orbiting spacecraft also confirm that if something is dropped or thrown it continues moving in a straight line at constant speed until it hits something.

One of the most important effects of Galileo's work was the realization that *no resultant force is required to maintain uniform motion* (since none is required to keep something at rest). Newton developed Galileo's ideas into the laws of motion.

> **Newton's first law of motion**
>
> Objects continue to move at constant velocity (which may be zero) until they are acted upon by a resultant force.

What does the first law tell us?

The consequences of the first law can often seem surprising.

- An object moving at constant velocity (remember this means constant speed and direction) is in equilibrium. All the forces that act upon it (if any) cancel out. Sometimes people find this a bit hard to take. 'Surely,' they argue, 'if a car is travelling forwards at a steady 70 mph the thrust from the engine must be *a bit bigger* than the total of drag and other frictional forces?' Not true – thrust and drag must balance. 'Well, what about lifting something, surely you need to pull up *a bit harder* than gravity pulls it down?' Not true – to lift something at a steady rate you must apply a lifting force exactly equal to the weight of the thing you are lifting. Of course, *to start it moving* you need a bigger force, but that too agrees with Newton's law: the object is *changing* its velocity as it starts to rise.

- Anything that is changing its velocity *must* have a resultant force acting on it. That could mean accelerating, decelerating, or changing its direction. For example, when you drop something it falls with an increasing velocity. There must be a resultant force acting on it. We call this force gravity. The Moon is changing its velocity. Its speed is constant but its direction of motion is continually turning toward the Earth. There must be a force acting on it in the direction of the Earth – gravity again.

- The first law is sometimes called the 'law of inertia', since it seems to encapsulate the tendency of matter to keep moving in the way it is already moving.

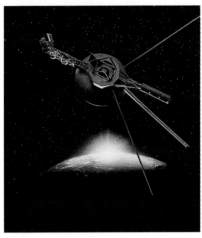

Voyager 1 was launched in 1977, and Voyager 2 in 1979. They have now passed and photographed the outer planets and are headed away from the solar system. They both carry specially coded audio recordings entitled 'Sounds of Earth' for the entertainment of any aliens who happen to find them.

Shut your eyes on board a cruising plane and you can almost imagine it is motionless. There is no way to distinguish uniform motion from rest, so there is no meaning to the idea of 'absolute rest'; all motion is relative.

There is no resultant force on this cruising airliner.

PRACTICE

1 A stone is dropped from the top of a sailing boat's mast to the deck below. Will it land in front of, behind, or level with the foot of the mast if:

  a The boat is at rest on the water?

  b The boat is moving forwards at a steady speed?

  c The boat is moving backwards at a steady speed?

  d The boat is accelerating in the forward direction?

2 What two forces act on a person in a stationary lift? How do these forces compare?

3 How do the forces described in question 2 change when the lift is:

  a Moving up at constant speed;

  b Moving down at constant speed;

  c Accelerating upwards;

  d Accelerating downwards;

  e Decelerating downwards;

  f Decelerating upwards?

4 A friend argues that he knows how to survive if a lift falls down its shaft. You have to wait until the moment just before the crash and then leap into the air. The lift will crash and you can step out of the wreckage. Is this possible?

5 A car is driving straight up a steep hill into a headwind at a steady speed. Draw a free-body diagram of the forces acting on the car, and use a vector diagram to show their resultant.

6 If uniform motion requires no resultant force to keep it going, why do we need to put petrol in cars?

7 Copernicus revolutionized cosmology by claiming that the planets orbit the Sun rather than the Earth. If all motion is relative isn't either point of view equally acceptable?

- the effects of a resultant force

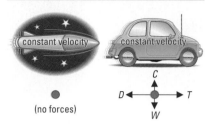

$W = C \quad T = D$
The resultant force is zero in both cases.
W = weight, C = contact force, T = thrust,
D = drag.

Using tickertape.

This bullfrog uses a resultant force to accelerate from rest as it jumps.

Using F = ma

- Draw a free-body diagram showing all forces acting on the object whose motion is to be analysed.
- Work out the resultant force F (this is equal to the vector sum of all the forces in the free-body diagram).
- Substitute values into F = ma.

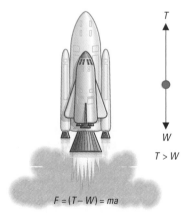

$F = (T - W) = ma$

Using a free-body diagram to find resultant force.

FORCE AND MOTION

A spacecraft drifting at constant velocity in deep space has no forces acting upon it. A car travelling along a motorway at a constant $100 \, \mathrm{km \, h^{-1}}$ is affected by many different forces, but both the car and the spacecraft have the same kind of motion – constant velocity in a straight line – and in both cases the *resultant force* is zero. The total frictional drag on the car is exactly equal to its forward thrust, and the weight of the car is balanced by an equal upward force from the road on the tyres.

If the spacecraft fires its rockets it will no longer continue at a constant rate; its velocity will increase and it will accelerate in the direction of the resultant force. If the car engine suddenly stops and the thrust disappears there will be an unbalanced or resultant force on the car which opposes its motion. It will decelerate until it stops moving. In both cases the effect of a resultant force is an acceleration in the direction of the force (in the case of the car a forward deceleration is the same as a backward acceleration).

- *The effect of a resultant force is to change the state of motion of the object on which it acts.*

Resultant force, mass, and acceleration

You could investigate the link between resultant force and acceleration by applying known forces to a vehicle and measuring its acceleration using a motion sensor: either light gates or tickertape. You should get the following results.

- Acceleration is directly proportional to resultant force (rider of constant mass).

  $a \propto F$ (constant mass)

- Acceleration is inversely proportional to rider mass (constant resultant force).

  $a \propto \dfrac{1}{m}$ (constant resultant force)

This is best summarized by:

  $a \propto \dfrac{F}{m}$

Definition: the newton

The unit of force (the **newton**) is defined so that the constant of proportionality in the equation above can be replaced by an equality. To do this 1 unit of resultant force must make a 1 kg mass accelerate at $1 \, \mathrm{m \, s^{-2}}$, so:

- 1 newton = the resultant force that accelerates a 1 kg mass at $1 \, \mathrm{m \, s^{-2}}$

This means that the 'newton' is simply a fancy way of saying $1 \, \mathrm{kg \, m \, s^{-2}}$!

Newton brought all this together in his second law of motion:

If a resultant force is applied to an object of constant mass, it will accelerate in the direction of the resultant force. The acceleration is directly proportional to the resultant force and inversely proportional to the mass (or inertia) of the object. If **SI** units are used throughout, this can be written as:

  $F = ma$

This is probably the most important single equation in physics, because it allows us to predict what happens next when things interact.

Worked example 1

5 seconds after launch the total vertical thrust of the space shuttle was $3.0 \times 10^7 \, \mathrm{N}$. Its mass at this time was $1.9 \times 10^6 \, \mathrm{kg}$. What was its acceleration?

Resultant vertical force = thrust − weight

  $F = 3.0 \times 10^7 \, \mathrm{N} - 1.9 \times 10^6 \, \mathrm{kg} \times 9.8 \, \mathrm{N \, kg^{-1}} = 1.14 \times 10^7 \, \mathrm{N}$

  $a = \dfrac{F}{m} = \dfrac{1.14 \times 10^7 \, \mathrm{N}}{1.9 \times 10^6 \, \mathrm{kg}} = 6.0 \, \mathrm{m \, s^{-2}}$

Linear momentum

So far we have restricted ourselves to resultant forces acting on objects of constant mass. In the real world this is often not the case – the rocket, for example, reduces in mass as its fuel is consumed and expelled. Newton's second law can be used to analyse these situations, but it needs to be recast in terms of a new quantity: **linear momentum**.

Imagine stepping in front of someone to block them. If they are walking, this involves less of an impact than if they are running, and a skinny child will be easier to stop than an adult rugby player running at the same speed. In general there will be a greater impact if the 'amount of motion' is greater. 'Amount of motion' depends on *both* mass and velocity and is called *linear momentum* – a sprinting adult has much more momentum than a toddling child and a cricket ball has more momentum than a table tennis ball moving at the same speed.

A dramatic change of momentum.

Definition: linear momentum

linear momentum (kg m s^{-1}) = mass (kg) × velocity (m s^{-1})

$$p = mv$$

Like velocity, momentum is a vector quantity.

Worked example 2

What is the linear momentum of a tennis ball (mass 0.058 kg) served at 35 m s^{-1}?

$p = mv = 0.058\,\text{kg} \times 35\,\text{m s}^{-1} = 2.0\,\text{kg m s}^{-1}$ (2 sig. figs)

The first law of motion could be recast as: *The linear momentum of an object is constant unless* it is acted upon by a resultant force'. The second law of motion describes what happens when there is a resultant force – the linear momentum changes. *The larger the resultant force, the greater the rate at which momentum changes.*

Internal forces

Complex systems have internal forces as well as external forces. Internal forces arise from interactions between objects within the system so they are action–reaction pairs. The vector sum of internal forces is always zero, so they do not affect the resultant force on the system.

The resultant force is the vector sum of the external forces acting on the system. For example, intermolecular forces on air molecules inside a tennis ball do not affect the motion of its centre of mass. This is determined by external forces such as gravity or impacts.

PRACTICE

1 A 900 kg car travelling at 25 m s^{-1} brakes hard and comes to rest in 5 s. What is the average braking force? (Ignore drag.)

2 Use the second law of motion to explain why (ignoring frictional forces) the free-fall acceleration of all objects dropped close to the Earth's surface is the same.

3 Draw free-body diagrams for each of the following and describe their motion in each case:

  a a bouncing ball (i) at the top of its bounce and (ii) at the instant when it is momentarily at rest during a collision with the ground;

  b a parachutist at the moment her parachute just opens;

  c a rain drop falling at its terminal velocity;

  d the Moon;

  e a ping-pong ball released under water (i) as it is released and (ii) a short time later before it reaches the surface of the water.

4 A child is tobogganing. If the total frictional force on the sledge is 100 N and the mass of the child and sledge is 45 kg, calculate the initial acceleration down a 30° slope.

5 The diagram shows a heavier child (50 kg) lifting a lighter child (40 kg) by hanging on one end of a rope which passers over a (reasonably smooth) tree branch. Calculate the initial acceleration of the two children*, assuming that friction on the branch can be neglected. In practice, of course, the friction is likely to be significant. What difference will it make to the initial acceleration?

(*Draw a free-body diagram for each child and apply the second law of motion to each of them.)

6 Estimate your linear momentum when you are running as fast as you can.

7 What velocity would a ball of mass 500 g need to equal the linear momentum of a 750 kg car travelling at 20 m s^{-1}?

3.22

- resultant force equals rate of change of linear momentum

In order to attack its prey effectively, this cheetah must be able to change its momentum rapidly.

NEWTON'S SECOND LAW OF MOTION

The second law of motion

> **Newton's second law of motion**
>
> The resultant force F exerted on a body is directly proportional to the *rate of change* of linear momentum p of that body.

$$\text{resultant force (N)} \propto \text{rate of change of linear momentum (kg m s}^{-2})$$

$$\propto \frac{\text{final linear momentum (kg m s}^{-1}) - \text{initial linear momentum (kg m s}^{-1})}{\text{time for which force acts (s)}}$$

The units of force are defined to make the constant of proportionality unity so we can write

$$F = \frac{mv - mu}{t} \quad (1) \qquad \text{or} \qquad F = \frac{\Delta p}{\Delta t}$$

Strictly speaking, F in the equation above is the average resultant force during time t. The instantaneous value of the force is given by

$$F = \frac{dp}{dt} \quad (2) \qquad \left(\text{remember that } \frac{\text{'d'}}{dt} \text{ means 'rate of change'} \right)$$

NB Momentum, force, and velocity are all vectors, so the momentum changes above are parallel to the resultant applied forces.

> **Worked example 1**
>
> What was the average resultant force needed to accelerate the space shuttle (mass 2.0×10^6 kg) from rest to 400 m s^{-1} in 22 s?
>
> $$F = \frac{mv - mu}{t} = \frac{2.0 \times 10^6 \text{ kg} \times 400 \text{ m s}^{-1}}{25 \text{ s}} = 3.2 \times 10^7 \text{ N}$$
>
> (This could be solved equally quickly using Newton's second law in the form $F = ma$.)

We can rearrange equation (1) to give

$$F = \frac{m(v - u)}{t} = m \frac{(v - u)}{t} = ma$$

$$F = ma \tag{3}$$

This is a special case of the general equation (1) and applies if the mass is constant.

Impulse and momentum change

Rearranging equation (1) gives us

$$Ft = mv - mu \tag{4}$$

The quantity on the left-hand side is called the **impulse**, the product of force and time, and is measured in N s. The quantity on the right-hand side is the change of momentum produced by this impulse. A word equation for this is impulse (N s) = change of momentum (kg m s^{-2}), or

$$F\Delta t = \Delta p$$

which is an alternative version of equation (4) and can be used even if the mass is not constant.

A large force acting for a short time causes the same *change of momentum* as a small force acting for a longer time. You use this every time you jump off something and bend your knees on landing. If you were to land rigidly, your momentum would fall to zero in a very short time. The large *rate of change of momentum* this involves would exert large forces on your body and probably cause injuries. Bending your knees allows the *same* change of momentum to occur over a *longer* time and so reduces the force on your body. A similar principle applies in the use of air bags and crumple zones to make cars safer.

Bend your knees or break your legs!

Worked example 2

A gymnast of mass 62 kg bounces vertically on a trampoline so that she approaches and leaves the trampoline with a speed of 8.0 m s^{-1}. Calculate (i) her change of momentum and (ii) the average resultant force exerted on her whilst in contact with the trampoline (contact time of 0.8 s).

(i) Take upwards as positive:

$$\Delta p = mv - mu = 62 \text{ kg} \times 8.0 \text{ ms}^{-1} - (-62 \text{ kg} \times 8.0 \text{ m s}^{-1}) = 990 \text{ kg m s}^{-1} \text{ (2 sig. figs)}$$

(ii) $F = \dfrac{\Delta p}{\Delta t} = \dfrac{990 \text{ kg m s}^{-1}}{0.8 \text{ s}} = 1200 \text{ kg m s}^{-1}$ (2 sig. figs)

(The total force exerted on the gymnast by the trampoline is larger than the resultant force calculated above because of the gymnast's weight.)

Force–time graphs

The term *impulse* is usually used when a force acts for a short time (in collisions, for example), but the equation works for constant forces acting for a longer period too. If the force changes with time then *total impulse is equal to the area under the force–time graph*. The force–time graph is for a short 'burn' of a solid fuel rocket.

Each vertical strip represents an impulse $F\delta t$ (force times short time interval) and results in a small change of momentum δp. The area under the whole graph is therefore the total impulse during the 'burn' and is equal to the change of momentum Δp of the rocket. This can be evaluated by counting squares. The velocity change of the rocket is then $\Delta v = \Delta p/m$ (ignoring the small change in rocket mass due to burnt fuel).

Conservation of linear momentum

When two things interact the force exerted on one is always equal and opposite to the force exerted on the other (Newton's third law). The interaction time is the same for both so they receive equal and opposite impulses.

The sketch graph shows how forces acting on two cars depend on time during a head-on collision. The area between each graph line and the time axis is the total impulse received by that vehicle during the collision. Since the forces are equal in magnitude but opposite in direction at all times, the two impulses have the same magnitude but opposite directions. This means that the change in momentum of the vehicles is also equal in magnitude and opposite in direction. There is no change in the total linear momentum of the system of two bodies.

This is an example of a conservation law – the two cars have interacted but the total linear momentum is the same before and after the collision. *All* interactions conserve linear momentum and *all* forces arise through interactions, so the total linear momentum of the universe is constant. For practical purposes we apply the law in a more restricted way to **closed systems**.

Law of conservation of linear momentum

The linear momentum of any closed system is conserved. Linear momentum is a vector quantity and each component is conserved independently.

Maths box

Counting squares is equivalent to integrating.

If force F acts for a short time δt, the momentum changes by δp:

$$\delta p = F\delta t$$

The total momentum change Δp during time t will be the sum of all small changes

$$\Delta p = \sum_0^t \delta p = \sum_0^t F\delta t$$

If $\delta t \to 0$, the summation becomes an integration

$\Delta p = \displaystyle\int_0^t F\delta t$ this is equal to the area under the graph between 0 and t.

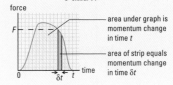

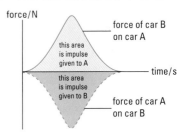

A closed system

A **closed system** is not acted on by any external resultant forces. Another way of looking at this is to say that it includes both ends of all interactions. For example, a falling stone increases momentum because it is acted on by a resultant force from the Earth. A closed system for a falling body must include the Earth. The increasing downward momentum of the stone is then balanced by an equal upward momentum imparted to the Earth. The huge mass of the Earth means that when the momentum of a human-sized object is transferred to it there is no noticeable change in its motion. This is one of the reasons that Newton's first law and the conservation of linear momentum took so long to be discovered.

PRACTICE

1 A 65 kg woman can sprint to 9 m s^{-1} in 5 s. Her 25 kg daughter can accelerate to 3 m s^{-1} in 3 seconds.
  a Calculate the change in momentum in each case.
  b Compare the *rates of change* of momentum. What are appropriate units for rate of change of linear momentum? **c** Show that the units you used in **b** are equivalent to newtons, the unit of force.

2 Give two reasons why an airbag can reduce a driver's injuries in a car crash.

3 A 72 kg man steps forwards out of a rowing boat with a horizontal velocity of 1.0 m s^{-1} relative to a quay. The mass of the boat is 200 kg. How fast does it move away from the quay?

NEWTON'S THIRD LAW OF MOTION

Everything pushes back

Clap your hands. Your left hand feels a force from your right and your right hand feels a force from your left. These forces are equal in magnitude but act in opposite directions along the same line of action. Newton's third law is deceptively simple – it states that every time a force is applied by one object on another the other object pushes back with an equal and opposite force. The third law is often quoted as 'every action has an equal and opposite reaction', but in this form it is often misapplied. Test this yourself with a simple example. A book rests on a table: write down the action and reaction pairs before reading on.

If you thought that the weight of the book and the contact force from the table are one such pair, you have made a very common error. Before you check the correct pairs in the diagram and note on the left, think about this more complicated example.

- Imagine you are standing on a flat horizontal surface, perhaps in an elevator. Your weight is W and the contact force from the floor is N. It is true that $W = N$ if the elevator is not accelerating, but what if it *is* accelerating?

- Let's say it accelerates upwards and so do you. Now there must be a resultant upward force *on you* to produce this acceleration. But only N and W act on you, so N must now be greater than W. If N and W are not equal, they cannot be an action–reaction pair.

This should convince you that the book's weight and the contact force from the table in the first example also cannot be an action–reaction pair in the sense of Newton's third law.

There is another reason why these two forces cannot be an action–reaction pair. They both act on the *same* body. If action–reaction pairs act on the same body and are always equal, then resultant forces would be impossible, and nothing would ever change its momentum!

So what does Newton's third law of motion really say?

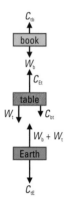

Free-body diagrams showing pairs of forces acting when a book rests on a table.

Key:
C_{tb} = contact force, table on book
W_b = weight of book
C_{Et} = contact force, Earth on table
W_t = weight of table
C_{bt} = contact force, book on table
C_{tE} = contact force, table on Earth

NB ($W_b + W_t$) acting on Earth is actually the weight of Earth in the gravitational field of the table and book.

> ### Newton's third law of motion
>
> - Forces never arise singly but always in pairs as the result of *interactions*.
>
> - When A interacts with B the force A exerts on B is always equal to the force B exerts on A but in the opposite direction along the same line of action.
>
> - Because these pairs of forces arise from an interaction they are always of the same type: for example, both gravitational or both electromagnetic.
>
> - Action–reaction pairs *always* act on different bodies, never on the same body.

Types of interaction

It is believed that all forces arise from four underlying interactions: electromagnetism, gravitation, and the strong and weak nuclear forces. The last two are extremely short-range and have no directly observable effects in everyday life. The first two account for all the forces we experience. What we 'feel' as contact when we touch something is transmitted through the electromagnetic field; electron orbits in atoms on the surface of our hand and on the object touched are distorted.

In principle this is no different from the apparent action-at-a-distance when magnets or charges exert forces on one another. The only real difference is that the distance involved is so small that it looks to us as if we 'really touch' things. In the end, every force we feel is transmitted through a field, so our simple concept of contact has to be modified.

Frictional forces

Frictional forces always oppose the relative motion of two objects. Surface friction arises when one object slides over another, and is caused

---

The book, the table, and the world

The book's weight arises because of a gravitational attraction to the Earth. There is an equal gravitational force on the Earth attracting it to the book. The force from the table on the book arises through the distortion of the bodies in contact, and is equal and opposite to the contact force from the book acting on the surface. The origin of these contact forces can be found in the elasticity of the materials. The book is effectively 'squashed' between the contact force from the table and its own weight. The repulsive forces between its particles, as they are pushed closer together, result in a force on the surface of the table.

by the microscopic roughness of the two surfaces and by the formation of temporary bonds between them at points where the stress is large. Fluid friction arises when a liquid or gas flows through a tube or when a solid body moves through the liquid. It is caused by cohesive forces in the liquid and by the sheer effort of pushing the liquid out of the way in order to move through it. (Friction and its effects on motion are discussed in spread 3.25.)

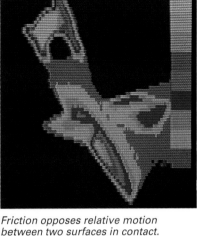

Friction opposes relative motion between two surfaces in contact. This thermogram of a wood saw demonstrates an effect of friction: heat.

Worked example

The diagram below shows a train pulling two carriages. It is moving along a horizontal track with an acceleration a. The engine has mass M and each carriage has mass m. The combined drag and frictional force on each carriage and on the engine is f. What is the thrust of the engine and the tension in each coupling? Identify any action–reaction pairs.

(i) To calculate the required thrust F from the engine, treat the whole train as a single object and draw a free-body diagram for it. Now write down the second law of motion for the train:

resultant force = total mass × acceleration

$$F - 3f = (M + 2m)a \quad \text{so} \quad F = (M + 2m)a + 3f$$

(ii) Now do the same for just the final carriage:

$$T_2 - f = ma \quad \text{so} \quad T_2 = ma + f$$

(iii) Now for the other carriage:

$$T_1 - (T_2 + f) = ma \quad \text{so} \quad T_1 = 2(ma + f)$$

If this is repeated for the engine alone it will give the same value for resultant force as in (i).

(iv) The thrust F is applied to the engine as a frictional force from the rails. It is a reaction to the engine's wheels pushing back on the track. Where each coupling connects to a carriage it exerts a force on the carriage (the tension in the coupling), equal but opposite to the force exerted on the coupling by the carriage.

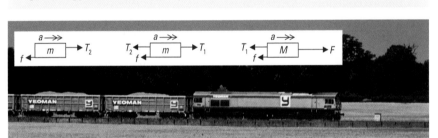

A photograph and free-body diagram of an engine pulling two trucks of equal mass.

PRACTICE

1 Identify two action–reaction pairs for a person sitting on the ground.

2 A helicopter is hovering some distance above the ground. Draw a free-body diagram showing the main forces acting on the helicopter and identify the reaction forces to each of these.

3 **a** Why is a jet engine useless in space?

  b Explain how it is possible for a rocket to accelerate in space and identify an important action–reaction pair essential for rocket propulsion.

4 Use Newton's third law to explain why your foot slips backwards when you start to run on an icy surface.

5 A beaker of water is placed on a top-pan balance and the balance reads 500 g. A 10 g mass is dropped into the water and sinks to the bottom of the beaker.

The balance now reads 510 g. If the mass had floated instead of sinking would the balance reading have been 500 g or 510 g? Explain.

6 A bird cage weighing 50 N contains a bird of weight 5 N. The cage is suspended from a newton-meter. What is the reading on the newton-meter if:

  a the bird is asleep on its perch?

  b the bird is flying across the cage (assume no air flow in or out of the cage as a result of flying)?

  c the bird has a seizure and falls from its perch (what is the reading while it is falling)?

7 Explain carefully how a rowing boat is propelled forwards. Identify important forces on the rower, the boat, and the water, and indicate which of these form action–reaction pairs.

INTERACTIONS AND COLLISIONS

When two things interact they exert equal and opposite forces on one another. When a force acts for a period of time it changes the momentum of the object on which it acts. When a force moves an object parallel to itself it does work on that object. To analyse an interaction you must consider momentum *and* energy transfers.

Worked example

A high-speed diesel locomotive was made to collide with a nuclear fuel container to demonstrate the strength and safety of these containers. The two bodies locked together after the collision and travelled some way down the track. The mass of the locomotive was M and the mass of the container was m. Calculate the velocity of the two just after they locked together.

The direction of motion of the locomotive is taken as positive.

Before collision: $p = Mu$

After collision: $p = (M + m)v$

Momentum is conserved: $Mu = (M + m)v$

$$v = \frac{Mu}{(M + m)}$$

If $M = 25\,000\,\text{kg}$, $m = 8000\,\text{kg}$ and $u = 25\,\text{m s}^{-1}$, then

$$v = \frac{25\,000\,\text{kg} \times 25\,\text{m s}^{-1}}{33\,000\,\text{kg}} = 19\,\text{m s}^{-1} \quad \text{(2 sig. figs)}$$

Kinetic energy and momentum

Momentum is conserved in all interactions, but what about kinetic energy? The kinetic energies before and after the collision described above are:

Before collision: $KE_1 = \frac{1}{2}Mu^2$

After collision: $KE_2 = \frac{1}{2}(M + m)v^2$

Substitute

$$v = \frac{Mu}{(M + m)}$$

to get

$$KE_2 = \frac{(M + m)M^2u^2}{2(M + m)^2} = \frac{M^2u^2}{2(M + m)} = \frac{M}{(M + m)}KE_1$$

Kinetic energy has been lost:

$$KE_1 - KE_2 = \left(1 - \frac{M}{(M + m)}\right)KE_1 = \left(\frac{M}{(M + m)}\right)KE_1$$

Elastic and inelastic collisions.

	Kinetic energy	Total energy	Momentum
Elastic collision	conserved	conserved	conserved
Inelastic collision	not conserved	conserved	conserved

A collision in which kinetic energy is not conserved is called **inelastic**. All large-scale collisions are inelastic to some extent so kinetic energy is not so useful as momentum (which is always conserved) in calculating the outcome of interactions. Total energy is *always conserved in any closed system*, so what must be happening is that some of the initial kinetic energy is converted to other forms in the collision. The equation above shows that this will amount to exactly 50% of the incident kinetic energy if the two masses are equal and stick together on impact.

Collisions and explosions

It is often useful to analyse an interaction from the position of the system's centre of mass. In a collision in which the two bodies stick together (perfectly inelastic) the centre of mass of the system must be that of the final composite body. From this point of view the linear momentum before and after the collision is zero and the kinetic energy before the collision is entirely converted to other forms (since the final composite body is at rest in this reference frame). Reverse the time order of an inelastic collision and it becomes a simple explosion. If the exploding object is at rest then the total linear momentum of all the fragments must be zero. The kinetic energy of the fragments has come from the chemical or nuclear energy of the explosives.

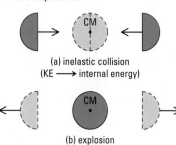

(a) inelastic collision
(KE ⟶ internal energy)

(b) explosion
(internal energy ⟶ KE)

An elastic collision

A perfectly **elastic** collision conserves kinetic energy as well as linear momentum.

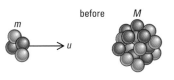

Rutherford estimated the size of a nucleus by analysing the mechanics of a head-on collision between an alpha particle and a gold nucleus. If we assume this is an elastic collision (which will be the case if the collision does not 'excite' the nucleus to a higher internal energy state), we can use conservation of both linear momentum and kinetic energy to work out the motion of the alpha particle and the nucleus after the collision. (Below, the alpha particle mass is m and the gold nucleus mass is M. Right is positive.)

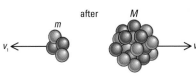

A head-on collision between an alpha particle and a gold nucleus.

$$\text{Conservation of linear momentum:} \quad mu = mv_1 + Mv_2 \qquad (1)$$

$$\text{Conservation of kinetic energy:} \quad \tfrac{1}{2}mu^2 = \tfrac{1}{2}mv_1^2 + \tfrac{1}{2}Mv_2^2 \qquad (2)$$

With two simultaneous equations we can solve for the two final velocities. From (2)

$$mu^2 = mv_1^2 + Mv_2^2$$

so

$$v_2^2 = \frac{m(u^2 - v_1^2)}{M} \qquad (3)$$

From (1)

$$m(u - v_1) = Mv_2 \qquad (4)$$

Square (4) and use (3) to substitute for v_2^2:

$$m^2(u - v_1)^2 = M^2v_2^2 = Mm(u^2 - v_1^2) = Mm(u + v_1)(u - v_1)$$

Simplifying this gives

$$m(u - v_1) = M(u + v_1)$$

$$mu - mv_1 = Mu + Mv_1$$

$$(M + m)v_1 = -(M - m)u$$

$$v_1 = \frac{-(M - m)u}{(M + m)} \qquad (5)$$

The minus sign here indicates that the alpha particle velocity is to the left (it rebounds). Expression (5) can be substituted back into (1) to find the velocity of recoil of the gold nucleus. This gives

$$v_2 = \frac{2mu}{(M + m)} \qquad (6)$$

Since $M = 49m$, $v_1 = -0.96u$ and $v_2 = 0.04u$.

Coefficient of restitution

Another way of looking at collisions is to consider the relative velocities of the objects before and after the collision using the **coefficient of restitution** C_R.

For a one-dimensional collision this is given by:

$$\text{coefficient of restitution } C_R = \frac{\text{separation velocity}}{\text{approach velocity}}$$

- A perfectly elastic collision has $C_R = 1$.
- A perfectly inelastic collision has $C_R = 0$.

Example: A squash ball approaches the court wall along a normal at a velocity of 20 m s⁻¹. It rebounds with a velocity of –12 m s⁻¹. What is the coefficient of restitution for this collision?

Approach velocity = 20 m s⁻¹

Separation velocity = 12 m s⁻¹

$C_R = \dfrac{12}{20} = 0.60$ (ignore sign, only relative velocity matters)

PRACTICE

1 In an experiment using a linear air track a 0.5 kg rider travelling to the right at 2 m s⁻¹ collides with a stationary rider of mass 1 kg. Find the velocities of the two riders after the collision if:

  a it is perfectly elastic;

  b it is totally inelastic.

2 A 1600 kg van moving at 20 m s⁻¹ crashes into the back of a small car of mass 700 kg travelling in the same direction at 15 m s⁻¹.

  a What is the velocity of the two vehicles immediately after the collision if they lock together?

  b How much kinetic energy is transferred to other forms in the collision?

3 A small space vehicle in deep space has mass 1000 kg and length 5 m. The lone astronaut exercises inside the capsule by pushing off from one end, drifting to the other, and stopping herself. She returns in a similar way and repeats the exercise 50 times. Her mass is 60 kg and her drift speed (relative to her original speed) is 4 m s⁻¹. To start herself she pushes against the end of the capsule for 0.6 s. To stop she pushes for 0.4 s.

  a What force and impulse does she apply to the spacecraft each time she starts to move?

  b How does the impulse in **a** affect the motion of the astronaut and the spacecraft?

  c How does the system conserve momentum?

  d Are the collisions elastic or inelastic (assume she stops briefly before changing direction)?

  e Describe the motion of the spacecraft during these daily exercises.

  f Describe the motion of the centre of mass of the system (astronaut plus spacecraft) during the exercises.

  g Calculate the average rate at which kinetic energy is converted to other forms. Will this have any effect on the temperature of the spacecraft?

O B J E C T I V E S

- surface friction
- the coefficient of friction
- modelling fluid friction

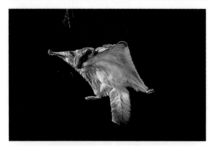

Evolution perfected gliding long before we did: the flying squirrel.

Friction welding.

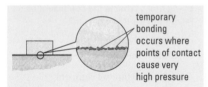

temporary
bonding
occurs where
points of contact
cause very
high pressure

Friction between surfaces.

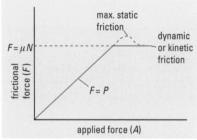

The graph shows how surface friction changes as you apply progressively greater forces. The bump is caused by small irregularities between the surfaces that temporarily bind them together. It is important to remember that the frictional force only has its limiting value when the surfaces are just about to slip or are slipping.

Friction and internal energy

Frictional forces often result in energy losses and inefficiency. It is wrong to say that the waste energy has 'gone to friction' since friction is a force, not a form of energy. It is much better to say it is transferred to internal energy by work done against frictional forces.

FRICTION

It's a drag, but we can't do without it

We rely on the ground pushing forwards on our foot to prevent it slipping back when we walk. We use the **friction** between a road surface and the tyres on our car to provide a force to turn us around a corner. If friction is unexpectedly reduced because of a wet road or ice, we may skid and crash. If table tops were friction-free it would be almost impossible for anything to remain on top of one. If fluid friction disappeared, divers would break their necks in swimming pools and parachutes would be useless.

Frictional forces prevent or oppose *relative motion*. The study of frictional forces is called **tribology**.

- **Static friction** acts between surfaces *at rest* when a force is applied to make them slide past one another.
- **Kinetic (or dynamic) friction** acts between surfaces *as they slide* over one another.
- **Fluid friction** opposes the motion of solids through liquids or gases, or liquids through pipes, creating 'viscous drag'.

Friction between surfaces

Microscopically all surfaces are rough. When placed in contact there will be regions where the irregularities interlock, and points where high pressure may result in temporary bonds. Any attempt to slide the surfaces over one another will require a certain amount of work to be done, lifting and deforming the surface. This will require a force and the work done, as the surfaces move, will increase their thermal energy.

Although the frictional force between real surfaces will vary from one place to the next as the local surface changes, it is possible to describe sliding friction by a fairly simple rule. This rule links the normal force between the two surfaces (that is, how hard they are pressed together) to the frictional force acting on them when they just begin to slip past one another (the limiting friction):

$$\left(\begin{matrix} \text{limiting friction} \\ \text{between surfaces} \end{matrix} \right) \propto \left(\begin{matrix} \text{normal contact force} \\ \text{between surfaces} \end{matrix} \right)$$

$$F \propto N \text{ or } F = \mu N$$

μ is called the **coefficient of friction** and is roughly constant for a particular pair of surfaces.

Worked example

What is the minimum horizontal force required to slide a 300 N packing case across a floor if the coefficient of friction between the case and the floor is 0.3?

To slide the case the applied force must just equal the limiting friction:

$F = \mu N = 0.3 \times 300 \text{ N} = 90 \text{ N}$

Think of sliding a heavy case across the floor. If you push it gently it doesn't move. Push it a bit harder and it still doesn't move. In both cases the static frictional force has balanced your applied force. Push harder still and it just begins to slide at a steady rate. Your force is now equal to the limiting frictional force.

As long as the case slides slowly the rule above is reasonably good, and the kinetic friction is equal to the limiting static friction. At higher relative speeds, the nature of the surfaces is changed by the heat that is generated, and the coefficient of friction changes.

Fluid friction

Fluids are often used as lubricants because they flow more easily than solids slide. However, there are internal forces between molecules in liquids and gases and these result in **viscosity**, a resistance to flow.

Definition: viscosity

In laminar or streamline flow through a pipe, adjacent layers of a liquid move parallel to one another. The molecules in the layer next to the surface stick to it so the velocity of this layer is zero. Pressure in the fluid provides a shear force (moving adjacent layers relative to each other) that makes layers further from the surface move faster. Higher viscosity means that the velocity gradient is lower; low viscosity allows the velocity to increase rapidly away from the surface.

$$\text{viscosity} = \frac{\text{shear stress}}{\text{velocity gradient}} = \frac{F/A}{\delta v/\delta y} \text{ (units are N s m}^{-2})$$

where F is the shear force applied to layers of area A separated by δy to produce a velocity difference of δv.

For objects moving slowly through a liquid (so that the liquid does not get turbulent) the frictional resistance is proportional to the velocity of relative motion. **Stokes's law** is a good description for a sphere falling slowly through a liquid (for example, ball bearings through glycerol):

total frictional drag, $F = 6\pi\eta rv$

where η is the viscosity of the fluid (a measure of its resistance to shear), r is the radius of the sphere, and v is its velocity.

When a plane flies or a torpedo is fired through water, there is a great deal of turbulence, and the frictional force is calculated as if the moving object is making a collision with the fluid. The analysis leads to a frictional force proportional to the velocity (v) squared, the cross-sectional area (A) of the projectile, and the density (ρ) of the fluid (see below).

Viscosity and temperature

As temperature increases so do molecular kinetic energies, making it easier for a liquid to flow. Viscosity falls with increasing temperature. This is important to bear in mind when choosing oil for a car engine. Car engines start cold but run hot so car engine oils are labelled according to their viscosity at low and high temperatures. An oil graded 5W-30 has the same high-temperature viscosity as one graded 10W-30, but is less viscous at low temperatures, making it more suitable for use in the winter or in a colder climate. Choosing the correct oil protects the engine at all temperatures.

Maths box: colliding with a fluid

Volume of fluid encountered in t seconds $= Avt$

Mass of fluid in this volume $= \rho Avt$

Momentum imparted to fluid $= \rho Av^2 t$

(This assumes fluid is accelerated to v in a totally inelastic collision with the projectile.)

Rate of change of fluid momentum = force on fluid from projectile $= \rho Av^2$

Frictional drag on projectile $= -\rho Av^2$

PRACTICE

1 A large trunk rests on a wooden floor. The coefficient of friction between the trunk and the floor is 0.35, and the trunk has mass 50 kg.

  a What is the minimum horizontal force that will just make the trunk slide?

  b Does it make any difference to the required force if it is applied at 45° to the horizontal?

  c What is the frictional force between the trunk and the floor if a horizontal force of 100 N is applied to the trunk?

2 A simple way to measure the coefficient of friction between two surfaces is to tilt one surface until a block of the other material just begins to slip down the slope. The coefficient of friction is then the tangent of the angle between the sloping surface and the horizontal. Prove this.

3 Use Stokes's law to determine appropriate units for viscosity.

4 In a famous experiment to measure the charge on tiny oil drops, Millikan worked out the radius of the oil drops by measuring their terminal velocity as they fell through the air and applying Stokes's law. A particular drop fell 3.0 mm in 75 s.

  a What is the terminal velocity of this drop?

  b Write down an equation for the weight of the drop in terms of its radius r, the gravitational field strength g, and oil density ρ. (Assume it is spherical.)

  c What can you say about the weight of the drop and the fluid friction when the drop falls at constant velocity?

  d Use your answer to c, your formula for weight, and Stokes's law to derive an expression for the radius of the drop.

  e Calculate the radius if the viscosity of air is 1.7 N s m^{-2} and the density of the oil is 800 kg m^{-3}.

5 A simple model of how viscosity varies with time is given by the equation:

$\eta = Ae^{-B/T}$ where A and B are constants and T is measured in kelvin.

  a Take natural logarithms of both sides of this equation and explain how a graph of $\ln \eta$ against $1/T$ could be used to
  i test the relationship, **ii** find A and B.

  b Use the data for water in the table below to test the proposed relationship.

Temperature/°C	20	30	40	50	60	70	80	90
η/cP	1.00	0.798	0.653	0.547	0.467	0.404	0.355	0.315

- application of Newton's laws of motion

- application of conservation of momentum

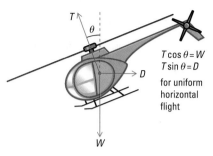

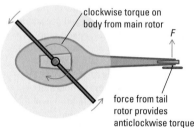

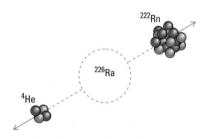

Helicopter motion.

Alpha decay.

Evolution beat us to jet propulsion: the nautilus.

NEWTON IN ACTION – 1

The examples below show how Newton's laws and momentum conservation can be applied to real situations.

The helicopter

Lift and thrust are provided by the main rotor. The rotor blades collide with the air and force it downwards. This force on the air has a reaction force on the blades which pushes them up. If the blades are tilted the air is pushed downwards and backwards. This pushes the helicopter upwards and forwards. In terms of momentum, the rotor increases the momentum of the air. The force on the air is equal to the rate of change of momentum of the accelerated mass of air.

Rotational stability is also important. The rotor shaft exerts a torque on the blades to keep them turning. There is an equal and opposite torque exerted on the helicopter. This would cause the helicopter to rotate. However, a small additional tail rotor is used to balance the turning effect of this torque, as shown in the diagram.

Alpha decay

Alpha decay occurs in some unstable heavy nuclei. Radium-226 is one example. This emits an alpha particle (helium-4 nucleus) and decays to radon-222. The energy released in the decay (that is the total kinetic energy of the two particles) is about 4.2 MeV. Linear momentum must be conserved in the decay, so the momentum of the alpha particle must be equal and opposite to the momentum of the radon nucleus. Since the radon nucleus is 55.5 times the mass of the alpha particle it moves off with a velocity 1/55.5 as large. This results in the alpha particle getting almost all the energy.

If the alpha particle has mass m and speed u, and the radon nucleus has mass M and speed v then:

$$Mv = mu$$

$$v = \frac{mu}{M} \quad \text{and} \quad u = \frac{Mv}{m}$$

so

$$KE_\alpha = \frac{1}{2}mu^2 = \frac{1}{2}m\left(\frac{Mv}{m}\right)^2 = \frac{M}{m} \times \frac{1}{2}Mv^2 = \frac{M}{m} \times KE_{\text{nucl}}$$

In this case

$$\frac{M}{m} = \frac{222}{4} = 55.5$$

So the kinetic energies are in the ratio 55.5:1. The alpha particle gets 98% of the energy.

The jet engine

Air entering the front of the engine is compressed, combusted with fuel, and hot gases are expelled at very high speed from the rear of the jet engine. The net effect is to increase the momentum of the air and burnt fuel behind the plane. The force on the expelled gases is equal to their rate of change of momentum, and equal and opposite to the force on the plane.

You might expect this to cause the plane to increase its momentum in the forward direction. This is not necessarily the case. The plane is moving through the air and its thrust may be balanced by drag. If this is so, we do not have to give up on momentum conservation. The increased backward momentum of the expelled gases exactly balances the increase in forward momentum of the air that is pushed out of the way by the plane.

Rockets

A rocket carries both fuel and oxidizer on board. These are burnt inside the rocket and allowed to escape at high velocity behind it. The force on the gases is equal to their rate of change of momentum and equal and opposite to the force on the rocket. This propels the rocket. It is strange to think that in deep space, the centre of mass of a rocket and its fuel doesn't move as the rocket and ejected gases separate in opposite directions. This is because no external force acts on the rocket and its fuel.

Rocket propulsion.

Changing mass

Some problems involve a change in mass of the material in motion, for example, continually loading a moving conveyor belt with soil or coal, a water jet striking a wall, or a body falling through a fluid (and accelerating the fluid as it goes). In these cases we can adapt the equation for Newton's second law:

$$F = \frac{\mathrm{d}p}{\mathrm{d}t} = \frac{\mathrm{d}(mv)}{\mathrm{d}t}$$

now becomes

$$F = v\frac{\mathrm{d}m}{\mathrm{d}t}$$

since v is now constant and only m changes.

Worked example

A jet of water issues from a stationary hose pipe of cross-sectional area $2\,\mathrm{cm^2}$, at velocity $20\,\mathrm{m\,s^{-1}}$. The density of water is $1000\,\mathrm{kg\,m^{-3}}$. Calculate the average force exerted on a person struck by the jet.

First we must make an assumption about the collision – is it elastic or inelastic? It is certainly inelastic, but to what extent? Experience tells us there will be a bit of splashing but nothing like a complete rebound, so we use the approximation that it is totally inelastic since this is closer to the actual effect. This assumption is important. If the water did bounce back elastically it would exert exactly double the force it exerts in a perfectly inelastic collision because its change of momentum is doubled.

Water velocity is v, pipe exit hole has area A. In a short time δt a 'tube' of water of length $v\delta t$ is ejected. In the same time a volume $vA\delta t$ hits the person and stops. The change of mass of moving water is $-\rho vA\delta t$ in time δt. Hence the force on the water jet is given by

$$F = v\frac{\mathrm{d}m}{\mathrm{d}t} = -\rho Av^2$$

The reaction to this force acts on the person and has magnitude ρAv^2.

Using the values above,

the force on the person $= 1000\,\mathrm{kg\,m^{-3}} \times 2.0 \times 10^{-4}\,\mathrm{m^2} \times (20\,\mathrm{m\,s^{-1}})^2 = 80\,\mathrm{N}$

A water cannon used for crowd control. What kind of collision does the water jet make with a person? How does the nature of the collision affect the force exerted on the person?

PRACTICE

1 Tilting a helicopter's main rotor forward to fly forwards could cause the helicopter to lose altitude. Why? How could this be avoided?

2 Air pressure is the average force exerted on a surface by molecules as they collide with it. Atmospheric pressure is about $10^5\,\mathrm{N\,m^{-2}}$, and nitrogen molecules (mass $5 \times 10^{-26}\,\mathrm{kg}$) have an average velocity at room temperature of about $600\,\mathrm{m\,s^{-1}}$.

  a What is the average linear momentum of a nitrogen molecule?

  b What is the change of momentum of a nitrogen molecule that makes an elastic collision with a solid surface normal to its path?

  c How many collisions per second like the one in **b** are needed to account for atmospheric pressure?

3 In nuclear reactors it is necessary to slow down fast neutrons (mass m) so they have a greater chance of splitting uranium nuclei and causing a chain reaction. They are slowed down when they collide with nuclei of a 'moderator'. Carbon (nuclear mass $12\,m$) and deuterium (nuclear mass $2\,m$) can both be used as a moderator.

  a Assume the collisions are perfectly elastic and calculate the fraction of the neutron's kinetic energy that is transferred to the moderating nucleus in each case.

  b The energy of a fast neutron has to be reduced by a factor of about 10^4. How many collisions does this take in each case?

An alpha particle (entering from bottom right) collides with a stationary helium nucleus in a cloud chamber.

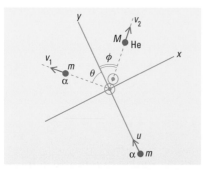

An alpha particle colliding with a helium nucleus.

If the Galileo spacecraft had not used gravity assists it would have needed to carry six times as much fuel to complete its mission to Jupiter and the Jovian moons!

NEWTON IN ACTION – 2

Collisions in two dimensions

The examples given so far have all involved motion along a line. However, many collisions and interactions occur in more than one dimension: a hockey player deflecting a ball into goal; an alpha particle scattered through some angle in Rutherford's experiment (see chapter 8); a pool player cutting the black ball into the top pocket.

Linear momentum is a vector quantity, and so the vector sum of all momenta after a collision or interaction must equal the original momentum vector. To solve problems in more than one dimension, it is best to resolve all momenta along a carefully chosen set of axes and then conserve each component of momentum independently.

The example below left shows an incident alpha particle (mass m) colliding with a helium nucleus (mass M) in a cloud chamber. In the analysis we have *not* assumed the masses of the alpha particle and the helium nucleus are equal, so the results would be true for any similar scattering event.

Resolving parallel and perpendicular to the incident alpha particle (along y and x as shown):

Resolving along y: $mu = mv_1 \cos \theta + Mv_2 \cos \phi$ (1)

Resolving along x: $0 = Mv_2 \sin \phi - mv_1 \sin \theta$ (2)

Assuming an elastic collision we also have:

$$\tfrac{1}{2}mu^2 = \tfrac{1}{2}mv_1^2 + \tfrac{1}{2}Mv_2^2 \qquad (3)$$

These three equations can be used to solve for up to three unknown quantities. For example, if we know the masses of the colliding particles and the initial velocity and scattering angle of the incoming particle we can calculate the final velocities of both particles, and the direction in which the second particle moves off. In particle physics, collisions like this can be used to compare the masses of different particles. In this case the fact that the angle between the tracks of the scattered particles is 90° tells us that their masses are equal (the proof of this requires some algebraic dexterity), and is evidence to support the idea that alpha particles are helium nuclei.

The pressure of light

In 1905 Einstein showed that energy has mass according to the equation:

$E = mc^2$

This means that electromagnetic radiation transfers mass as well as energy.

It also implies that radiation will carry momentum, and so exerts a force on a surface when it is reflected or absorbed, and an atom will recoil when it emits radiation. If a burst of radiation has total energy E it will have a mass:

$$m = \frac{E}{c^2}$$

Since it travels at the speed of light it will have a linear momentum p:

$$p = mc = \frac{E}{c}$$

This means that light reflecting from a mirror will exert a force on the mirror equal and opposite to the rate of change of linear momentum of the light itself. The force exerted per square metre of reflector is called **radiation pressure**, and for normal incidence on a perfect reflector is given by:

$$P = \frac{I}{c}$$

where I is the intensity of radiation (power per unit area) falling on the reflector, and c is the speed of light.

This rather surprising result can also be derived from classical electromagnetism, and experiments to measure radiation pressure on a surface had been carried out successfully even before Einstein's theory was published.

Sailing

How is it possible for any component of a sailing boat's velocity to be directed into the wind if it is using the wind to provide a motive force? To answer this we need to draw a free-body diagram showing all the forces which act on the boat. In particular there will be a force on the keel which opposes sideways drift. The resultant of the force on the keel and the force of the wind on the sail is in the forward direction.

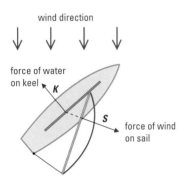

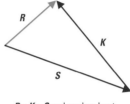

The forces acting on a sailing boat.

Arthur Compton (1892–1962) carried out an experiment to scatter X-rays from electrons in 1923. He explained his results by treating the photon and electron like billiard balls colliding, and applied the laws of conservation of energy and momentum. This was the first hard evidence for wave–particle duality of light.

PRACTICE

1 A water droplet of mass 0.04 g breaks into three smaller drops that move apart symmetrically. Two of the drops have mass 0.01 g; the third is 0.02 g. Find the ratio of the velocities of the three drops.

2 **a** What is the average force due to radiation pressure on a 1 m² plane mirror held up normal to the light from a 100 W light bulb? The distance from the bulb to the centre of the mirror is 5 m (ignore the effects of curvature and assume 100 W is the power radiated).

  b How would your answer to **a** be affected if, instead of a perfect reflector, a perfect absorber of the same size was held up normal to the incident light?

3 It has been suggested that radiation pressure could be used to propel spacecraft of the future. They would use huge low-mass reflective sails. (The Earth orbits at a distance of 1.5×10^{11} m from the Sun, and the normal intensity of radiation at this distance is 1400 W m⁻².)

  a How large a sail would be needed to produce a force of 0.1 N at the same distance from the Sun as the Earth's orbit?

  b What velocity could a 50 kg unmanned craft reach in one year under the influence of this resultant force?

4 Saloon cars are designed with a rigid passenger compartment and a reasonably collapsible engine compartment. How does this help to reduce the risk of serious injuries in an accident?

5 Space missions to the outer planets use the gravity of closer planets to provide a slingshot effect. This is called 'gravity assist'. *Voyager*, for example, was accelerated as it passed behind Jupiter.

  a How is it possible for a spacecraft to be accelerated by passing in and out of Jupiter's gravitational field?

  b Where does the craft's increased linear momentum and kinetic energy come from?

6 A sailing boat sails north-east on a day when a steady wind blows from the north. Find the direction of the force on the sail if it is 1.2 times the magnitude of the normal force of water on the keel. Ignore other forces.

7 Electromagnetic waves transmit energy and momentum. The momentum of a photon is related to its energy by $p = E/c$, where c is the speed of light. Show that light reflecting normally from a mirror exerts a pressure $P = 2I/c$ on the mirror, where I is the intensity of light hitting the mirror (W m⁻²). How would the radiation pressure on a matt black surface differ from this?

3.28

O B J E C T I V E S

- angular displacement and velocity
- period and frequency of rotation

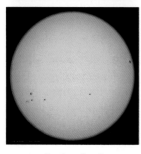

A sequence of solar disc images. Sunspots reveal the rotation of the Sun. More detailed analysis shows that the equator (period of about 27 days) revolves faster than the poles (period of about 34 days). This is possible because the Sun is not a rigid body.

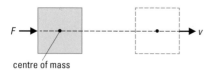

If the line of action passes through the centre of mass the object translates but does not rotate.

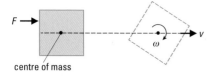

If the line of action does not pass through the centre of mass the object translates and rotates. There is a resultant force and a resultant moment.

ROTATION

Rotational motion is found everywhere in the universe, from spiral galaxies to atoms, from spin dryers to turbine blades, and from spin bowling to pirouetting ice skaters.

Definitions

Period of rotation (T): the time (s) to complete one revolution about some axis.

Frequency (f): the number of revolutions per second (Hz). (1 Hz = 1 s^{-1}.)

Angular displacement (θ): the angle (in radians) turned through in some time.

Angular velocity (ω): the rate of change of angular displacement (rad s^{-1}).

These terms are all related to one another.

$$f = \frac{1}{\text{time for 1 revolution}} \quad \text{so} \quad f = \frac{1}{T}$$

$$\omega = \frac{d\theta}{dt} \text{ (instantaneous angular velocity, like } v = \frac{ds}{dt} \text{ in linear mechanics)}$$

$$\omega_{av} = \frac{\Delta\theta}{\Delta t} = \frac{\text{change in angular displacement (rad)}}{\text{time taken (s)}}$$

If the last equation is applied to a single rotation, $\Delta\theta = 2\pi$ and $\Delta t = T$:

$$\omega_{av} = \frac{2\pi}{T} = 2\pi f$$

These are particularly useful relations, especially for uniform circular motion where instantaneous and average angular velocities are equal.

Worked example 1

The Earth rotates once in about 24 h. Calculate its rotation frequency, and its angular velocity.

$$f = \frac{1}{T} = \frac{1}{24 \times 3600\,\text{s}} = 1.2 \times 10^{-5}\,\text{Hz}$$

$$\omega = \frac{2\pi}{T} = \frac{2\pi}{24 \times 3600\,\text{s}} = 7.3 \times 10^{-5}\,\text{rad s}^{-1}$$

Angular velocity and tangential velocity

When something moves in a circle, its instantaneous linear velocity is always parallel to a tangent to the circle; this is called its **tangential velocity**. It has no **radial velocity**. If it moved along a spiral path it would have both a tangential and a radial velocity at all times.

Angular velocity is related to tangential velocity. This relation is derived below for an object moving in uniform circular motion with constant angular velocity ω: in a short time δt the object will increase its angular displacement by

$$\delta\theta = \omega\delta t$$

If the object moves through an angle $\delta\theta$, it moves a distance

$$\delta s = r\delta\theta$$

'along the circumference'. This means that the object has a speed

$$v = \frac{\delta s}{\delta t} = \frac{r\delta\theta}{\delta t} = r\omega$$

If δt is made very short, this gives the instantaneous tangential velocity:

$$v = r\omega$$

Worked example 2

What is the tangential velocity of a point on the equator of Mercury which has a rotation period of 59 days and an equatorial radius of 2500 km?

$$v = r\omega = \frac{2\pi r}{T} = \frac{2\pi \times 2.5 \times 10^6\,\text{m}}{59 \times 24 \times 3600\,\text{s}} = 3.1\,\text{m s}^{-1}$$

Torque and angular acceleration

Resultant forces cause linear accelerations; resultant moments or torques cause angular accelerations.

Angular acceleration = rate of change of angular velocity

$$\alpha = \frac{\delta\omega}{\delta t} \quad (\text{rad s}^{-2})$$

For uniform circular motion:

$$\alpha = \frac{\text{tangential acceleration}}{\text{radius}} = \frac{a}{r} \quad (\text{see previous page})$$

Resistance to twist

The **moment of inertia** of an object is its 'reluctance' to change its state of rotational motion. It is the rotational equivalent of mass in linear mechanics. Unlike mass it depends on both the body *and* the axis about which it is rotated, so the same thing has an infinite number of moments of inertia.

It is easy to 'feel' this 'reluctance' to rotate. Stand with your feet together and your arms by your side and twist back and forth about your vertical axis. Now stretch out your arms and do the same thing, trying to keep your arms moving rigidly with your body. The total mass in motion is the same as before but now it takes more effort, or greater torque, to reverse the motion each time. (The effect is dramatically increased if you hold a 1 kg mass in each hand.) With arms outstretched your body has a greater reluctance to change its rotational motion; it has a greater moment of inertia.

This idea of 'reluctance to change rotational motion' can be developed into a physical definition of moment of inertia. If a resultant torque *T* produces an angular acceleration α then the moment of inertia *I* is the ratio of torque applied to angular acceleration produced:

$$I = \frac{T}{\alpha} \quad \text{the units of moment of inertia are kg m}^2.$$

The moment of inertia depends on the distribution of mass about the axis of rotation.

Moments of inertia.

Object	Axis of rotation	Moment of inertia (I)	Radius of gyration (k)
point mass	perpendicular distance *r* away	mr^2	r
rod	perpendicular to rod, at one end	$\frac{1}{3}mr^2$	$r/\sqrt{3}$
rod	perpendicular to rod, through centre	$\frac{1}{12}mr^2$	$r/2\sqrt{3}$
solid disc/cylinder	central axis of symmetry	$\frac{1}{2}mr^2$	$r/\sqrt{2}$
hoop/cylindrical shell	central axis of symmetry	mr^2	r
solid sphere	diameter	$\frac{2}{5}mr^2$	$r\sqrt{2}/\sqrt{5}$
spherical shell	diameter	$\frac{2}{3}mr^2$	$r\sqrt{2}/\sqrt{3}$

Maths box: calculating moments of inertia

Mathematically, the moment of inertia is defined as:

$$I = \sum_{i=1}^{i=N} m_i r_i^2$$

where the body is made up from N small particles. m_i is the mass of the *i*th particle of N particles in the body. r_i is the perpendicular distance of the *i*th particle from the rotation axis. The moment of inertia is the sum of mass times the distance squared for all particles. This explains why moving mass away from the axis (e.g. when a figure skater spreads her arms) dramatically increases moment of inertia.

The final expression for moment of inertia will obviously depend on the total mass of the body, and will have dimensions of M L². For this reason moments of inertia are often quoted in the form:

$$I = mk^2$$

where m is the total mass of the object and k is called its **radius of gyration**.

The physical significance of k is that it is the distance from the axis that a single point mass m would orbit with the same moment of inertia.

Angular acceleration

If the motion is purely circular, the angular acceleration is very simply related to tangential acceleration:

$$\alpha = \frac{d}{dt}\left(\frac{v}{r}\right) = \frac{1}{r}\frac{dv}{dt} = \frac{a}{r} \quad \begin{array}{l}(\text{since } r \text{ is}\\ \text{constant})\end{array}$$

PRACTICE

1 An ice skater completes 4 revolutions in 1.2 s. What is her average angular velocity and her rotation frequency?

2 A pram has wheels of diameter 30 cm. Calculate their angular velocity and frequency of rotation when the pram is pushed along at a steady 2 m s^{-1}.

3 A flywheel is connected to a shaft of diameter 20 mm. The moment of inertia of the flywheel and shaft is 1.5 kg m^2. The flywheel is made to spin by winding a long string around the shaft and then pulling it with a steady force of 4.0 N.

a What is the turning effect of the applied force about the centre of the shaft?

b What is the torque applied to the flywheel and shaft?

c What is the angular acceleration of the flywheel and shaft?

d What is the final angular velocity of the flywheel if the string becomes detached after 6 s?

UNIFORM CIRCULAR MOTION

When the gymnast releases her grip on the parallel bars, her initial motion is tangential, not radial. Of course, gravity will cause her centre of mass to follow a parabolic trajectory.

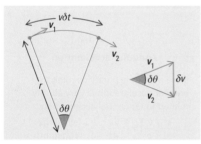

Velocity changes in uniform circular motion.

Useful formulae

Period of rotation: $T = \dfrac{2\pi}{\omega}$

Tangential speed: $v = \dfrac{2\pi r}{T} = r\omega$

Centripetal acceleration: $a = \dfrac{v^2}{r} = r\omega^2$

Centripetal force: $F = \dfrac{mv^2}{r} = mr\omega^2$

Helpful hint

The formulae for centripetal force are simply mathematical expressions for the magnitude of the resultant force needed to maintain uniform circular motion. Centripetal force is not a force in its own right, and does not act as an additional force on things in circular motion. Gravity is the centripetal force on the Moon, and tension in the string is the centripetal force on a stone swung in a circle.

Think of anything that moves in a circle; the Earth around the Sun or clothes in a tumble drier. What makes them follow a circular path? Newton's first law states that the natural state of motion is in a straight line at constant speed, so what turns these objects from their straight-line paths?

The Earth is pulled towards the Sun by gravitational attraction. The clothes are prevented from going straight by the barrel of the drier. In both cases, the force which changes the direction of the moving object is directed in *towards the centre of rotation*. This is called a **centripetal** (centre-seeking) force.

If a centripetal force is removed, the object concerned is free to continue moving in a straight line. For example, if a stone is swung in a circle on the end of a string and the string breaks, the stone flies off along a tangent (it does not travel radially outwards). A Grand Prix car hitting oil at a hairpin goes off the track in a straight line.

A centripetal force acts at right angles to velocity, so it has no component parallel to motion. It therefore leaves the tangential velocity and kinetic energy unchanged. It does no work on the object; it merely changes the direction of its velocity vector.

Centripetal force and acceleration

Resultant forces cause acceleration. It follows that objects moving in circular motion are accelerating in the direction of the centripetal force, that is, towards the centre of rotation. The magnitude of the acceleration and force can be calculated from the rate of change of the velocity vector. The diagrams show two positions during uniform circular motion separated by a short time. The right-hand diagram shows the velocity change that occurs in this time. If the time is made extremely short the velocity change is perpendicular to the tangential motion, in other words radially inwards or centripetal.

Tangential speed is v, so vectors $\mathbf{v}_1$ and $\mathbf{v}_2$ have this same magnitude. From the space diagram:

$$\delta\theta = \frac{v\delta t}{v} \text{ (definition of angle in radians)}$$

From the vector diagram:

$$\delta\theta = \frac{\delta v}{v} \text{ (small-angle approximation)}$$

Equating these gives:

$$\frac{\delta v}{\delta t} = \frac{v^2}{r}$$

In the limit of small times:

$$a = \frac{\mathrm{d}v}{\mathrm{d}t} = \frac{v^2}{r} \text{ (centripetal acceleration)}$$

Using $F = ma$ and $v = r\omega$ produces the useful formulae on the left.

Worked example 1

What is the centripetal acceleration of a 40 kg child sitting 2 m from the centre of a roundabout which turns once in 5.0 s? What is the resultant horizontal force acting on the child?

$$a = r\omega^2 = r\left(\frac{2\pi}{T}\right)^2 = \frac{4\pi^2 r}{T^2} = \frac{4\pi^2 \times 2.0\,\text{m}}{(5.0\,\text{s})^2} = 3.2\,\text{m s}^{-2}$$

$$F = ma = 40\,\text{kg} \times 3.2\,\text{m s}^{-2} = 130\,\text{N}$$

This force would arise from friction where the child sits and from the child holding onto part of the roundabout.

Worked example 2

What is the period of revolution of a spy satellite in a low Earth orbit (distance 7100 km from the centre of the Earth) where the gravitational field strength is 8.0 N kg⁻¹?

$$F = mg = mr\omega^2 \text{ so } \omega = \sqrt{\frac{g}{r}}$$

and

$$T = 2\pi\sqrt{\frac{r}{g}} = 2\pi\sqrt{\frac{7.1 \times 10^6 \text{ m}}{8.0 \text{ N kg}^{-1}}} = 5900 \text{ s} = 1 \text{ h } 39 \text{ min}$$

The bucket trick.

A party trick

A bucket of water can be swung in a vertical circle so that the water remains in the bucket even though it is upside down at the top of the circle. How is this possible? It is possible to perform a similar trick on a smaller scale by rapidly rotating a mug half-filled with water.

Water falls out of a stationary inverted bucket because of its own weight. This gives it a downward acceleration of 9.8 m s⁻². When the bucket is moving in a vertical circle it has a centripetal acceleration, and this is also downwards when the bucket is at the top of its motion. If the downward acceleration of the bucket is at least equal to the free-fall acceleration then the water will not fall out. The condition for this is:

$$a = r\omega^2 \geq g = 9.8 \text{ m s}^{-2}$$

If the bucket revolves faster than the minimum rate, its centripetal acceleration is greater than free fall and so it actually pushes down on the water at the top of the motion to provide the water with sufficient centripetal acceleration to move in the same circular motion as the bucket.

The conical pendulum

A mass attached to the end of a light string can be made to move in a horizontal circle, so that the string traces out the surface of a cone. This is called a conical pendulum. Since the mass is moving in uniform circular motion, the resultant of all forces acting on it must act toward the centre of the circle and equal $mr\omega^2$. The only forces are tension S in the string and weight of the mass. If the string is at an angle θ to the vertical:

Resolving vertically: $S \cos\theta - mg = 0$ vertical equilibrium

Resolving horizontally: $S \sin\theta = ma = mr\omega^2$ horizontal centripetal acceleration

S and m can be eliminated from the equation to give:

$$\tan\theta = \frac{r\omega^2}{g} = \frac{v^2}{rg}$$

> ### Centrifugal force?
>
> If you are in a car that suddenly turns, you feel as if you are being thrown *outwards*, away from the centre of the curve. It is common to explain this by saying there is a **centrifugal** force that acts *away* from the centre. However, if this situation is viewed from above a different explanation is possible. The car has turned suddenly because of an inward force from the road as its wheels turn. You, on the other hand, tend to carry on moving with straight-line motion and constant velocity until the side of the car 'hits you' and makes you corner too (it applies a *centripetal* force to you). Centrifugal force is an 'imaginary force' introduced to account for the effects of inertia in an accelerated reference frame.

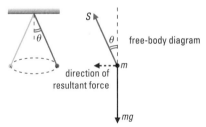

A conical pendulum.

PRACTICE

1 What provides a centripetal force for each of the following things?

  a The rings of Saturn;

  b the blades of a propellor;

  c an electron in a hydrogen atom.

2 Calculate the centripetal acceleration and force on 1 kg of matter at the surface of a neutron star of mass 4×10^{30} kg and radius 10 km rotating once every 50 ms. What provides this force?

3 A passenger is riding in a roller-coaster car on a track that loops the loop. Draw free-body diagrams to show the forces acting on her when she is at the top and bottom of the loop. Is the resultant force always toward the centre of the loop?

4 High-speed sanding discs and grinding wheels sometimes explode. Explain why this happens and discuss the stresses and strains in the wheels when they are rotating.

5 The Earth has a mass of 6.0×10^{24} kg, and it orbits the Sun at a mean distance of 1.5×10^{11} m once a year.

  a What is the Earth's angular velocity about the Sun?

  b What is the Earth's centripetal acceleration?

  c What is the magnitude of the gravitational force between the Earth and Sun?

  d What is the minimum diameter steel cable that could withstand this force? (Take the tensile strength of steel as 1×10^9 Pa.)

3.30

OBJECTIVES

- applying equations for angular velocity and angular acceleration

- rotational kinetic energy

- conservation of angular momentum

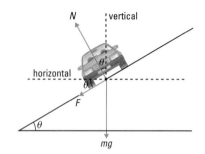

A bank on a corner means that it can be negotiated at a higher speed.

'Artificial gravity' can be experienced at a fairground.

Rotational kinetic energy

When a rigid body rotates with angular velocity ω about an axis through its centre of mass, then the ith particle (of n) in the body has an instantaneous tangential velocity and kinetic energy (KE) given by

$$v_i = r_i\omega \quad KE_i = \tfrac{1}{2}m_iv_i^2 = \tfrac{1}{2}m_ir_i^2\omega^2$$

ω has the same value for all particles

The **rotational kinetic energy (RKE)** of the whole body is the sum of all these individual kinetic energies:

$$RKE = \sum_{i=1}^{i=n} KE_i = \sum \tfrac{1}{2}m_ir_i^2\omega^2$$

$$= \tfrac{1}{2}\left(\sum_{i=1}^{i=n} m_ir_i^2\right)\omega^2 = \tfrac{1}{2}I\omega^2$$

I is called the **moment of inertia** of the body.

EXAMPLES OF ROTATION

Banking

Olympic cycle tracks are steeply banked, as are some Grand Prix track corners (for example, at Monza). This is so that a component of the normal contact force acts towards the centre of the turn and helps provide the centripetal force. This means that the curve can be negotiated at a higher speed before limiting friction is reached. The resultant force is horizontal and must equal mv^2/r.

Resolving horizontally: $N\sin\theta + F\cos\theta = \dfrac{mv^2}{r}$

Resolving vertically: $N\cos\theta - mg = 0$

These can be solved when the limiting friction is reached: $F = \mu N$

$$v_{max} = \sqrt{\frac{rg(\sin\theta + \mu\cos\theta)}{\cos\theta}}$$

If $\theta = 0$ (no banking),

$$v_{max} = \sqrt{rg\mu}$$

Since $\cos\theta$ approaches 0 as θ increases, it is clear that a faster speed is possible if the track is banked.

On a flat track it is not possible to corner if there is no friction ($F = 0$). This is not true on a banked track. Cornering is still possible, but only at one particular speed.

Solving the same pair of equations with $F = 0$ gives

$$\tan\theta = \frac{v^2}{rg} \text{ and so } v = \sqrt{rg\tan\theta}$$

In this case, the horizontal component of N is entirely responsible for the centripetal force.

Artificial gravity

Although the apparent weightlessness experienced inside a space station would be fun for a while, it is also rather inconvenient and can cause medical problems for astronauts who spend too long there. One possible solution is to create an artificial gravity by building the space station like a large doughnut and spinning it about a central axis. The outer wall provides a centripetal force on astronauts that feels like the reaction from the ground when they are standing in a gravitational field. The value of the artificial gravity is fixed by the radius and rate of rotation of the space station. To create an artificial gravity equivalent to a field strength g the centripetal acceleration at the circumference of the station must be g.

$$a = r\omega^2 = g$$

For a space station of radius 100 m this gives

$$\omega = \sqrt{\frac{g}{r}} = \sqrt{\frac{9.8\,\text{m s}^{-2}}{100\,\text{m}}} = 0.31\,\text{rad s}^{-1}$$

Its period of revolution would be

$$T = \frac{2\pi}{\omega} = 20\,\text{s}$$

An alternative to building a large space station is to 'catch' an asteroid, drill circular tunnels around a central axis, and then use rockets attached to the outside surface of the asteroid to give it a suitable angular velocity. People could live on the inside surface of the tunnels (which could be very large). The value of 'g' in a tunnel would vary depending on its distance from the axis of rotation.

Rotational kinetic energy

Rotating masses can store very large amounts of kinetic energy, and can be used to power vehicles for short journeys. Very large flywheels are also used to help stabilize ships. If something is *rolling* it will have both **translational kinetic energy** (of its centre of mass) and **rotational kinetic energy** (**RKE**) (about its centre of mass). The RKE is equal to the sum of the kinetic energies of all the particles in the body because of their tangential velocities about the axis. The expression for RKE is similar in form to that for KE; however, moment of inertia replaces mass and angular velocity replaces velocity:

$$RKE = \frac{1}{2}I\omega^2 \quad (\text{whereas } KE = \frac{1}{2}mv^2)$$

The units are, of course, joules (J).

Worked example

(a) How much energy is stored in an 800 tonne flywheel of radius 6.0 m rotating at 18 rad s⁻¹?

(b) How long would it take to reach this angular velocity from rest using a 100 kW motor?

(a) $RKE = \frac{1}{2}I\omega^2$ and $I = \frac{1}{2}mr^2$ (disc)

$RKE = \frac{1}{2} \times \frac{1}{2} \times 8.0 \times 10^5 \text{ kg} \times 6.0^2 \text{ m}^2 \times 18^2 \text{ rad}^2\text{ s}^2 = 2.3 \times 10^9 \text{ J}$

(b) $P = \dfrac{RKE}{t}$ $\therefore t = \dfrac{RKE}{P} = \dfrac{2.3 \times 10^9 \text{ J}}{10^5 \text{ W}} = 2.3 \times 10^4 \text{ s} = 6.4 \text{ h}$ (2 sig. figs)

Angular momentum

Angular momentum plays the same role in rotation as linear momentum does in translation. The angular momentum (L) of a rigid body is given by

$L = I\omega$

(As with RKE, I replaces m and ω replaces v in the linear equation $p = mv$.) The units of angular momentum are kg m² s⁻¹ (or J s).

Resultant forces change linear momentum, and resultant torques change angular momentum, so there is a rotational equivalent of Newton's second law:

resultant torque ∝ rate of change of angular momentum
(about some axis)

$$\boldsymbol{T} = \frac{dL}{dt}$$

which becomes $\boldsymbol{T} = I\alpha$ (like $F = ma$) if the body has constant I.

Conservation of angular momentum

The angular momentum about a particular axis is constant if there is no resultant torque acting about that axis.

Comets and planets *conserve angular momentum* about the Sun, since the gravitational force acts toward the Sun and therefore generates no resultant torque about it.

Spinning ice-skaters and somersaulting high divers use conservation of angular momentum to control their rotation rates. The principle applies because there are no significant torques acting around their axis of rotation. They reduce their moment of inertia by bunching up close to the axis, and increase it by spreading out away from the axis. This has the following result:

$I_1\omega_1 = I_2\omega_2$ (initial and final angular momenta are equal)

so

$$\frac{\omega_2}{\omega_1} = \frac{I_1}{I_2}$$

Their angular velocity is inversely proportional to their moment of inertia, so their rotation rate increases dramatically as their mass is distributed close to the axis. (The reason there is such a marked change is because I depends on the *squares* of distances of masses from the axis.)

By bringing her arms and legs closer to her axis of rotation the skater reduces her moment of inertia. To conserve angular momentum about her vertical axis, her angular velocity must increase.

OBJECTIVES

- use of analogy

- linear motion as a model for rotational motion

- worked examples

The sycamore seed converts gravitational potential energy into both translational and rotational kinetic energy. The translational velocity is therefore less than if it fell without turning. It takes longer to fall and so is likely to travel further from the parent tree.

Linear equation	Rotational equation
$a = \dfrac{v - u}{t}$	$\alpha = \dfrac{\omega_2 - \omega_1}{t}$
$s = ut + \frac{1}{2}at^2$	$\theta = \omega_1 t + \frac{1}{2}at^2$
$s = \dfrac{u + v}{2}t$	$\theta = \dfrac{(\omega_1 + \omega_2)}{2}t$
$v^2 = u^2 + 2as$	$\omega_2{}^2 = \omega_1{}^2 + 2\alpha\theta$
$KE = \frac{1}{2}mv^2$	$RKE = \frac{1}{2}I\omega^2$
$W = Fs$	$W = T\theta$
$P = Fv$	$P = T\omega$
$p = mv$	$L = I\omega$ (angular momentum)
$F = ma$	$T = I\alpha$
$F = \dfrac{d(mv)}{dt}$	$T = \dfrac{d(I\omega)}{dt}$
$Ft = \Delta(mv)$	$Tt = \Delta(I\omega)$

The angular velocity of air increases as it is drawn toward the centre of a cyclone.

The formulae for rotational kinetic energy and angular momentum used in spread 3.30 have the same mathematical form as their counterparts in translational motion. This is not just a coincidence. It is because they both derive from Newton's laws of motion and relate quantities that are defined in a similar way. The great advantage of this is that once you know how to solve problems in linear motion, you can solve problems in rotational motion using exactly the same techniques and simply substituting the rotational terms in place of their translational counterparts.

Rotation and translation

Linear velocity is the rate of change of linear displacement, so if I move 6 m in 2 s I have an average linear velocity of 3 m s⁻¹. Rotational (angular) velocity is the rate of change of angular displacement, so if I turn through 6 rad in 2 s I have an average angular velocity of 3 rad s⁻¹.

Since the mathematical definitions of linear and angular displacements, velocities, and accelerations have the same form it is possible to make an analogy between rotation and translation.

$$v = \frac{ds}{dt} \quad \text{(definition of linear velocity)}$$

$$\omega = \frac{d\theta}{dt} \quad \text{(definition of angular velocity)}$$

To make the analogy we simply swap terms according to the following rules:

$$s \rightarrow \theta \quad u, v \rightarrow \omega_1, \omega_2 \quad a \rightarrow \alpha \quad t \rightarrow t$$

The analogy can be extended further if force and torque and mass and moment of inertia are also paired. (This is justified by the rotational equivalent of Newton's second law, and the expressions for RKE and angular momentum discussed in spread 3.30).

$$F \rightarrow T \quad m \rightarrow I$$

Worked example 1

A bicycle wheel of radius 0.32 m has a moment of inertia of 0.035 kg m² about its axle. When the wheel is lifted from the ground and set spinning at a frequency of 10 Hz it comes to rest after 3.0 min. (a) What is the angular deceleration of the wheel? (b) What is the average frictional torque acting on it? (c) What is the rate at which it dissipates energy when rotating at 10 Hz?

(a) $\omega_1 = 2\pi \times 10\,\text{Hz} = 62.8\,\text{rad s}^{-1} \quad \omega_2 = 0\,\text{rad s}^{-1}$

$$a = \frac{\omega_2 - \omega_1}{t} = \frac{-62.8\,\text{rad s}^{-1}}{180\,\text{s}} = -0.35\,\text{rad s}^{-2}$$

(b) $T = I\alpha = 0.035\,\text{kg m}^2 \times (-0.35\,\text{rad s}^{-2}) = -0.012\,\text{N m}$ $(1\,\text{N m} = 1\,\text{kg m}^2\,\text{s}^{-2})$

(c) $P = \dfrac{dW}{dt}$ but $W = T\theta$ so $P = T\dfrac{d\theta}{dt} = T\omega = -0.012\,\text{N m} \times 62.8\,\text{rad s}^{-1} = 0.75\,\text{W}$

Note that $P = T\omega$ is directly analogous to $P = Fv$ for translation.

Worked example 2

How long will it take for a steel ball bearing of mass 33 g and radius 0.01 m to roll 25 cm from rest down a ramp inclined at 10° to the horizontal?

The ball bearing falls through a vertical height

$$h = 0.25\,\text{m} \times \sin 10° = 0.043\,\text{m}$$

$$\Delta E = -\Delta GPE = mgh$$

This equals the sum of rotational and translational kinetic energies at the bottom:

$$\frac{1}{2}mv^2 + \frac{1}{2}I\omega^2 = mgh$$

Worked example 2 *(continued)*

The moment of inertia of a sphere is $I = \frac{2}{5}mr^2$ and $r\omega = v$, so:

$$\frac{1}{2}mv^2 = \frac{1}{2} \times \frac{2}{5}mv^2 = \frac{1}{5}mv^2 = mgh$$

$$v^2 = 5gh = 5 \times 9.8\,\text{N kg}^{-1} \times 0.043\,\text{m} = 2.11\,\text{m}^2\,\text{s}^{-2}$$

$$v = \sqrt{2.11\,\text{m}^2\,\text{s}^{-2}} = 1.45\,\text{m s}^{-1}$$

This is the final velocity so the average velocity is $0.73\,\text{m s}^{-1}$

Time to roll 25 cm is $t = \dfrac{0.25\,\text{m}}{0.73\,\text{m s}^{-1}} = 0.34\,\text{s}$

Worked example 3

Pluto's distance from the Sun varies from a minimum of 4.43×10^9 km to a maximum of 7.38×10^9 km. What is the ratio of its maximum to minimum tangential velocities?

The only significant force on Pluto is from the Sun. This acts through the centre of mass of the Sun so there is no resultant torque on Pluto about the Sun. Its angular momentum is therefore constant and so is equal in both positions:

$$mv_1r_1 = mv_2r_2 \quad \text{so} \quad \frac{v_1}{v_2} = \frac{r_1}{r_2} \quad \text{and} \quad \frac{v_{\max}}{v_{\min}} = \frac{r_{\max}}{r_{\min}} = \frac{4.43 \times 10^9\,\text{km}}{7.38 \times 10^9\,\text{km}} = 0.60$$

Worked example 4

A turntable with moment of inertia $0.05\,\text{kg m}^2$ and radius 20 cm is rotating freely at 45 revolutions per minute when a disc of moment of inertia $0.02\,\text{kg m}^2$ is dropped on top of it so that its centre coincides with the axis of rotation. What is the new rate of revolution? Angular momentum is conserved, since no external resultant torque acts.

Initial angular velocity $\omega_1 = \dfrac{45 \times 2\pi}{60\,\text{s}} = 4.7\,\text{rad s}^{-1}$

Initial angular momentum $= I_1\omega_1$

Final angular momentum $= I_1\omega_2 + I_2\omega_1 = (I_1 + I_2)\omega_2$
(ω_2 is the final angular velocity of both discs.)

$$I_1\omega_1 = (I_1 + I_2)\omega_2$$

$$\omega_2 = \frac{I_1\omega_1}{(I_1 + I_2)} = \frac{0.05\,\text{kg m}^2 \times 4.7\,\text{rad s}^{-1}}{(0.05\,\text{kg m}^2 + 0.02\,\text{kg m}^2)} = 3.4\,\text{rad s}^{-1}$$

$$\frac{3.4\,\text{rad s}^{-1} \times 60\,\text{s}}{2\pi} = 32\,\text{r.p.m.}$$

How can a cat turn over if no external resultant torque is applied to it? Try to do a similar thing by sitting on a rotating chair and trying to turn around without pushing on anything outside. Think about how your moment of inertia changes as you turn.

PRACTICE

1 A wooden disc of radius 30 cm and mass 1.0 kg is spun at 50 r.p.m..

   a What is its rotation frequency and angular velocity?

   b What is its rotational kinetic energy?

   c What is its angular momentum?

   d If it slows to 40 r.p.m. in 30 s what is its average angular deceleration?

   e What average frictional torque would slow the disc as in **d**?

2 Explain why comets in very eccentric orbits move very slowly far from the Sun and very fast close to it. Give three explanations, one in terms of energy, one in terms of angular momentum, and one in terms of forces acting on the comet around its orbit.

3 A pirouetting skater halves his moment of inertia by pulling in his arms and legs closer to his axis of rotation.

   a By what factor does his angular velocity increase?

   b What happens to his rotational kinetic energy?

   c Explain how conservation of energy and angular momentum apply to this example.

4 A roundabout has moment of inertia $1400\,\text{kg m}^2$ and radius 2.0 m. A 50 kg child runs towards the roundabout, parallel to a tangent, and leaps onto it while running at $7\,\text{m s}^{-1}$.

   a Ignoring friction calculate the final angular velocity of the roundabout and child.

   b Compare the final rotational kinetic energy of the system with the initial kinetic energy of the running child, and comment on this.

A multiflash image of a child in a swing. You probably know that you can increase your amplitude of swing if you tuck your legs back underneath you as the swing moves backwards, and point them straight in front as the swing moves forwards. How is this consistent with the laws of conservation of energy and angular momentum?

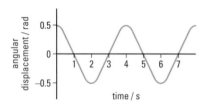

A graph of angular displacement against time for the child on the swing.

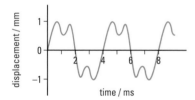

A graph of displacement against time for an oscillating guitar string.

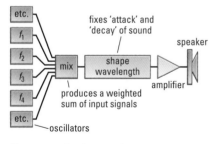

How a synthesizer works.

Sinusoidal curves

'Sinusoidal' is used to describe the shape of any sine or cosine curve regardless of whether it is zero at $t = 0$.

OSCILLATORS

The speed at which we walk, the sound of a musical instrument, the ticking of a clock, the colours of street lamps and stars, the buildings that will survive or be destroyed by an earthquake, and the speed of sound are all determined, to some extent, by periodic motions or **oscillations**. The essential characteristic of an oscillation is that it repeats itself. A child on a swing is a good example of oscillation. A complete oscillation would start at any point on the arc and continue until the child moved through the same point in *the same direction* a second time. (For example, from A to E *and back* or from C to A to E *and back* to C in the figure in Practice question 2.)

Definitions

Amplitude (A): the maximum displacement from equilibrium (m).

Period (T): the time for one complete cycle of the oscillation (s).

Frequency (f): the number of complete oscillations (cycles) per second (Hz).

Period and frequency are inversely related: $f = \dfrac{1}{T}$

Simple harmonic oscillators

The graphs show how the displacements of two oscillators vary with time. The first is for the swing above left (angular displacement is used here); the second is for a guitar string as it produces a note.

The graph for the child on the swing is the simplest shape, and looks rather like a cosine curve. A similar graph also results if a microphone detects the sound of a single pure frequency from, say, a signal generator connected to a loudspeaker. Oscillators that produce these regular 'sinusoidal' displacements are called **simple harmonic** and are analysed in detail later. The guitar string has a more complex vibration.

Simple harmonic motion (SHM) is important because all oscillations are approximately simple harmonic if their amplitude is small. More complex oscillations like the signal from the guitar string can be **synthesized** (built up) by combining a number of simple harmonic oscillations of different frequencies. This is how an electronic synthesizer works. It adds several simple harmonic oscillations electronically to mimic the sound of a particular instrument. It also shapes the overall waveform by imposing an **envelope** on the combined signal. This envelope determines the **attack** and **decay** of the sound generated, that is, how quickly the sound cuts in and how long it takes to fade out.

Phase

Imagine two swimmers treading water at different distances from a beach as regular waves come in at right angles to the beach (see the diagram at the top of the next page). They will both bob up and down with the same amplitude and frequency (equal to the amplitude and frequency of the incoming waves), but they are unlikely to bob up and down at the same time because the waves reach one swimmer before the other. Their oscillations are **out of phase**.

Phase measures the position of an oscillator within its cycle of oscillation. Cyclic motion is related to circular motion (see spread 3.28), so phase is related to the angle, and changes by 360° or 2π radians in one cycle. For example, if one swimmer is at the top of a wave crest while the other is at the bottom of the next trough, they are π (180°) out of phase. If both swimmers are on the crests of different waves at the same time their phase difference will be a multiple of 2π, but this is the same as being *in phase* – the oscillation repeats when the phase changes by any integral multiple of 2π.

Navigating with pendulums

Simple harmonic oscillators are **isochronous**, that is, their period is independent of amplitude. Galileo discovered that the period of a simple pendulum is almost isochronous. This is because its oscillations approximate very closely to simple harmonic motion, especially for small amplitudes. He also showed that the period of a simple pendulum is proportional to the square root of its length. His work was developed by a Dutch scientist, Christiaan Huygens, who invented the **clock escapement** in 1656. This enabled the accuracy of time-keeping to improve from an error of 5 min per day to just 10 s per day. Following this success, Huygens hoped to develop a marine chronometer that would help sailors to navigate.

An accurate clock would allow sailors to determine their longitude. It is easy to see how this works. If you move west around the Earth the Sun will rise later, reach its highest point later and set later. You would have to go right around the world to get a delay of 24 h so there is an additional 1 h delay for every 15° west (or 4 min per degree of longitude). If the Sun is at its highest point in port at midday, and reaches the highest point where the ship is located at 12:30, then the ship's longitude is 7.5° west of the port.

Unfortunately pendulum clocks have a number of problems that make them unsuitable for accurate marine navigation.

- They are affected by the movement of the ship on a rough sea.
- They are not accurately isochronous (but this can be allowed for in their design).
- Their length, and therefore their time period, is affected by temperature (thermal expansion).
- Their time period depends on the gravitational field strength, and this varies from place to place on Earth, especially with latitude.

The problem was eventually solved in 1762 by an Englishman, John Harrison, whose prize-winning design did not use pendulums. Nowadays, traditional navigation is reinforced by the global positioning system, a network of geostationary satellites that can be used to pinpoint a position anywhere on Earth (see spread 5.6).

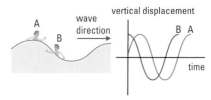

A and B are out of phase.

Phase difference

A phase difference of, say, 3π is indistinguishable from one of π, so in most cases we only use values between 0 and 2π to describe phase difference. The range of values used to compare phases can be restricted still further by realizing that a phase advance of x radians is equivalent to a phase delay of $(2\pi - x)$ radians. This means we only really need to use values between 0 and π.

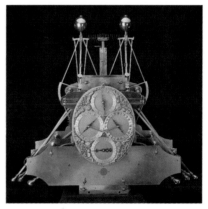

The first of John Harrison's four attempts to win the 'Longitude Prize' – the first accurate global positioning system?

PRACTICE

1 a What is the period of a 50 Hz oscillation?

  b What is the frequency of a swing that moves from one extreme to the centre of its motion in 0.7 s?

  c What is the fundamental (lowest) frequency of the guitar note in the graph on the previous page?

2 Refer to the following diagram.

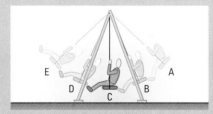

  a Sketch the graph for one cycle of the swing's motion and label points that could correspond to A, B, C, D, and E in the diagram.

  b Where does the swing have maximum velocity, maximum kinetic energy, maximum gravitational potential energy, maximum acceleration, zero velocity?

  c If nobody pushes the swing it eventually stops swinging. Why?

3 Sketch graphs of displacement versus time for each of the following oscillations and suggest suitable values for amplitude and frequency in each case.

  a Your arm swinging freely as you walk. (Use angular displacement.)

  b A perfectly elastic ball bouncing vertically on a rigid solid surface.

  c The free end of a plastic ruler held over the edge of a table, bent downwards, and then released to vibrate vertically.

4 Globally the Earth has two high tides per day. (Local geography can change this.) How would you expect the phase of the tides to vary with longitude? Try to give a quantitative answer – perhaps a formula – taking the phase to be zero at the Greenwich meridian (0° longitude). The times of actual high tides do not conform to such a simple rule. Can you suggest why?

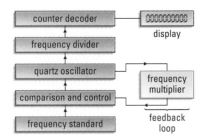

Block diagram showing the essential components of an atomic clock.

Measuring time

Any measurement of time is a comparison of change between two systems; for example, your heart might beat 70 times while the second hand of your watch moves round once. Clocks are simply devices chosen as a standard because they undergo cyclic changes in an apparently regular way. Many such cyclic motions have been used as a standard: the rotation of the Earth on its axis for days and hours, and the orbit of the Earth around the Sun for years, are the most important regular changes in our everyday life. However, these are not regular enough when measured against reproducible atomic changes to serve as a scientific measure of time.

Unfortunately, a simple pendulum is not accurately isochronous and is prone to disturbances from vibration and thermal expansion.

- **Quartz crystal clocks** replaced pendulum clocks in the 1930s. These depend on the vibrations of a quartz crystal, which operates tiny electrical signals which can be used to control the frequency of an electronic oscillator to an accuracy of about one part in 10^8.

- The quartz clock was replaced by the **atomic clock** in the 1950s. This clock relies on the ability of atoms to absorb and emit radiation only at sharply defined frequencies. The radiation from caesium atoms is used to control the vibrations of a quartz crystal, which in turn controls the vibrations of the electronic oscillator.

Defining the second

The second is now defined as the duration of 9 192 631 770 periods of the radiation corresponding to the transition between the two hyperfine levels of the ground state of a caesium-133 atom. Clocks based on the caesium atom set the present standard for time measurement, and are accurate to about 1 part in 10^{13}, or one second in 300 000 years.

Simple harmonic motion and circular motion

One of the most familiar periodic motions is uniform circular motion. The Earth itself (almost) does this as it orbits the Sun. Imagine watching the orbital motion 'edge-on', so it is restricted to movement along one line in your field of view. The Earth would move swiftly across the centre of its motion and slow down as it approached each end before accelerating back toward the centre, and so on. What you are looking at is a projection of its uniform circular motion onto one diameter, and this would seem very similar to **simple harmonic motion** (see spread 3.34). In fact, if the circular motion is uniform, its projection onto a diameter *is* simple harmonic. This is shown in the diagram.

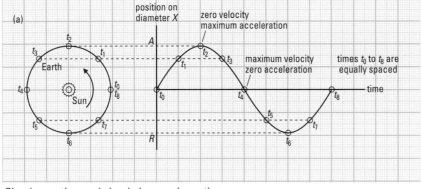

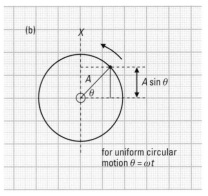

Circular motion and simple harmonic motion.

Pulsars

The discovery of **pulsars**, rapidly rotating neutron stars that emit regular pulses of radio waves, provides an extraterrestrial timekeeper whose accuracy, in some cases, rivals that of terrestrial atomic clocks. One day it is conceivable that we may use these cosmic pacemakers as a standard.

The relation between the two motions is summarized in the tables on the next page. Equations for displacement, velocity, and acceleration along the diameter are simply projections of the displacement, velocity, and acceleration of the circular motion. Both motions have a common period T related to the angular velocity by:

$$T = \frac{2\pi}{\omega}$$

giving

$$f = \frac{1}{T} = \frac{\omega}{2\pi} \quad \text{or} \quad \omega = 2\pi f$$

Circular and simple harmonic motion.

Circular motion	Simple harmonic motion
radius (A)	amplitude (A)
angular displacement (θ)	phase (ωt)
angular velocity (ω)	angular frequency (ω)

Quantities in circular and simple harmonic motion.

Quantity	Circular motion	Simple harmonic motion
displacement	A at angle $\theta = \omega t$	$x = A\cos\theta = A\cos\omega t$
velocity	ωA along tangent	$v = -\omega A\sin\theta = -\omega A\sin\omega t$
acceleration	$\omega^2 A$ centripetal	$a = \omega^2 A\cos\theta = \omega^2 A\cos\omega t$

Graphical representation

A simple pendulum can create its own displacement–time graph if it swings from a linear potentiometer (see spread 4.10) that is connected to a chart-recorder or datalogging equipment. If it is started from one amplitude ($x = A$) the graph will have a maximum value at $t = 0$ and be a cosine curve. Since velocity is rate of change of displacement, a graph of velocity versus time can be derived from the displacement graph. A graph of acceleration versus time can be derived from the velocity graph in a similar way.

A position sensor interfaced to a computer can be used to collect data on mechanical oscillations.

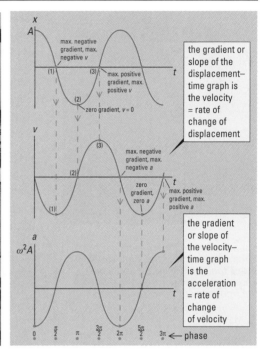

Graphical representation of a simple pendulum.

PRACTICE

1 After exercise your heart rate has risen to 180 beats per minute. At what frequency is it beating?

2 A pair of clocks are each driven by electronic oscillators. If the oscillators keep time with one another to an accuracy of one part in 10^8, what is the maximum difference between the two clocks after one year?

3 What is the frequency of radiation emitted in a transition between the hyperfine levels of the ground state of caesium?

4 A mass on a spring oscillates vertically with simple harmonic motion of amplitude 0.04 m and frequency 1.5 Hz.

   a Write down an equation for its displacement (x) from equilibrium as a function of time (t).

   b What is its maximum velocity and when does it occur?

   c What is its maximum acceleration and where does it occur?

5 Sketch a graph to show how the distance of a pendulum bob from its equilibrium position varies with time. Assume the motion is approximately linear and has amplitude 0.05 m and period 2.0 s. Label your graph to show significant values and indicate the positions of maximum and minimum kinetic energy.

6 The comparison of circular and simple harmonic motion above assumed that the object moving in circular motion started at the top of the circle at $t = 0$.

   a How would the equation for displacement in simple harmonic motion have differed if it had started at the right-hand end of the horizontal axis?

   b It is obvious that the projection of circular motion onto a diameter is *always* simple harmonic, regardless of the starting position of the object. What does this tell you about the equations for displacement, velocity, and acceleration for something moving with simple harmonic motion?

SIMPLE HARMONIC MOTION

Trees bend in a steady wind until their internal restoring force balances the force of the wind. If the wind suddenly drops the tree will sway (oscillate) backwards and forwards. What affects the frequency of these oscillations?

Maths box

The question that must be answered is why does the displacement vary sinusoidally in the first place? The equation

$$a = -\omega^2 x \quad \text{or} \quad \frac{d^2x}{dt^2} = -\omega^2 x$$

is a **second-order differential equation** that defines simple harmonic motion. Solutions to this equation are expressions for x as a function of time which satisfy the equation. (To 'satisfy the equation' the second derivative of the expression for x must equal minus a constant times itself). $x = A \cos \omega t$ is one such solution, but the general solution of the equation is

$$x = A \cos (\omega t + \phi)$$

where ϕ is a constant phase whose value is determined by the position of the oscillator at $t = 0$. For example, if $x = 0$ at $t = 0$, $\phi = -\pi/2$ and $x = A \cos (\omega t - \pi/2) = A \sin \omega t$.

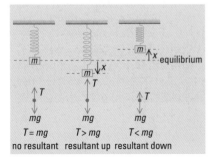

Forces in SHM.

Equilibrium and restoring forces

Hold a ruler flat on a table so that its end extends about 10 cm beyond the edge of the table. It will stay there indefinitely in equilibrium. Now push down the free end a little so that the ruler bends at the table edge. Feel it push up. Release it and it moves rapidly back up towards equilibrium, passes that position, bends upwards slightly, and then moves back down again. It completes a number of fast oscillations before eventually stopping in its equilibrium position. Now *lift* its free end slightly and feel it pull back down. If released, it again oscillates for a while before stopping at equilibrium.

This example illustrates the essential physics of all mechanical oscillators.

• They have an equilibrium position.

• If displaced in either direction from equilibrium, a resultant force acts toward equilibrium. This is sometimes called a **restoring force**. (For angular oscillations it is more convenient to consider the angular displacement and **restoring moment**.)

With the same length of ruler extending beyond the table edge as before, attach a piece of Plasticine to the free end and ping the ruler so that it oscillates. The sound that it makes is at a lower frequency, because it is taking longer to complete each cycle. This demonstrates the factors that control the period and frequency of a mechanical oscillator.

• The size of the restoring force affects the time period. In this case the force is determined by the ruler's elasticity. Larger forces reduce the time period (increase the frequency).

• The inertia (mass) of the oscillator also affects the time period. Greater inertia increases the time period.

In this case the inertia increased while the forces remained unchanged. The acceleration at each point in the oscillation was reduced, and the cycle took longer to complete. This argument is based on Newton's second law of motion in the form $F = ma$.

Definition of simple harmonic motion

So far simple harmonic motion (SHM) has been described rather than defined. Recall the equations for SHM:

$$x = A \cos \omega t$$
$$v = -\omega A \sin \omega t$$
$$a = -\omega^2 A \cos \omega t$$

The acceleration is simply $-\omega^2$ times the displacement:

$$a = -\omega^2 x = -(2\pi f)^2 x$$

This is the defining equation for SHM. If the oscillator has constant mass m then it can be used to find the characteristic force law that will produce SHM:

$$F = ma = -m\omega^2 x$$

SHM will always occur if the force acting on the oscillator is proportional to displacement from some equilibrium position ($x = 0$), and always directed back towards equilibrium. (This is the significance of the minus sign.)

The force law

Anything that moves with an acceleration proportional to displacement from a fixed point and always directed towards that point is undergoing simple harmonic motion.

If $a \propto -x$, then the motion is SHM and $a = -\omega^2 x$.

advanced **PHYSICS**

For an object of constant mass, this will occur if it experiences a restoring force (towards equilibrium) proportional to its displacement from equilibrium:

$$F \propto -x \quad \text{giving} \quad F = ma = -m\omega^2 x$$

The beauty of all this is that it gives a foolproof strategy for analysing oscillators.

Worked example

A mass m is suspended from a spring with spring constant k. It is pulled down a distance x below equilibrium and released. The spring obeys Hooke's law throughout the subsequent oscillations. a) Show that the oscillations are simple harmonic. b) What is their time period?

Step 1: The weight of the mass is supported by tension in the spring when it is at equilibrium. If it is displaced a distance x below equilibrium the spring tension increases by an amount kx (from Hooke's law).

Step 2: There is a restoring force kx acting on the mass: $F = -kx$

Step 3: The motion is simple harmonic because $F \propto -x$ (answers part (a)).

Step 4: $\omega^2 = \dfrac{k}{m}$ so $f = \dfrac{1}{2\pi}\sqrt{\dfrac{k}{m}}$ then period $= \dfrac{1}{f}$

Hence, the period is $T = 2\pi\sqrt{\dfrac{m}{k}}$

Solving SHM problems

Step 1: Identify the resultant force F on the system.

Step 2: Determine how this varies with displacement x.

Step 3: If F is proportional to $-x$ then SHM will occur and $F = -kx$ (k is a constant).

Step 4: ω and f can be derived from the constant of proportionality in $F = -kx$, since $F = m\omega^2 x = -kx$ so

$$\omega^2 = \frac{k}{m} \text{ and } f = \frac{\omega}{2\pi}$$

PRACTICE

1 Identify the restoring forces for each of the following:

   a A cork bobbing up and down in water.

   b A sodium ion vibrating in the cubic sodium chloride lattice.

   c A mass bouncing up and down on a spring.

   d A boxer's brain slopping back and forth in his skull after he is punched on the nose.

2 Show that the following oscillators are simple harmonic:

   a A mass m suspended between two stretched horizontal springs, each of which obey Hooke's law as they are stretched and compressed. (Assume the mass is resting on a smooth horizontal surface so the springs do not sag.)

   b A block of wood bobbing up and down in water. (Assume the block is rectangular, but the result is true for all shapes.)

3 A friction-free trolley of mass 2 kg is tethered to rigid supports at each end by two identical springs of spring constant $20\,\text{N}\,\text{m}^{-1}$ each. The springs obey Hooke's law and remain in tension as the trolley is displaced 2 cm to one side and released.

   a What is the effective spring constant of the system tethering the trolley?

   b What is the frequency of the resulting simple harmonic motion?

4 A trolley is tethered between two rigid supports by a pair of similar springs. When displaced from equilibrium it undergoes a series of simple harmonic oscillations of amplitude A and period T. For the questions that follow assume that the springs remain in tension and do not pass their limit of proportionality as the trolley moves. What happens to A and T if:

   a The trolley is initially pulled back twice as far?

   b A second mass equal to the mass of the trolley is placed on the trolley before it is displaced a distance A?

   c Instead of being placed on the trolley at the beginning, the mass is dropped onto the moving trolley as it passes its equilibrium position?

   d As in c, but the mass is dropped on at one extreme of the trolley's motion?

   e The two springs are replaced by a pair which each have half the spring constant of the originals?

5 An atom of mass 5×10^{-26} kg is in a cubic lattice with all bonds between adjacent pairs of atoms having a spring constant about $100\,\text{N}\,\text{m}^{-1}$. Estimate the natural frequency of oscillation of the atom. If it is able to absorb radiation at this frequency, what part of the electromagnetic spectrum does it absorb?

6 During an earthquake the floor of a building vibrates vertically with an amplitude of 5 cm. At what frequency would objects resting on the floor just begin to lose contact with it at one point in the oscillation? Where in the oscillation do they lose contact?

OBJECTIVES

- energy in simple harmonic motion
- damping
- forced oscillations and resonance

Maths box

This page shows how energy varies with *time*. The expressions below give the variation with position (see graph):

$F = -kx$ from the definition of SHM

$$F = -\frac{dPE}{dx}$$

$$PE = \int_0^x F\,dx = \int_0^x kx\,dx = \frac{1}{2}kx^2$$

$$TE = \frac{1}{2}kA^2$$

$$KE = TE - PE = \frac{1}{2}kA^2 - \frac{1}{2}kx^2$$

$$= \frac{1}{2}k(A^2 - x^2)$$

Substituting $k = m\omega^2$ leads to a new formula for velocity:

$$KE = \frac{1}{2}mv^2 = \frac{1}{2}m\omega^2(A^2 - x^2)$$

giving $v = \omega\sqrt{A^2 - x^2}$

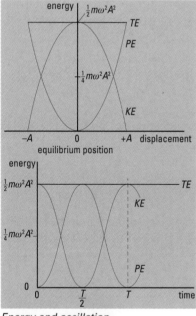

Energy and oscillation.

Shock absorbers on a bicycle.

Oscillator energy

Mechanical oscillators continually transfer energy back and forth between kinetic and potential. Ignoring frictional forces, the total energy of the oscillator is conserved:

$$TE = KE + PE$$

It is convenient to define the potential energy for the oscillator at its equilibrium position as zero. Then the total energy will equal the maximum kinetic energy, which it has when it has maximum velocity as it passes through the equilibrium position.

$$TE = \frac{1}{2}mv^2_{\text{max}}$$

$$v = \omega A \cos \omega t$$

$$\therefore v_{\text{max}} = \omega A \quad \text{when } \cos \omega t = 0$$

$$TE = \frac{1}{2}m\omega^2 A^2$$

The values of KE and PE at any time in the motion are

$$KE = \frac{1}{2}m\omega^2 A^2 \cos^2 \omega t$$

$$PE = TE - KE = \frac{1}{2}m\omega^2 A^2(1 - \cos^2 \omega t) = \frac{1}{2}m\omega^2 A^2 \sin^2 \omega t$$

These are plotted as a function of time on the sketch graph left. The horizontal line, representing total energy, is the sum of values from the two lines (KE and PE) beneath it. Beware of oscillators which move vertically, like a mass on a spring, since these may involve more than one kind of potential energy. The PE line on this graph is total potential energy – for a mass at the upper amplitude on the end of a spring this is stored partly as elastic potential energy in the spring, and partly as gravitational potential energy.

Total energy is proportional to amplitude squared, so doubling the amplitude of oscillation *quadruples* the total energy. This also applies to waves where intensity is proportional to amplitude squared.

Damping

When a bell rings it is transferring energy stored in its oscillation to sound by moving the air around it. It does work against frictional forces and is said to be **damped**. Whenever frictional forces act on an oscillator its total energy will diminish with time so that its amplitude decays to zero. The heavier the damping (larger frictional forces) the greater the rate of decay. For many oscillations the damping forces are roughly proportional to velocity, and this leads to an *exponential* decay of amplitude.

A simple test for exponential decay is to look for a constant proportion in successive amplitudes. That is if the second amplitude is 0.9 of the first, the third will be 0.9 of the second (0.9^2 of the first) and so on. The **constant proportion property** is common to all exponential changes. (It is also used in the time constant for resistance–capacitance circuits (see chapter 4) and the half-life in radioactive decay.)

Damping is deliberately introduced into some systems to prevent continuous oscillations. For example, some compass needles are damped by filling the compass with a liquid. A light-beam galvanometer relies on electromagnetic damping to prevent it swinging back and forth past the correct reading, and a car's shock absorbers have hydraulic pistons to prevent the car bouncing after hitting a bump. Shock absorbers can be tested by pushing down on the car and then releasing it. It should return to its normal height with negligible overshoot. A system like this, which returns to equilibrium in the shortest possible time (after about a quarter of a damped cycle) is said to be **critically damped**.

Forced oscillations and resonance

If an undamped oscillator is displaced and then left to oscillate, it is a **free oscillator**. If some external agent continues to shake it, it is a **forced** (or **driven**) **oscillator**. Its response to being forced depends on the frequency of the forcing oscillator (sometimes called the **driver**). The response is greatest if an oscillator is driven at its own natural frequency. This is called **resonance**.

Try shaking the string of a simple pendulum sideways at a regular rate. If it is shaken too quickly the pendulum is unable to respond, and hardly moves one way before changing direction. If it is shaken too slowly the whole string moves sideways with the same amplitude as the driver, but this oscillation never grows. It is easy to 'feel' where the resonance is: if the top of the string is moved even very slightly back and forth at this frequency, then the resulting oscillation grows and quickly attains a large amplitude.

At resonance the driver applies forces which continually supply energy to the oscillator. This is why its amplitude increases. In fact, it would continue to increase indefinitely unless some other process transferred energy away from the oscillator. In severe cases this limit is reached when the driven oscillator destroys itself.

The size of the resonant amplitude is limited by damping forces. At resonance the rate of energy supply from the driver exactly equals the rate of loss as work is done against damping forces. Increasing the damping reduces the sharpness and strength of the resonance. The sharpness of resonance is measured by its **Q-factor**. Although this does have a precise mathematical definition, it is roughly the number of free oscillations the oscillator will complete before decaying to zero. If damping is light Q will be large; if it is heavy Q will be small.

Resonances in structures such as buildings and bridges can cause large stresses and could even cause the structure to collapse (for example, the Tacoma Narrows bridge collapse, or a wine glass shattered by an opera singer). Unwanted resonances in structures and machinery can be made less significant by damping, or may be moved to a different frequency by changing the mass of the oscillator or the stiffness of its supports (for example, the Millennium Bridge in London). The latter method is used to move the resonance away from the frequency of the driving vibrations.

Resonances can also be useful, for example, in microwave cooking (resonance of water molecules) or radio and TV tuning circuits (electrical resonance with the received carrier frequency).

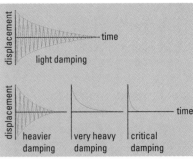

Damped oscillations. Critical damping stops the oscillations in the minimum time.

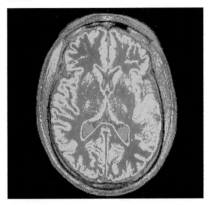

An image of a transverse cross-section through the brain created by nuclear magnetic resonance imaging (NMR).

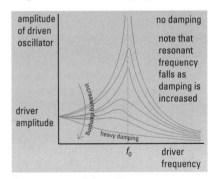

Resonance and damping.

PRACTICE

1 A 500 g mass oscillates with a vertical amplitude of 3 cm on the end of a spring of spring constant 25 N m⁻¹. What is its total energy and its maximum velocity?

2 The amplitude of the oscillator in question 1 above is now reduced to 2 cm. How does this affect its maximum kinetic energy, velocity, and acceleration?

3 As the engine of a small boat starts up, a passenger notices that all the seats vibrate quite violently for a second or two and then stop once the engine is up to full speed. Explain this.

4 Generally big musical instruments produce low notes and small instruments produce high notes. Why?

5 If the supports of a suspension bridge deck are stiffened, what effect will this have on its resonant frequency?

6 Assume different people move their legs through the same angle when walking at a comfortable unhurried pace, and that their legs act like freely pivoted simple pendulums.

   a Show that comfortable walking pace is proportional to the square root of leg length.

   b Explain why fast walking is more of an effort and difficult to maintain.

7 Is it true that taller buildings will always be more susceptible to damage by earthquakes? Explain.

8 After 1 s the amplitude of a lightly damped oscillator has fallen to 95% of its original value.

   a What fraction of the incident energy has been dissipated?

   b If the decay of amplitude is exponential, what will the amplitude be after 5 s?

- application of simple harmonic motion to molecular structures and pendulums

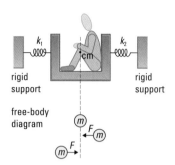

Massing an astronaut. If both springs obey Hooke's law,
$$F = -(k_1 + k_2)x = -k_{tot}x$$
Therefore SHM occurs and $\omega^2 = k_{tot}/m$

SIMPLE HARMONIC MOTION IN ACTION

Massing an astronaut

It is important to monitor the health and physical condition of an astronaut in space. One thing doctors need to know is how an astronaut's body mass changes during a mission. On Earth the astronaut could simply be 'weighed', but this cannot be done in the apparently weightless environment of a free-falling orbital spacecraft. Instead astronauts sit on a specially mounted chair attached to the craft via springs. Once they are displaced, the period of their oscillations is measured and is used to determine their mass. The method works because the inertia of a body is unaffected by gravity or free fall. Greater inertia (mass) produces a longer oscillation period. The arrangement is equivalent to the mass–spring oscillator analysed in spread 3.34, so the time period is given by:

$$T = 2\pi \sqrt{\frac{m}{k_{tot}}} \quad \text{and} \quad m = \frac{k_{tot}T^2}{4\pi^2}$$

Note that the spring constant used in the formulae above is the sum of the spring constants for the springs on either side of the chair ($k_{tot} = k_1 + k_2$), since both apply a restoring force when it is displaced.

The analysis above applies to any system where the restoring forces arise from the elasticity of the supports. For example, the resonant frequency of ions in a lattice can be estimated from the interatomic force constants and ionic mass. (Conversely, we can use the measured frequency to estimate bond stiffness.) The swaying of buildings during an earthquake depends on the stiffness of the building and its mass distribution.

Worked example

HCl gas absorbs electromagnetic radiation very strongly in the infrared region of the spectrum, at a wavelength of $3.5\,\mu m$. This frequency corresponds to a natural frequency for vibration of the molecule, so the incident waves cause resonance. The chlorine atom is about 35 times the mass of a hydrogen atom, so a simple model of the vibrating molecule assumes the chlorine atom is fixed and the hydrogen atom behaves like a mass on a spring obeying Hooke's law. Estimate the 'spring constant' for the interatomic bond. (The mass of the hydrogen atom is $1.7 \times 10^{-27}\,kg$.)

$$f_0 = \frac{1}{2\pi} \sqrt{\frac{k}{m_H}} \quad \text{natural frequency of a mass–spring oscillator}$$

$$f_d = \frac{c}{\lambda} = \frac{3 \times 10^8\,m\,s^{-1}}{3.5 \times 10^{-6}\,m} = 8.6 \times 10^{13}\,Hz \quad \text{electromagnetic waves act as driving oscillator}$$

$$\therefore k = 4\pi^2 f^2 m_H = 490\,N\,m^{-1}$$

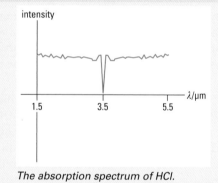

The absorption spectrum of HCl.

Absorption in practice

In practice an absorption band – a range of absorption frequencies – rather than a line would be observed, because the molecules can also rotate, and different states of rotation have slightly different energies, so a range of photon energies and frequencies can be absorbed.

SHM and the speed of sound

The speed of sound in a material depends on how rapidly a disturbance at one place can be communicated to the next. This in turn depends on the masses and separations of particles in the medium, and on the stiffness of the bonds that connect them.

The simple pendulum

A simple pendulum is a point-like mass suspended from a light inextensible string. Does it undergo simple harmonic motion and what is its period of oscillation?

There are a number of ways to analyse the simple pendulum. Here we focus on the motion along the circular arc in which it swings.

Step 1: The resultant force is the component of its weight along the tangent to the arc (see the free-body diagram on the next page), and is always directed toward equilibrium.

Step 2: For an angular displacement θ the linear displacement along the arc is

$$x = l\theta \quad \text{(definition of angle in radians)}$$

and the resultant force is given by

$$F = -mg \sin \theta$$

Step 3: F is not proportional to $-x$, so this is *not* SHM.

However, if the angle of swing is small, the small-angle approximation can be used:

$$\sin \theta \approx \theta = \frac{x}{l}$$

If this is substituted into the force equation F is proportional to $-x$:

$$F = -\frac{mgx}{l}$$

So a simple pendulum approximates to SHM for small angles of swing.

Step 4:

$$F = -\frac{mgx}{l} = -m\omega^2 x \quad \text{so} \quad \omega = \sqrt{\frac{g}{l}}$$

giving

$$f = \frac{1}{2\pi}\sqrt{\frac{g}{l}} \quad \text{and} \quad T = 2\pi\sqrt{\frac{g}{l}}$$

Compound pendulums

A **compound pendulum** is any rigid object that oscillates in a vertical plane about a fixed axis. (The simple pendulum is the simplest case of compound pendulum, in which all the mass is concentrated at a point.)

Step 1: Identify the resultant torque, $\boldsymbol{T}$ ($\boldsymbol{T}$ is used for torque to avoid confusion with T for period). In the case of both the simple and the compound pendulum this arises because the weight of the pendulum generates a turning effect about the pivot when the pendulum is displaced from equilibrium.

Step 2: This turning effect is given by the product of the weight of the pendulum and the horizontal displacement of its centre of mass from equilibrium:

$$\boldsymbol{T} = -mgl \sin\theta$$

Step 3: This torque is not proportional to angular displacement, but to the sine of the angular displacement, so in general the motion is not SHM. However, if the amplitude of angular displacement is small then the sine of the angle is approximately equal to the angle in radians and:

$$\boldsymbol{T} = -mgl\theta$$
$$\boldsymbol{T} \propto -\theta$$

so SHM occurs

Step 4:

$$a = \frac{\boldsymbol{T}}{I} = -\frac{mgl\theta}{I} = -\omega^2\theta \qquad \omega^2 = \frac{mgl}{I} \qquad T = \frac{2\pi}{\omega} = 2\pi\sqrt{\frac{I}{mgl}}$$

For a compound pendulum I (moment of inertia) depends on the mass distribution from the pivot. For a simple pendulum there is one mass point distance l from the pivot so:

$$I = ml^2 \qquad T = 2\pi\sqrt{\frac{l}{g}}$$

which is identical to the result obtained by treating it as one-dimensional oscillation.

Rotational oscillations

A **torsional pendulum** is a mass suspended from a string or wire, twisting back and forth around the wire. The best way to analyse rotational oscillations is by using torques and angular displacements in place of forces and linear displacements. The form of the equations is exactly the same as for linear motion but with the following substitutions:

$$F \to \boldsymbol{T}$$
$$x \to \theta$$
$$m \to I \quad \text{(moment of inertia)}$$
$$a \to \alpha \quad \text{(angular acceleration)}$$

For linear SHM: $\quad F \propto -x \quad$ For rotational SHM: $\boldsymbol{T} \propto -\theta$

Equation of motion: $a = -\omega^2 x \quad$ Equation of motion: $\alpha = -\omega^2\theta$

In both cases: $\quad T = \frac{2\pi}{\omega}$

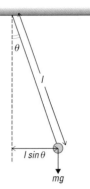
A simple pendulum (point mass, light inextensible string).

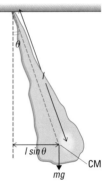
Compound pendulum (distributed mass).

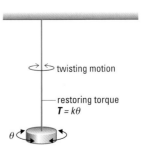
A torsional pendulum.

PRACTICE EXAM QUESTIONS

1 A person is standing on the floor of a lift which is accelerating uniformly upwards.

    a Draw a free body diagram showing all the forces acting upon the person.

    b What are the Newton's Third Law reactions to each of these and on what does each act?

    c Which, if any, of the forces in **a** change as the acceleration changes? Give brief reasons.

2 **a** Explain what is meant by the moment of a force.

    b State the conditions for the equilibrium of a rigid body under a system of coplanar forces.

    c Masses of 2.0 kg and 3.0 kg are hung from points B and C of a string ABCD as shown in Figure 3.1. A and D are attached to a fixed support. AB makes an angle of 30° with the vertical and CD makes an angle θ with the vertical. BC is horizontal.

    Calculate

    i the tension T_1 in AB,

    ii the tension T_0 in BC,

    iii the angle θ.

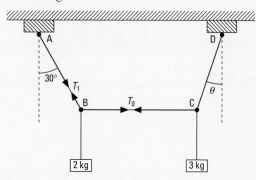

Figure 3.1

3 **a** A tug is towing two barges at constant speed as shown in Figure 3.2. The tension in cable CD is 12 000 N. Calculate the tension in each of the cables, AC and BC.

    b The speed of the tug is 1.8 m s⁻¹ and 25% of the power of the propulsion system is used to drive the tug and barges. Calculate the power of the propulsion system.

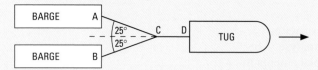

Figure 3.2

4 Figure 3.3 shows a straight, horizontal, uniform swimming bath spring board of length 4.00 m and of weight 300 N. It is freely hinged at A and rests on a roller B, where AB is 1.60 m. A boy of weight 400 N stands at end C.

    a **i** Copy Figure 3.3 and show on it the directions of the forces acting on the board at A and B.

    ii Calculate the magnitudes of the forces at A and B.

    b With the boy still on the board, a girl of weight 630 N also stands on the board. How far is she from A if the force at B is doubled?

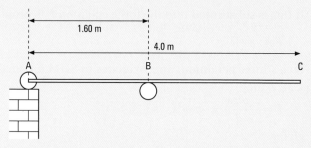

Figure 3.3

5 This question is about the stopping distances of vehicles given in the British Highway Code.

The graphs in Figure 3.4 summarise the information relating to thinking distance and braking distance.

Thinking distance is the distance the vehicle travels between the driver seeing an incident and beginning to apply the brakes. Braking distance is the distance the vehicle travels while it decelerates.

    a Suggest why the graph of thinking distance against speed is a straight line.

    b **i** The graph of braking distance against speed is curved. Use information from the graph to test whether braking distance is proportional to (speed)².

    ii Suggest a *physical* explanation for the shape of this graph.

    c **i** What is the total stopping distance for a vehicle initially travelling at 30 m s⁻¹?

    ii This is the minimum stopping distance. Suggest one possible reason why the stopping distance may be significantly greater than this. Explain your answer.

    d Use the graph to calculate the mean braking force acting on a vehicle of mass 1000 kg as it decelerates from 30 m s⁻¹ to rest.

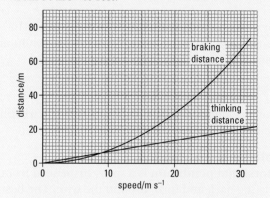

Figure 3.4

6 A car travelling at 20 m s⁻¹ collides head on with a massive wall and stops virtually instantly. A passenger of mass 75 kg, who is wearing a seat-belt, is brought to rest in 0.2 s. Find:

    a the force (assumed constant) exerted by the seat-belt on the passenger;

    b the energy absorbed in the seat-belt system as a result.

7 A catapult fires an 80 g stone horizontally. The graph in Figure 3.5 shows how the force on the stone varies with distance through which the stone is being accelerated horizontally from rest.

a Use the graph to estimate the work done on the stone by the catapult.

b Calculate the speed with which the stone leaves the catapult.

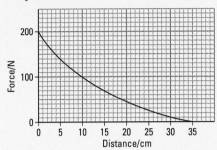

Figure 3.5

8 a Peregrine falcons dive vertically on their prey with speeds of up to 30 m s⁻¹. Through what height must a falcon dive from rest to achieve this speed?

b State *one* assumption that you make in your calculation.

c Calculate a value for the kinetic energy of a falcon of mass 0.80 kg just before it strikes its prey.

9 A glider travels in a straight line along an air track. Figure 3.6 is a graph of its velocity plotted against time.

a i Calculate the acceleration of the glider in each of the time intervals AB, BC, CE and EF of its motion.

ii Plot a graph of the acceleration of the glider against time.

b Obtain a value of the displacement of the glider from its starting time A to time F.

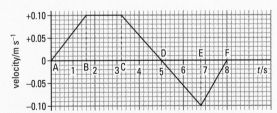

Figure 3.6

10 A typical take-off speed for a flea of mass 4.5×10^{-7} kg, jumping vertically, is 0.80 m s⁻¹. It takes such a flea 1.2 ms to accelerate to this speed.

a i Show that the average acceleration is 670 m s⁻².

ii Calculate the force required to produce this acceleration.

iii Calculate the *average* power developed during the acceleration.

b Assuming that the effect of air resistance is negligible, estimate the height to which the flea jumps. (Acceleration of free fall, $g = 10$ m s⁻².)

11 A transport plane is towing a glider, which in turn is towing a second glider. They accelerate from rest on a level runway. The plane and gliders remain in a straight line and the tow ropes are horizontal. The gliders *each* have a mass of 1000 kg. Assume the force opposing the motion of each glider is constant at 1500 N. The plane reaches a speed of 40 m s⁻¹ after accelerating uniformly over a distance of 320 m.

a Calculate the acceleration of the plane.

b During the acceleration of the plane, calculate

i the tension in the tow rope between the two gliders.

ii the tension in the tow rope between the plane and the first glider.

12

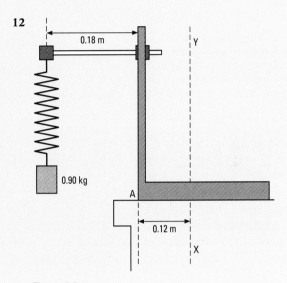

Figure 3.7

The mass of a retort stand and clamp is 1.6 kg and their combined centre of mass lies along the line XY. A spring which has negligible mass is attached to the clamp and supports a mass of 0.90 kg as shown in Figure 3.7. The spring requires a force of 6.0 N to stretch it 100 mm.

a Calculate the extension of the spring.

b Show that this arrangement will not tip (i.e. will not rotate about A) when the 0.90 kg mass is at rest in its equilibrium position.

c If the mass is lifted up and released, it will vibrate about the equilibrium position. Explain, without calculation, why the stand will tip if the amplitude exceeds a certain value.

13 a State the difference between a vector and a scalar quantity.

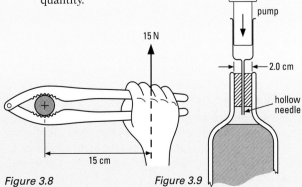

Figure 3.8   *Figure 3.9*

b A device for removing tightly fitting screw tops from jars and bottles is shown in Figure 3.8.
In one case a constant force of magnitude 15 N has to be applied, as shown in Figure 3.8. The force is applied for one complete turn to remove the top.

  i Calculate the torque which has to be applied to remove the top.

  ii Calculate the work done when opening the bottle.

c Tight-fitting corks can be removed by pumping air into the bottle using the device shown in Figure 3.9. Assuming that the force which has to be overcome to remove the cork is 30 N, calculate the pressure inside the bottle when the cork begins to move. Atmospheric pressure = 1.0×10^5 Pa.

14 This question is about the path followed by an object moving freely under gravity.

A cricketer throws a cricket ball at an angle of $45°$ above the horizontal. The graph, Figure 3.10, shows the variation of height, h, with horizontal distance covered, x. Assume that the ground is horizontal, and that air resistance is negligible.

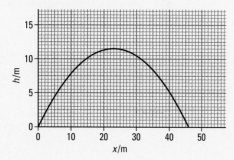

Figure 3.10

a The horizontal component of the ball's velocity is $15\,\text{m s}^{-1}$.

Use the graph to calculate the time for which the ball is in the air.

b **i** Write down the ball's vertical component of velocity at the moment of release. Justify your answer.

  ii Use the graph to calculate the time the ball takes to reach its maximum height. Explain your answer.

  iii The maximum height reached is 11.5 m. Use this value to calculate the deceleration of the ball during this time.

c Copy the graph and sketch on your copy the path for a cricket ball thrown at the same angle, but with half the initial velocity.

15 a Describe briefly how you would demonstrate the phenomenon of mechanical resonance in the laboratory. Sketch a graph showing how the amplitude of a damped oscillatory system varies with the frequency of the driving force.

b Standing on one leg, a student finds that his other leg swings naturally forwards and backwards 10 times in 15 s. Calculate the length of a simple pendulum with the same period as the student's leg. Comment on the result.

c During a single stride the horizontal push F of the ground on a walker's foot varies with time t approximately as shown in Figure 3.11. F is taken to be positive when it is in the direction in which the walker is moving.

  i What physical quantity is represented by the area between the graph line and the time axis?

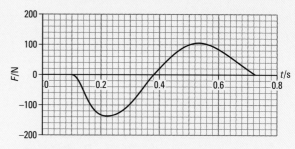

Figure 3.11

  ii Estimate the size of this quantity for the part of the graph for which F is positive. Explain how you made your estimate.

  iii The area above the time axis is the same as the area below it. Explain what this tells you about the motion of the walker.

16

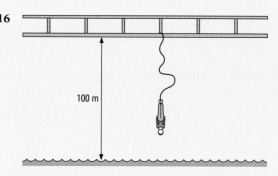

Figure 3.12

This question is about the sport of bungee jumping.
A girl of mass 50 kg jumps from a bridge 100 m above a river. Attached to her ankles is an elastic rope of natural length 50 m. The rope extends as she falls and brings her to a momentary halt 10 m above the water surface.

a Describe the energy transfers which occur from the time the girl jumps until she first comes to rest.

b At what point during the jump does the elastic rope exert the greatest force on the girl? Explain your answer.

c A relative of the girl, observing the jump, thought this activity was far too dangerous, since the decelerating force on the girl was so large.

  i Estimate the energy stored in the rope when the girl first comes to rest 10 m above the water. ($g = 9.8\,\text{N kg}^{-1}$)

  ii Hence estimate the mean force exerted by the rope on the girl whilst bringing her to rest. Assume the rope obeys Hooke's Law.

d The organisers of the event have ropes with different values of forces constant (k) where k is defined by Hooke's Law.

force = $k \times$ extension

They match the rope to the weight of the person.

Discuss what would happen if the force constant of the rope were:

i too large,

ii too small.

17

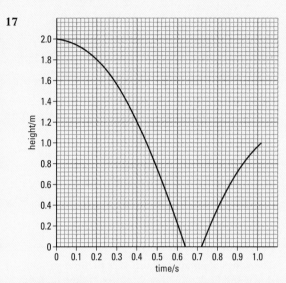

Figure 3.13

A rubber ball of mass 0.120 kg is dropped from a height of 2.00 m (measured from the bottom of the ball) on to a flat horizontal patch of hard soil.

a Calculate the speed of the ball when it hits the ground.

The rubber ball loses speed each time it bounces. Figure 3.13 shows how the height of the bottom of the ball varies with time during the first second of its motion.

b Use the graph to show that the speed of the ball as it leaves the ground is approximately 4.7 m s^{-1}.

c Calculate the average force exerted by the ground on the ball while it is in contact with the ground.

18 A snooker ball *A* of mass 0.2 kg travelling at 2.5 m s^{-1} collides perfectly elastically and head-on with a second, identical, stationary ball *B*. They remain in physical contact for 50 μs.

a What does perfectly elastically mean?

b Find the speed and direction of motion of *A* and of *B* after the collision.

c Find the average force exerted by *A* upon *B* during the collision.

*19 Figure 3.14 shows a view of the end of a short solid cylindrical rod radius *R* and length *L* made from a metal of density ρ.

Figure 3.14

a i Write down the formula for the mass of the hollow cylinder, shown shaded, of radius *r* and thickness δ*r*.

ii Write down the moment of inertia δ*I* of the shaded element of the hollow cylinder.

*Pre-U-style question

iii Use integration of the equation from **ii** to show that the moment of inertia of the solid cylinder is given by:

$$I = \frac{1}{2}\pi L\rho R^4$$

and that this is equivalent to:

$I = \frac{1}{2}Mr^2$ where *M* is the mass of the cylinder.

b Figure 3.15 shows a cylinder of mass 0.028 kg radius 8.0 × 10^{-3} m on a table.

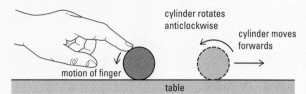

Figure 3.15

When a finger provides an anticlockwise torque momentarily as shown in the diagram, the cylinder moves forwards and spins anticlockwise. In one instance the ball was initially spinning with angular velocity 70 rad s^{-1} and linear velocity 0.25 m s^{-1}.

Neglect air resistance.

The frictional force between the cylinder and the table is 0.20 *Mg* and acts to reduce the angular speed and the linear speed.

i Calculate the distance travelled before the forward motion ceases.

ii Calculate the time taken for the cylinder to stop moving forwards.

iii Calculate the angular speed when the cylinder stops moving forwards.

iv Describe the motion after the cylinder stops moving forwards.

20 This question is about the performance of jet engines. A simplified diagram of a jet engine is shown in Figure 3.16.

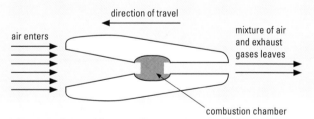

Figure 3.16

a When an aircraft is cruising at a speed of 250 m s^{-1}, air enters each of its jet engines at a rate of 220 kg s^{-1}.

Fuel is burnt in this air, in the combustion chamber, and the air and exhaust gases leave the engines at a speed of 420 m s^{-1}, relative to the aircraft.

i Show that the momentum of the air entering each jet engine every second is 55 000 kg m s^{-1}.

ii Calculate the momentum of the air and exhaust gases leaving the jet engine every second. State any assumptions that you make.

iii Calculate the forward thrust produced by each jet engine, under these circumstances. Make your reasoning clear.

b When the aircraft is cruising at a constant speed of 250 m s⁻¹, in level flight, the two jet engines are producing a constant forward thrust of about 74 000 N.

   i Explain this statement in terms of Newton's laws of motion.

   ii Describe and explain two ways in which the speed of the aircraft could be increased.

21 a For a body undergoing simple harmonic motion, the displacement s and velocity v at time t are given by the equations:

$s = A \sin (2\pi ft)$
$v = 2\pi fA \cos (2\pi ft)$

Write down an expression for the maximum speed of the oscillator in terms of f, the frequency of the oscillations, and A, the amplitude.

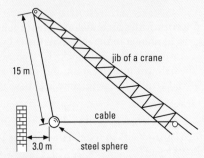

Figure 3.17

b Figure 3.17 shows a system used for demolishing buildings.

A 2500 kg steel sphere is suspended by a steel cable. The distance from the point of suspension to the centre of mass of the sphere is 15 m. The sphere is pulled a distance of 3.0 m to one side and then released.

When the wall is not in the way, the system performs simple harmonic motion and behaves like a simple pendulum of period T given by

$$T = 2\pi \sqrt{\frac{l}{g}}$$

where l is the length of the pendulum and g is the acceleration of free fall (10 m s⁻²).

   i Calculate the frequency of oscillation of the system.

   ii The suspension cable is vertical when the sphere hits the wall. Calculate the speed of the sphere when it hits the wall.

   iii Determine the kinetic energy of the sphere when it hits the wall.

   iv State with reasons whether using double the mass or double the initial displacement would provide the greater increase in kinetic energy (assume that the motion is still simple harmonic).

22 This question is about some of the physics involved in the design and operation of aircraft.

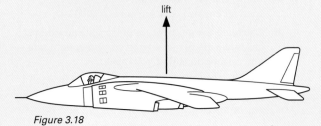

Figure 3.18

a i Figure 3.18 shows an aircraft with an arrow indicating the lift force. Make a sketch of the diagram and on it indicate and label the other principal forces acting on the aircraft when it is in level flight.

   ii Explain how the design of the aircraft provides the lift force.

b An aircraft has a mass of 11 000 kg when fully loaded. The take-off speed is 180 km h⁻¹ and is achieved using 750 m of runway starting from rest.

   i Assuming that the acceleration is uniform, determine the time taken to reach the take-off speed.

   ii Calculate the resultant force on the aircraft while accelerating along the runway.

   iii Calculate the average power used to accelerate the aircraft on the runway.

   iv Explain why the power developed by the engines is much greater than the answer to **iii**.

c Jet aircraft cruise at a height of 10 000 m.

Why is it an advantage to cruise at 10 000 m rather than at a much lower height?

d The wings of an aircraft are seen to move up and down and undergo other vibrations when in flight. State and explain one property of the material used for the manufacture of the wings of an aircraft which is desirable in view of these vibrations.

23 This question is about a toy consisting of a plastic head attached to a light spring (Figure 3.19). When the spring is compressed and released the toy jumps into the air. Figure 3.20 shows the variation of compression with applied force for this spring.

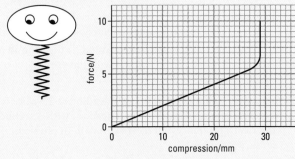

Figure 3.19   Figure 3.20

a Why is there a change in the gradient of the graph?

b The spring is compressed by pushing the head down 20 mm. The toy, of mass 0.030 kg, is then released.

   i Use the graph to calculate the energy stored by the spring when it has been compressed by 20 mm.

   ii What has happened to this energy at the time the spring has just returned to its uncompressed length?

   iii Estimate the maximum distance through which the toy will eventually rise.

   [Earth's gravitational field strength = 9.8 N kg⁻¹]

   iv In practice, the toy does not rise as far. Suggest one reason for this.

c A student suggests that, when the spring is compressed by only 10 mm, the toy will only reach half the estimated maximum height.

Without detailed calculation, comment on this suggestion.

24 The shot-putter shown in Figure 3.21 throws the shot forwards with a velocity of 12 m s⁻¹ with respect to his hand, in a direction 54° to the horizontal. At the same time, the shot-putter's body is moving forward horizontally, with a velocity of 3.0 m s⁻¹.

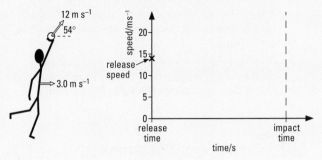

Figure 3.21   *Figure 3.22*

a Draw a vector diagram to show the addition of the two velocities of the shot at the moment of release. Hence, or otherwise, show that the vector sum of the two velocities has a magnitude of approximately 14 m s⁻¹.

b At the moment of release, the shot is 2.3 m above the ground. The shot has a mass of 7.3 kg.

   i Calculate the kinetic energy of the shot at the moment of release.

   ii Calculate the potential energy, relative to the ground, of the shot at the moment of release. The acceleration of free fall, g, is 9.8 m s⁻².

   iii Use the law of conservation of energy to calculate the speed of the shot as it hits the ground. (Neglect air resistance.)

   iv Copy Figure 3.22 and sketch the variation of the shot's speed with time, up to the moment when it touches the ground.

25 a State the conditions for a system of coplanar forces to be in equilibrium.

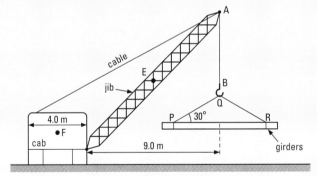

Figure 3.23

b Figure 3.23 shows a crane being used to lift a load of girders, each of which has a mass of 500 kg.

The jib of the crane has a mass of 2500 kg and the cab has a mass of 20 000 kg. The centres of mass (gravity) of the jib and the cab are at their mid-points, E and F respectively. The hook and cable have negligible mass. Gravitational field strength, g = 10 N kg⁻¹.

   i Calculate the tension in the cable AB when a single girder is lifted.

   ii Determine the corresponding tension in the cable PQR, when a single girder is lifted.

   iii For the jib in the position shown, determine the tension in AB which will just topple the crane.

   iv For the jib in the position shown, determine the maximum number of girders which can be lifted without the crane toppling over.

26 a Figure 3.24 shows a body moving with uniform speed in a horizontal circle.

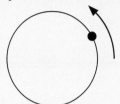

Figure 3.24

   i Copy the diagram and show on it the direction of the resultant force P acting on the body.

   ii Show on your diagram the path the body would follow if the force ceases to exist when it is at the point shown. Label this path D.

b In a fairground ride called a 'rotor', a person of mass 60 kg stands against a wall, as shown in Figure 3.25, and the wall is rotated. When it is spinning at a suitable speed the floor is dropped so that the person is left 'stuck to the wall'.

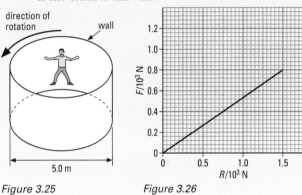

Figure 3.25   *Figure 3.26*

Figure 3.26 shows the variation in frictional force, F, with normal reaction, R, between the person and the wall.

Determine:

   i the normal reaction when the frictional force is equal to the weight of a person of mass 60 kg;

   ii the minimum angular speed, in rad s⁻¹, at which such a person must be rotated to remain in position when the floor is dropped.

   Acceleration of free fall, g = 10 m s⁻².

27 A metre rule is clamped to the top of a bench so that most of its length overhangs and is free to vibrate in a vertical plane with simple harmonic motion.

a Explain what is meant by simple harmonic motion (SHM).

b The metre rule oscillates with frequency 5.0 Hz. Calculate the angular frequency of the SHM.

c A small mass rests on the end of the metre rule and oscillates with it (Figure 3.27). The rule is oscillated with different amplitudes. As the amplitude is increased, it is observed that at one point in the cycle of motion of the rule and mass, the mass loses contact with the rule.

   i Explain why the mass loses contact with the rule.

   ii State at which point in the cycle of motion this loss of contact first occurs as the amplitude is increased.

iii Calculate the minimum amplitude of oscillation at which the loss of contact occurs.

d In practice, all vibrating systems are subject to damping.

   i Describe what difference you would observe between heavily and lightly damped oscillations.

   ii State a cause of damping.

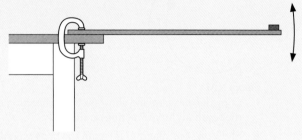

Figure 3.27

28 a Two wheels A and B are both made from the same material of uniform density and have the same mass (Figure 3.28).

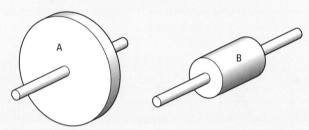

Figure 3.28

   i Define moment of inertia.

   ii State which of wheels A and B has the greater moment of inertia. Give a reason for your answer.

b Wheel A has a mass of 5.0 kg and a moment of inertia of 0.25 kg m². It is accelerated from rest by a constant torque of 2.4 N m applied for 4.0 s. Calculate

   i the angular velocity reached by the wheel,

   ii the rotational kinetic energy stored in the rotating wheel.

c The spinning wheel slows down under the action of friction and air resistance. Both produce torques proportional to the angular velocity of the wheel. Sketch a graph to show how the angular velocity varies with time as the wheel slows down to a stop. What is the shape of the graph?

29 The suspension of a car is tested by dropping the car from a low height on to a rigid concrete surface. The displacement–time graph for the resulting vertical oscillations of the car is shown in Figure 3.29.

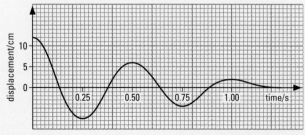

Figure 3.29

a **i** What is the frequency of the oscillation of the car?

   ii State how the results show that oscillations are damped.

b The effective oscillating mass of the car is 750 kg. The car has an identical spring at each of the four wheels. Determine the spring constant, in N m⁻¹, of each spring.

c As a warning for speeding drivers approaching a roundabout it is suggested that the road be made so that it rises and falls as shown in Figure 3.30.

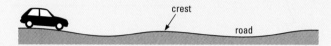

Figure 3.30

Resonant oscillations are produced when the speed of the car is 110 km h⁻¹.

   i State the condition for resonant oscillations to occur.

   ii Estimate the distance required between the crests to produce resonance.

   iii Sketch a graph showing how you would expect the amplitude of oscillation of the car to vary with speed of approach to the roundabout.

30 a **i** Use the continuity equation to show that an incompressible liquid moving from a wider pipe to a narrower pipe must increase its velocity.

   ii State Bernoulli's relation for the flow of an incompressible inviscid fluid along a horizontal stream line.

b Figure 3.31 shows a cross-section through a simple laboratory filter pump.

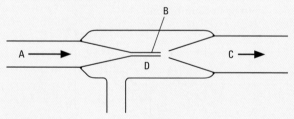

Figure 3.31

The pressure at C is atmospheric pressure (100 kPa) and the water at C is moving very slowly.
The pressure at D is 45 kPa. Nozzle B has a diameter of 2.0 mm.

   i Explain how a flow of water from A to C through the pump produces a partial vacuum at D.

   ii Calculate the velocity of the water emerging from the nozzle B.

   iii Calculate the rate, in m³ s⁻¹, at which water flows through the pump.

(density of water = 1000 kg m⁻³)

31 An orbiting satellite carries two sensors each of mass 2.5 kg at the ends of telescopic booms (Figure 3.32). When the booms are fully extended the centre of mass of each sensor is 2.0 m from the axis of the satellite, which spins at 0.10 rad s^{-1} about this axis, as shown in the diagram. The mass of each boom is negligible compared with that of the sensor. A motor can be used to retract the booms so that the sensors lie on the axis of the satellite and make negligible contribution to the moment of inertia of the satellite. Under these conditions the moment of inertia of the satellite about its axis is 100 kg m^2.

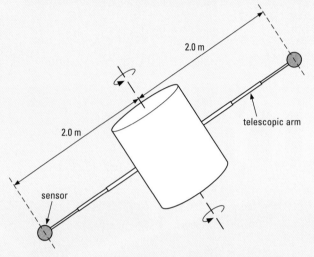

Figure 3.32

a When the booms are fully extended,

   i show that the moment of inertia of the satellite is 120 kg m^2;

   ii calculate the rotational kinetic energy of the satellite when spinning at 0.10 rad s^{-1}.

b With the satellite spinning initially at 0.10 rad s^{-1}, the booms are fully retracted. Calculate

   i the new rotational speed of the satellite, explaining your reasoning carefully,

   ii the rotational kinetic energy of the satellite.

c With reference to your answers to **a ii** and **b ii**, state whether or not the law of conservation of energy can be applied in this situation, and explain your reasoning.

32 a When a solid block is placed in water, it experiences an upthrust.

   i What is meant by an upthrust? How can it be calculated?

   ii By reference to the mean density p of the material of the block and the density p^w of water, state the condition of the block

      1 to sink in water,

      2 to float in water.

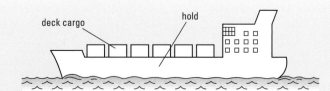

Figure 3.33

b A container ship is a cargo vessel which is designed to carry cargo in its holds and on deck, as illustrated in Figure 3.33.

   i Explain how the stability of a ship is maintained.

   ii Hence discuss the importance of the correct loading of the container ship with regard to light and to heavy containers.

c The hull of the ship in **b** may be assumed to have a uniform horizontal cross-sectional area in the region above and below the water line. When in sea-water of density 1.06×10^3 kg m^{-3}, the ship floats with 3.0 m of its hull below water. The ship then travels into a river estuary where the density of the water is 1.01×10^3 kg m^{-3}.

   i Explain why the level at which the ship floats will change.

   ii Calculate the change in the submerged depth of the hull.

* 33 a A rod has a mass per unit length m and a length L. It can rotate about an axis through one end.

Write down an expression for the moment of inertia of a length δx of the rod that is a distance x from one end of the rod. Hence show that the moment of inertia of the rod is $ml^2/3$, where m is the total mass of the rod and l is its length.

b The mass per unit length of a wind turbine blade varies along its length. To estimate its moment of inertia a student suggests a simplified model for a blade of total length L, shown in Figure 3.34.

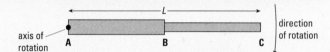

Figure 3.34

The model consists of two rods, **AB** and **BC** of equal length. **AB** has a mass $2M$ and **BC** has a mass M so that the total mass of the model blade is $3M$.

   i Deduce an expression for the moment of inertia of the student's model for the turbine blade. (Hint: You may wish to consider the model blade as a single rod of length L with another rod of half the mass and length attached to it to form **AB**.)

   ii Use the expression for the student's model to calculate the moment of inertia of the three blades of a wind turbine when each blade has a total mass of 21 000 kg and length 55 m.

   iii Calculate the rotation kinetic energy of the blades when they rotate once every 7.0 s.

*Pre-U-style question

4

ELECTRICITY

Electricity has been known about since ancient times, but it was only in the mid-nineteenth century that the phenomenon was studied systematically and a theory developed. Michael Faraday (1791–1867) was the first to produce an electric current from a magnetic field and showed that there was a relationship between electricity and chemical bonding. He also invented the first electric motor. James Clerk Maxwell (1831–1879) built upon the foundations laid by Faraday, and in integrating electricity and magnetism he formulated classical electromagnetic theory. Einstein described the influence of Maxwell's work as 'the most profound and the most fruitful that physics has experienced since the time of Newton'.

In this chapter you will be introduced to the basic ideas of electric charge and current, which quickly lead on to an examination of simple electric circuits. Once this foundation has been laid, you will be able to look more closely at the physics of resistors, cells, and capacitors in both d.c. and a.c. circuits, in preparation for further topics in electromagnetism developed in chapter 5.

Testing power transmission lines with artificial lightning. The amount of electricity in a typical lightning bolt is staggeringly large: the potential difference between a cloud and the ground can be several hundred million volts, peak currents of about 20 000 amperes can flow, and the temperature along the path of the strike can rise to 30 000 K.

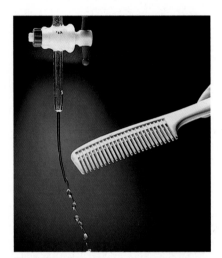

Electrostatic attraction between water and a charged comb. It is unusual for **macroscopic** objects (that is, objects on the scale of tables, chairs, etc.) to have an electric charge. They are neutral because they contain equal numbers of protons and electrons. When you rub an object, friction can give it a negative charge by adding more electrons, or a positive charge by removing some electrons. The term 'negative' was coined to refer to the charge on a piece of amber rubbed with fur. A glass rod rubbed with silk becomes positively charged. The electrical force is very strong; moving even a relatively small number of electrons will produce a big electrical effect.

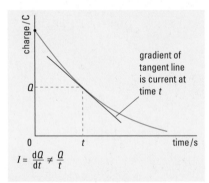

$$I = \frac{dQ}{dt} \neq \frac{Q}{t}$$

The gradient of the tangent line gives the current at that instant. This is not the same as dividing the values of Q and t on the graph.

Helpful hint

See the Mathematics toolbox for the connection between gradients of graphs, rates of change, and differential calculus.

ELECTRIC CHARGE AND CURRENT

Imagine a world without electricity. There would be no microwaves, no television, no hospitals or life-support systems, and no computer games. In the classic science fiction movie, *The Day the Earth Stood Still*, an alien spacecraft neutralizes all electricity on Earth. As a result, society collapses, and technology is reduced to that of the 16th century.

Electricity has become so vital in our everyday life that we no longer notice its existence. In the future this influence can only grow. An understanding of the basic properties of electric current is important for its safe and effective use.

Electric charge

Electric charge is an important property of nature.

• Some particles, for example the **proton**, have a positive charge.

• Some particles, like the **electron**, have a negative charge.

Not every particle has a charge.

• **Neutrons** have no overall charge, because they contain three **quarks** (see below) whose charges cancel each other out.

• **Photons**, which are particles of electromagnetic radiation, have no electric charge at all.

Although physicists are still not sure what electric charge *is*, they understand what it *does*. Objects that have an electric charge exert **electrostatic** forces on each other.

• Two **like** charges (charges of the same sign) repel (push each other apart).

• Two **unlike** charges (charges of different signs) attract (pull each other together).

The unit of charge

The unit used to measure electric charge is the **coulomb** (C). Physicists used to think that electric charge could only be found in whole number multiples of a fundamental amount of charge,

$e = 1.602\,192 \times 10^{-19}\,\text{C}$.

Charge on the proton $= +e$

Charge on the electron $= -e$

However, in the late 1960s, small particles called quarks were discovered with electric charges of either $+\frac{2}{3}e$ or $-\frac{1}{3}e$. These, along with $-e$, seem to be the basic quantities of electric charge.

The coulomb is not an SI base unit. This is because it is easier to define the unit of electric current, which is linked to electric charge, and then derive the coulomb from that.

Current and charge

An electric current is a transfer of charge from one place to another. In a metal, the movement of free electrons produces the charge transfer.

• The size of the current is the rate of charge transfer per second.

$$\text{Average current (A)} = \frac{\text{amount of charge transferred (C)}}{\text{time taken (s)}}$$

or

$$I = \frac{\Delta Q}{\Delta t} \qquad 1\,\text{A} = 1\,\text{C s}^{-1}$$

• The unit of current is the ampere (A), which is defined in terms of magnetic forces between currents.

• The coulomb (C) is defined as the charge passed by a current of 1 A in 1 s.

- The rate of charge transfer does not have to be constant. In many situations it is continually changing with time. If this is so, the size of the current at any time is the gradient of the graph of charge against time.

$$I = \frac{dQ}{dt}$$

Worked example

A TV tube produces a beam of electrons which strikes the fluorescent screen at the front of the tube. The beam current is 10 mA. How much charge strikes the screen in 10 min?

$$\Delta Q = \Delta It = (10 \times 10^{-3}\,A) \times (10 \times 60\,s)$$

$$\therefore \quad \Delta C = 6\,C$$

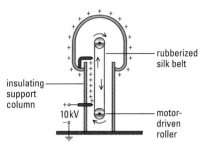

Van de Graaff generators are still used to provide the charges used in modern particle accelerators and in some nuclear physics experiments.

Examples of charge transfer

The very strong forces that exist between objects of opposite charge often cause a transfer of charge from one object to the other. This charge transfer discharges (neutralizes) the objects by bringing the opposite charges together.

- The dome of a Van de Graaff generator is charged by the friction of a rotating belt. The force due to the large positive charge on the belt nearby **ionizes** atoms in the air (this means it gives them a charge by adding or removing some electrons). If a second sphere is brought close then a charge transfer results, often producing a spark between the sphere and the dome.

- A thundercloud has a large negative charge on its lower surface. A lightning strike begins when electrons detach themselves from the cloud and move down towards the ground.

  The electric forces due to these charges ionize the air molecules as they go. This produces a 'tube' of ionized gas along the path that the charge has taken. As the jagged path or **stepped leader** gets close to the ground, it can draw weakly glowing threads of ionized gas upwards from the ground.

  If one of these threads makes contact with the stepped leader, a complete path of charged particles connects the cloud to the ground. Charge transfer takes place along this path – with a current of about 10 000 A.

- Io, one of the moons of Jupiter, is the most volcanically active body in the solar system. Its volcanoes are constantly spewing out oxygen and sulphur atoms. These then settle on its surface.

  Jupiter's immensely strong magnetic field strips these atoms from the surface of Io at the rate of 900 kg per second. The magnetic field ionizes the atoms, and then accelerates them to up to 1/10 of the speed of light. Some of the ions spiral towards Jupiter along the magnetic field lines. This causes a current of 3×10^6 A.

A lightning strike is formed when the stepped leader makes contact with ionized gas from the ground. Charge drains downwards, emptying the lower regions of the stepped leader first. This means that the flash of light moves upwards along the stepped leader. Once the path of ions is fully formed, several lightning strikes can flow along the same path.

A current flows continually between Io and Jupiter. This is because of Jupiter's enormous magnetic field, which revolves with the planet.

PRACTICE

1 The forces acting in the top diagram on the previous page are equal in size and opposite in direction. What can be deduced about the sizes of the charges on the two objects?

2 If a lightning strike has an average current of 10 000 A and carries a charge of 20 C to Earth, how long does the strike last for?

3 A copper wire carries a current of 1 A. How many electrons per second pass a given point in the circuit?

4 Copper has a density of 8930 kg m^{-3}. Calculate the mass of a copper wire that has a cross-section of 1 mm^2 and is 10 cm long.

Each copper atom has the mass of approximately 64 protons (the mass of one proton = 1.67×10^{-27} kg). How many copper atoms are there in the wire? If each copper atom has 64 electrons, how many electrons are there in the wire?

- charge carriers

- $I = nAqv$

- conventional current

- thermal motion of electrons

Conventional current

In a simple circuit the charge carriers are electrons. They move round the circuit from the negative side of a cell towards the positive side. In a more complicated case, such as the current in a fluorescent tube, there are two types of charge carrier in motion at once – positive and negative ions that travel in opposite directions.

By convention, diagrams always show current as moving *from positive to negative* (as if the charge carriers were positive). While this is not true in a wire, in general there is no single direction in which all the charge carriers are moving. The use of a single conventional current direction (positive to negative) avoids having to draw diagrams showing many different current directions at the same time.

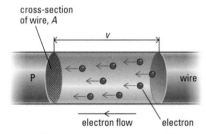

cross-section
of wire, A

P   wire

electron flow   electron

Any charge carrier that starts its journey more than v away from the shaded section will not arrive within one second.

Average drift speed

The **average drift speed** is the average speed attained by a charge carrier in a material under the influence of an electric field (for example, from a cell) and of collisions between the atoms of the material.

Electrical conduction and charge carrier density (number n per m³)

Type of material	$n\,\mathrm{m^{-3}}$
conductor, e.g. copper	$\sim 10^{28}$
insulator	$\sim 10^{7}$
semiconductor, e.g. silicon (in pure form)	$\sim 10^{16}$

THE MICROSCOPIC NATURE OF CURRENT

Benjamin Franklin discovered the relationship between static electricity and electric current in the 1740s. He thought that current was the flow of a positively charged 'electrical fluid' through a conductor. Transferring the fluid from one material to another would leave one of the materials with an excess of the fluid, making it positively charged. The other would have a shortage and would be negatively charged. Today we know that there are both positively and negatively charged particles, and it is their movement that causes electric current.

Charge carriers
The moving charged particles that make up an electric current are called **charge carriers**.

Different charge carriers.

Charge carriers	Conductor	Charge	Example
ions	liquid	positive or negative	car battery
ions	gas	positive or negative	fluorescent tube
free electrons	metal or gas	negative	wires

Not all the charged particles within a material are free to move. Even in a good conductor like copper, only about 3% of the electrons take part in conduction. The more free electrons there are in a metallic conductor (for example), the lower its resistance. The more highly ionized a liquid or gas, the more it will conduct.

Current and charge carrier flow
The size of a current in a material depends on the number of charge carriers taking part, and the speed at which they are moving:

$$\text{current} = \text{rate of flow of charge}$$
$$= (\text{number of charge carriers per second})$$
$$\times (\text{charge on the carriers})$$

Take electron motion in a wire as an example, and assume that all the charge carriers are moving in the same direction at a constant average speed v.

The number of charge carriers passing a point (P) in the wire *per second* is equal to the number contained in a volume of wire with length v.

$$\therefore \quad \text{number of charge carriers per second} = n \times A \times v$$

n is the number of charge carriers in every cubic metre of material (m^{-3}), A is the cross-sectional area of material (m^2), and v is the average drift speed at which they are moving (m s^{-1}).

$$\therefore \quad \text{current in wire } I = (nAv) \times (q)$$

q is the charge on each charge carrier (C). In the case of a metal this would be e (the charge on an electron).

$$\therefore \quad I = nAqv$$

Worked example

The number of free electrons per cubic metre in a copper wire is approximately 8×10^{28}, that is, about 1 or 2 per atom. A typical diameter of wire is 1 mm. If the current is 1 A, what is the average drift speed of the electrons along the wire?

$$I = nAqv$$
$$\therefore \quad v = \frac{I}{nAq} = \frac{I}{n\pi r^2 q}$$
$$= \frac{1\,\text{A}}{(8 \times 10^{28}\,\text{m}^{-3}) \times (\pi \times (0.5 \times 10^{-3})^2\,\text{m}^2) \times (1.6 \times 10^{-19}\,\text{C})}$$
$$= 9.9 \times 10^{-5}\,\text{m s}^{-1}$$

NB This is a very small number – it would take an individual electron many minutes to travel round even a small circuit.

Resistance

The charges on the plates of a cell will attract and repel the charge carriers in a wire. These forces make the charge carriers accelerate until they collide with atoms or ions within the material, causing them to transfer energy to the material, and increasing its temperature. The charge carriers then continue to accelerate as before. As a result of this cycle of acceleration and collision, the charge carriers settle into a motion with a constant average drift speed. They are continually gaining energy from the cell and losing it to the material. This is a microscopic view of resistance in a circuit.

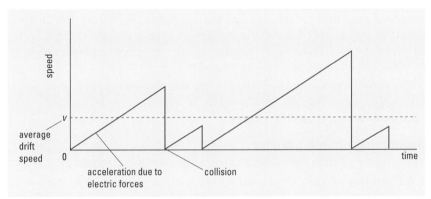

Although we talk about the charge carriers having an average drift speed, in reality their motion is a complex pattern of acceleration because of the force of the cell and sudden deceleration because of collisions with metal ions in the lattice structure of the wire.

Thermal motion

At room temperature free electrons in a circuit have far more thermal than electrical energy.

- This makes them move about very quickly (around $10^6 \, \text{m s}^{-1}$). The collisions between electrons and metal ions are random, and the thermal motion does not carry the electrons any distance along the wire.

- If there is a temperature difference between the ends of the wire, the free electrons at the hotter end will be moving faster than those at the colder end. Slowly moving electrons colliding with faster ones will tend to speed up, and there will be a transfer of energy along the wire. This process happens randomly, because there is no overall force acting on the electrons, and an individual electron will remain in the same region of the wire. *Energy* is transferred, but *charge* is not.

If the wire is connected to a cell, the free electrons gain an additional **drift motion** that carries them along the wire, and so charge is transferred. The direction of this drift is not random, because they are being directed by the lines of force from the cell.

> **Instantaneous current**
>
> The worked example shows how slowly electrons drift along a wire, but there is no time delay between pressing a light switch and the light coming on. This is because all the charge carriers start to move at virtually the same instant. Like queuing people jostling through a turnstile, the movement is continuous, although it may take an individual a long time to reach the turnstile.

> **Electron speeds**
>
> A typical drift speed in a copper wire carrying current is about $10^{-4} \, \text{m s}^{-1}$.
>
> Thermal speed (the motion due to thermal energy) in a copper wire at room temperature is about $10^6 \, \text{m s}^{-1}$.

PRACTICE

1. A copper printed circuit board track is 3 cm long, 4 mm wide, and 0.1 mm thick. The track carries a current of 100 mA. What is the average drift speed of the electrons in the track?
   (Take $n_{Cu} = 8 \times 10^{28} \, \text{m}^{-3}$.)

2. A copper wire carries current from the brake pedal switch to the brake light on a car. The wire has a cross-section of $1.0 \, \text{mm}^2$, and is 6.0 m long. If the current in the wire is 2.0 A when the light is on, estimate how long it takes an electron in the wire to travel from one end to the other.

3. Two copper wires of diameter 1.0 mm and 2.0 mm are joined end to end. As they are connected in series, the same current goes through each wire when there is a potential difference across the total length. What is the ratio of the drift speed of the electrons in the two wires?

4. Silicon is a semiconducting material and has fewer charge carriers per unit volume than a metal. A piece of copper and a piece of silicon have exactly the same dimensions and are carrying the same current (100 mA). Given that the average drift speed of the electrons in the copper is $2 \times 10^{-5} \, \text{m s}^{-1}$, what is the average drift speed of electrons in the silicon?
   (Take $n_{Cu} = 8 \times 10^{28} \, \text{m}^{-3}$ and $n_{Si} = 3 \times 10^{18} \, \text{m}^{-3}$.)

O B J E C T I V E S

- maintaining charge transfer
- comparison with fluid flow
- electromotive force (e.m.f.) and voltage drop
- series and parallel circuits
- the definition of resistance

Lightning is responsible for transferring negative charge to the ground. This must be balanced by another charge transfer back into the atmosphere.

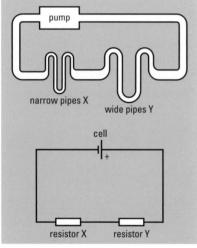

A simple water circuit can be compared to an electric circuit. The pump plays the role of the cell, the narrow pipe X that of the larger resistance X, and the wide pipe Y that of a smaller resistance Y.

> **Electromotive force (e.m.f.) and voltage drop**
>
> These, along with potential difference, will be more precisely defined in spread 4.4.

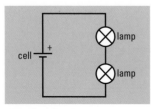

A series circuit.

ELECTRIC CIRCUITS

At any moment there are about 2000 thunderstorms taking place on Earth. In total these are responsible for transferring negative charge down to the ground at a rate of 1800 A. At this rate the ground gains -1.6×10^8 C per day. If this gain in charge were not balanced by an equivalent loss, the ground would soon become so charged that lightning strikes would not be possible due to repulsion.

During fair weather, charge is pumped back up into the atmosphere. This is a much more gentle affair. The current is spread over the whole surface of the Earth rather than localized in lightning flashes. The current density is only a few picoamperes per square metre, and the charge carriers are ions (produced by cosmic rays in the upper atmosphere, or by radioactivity near the Earth's surface).

The combination of current during fair weather and lightning means that charge is continually circulating in the atmosphere. This is an example of a simple electric circuit.

Maintaining charge transfer

There are two basic requirements for continual charge transfer.

1 **A complete circuit** Separated positive and negative charges, for example on the plates of a cell, exert electric forces on charge carriers, producing a current. The circuit provides a path for the charge carriers to move in response to these forces.

2 **A process that continually separates positive and negative charges** On its own, the current in a circuit would rapidly neutralize the separated charges. To maintain a current the charges must be separated again. This is a process that requires a transfer of energy. In a simple circuit, it is a chemical reaction in the cell that charges the plates. In a thundercloud, the process may be caused by friction between large falling water droplets and smaller ones that are rising.

Electromotive force (e.m.f.) and voltage drop

The flow of charge in an electric circuit is often compared to the flow of water in a water circuit such as a central heating system. Although it is useful to think about electricity in this manner, the comparison should not be taken too literally.

Water will not flow along a pipe unless there is a pressure difference. A metal wire containing electrons is like a pipe full of water. Without some 'electrical pressure difference' the electrons will not flow. In strict terms, the 'electrical pressure difference' is due to the complicated forces between the charges on the plates of the cell and the electrons in the circuit.

In any circuit there must be a device that provides this 'electrical pressure difference'. Often it is a cell; the amount of 'electrical pressure difference' that it can provide is the electromotive force (e.m.f.) of the cell. The **volt**, V, is the unit of e.m.f. A cell will have its e.m.f. written on its casing. For example, a PP3 cell has an e.m.f. of 1.5 V.

A voltmeter connected across a lamp will measure the **voltage drop** across the lamp. This is the 'electrical pressure difference' required to maintain the current through the lamp. Voltage drop is also measured in volts.

Series and parallel circuits

The are two types of electric circuit: **series** and **parallel**.

- In a series circuit the current is not diverted, and is the same in each component. However, the e.m.f. of the cell is shared between the components.

- A parallel circuit has at least two possible branches. The current can be different in each branch, but it adds up to the current that is drawn from the cell. The voltage drop across each branch is the same as the e.m.f. of the cell.

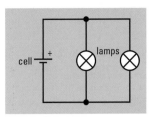

A parallel circuit.

Current and voltage in series and parallel circuits.

Circuit	Current	Voltage
series	same	splits
parallel	splits	same

Resistance

The e.m.f. provided by a cell drives the current through the components in the circuit. The amount of current depends on the total **resistance** of the circuit. The *lower* the resistance, the *greater* the current that can be maintained by a given e.m.f.

Whenever a current passes through a conductor a voltage drop can be measured across it. The better the conductor, the smaller the voltage drop will be for a given size of current.

Definition: resistance

The **resistance** of a conductor is the ratio between the voltage drop across it and the current through it:

$$\text{resistance } (\Omega) = \frac{\text{voltage drop across conductor (V)}}{\text{current through conductor (A)}}$$

$$R = \frac{V}{I}$$

The unit of resistance is the **ohm**. $1\,\Omega = 1\,\text{V A}^{-1}$.

For some materials (all metals, for example) the size of the voltage drop produced is *directly proportional to the current*. Such materials are called **ohmic conductors** and they have a constant resistance provided the temperature remains constant. Resistors designed to be ohmic conductors are used to set the current in a circuit.

Worked example

A sample of an ohmic conductor has a voltage drop of 9 V measured across it. The resistance of the sample is known to be 3 kΩ. What is the current in the sample?

Definition of resistance:

$$R = \frac{V}{I}$$

$$\therefore \quad I = \frac{V}{R} = \frac{9\,\text{V}}{3\,\text{k}\Omega} = \frac{9\,\text{V}}{3000\,\Omega}$$

$$= 3\,\text{mA}$$

Ammeters and voltmeters

Ammeters are used to measure current. They must always be connected into the circuit *in series*. The current has to pass through the meter if it is to be measured correctly. We say that we have measured the current *through* the component.

Electromotive force (e.m.f.) and voltage drop are measured by **voltmeters**. The voltmeter must be connected *in parallel* with the component. We say that we have measured the voltage drop *across* the component.

Superconductors

There are some materials called **superconductors** that have no resistance to current at very low temperatures. There is no voltage drop across a superconductor.

Helpful hint

Ohm's law is often stated as: the voltage drop across an ohmic conductor is proportional to the current through it provided the temperature is constant.

This really defines an ohmic conductor, rather than a law of nature.

PRACTICE

1 If you place two electrodes on your skin a couple of centimetres apart, the resistance measured between them will be about 100 kΩ. If you were unfortunate enough to receive an electric shock between those points with an e.m.f. of 230 V, what would be the current?

2 What is the **ionosphere**? Why is it such a good conductor of electricity?

3 The voltage drop between the top of the ionosphere and the ground is about 300 000 V. Given that the constant current between the ground and the ionosphere is 1800 A, what is the effective resistance of the atmosphere?

4 The filament of a lamp is made of metal (for example, tungsten), yet lamps are not classed as ohmic conductors. Why not?

ENERGY TRANSFORMATIONS IN CIRCUITS

Many people think that moving electrons carry energy from the cell to different parts of the circuit. This is not true. An individual electron takes many minutes to move through a circuit. Yet a lamp emits light as soon as you close a switch.

Think of a bike chain that links the pedals to the rear wheel. Energy is transferred to the rear wheel as the pedals are pushed. However, does the chain really *carry* the energy? It is the motion of the chain that allows the energy to be transferred. Similarly, it is the motion of the charge carriers through a circuit that transfers energy.

A bicycle chain allows energy to be transferred from the rider to the rear wheel.

Energy transfers and transformations

The charges on the plates of a cell exert forces on the electrons in a circuit as soon as the circuit is connected. All the electrons in the circuit will accelerate, because the electric forces have a very long range.

Whenever a force acts on a moving object, work is done. The cell therefore starts to transfer energy to all the electrons in the circuit at once (not just the ones that are near to it). As the electrons move through the circuit they lose energy by colliding with atoms or ions, but this is immediately replaced by the cell – the electrons start to accelerate again the instant after the collisions have taken place.

Inside the cell, chemical energy is transformed into the **electrical potential energy** (or **electrical energy** for short) of the electrons. This is transformed into the kinetic energy (KE) of the electrons by the forces acting on them, in the same way as gravitational potential energy is transformed into kinetic energy when an object falls.

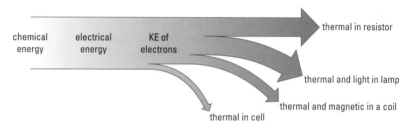

Chemical energy is transformed in the circuit to other useful forms of energy.

Potential difference, electromotive force (e.m.f.), and voltage drop

In order to keep track of the energy in a circuit, it is helpful to know the amount of energy converted in conducting a certain amount of charge.

Definition: potential difference (p.d.)

$$\text{p.d.} = \frac{\text{energy converted when charge } Q \text{ moves between two points in a circuit}}{\text{amount of charge, } Q}$$

$$\text{p.d. (V)} = \frac{\text{energy (J)}}{\text{charge (C)}}$$

$$V = \frac{W}{Q}$$

$$1\,\text{V} = 1\,\text{J}\,\text{C}^{-1}$$

Electromotive force or e.m.f.?

In the early days of research into electric circuits, the e.m.f. was thought to be a genuine force (which should be measured in newtons). The modern understanding is that the e.m.f. is an energy conversion, so we tend to drop the full name and just refer to it by the abbreviation, to avoid confusion. **Voltage** is the commonly used term for potential difference.

Some components convert chemical energy into electrical energy (for example, a cell), while others convert electrical energy into light and heat (for example, a lamp). It is helpful to distinguish between these two types of p.d. by introducing the terms **e.m.f.** and **voltage drop**. The first of these is in common use, but there is no generally agreed term to represent a p.d. that is a conversion of electrical energy into thermal energy.

$$e.m.f. = \frac{\text{amount of energy converted into electrical energy (J)}}{\text{charge passed (C)}}$$

$$\text{voltage drop} = \frac{\text{amount of electrical energy converted into other types (J)}}{\text{charge passed (C)}}$$

> **Electromotive force (e.m.f.) and voltage drop**
>
> Both e.m.f. and voltage drop are potential differences and so they are measured in volts using a voltmeter.

Worked example 1

In this circuit, the two lamps are identical. While 3 C of charge move round the circuit, the cell converts 6 J of chemical energy into electrical energy. What is the e.m.f. of the cell? What is the voltage drop across each lamp?

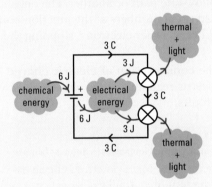

$$\text{e.m.f. of cell} = \frac{6\,J}{3\,C} = \frac{2\,J}{1\,C} = 2\,V$$

As the lamps are identical, they each convert the same amount of electrical energy into light and heat while the 3 C of charge is passing through them. They each convert half of the available energy.

$$\therefore \text{voltage drop across each lamp} = \frac{3\,J}{3\,C} = \frac{1\,J}{1\,C} = 1\,V$$

The important point here is that both lamps have the same amount of charge passing through them.

Worked example 2

In this circuit R_1 has twice the resistance of R_2. Again, the cell converts 6 J of energy while 3 C of charge is moving round the circuit, so the e.m.f. of the cell is 2 V. What is the voltage drop across R_1 and R_2?

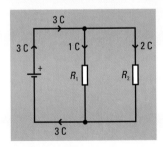

Now, the resistor with the smaller value will pass more of the charge, so that 2 C pass through R_2 and 1 C passes through R_1. The energy converted by R_2 is twice that converted by R_1.

$$\therefore \text{voltage drop across } R_1 = \frac{2\,J}{1\,C} = 2\,V$$

$$\text{voltage drop across } R_2 = \frac{4\,J}{2\,C} = \frac{2\,J}{1\,C} = 2\,V$$

- Kirchhoff's first law
- Kirchhoff's second law

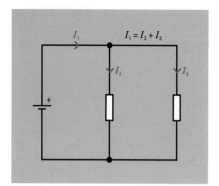

Kirchhoff's first law.

KIRCHHOFF'S LAWS

Solving problems to do with electric circuits can at first seem to be a confusing collection of 'tricks' applied in a jumbled manner. Actually, there is a very systematic way of going about solving any particular problem. There are two basic rules that apply in all electric circuits – **Kirchhoff's laws**. Applying these rules can solve any problem.

Kirchhoff's first law

The total current arriving at a junction in a circuit must equal the total current leaving the junction.

This simple rule makes sense in terms of charge carriers like electrons. Electrons cannot be created or destroyed inside a material. As they move round a circuit the total number must remain the same. Electrons arriving at a branch point must follow one path or another. The only other possibility is for them to build up in numbers at the junction point. This accumulation of charge would repel other electrons, stopping the current altogether.

Kirchhoff's first law expresses the **conservation of electric charge** – a fundamental law of nature that governs reactions at the subatomic level, as well as simple charge flow in circuits.

Kirchhoff's second law

The sum of the potential differences round any closed loop in a circuit must be zero.

This is a little more subtle than the first law. It is the application of Kirchhoff's second law that students often find difficult. The following examples illustrate the general ideas.

Worked example 1

In this simple circuit, the conventional current would be in the direction shown.

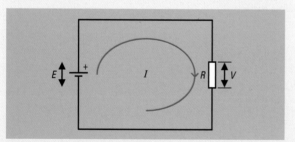

There is a negative potential difference (voltage drop) across the resistor, as electrical energy is being converted into thermal energy while the current is passing through it. There is a positive potential difference (e.m.f.) across the cell, as electrical energy is being produced from chemical energy.

Kirchhoff's second law implies that

$$E - V = 0$$

Using the resistance equation

$$V = I \times R$$

we get

$$E - IR = 0 \quad \text{or} \quad I = \frac{E}{R}$$

Worked example 2

In this circuit one of the cells is reversed.

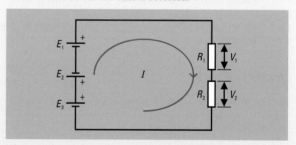

In passing through the reversed cell, positive charge will have to do work against the charge on the terminals of the cell; in other words, electrical energy will be lost. In this case the potential difference is still an e.m.f., but it is negative.

Applying Kirchhoff's second law gives:

$$E_1 - E_2 + E_3 - V_1 - V_2 = 0$$

$$\therefore \quad E_1 - E_2 + E_3 = I \times R_1 + I \times R_2$$

so

$$I = \frac{E_1 - E_2 + E_3}{R_1 + R_2}$$

Worked example 3

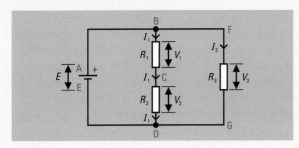

Kirchhoff's second law refers to any closed loop in a circuit. In this case there are three separate loops to look at. They are distinguished by the letter labels.

For loop ABCDE:

$$E - V_1 - V_2 = 0$$

$$\therefore E = I_1 R_1 + I_1 R_2$$

For loop ABFGDE:

$$E - V_3 = 0$$

$$\therefore E = I_2 R_3$$

The awkward one is the third possible loop – BFGDC – which misses out the cell altogether. From F to G we are passing through the resistor in the same direction as the conventional current, so there is a positive p.d. However, along the section DCB we are passing through R_1 and R_2 in the opposite direction to the conventional current. If we were to force a positive charge to do this, we would have to give the charge energy to overcome the repulsion of the cell. We must count this as a positive p.d. as we would be increasing the electrical energy by pushing a charge in this direction.

For this loop:

$$-V_3 + V_2 + V_1 = 0$$

$$\text{i.e.} \quad V_1 + V_2 = V_3$$

$$\text{so} \quad I_1 R_1 + I_1 R_2 = I_2 R_3$$

Worked example 4

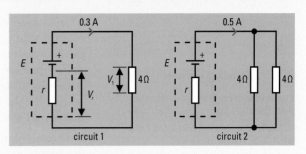

This example applies Kirchhoff's second law to solve a more realistic problem. See spread 4.11 for a definition of the e.m.f. and internal resistance of a cell.

A current of 0.3 A is drawn from a cell that is connected to a 4 Ω resistor. A second 4 Ω resistor is connected in parallel with the first. The current drawn from the cell is now 0.5 A. What are the e.m.f. and internal resistance of the cell?

Applying Kirchhoff's second law to circuit 1 (there is also a p.d. across the internal resistance):

$$E - V_r - V_1 = 0$$

$$\therefore E = V_r + V_1$$

$$E = (0.3\,\text{A}) \times r + (0.3\,\text{A}) \times (4\,\Omega)$$

$$E = 0.3r + 1.2$$

Applying Kirchhoff's second law to circuit 2 (treating the two 4 Ω resistors in parallel as a single 2 Ω resistor):

$$E = (0.5\,\text{A}) \times r + (0.5\,\text{A}) \times (2\,\Omega)$$

$$E = 0.5r + 1.0$$

Eliminating E between the two equations:

$$0.3r + 1.2\,\text{V} = 0.5r + 1.0\,\text{V}$$

$$\therefore \quad 1.2\,\text{V} - 1\,\text{V} = 0.5r - 0.3r$$

$$\text{so} \quad r = \frac{0.2\,\text{V}}{0.2\,\text{A}} = 1\,\Omega$$

The value of r can now be inserted into either equation:

$$E = (0.3\,\text{A}) \times (1\,\Omega) + 1.2\,\text{V} = 1.5\,\text{V}$$

or

$$E = (0.5\,\text{A}) \times (1\,\Omega) + 1.0\,\text{V} = 1.5\,\text{V}$$

This gives the same answer in either case.

PRACTICE

1 What are the currents labelled in the diagram? What is the p.d. across **a** AB, **b** BC, **c** CD, **d** AE, **e** EF, **f** FG, **g** AF, **h** BG, and **i** EC?

2 A power supply of e.m.f. E and internal resistance R is connected to a 10 kΩ resistor. The current drawn from the power supply is 0.91 mA. A voltmeter of internal resistance 10 kΩ is connected in parallel with the 10 kΩ resistor, and reads 8.3 V. Calculate the e.m.f. and the internal resistance of the power supply.

3 Two identical cells of e.m.f. E and internal resistance r are connected in series. A 7 Ω resistor is connected across the combination and draws a current of 0.333 A. The two cells are now connected in parallel; the 7 Ω resistor now draws a current of 0.375 A from the combination. Calculate the e.m.f. and the internal resistance of the cells.

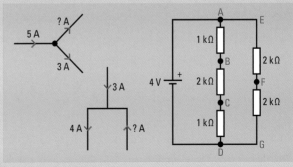

4 A standard AA cell has an e.m.f. of 1.5 V and an internal resistance of 1 Ω. A 3 V supply is required to light a lamp. This can be done either by connecting two AA cells in series, or by having four cells grouped so that two series pairs are in parallel with each other. Calculate the e.m.f. and internal resistance of each arrangement. Which is the better way of supplying the lamp?

Standard resistors

The standard resistor values that are manufactured are:

1, 1.2, 1.5, 1.8, 2.2, 2.7, 3.3, 3.9, 4.7, 5.6, 6.8, 8.2, 9.1 (20% tolerance)

in the appropriate multiples of ten, for example $0.18\,\Omega$ to $1.8\,M\Omega$ (20% tolerance).

The value of a resistor can often be read from a colour-coded sequence of stripes on the body of the component. The last stripe in the sequence represents the tolerance of the resistance, that is, the precision with which it has been manufactured.

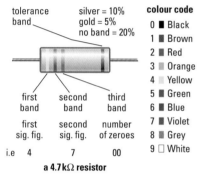

		colour code	
tolerance band	silver = 10% gold = 5% no band = 20%	0 ■	Black
		1 ■	Brown
		2 ■	Red
		3 ▨	Orange
		4 ▢	Yellow
first band	second band	third band	5 ■ Green
		5 ■	Green
first sig. fig.	second sig. fig.	number of zeroes	6 ■ Blue
		7 ■	Violet
i.e 4	7	00	8 ▨ Grey
		9 ▢	White

a 4.7 kΩ resistor

The resistor colour code chart.

Helpful hint

When you connect equal resistors in parallel, the resistance is the value divided by the number of resistors. For example, three $15\,\Omega$ resistors in parallel have a total resistance of $(15\,\Omega)/3 = 5\,\Omega$.

Unfortunately, it is impossible to buy off the shelf all the values of resistor that you might need in a circuit. Manufacturers concentrate their production on a set of standard resistor values. Normally it is possible to design a circuit to use only the standard values, but occasionally it is necessary to combine resistors to make up a special value. Alternatively, you might need to know the value of a combination of resistors to calculate the total current drawn from a cell by a circuit.

Of course, resistors are not the only objects that have resistance. There are many other situations in which resistances combine, for example the heating elements in a cooker or in the rear window de-mister of a car.

Resistances can be connected in **series** or **parallel** to combine their values.

Series

Any set of resistors in series can be replaced by a single resistor of value R_{tot} provided

$$R_{tot} = R_1 + R_2 + R_3 + \dots \text{ etc.}$$

Parallel

Any set of resistors in parallel can be replaced by a single resistor of value R_{tot}:

$$\frac{1}{R_{tot}} = \frac{1}{R_1} + \frac{1}{R_2} + \frac{1}{R_3} + \dots \text{ etc.}$$

If there are only two resistors, the equation reduces to a simpler form:

$$R_{tot} = \frac{\text{product of values}}{\text{sum of values}}$$

Worked example 1

What value of resistor could be used to replace a set of $3\,\Omega$, $4\,\Omega$, and $5\,\Omega$ resistors connected in parallel?

$$\frac{1}{R_{tot}} = \frac{1}{3\,\Omega} + \frac{1}{4\,\Omega} + \frac{1}{5\,\Omega} = \frac{(4 \times 5) + (3 \times 5) + (3 \times 4)}{3 \times 4 \times 5}\,\Omega^{-1} = \frac{47}{60}\,\Omega^{-1}$$

$$\therefore\ R_{tot} = \frac{60}{47}\,\Omega = 1.3\,\Omega \ \ (2 \text{ sig. figs})$$

NB The final answer is smaller than the value of any of the individual resistors.

People often forget to invert the answer at the last but one step. For example, they will quote 47/60 as being R_{tot}, rather than 60/47.

Worked example 2

Calculate the value of the single resistor that could replace the network on the right.

The place to start is the parallel combination of R_1 and R_2.

$$\frac{1}{R_{tot}} = \frac{1}{10\,\Omega} + \frac{1}{10\,\Omega} = \frac{1}{5\,\Omega}$$

$$\therefore\ R_{tot} = 5\,\Omega$$

This pair is in series with $R_3 = 5\,\Omega$, so the total resistance in this arm of the circuit $= 10\,\Omega$. R_5 and R_6 are in series, so they have a total resistance of $10\,\Omega$.

Now that we have the total resistance of the three arms of the circuit, we can combine all three in parallel to get the final answer:

$$\frac{1}{R_{tot}} = \frac{1}{10\,\Omega} + \frac{1}{10\,\Omega} + \frac{1}{10\,\Omega} = \frac{3}{10\,\Omega}$$

$$\therefore\ R_{tot} = 3.33\,\Omega \ \ (3 \text{ sig. figs})$$

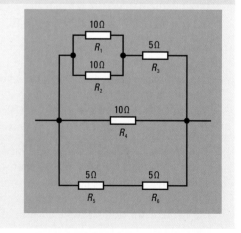

Maths box 1: derivation of resistors-in-series equation

Consider a circuit in which three resistors R_1, R_2, and R_3 are connected to a cell of e.m.f. E.

As the resistors are in series, the current through each one is the same $= I$.

$\therefore$ the voltage drop across resistor $R_1 = V_1 = I \times R_1$

$\therefore$ the voltage drop across resistor $R_2 = V_2 = I \times R_2$

and the voltage drop across resistor $R_3 = V_3 = I \times R_3$

By Kirchhoff's second law:

$E - V_1 - V_2 - V_3 = 0$

$\therefore \quad E = V_1 + V_2 + V_3 = I \times R_1 + I \times R_2 + I \times R_3$

$\quad\quad\quad = I \times (R_1 + R_2 + R_3)$

Now consider a circuit in which there is only a single resistor of value R_{tot}. The current drawn from the cell will be

$I = \dfrac{E}{R_{\text{tot}}}$

$\therefore \quad E = I \times R_{\text{tot}}$

and this current will be the same as that in the previous circuit, provided

$R_{\text{tot}} = R_1 + R_2 + R_3$

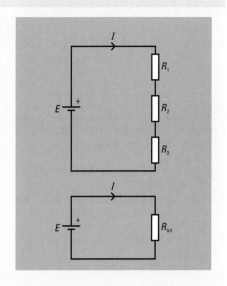

Maths box 2: derivation of resistors-in-parallel equation

Proving the parallel formula follows similar lines to the series case. The starting point is a circuit in which three resistors, R_1, R_2, and R_3, are connected in parallel to a cell of e.m.f. E.

The current in each resistor is:

$I_1 = \dfrac{E}{R_1} \qquad I_2 = \dfrac{E}{R_2} \qquad I_3 = \dfrac{E}{R_3}$

The total current drawn from the cell is the sum of these currents, by Kirchhoff's first law, that is:

$I = I_1 + I_2 + I_3 = \dfrac{E}{R_1} + \dfrac{E}{R_2} + \dfrac{E}{R_3}$

$\quad\quad = E\left(\dfrac{1}{R_1} + \dfrac{1}{R_2} + \dfrac{1}{R_3}\right)$

In a circuit with a single resistor of value R_{tot}:

$I = \dfrac{E}{R_{\text{tot}}} = E \times \dfrac{1}{R_{\text{tot}}}$

$\therefore$ the currents will be the same in each case, provided

$\dfrac{1}{R_{\text{tot}}} = \dfrac{1}{R_1} + \dfrac{1}{R_2} + \dfrac{1}{R_3}$

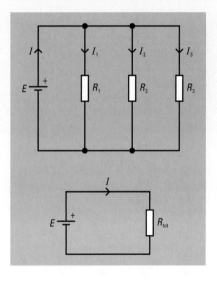

PRACTICE

1 A connecting lead that is used in a laboratory consists of 60 strands of thin wire. If each strand has a resistance of $3.0\,\Omega$, what is the total resistance of the wire?

2 What is the smallest number of resistors that you need to provide a resistance of:

a $2\,\Omega$, given $3\,\Omega$ resistors only;

b $7\,\Omega$, given $4\,\Omega$ resistors only.

In each case draw a diagram to show how you would connect them together.

3 An electric blanket contains three heating elements of $480\,\Omega$ resistance each. The blanket has low, medium, and high settings. How are the elements connected in each case? What is the total resistance of the blanket for each setting?

If the mains is $230\,\text{V}$, what current is drawn by the blanket on each setting?

4 For the following circuit, calculate:

a the voltage drop across each resistor;

b the current through each resistor;

c the total current drawn from the cell.

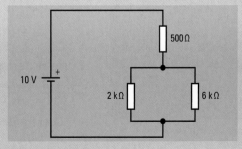

FACTORS THAT INFLUENCE RESISTANCE

Electrical energy is converted into thermal energy whenever current passes through a conductor. This may be necessary to reduce the current in a circuit, but it may also be a wasteful loss of energy needed elsewhere in the circuit. Engineers would like the option of using materials that do not have any resistance to current.

Such materials, called **superconductors**, do exist, but they only work at extremely low temperatures (a few kelvin). The race is on to develop the first *room-temperature* superconductor. This would revolutionize the design of electromagnets and power transmission cables. In the meantime, engineers need to be able to control the resistance of a conductor by choosing the right material and the right size of sample.

Resistivity

The resistance of a conductor depends on:

- the nature of the material;
- the size of the sample (length and cross-sectional area);
- the temperature of the sample.

In many cases the change in resistance with temperature is so small that it can be ignored except under extreme conditions.

Written mathematically:

$$R \propto l \text{ and } R \propto 1/A$$

where l is length and A is cross-sectional area. Putting these two relationships together gives:

$$R = \frac{\rho l}{A}$$

where the constant of proportionality, ρ, is the **resistivity** of the material (Ω m). The resistivity (which depends on temperature) is the only factor in this equation governed by the nature of the material. Tables of resistivity values are available for a variety of materials.

> ### Resistance and size
>
> A simple argument shows how the resistance of a sample of material depends on its size. Imagine connecting two pieces of material of the same length (l) and cross-sectional area (A) together:
>
> - Linking the pieces end to end is like connecting resistors in series – the resistance will increase, therefore resistance will be proportional to length.
>
> - Linking the two pieces side by side is like connecting resistors in parallel – the resistance will decrease, therefore resistance will be inversely proportional to area.

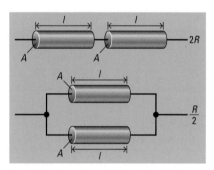

Samples in series and parallel illustrate how the resistance of a sample of material depends on its dimensions.

> ### Gold-plated connectors
>
> You can see from the table that silver has a lower resistivity than gold. However, when you buy a high-quality connection lead for a hi-fi, the plugs are always gold-plated. This is because silver-plated plugs would tarnish in the atmosphere, and the silver oxide formed has a higher resistivity than gold.

Resistivity of various materials.

Material	Resistivity, $\rho/10^{-8}\,\Omega\,\text{m}$	Temperature coefficient, $\alpha/10^{-3}\,\text{K}^{-1}$
copper	1.7	4.3
lead	21.0	4.3
gold	2.4	4.0
silver	1.6	4.0
nichrome	130.0	0.17
constantan	47.0	0.02

> **Worked example**
>
> Power cables capable of supplying large currents (30 A) have solid copper cores of cross-sectional area equal to 1.5 mm². What is the resistance of a 10 m length of such a cable?
>
> $$R = \frac{\rho l}{A}$$
>
> $$= \frac{(1.7 \times 10^{-8}\,\Omega\,\text{m}) \times (10\,\text{m})}{(1.5 \times 10^{-6}\,\text{m}^2)}$$
>
> $$= 0.11\,\Omega \text{ (2 sig. figs)}$$

Current density, resistivity, and conductivity

Sometimes it is convenient to work with the **current density**, for example if conductors of different area are connected in series.

$$\text{Current density } j = \frac{1}{A} = nqv$$

(The last term is obtained by dividing $I = nAqv$ by the area A.)

High-quality connecting leads have gold-plated plugs.

Combining this with the equation $V = IR$ gives:

$$V = I \times R = jA \times \frac{\rho l}{A} = j\rho l$$

The **conductivity** of a material, $\sigma = 1/\rho$, is another useful quantity. Using this, the conductance G of a sample of material can be defined:

$$G = \frac{\sigma A}{l} = \frac{1}{R}$$

The unit of conductance is the **siemens** (S), which is equivalent to Ω^{-1}.

As $R = \dfrac{V}{I}$ it follows that $G = \dfrac{I}{V}$

Variation of resistance with temperature

The ions in a metal lattice vibrate more quickly as the temperature increases. This makes it more likely that an electron will interact with an ion and lose energy. In practical terms, the resistance of the metal increases, and for pure metals it increases linearly with temperature:

$$R_\theta = R_0 (1 + \alpha\theta)$$

R_θ is the resistance at temperature $\theta\,°C$ (Ω), R_0 is the resistance at $0\,°C$ (Ω), α is the temperature coefficient of resistance $(°C^{-1})$, and θ is the temperature $(°C)$.

Some materials, most notably semiconductors, have a *negative* temperature coefficient of resistance. They experience the same increase in ionic vibration, but as they get warmer the number of charge carriers increases, so overall their resistance drops.

Superconductors

In 1911, H. K. Onnes discovered that as mercury is cooled to $4.15\,K$ its resistivity suddenly drops to a very low value. The mercury enters a **superconducting** state. The resistivity of a superconductor is at least 10^{12} times less than that of the same material at room temperature. An electric current induced in a superconductor has been observed to continue for several years without any applied potential difference. This makes them very useful in particle accelerator magnets, MRI scanners and as technology develops, motors and generators.

A variety of different materials turn into superconductors below their **critical temperatures**. Recently, high-temperature superconducting materials have been discovered, but their critical temperatures are still about $-150\,°C$. This temperature is above the boiling point of nitrogen $(-196\,°C)$. Consequently liquid nitrogen can be used as a coolant that keeps such materials in a superconducting state.

To understand how superconductors work in detail requires a knowledge of quantum mechanics, but in simple terms, at very low temperatures free electrons in the material join together to form **Cooper pairs**. These collide with ions or atoms in the material, but unless there is enough energy in the collision to split the pair, they will not lose energy. This reduces the resistivity of the material to zero.

> **Resistivity and n**
>
> The resistivity of a material is heavily influenced by the number of charge carriers $n\,m^{-3}$ (spread 4.2). The large range of n values explains the large range of resistivity in different materials.

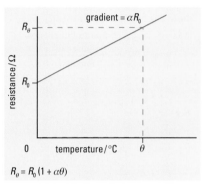

$$R_\theta = R_0 (1 + \alpha\theta)$$

The graph of resistance versus temperature for a normal conductor.

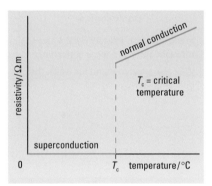

Below the critical temperature, the resistivity of a superconductor is essentially zero.

> **Helpful hint**
>
> The critical temperature is sometimes called the transition temperature.

PRACTICE

1 A rectangular copper track on a printed circuit board measures 3 cm long, 5 mm wide, and 0.1 mm thick. Calculate the resistance of the track.

2 Calculate the resistance per metre length of constantan wire of diameter 0.4 mm. What length of constantan would be required to make a resistance of $1.5\,\Omega$?

3 A small rectangular block of metal has length l, width w, and thickness t. Current can be passed through the block in any of three directions – parallel to each edge. Given that the resistivity of the material is ρ show that the resistance is the same for all three current directions.

4 In an experiment the current through a 12 V lamp is measured as a function of the p.d. across it, with the following results:

p.d./V	0.00	2.00	4.00	6.00	8.00	10.00	12.00
Current/A	0.00	1.36	2.07	2.64	3.13	3.58	4.05

Plot a graph of p.d. against current for the lamp. Using the graph, estimate the resistance of the lamp's filament at 1.0 A, 2.0 A, and 4.0 A. If the filament's wire is a material similar to nichrome, and the change in resistance between 1.0 A and 4.0 A is entirely due to the increase in temperature of the filament, estimate the temperature increase of the filament.

Definition: electrical power

Power is the rate of energy conversion:

$$\text{power (W)} = \frac{\text{amount of energy converted (J)}}{\text{time taken (s)}}$$

$$P = \frac{ItV}{t} = IV$$

but

$$V = IR$$

so

$$P = I^2R = \frac{V^2}{R}$$

Don't confuse

In the equation $W = QV = ItV$, W stands for work done or energy converted.

W is also used to stand for watts, the unit of power.

A good question in a physics quiz would be 'what are you paying for when you pay your electricity bill?' The automatic answer would be 'electricity'. However, this is not correct. The same amount of charge leaves your house as arrives at it. You are paying for the *energy transformations* that take place while there is a current in the appliance you are using. Electric current provides the means for bringing about an energy conversion in a particular place, for example, inside a kettle used for boiling water.

Useful equations

An energy conversion takes place when a charge passes through a potential difference:

Energy converted = charge × potential difference
$$W = Q \times V$$

The amount of charge conducted is the current multiplied by the time, so:

$$W = I \times t \times V$$

Sometimes the amount of energy converted is not the only important factor. The *rate* of energy conversion can be critical. A light bulb and a kettle can convert the same amount of energy, but boiling water by the heat from a light bulb would not be possible; the rate of energy supply will not be great enough to overcome thermal losses. The kettle is more appropriate because, among other things, it can convert energy at a faster rate. It is more **powerful** than the bulb.

Fuses

Fuses are safety devices that are used in the UK in combination with the earthing system on domestic mains equipment. They are constructed from thin pieces of wire that melt when their **current rating** is exceeded. An **anti-surge fuse** will tolerate short bursts of current greater than its rated value. These are useful in devices that produce a surge of power when they are switched on.

More modern circuit breakers and earth leakage current detectors are capable of cutting off the current supply more quickly than the comparatively slow-melting fuses.

Worked example 1

A 100 W light bulb in a table lamp is being used on the domestic 230 V power supply. What rating of fuse would be the safest to use in the plug?

$$P = IV$$
$$\therefore \quad I = \frac{P}{V} = \frac{100\,\text{W}}{230\,\text{V}} = 0.43\,\text{A} \quad (2 \text{ sig. figs})$$

Fuses that fit in plugs come in a variety of ratings: 1 A, 3 A, 5 A, and 13 A. The best fuse has a rating just greater than the normal current required by the appliance. In this example a 1 A or 3 A fuse would be best.

Worked example 2

What is the maximum power appliance that can safely be used with a 13 A fuse?

$$P = IV$$
$$= (13\,\text{A}) \times (230\,\text{V})$$
$$= 2990\,\text{W}$$

So a 2.5 kW appliance could be used. A more powerful device would blow the fuse during normal operation.

Energy units

The standard SI unit of energy, the joule, is too small to be convenient in the recording of domestic energy conversions. The electricity supply companies use a larger unit, the **kilowatt hour**, for billing.

Fuses come in a variety of ratings, shapes, and types.

Definition: the kilowatt hour

1 kilowatt hour (kWh) = amount of energy converted when a 1 kW
appliance runs for one hour

= (1000 W) × (1 × 60 × 60 s)

= 3.6 MJ

Worked example 3

While on holiday, a couple decide to leave three 100 W lights on automatic timer. The timers are set to turn the lights on from 6.00pm until 11.00pm every night. How much will it cost to leave the light on during a two week holiday?

The price of 1 kWh = 6.0p   6.00pm–11.00pm = 5 h

∴ over two weeks = 5 h × 14 = 70 h

total power of 3 lamps = 3 × 100 W = 300 W = 0.3 kW

∴ total number of kWh = (0.3 kW) × (70 h) = 21 kWh

∴ cost = (21 kWh) × (6.0 p) = 126.0 p = £1.26

Capacity of a cell

An electrical cell will convert chemical energy into electrical energy when charge flows through it. The cell has a limited supply of energy, and so will run down after a while. We could record the amount of energy stored, but it is more convenient to label the cell with the amount of charge it can conduct before it runs out of energy. The **capacity** of a cell is often quoted in **ampere hours** (A h).

A 1 A h cell will have enough energy to supply a current of 1 A for 1 h, or a 0.5 A current for 2 h, etc. Typical battery capacities are shown in the table on the right.

Typical battery capacities.

Cell	e.m.f./V	Capacity/A h
torch battery	1.5	8
car battery	12	60
watch battery	1.25	0.1

Worked example 4

A 1.5 V cell has a capacity of 1.2 A h. How much energy can it supply?

total charge that the cell can conduct before energy supply runs out

= (1 A) × (1.2 × 60 × 60 s) = 4320 C

energy = charge × p.d.

= (4320 C) × (1.5 V) = 6480 J

= 6.5 kJ (2 sig. figs)

The equivalent amount of energy supplied by the mains would cost about 0.01p. Cells are an expensive energy source, and we pay for the convenience of having portable appliances.

Don't confuse

The capacity of a cell must not be confused with the capacity of a capacitor. The cell's capacity tells us how much charge it can pass before it runs out of energy. The capacity of a capacitor is a measure of the charge it can store at a given voltage.

PRACTICE

1 If it takes 2.58 MJ of energy to boil one litre of water from a temperature of 25 °C, estimate how long it would take a 2.5 kW kettle to boil the water. How long would it take a 100 W bulb if all the energy could be transferred to the water?

2 Electric cookers are connected to a household fuse box on their own circuit. Typically, a 30 A fuse is used. What is the maximum power of the cooker? The same is true for electric showers which typically use 15 A fuses. What is the maximum power of shower that can be used with this fuse?

3 A 2 kW kettle runs off the 230 V domestic mains supply. What is the resistance of the heating element in the kettle?

4 My typical quarterly electricity bill shows that I have used 990 units. How much energy is this in MJ? If I am mostly using electricity between 8.30am and 10.30pm and there are 90 days in a billing period, what is my average power consumption when I am using electricity?

5 I use a tablet computer. It runs off 2 AA cells, each of 1.5 V e.m.f. and 2 A h capacity connected in series. The computer also has a mains adaptor that provides 40 mA at 9 V. Assuming that the computer uses the same power on battery as it does on mains, how long can I use it for before I have to change the batteries?

OBJECTIVES

- ohmic and non-ohmic conductors
- variation of resistance with current
- semiconductor diodes
- thermistors and light-dependent resistors

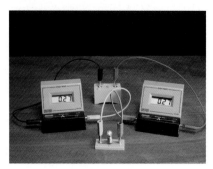

A circuit for obtaining the characteristics of an unknown component.

COMPONENT CHARACTERISTICS

Imagine that you have been set a puzzle. In front of you is a box with two terminals connected to it. You are told that the box contains either a lamp or a resistor, and that you can take whatever electrical measurements you wish in order to determine which is inside the box.

You might connect the terminals to some power supply and measure the current drawn by the box, and the voltage drop across it. From these measurements you can determine the resistance of the 'contents'. On its own, however, one such measurement is not enough. Both lamps and resistors have resistance.

In order to be sure of what was in the box, a range of different potential differences would have to be tried, and the current drawn from the box measured in each case. From these measurements a graph of current drawn against voltage drop could be plotted. Each type of component has a characteristic shape of graph when I is plotted against V. In fact, such graphs are commonly called **characteristics**. A lamp's graph is curved, while a resistor's graph is a straight line. A wide enough set of p.d. readings would make the difference quite clear.

Ohmic and non-ohmic conductors

An **ohmic conductor** has a constant resistance independent of the current that is passing through it. In electronics, resistors are designed to be as nearly ohmic as possible. They are very good at dissipating thermal energy generated by the current passing through them.

- A graph of current against resistance for an ohmic conductor is a straight line passing through the origin.
- The gradient of the line is 1/resistance of the conductor.

There are many other conductors that are **non-ohmic**. This may be because they do not dissipate heat as well as resistors, and so their temperature rises and alters their resistance (for example, a filament lamp). On the other hand, there are semiconductor devices in which the number of charge carriers changes with temperature.

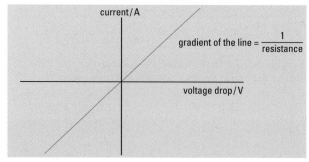

The characteristic line of an ohmic conductor, for example a resistor.

Variation of resistance with current

The filament of a lamp gets hotter as the current increases. This increases the lamp's resistance. Consequently, a graph of current against voltage drop for the lamp shows that it is not an ohmic conductor. If the lamp filament could be maintained at a constant temperature, then its behaviour would be ohmic.

At any current, the resistance of the lamp is given by V/I. Note that this is not the same as 1/gradient of the characteristic line at this current. *It is only for ohmic conductors* that the gradient is related to the resistance.

A **semiconductor diode** is also a non-ohmic conductor. Once a silicon diode is conducting, it will adjust its resistance to maintain a voltage drop of about 0.6 V across its terminals. If a diode is connected across a supply of e.m.f. greater than 0.6 V, then a very large current will pass through the diode, which will probably destroy it. For this reason, diodes must have

Don't confuse

The gradient of the characteristic line of a non-ohmic conductor is not related to the resistance. Resistance is defined as the *ratio* of V to I, *not the rate of change* of V with I. For an ohmic conductor with a straight line characteristic, the ratio of V to I is the same as 1/gradient.

protective resistors in series with them to limit the amount of current. This also applies to **light-emitting diodes** (**LEDs**), which are often used as indicator lamps in circuits.

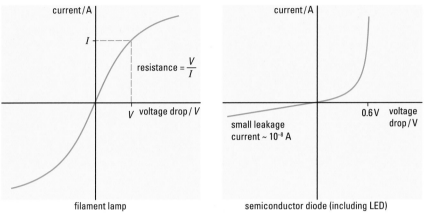

Lamps and diodes have curved characteristics, although for different reasons.

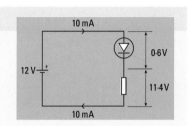

If the diode is to conduct, the anode must be connected to a potential that is more positive than the cathode.

	anode	cathode
conduct	+5 V	0 V
conduct	−2 V	−4 V
not conduct	0 V	+5 V

A diode will only conduct current in one direction.

Worked example 3

An LED has been mounted on the dashboard of a car and is to be used as an indicator for the car alarm. The car battery can supply 12 V and the LED requires 10 mA to run correctly. What value of protective resistance is required?

The protective resistance must drop all but 0.6 V of the e.m.f. supplied by the battery.

$$\therefore \ R = \frac{(12\,\text{V} - 0.6\,\text{V})}{(10 \times 10^{-3}\,\text{A})} = 1.14\,\text{k}\Omega$$

Variation of resistance with temperature and light

Thermistors and **light-dependent resistors** (**LDRs**) are semiconductor devices with resistances that change dramatically under different conditions. Unlike a device made from a metallic conductor, a thermistor's resistance *falls* as the temperature *rises*. As the temperature increases, the number of charge carriers in the material increases, reducing the resistivity.

Similarly, the energy from light falling on the surface of an LDR is used by the semiconductor material to make more charge carriers available, so its resistivity falls as the level of light rises.

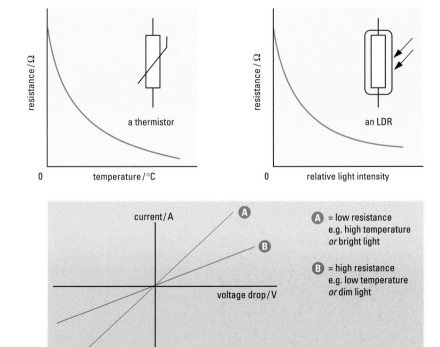

Thermistors and light-dependent resistors change their resistance under different conditions.

The characteristics of a thermistor or LDR are ohmic at a fixed temperature/ light level.

4.10

OBJECTIVES

• variable resistors

• controlling current and voltage

• potential dividers

• measuring the e.m.f. of a cell

Variable resistors are used to control the volume in hi-fi amplifiers.

VARIABLE RESISTORS AND POTENTIAL DIVIDERS

Variable resistors

There are many situations in which a device that provides a variable resistance can be very useful. These include:

• volume controls on hi-fi amplifiers;

• dimmers on light switches;

• adjusters for the electronic control of car accelerators ('drive by wire');

• thermostat or light-dependent switch settings.

In all of these cases a **variable resistor** is being adjusted to control a voltage drop or the current in a circuit.

Variable resistors come in a variety of different forms. However, they all work in the same way. They are made from a piece of resistive material with a connection that can be moved along its length. The amount of material that the current has to pass through can be changed by altering the position of this contact. Increasing the amount of material in the conduction path increases the resistance. Variable resistors are sometimes called **potentiometers**, or **pots** for short.

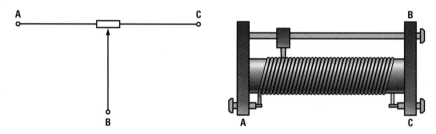

Variable resistors, like the one shown on the right, are represented by the symbol on the left.

A connection made between A and C provides a fixed resistance that is determined by the length of the resistance material. B is connected to the **slider** or **wiper** of the variable resistor, and connecting between A and B or B and C gives an adjustable resistance. Some circuits use all three connections, which splits the variable resistor into two resistances.

Controlling current and potential difference

Variable resistors can control either current or potential difference. For example, there are two circuits in the diagram on the left for adjusting the brightness of a lamp.

In the first circuit the variable resistor will adjust the total resistance in the circuit, and hence the current in the lamp.

The second circuit is more subtle. The e.m.f. of the cell produces a p.d. across each part of the variable resistor. The position of the sliding contact controls the ratio of V_1 to V_2 ($V_1 + V_2 = E$). As the lamp is connected in parallel with the lower section of the variable resistor, it will have the p.d. V_2 across it. Moving the sliding contact all the way down to A reduces the p.d. across the lamp to zero. If the contact is moved up to C, then the full e.m.f. of the cell is across the lamp. So, the p.d. across the lamp can be adjusted from zero to the full e.m.f. The advantage of this over the first circuit is that with p.d. set at zero there is no current in the lamp. With the first circuit there is always some current in the lamp, and unless the resistance is very high, it may not be possible to turn the lamp off completely.

circuit 1

circuit 2

Circuit 1 adjusts the brightness of the lamp by controlling the current in it (note the alternative symbol for the variable resistor). Circuit 2 works by adjusting the p.d. across the lamp. A similar circuit can be used to control the volume in an audio amplifier, with the cell being replaced by a source, such as a CD player, and the lamp by the amplifier.

Potential divider circuits

A **potential divider** circuit is formed whenever two resistances are connected in series across a source of e.m.f. The e.m.f. produces a voltage drop across each component in the ratio of their resistances, and the potential is divided between the resistances. In general, if a further circuit is connected in parallel with one of the components, it will have the same p.d. across it – it is being 'fed' part of the divided p.d.

$$\text{p.d. across external circuit} = \frac{R_2}{R_1 + R_2} \times E = V_{\text{out}}$$

NB This equation only works if the current drawn by the external circuit is small compared with I (in other words, the resistance of the external circuit is very much bigger than R_2).

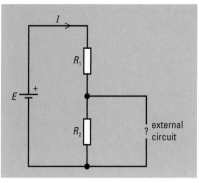

A potential divider circuit.

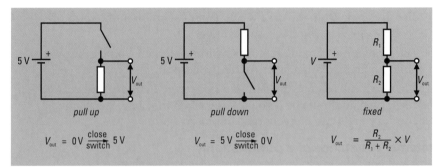

Examples of potential dividers.

Uses of potential dividers

Circuits such as (a) and (b) opposite are used to control systems based on temperature (a) or illumination (b).

For example, in circuit (a), V_{out} from the potential divider could be passed to a p.d.-sensing circuit. As the temperature sensed by the thermistor drops, so its resistance increases and V_{out} increases in turn. Once it passes a threshold value, the p.d.-sensing circuit switches on a heater. So, the complete circuit acts as a thermostatically controlled heating unit. The temperature can be set from the value of R_1.

A circuit like (b) can be used to switch on a lamp when the illumination drops to a set level. As it gets darker, the resistance of the LDR increases and with it V_{out}. Once again, when the p.d. rises above a set level, the sensing circuit triggers something to happen; in this case, a light to come on. In a typical application, R_1 is a variable resistor used to adjust the light level at which the lamp is triggered to come on.

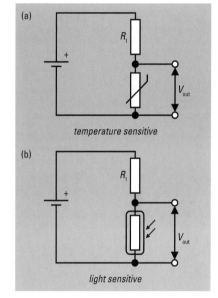

temperature sensitive

light sensitive

PRACTICE

1 A $2\,\text{k}\Omega$ variable resistor is connected across a 10 V supply. What is the p.d. between the sliding contact and the negative side of the supply when the slider is **a** halfway along, **b** two thirds of the way along, and **c** 5/6 of the way along?

2 Two resistors of $1.2\,\text{k}\Omega$ and $800\,\Omega$ are connected in series with a 24 V battery. What is the p.d. across the $800\,\Omega$ resistor if **a** there is a resistor of $600\,\Omega$ in parallel with it, and **b** there is a voltmeter of $1\,\text{M}\Omega$ connected across it?

3 In circuit 2 on the opposite page, the cell is replaced by a battery of e.m.f. 10 V and negligible internal resistance. The variable resistor is $4\,\text{k}\Omega$.

What is the p.d. across the lamp when the variable resistor is set to its midway point? (Assume that the lamp has a fixed resistance of $600\,\Omega$.)

4 A standard Weston cell has an e.m.f. of 1.0186 V at 293 K. Such a cell is used to calibrate a potential divider made from 1 m of nichrome wire and a 2 V lead–acid cell. The balance length for the standard cell is 0.51 m. What is the e.m.f. of a test cell which has a balance length of 0.75 m? Why is the calibration of the potentiometer wire not exactly $2\,\text{V}\,\text{m}^{-1}$?

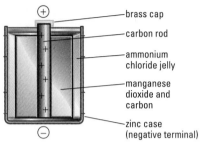

A dry cell. The zinc that will be dissolved also acts as a container for the cell. The hydrogen is removed chemically.

Different types of cell.

Type of cell	e.m.f. Internal resistance	Behaviour
dry cells (torches, etc.)	1.5 V ~ 1 Ω	e.m.f. drops steadily with use. Case part of reaction ∴ can leak
alkaline (radios, toys, etc.)	1.5 V 0.5 Ω	can supply large currents, longer life
lithium-ion cell	3.2 V, ~6 mΩ	rechargeable cell used in mobile phones, digital cameras, etc.
lead–acid (car battery)	2.0 V ~ 0.01 Ω	constant at 2.0 V until needs recharging
NiCad (hand tools, etc.)	1.25 V very low internal resistance	e.m.f. starts at 1.3 V, then drops to 1.25 V and remains steady
solar cell (watches, calculators, etc.)	0.5 V in bright light	relies on the properties of semi-conductors

Key:

	primary cells		secondary cells

	solar cells

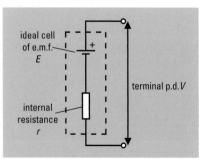

A real cell can be modelled as an ideal cell with a resistor connected in series.

CELLS

Cells

There are a wide variety of cells on the market to fulfil various uses. Selecting the right cells to use is a matter of matching cost and properties to the application. You would not want to carry a car battery about to power an ipod, and dry cells can not provide the 100 A required to turn over a car's starter motor.

The basic design of the dry cell is shown in the diagram on the left.

- The positive terminal is the central carbon rod, while the case forms the negative terminal.

- A chemical reaction removes electrons from the carbon, and also dissolves zinc from the case which enters solution in the form of positive ions, leaving the case negative.

- The manganese dioxide removes hydrogen which would otherwise build up an insulating layer round the carbon.

All chemical cells can be divided into two types. **Primary cells** cannot be recharged and must be thrown away when they have discharged. **Secondary cells** can be recharged and used again.

Internal resistance of a power supply

- A cell converts chemical energy into electrical energy.

- A dynamo converts mechanical energy into electrical energy.

- A solar cell converts light energy into electrical energy.

In each case above the conversion is not 100% efficient. The amount of electrical energy produced by a cell is always less than the amount of chemical energy that is converted. This is due to the complex chemical processes that occur within the cell, but a simple model can be used to explain this inefficiency. The behaviour of a real cell is like that of an **ideal cell** (one that does not produce thermal energy) in series with a small resistor. Current drawn from the cell has to pass through the resistor which warms up. This represents the inefficiency of the cell.

Terminal p.d. and e.m.f.

If the current drawn from a *real cell* is I, then the voltage drop across its internal resistance will be Ir. This represents the amount of electrical potential energy converted into thermal energy by the resistor per coulomb of charge that passes through it.

The e.m.f., E, of the *ideal cell* represents the amount of electrical energy that is produced by the cell per coulomb of charge. The net amount of energy available to this coulomb of charge when it passes out of the real cell into the external circuit is $E - Ir$. This is the **terminal p.d.** of the cell, V.

$$V = E - Ir \qquad \text{or} \qquad IR = E - Ir$$

where R is the resistance of the load connected to the cell and $V = IR$. The terminal p.d. of a cell is *reduced* as the current drawn from it is *increased*.

A typical cell used in a torch has an e.m.f. of 1.5 V and an internal resistance of about 1 Ω, while a typical car battery (6 cells) has an e.m.f. of 12 V and an internal resistance of 0.05 Ω. All power supplies have internal resistance, but to deliver large current this should be as low as possible. Laboratory power supplies can produce in excess of 1000 V, and sometimes use resistors (of a few MΩ) connected in series to reduce the current.

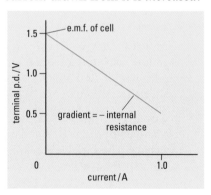

The graph of terminal p.d. against current drawn from a cell.

Combinations of cells

Sometimes the p.d. required in a circuit is greater than that of a single cell (typically 1.5 V). In such cases it is convenient to combine cells together in series to form a **battery**.

The combination of cells provides a total p.d. equal to the sum of the e.m.f.s of the individual cells. However, the internal resistances of the cells also act in combination. So the battery appears to the external circuit as if it were a single cell of e.m.f. = $E_1 + E_2 + E_3 + \ldots$ and internal resistance $r = r_2 + r_2 + r_3 + \ldots$

In a different situation, the p.d. required might not be as important as keeping the internal resistance of the cell down to the smallest possible value. Consequently it can be convenient to combine identical cells in parallel.

The total current drawn from the combination is partly supplied by each cell. With three identical cells in parallel the total current I would be divided, so that $I/3$ comes from each cell. This makes the p.d. dropped across the internal resistance $Ir/3$ in each case. As a result, the combination of cells presents to the external circuit as a single cell of e.m.f. and internal resistance $r/3$. With n identical cells in parallel, the internal resistance appears to be r/n.

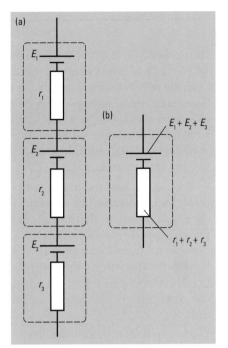

Cells in series. The series combination of cells in (a) acts like a single cell with an e.m.f. E equal to the sum of the separate e.m.f.s. The same is true for the internal resistance r, as shown in (b).

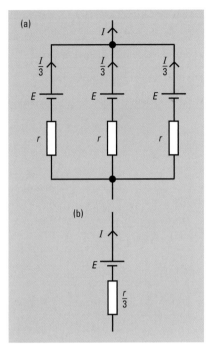

Cells in parallel. The combination (a) appears to the external circuit as a single cell with a lower internal resistance, (b).

PRACTICE

1 A cell of e.m.f. 1.5 V and internal resistance 0.5 Ω is connected in series to each of the following in turn: **a** a 1 Ω resistor; **b** a 20 Ω resistor; **c** a 50 Ω resistor. Calculate the current in the circuit and the p.d. across the external resistor in each case.

2 A cell has an internal resistance of 1 Ω and an e.m.f. of 1.5 V. Sketch a graph showing how the terminal p.d. varies with current for this cell. On the same diagram, sketch graphs showing the variation of terminal p.d. with current for a cell of: **a** twice the internal resistance but the same e.m.f.; **b** twice the e.m.f. but the same internal resistance.

3 A cell of e.m.f. E and internal resistance r draws a current of 1 A when connected to a 2 Ω resistor. The same cell provides a current of 0.5 A when connected to a 7 Ω resistor. Calculate the e.m.f. and internal resistance of the cell.

4 **a** A selection of identical cells of e.m.f. 1.5 V and internal resistance 0.2 Ω are available. Draw a diagram showing how the cells should be combined to provide an internal resistance of 0.01 Ω. What would be the combined e.m.f. of the cells in this diagram?

   b The same selection of cells now have to be combined to provide an e.m.f. of 4.5 V and an internal resistance of 0.5 Ω. Draw another diagram showing this combination, and demonstrate how the combination provides the required e.m.f. and internal resistance.

Real-life examples of batteries

A car battery needs to supply up to 100 A at 12 V to the starter motor, so the internal resistance of the battery needs to be very low or the p.d. will drop too much.

High voltage power supplies, providing em.f.s in the 1000s of volts, need to have large internal resistances as a form of protection. If the terminals of the supply are accidentally shorted together, the internal resistance ensures that the current is not fatally high.

4.12

O B J E C T I V E S

- a.c. movement of charge carriers
- r.m.s. current and power
- r.m.s. voltage

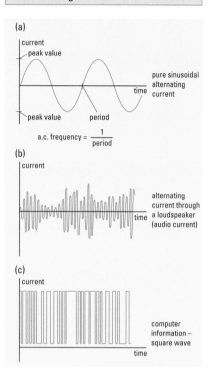

Various forms of a.c. current.

Electron movements

NB The answer in the example is a very small distance! The electrons that are moving inside the generator are very far away from the electrons moving inside one of your light bulbs. The energy is not carried by electrons! The energy is in the alternating electric field set up inside the wires that drives the motion of the electrons wherever they are.

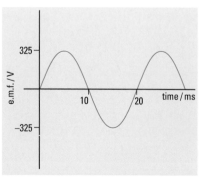

The variation of UK mains e.m.f. with time.

ALTERNATING CURRENT

How could you tell that the light bulb in your room is flashing on and off 100 times a second? One way would be to have an electric motor spinning at a rate of 100 times a second. In the flashing light of the mains lamp, the motor would appear not to be moving. (This is the same effect that allows stroboscopes to freeze motion.) In certain situations this could be rather dangerous, which is why in areas where moving equipment is being used, for example in a factory, people prefer to use fluorescent lamps that maintain a steady light even when run off the mains supply.

The reason that the mains lamp flashes on and off is that it is being powered by **alternating current** (**a.c.**). The diagram on the left shows a variety of different alternating currents. When the graph of the current goes below the time axis, this means that the current has reversed direction. The mains supply is very close to being a pure **sinusoidal** wave form (i.e. having the form of a sine wave). A lamp will glow at its brightest when the current is a maximum (in either direction) and will go out when the current is zero (in practice it does not completely cool down in time).

Although all the examples in the diagram can be called a.c., the rest of this section will discuss only the simple sinusoidal a.c.

Alternating current and charge carriers

Putting an alternating e.m.f. across a conductor will drive the charge carriers back and forth in the direction of the e.m.f. The current equation $I = nAqv$ will still apply, but v will be continually changing. It will climb to a maximum value as E reaches a maximum, then as E decreases the resistance in the wires will slow the charge carriers down to rest. The e.m.f. will then drive them back in the opposite direction. Under the influence of an alternating e.m.f., charge carriers are doomed to move back and forth along the same piece of conductor.

Worked example 1

A copper wire of $1\,mm^2$ diameter carries a.c. of 10 A maximum. How far does an electron move in the wire during half a cycle? (Assume there are 1×10^{29} charge carriers in $1\,m^3$.)

To calculate this correctly we would have to plot the velocity–time graph for the electron and find the area beneath the curve. We can make an approximation by assuming that the electron moves at an average velocity equal to half the maximum velocity.

$$I = nAev$$

$$\therefore \quad v = \frac{I}{nAe} = \frac{10\,A}{(10^{29}\,m^{-3}) \times (1 \times 10^{-6}\,m^2) \times (1.6 \times 10^{-19}\,C)}$$

$$= 6.25 \times 10^{-4}\,m$$

In half a cycle, which is 1/100 s, the distance moved will be:

$$x = (3.13 \times 10^{-4}\,m\,s^{-1}) \times (0.01\,s)$$

$$= 3.13 \times 10^{-6}\,m = 0.003\,13\,mm$$

Root mean square current and power

Alternating current is difficult to use in calculations as the size of the current is continually changing. If the e.m.f. producing the current varies as shown on the left, then the variation in the current will be:

$$I = I_0 \sin(2\pi ft) = I_0 \sin(2\omega t)$$

where $\omega = 2\pi f$ and $f = 50\,Hz$ for UK mains $\qquad$ (1)

In many situations, the instantaneous value of the current is not as important as the average effect that the current has over a period of time. One example is the heat dissipated by a resistor. The instantaneous rate at which a resistance produces thermal energy is I^2R, where I is the size of the current at that instant. The direction of the current in the resistor is unimportant. Only the variation in the size of the current affects the power.

The average power produced in the resistor over one cycle is:

$$\langle P \rangle = \frac{1}{2} I_0^2 R \qquad \text{(where } I_0 \text{ is the peak current)}$$

$$= \left(\frac{I_0}{\sqrt{2}} \right) \times \left(\frac{I_0}{\sqrt{2}} \right) \times R = \left(\frac{I_0}{\sqrt{2}} \right)^2 R$$

This average power is the same as the constant power produced by a d.c. current of $I_0/\sqrt{2}$. In other words, a lamp powered by an a.c. current that varies between I_0 and $-I_0$ would glow with the same brightness as an identical lamp carrying a d.c. current of $I_0/\sqrt{2}$. This d.c. current is known as the **r.m.s. equivalent current** to the a.c.

The r.m.s. equivalent current is defined as *that d.c. that will provide the same power in the resistor as the a.c. does on average*.

$$I_{\text{r.m.s.}}^2 R = \frac{1}{2} I_0^2 R$$

$$\therefore \quad I_{\text{r.m.s.}} = \sqrt{\frac{1}{2} I_0^2} = \frac{I_0}{\sqrt{2}}$$

Using r.m.s. values of current, voltage drop, and e.m.f., a.c. can be treated as d.c. provided the circuit only contains components with resistance and not capacitance or inductance. The ordinary d.c. rules apply:

$$I_{\text{r.m.s.}} = \frac{V_{\text{r.m.s.}}}{R}$$

$$\text{average power} = I_{\text{r.m.s.}} V_{\text{r.m.s.}} \qquad I_{\text{r.m.s.}}^2 R = \frac{V_{\text{r.m.s.}}^2}{R} = \frac{1}{2} I_0 V_0$$

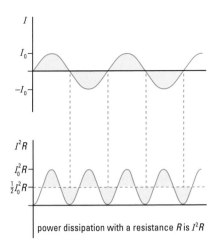

power dissipation with a resistance R is I^2R

The average value of a sine curve is zero as there are equal areas above and below the x axis. The average value of a sine^2 curve is 1/2 because there are equal areas above and below the 1/2 line.

Worked example 2

Domestic a.c. has a peak value of 325 V. What is the r.m.s. voltage? What is the r.m.s. current in a 100 W light bulb?

$$V_{\text{r.m.s.}} = \frac{V_0}{\sqrt{2}} = \frac{325 \, \text{V}}{\sqrt{2}} = 230 \, \text{V}$$

So the mains supply is referred to as being 230 V.

$$\langle P \rangle = I_{\text{r.m.s.}} V_{\text{r.m.s.}}$$

$$\therefore \quad I_{\text{r.m.s.}} = \frac{\langle P \rangle}{V_{\text{r.m.s.}}} = \frac{100 \, \text{W}}{230 \, \text{V}} = 0.43 \, \text{A}$$

Root mean square

The letters r.m.s. stand for **root mean square**. The root mean square of a set of current values is obtained by squaring each current value, taking the average of the results, and finally taking the square root of the average ($\sqrt{\text{average (current}^2)}$). (See Practice question 1.)

PRACTICE

1 Calculate the r.m.s. value of the following set of numbers:

(12, 2, –10, 6, 4, –7, 8, 2, –3, 1)

2 The wires that make up the cables that carry power from pylon to pylon across the countryside generally carry an r.m.s. current of 42 A. If each wire has a cross-sectional area of $5 \times 10^{-5} \, \text{m}^2$, estimate the distance travelled along the wire by an electron during half a cycle.
(a.c. supply has a frequency of 50 Hz, and n for aluminium is $4.3 \times 10^{28} \, \text{m}^{-3}$.)

3 Two light bulbs are glowing at the same brightness. One is supplied with a.c. and the other with d.c. Both bulbs have a resistance of $4 \, \Omega$ (assumed constant) and the d.c. bulb is drawing 3 A at 12 V. What is the peak value of the current in the a.c. lamp?

4 Household mains is 13 A r.m.s. at 50 Hz.

a Calculate the peak value of the current.

Using equation (1) (opposite page) calculate:

b the value of the current when $t = 5 \, \text{ms}$;

c the value of the current when $t = 17 \, \text{ms}$.

If the frequency were to be changed to 75 Hz, what would be the earliest moment after $t = 0$ at which the current would be (momentarily) zero?

5 Use a calculator to produce a table of values of I for an a.c., given that I_0 is 10 A and $f = 50 \, \text{Hz}$. Plot a graph of I against time for one complete cycle. Draw a new set of axes beneath the graph of I so that the horizontal time scales line up. On this second set of axes draw a graph of I_2 against time. By counting squares, or otherwise, estimate the total area under the I_2 curve over one cycle. Estimate the average value of I_2 over one cycle. Does this agree with the r.m.s. value of the current?

OBJECTIVES

- diodes and light-emitting diodes (LEDs)
- laser diodes
- photodiodes
- thermistors and light-dependent resistors (LDRs)

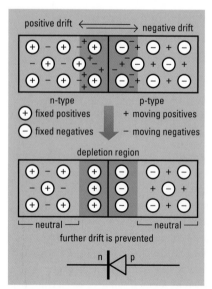

Modern electronic devices can be made remarkably small (and tasty).

Semiconductor materials

The way in which electric current passes through a semiconductor is discussed fully in chapter 10. Semiconductors contain two types of charge carrier: electrons (which are negative) and holes (which are positive). Doping is the process by which impurities can be added to a semiconductor crystal to control the number of electrons and holes which are free to move through the crystal. In a p-type crystal an impurity is added that donates free electrons, leaving a fixed positive ion. In an n-type semiconductor crystal the donor generates holes and fixed negative ions.

A junction diode at manufacture (top) and shortly afterwards (bottom).

Semiconducting materials are useful because of the wide variety of electronic devices that can be made from them. The other key point is that the devices can be miniaturized to a remarkable extent, enabling highly complicated systems to be packaged into sizes of the order of a few millimetres. The range of devices mentioned in this section represents a small fraction of the specialist equipment that is available.

Thermistors and light-dependent resistors

The **thermistor** is a temperature-dependent resistor. Despite the name, thermistors are actually made from an intrinsic semiconductor such as silicon. An ordinary metallic conductor's resistivity increases with temperature, because of the increased atomic vibrations. The same thing happens in a semiconductor, but at higher temperatures more charge carriers are made available. The second effect is more significant than the first, and the resistivity drops as temperature increases.

A light-dependent resistor (LDR) works in a similar manner. Energy from photons falling on the material is used to free charge carriers, and so as the light level increases, the resistivity falls.

Diodes

Diodes are formed from a block of semiconductor material, half of which has been doped to make it p-type, and the other half n-type. With the diode *not* connected to a circuit the free motion of electrons and holes creates a **depletion region** either side of the junction. Here the free electrons and holes have combined to cancel each other out. This makes the depletion region almost free of charge carriers, but fixed charges are still present on the impurity atoms that have been added (positive in n-type, negative in p-type). As there are no free charge carriers to cancel these fixed charges, the depletion region has a net charge, and repulsion from this charge stops further wandering of the charge carriers across the junction.

Because of the charge the depletion region has a p.d. across it (0.6 V for silicon). If the diode is connected to an external e.m.f. greater than this, and in the opposite direction, a current is produced and the diode is said to be **forward biased**. If the e.m.f. is in the same direction as the internal p.d., the depletion region increases in size as there is no current – the diode is **reverse biased**.

Note that the p.d. across the diode remains about 0.6 V regardless of the current. This is why protective resistors are often used to limit the current in the diode.

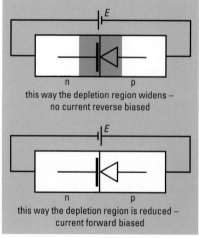

A reverse-biased and forward-biased diode.

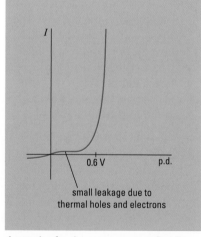

A graph of p.d. versus current flowing through the diode shows that the p.d. remains around 0.6 V even though the current flowing through the diode may increase significantly.

Light-emitting diodes (LEDs)

Light-emitting diodes (**LEDs**) behave just like ordinary junction diodes, but in addition they are extremely efficient converters of electrical energy into light. When electrons and holes combine in the junction of an ordinary diode, the energy released increases the lattice vibrations of the atoms in the semiconductor, and the diode warms up. In an LED, the materials are chosen so that a great deal of the energy is released as light (photons). The junction is deliberately designed to be close to the surface of the material so that the emitted light can escape from the diode. The material used determines the colour of the resulting light.

Colours associated with various semiconductor materials.

Semiconductor	Colour of light	Wavelength
gallium aluminium arsenide	infrared	880 nm
gallium aluminium arsenide	red	645 nm
aluminium indium gallium phosphate	yellow	595 nm
gallium phosphate	green	565 nm
gallium nitride	blue	430 nm

Advantages of LEDs

Uses of LEDs include: household lighting, automotive lighting, garden and other solar-powered lighting, indicators, remote controls (IR), and alarms.

Advantages of using LEDs for these applications include:

- long life (50 000 hours +);
- low power (efficiency);
- no heat production (safety);
- fast switching;
- compact;
- easily made waterproof.

Laser diodes

These extremely small and efficient lasers have revolutionized the communications industry. They are used in fibre-optic communications, computer links, and compact disc players as well as in more mundane applications such as the laser pointers used by lecturers. In essence they work in the same way as LEDs. With an LED, light is produced in all directions. In a laser diode two faces of the semiconductor wafer are turned into mirrors so that they reflect light back along the junction plane. The photons passing along the junctions stimulate electrons to emit light of the same wavelength and in the same direction. This increases the amount of light that is produced and also directs it along the plane of the junction. One of the mirrors is designed to let some light through, and the result is a laser beam emerging from this face of the wafer. It is now possible to miniaturize these devices to the extent that one million such lasers can be packed onto a chip with an area of 0.5 cm².

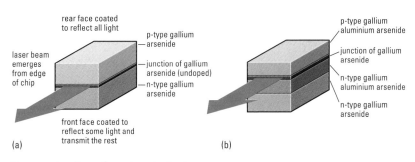

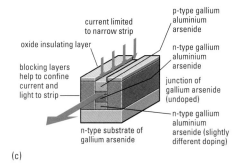

The construction of semiconductor lasers.

Photodiodes

When a diode is reverse biased a small current (typically 10^{-8} A) arises because of thermal production of electron–hole pairs. If light of a specific frequency is shone onto the diode's depletion region, the energy of the photons can be absorbed and used to produce extra electron–hole pairs. The result is an increase in the reverse-bias current that is proportional to the light intensity. Among other applications, photodiodes are used as detectors for infrared remote control units. A variation on this device is the **solid-state particle detector** used in many particle physics experiments. In this context some of the energy from a high-energy subatomic particle passing through the detector is used to create the electron–hole pairs. Strips of these detectors close to one another are used to record the direction in which particles are travelling.

A silicon strip detector used at the Stanford Linear Accelerator Center, California, USA.

ELECTRICITY IN THE HOME

Every year in the UK about 2000 people are treated for electric shocks, and 20% of those people die. Furthermore, about four times that number die each year due to fires started by electrical faults.

In the course of a year a typical household will use about 1.2×10^4 MJ of electrical energy. With this much energy conversion going on, it is not surprising that there are some accidents. However, most of the accidents that happen are avoidable. The electric circuits and appliances in the home have been carefully designed to be as safe as possible, and an understanding of these design features is important for their correct operation.

Effects of electric shock

Electric shock is a general term for a variety of effects that are produced by the passage of a current through the body. The severity of the effect depends on the size of the current, the duration of the shock, and also on the frequency of the a.c. Nerves that control muscles are excited by the passage of a.c. This can result in muscle contractions that prevent a person from letting go of the source of the current. At 50 Hz any current above about 15 mA will cause this effect. However, at 5 Hz and 1000 Hz the current is greater – about 30 mA.

There are four ways in which a person can suffer if they have an electric shock.

- **Burns** A current passing through the skin can cause local heating, resulting in burns at the point of contact with the source of the shock.
- **Fibrillation** A shock can cause the heart to start beating irregularly (**fibrillation**). This is the most common cause of death due to electric shocks.
- **Muscular contraction** This can happen suddenly and violently enough to throw the victim across the room.
- **Clinical shock** When a person goes into shock, the pulse becomes weak and the breathing irregular. This can be lethal if not treated properly.

The current that will cause these different effects depends on the resistance of the body. This varies widely between individuals, and depends on which part the current is passing through, as well as other factors such as perspiration. The severity of the shock also depends on how long it lasts. The graph on the left shows how different durations of current affect the heart. A surprisingly small current can have severe effects.

Why is 230 V used?

Before 1948 (when the industry was nationalized) there were several companies supplying electricity using different voltages from 100 V to 250 V. The national voltage is now set to 230 V. The American system uses 110 V. It would seem that the UK uses a more dangerous setting. However, there are other factors that need to be taken into account.

- **Current levels** A domestic appliance like a kettle has a power of 2 kW. At 230 V it will draw about 9 A, while at 110 V the current would be 18 A. This is high enough to cause significant heating in the wires, leading to a fire risk. To counter this the wires have to be much thicker and more awkward to use, making them more expensive.
- **Voltage drop** A voltage drop across wires supplying current is inevitable, as they have resistance. With a significant distance between the substation and the user, it is important to minimize the loss. The lower the supply voltage, the higher the current for the same power and the more significant the voltage drop will be.

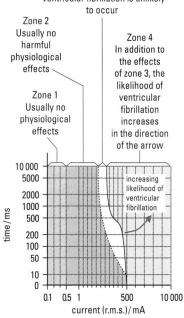

Zone 3
Usually no damage to the body. In this zone, muscular contractions are likely – if current flows through the chest, this may lead to difficulties in breathing, and there is also a risk of short-term heart stoppage, although ventricular fibrillation is unlikely to occur

Zone 2
Usually no harmful physiological effects

Zone 4
In addition to the effects of zone 3, the likelihood of ventricular fibrillation increases in the direction of the arrow

Zone 1
Usually no physiological effects

increasing likelihood of ventricular fibrillation

time / ms — current (r.m.s.) / mA

An electric shock can seriously affect the function of the heart.

Safety features

In the UK the problems of a higher voltage are countered by a combination of safety features to protect the user.

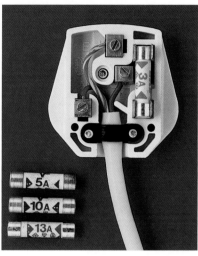

- **Double insulation** Many modern appliances use two layers of insulation to protect the user from the electric supply. An inner layer (the **functional insulation**) is made of a material that is chosen for its lack of electric and thermal conductivity. The outer insulation (the **protective insulation**) normally forms part of the case of the appliance (for example, the plastic case of a kettle).

- **Fuses** Fuses are thin pieces of wire designed to melt when the current rises above a set value. They come in many different ratings and in casings of various shapes, so they can fit in plugs or fuse boxes. All UK 13 A plugs contain a fuse. The size of the fuse depends on the power rating of the appliance. The maximum power appliance that can be used is:

$$P = I_{\text{r.m.s.}}.V_{\text{r.m.s}} = (13\,\text{A}) \times (230\,\text{V}) = 3.0\,\text{kW}$$

A standard 13 A plug used in the UK.

Most household appliances are much less powerful than this, and so lower-rated fuses should be used in the plugs. More powerful appliances have fuses in the fuse box and are wired directly to them – not via a plug and socket.

- **Earthing** Some appliances which have metal cases are **earthed**. This means that the case is connected via a low-resistance path to earth (usually a metal spike driven into the ground). The third pin in a UK 13 A plug provides this connection, and if a fault causes the case to become live, then a large current will pass from the appliance to ground, and the fuse connected to the live wire will blow. Without the combination of fuse and earth a person touching the appliance could receive a fatal current, which would still be much less than the fuse rating.

- **Residual current devices (RCDs)** These devices operate faster than fuses and can cut off the current within 40 ms. They work by sensing the difference between the currents in the live and neutral wires. Under normal circumstances these currents should be the same (but in opposite directions) causing no net magnetic flux through the core. However, if a person is receiving a shock then some of the current will be diverted to earth. The net flux will induce a current in the secondary coil, activating the relay to switch off the current.

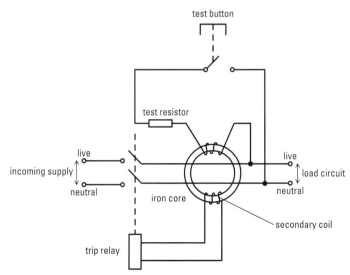

The circuit used inside an RCD.

An RCD.

4.15

OBJECTIVES

• $Q = CV$

• definitions of capacitance and the farad

• manufacture of capacitors

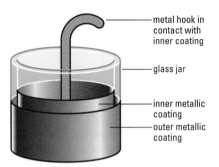

metal hook in contact with inner coating

glass jar

inner metallic coating

outer metallic coating

The Leyden jar was discovered by accident at the University of Leyden in 1745. Connecting the jar to a source of static electricity would cause it to 'accumulate' electrical effects. A person touching the jar could still receive a shock some time after the jar had been disconnected from the static source.

INTRODUCTION TO CAPACITORS

Often when the power to an electrical device, such as a radio or stereo, is turned off, the power lamp will fade out slowly. With a stereo the sound can continue playing for some seconds after the power is turned off. This indicates that electrical energy is being stored inside the device and continues to supply the electronics for a while after the external power has been removed.

The components responsible for the storage of electrical energy in a circuit are called **capacitors**. The modern devices are a development of the original Leyden jars used by Faraday (among others) to experiment on the connection between static electric charges and electric currents.

Capacitance

Any arrangement of two conductors isolated from one another by an insulator will form a capacitor. In the lab this may be two sheets of metal with small pieces of plastic between. In this case air forms the insulator. A commercial capacitor can be made by using sheets of wax paper to separate two sheets of thin metal foil. Very large areas of foil can be rolled up and contained in small cylinders. The two conductors forming the capacitor are always called the capacitor **plates**. Capacitors store energy by keeping electric charges on their plates. They can also be viewed as devices for storing charge.

When a capacitor is **charged up**, one of the plates will have a positive charge and the other an equal negative charge. A voltmeter connected between the plates will record a p.d. The ratio between the charge and the p.d. is a constant for the particular capacitor, and is called the **capacitance**.

Definition: capacitance

$$\text{capacitance} = \frac{\text{size of charge on plates (C)}}{\text{p.d. between plates (V)}}$$

$$C = \frac{Q}{V} \qquad \therefore \qquad Q = CV$$

The capacitance is determined by the size of the plates, the nature of the insulator, and the way in which the plates are arranged. The unit used to measure capacitance is the **farad**.

$$1 \text{ farad, F} = 1 \text{ C V}^{-1} = 1 \text{ A}^2 \text{ s}^4 \text{ m}^{-2} \text{ kg}^{-1}$$

In practice this unit is far too large for most capacitors used in modern electric circuits. It is normal to see capacitor sizes expressed in **microfarads** (10^{-6} F) or **picofarads** (10^{-12} F).

Worked example

A capacitor of capacitance 250 µF is allowed to charge up until the p.d. between its plates is 10 V. How much charge accumulates on the plates during the charging process?

$$Q = C \times V = (250 \,\mu\text{F}) \times (10 \,\text{V}) = 2500 \,\mu\text{C} = 2.5 \,\text{mC}$$

One plate has a charge of +2.5 mC; the other has a charge of –2.5 mC.

Net charge on a capacitor

When we say the charge on the capacitor is Q, we mean that one plate will have a charge of $+Q$ and the other $-Q$. This means that the net charge stored is actually zero. However, because the charges are kept separated we say that the capacitor is charged up.

Manufacture of capacitors

The following are different types of capacitor.

a Paper or plastic capacitors The conductors are metal foil strips; the insulator is waxed paper or plastic.

b Mica capacitors The sheets of metal foil are separated by strips of mica, a natural mineral that is easily split into very thin sheets (like slate).

c Electrolytic capacitors Metal sheets (aluminium) are separated by paper that has been soaked in a chemical. The paper does not insulate the two plates, but as the capacitor is charging up a thin layer of aluminium oxide is formed on the positive plate, providing an insulating layer between the plates. The layer is very thin (0.01 mm) and so the capacitance is very high. **Warning:** because of the nature of the chemical reaction, electrolytic capacitors are labelled with positive and negative ends. It is very dangerous to connect such capacitors the wrong way round. At best this will destroy the capacitor, at worst it can cause a dangerous explosion.

d Variable capacitors The insulator is air, and the capacitance can be changed by altering the overlapping area between the plates.

Different types of capacitor.

Uses of capacitors

A capacitor's ability to store energy (and hence charge) is exploited in several simple situations, as described below.

- In a power supply circuit designed to convert a.c. into d.c. (spread 5.28), the capacitor's store of charge helps to 'smooth over' the variations in rectified a.c.;

- In a camera flash gun, a rapid burst of energy is required to produce a bright light over a short time period. Cells in the flash gun charge the capacitor over a few seconds. When the flash is triggered, the capacitor discharges by releasing its stored charge through the lamp in a very short period (hence the time constant of the discharge needs to be very short, see spread 4.16). If the cells in the flash gun supplied the charge directly, the internal resistance of the cells would cause such a high current to reduce the p.d. supplied to the lamp, hence the advantage of the capacitor. Also, flash guns typically require p.d.s of several hundred volts to operate, which is more than can be provided by easily portable cells. Using more than one capacitor, and a circuit known as a charge pump, charge can be transferred to the main capacitor in several steps, building up a very high voltage from small cells.

- In filter circuits, a capacitor can be used to select ranges of a.c. frequencies that are allowed to pass through (spread 5.26).

PRACTICE

1 In a typical defibrillating device, used to restore normal heart function, a 16 μF capacitor is charged to 6 kV and then discharged in 2 ms. How much charge is passed through the body and what is the average current?

2 A capacitor stores charge of ± 0.025 μC on its plates. The p.d. between the plates is 250 V. What is the capacitance of the capacitor?

3 A 100 μF capacitor is charged from a battery of e.m.f. 6 V. What is the charge on the plates of the fully charged capacitor? How much energy has been converted by the cell in the charging process?

4 A capacitor is charged from a cell through a resistor. The charging current is recorded over a period of time.

time/s	0.0	1.0	2.0	3.0	4.0	5.0
current/μA	15.0	5.5	2.0	0.7	0.3	0.1

a Plot a graph of charging current against time.

b Use the graph to estimate the charge on the plates of the fully charged capacitor.

c If the cell has an e.m.f. of 1.5 V, calculate the capacitance of the capacitor.

CHARGING AND DISCHARGING CAPACITORS

Electronic flash guns for cameras have to be left for a short period of time between flashes. The capacitor inside needs to be charged up again by the cell in the flash gun. The time taken for this depends on the rate at which the charge is flowing, which in turn is determined by the resistance of the circuit.

It can take a couple of seconds before the capacitor is fully charged. However, the discharge through the lamp is very rapid – of the order of a few microseconds.

Charging a capacitor

Charge starts to flow in the wires the instant that the cell is connected. As there is insulation between the plates of the capacitor, electrons are forced to accumulate on the plate connected to the negative side of the cell. Electrons move away from the other plate, partly because of the repulsion from the negative charge nearby and partly because of the attraction of the positive side of the cell.

The initial current in the circuit is determined by the resistance R (which may just be the resistance of the wires). Surprisingly, the uncharged capacitor does not resist the current arriving at one plate or leaving the other.

$$\text{initial charging current} = \frac{\text{e.m.f. of cell}}{\text{resistance in circuit}}$$

On a graph of capacitor charge against time, the initial current is the gradient of the graph at the origin.

As charge builds up on the plates, it repels the charge moving through the wires, and the current drops as the charge on the plates increases. Charging will stop when the p.d. between the capacitor plates is equal to the e.m.f. of the cell.

$$\text{maximum charge on capacitor} = \text{capacitance} \times \text{e.m.f. of cell}$$

Time constant

In principle, a capacitor can never charge up fully, because the rate of charging decreases as the charge increases. In practice, after a finite time the charging current becomes too small to measure, and the capacitor is effectively fully charged. The time taken to charge a capacitor in a given circuit is determined by the **time constant** of the circuit. The bigger the capacitance, the longer it takes to charge the capacitor. The larger the resistance, the smaller the current, which also increases the charging time.

$$\text{time constant, } \tau = \text{resistance} \times \text{capacitance} = RC$$

The **half-time** is the time taken to halve the charging current:

$$\text{half-time, } t_{\frac{1}{2}} = \text{time constant} \times \ln 2 = RC \ln 2 = 0.693RC$$

After every time interval $t_{\frac{1}{2}}$, the amount of extra charge still needed to fully charge the capacitor halves.

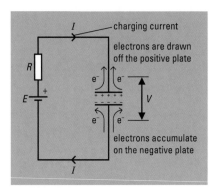

The process of charging a capacitor, showing conventional current and electron flow.

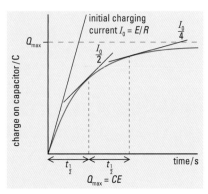

Graph of charge versus time for a charging capacitor.

Opposite sign

The charge on each plate at any moment is equal in size, but of opposite sign.

Equipotentials

The connecting wire, the capacitor plate, and one of the cell's terminals form a complete conducting path. Conductors always tend to form **equipotentials** (see spread 5.9). When the cell is first connected, there is a p.d. between the capacitor plate and the terminal of the cell. The charge carriers in the wire redistribute themselves until an equipotential is formed.

Worked example 1

A 250 μF capacitor is being charged from a 6 V cell of negligible internal resistance via a 10 kΩ resistor. What is the initial charging current? What is the charging current after 10.4 s?

$$\text{initial charging current} = \frac{E}{R} = \frac{6\,\text{V}}{10\,\text{k}\Omega} = 0.6\,\text{mA}$$

$$\text{time constant} = RC = (250 \times 10^{-6}\,\text{F}) \times (10\,000\,\Omega) = 2.5\,\text{s}$$

$$\therefore t_{\frac{1}{2}} = 0.693 \times 2.5\,\text{s} = 1.73\,\text{s (3 sig. figs)}$$

10.4 seconds corresponds to $10.4/1.73 = 6$ half-times

$$\therefore \text{charging current after } 6t_{\frac{1}{2}} = \frac{I_0}{2^6} = \frac{0.6\,\text{mA}}{64} = 9.4\,\mu\text{A (2 sig. figs)}$$

Discharging a capacitor

A charged capacitor that is isolated from a circuit should hold its charge indefinitely. However, leakage between the plates (due to the imperfect nature of the insulators) will eventually completely discharge it.

A capacitor can be deliberately discharged by connecting its plates together via a resistor. At any time:

$$\text{discharge current} = \frac{\text{p.d. between plates}}{\text{resistance}}$$

On a graph of charge against time the discharge current is given by the gradient at any point. The gradient is:

$$\frac{\delta Q}{\delta t} = I = \frac{V}{R} = \frac{Q}{CR} \qquad \left(V = \frac{Q}{C}\right)$$

So the gradient of the curve is proportional to its value at any point. There is only one type of curve in mathematics that has this property – the exponential curve (see Mathematics toolbox). For a discharging capacitor the equation of the curve has the form:

$$Q = Q_0 e^{-t/RC}$$

where Q is the charge at elapsed time t and Q_0 is the initial charge.

The time constant determines the time taken to halve the amount of charge on the plates.

time constant, $\tau = RC$ time to halve charge, $t_{\frac{1}{2}} = RC \ln 2$

Substituting $Q = CV$ into the charge equation gives:

$$V = V_0 e^{-t/RC}$$

where V is the p.d. between the plates at time t and V_0 is the initial p.d. At any time the current in the resistor = V/R, so dividing both sides by R gives:

$$\frac{V}{R} = \frac{V_0}{R} e^{-t/RC} \qquad \text{or } I = I_0 e^{-t/RC}$$

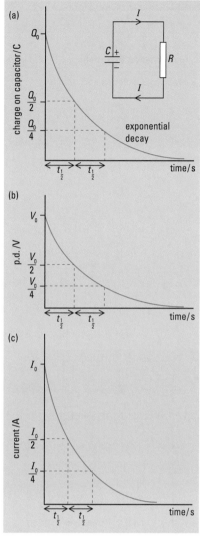

Graphs of (a) charge; (b) p.d.; (c) current versus time for a discharging capacitor.

Worked example 2

A 1000 µF capacitor has been charged to a p.d. of 25 V. The capacitor is being discharged through a 470 kΩ resistor. After 100 s, what is the charge on the plates, and the p.d. between them?

initial charge on plates, $Q = CV = (1000\,\mu F) \times (25\,V) = 25\,mC$

after 100 seconds, $Q = (25\,mC) \times e^{\frac{-100\,s}{(470\times10^3\,\Omega)\times(1000\times10^{-6}\,F)}} = 20.2\,mC$ (3 sig. figs)

p.d. between plates $= (25\,V) \times e^{\frac{-100\,s}{(470\times10^3\,W)\times(1000\times10^{-6}\,F)}} = 20.2\,V$ (3 sig. figs)

Exponential curves

All exponential curves have a constant ratio property. This means that equal time intervals produce the same fractional change in charge/voltage/current.

Helpful hint

As a rule of thumb, you can assume a capacitor to be fully discharged if $t \geq 5CR$.

PRACTICE

1 A 50 µF capacitor is being charged from a 6 V battery via a 100 kΩ resistor. What is the initial charging current? After a period of time the charging current is 30 µA. What are the p.d.s across the resistor and the capacitor at that moment? How much charge has been stored on the capacitor up to that time?

2 A 20 µF capacitor is charged to 20 V and isolated. After 6 min the p.d. between its plates has fallen by 6 V. What is the average leakage current between the plates? If the leakage continues at the same rate, how long would it take to fully discharge the capacitor?

With reasons, say if you would expect it to take less or more time than this in practice.

3 A memory chip in a computer runs off a 5 V power supply and draws a current of 80 mA. The chip will 'forget' its contents if the power supply drops below 3 V. In order to protect the data against power supply faults, a 1 F capacitor is connected across the power supply output. Estimate how long the capacitor could keep the memory chip going. You may assume that the only resistance in the circuit is that of the chip.

- charging and discharging – mathematical treatment

- capacitors in series and parallel

FURTHER WORK ON CAPACITORS

Combining capacitors

If you continue to charge up a capacitor, eventually the p.d. between its plates will be sufficient to break down the insulation. The resulting current can cause a rapid temperature rise, and even an explosion as the material inside the capacitor expands. All capacitors have a voltage rating that should not be exceeded, and because the energy stored in a capacitor depends on V^2 it is desirable to have as large a rating as possible.

One way of adjusting the voltage rating and capacitance is to connect capacitors together in series or in parallel. (See question **3**)

Series

A set of capacitors in series can be replaced by a single capacitor of the right value. The value of the replacement capacitor can be calculated from the equation:

$$\frac{1}{C_{\text{tot}}} = \frac{1}{C_1} + \frac{1}{C_2} + \frac{1}{C_3} + \dots \text{ etc.}$$

NB Capacitors in *series* add like resistors in *parallel*.

The total p.d. across the combination is the sum of the potential differences across the individual capacitors, and the voltage rating of the set is the sum of the individual p.d. ratings.

Maths box 1: derivation of the capacitors-in-series equation

A circuit has three capacitors, C_1, C_2, and C_3, connected in series to a cell of e.m.f. E. Even though the capacitor values may not be the same, they will each accumulate the same charge on their plates.

By Kirchhoff's second law:

$$E - V_1 - V_2 - V_3 = 0$$

$$\therefore\ E = V_1 + V_2 + V_3 = \frac{Q}{C_1} + \frac{Q}{C_2} + \frac{Q}{C_3}$$

A single capacitor of value C_{tot} in a circuit must charge to a p.d. of $E = Q/C_{\text{tot}}$, so

$$\frac{Q}{C_{\text{tot}}} = \frac{Q}{C_1} + \frac{Q}{C_2} + \frac{Q}{C_3} \qquad \therefore \qquad \frac{1}{C_{\text{tot}}} = \frac{1}{C_1} + \frac{1}{C_2} + \frac{1}{C_3}$$

Parallel

Connecting capacitors together in parallel involves linking together their plates. This is like adding the areas of the plates to form one large plate of the same total area. In parallel, simply add the capacitor values:

$$C_{\text{tot}} = C_1 + C_2 + C_3 + \dots \text{ etc.}$$

In this case the p.d. across each capacitor is the same, so the voltage rating of the set is equal to that of the capacitor with the smallest individual rating.

Maths box 2: derivation of the capacitors-in-parallel equation

Three capacitors, C_1, C_2, and C_3 are connected in parallel to a cell of e.m.f. E. The p.d. across each plate must be the same:

$$\frac{Q_1}{C_1} = \frac{Q_2}{C_2} = \frac{Q_3}{C_3} = E$$

If an equivalent circuit consists of one capacitor of value C_{tot}, the charge drawn from the cell Q must be the same as the total charge on the plates of the other capacitors:

$$Q = Q_1 + Q_2 + Q_3$$

$$\therefore\ C_{\text{tot}}E = C_1E + C_2E + C_3E$$

so

$$C_{\text{tot}} = C_1 + C_2 + C_3$$

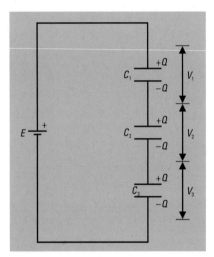

Capacitors in series.

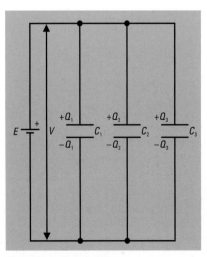

Capacitors in parallel.

Worked example

What value of capacitor could be used to replace a set of $10\,\mu F$, $20\,\mu F$, and $30\,\mu F$ capacitors connected in series?

$$\frac{1}{C_{tot}} = \frac{1}{C_1} + \frac{1}{C_2} + \frac{1}{C_3} \dots \text{etc.}$$

$$= \frac{1}{10\,\mu F} + \frac{1}{20\,\mu F} + \frac{1}{30\,\mu F}$$

$$= \frac{(20 \times 30) + (10 \times 30) + (10 \times 20)}{10 \times 20 \times 30}\,\mu F^{-1} \quad \text{so } C_{tot} = \frac{6000\,\mu F}{1100} = 5.45\,\mu F$$

NB The final answer is smaller than the value of any of the individual capacitors.

Maths box 3: discharging a capacitor

A capacitor of initial charge Q_0 is being discharged via a resistance R (which includes any resistance in the wires).

Applying Kirchhoff's second law round the loop of the circuit in the direction of the discharge current:

$$V - IR = 0 \quad \text{so} \quad V = IR$$

At any moment the charge on the capacitor is $Q = CV$, so

$$I = \frac{V}{R} = \frac{Q}{CR}$$

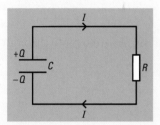

In time δt an amount of charge equal to $I\delta t$ moves round the circuit, decreasing the change on the plates by the same amount.

The rate of change of charge on the plates is therefore

$$-\frac{dQ}{dt} = I \quad \text{so} \quad \frac{dQ}{dt} = -\frac{Q}{CR}$$

This differential equation can be solved by separating the variables and integrating:

$$\int_{Q_0}^{Q} \frac{dQ}{Q} = -\int_0^t \frac{dt}{CR} \quad \therefore \quad [\ln(Q)]_{Q_0}^{Q} = \left[-\frac{t}{CR}\right]_0^t$$

so

$$\ln\left(\frac{Q}{Q_0}\right) = -\frac{t}{CR}$$

Taking the antilog of both sides and tidying up we get

$$Q = Q_0 e^{-t/CR}$$

Maths box 4: charging a capacitor

In this circuit, the cell has an e.m.f. E. The internal resistance of the cell, resistance of the wires, and any resistors in the circuit are included in the value of R.

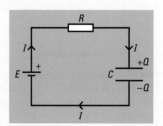

Applying Kirchhoff's second law we have

$$E - V - IR = 0 \quad \therefore \quad E = V + IR$$

V and I will be changing as the capacitor charges. At any moment the charge on the capacitor is

$$Q = CV$$

Also $\quad I = \frac{dQ}{dt}$

(This time it is positive, because the charge on the plates is *increasing* with time.)

$$\therefore \quad E = \frac{Q}{C} + R\frac{dQ}{dt} \quad \text{and hence} \quad \frac{dQ}{dt} = \frac{E}{R} - \frac{Q}{CR} = \frac{EC - Q}{CR}$$

Solving this differential equation shows how the charge Q varies with time t.

If we introduce a 'dummy variable' $U = EC - Q$, then as E and C are constants:

$$\frac{dU}{dt} = -\frac{dQ}{dt} \quad \therefore \quad \frac{dU}{dt} = \frac{U}{CR}$$

which can be solved in the same way as the discharge equation, and:

$$U = U_0 e^{-t/CR}$$

with U_0 being the value of U when $t = 0$.

Now we return the equation to the variable Q, remembering that when $t = 0$, $Q = 0$:

$$EC - Q = ECe^{-t/CR}$$

so

$$Q = EC\,(1 - e^{-t/CR})$$

PRACTICE

1 Three capacitors of values $2\,\mu F$, $3\,\mu F$, and $6\,\mu F$ are connected in series and then in parallel. What is the equivalent capacitance in each case?

2 You are given a set of three $2\,\mu F$ capacitors and asked to connect them to make up the following equivalent capacitances (in turn): $1.5\,\mu F$, $3\,\mu F$, $0.75\,\mu F$, $4\,\mu F$, and $6\,\mu F$. Which value *cannot* be made?

3 A $200\,\mu F$ capacitor has a voltage rating of $60\,V$. What is the maximum energy that it can store safely?

If two $200\,\mu F$ capacitors are connected in series, what is the capacitance and the voltage rating of the combination? What is the maximum energy that can be stored safely in the combination? If two $200\,\mu F$ capacitors are connected in parallel, what is the capacitance and the voltage rating of the combination? What is the maximum energy that can be stored safely in the combination? How would you connect four $200\,\mu F$, $60\,V$ capacitors to produce a combination of $200\,\mu F$ capacitance and $120\,V$ rating?

ENERGY STORED IN A CAPACITOR

When a cell is used to charge a capacitor a certain amount of chemical energy is converted into electrical energy. Some of this appears as thermal energy in the resistance of the charging circuit, but this only accounts for half of the energy converted by the cell. The rest is stored in the capacitor. This is shown by the ability of a capacitor to provide energy to drive a current as it is discharging.

The stored energy inside the capacitor is a form of **potential energy**.

- Any situation in which opposite charges are held apart from each other against the action of their electrostatic attraction will store electrical potential energy.
- This is similar to the storage of gravitational potential energy when two masses are held apart against their gravitational attraction.

Calculating the energy stored in a capacitor

Imagine a small amount of charge δq arriving at the positive plate of a capacitor which is already charged to q. This extra positive charge would repel an equal amount of charge away from the negative plate, and the charges on the two plates would become:

$+$ve plate charge $= q + \delta q$

$-$ve plate charge $= -q - \delta q$

This is equivalent to a charge δq moving from the negative plate through the insulation and onto the positive plate.

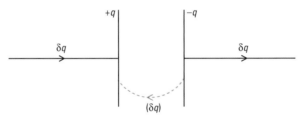

The charge flow round the circuit is equivalent to a charge jumping between the plates in the opposite direction.

Physically, this does not happen. The charge is conducted through the external circuit (and in the other direction). The two situations are equivalent, but it is easier to calculate the work done in charging the capacitor if we imagine that the charge moves between the plates rather than through the circuit.

If the charge δq moves between the plates it must cross a p.d. $= V$.

$\therefore$ work done in moving the charge $= \delta q \times V$

Increasing the charge on the plates also alters the p.d. between them. A small charge δq was specified so that the change in the p.d. is small enough to be neglected.

$\therefore$ energy stored in capacitor when charge is increased from q to $q + \delta q$ $= V\delta q$

On a graph of p.d. against Q, this stored energy is represented by the area of the small strip A.

If we charge the capacitor from zero to a total charge Q, the energy stored will be the sum of all the small strips that lie between the origin and Q. This is the total area under the line of the graph.

$$\text{area under line} = \text{area of triangle} = \frac{1}{2} \times \text{base} \times \text{height}$$

$$\text{so work done} = \frac{1}{2}QV = \frac{1}{2}CV^2 = \frac{1}{2}\frac{Q^2}{C}$$

Charge carriers

We are assuming that the small charge δq is positive, that is, that the charge carriers in the circuit that is charging the capacitor are positive. In practice they will almost certainly be electrons, so δq will be a negative charge arriving at the negative plate and pushing electrons away from the other plate, making it positive.

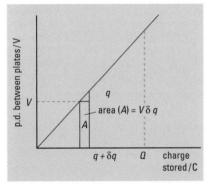

The energy stored is equal to the area under the line.

Total energy stored

Using calculus, energy stored in charging from q to $q + \delta q$ is $V\delta q$, so the total energy stored in charging from 0 to Q is

$$\int_0^a V dq = \int_0^Q \frac{Q}{C} dq = \frac{1}{2}\frac{Q^2}{C} = \frac{1}{2}QV$$

Using the capacitor equation $Q = CV$ to substitute for Q or for V, a set of alternative equations can be derived. These can be used to solve a variety of problems.

Worked example

A 10 000 μF capacitor is charged until the p.d. between its plates is 230 V. How much energy is stored in the capacitor?

$$\text{energy stored} = \tfrac{1}{2}CV^2 = \tfrac{1}{2} \times (10\,000 \times 10^{-6}\,\text{F}) \times (230\,\text{V})^2 = 265\,\text{J}$$

At normal listening levels this is enough energy to keep a stereo playing for about 5 seconds after the power has been turned off.

Energy losses in a circuit

In the simple circuit on the right, the capacitor will charge until the p.d. between its plates is equal to the e.m.f., E, of the cell, which is assumed to have negligible internal resistance. A total charge Q will have moved round the circuit where $Q = CE$.

$$\text{chemical energy converted by cell} = QE$$

$$\text{energy stored in capacitor} = \tfrac{1}{2}QE$$

$$\therefore \quad \text{energy dissipated as heat in the resistance} = \tfrac{1}{2}QE$$

At first sight this is a very puzzling result, as it is independent of the size of the resistance. As the capacitor charges up the p.d. across its plates must increase, and the p.d. across the resistor decreases.

At any moment $V_C + V_R = E$, and $Q = CV$, so the p.d. across the resistor falls linearly as the charge on the capacitor plates increases. The area of the rectangle (right) represents the energy converted by the cell. This is broken into two equal triangles which represent the energy stored by the capacitor and the heat dissipated by the resistance. What is *not* shown on this graph is *how long* the capacitor takes to charge. This is determined by the resistance.

A large resistance reduces the current and increases the time needed to charge the capacitor. The opposite is true of a small resistance. Also, a large current flowing for a short time produces the same amount of heat as a small current flowing for a longer time.

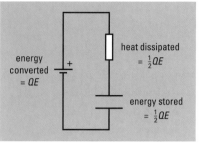

Irrespective of the size of the resistance, it will always dissipate half the energy converted by the cell.

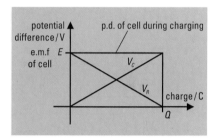

The area of the rectangle is the energy converted by the cell. This is split equally between the capacitor and the resistor as shown by the two triangles. V_C is p.d. across capacitor during charging; V_R is p.d. across resistor during charging. Q is the total charge passed in charging the capacitor.

PRACTICE

1 A typical electronic flash gun has to provide 2 kW for about 2 ms. This is done by discharging a 50 μF capacitor. To what voltage must the capacitor be charged? If the capacitor is replaced by one of 250 μF, to what voltage would it have to be charged? What would be the disadvantage of using the higher-value capacitor?

2 A 20 μF capacitor is charged to 10 V and isolated. A second, uncharged, 20 μF capacitor is now connected to the first so that the plates of one are connected to the plates of the other. The capacitors rapidly come to the same p.d. between their plates. What is the charge on each capacitor now? What is the total energy stored in the two capacitors? Comment on your answer.

3 Repeat question 2 with the second capacitor rated at 30 μF.

4 A 50 pF capacitor is charged to 20 V and isolated. What is the charge on its plates and what energy does it store? The plates are then pulled apart until the capacitance decreases to 25 pF. In the process of doing this no charge is lost from the plates. What is the p.d. between the plates now? What is the energy stored by the capacitor now? What was the average force required to separate the plates if the distance that they moved was 2.8 cm?

5 A 100 μF capacitor is charged to 10 V, and the charge on the plates is 1 mC. What is the energy stored? If you move a 1 mC charge through a p.d. of 10 V how much energy is converted? Why is this not the same as the energy stored on the capacitor? A thundercloud acts as a parallel-plate capacitor with the base of the cloud as one plate and the ground as the other. The p.d. between the ground and the cloud is typically 10^8 V. A flash of lightning discharges the cloud and carries 20 C of charge to ground. How much energy is there in the flash?

PRACTICE EXAM QUESTIONS

1

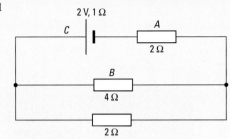

Figure 4.1

In Figure 4.1, *C* is a cell of e.m.f. 2 V and internal resistance 1 Ω. Calculate

a the current through *A*,

b the current through *B*,

c the p.d. between the terminals of the cell.

2

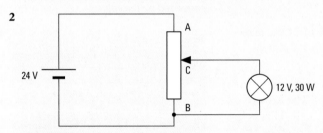

Figure 4.2

In Figure 4.2, AB is a variable resistor having a total resistance of *R*. C is a moveable contact. The resistor is used as a potential divider which, when connected across a cell of emf 24 V and no internal resistance, provides a p.d. to a bulb rated 12 V, 30 W. When the resistance of AC is one third that of AB the bulb operates under the rated conditions (i.e. normally).

a Calculate the current through the bulb.

b What is the p.d. across BC?

c *In terms of R* find an expression for the current through

  i BC,

  ii AC.

d Hence calculate the value of R.

3

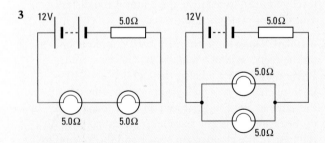

Figure 4.3   *Figure 4.4*

In the circuits in Figures 4.3 and 4.4 the battery has negligible internal resistance, and the bulbs are identical.

a For the circuit shown in Figure 4.3 calculate

  i the current flowing through each bulb,

  ii the power dissipated in each bulb.

b In the circuit shown in Figure 4.4 calculate the current flowing through each bulb.

c **i** Explain how the brightness of the bulbs in Figure 4.3 compares with the brightness of the bulbs in Figure 4.4.

  ii Explain why the battery would last longer in the circuit shown in Figure 4.3.

d One of the bulbs in Figure 4.4 develops a fault and no longer conducts. Describe and explain what happens to the brightness of the other bulb.

4 This question is about two low-voltage filament lamps.

A student is provided with two filament lamps, A and B. Each is rated 12 V, 60 W, but lamp A has a *carbon filament*, and lamp B a *tungsten* filament.

a What is the resistance of each filament lamp under normal operating conditions?

The graph in Figure 4.5 shows the relationship between the potential difference *V* and the current *I*, for the two filament lamps.

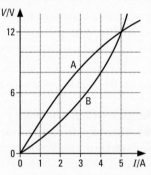

Figure 4.5

b Describe how the resistance of lamp A and lamp B vary with the current.

c Figure 4.6 shows the two filament lamps connected in series across a cell of emf *E* = 12 V, and internal resistance *r*. Both filaments glow dimly, though the carbon filament is the brighter, and the reading on the ammeter is 2 A.

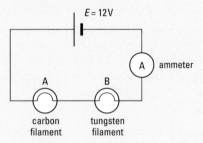

Figure 4.6

  i Explain why each filament glows only dimly.

  ii Suggest why the carbon filament is the brighter of the two.

  iii Use the graph, and information you have been given, to calculate the value of the internal resistance *r* of the cell.

5 A car battery has an e.m.f. of 12.0 V. When a car is started the battery supplies a current of 105 A to the starter motor. The terminal potential difference between the battery terminals drops at this time to 10.8 V due to the internal resistance of the battery.

a Explain briefly what is meant by:

  i an e.m.f. of 12.0 V;

  ii the internal resistance of the battery.

    advanced **PHYSICS**

b Calculate the internal resistance of the battery.

c The manufacturer warns against short-circuiting the battery.

 i Calculate the current which would flow if the terminals were to be short-circuited.

 ii Explain briefly why the manufacturer provides this warning. Justify your explanation with an appropriate calculation.

d When completely discharged, the battery can be fully recharged by a current of 2.5 A supplied for 20 hours.

 i How much charge is stored by the battery?

 ii For how long could the motor be operated on a fully-charged battery? Assume that the motor could be operated continuously.

6 a State Kirchhoff's laws.

b

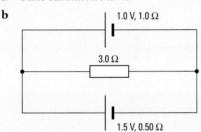

Figure 4.7

 i Calculate the current through the 3.0 Ω resistor in the circuit in Figure 4.7.

 ii Calculate the pd across the 3.0 Ω resistor.

 iii Why is the pd across the terminals of the 1.0 V cel higher than its emf?

7 a Define the term resistivity.

b The resistivity of copper is 1.7×10^{-8} Ωm. A copper wire is 0.6 m long and has a cross-sectional area of 1 mm². Calculate its resistance.

c Two such wires are used to connect a lamp to a power supply of negligible internal resistance. The potential difference across the lamp is 12 V and its power is 36 W. Calculate the potential difference across each wire.

d Draw a circuit diagram of the above arrangement. Label the potential differences across the wires, lamp and power supply.

8 A student's notes contain the two statements below in the section on electrical conduction.
(1) 'Resistivity depends only on the material and the temperature, whereas resistance depends also on the shape and size of the sample under test.'
(2) 'The resistivity of copper is very much smaller than that of glass.'

a Discuss experiments you might perform in order to verify each of these statements.

b Explain how you would expect the resistivity of a metal such as copper to vary with temperature.

c Consider the resistances of various arrangements of pieces of copper wire, each of length *l* and area of cross-section *A*. Using the idea of resistivity as in statement (1), show how statement (1) relates to the usual formulae for the connection of resistors in series and in parallel.

9 An electricity supply cable is made up of 7 strands of steel and 54 strands of aluminium as illustrated in Figure 4.8. All strands have a diameter of 3.0 mm.

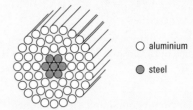

Figure 4.8

a Calculate the total resistance of the aluminium strands for a 1 km length of electricity supply cable. The resistivity of aluminium is 2.5×10^{-8} Ω m.

b The central group of steel strands has a resistance of 4.0 Ω for a 1 km length. By considering the steel strands and the aluminium strands to be two resistors in parallel, show that the presence of the steel does not significantly affect the resistance of the cable.

c When the cable is in use, the resistance increases. Explain briefly, in terms of electron motion, why this happens.

d By referring to the non-electrical properties of steel and aluminium, state and explain one reason why steel is used in this cable.

10 This question is about the behaviour of a lamp in a simple circuit.

a A student has a lamp which he wishes to light to 'normal brightness'. The rating given on the lamp is '4.5 V, 1.35 W'.

 i Explain what is meant by a rating of '4.5 V, 1.35 W'.

 ii Use this information to calculate the resistance of the lamp filament, under normal operating conditions.

The only apparatus available to the student is a 6.0 V d.c. supply, of negligible internal resistance, some resistors, and a digital voltmeter. He connects the circuit shown in Figure 4.9, with which he intends to produce a voltage difference of 4.5 V across the points YZ.

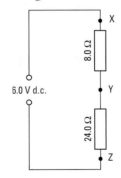

Figure 4.9

b Explain the reasoning behind this idea.

c To check, he connects the digital voltmeter across YZ and finds that the reading is 4.5 V. He then *disconnects* the voltmeter, and connects the lamp in its place.

 i Draw a diagram of the circuit showing the lamp connected as described and explain why the lamp, which is not defective, does not light.

ii He checks the voltage difference across the lamp with the digital voltmeter and finds that the reading is only 1.5 V. Use this fact to show that the resistance of the lamp filament is 3.0 Ω, when connected as described.

iii How do you account for two different values of the resistance of the lamp calculated in part **a ii** and part **c ii**?

11 Figure 4.10 shows a circuit containing the components for which the characteristics are shown in Figures 4.11 and 4.12. The diodes are identical.

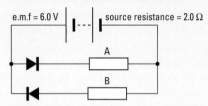

Figure 4.10

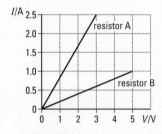 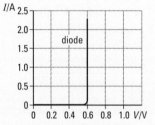

Figure 4.11   Figure 4.12

Determine

a the potential difference across a diode when conducting.

b i the current through resistor A,

ii the current through resistor B.

12

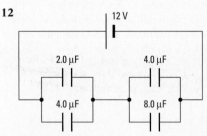

Figure 4.13

Four capacitors are connected to a 12 V battery, as shown in Figure 4.13.

Calculate

a the effective capacitance of the combination of capacitors,

b the total energy stored in the capacitors.

13 a A capacitor of capacitance **C** is charged so that the potential difference between its terminals is **V**. Show that the energy stored by the capacitor is given by:

$$\text{energy} = \tfrac{1}{2}CV^2$$

b A capacitor whose capacitance can be varied is set to its maximum value of 500 pF and connected to a 12 V battery so that it becomes fully charged. The capacitor is now removed from the battery and its terminals are isolated.

$1\ \text{pF} = 10^{-12}\ \text{F}$

i Calculate the charge stored by the capacitor.

The capacitance is now changed to a value of 250 pF. If no charge leaves the capacitor in this process, calculate

ii the change in potential difference across the capacitor,

iii the change in energy stored by the capacitor.

14 This question is about the transfer of energy in a circuit containing a capacitor, a resistor and two centre-zero galvanometers.

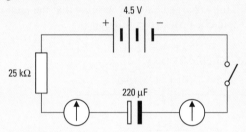

Figure 4.14

In the circuit shown in Figure 4.14, the battery and galvanometers have negligible resistance. Initially the capacitor is uncharged.

a Describe what you would observe on the two galvanometers after the switch has been closed.

b Now assume that the switch has been closed for several minutes. Show that the charge stored by the capacitor is about 10^{-3} C.

c As charge moves around the circuit, some but not all of the energy transferred from the battery is stored as potential energy in the capacitor.

i How much energy is transferred to each coulomb of charge flowing through the battery?

ii Hence calculate the work done by the battery during the charging of the capacitor.

iii Calculate the potential energy stored in the fully charged capacitor.

d Account for the difference in your answers to **c ii** and **iii**.

15 This question is about the use of a large value capacitor to provide a back-up energy source for a solar-powered wristwatch.

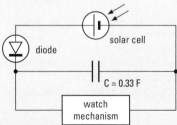

Figure 4.15

Figure 4.15 shows the basic circuit.

a In direct sunlight, the solar cell supplies power to the watch mechanism and charges the capacitor C to a voltage $V = 2.4$ V.

i Calculate the charge on the charged capacitor.

ii Calculate the energy stored in the charged capacitor.

b In conditions of poor light, the voltage produced by the solar cell drops to zero and the capacitor acts as a back-up power supply, discharging through the watch mechanism. The watch mechanism will cease to function if the voltage across it falls below 1.0 V.

   i What is the purpose of the diode in the circuit?

   ii Calculate the charge which will have flowed through the watch mechanism when the voltage across the capacitor has fallen to 1.0 V.

   iii The watch mechanism is designed to draw a constant current of 1.0 µA, as long as the voltage across it is greater than 1.0 V. Use this fact, and your answer to **b ii**, to estimate how long the capacitor can 'back-up' the watch mechanism. Express your answer in hours.

c Calculate the average power delivered by the capacitor to the watch mechanism during 'back-up'.

16 a An engineer required a 20 µF capacitor which operated safely using a voltage of 6.0 kV.

   Ten similar capacitors connected in series were found to provide a suitable arrangement when operating at their maximum safe voltage.

   i State the maximum safe working voltage of each capacitor.

   ii Determine the capacitance of each capacitor.

   iii Calculate the total energy stored by the combination of capacitors when fully charged to 6.0 kV.

b The circuit shown in Figure 4.16 was used in an experiment to investigate the charging of the series combination. The graph in Figure 4.17 shows how the potential difference V_R across the resistor varied with time, t, after the switch was closed.

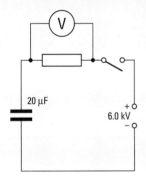

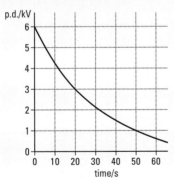

Figure 4.16　　　　*Figure 4.17*

   i Use the graph to determine the potential difference across the capacitor 30 s after charging commenced.

   ii At what time after charging commenced were the potential differences across the capacitor and the resistor equal?

   iii Determine the resistance of the resistor used in the experiment.

Calculus is not required to answer this question.

17 The diagram shows a parallel-plate capacitor being used as part of the mechanism in an electronic balance. A weight placed on the balance plate causes one of the plates of the capacitor to move closer to the other. When the balance is turned on the plates are charged up and then isolated. A sensitive, high-resistance voltmeter is used to monitor the p.d. between the plates. As the plates are moved together, the capacitance changes. This causes the p.d. between the plates to change. The p.d. becomes a measure of the weight placed on the balance.

a Why must a high-resistance voltmeter be used?

b What is the purpose of the spring?

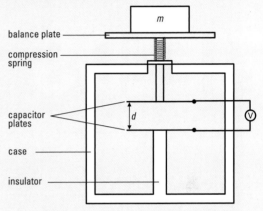

Figure 4.18

It can be shown that the p.d. between the capacitor plates is given by:

$$V = \frac{Qd}{\varepsilon_0 A} - \frac{mgQ}{\varepsilon_0 Ak}$$

where Q is the charge on the plates, A is the area of each plate, d is the separation of the plates when there is no mass on the balance plate, and ε_0 is a constant called the **permittivity of free space** ($\varepsilon_0 = 8.85 \times 10^{-12}\,\text{F m}^{-1}$).

c Show that this equation is dimensionally correct.

d Sketch a graph showing how V varies with the mass placed on the balance.

e The sensitivity of the balance means how well it can measure small differences in mass. Add to your sketch graph a second line for a more sensitive balance.

f What mechanical change could you make to produce a more sensitive balance?

g What would be the disadvantage of doing this?

h How might you alter the capacitor in order to make the balance more sensitive?

5

FIELDS

The history of field ideas is a very interesting one. James Clerk Maxwell's (1831–1879) theory of electromagnetism brought about the first real understanding of the nature of the electromagnetic field, but Maxwell himself thought in very mechanical terms unrelated to the field ideas developed from his mathematical treatment. Fields are, by their nature, abstract things and seem quite unlike matter. This is why students find them difficult to understand – so did the pioneering physicists who worked in this area. Fortunately we have many technological aids to help us grasp the reality of fields, levitating frogs among them!

Classical physics (that is, physics prior to quantum theory and relativity) recognized the existence of particles of matter and fields of force. The particles interacted with one another by means of forces that were described in terms of non-material fields. In modern physics the merging of quantum theory and relativity has led to a blurring of the distinction between field and particle. Photons are regarded as particles, yet they are also connected with the electromagnetic field. Similarly electrons can be linked to an electron field by their quantum nature.

In this chapter you will be gently introduced to the idea of the field with examples from electricity and gravitation. You will then see how powerful the idea is in explaining a variety of electromagnetic phenomena and practical applications.

A (live) frog suspended in a magnetic field of about 16 tesla (over 300 000 times stronger than that of the Earth) produced by a so-called Bitter solenoid in the University of Nijmegen's High Field Magnet Laboratory.

THE GRAVITATIONAL FIELD

Why things fall

Physics attempts to explain phenomena in the simplest and most general way. Here are four attempts to explain why things fall.

Aristotle (384–322 BC) Things try to reach their natural resting place. For solid 'earthy' bodies this would be the Earth's centre, so they fall towards it.

Newton (1642–1727) All masses exert attractive forces on all other masses. Near the Earth this force is almost entirely an attraction to the Earth, so if something is dropped it falls towards the centre of the Earth.

Newtonian field theory All masses create a gravitational field in the space surrounding them. This field exerts a force on any mass placed in it, directed toward the centre of the mass which created the field. Near the Earth this is the centre of the Earth.

Einstein (1879–1955) The gravitational field is really a distortion of the geometry of space and time. A free object moves along the shortest path through space–time. These paths are curved near masses and tend to bring them together.

Einstein's theory is beyond the scope of this book, but that is not a big problem because just about every effect of gravity that has any immediate bearing on human beings is accurately modelled by the Newtonian field theory.

Gravity is mysterious. Something in the space surrounding a star or planet (or anything with mass) 'grabs hold of' other masses, bungee jumpers, for example, and accelerates them.

Jargon buster: fields

The term **field** is used whenever a value can be associated with all points in some region of space. For example, you could talk about the **temperature field** in your body since there is a particular value for temperature at each position. This would be an example of a *scalar field* since temperature is a scalar quantity.

A field theory

It may not be immediately obvious how Newtonian field theory differs from Newton's original theory, but there is a very important distinction. The field exists in space and exerts a force on things *where they are*. Newton's original theory of gravity involved action at a distance. This involves forces being exerted by one object on another with no physical connection between them. Newton himself was well aware that this seemed rather spooky. Nowadays the gravitational field is often discussed as though it really exists, but it is worth remembering that there is no way to directly observe it; its strength is always determined by making measurements of gravitational effects on masses, so we should be careful about claiming that it is *really* there! The modern view is that the gravitational field will eventually be described by quantum field theory in which the force is transmitted between masses when they exchange '**virtual gravitons**'. No one has yet observed a graviton.

Representing the gravitational field

What properties can a field have? The primary effect of a field is to exert forces on particular kinds of objects, so it must have strength and direction at each point in space. This is described by the **gravitational field strength** g. When masses are moved in a gravitational field their potential energy in the field changes, so an alternative and complementary description uses the way potential energy changes in the field. This is **gravitational potential**.

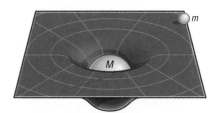

the trajectory of a free body is the shortest path through curved spacetime

m

M

Einstein described gravity as a distortion in space–time geometry. In this view projectiles follow the shortest possible path (a geodesic) through curved space–time.

Definition: gravitational field strength

Gravitational field strength is the gravitational force exerted per unit mass at a point in the field, and like force it is a vector quantity:

$$g = \frac{F}{m}$$

The units of g are $N\,kg^{-1}$.

Field strength is a property of the field. It is independent of the test mass.

Field lines

The idea of a gravitational field is like Faraday's idea of an electric or magnetic field, and it can be pictured in a similar way. Field lines point in the direction a free body would begin to accelerate if released. Their density is a measure of the field strength which is equivalent to the free-fall acceleration at each point.

$$a = \frac{F}{m} = \frac{mg}{m} = g$$

Although the possibility of gravitational repulsion has not been ruled out, none has so far been discovered, so all gravitational field lines come from infinity and end on masses.

> **Definition: gravitational potential**
>
> **Gravitational potential** is the gravitational potential energy (GPE) per unit mass at a point in the field:
>
> $$V_g = \frac{GPE}{m}$$
>
> The units of V_g are $J\,kg^{-1}$.
>
> Potential is also a property of the field. Potential is a scalar quantity.

Equipotential surfaces

A table top is an example of an **equipotential surface**, since it takes a fixed amount of energy to lift a 1 kg mass from the floor to the table top, but no further work must be done (against gravitational forces) to move it across the surface of the table. The table top (or equipotential surface) is horizontal because the gravitational field strength is vertical. (*Equipotential surfaces are always perpendicular to field lines.*)

If you lift something in a gravitational field you must apply a force to oppose gravity, and the work you do is stored as gravitational potential energy (GPE). If something falls, gravity does work on it, reducing its GPE. If you move something perpendicular to the gravitational field lines then no work is done by the field on the mass or by you against forces from the field. This leaves the GPE of the mass unchanged, and the potential of the field is constant. Equipotential surfaces are perpendicular to the field lines. Since GPE changes more rapidly in a strong field than in a weak field the equipotentials (which are standardly drawn at regular intervals of field strength) will be closer together where the field is strongest (that is, where the density of field lines is greatest).

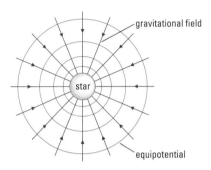

An equipotential line is always perpendicular to the gravitational field lines.

> **Zero potential**
>
> It is often convenient to measure potential and GPE relative to a local zero, for example, at the ground or on a laboratory bench. For theoretical work the zero of potential is defined to be at infinity.

PRACTICE

1 What is the gravitational field strength at the centre of the Earth? Justify your answer.

2 The gravitational potential is not zero at the centre of the Earth. Why not? Think of the work done in moving a 1 kg mass from infinity to the centre of the Earth.

3 Earth, Mars, and Jupiter have gravitational field strengths at their surfaces of $9.8\,N\,kg^{-1}$, $3.7\,N\,kg^{-1}$, and $25\,N\,kg^{-1}$ respectively.

  a What is the weight of a 75 kg man on the surface of each planet? (You could not stand on the surface of Jupiter because it is gaseous.)

  b How much work must be done to lift the man 10 m above the surface in each case?

4 Less fuel is needed to escape from Mars than from the Earth. What does this tell you about the gravitational potentials at the surfaces of Mars and Earth?

5 Stars often form in binary systems that revolve about their common centre of mass. Sketch field lines and equipotentials for a binary system in which both companions are approximately the same mass.

- the inverse-square law of force
- spheres, points, and shells
- 'big G'

Attractive forces

The negative sign in the equation for gravitational force is merely a mathematical way of describing an attractive force. The force on m_2 is in the opposite direction to the displacement of m_2 from m_1.

Schiehallion, a mountain in the Scottish Highlands, was chosen by the Astronomer Royal, Nevil Maskelyne, in 1774 because of its symmetrical shape for an ingenious experiment to measure the density of the Earth. The inclination from the vertical of a suspended test weight either side of the mountain due to its gravitational field enabled Maskelyne to calculate a value for G: $5.9 \times 10^{-11} \, N \, m^2 \, kg^{-2}$, from which he calculated that the mass of the Earth is $4.9 \times 10^{24} \, kg$. Accepted values are now $G = 6.6732 \times 10^{-11} \, N \, m^2 \, kg^{-2}$ and $M_E = 5.983 \times 10^{24} \, kg$.

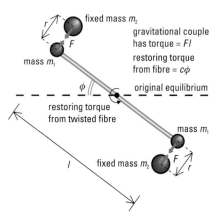

Cavendish's experiment balanced a gravitational couple against the restoring torque from a twisted fibre.

NEWTON'S LAW OF GRAVITATION

The law

The law of gravitation is attributed to Newton, but his claim to priority was contested by Robert Hooke. In an ingenious experiment, Hooke took a balance high up inside Westminster Abbey and weighed an object in one pan. He then repeated the measurement, but this time the same object was suspended from the pan so it was closer to the Earth and therefore subjected to a stronger gravitational force. His experiments were inconclusive. His arguments with Newton were bitter.

The inverse-square law of gravitation

All masses exert an attractive force on all other masses.

Two point masses m_1 and m_2 separated by a distance r attract each other with a force F directly proportional to the product of the masses and the inverse square of their separation:

$$F \propto \frac{m_1 m_2}{r_2}$$

$$F = -\frac{G m_1 m_2}{r_2}$$

'Big G' is the **universal constant of gravitation** and is a measure of the strength of gravitational forces. It is equal to the force in newtons exerted on a pair of 1 kg masses separated by 1 m.

$$G = 6.673 \times 10^{-11} \, \text{N m}^2 \, \text{kg}^{-2}$$

This tiny value explains why gravitational forces between small objects are usually completely insignificant (unlike the electromagnetic forces which hold us together).

Worked example 1

What (approximately) is the gravitational force between two people standing 2 m apart?

Take $m_1 = m_2 = 70 \, \text{kg}$; then

$$F = -\frac{G m_1 m_2}{r^2} = -\frac{6.7 \times 10^{-11} \, \text{N m}^2 \, \text{kg}^{-2} \times 70 \, \text{kg} \times 70 \, \text{kg}}{(2 \, \text{m})^2} = 8 \times 10^{-8} \, \text{N}$$

Measuring G

In principle it is very easy to measure G. Measure the force needed to hold two known masses a measured distance apart and then substitute these values into Newton's equation. In practice it is very difficult to do this accurately because the forces are so small.

The first direct measurement of G was made by Henry Cavendish in 1798 using a torsion balance. Two small masses are attached to the ends of a light horizontal rod which is suspended at its centre by a fine fibre. Two large lead spheres are then placed close to the small masses in such a way as to make the balance rotate to a new equilibrium position. In this position the gravitational torque is balanced by the torsional torque in the twisted fibre. The angle of rotation is a measure of this torque and of the force of gravity between the pairs of masses. To calculate its value, Cavendish needed to know the torsional spring constant for the fibre. This can be found by measuring the time period of free oscillations for the rod and small masses and using the equations of simple harmonic motion (see spreads 3.34 and 3.36).

Points, spheres, and shells

Newton's law gives the force between two point masses. However, real objects are extended in space and every point mass in one body attracts every point mass in all other bodies. In principle this means a huge and often very complicated calculation would need to be carried out to calculate the resultant gravitational force on just about anything. This problem is avoided in the following ways.

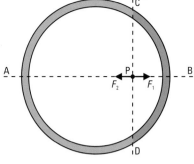

- If the objects are small compared to the distance that separates them they can be approximated as point masses at their centre of mass. The trajectory of a spacecraft is calculated using this assumption. The same assumption is made when calculating someone's weight, the radius of the Earth being much greater than the size of a person.

- Uniform spherical masses can be treated like point masses at their centre of mass. The Sun, Earth, and other planets and moons are very nearly spherical and can usually be treated like this. This is accurate even if their density is not uniform as long as it is spherically symmetric. This result is true only for points outside the surface of the sphere.*

- The gravitational effects of a spherically symmetric hollow shell cancel out at all points inside the sphere. This is useful for calculating gravitational forces inside the Earth (see spread 5.3) or inside a large gas cloud in space.*

(*These rigorous results can be proved using calculus. If you think your mathematics is up to it you might try to do this. Newton struggled with this problem and it delayed the publication of his theory.)

The gravitational force acting on a mass at P is zero. This is because the force F_2 due to the mass in the blue region DAC is balanced by F_1 due to the green region DBC. Although there is more mass in the blue region (as it is bigger) much of it is further away from P than that in the green region.

Worked example 2

The Earth has mass 6.0×10^{24} kg. The Moon has mass 7.4×10^{22} kg, and orbits at a mean distance of 3.8×10^8 m. How far from the Earth will the attraction to the Moon be equal and opposite to the attraction to the Earth?

If a mass m is placed at this point the magnitude of gravitational forces from Earth and Moon will be equal. Let the distance from Earth be x and the distance from the Moon be y.

$$\frac{GM_{\mathrm{E}}m}{x^2} = \frac{GM_{\mathrm{M}}m}{y^2} \qquad \frac{x^2}{y^2} = \frac{M_{\mathrm{E}}}{M_{\mathrm{M}}} \qquad \frac{x}{y} = \sqrt{\frac{M_{\mathrm{E}}}{M_{\mathrm{M}}}} = \sqrt{\frac{6.0 \times 10^{24}\,\mathrm{kg}}{7.4 \times 10^{22}\,\mathrm{kg}}} = 9.0$$

Now, $x = 9y$, so this point is $\frac{9}{10}$ of the way from the Earth to the Moon. Hence

$$x = 0.9 \times 3.8 \times 10^8\,\mathrm{m} = 3.4 \times 10^8\,\mathrm{m}$$

PRACTICE

1 Show that $\mathrm{N\,m^2\,kg^{-2}}$ are appropriate units for G.

2 What is the gravitational attraction between a spherical asteroid of mass 10^{13} kg and radius 1 km and a 70 kg human standing on its surface?

3 a At what altitudes above the Earth (in terms of the Earth's radius) would your weight be **i** half its surface value, and **ii** one percent of its surface value?

   b True or false: 'Astronauts are weightless because they are outside the Earth's atmosphere.' Explain.

4 Refer to the diagram of Cavendish's apparatus. Derive an equation for G in terms of the angle ϕ that the balance rotates, the torsional spring constant c, the length of the rod l, the masses m_1 and m_2, and the distance between the pairs of masses r.

5 If gravitational forces are so weak how can they be responsible for the large-scale structure of the universe?

6 The force between two 1 kg spheres separated by 1 m is tiny (given by G). Now scale up the system by a linear factor of 10^9. By what factor do each of the following increase?

   a the radius of each sphere;

   b the volume of each sphere;

   c the mass of each sphere (assume density does not change);

   d the gravitational attraction between the spheres.

   How does this back up your answer to question 5?

7 Estimate the difference in weight between 1 kg close to the floor, and the same mass close to the roof of Westminster Abbey.

8 How much would a 75 kg person weigh:

   a on Earth (take g as $9.8\,\mathrm{N\,kg^{-1}}$)?

   b on the Moon (mass of Moon is 7.4×10^{22} kg and its radius is 1.7×10^6 m)?

   c at the surface of Jupiter where free-fall acceleration is $25\,\mathrm{m\,s^{-2}}$?

9 Discuss how you would expect a material which has negative mass to behave. (There is more than one possibility here: mass determines both gravitational effects and inertia, and negative mass may affect either or both of these.)

5.3

O B J E C T I V E S

- radial force and field strength
- the Earth's field
- uniform field approximation

Don't confuse

A lot of people confuse field strength and force, especially since their formulae are so similar. Field strength measures a property of the field *out there* at a point in space and can result in *any* value of force. For example, at the Earth's surface a 1 kg mass weighs 9.8 N, whereas 500 kg weighs 4900 N at the same point.

The symbol g

The same symbol g is often used in two different ways: as the field strength at the Earth's surface, and as the symbol for field strength anywhere in space. Confusion can arise because the surface value is usually quoted as a magnitude only, that is, $g = 9.8\,\text{N kg}^{-1}$ rather than $-9.8\,\text{N kg}^{-1}$. When solving problems it is wise to check that the sign of your answer makes sense in the context of the question! Often it is simpler to use the magnitude of g and work out directions by inspection.

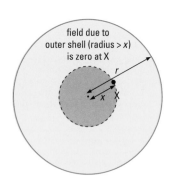

field due to
outer shell (radius > x)
is zero at X

Working out gravitational field strength inside a solid sphere.

FORCE AND FIELD STRENGTH –
1. GRAVITATION

Radial field strength

The gravitational force on a test mass m a distance r from the centre of a spherical symmetric mass M is

$$F = -\frac{GmM}{r^2} \quad \text{(the negative sign indicates attraction)}$$

The gravitational field strength is given by

$$g = \frac{F}{m} = \frac{-GM}{r^2} \tag{1}$$

This formula applies outside a sphere or point mass. It is an inverse-square law, which means if the distance is increased by some factor then the strength of the field is reduced by that factor squared. For example, the radius of the Moon's orbit is about 60 times the radius of the Earth and the strength of the Earth's field at this distance is about $0.0027\,\text{N kg}^{-1}$, which is about 3600 (that is, 60^2) times smaller than the field strength at the Earth's surface (about $9.8\,\text{N kg}^{-1}$).

Worked example

The gravitational field strength at the Earth's surface is $-9.8\,\text{N kg}^{-1}$, so the mass of the Earth is:

$$M = -\frac{gr^2}{G} = \frac{-(-9.8\,\text{N kg}^{-1}) \times (6.4 \times 10^6\,\text{m})^2}{6.7 \times 10^{-11}\,\text{N m}^2\,\text{kg}^{-2}} = 6.0 \times 10^{24}\,\text{kg}$$

Inside a solid sphere

The inverse-square law describes how g falls with distance as you leave the surface of the Earth, but what if you journey to the *centre* of the Earth? How does the gravitational field strength vary *inside* a solid sphere? It will be zero *at* the centre of the sphere, since any other value would have to point in a particular direction and that would violate the symmetry of the sphere. This suggests the field strength falls from its surface value to zero as we burrow into the sphere. The way it changes will depend on how the density of the sphere changes the deeper we go. The simplest model is a uniform sphere of constant density throughout.

Imagine a test mass m placed at a distance x from the centre of the sphere where x is less than the radius r. The spherical shell of matter outside x will have no effect on m, so the force (F_x) at radius x is due entirely to the inner sphere:

$$F_x = -\frac{GM_x m}{x^2} \quad \text{(magnitude)}$$

$$g_x = -\frac{GM_x}{x^2}$$

(M_x is the mass of a sphere of radius x.) At first glance this looks like another inverse-square law, but the mass of the inner sphere, M_x, depends on x:

$$M_x = \frac{4}{3}\pi x^3 \rho \quad \text{(the volume of the inner sphere times its density, } \rho\text{)}$$

Therefore

$$g_x = -\frac{4}{3}G\pi\rho r \tag{2}$$

g_x increases linearly from zero at the centre to its surface value:

$$g_r = -\frac{4}{3}G\pi\rho r$$

The Earth's gravitational field

The diagram below shows how the magnitude of g varies along a line passing through the centre of the Earth (ignoring the contributions of other bodies such as the Sun and Moon). The reason it differs from the theoretical formulae derived above is because its density increases toward its centre.

Our experience of the Earth's field is usually limited to a space measured in tens of metres. This is tiny compared to the Earth's 6400 km radius, so we do not experience the inverse-square law or the divergence of the Earth's field lines. Locally, the field is almost uniform and can be represented as parallel equally spaced field lines. This is the region in which the expression $\Delta GPE = mgh$ can be used to calculate changes of gravitational potential energy.

Gravitation has infinite range, so every object in the universe is affected by the gravity of every other object. However, changes in the gravitational field travel at the speed of light, so what we do now may not affect beings in a distant galaxy for millions or even billions of years.

A view of the Earth–Moon system from the Galileo *space probe. The Earth and Moon form a binary system bound together by their mutual gravitational attraction. They orbit around their common centre of mass which, because of the Earth's much greater mass, is actually inside the Earth.*

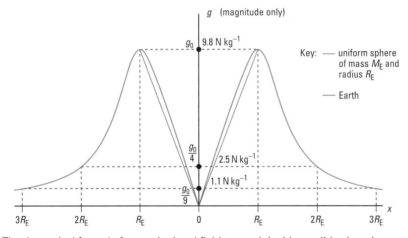

The theoretical formula for gravitational field strength inside a solid sphere is inaccurate when applied to the Earth, because the Earth's density varies.

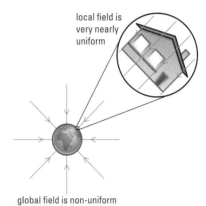

In everyday life we experience the Earth's gravitational field as practically uniform.

PRACTICE

1 What is the gravitational field strength at the surface of Jupiter (mass 1.9×10^{27} kg, radius 7.1×10^7 m)? How would your mass and weight on Jupiter compare to their values on Earth?

2 Show that the expression for g obtained by putting $x = r$ in equation (2) is equivalent to that obtained by putting $x = r$ into (1).

3 Sketch how g would vary with radius from the centre of the Earth if the Earth had a large spherical cavity at its centre.

4 a How far from the Moon is the **Lagrange point**, where the Earth and Moon's gravitational field strengths are equal in magnitude but opposite in direction? ($M_E = 6.0 \times 10^{24}$ kg; $M_M = 7.4 \times 10^{22}$ kg; mean radius of Moon's orbit $= 3.8 \times 10^8$ m.)

 b Explain why the Moon's gravity has a more significant effect on the Earth's tides than the Sun's gravity. (The Sun has a mass of 2.0×10^{30} kg and is 1.5×10^{11} m from the Earth.)

5 If the universe was infinite in extent and of uniform density throughout, what would the gravitational field strength at any point be?

6 If the Sun collapsed to a radius of 3 km it would become a black hole.

 a Calculate the gravitational field strength at this radius.

 b What would a 75 kg person weigh here?

 c How much does the gravitational field strength change over the length of a human (about 2 m) at this radius? (Assume the human is falling towards the hole on a radial trajectory.)

 d Would this difference in field strength (a **tidal effect**) have any noticeable effect on the person?

7 The value of g decreases as we go deeper into the Earth. How far vertically must we go to reduce it by 0.1%? (Assume that the Earth has constant density and say whether in reality we would have to go deeper or less deep.)

8 The Earth is not a uniform sphere. It is an oblate spheroid bulging at the equator. It is not of constant density – its density increases toward its centre – and it spins. Explain whether any or all of these affect the actual value of g at the Earth's surface, and the value of g we would measure using a spring balance to weigh a 1 kg mass.

OBJECTIVES

- potential in a radial field
- link between field and potential
- energy changes

Glen Canyon Dam, Arizona, USA, on the Colorado river. The hydroelectric station at its base has a maximum power generation of 1356 MW.

Zero at infinity

GPE and V_g are always measured relative to some reference position where their values are defined to be zero. For example, if you lift a case from the floor to a table and then let it fall back, the obvious choice for the zero position is the floor, even if you happen to be at the top of a twenty storey building. But the floor of your building obviously won't do for a universal standard; it is too arbitrary.

Gravitational potential energy and potential are defined to be zero at infinity.

Why infinity? The answer is because it gives an idealized reference place which is the same for all gravitational fields. The field strength due to all bodies falls to zero at infinity so it is a reasonable choice for our reference of potential.

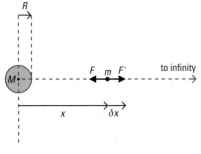

Calculating gravitational potential.

POTENTIAL ENERGY AND POTENTIAL

Energy in the gravitational field

Hydroelectric power harnesses gravitational potential energy (GPE) and converts it to electricity. Water is moving from a place (high up) where it has a large amount of GPE to a place (low down) where it has much less. The GPE is transferred by gravitational forces that do work (force times distance fallen) on the water as it falls. The water in turn does work on the blades of a turbogenerator which generates electricity by electromagnetic induction (see spread 5.23). The amount of work done depends on the quantity of water that falls and the distance through which it falls, but the work done per kilogram depends only on the displacement of the water in the gravitational field.

The change in GPE per kilogram between two points in a gravitational field is called the gravitational potential difference and is a *property of the field* independent of the mass that moves between those two points. (This is exactly analogous to potential difference or voltage drop in an electric circuit, which measures the work done per unit charge between two points.)

Definitions

Gravitational potential energy (*GPE*): the energy stored (J) as a result of a body's position in the gravitational field.

Gravitational potential (V_g): the GPE per unit mass at a point in the field:

$$V_g = \frac{GPE}{m}$$

Gravitational potential difference (ΔV_g) is the difference in gravitational potential between two points in a gravitational field and is equal to the work done per unit mass in moving it between the two points.

$$\Delta V_g = V_g{}^2 - V_g{}^1 = \frac{\Delta GPE}{m} = \frac{W}{m}$$

This leads to the very useful equation

$$W = m\Delta V_g$$

Maths box 1: calculating gravitational potential

Once the zero is defined, gravitational potential is calculated in the following way.

Calculate how much work must be done to lift a test mass m from the point at which you want to know the potential to infinity. Since all known gravitational forces are attractive this will be a positive amount of work. You have to *increase* the GPE of the mass to take it to infinity where its GPE is zero. Its original position must have been one of *negative* GPE (equal to minus the work you had to do), and therefore negative gravitational potential. All values of GPE and V_g are negative.

If a test mass m is a distance x from a mass M the gravitational force on it is

$$F = -\frac{GMm}{x^2}$$

To lift it you must apply a force F' equal and opposite to F. The work done by F' in moving a distance δx is

$$\delta W = F'\delta x = \frac{GMm\delta x}{x^2}$$

The total work done moving m from r to ∞ is

$$W = \int_r^\infty \frac{GMm}{x^2}\,dx = \left[-\frac{GMm}{r}\right]_r^\infty = 0 - \left[-\frac{GMm}{r}\right] = \frac{GMm}{r}$$

$$GPE = 0 - W = -\frac{GMm}{r} \qquad V_g = -\frac{GM}{r} \ \ \text{J kg}^{-1}$$

As the calculation shows, the gravitational potential at any point can be defined as:

$$V_g = -[\text{work done in bringing unit mass from infinity to a point}]$$

NB Do not confuse this relation for potential, $1/r$, with the relation for field strength, $1/r^2$.

Maths box 2: when to use '*mgh*'

When something moves from A to B in a gravitational field the change in its GPE is given by

$$\Delta GPE = m\Delta V_g = m(V_B - V_A)$$
$$= \frac{-GMm}{r_B} - \frac{-GMm}{r_A} \quad \text{(in the field of a central mass } M\text{)}$$
$$= GMm\left(\frac{1}{r_A} - \frac{1}{r_B}\right) = GMm\left(\frac{r_B - r_A}{r_A r_B}\right)$$

If r_A and r_B are very close to some value r but have a small separation h then their product is approximately r^2 and their difference h:

$$\Delta GPE \to \frac{GMmh}{r^2} \quad \text{and since} \quad g = -\frac{GM}{r^2} \quad \text{we have} \quad \Delta GPE = -mgh$$

In practice, everybody uses the magnitude of g so that

$$\Delta GPE = mgh \quad \text{when} \quad h \ll r$$

This justifies using the equation near the surface of the Earth for vertical displacements that are small compared with R_E (6400 km).

The Earth's potential

Outside its surface the Earth's potential varies as $-1/r$, as in the formula above. The negative sign is important. It means work must be done to increase the GPE of a mass in order to move it away from the Earth. You experience this if you walk up a hill. Because of this the majority of the mass of a rocket when it is launched is its fuel – the 'energy tax' it must pay to climb out of the Earth's **'gravitational well'**.

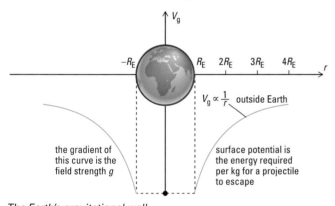

The Earth's gravitational well.

Maths box 3: field and potential

GPE changes when work is done by or against gravitational forces. The stronger the forces the more work is done per metre moved, so GPE changes rapidly from place to place where the forces are strong. This implies a link between gravitational potential and gravitational field strength.

Let a test mass move through a distance δx in a gravitational field. The work done by the field on the mass is

$$\delta W = F\delta x = mg\delta x$$

Since this is work done by the field it must reduce the GPE of m in the field so

$$\delta GPE = -mg\delta x$$

Dividing by m gives

$$\delta GPE = -g\delta x \quad \text{or} \quad g = -\frac{\delta V_g}{\delta x}$$

In the limit $\delta x \to 0$, $g = -\dfrac{dV_g}{dx}$

In words this means: gravitational field strength = negative potential gradient.

This is a very important result. (A similar result holds in electrostatics.) The negative sign indicates that gravitational field strength always points towards lower potentials.

A good way to think about this is in terms of cars parked on hills; the tendency is always to slip down the hill, never up. Also, the force tending to make the car slip depends on the steepness of the slope, that is, the rate at which GPE increases.

Summary of equations

Equation:	Magnitude of:
$F = \dfrac{-Gm_1m_2}{r^2}$	force between two point masses (N)
$g = \dfrac{F}{m}$	gravitational field strength (N kg^{-1})
$V_g = \dfrac{GPE}{m}$	gravitational potential (J kg^{-1})
$W = \Delta GPE = m\Delta V_g$	work done moving mass in a gravitational field (J)
$\Delta GPE = mgh$	change in GPE in a uniform field (g constant)
$g = \dfrac{-GM}{r^2}$	gravitational field strength distance r from a spherical mass M
$V_g = \dfrac{-GM}{r}$	gravitational potential distance r from a spherical mass M

PRACTICE

1 Calculate the gravitational potential at the Earth's surface ($M_E = 6.0 \times 10^{24}$ kg, $RE = 6400$ km).

2 What is the change in gravitational potential per vertical metre near the Earth's surface? Comment on your answer.

3 Calculate the change in GPE of an 80 kg astronaut when she goes from the surface of the Earth to the orbiting Hubble Space Telescope, about 300 km above the surface:

  a assuming an approximately uniform field;

  b using the equations for potential in a radial field.

4 What is the overall change of GPE and gravitational potential for an astronaut between leaving home and returning home after a space mission? Is this result true for any closed loop in a gravitational field?

Changes in gravitational field strength before an eruption can be monitored by sensitive gravimeters and may give an early warning of potential disasters.

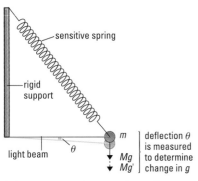

Laboratory gravimeters depend on accurate measurements of free fall using the 'suvat' equations. Portable gravimeters such as this used for prospecting usually balance the moment of a mass fixed on a pivoted rod against the restoring moment of a weak spring attached to the same rod.

A commercial gravimeter from the 1990s.

THE EARTH IN SPACE

Weighing the Earth

Once *G* has been measured, Newton's law can be used to weigh the Earth (and hence find its mass). It is a particularly simple experiment. Suspend a known mass from a spring balance and measure its gravitational attraction to the Earth. This should be

$$F = \frac{GM_Em}{R_E^2} \quad \text{so} \quad M_E = \frac{FR_E^2}{Gm} \quad \text{(magnitudes only)}$$

Try it! ($R_E = 6.37 \times 10^6$ m, $G = 6.67 \times 10^{-11}$ N m^2 kg^{-2}.)

Measuring *g*

The value of *g* at the surface of the Earth is not constant. It varies by about 0.5% and is strongest at the poles. This is because the Earth is neither perfectly spherical nor of uniform density. It is flattened because of its rotation, and local geology and topology will affect the magnitude and direction of *g* from place to place. Blips or anomalies in the otherwise smooth field indicate different geological formations beneath the surface and are used to prospect for minerals and oil. Some recent research has used variations in *g* caused by magma moving beneath the surface to try to predict volcanic eruptions. For these purposes a gravimeter must be capable of detecting variations as small as one ten-millionth of *g*.

Escape into space

Throw a ball vertically upwards and it will eventually come back down. Throw it harder and it goes higher and stays up longer, but it still comes back down. Ignoring friction from the air, the only energy changes taking place during its up and down motion are kinetic energy (KE) to gravitational potential energy (GPE) on the way up and GPE to KE on the way down. The result is that it hits your hand moving down at the same velocity with which you threw it. Throughout its motion its total energy (TE) remains constant:

$$TE = KE + GPE$$

The GPE of the ball near the Earth is negative. The GPE at infinity is zero. The reason the ball fails to escape is that it wasn't given enough KE at the start to make its total energy greater than or equal to zero. How fast *would* it have to be thrown (again ignoring friction) to escape?

Maths box: calculating escape velocities

If an object escapes to infinity its GPE there will be zero (by definition). If the object only just gets there its KE will be zero as well. Hence the condition for just escaping is

$$TE = KE + GPE = \frac{1}{2}mv^2 - \frac{GM_Em}{R_E} = 0$$

$$\frac{1}{2}mv^2 = \frac{GM_Em}{R_E}$$

$$v = \sqrt{\frac{2GM_E}{R_E}}$$

$$v_{esc} = \sqrt{\frac{2GM_E}{R_E}}$$

(Notice that *m* has cancelled – escape velocity is the same for all masses.)

$$v_{esc} = \sqrt{\frac{2 \times 6.7 \times 10^{-11}\,\text{N m}^2\,\text{kg}^{-2} \times 6.0 \times 10^{24}\,\text{kg}}{6.4 \times 10^6\,\text{m}}} = 11\,000\,\text{m s}^{-1} \text{ for Earth}$$

$$v_{esc} = \sqrt{\frac{2 \times 6.7 \times 10^{-11}\,\text{N m}^2\,\text{kg}^{-2} \times 7.4 \times 10^{22}\,\text{kg}}{1.7 \times 10^6\,\text{m}}} = 2400\,\text{m s}^{-1} \text{ for the Moon}$$

The equation for escape velocity can be written in terms of surface gravity:

$$g_0 = \frac{GM}{R^2} \quad \therefore \quad G = \frac{g_0R^2}{M} \qquad v_{esc} = \sqrt{\frac{2GM}{R}} = \sqrt{2g_0R}$$

No escape!

Back in the eighteenth century, John Mitchell (who discovered the inverse-square law between magnetic poles) and Pierre Laplace (who did major work in celestial dynamics and probability theory) both speculated that if a star was sufficiently massive, its escape velocity would equal or exceed the speed of light. Such a star would neither reflect nor radiate light; it would be truly black. In the twentieth century Einstein showed that nothing can exceed the speed of light and massive bodies can never reach the speed of light. There would be no escape from such a black hole. Although the full treatment of black holes needs Einstein's general theory of relativity, many important results can be derived from Newtonian gravitation, in particular the radius at which a body of mass M becomes a black hole when it is squashed. This is known as the **Schwarzschild radius** (R_s).

For the body to become a black hole the escape velocity must equal or exceed the speed of light:

$$v_{esc} = c = \sqrt{\frac{2GM}{R_s}}$$

$$\therefore \quad R_s = \frac{2GM}{c^2}$$

For the Earth $\quad R_s = \dfrac{2 \times 6.7 \times 10^{-11}\,\text{N m}^2\,\text{kg}^{-2} \times 6.0 \times 10^{24}\,\text{kg}}{(3.0 \times 10^8\,\text{m s}^{-1})^2} = 8.9 \times 10^{-3}\,\text{m}$

For the Sun $\quad R_s = \dfrac{2 \times 6.7 \times 10^{-11}\,\text{N m}^2\,\text{kg}^{-2} \times 2.0 \times 10^{30}\,\text{kg}}{(3.0 \times 10^8\,\text{m s}^{-1})^2} = 3000\,\text{m}$

The idea of squashing the Earth to about the size of a table-tennis ball should convince you that it is fairly unlikely to end up as a black hole. The same is true for our Sun, whose eventual collapse will turn it into a white dwarf star. Stars collapse when the internal pressure of their nuclear reactions can no longer oppose the force of their own gravity. No known physical forces are strong enough to stop the collapse of stars of more than about three solar masses, so it is thought that there are many black holes in the universe.

This Hubble Space Telescope image shows a massive black hole consuming the galaxy NGC4261.

> **A conservative field**
>
> Potential is a property of the field, so the work done by or against gravitational forces when something moves from one place to another is independent of the route taken.
>
> If an object is moved around a closed loop there is no overall change in its GPE: all the work that is done in moving it against gravitational forces is returned as work done by gravity on it. Since no energy is dissipated in the gravitational field it is called a **conservative field**.

PRACTICE

1 The asteroid Ceres has a mass of 7.0×10^{20} kg and a radius of 550 km.

 a What is the gravitational field strength at the surface of Ceres?

 b What is the gravitational potential at the surface of Ceres?

 c What is the minimum escape velocity from the surface?

 d What is the minimum energy required to escape from Ceres?

2 The accuracy of the experiment to 'weigh the Earth' will change slightly with latitude. Why?

3 Rockets are usually launched as close to the equator as possible and towards the east. Why?

4 Draw a free-body diagram for a 1 kg mass suspended from a spring balance at the Earth's equator.

 a What is the resultant force on this mass?

 b Explain why the measured value of gravitational field strength at the equator is less than it would be if the Earth did not spin on its axis.

 c Calculate the proportional reduction in the measured value of g as a result of the Earth's rotation.

5 Will the variation of g across the Earth's surface affect the values for isotopic mass measured using a mass spectrometer?

6 The energy and mass of a photon of electromagnetic radiation are proportional to its frequency.

 a What will happen to the path of a light ray grazing the edge of a massive galaxy?

 b What will happen to the frequency and wavelength of light emitted from a massive star?

7 Does the Sun's gravity make a significant contribution to the work done in moving a spacecraft from the surface of Mars to the surface of the Earth? (Mass of Mars = 6.5×10^{23} kg, mean orbital radius of Mars = 2.1×10^8 km; mass of Earth = 6.0×10^{24} kg, mean orbital radius of Earth = 1.5×10^8 km; mass of Sun = 2.0×10^{30} kg.)

- gravity as centripetal force

- conservation laws

- applications

Maths box: analysing a circular orbit

Force:

Gravity provides a centripetal force:

$$\frac{GMm}{r^2} = \frac{mv^2}{r}$$

$$v = \sqrt{\frac{GM}{r}}$$

so the time period is

$$T = 2\pi\sqrt{\frac{r^3}{GM}}$$

Notice that the orbital speed and period are both independent of satellite mass.

Energy:

$$GPE = -\frac{GMm}{r}$$

$$KE = \tfrac{1}{2}mv^2 = \frac{GMm}{2r} = -\tfrac{1}{2}GPE$$

(using the result for v above)

$$TE = GPE + KE = -\frac{GMm}{r} + \frac{GMm}{2r}$$

$$= -\frac{GMm}{2r}$$

Notice that the total energy is negative (an orbit is a bound state).

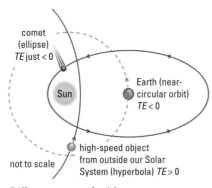

Different types of orbit.

ORBITS

Gravitational trajectories

The shape of a projectile's trajectory depends on the field in which it moves and its total energy. (The total energy, TE, is the sum of gravitational potential energy (GPE) and kinetic energy (KE), and is constant.)

- If the total energy in a radial field is negative the mass is in a '**bound state**' (it cannot escape from the field) and its orbit is an ellipse. Circular orbits are a special case of elliptical orbits.

- If the total energy is zero the particle escapes and its path is a parabola.

- For positive total energy its path is a hyperbola. (In a uniform field the path of a projectile is always a parabola.)

Most planetary orbits are near-circular, and so are the orbits of some moons and Earth satellites. Circular orbits are the easiest to analyse mathematically.

Force, energy, and angular momentum

The fundamental physical principles are the same whatever trajectory is followed.

- The only force exerted on the moving mass is gravitational and acts towards the centre of mass of the central body (for example, the Sun).

- Total energy is conserved: all that happens is a transfer of KE to GPE and vice versa as distance from the central mass changes.

- The force on the moving mass has no moment about the centre of mass of the central body, so angular momentum about this point is conserved.

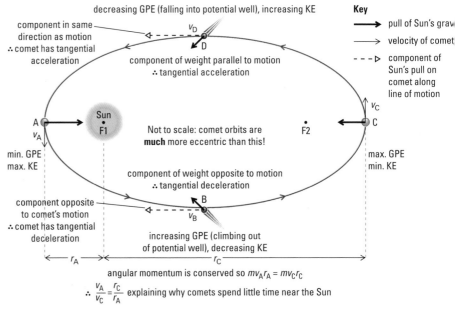

The orbit of a comet.

Artificial Earth satellites

The period of a satellite's orbit depends on the radius of its orbit. **Geostationary** satellites are put into equatorial orbits at an altitude (above the Earth's surface) of about 36 000 km where they have a period of 24 h. They are called geostationary because they remain above the same point on the rotating Earth. *Meteosat* is a European satellite in geostationary orbit above the equator where it intersects the Greenwich Meridian. Detectors on board the satellite take photographs in two ranges of infrared as well as with visible light. One infrared detector gives a thermal image of the Earth, the other measures the density of water vapour in the atmosphere and provides information about humidity.

A Meteosat image of Europe.

Earth observation or remote-sensing satellites are usually placed in lower, near-polar orbits. A typical period might be 100 min. As they orbit, the Earth rotates at a different rate beneath them and they effectively scan across its entire surface, several times a day. The **GPS** or '**global positioning system**' operated by the USA consists of 24 satellites placed so that several are always in line-of-sight contact with each point on the Earth. Small, relatively cheap receivers can now be purchased which can locate their own position to within a few metres anywhere on Earth. A related GPS system is being tested and will eventually replace the ground-based microwave beams that are currently used to guide aircraft during take-off and landing.

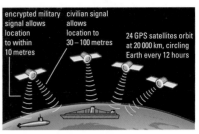

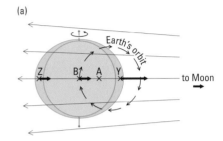

A GPS receiver compares transmission time with reception time to estimate a satellite's distance. Signals from four satellites give enough data to fix one's position.

Tides

There are approximately two tides per day everywhere on Earth. These are mainly caused by the interaction of the Earth's oceans with the Moon's gravitational field, but there is a second smaller influence from the Sun. The Earth's atmosphere and the crust also respond to these tidal forces. To simplify the explanation we shall focus on the ocean tides and consider only the Earth–Moon system.

Although it is often convenient to think of the Earth as fixed in space and to suppose that the Moon moves in a circular orbit around it, this is not correct. Both bodies orbit around their mutual centre of mass, which is actually *inside* the Earth because of its much greater mass. Gravitational attraction supplies the necessary centripetal forces. However, the Earth is an *extended* body, so the effect of the Moon's gravity varies across its diameter. On the side closest to the Moon the gravitational force is a bit bigger and on the side farther from the Moon it is a bit smaller than the value at the centre of mass itself. From the Earth's reference frame this is equivalent to a stretching force along the diameter that creates two tidal bulges on opposite sides of the Earth. As the Earth rotates each point on the surface will pass through these two 'high tides' in just over 24 hours.

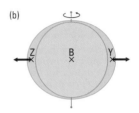

Tides. (a) A = centre of mass of Earth–Moon system. B = centre of mass of Earth. The Moon's gravitational field is stronger at Y than at B. This creates a bulge at Y. The Moon's gravity is weaker at Z than at B. This creates a bulge at Z. (b) In the Earth's reference frame there is a stretching force applied along YBX.

PRACTICE

1 Calculate the radius of a geostationary orbit, and of an orbit with period 100 min.

2 The tides are gradually slowing the Earth's rotation. How?

3 Use conservation of angular momentum to prove that a line joining the Sun to any planet sweeps out equal areas in equal times all around the orbit. (This was discovered by Kepler and is known as **Kepler's second law**.)

4 Find the ratio of r^3/T^2 for the Moon and for a geostationary satellite. Comment on your result. Do the same calculation and comparison for the orbits of the Earth and Jupiter around the Sun. Can you suggest a general rule? (Hint: use more data or prove it for circular orbits from Newton's law.)

5 Compare the energy needed to escape the Earth's gravity:

   a going up from the pole;

   b going initially east from the equator (with the Earth's rotation);

   c going initially west from the equator (against the Earth's rotation);

   d going from a low Earth orbit at 500 km above the surface. (Assume the craft continues in the direction it is orbiting.)

6 It is technologically challenging and very expensive to put satellites into orbit. Why bother? What are satellites able to do and what are their advantages over Earth-based systems? What are the problems involved in communicating with satellites, and how might these be overcome?

7 One way to study the interior structure of the Sun is to measure how it vibrates. The *SOHO* mission, launched in 1995, placed a spacecraft at the point where the Sun and Earth's gravitational fields balance each other, and it monitors both solar vibrations (**helioseismometry**) and the **solar wind** from the Sun.

   a What is the advantage of placing the spacecraft here?

   b Find out more about the solar wind.

   c The Sun is 3.327×10^5 times more massive than the Earth and at a distance of 1.496×10^8 km from it. How far from the Sun will the spacecraft be?

O B J E C T I V E S

• definition of field strength

• force of a charge in a field

• field of a point charge

• Coulomb's law

• the field of many charges

FORCE AND FIELD STRENGTH – 2. ELECTRICITY

Can you move an object without touching it? If I were to hang a piece of wire in the air and by using an electronic circuit cause electrons in it to accelerate back and forth, any other piece of wire in range would have some of its electrons moved about in response to those in the hanging wire. In this way I could transmit a TV picture to you. This is not magic, it is the action of electric fields over long distances giving rise to forces.

Earlier in this chapter the idea of a gravitational field was developed from Newton's law of gravity. Here, the electric field is defined, and used to obtain **Coulomb's law** for static charges.

Charges and fields

Electric charge is one of the fundamental properties of nature. Two objects of the same charge will push each other apart (repel) while objects of opposite charge pull each other together (attract).

Surrounding every charged object is an **electric field** – a region of space in which another charged object will experience a force. The electric field is characterized by *strength* and *direction*. The strength of the field at a point in space determines the size of the force that a charge will experience if it is placed at that point. **Electric field strength** is defined in a similar manner to gravitational field strength.

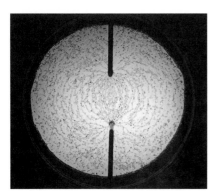

The electric field between two point charges can be illustrated by using a fine plastic dust which draws out the field, in a similar way to iron filings in a magnetic field.

Definition: electric field strength

electric field strength at a point, E (N C^{-1}) = $\dfrac{\text{force on charge } Q \text{ (N)}}{\text{size of charge } Q \text{ (C)}}$

$$E = \frac{F}{Q}$$

NB The units N C^{-1} are the same as V m^{-1}, which are also used to measure electric field strength.

Electric field strength is a **vector quantity**. The direction of the field at a point is defined to be the direction of the force acting on a small positive charge placed at that point. When a field-line diagram (such as the diagram below left) is drawn to illustrate the nature of an electric field, the arrows on the field lines are always drawn to show the direction of the force on a positive charge. A negative charge would experience a force in the opposite direction.

E defined strictly

E is defined strictly as:

$$E = \lim_{\delta q \to 0} \left(\frac{F}{\delta q} \right)$$

When a charge is placed into an electric field it will experience a force. It will also exert a force on the other charge that is producing the field which may cause it to move. So placing a charge into a field in order to measure its strength alters the field. In order to get round this we specify a small charge (δq) so the effect is small, and take the limit of a series of measurements as we use smaller and smaller charges.

The field of a point charge

The simplest electric field belongs to a single point charge (that is, a charge that is very small compared with any distance that we may be measuring), like an electron. As one would expect:

• The strength of the field is proportional to the size of the charge.

• It also varies with distance from the charge.

As the charge only occupies a point in space, the field must look the same along any line that is drawn radially from that point. This is shown in the diagram on the left. The field lines spread out as they get further from the charge, which indicates that the field is getting weaker.

At any distance r from the charge, we can imagine the field lines passing through the surface of a sphere of radius r. Hence the density of lines passing through the surface of the sphere is given by

$$\text{density of lines} = \frac{\text{number of lines}}{4\pi r^2} \propto \text{field strength at that distance}$$

$$\therefore \text{field strength} \propto Q_1 \quad (\text{the size of the charge})$$

$$\propto \frac{1}{4\pi r^2}$$

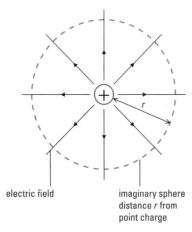

electric field

imaginary sphere distance r from point charge

The field of a point charge is radial.

Combining these gives

$$E = \frac{Q_1}{4\pi\varepsilon_0 r^2}$$

The constant of proportionality is gathered into ε_0, the **permittivity of free space**; $\varepsilon_0 = 8.854 \times 10^{-12}\,\text{F m}^{-1}$.

Coulomb's law

Any charge, Q_2, placed into an electric field will experience a force F:

F = size of charge × strength of field

$\quad = Q_2 E$

If the field is generated by a point charge, then:

$$E = \frac{Q_1}{4\pi\varepsilon_0 r^2}$$

$$\therefore \quad F = \frac{Q_1 Q_2}{4\pi\varepsilon_0 r^2}$$

This is Coulomb's law for the electric force between two point charges. Note the similarity to Newton's law of gravity. For gravity, the force is the *mass* multiplied by the *gravitational field strength*.

Worked example

What is the size of the electric field a distance of 3 m from a +3 nC point charge?

$$E = \frac{Q_1}{4\pi\varepsilon_0 r^2} = \frac{(+3 \times 10^{-9}\,\text{C})}{4 \times \pi \times (8.854 \times 10^{-12}\,\text{F m}^{-1}) \times (3\,\text{m})^2}$$

$$= 3.0\,\text{N C}^{-1}\ (2\ \text{sig. figs})$$

What would be the size of the force on +2 nC placed 3 m from a +3 nC point charge?

Either

$$F = Q \times E = (2 \times 10^{-9}\,\text{C}) \times (3.0\,\text{N C}^{-1}) = 6.0 \times 10^{-9}\,\text{N}\ (2\ \text{sig. figs})$$

or

$$F = \frac{Q_1 Q_2}{4\pi\varepsilon_0 r^2} = \frac{(+3 \times 10^{-9}\,\text{C}) \times (+2 \times 10^{-9}\,\text{C})}{4 \times \pi \times (8.854 \times 10^{-12}\,\text{F m}^{-1}) \times (3\,\text{m})^2} = 6.0 \times 10^{-9}\,\text{N}\ (2\ \text{sig. figs})$$

The field of more than one charge

The total force exerted on a charge by a collection of other charges can be calculated by adding up each of the 'one-on-one' forces. This is because electric fields obey the **principle of superposition**. The field at any point due to a collection of charges is the sum of the separate fields due to each charge. This seems so obvious that it does not appear to be worth stating as a principle. However, there are some examples in physics (especially in quantum theory) where fields do not add this simply!

In calculating the size of the field, or the force, one must remember that they are vector quantities.

Charging large objects

Subatomic particles may carry a charge as one of their basic properties. However, larger objects tend to be uncharged and have to be charged by artificial means.

- One way of doing this is by friction, for example, rubbing an object with a cloth.

- Another way is to transfer charge from an object that has already been charged by other means, for example, connecting a conductor to a high-voltage power supply.

In both cases, it is the negatively charged electrons that can be stripped away from a material (leaving it positively charged) or can be added to it (making it negative). The positively charged protons are locked up in the nuclei of the material's atoms, and cannot be moved.

Permittivity of free space

The permittivity of free space is a fundamental constant that determines the strength of the electric field in our units of measurement. $1/4\pi\varepsilon_0$ plays the same role in Coulomb's law as G does in Newton's law of gravity. See spread 5.11 for a method of measuring the value of ε_0.

Fields in a vacuum

Coulomb's law assumes that the point charges are in a vacuum (air behaves enough like a vacuum for the results to be accurate). In other materials ε_0 is replaced by $\varepsilon_r \times \varepsilon_0$ (see spread 5.12).

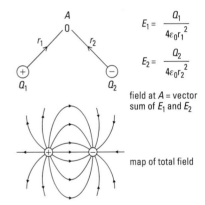

$$E_1 = \frac{Q_1}{4\varepsilon_0 r_1^2}$$

$$E_2 = \frac{Q_2}{4\varepsilon_0 r_2^2}$$

field at A = vector sum of E_1 and E_2

map of total field

The total field of two charges is the vector sum of the separate fields.

PRACTICE

1 A hydrogen atom consists of a proton with an electron in orbit. Both may be taken as point charges. The radius of the electron's orbit is 5.29×10^{-12} m. What is the strength of field felt by the electron and what is the force acting on it?

2 In fine weather the electric field near the surface of the Earth is $100\,\text{V m}^{-1}$ downwards. A drop of water of mass 25 mg is suspended in the electric field. What is the charge on the drop? (Take g as $9.81\,\text{N kg}^{-1}$.)

OBJECTIVES

- potential in a field
- the potential of a point charge
- electrical potential energy
- electrical potential energy in a system of two point charges

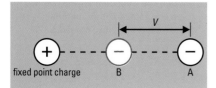

When a charge moves from A to B in the field of a fixed charge, it moves through a potential difference V.

Places of zero potential

Zero potential can be set anywhere. In a circuit it often makes sense to say that one side of a cell is at zero potential and to measure other potentials in the circuit relative to that. After all, it is only the *potential difference* that counts. In the circuit below, the 6 V lamp will still operate as normal!

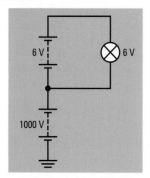

With electric fields around charges, it is convenient to place the zero at infinity. The field of a point charge decreases with distance and so would also reach zero at infinity.

ELECTRIC POTENTIAL AND POTENTIAL ENERGY

Electric and gravitational fields are rather mysterious. You can't see them, touch them, or smell them, but you can tell that they are there because they exert forces on objects. Both fields also act as sources of energy. If two charges are attracted towards each other they will accelerate, and their kinetic energy will increase.

This energy must come from somewhere: as the charges move, the electric field and the energy within it changes, and the field energy is converted into kinetic energy. **Electric potential** and **electric potential energy** help us keep track of the amount of energy in electric fields.

The potential of a point charge

When an electric charge moves through a conductor, electric potential energy is converted into thermal energy, and there is a p.d. across the conductor. One would expect a similar energy conversion whenever one charge moves through the electric field of another. In many cases the situation is far simpler than in an electric circuit, which has complicated patterns of charge on cell plates or on the ends of components. An especially simple case is moving one point charge towards or away from another that is fixed in place.

In an electric circuit, the p.d. between two points in the circuit is defined as:

$$\text{p.d.} = \frac{\text{energy converted when charge } Q \text{ moves between two points in a circuit}}{\text{amount of charge } Q}$$

This definition can be extended to include all circumstances in which a charge is moving through an electric field:

$$\text{p.d. between the points} = \frac{\text{amount of electrical energy converted}}{\text{size of moving charge}}$$

In a circuit the difference in potential between two points is fixed by the cells and components. When we are working with fields due to many large charges, the value of the potential at a single point is often useful. A point infinitely far from the positive charge is defined as having zero potential, and from this the value of the potential anywhere else in the field can be defined.

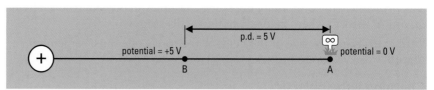

If the p.d. between A and B is 5 V then as the potential at A is 0 V, the potential at B must be 5 V. Hence the potential at a point in space can be defined in terms of the work done in bringing a unit positive charge from infinity to that point in space.

The potential in the field of a point charge can be obtained from Coulomb's law by integration in exactly the same way as gravitational potential. As a result, the potential of a point charge, V, is given by:

$$V = \frac{Q_1}{4\pi\varepsilon_0 r}$$

where r is the distance from the charge (m) and Q_1 is the size of the charge (C).

NB The *sign* of the potential is determined by the sign of the charge. As potential is a **scalar** quantity, the sign is *not* telling us about direction. The significance of the sign is simply to indicate if the potential is higher or lower than the zero at infinity.

Electric potential energy

If the definition of the p.d. between two points in an electric field is rearranged, then:

electric potential energy change between two points = $Q_2 \times$ p.d.

where Q_2 is the size of the moving charge.

$$\Delta U = Q_2 \times \Delta V$$

So, if the change in electric potential energy is the p.d. multiplied by the charge, the amount of electric potential energy that a charge Q_2 has at a point in an electric field = potential at that point $\times Q_2$.

$$\therefore\ U = \frac{Q_1 Q_2}{4\pi\varepsilon_0 r}\quad \text{if the field is due to a point charge } Q_1.$$

U is the electric potential energy in a system of two charges Q_1 and Q_2 (J).

NB The relationship between the *electric potential energy* and the *potential* is very similar to the relationship between the *force* and the *field*:

$$\text{force}\ =\ \text{size of field} \times Q$$
$$\text{electric potential energy}\ =\ \text{size of potential} \times Q$$

Helpful hint

If the two charges are separated by an infinite distance, the electric potential energy in the system will be zero. This will be the *largest* that the electric potential energy can be if the charges are *of opposite sign* (attractive force). On the other hand, if the charges are of the *same sign* (repulsive force), then zero is the *smallest* value the electric potential energy can be.

Worked example

What is the potential in the field of a $5\,\mu$C point charge at a distance of 30 cm from the charge?

$$V = \frac{Q_1}{4\pi\varepsilon_0 r^2}$$

$$= \frac{5 \times 10^{-6}\,\text{C}}{4 \times \pi \times (8.854 \times 10^{-12}\,\text{F m}^{-1}) \times (30 \times 10^{-2}\,\text{m})} = 0.149\,\text{MV}$$

Note: this is a large potential. In a circuit $5\,\mu$C would be a big charge!

A second charge of $+3\,\mu$C is placed in the field 30 cm from the first charge. What is the potential energy in the system?

Either

$$U = QV\ =\ (3 \times 10^{-6}\,\text{C}) \times (0.149\,\text{MV})\ =\ 0.45\,\text{J}$$

or

$$U = \frac{Q_1 Q_2}{4\pi\varepsilon_0 r^2}$$

$$= \frac{(5 \times 10^{-6}\,\text{C}) \times (3 \times 10^{-6}\,\text{C})}{4 \times \pi \times (8.854 \times 10^{-12}\,\text{F m}^{-1}) \times (30 \times 10^{-2}\,\text{m})} = 0.45\,\text{J}$$

NB Potential energy is a property of the set of charges – they have a mutual potential energy. The same result would be obtained by calculating the potential of the $3\,\mu$C charge and placing the $5\,\mu$C charge into the field.

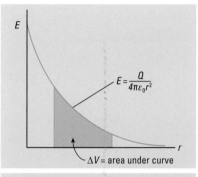

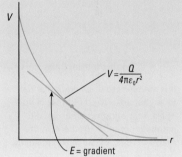

Field and potential for a point charge

PRACTICE

1 An electron starts from rest and is accelerated through a p.d. of 1000 V. At what speed is it moving when it emerges from the p.d.?

2 What is the electric potential in the field of a $+5$ nC point charge at a distance of 0.75 m from the charge? A charge of $+2$ nC is placed into the field at this point. What is the potential energy of the system? If the $+2$ nC charge is now moved through the field to a point 0.25 m from the first charge, how much work has been done by the force moving the charge?

3 A hydrogen atom consists of a proton and an electron orbiting at a distance of 5.29×10^{-12} m. What is the electric potential energy in this system? If the hydrogen atom is ionized it can be assumed that the electron has been removed to infinity. How much energy is required to do this?

(Hint: you will need to consider the maths box 'Analysing a circular orbit' on spread 5.6.)

4 An α particle has a charge equal to two proton charges. The nucleus of a gold atom has a charge equal to 79 proton charges. An α particle is fired on a direct line towards a gold nucleus with an initial kinetic energy of 1.6×10^{-12} J. How close can it get to the gold nucleus before being brought to rest? (Assume that the gold nucleus and the α particle are point charges. Assume the gold nucleus does not recoil.)

FIELD LINES AND EQUIPOTENTIALS

Equipotential lines

Contour lines on a map connect points of equal height. They are useful as they provide some information about the third dimension on an illustration that would otherwise be two-dimensional. Electric fields are very difficult to represent in a diagram. Both strength and direction must be indicated at every point in the field. As an alternative to field-line diagrams, 'contour maps' of the electric field can be drawn using **equipotential lines**. An equipotential line connects points in space where the potential of an electric field is the same.

For a point charge

$$V = \frac{Q}{4\pi\varepsilon_0 r}$$

so all points at the same distance r from the charge will have the same potential – the equipotential lines form circles centred on the charge.

If there are two or more charges present, then the potential at any point is the sum of the potentials due to each charge. Potentials are scalar quantities, so there is no direction to worry about when adding them. However, potential can be either positive or negative, depending on the sign of the charge. This needs to be taken into account in the sum.

Field lines and equipotential lines must always cross at 90°. Moving a charge *along* an equipotential is a process that, by definition, does no work. However, as the charge is moving through a field, it must be experiencing a force, and one would normally expect work to be done. The only way of reconciling these statements is if the force is always at 90° to the equipotential line. Work is only done if the charge has some component of its motion parallel to the force.

<table>
<tr><td>

Equipotential lines

It is standard to draw equipotential lines that are a fixed number of volts apart, so diagrams can be used to estimate potential gradient. In the diagram right, distance A represents the same difference in potential as distance B.
</td></tr>
</table>

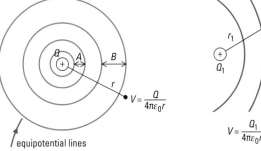

The equipotential lines of a point charge.

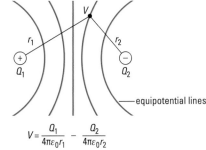

Equipotential lines in the field of two opposite charges.

$$V = \frac{Q_1}{4\pi\varepsilon_0 r_1} - \frac{Q_2}{4\pi\varepsilon_0 r_2}$$

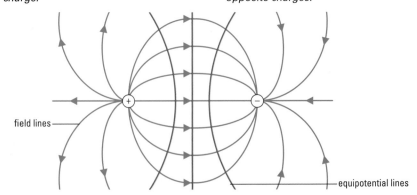

Field lines and equipotential lines cross at 90°.

Field and potential gradient

On a contour map, the steepness of the slope at any point can be estimated from the contour lines – the closer they are together, the steeper the slope. The same basic idea is true on an equipotential map – the closer the lines are together, the stronger the field is at that point.

If the equipotential lines are close together, the electric potential energy must be changing by large amounts in small distances, and there must be a large force acting. The exact relationship can easily be derived for the field of a point charge.

If a test charge δq is moved a small distance δr along a radius then

work done on test charge = force acting × distance moved

$$= F\delta r$$

In an electric field, the force acting is equal to the charge times the field strength:

$$F = E\delta q$$

$\therefore$ work done $= E\delta q\delta r$

Now, the work done on the charge is equal to the decrease in electric potential energy (remember that the potential energy is the charge times the potential), so

$$\delta q\delta V = -E\delta q\delta r$$

$$\therefore E = -\frac{\delta V}{\delta r}$$

The strength of the field is equal to the **potential gradient**. The negative sign indicates that the direction of the field is opposite to the direction in which the potential is increasing. (NB This relationship between field strength and potential gradient is also true for gravitational fields.)

Conservative fields

Like gravitational fields, electric fields are **conservative**. The potential difference between two points in an electric field does not depend on the route taken between the points. So if a charge is moved round a closed path no net work is done by the electric force. This can be demonstrated in the field of a point charge.

Consider this simple path ABCD around which a test charge, δq, is moved.

- Along AB, work is done by the field, as the test charge is moving in the direction of the force.

- Across BC, no work is done, as the charge is moving along an equipotential line.

- Back down CD, work must be done, as δq is being pushed against the force. This work done compensates exactly for the work done by the field along AB.

- Finally, by returning to the starting point along DA, the test charge moves along an equipotential line, and so no work is done.

In moving the charge round this closed path, no overall work is done.

The fact that electric fields are conservative has implications in circuit theory. The p.d. between two points is independent of the path taken – *that is why components in parallel have the same p.d.* Kirchhoff's second law is really a statement of what happens in conservative fields when non-electric forces act as well.

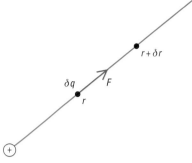

Moving a test charge in the field due to a point charge.

Very small charges

In this derivation, we need a very small test charge so that the original charge is not disturbed by its presence. Also, δr is so small that the force hardly changes over this distance. In calculus terms, we let $\delta r \to 0$ producing:

$$E = -\frac{dV}{dr}$$

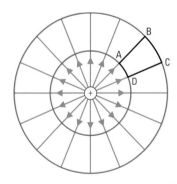

In following this closed loop in the field of a point charge, no net work is done. Therefore the field is conservative.

PRACTICE

1 The electric field near the surface of the Earth is $100\,\mathrm{N\,C^{-1}}$ downwards. Sketch the equipotential lines at 1 m intervals near the surface of the Earth. Take the surface as being at a potential of 0 V.

2 Two point charges, one of +4 nC and one of –2 nC, are placed 1 m apart. Locate all the points along a line through the centre of the two charges at which the total potential is zero. Consider points either side as well as between the charges.

3 Calculate the potential 1.00 m from a +2 nC charge. Calculate the potential 1.10 m from the charge. Use these figures to estimate the average potential gradient between 1.00 and 1.10 m. Calculate the field strength 1.05 m from the charge. Compare this answer to the average potential gradient.

Faraday cages prevent electromagnetic radiation getting in or out. Exposure to electromagnetic radiation is now the subject of strict regulation. The first edition of the Apple iMac™ had a metal Faraday cage, which can be seen through the translucent case.

Electromagnetic radiation

Electromagnetic radiation, such as light, microwaves, or radio, is composed of electric and magnetic fields. Compared with the electrical part of the wave, the magnetic field strength is very weak, so electromagnetic waves interact with conducting objects in a similar way to static electric fields (especially at low frequencies).

CONDUCTORS IN ELECTRIC FIELDS

In the early 1990s, new international regulations set a limit on the amount of electromagnetic radiation that could be produced by electronic devices such as computers, monitors, and CD players. The regulations arose because people were finding that a computer, when turned on, seriously interfered with radio reception in the same room.

There were also concerns about the long-term health risks associated with exposure to this electromagnetic radiation. As a result of these regulations, manufacturers had to become more careful in the design of their equipment. Electronic devices in which current changes very rapidly (such as computers) cannot help producing some radiation, so part of the answer lies in screening the devices so that the radiation does not escape. This is the same problem that manufacturers of microwave ovens face.

Screening can easily be constructed. It relies on the behaviour of conductors in electric fields.

Equipotential surfaces

Every point inside a perfect conductor must have exactly the same potential. This is because the charge carriers are free to move about inside the material. If there were points of different potential the charge carriers would move in order to balance them out. A conductor forms an **equipotential volume** and has an **equipotential surface**. Conductors may be charged and yet still be an equipotential surface.

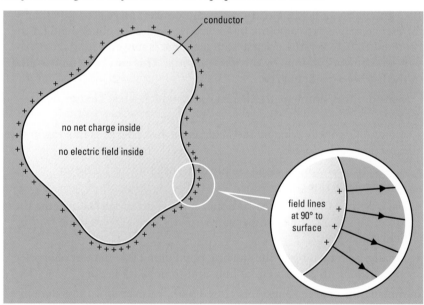

Charges distribute themselves on the surface of a conducting object.

There are several interesting consequences of this.

- **There can be no electric fields inside a perfect conductor.**

  Electric fields connect points of different potential. If all points in the conductor are at the same potential, then the net field inside must be zero.

- **A charged conductor carries its charge in a thin layer on the surface. There can be no net charge in the volume of a conductor.**

  A charge inside the conductor would set up an electric field. In response the charge carriers would move in order to cancel it.

- **Field lines touch conductors at 90°.**

  Field lines are always at 90° to equipotentials, so they must leave the surface of the conductor at 90°. If the field line was at an angle to the conductor's surface, then charge carriers would move about on the surface to cancel the effect.

Induced charges

A conducting sphere placed near a point charge will become **polarized**. Charge carriers (electrons) will move round the sphere until an equipotential surface is formed.

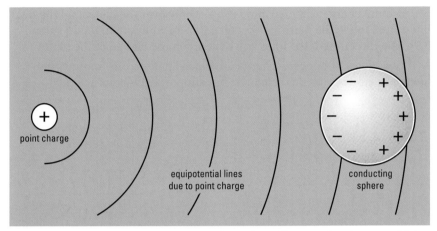

point charge

equipotential lines
due to point charge

conducting
sphere

The potential at any point on the sphere is found by adding together the potential due to the point charge and the potential due to the charges on the sphere. The total will be the same for every position on the sphere.

The electrons will not end up evenly arranged round the sphere, because the potential at any point is the sum of the potential of the point charge and the charge on the surface of the conductor. Parts of the sphere will carry a net positive charge and parts will be net negative. The arrangement of charge on the surface of the conductor has been **induced** by the presence of the point charge.

If the conducting sphere is connected to a fixed potential (earth, or one side of a power supply) then the sphere will have to gain a net charge in order to remain an equipotential. In the diagram on the right the sphere is connected to earth, and so is fixed at a potential of 0 V. The instantaneous effect of introducing a point charge is to alter the potential of the sphere. This temporary p.d. between the earth and the sphere causes a current. The electrons distribute themselves on the sphere to cancel the potential due to the point charge. The sphere now has a net charge, but it is still at 0 V.

The Faraday cage

Faraday cages are simple devices that use the properties of conductors to prevent electromagnetic fields getting into or out of the cage. They can be used to screen electronic and electrical equipment.

A Faraday cage is literally a cage constructed from a mesh of metal. If the cage is placed in the presence of an electric field it forms an equipotential – charges will be induced on the surface of the metal mesh. The total electric field inside the mesh is the sum of the external field and the field due to the induced charges. These two fields will always cancel each other inside the cage. Any equipment placed inside the cage will be screened.

Conversely, if the equipment inside the cage produces electric fields, then the charges induced on the cage will cancel the field trying to get out. The glass window of a microwave oven has a fine mesh built in to make a complete Faraday cage together with the walls of the oven. This prevents microwaves from escaping and becoming a health hazard.

Polarized objects

A polarized object is electrically neutral, but the positive and negative charges within it have been separated. A charged object will carry an excess of electrons if it is negative, or too few if it is positive.

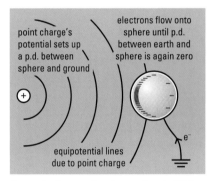

point charge's
potential sets up
a p.d. between
sphere and ground

electrons flow onto
sphere until p.d.
between earth and
sphere is again zero

equipotential lines
due to point charge

A conductor can be charged, yet still be at earth potential.

The Earth's potential

The Earth is like a giant capacitor plate, and charge can flow to and from it without altering the total charge by any measurable amount. Effectively, the potential of the Earth is fixed. This is why it is a good choice for setting the standard of 0 V in electric circuits.

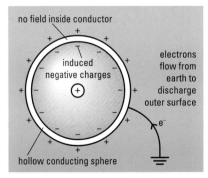

no field inside conductor

induced
negative charges

electrons
flow from
earth to
discharge
outer surface

hollow conducting sphere

A charge placed inside a hollow conductor will not have a field, as induced charges on the inner surface will produce a field which cancels that of the charge.

PARALLEL-PLATE CAPACITORS

If a single large metal plate is charged, the electric field lines near the plate will be uniform and parallel over most of the surface. This is because the charge will distribute itself evenly over the plate, and the field lines must touch the plate at 90°, since it is an equipotential (see spread 5.9). The direction of the lines will depend on the charge of the plates.

If two plates of opposite charge are placed near to one another, the fields will cancel outside the plates, but support each other inside.

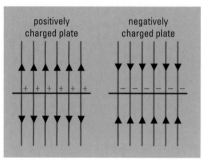

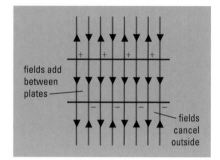

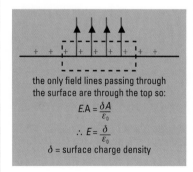

The field between two charged plates. Small plastic filings become polarized in electric fields, and line up along field lines.

The field between two plates.

In a parallel-plate capacitor, the field is only found between the plates. In practice, the edges of the plates spoil the uniform distribution of the charge. At these points the field stops being uniform.

If the positive plate is at potential V_1 and the negative plate is at potential V_2, then the electric field strength (see spread 5.7) between the plates is

$$E = -\frac{\delta V}{\delta r} = -\frac{(V_2 - V_1)}{d} = \frac{(V_1 - V_2)}{d} = \frac{V}{d}$$

where $V_1 - V_2 = V$, the p.d. between the plates, and d is the distance between the plates.

The strength of the electric field also depends on how closely packed the charges are on the plates. (Think of this in terms of drawing a field line to each charge.) The charge density on the plates is the total charge, Q, divided by the area of the plates, A. The electric field strength should be proportional to the charge density:

$$E \propto \frac{Q}{A} = K\frac{Q}{A}$$

where K is a constant of proportionality to be determined later. So

$$E = \frac{V}{d} = K\frac{Q}{A}$$

$$\therefore \quad Q = \frac{A}{Kd} V$$

showing that the capacitance of a parallel-plate capacitor, C, is related to A and d:

$$C \propto \frac{A}{d}$$

This is normally written as

$$C = \frac{\varepsilon_0 \varepsilon_r A}{d}$$

where ε_0 is the **permittivity of free space** = $8.854 \times 10^{-12}\,\mathrm{F\,m^{-1}}$ and ε_r is the **relative permittivity** of the material between the plates, which measures how strong an electric field can be in a material rather than in a vacuum.

NB Relative permittivity is also known as the **dielectric constant** of the material.

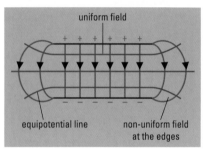

The field and equipotential lines of a real parallel-plate capacitor.

Gauss's law

The way in which the field near a capacitor plate depends on the charge density of the plate can be justified using **Gauss's law**. For any volume that surrounds a charge Q:

$$\int E\,dA = \frac{Q}{\varepsilon_0}$$

where the integral is taken over the surface area of the surrounding volume. This can be applied to a small volume that cuts through one of the plates.

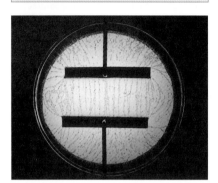

the only field lines passing through the surface are through the top so:

$$E.A = \frac{\delta A}{\varepsilon_0}$$

$$\therefore E = \frac{\delta}{\varepsilon_0}$$

δ = surface charge density

Experimental techniques for investigating parallel-plate capacitors

An experimental parallel-plate capacitor can be made in the lab with two sheets of aluminium spaced by thin strips of plastic. The distance between the plates can be varied by stacking strips of known thickness. (The strips can be measured with a micrometer screw gauge.) The area is most easily adjusted by altering the extent to which the plates overlap each other. (It is a reasonable approximation to worry only about the overlap area of the plates.)

The capacitance can be measured by connecting the plates into a circuit which allows them to charge and discharge rapidly. The sequence taken by the circuit is:

1 With the reed switch in position A, the plates charge to the p.d. set by the potentiometer.

2 With the reed switch in position B, the plates discharge through the ammeter.

The reed switch is continually driven between A and B by the variable-frequency a.c. supply. If the capacitor has time to charge fully while the switch is in position A, the charge on the plates will be CV. This amount of charge then passes through the ammeter when the switch flips over to B. If the switch flips f times a second, the amount of charge passing through the ammeter will be fCV. Provided f is of the order of a few hundred Hz (typically 200–400 Hz) the ammeter will read a steady current:

$$I = fCV = f\frac{\varepsilon_0\varepsilon_r A}{d} V$$

The graph of I against $1/d$ should be a straight line (with V, A, and f constant), and the graph of I against A should also be a straight line (with V, d, and f constant).

NB The same technique can be used for measuring ε_0 if I is plotted against V (with A and d constant). The gradient of the line is used to obtain ε_0 (ε_r is taken to be 1.00 for air). On the other hand, if I is measured for the capacitor with and without a dielectric between the plates, the ratio of the currents gives the ratio of ε_0 to $\varepsilon_r\varepsilon_0$.

Investigating a parallel-plate capacitor.

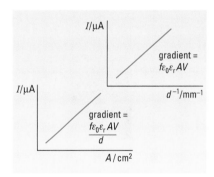

Analysing the data graphically.

PRACTICE

1 Two aluminium plates of area $16\,\mathrm{cm}^2$ are separated by a distance of 8 mm. The space between the plates is filled with air. (Take ε_r for air to be 1.00.) What is the capacitance of this arrangement?

2 A parallel-plate capacitor is to be made by sandwiching a 0.1 mm thick piece of mica between two metal plates. If mica has a relative permittivity of 6.00, what area of plates is required to achieve a capacitance of 250 nF?

3 Sparks will occur in air at the breakdown point, which is when the electric field strength has risen to $300\,\mathrm{V\,m}^{-1}$. Two metal plates of area $100\,\mathrm{cm}^2$ are separated by small plastic spacers of 1.0 cm thickness. What is the maximum p.d. between the plates before

sparking starts to occur? What will be the charge on the plates at this maximum p.d.?

4 An air-filled parallel-plate capacitor of $100\,\mu\mathrm{F}$ capacitance is charged to a p.d. of 10 V and then isolated. What is the energy stored in the capacitor? A material of dielectric constant 3.00 is pushed between the plates. What is the capacitance of the capacitor now? What is the p.d. between the plates now? What is the energy stored in the capacitor after the material has been inserted? Comment on the difference between this answer and the initial energy stored.

- polarizing dielectric materials with electric fields

- capacitors containing dielectrics

- forces on insulators

INSULATORS

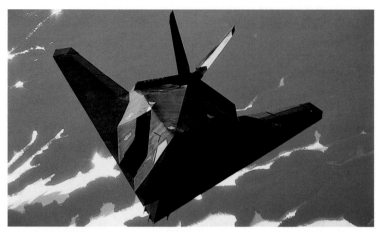

Many modern aircraft, such as this stealth bomber, use composite materials extensively as part of their structure.

Increasingly we are using composite materials to build cars and aeroplanes, because they are lightweight and strong. They also tend not to conduct electricity, which can lead to problems. When lightning strikes a metal aeroplane the damage is quite small as the charge is conducted through the metal, and the lightning bolt continues from the aeroplane to earth. If lightning strikes an aeroplane with a composite plastic body (such as a stealth bomber) it can do a lot of damage. As the material does not conduct, the charge does not spread out, so the place where the lightning has struck heats up to such an extent that it melts a hole in the structure. Aeroplane manufacturers have to consider the electrical properties of their insulating materials very carefully.

It would be easy to think that because an insulator does not conduct electricity, it has no interesting behaviour in electric fields. However, insulators contain charges and sometimes electrical effects are easier to see in them than in conductors. Insulators are always used in static electric experiments as the charge tends to spread out too much in conductors.

Polarizing dielectric materials

An insulating, or **dielectric**, material has no freely available charge carriers. The electrons are tightly bound to the atoms in the material. However, in the presence of an electric field the motion of the electrons round the atom can be disturbed. Normally the electrons circulate in a symmetrical manner so that the charge of the nucleus is screened. An electric field can exert a force on the electrons that disturbs the symmetrical nature of their motion, and the atom becomes polarized.

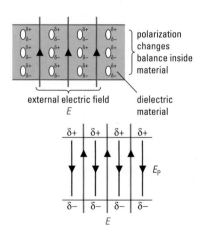

Polarizing an atom.

If an electric field is generated through the volume of a dielectric, the polarized atoms (or molecules) will tend to line up along the field lines. This has two effects.

- The end surfaces of the material become charged.

- A second electric field is set up in the material due to the charges on the polarized atoms.

If the applied field is not too strong, the size of the field due to the polarized atoms, E_p, will be proportional to the size of the externally applied field E:

$$E_p = \alpha E \quad (\alpha \text{ measures the extent of the polarization})$$

There are two electric fields inside the material – the externally applied field and the field due to the polarized atoms. Therefore, the total field in the material is given by

$$E_{tot} = E - E_p = (1 - \alpha)E$$

This reduction in the total field in the material helps to increase the capacitance of a parallel-plate capacitor.

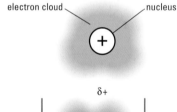
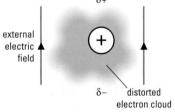

Applying an electric field to a dielectric.

Dielectrics between capacitor plates

Consider two capacitors with the same charge Q. Capacitor 1 has air between its plates and capacitor 2 has a dielectric material. Both capacitors have the same plate area and the same distance between the plates.

Measuring the p.d. between the plates of both capacitors would show that capacitor 2 is able to store charge at a lower p.d. The effect of placing the dielectric between the plates of capacitor 2 is to raise its capacitance. The amount by which it increases is related to the degree to which the atoms in the dielectric can be polarized.

The strength of the electric field between the plates of a capacitor is given by

$$E = \frac{V}{d}$$

where V is the p.d. and d is the distance between the plates.

$$\therefore \quad E_1 = \frac{V_1}{d} \text{ and } E_2 = \frac{V_2}{d}$$

Without the dielectric $E_1 = E_2$, but with the dielectric in place in capacitor 2

$$E_2 = (1 - \alpha) E_1$$
$$\therefore \quad V_2 = (1 - \alpha) V_1$$

As the charge is the same on both capacitors, $C_1V_1 = C_2V_2$.

$$\therefore \quad C_2 = \frac{C_1}{(1 - \alpha)}$$

For the air-filled capacitor

$$C_1 = \frac{\varepsilon_0 A}{d}$$

and for the capacitor with the dielectric

$$C_2 = \frac{\varepsilon_0 \varepsilon_r A}{d}$$

where ε_r is the relative permittivity of the dielectric. Hence

$$\therefore \quad \frac{\varepsilon_0 \varepsilon_r A}{d} = \frac{1}{(1 - \alpha)} \cdot \frac{\varepsilon_0 A}{d}$$

$$\therefore \quad \varepsilon_r = \frac{1}{(1 - \alpha)}$$

A material that could not be polarized at all would have $\alpha = 0$ and so $\varepsilon_r = 1$. On the other hand, a perfectly polarizable material would have $\alpha = 1$ and so ε_r would be infinite!

Forces on insulators

If you hold a charged rod near a stream of water, you can deflect the path of the water. A charged plastic comb will pick up small pieces of paper due to the same effect. Yet neither the water nor the pieces of paper were charged, so how can the charged objects exert a force on them?

As we have seen, in the presence of an electric field, insulating materials can become polarized with the opposite charge facing the source of the field. This always gives rise to an attractive force.

One might think that as the polarized object has both positive and negative charges the effect of the repulsive and attractive forces would cancel. However, one of the charges is always nearer to the source of the field than the other, so the force on this charge is always greater.

A very important example of the polarization force is the van der Waals force between neutral molecules. Without this force there would be no attraction between gas molecules, and gases could not be turned into liquids. The force arises when one molecule polarizes another (by altering the motion of the electrons round the molecule) and the second molecule polarizes the first in turn! The details of the effect are very complicated and need quantum theory to be explained thoroughly.

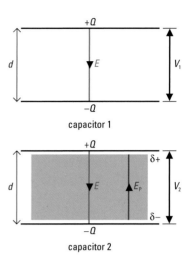

Capacitor 2 has an increased capacitance.

The stream of water is being deflected by the charged comb.

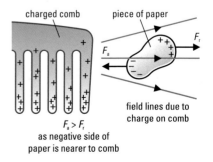

A charged plastic comb can be used to pick up small pieces of paper.

CONTRASTING ELECTRIC AND GRAVITATIONAL FIELDS

Physicists are ambitious people. By and large, they like things to be simple. Some of the processes that go on in the universe seem to be a long way from being simple. However, in most cases complex effects are due to very simple things going on at a fundamental level.

Many physicists believe that the whole of the universe can be explained by the action of a few simple laws. One of the most ambitious things that physicists want to do is to show that the **four fundamental forces** of nature are actually different aspects of a single more basic force, in the same way that friction can be explained by the electric force.

Some progress has been made along this path. However, one of the hardest problems that physicists face is how to include gravity. Gravity is a very mysterious force that we only partly understand at a fundamental level. It has many features in common with the electric force, but some important differences that are hard to reconcile.

Points of comparison

- Both electric and gravitational fields are inverse-square law fields; that is, the strength decreases as $1/\text{distance}^2$ as you move away from a point charge or mass.

- In both cases the force acting on a 'charge' is the strength of the field multiplied by the size of the 'charge':

  gravitational force = mass × g

  electric force = charge × E

- There are two types of electric charge: positive and negative. Mass can only be positive. This is important because it is possible to screen oneself from electric forces by inducing an opposite charge. It is not possible, however, to screen the effects of gravity.

- In both cases the potential in the field is zero at infinity. However, for gravity this is the largest potential in the field – at every other point the potential is negative. In the electric field of a positive charge, the potential is positive everywhere except at infinity, where it is zero. In the electric field of a negative charge, the potential is negative everywhere except at infinity, where it is zero.

- At the level of fundamental particles there seem to be only three values of charge that particles can have: $-e$ (-1.6×10^{-19} C), $+\frac{2}{3}e$, and $-\frac{1}{3}e$. There does not seem to be any regularity to the values of mass that the fundamental particles can have.

- The electric force is very much stronger than the gravitational force. When an apple is hanging from a tree it is the electric forces between the molecules of the stalk and the molecules of the branch that hold it to the branch. It requires the gravitational attraction of the whole of the Earth to exert enough force to break that attraction, when the apple is heavy enough to fall.

- The ratio of the force of electric repulsion to the force of gravitational attraction between two electrons is of the order of 10^{42}!

- Most objects in the universe are electrically neutral. This is because they have equal numbers of positively and negatively charged objects within them. The balance is maintained to a very high precision. Any object that is slightly charged will attract other objects of opposite charge, and so become neutral. For this reason, the electric force is hardly seen in the universe at large, though it is very strong compared with gravity. When we look at the motion of stars in galaxies, we see no evidence of any electric forces at work. Gravity may be a weaker force, but because it can't be screened out it has a very big effect in the universe.

A gravity shield!

In his novel '*First men on the Moon*' H.G. Wells' hero invents a substance 'Caverite' that is able to shield a spacecraft from the force of gravity. It's a nice idea, but such a material is completely impossible according to our current knowledge.

Attraction and repulsion

Gravity can only ever be an attractive force between masses. The electric force is attractive between opposite charges, and repulsive between like charges.

Force laws and fields

Comparing gravitational and electric fields.

	Force	Field	Potential	Constant
Gravitational	$-\dfrac{k_g m_1 m_2}{r^2}$ Newton's law	$-\dfrac{k_g m_1}{r^2}$	$-\dfrac{k_g m_1}{r}$	$k_g = G$
Electric	$\dfrac{k_q Q_1 Q_2}{r^2}$ Coulomb's law	$\dfrac{k_q Q_1}{r^2}$	$\dfrac{k_q Q_1}{r}$	$k_q = \dfrac{1}{4\pi\varepsilon_0}$

NB The equations have been written in this form to draw out the comparison between them.

Fields due to spherical objects

The gravitational and electric fields produced by solid spheres of mass and charge are very similar in the way that they vary with distance from the centre of the sphere. In the diagrams below, the variation of potential with distance is shown for a solid sphere (of charge or mass) with a uniform density and also a thin 'shell'. The latter is easy to imagine in the electrical case – any charged conducting sphere will have all its charge on the outer surface and so naturally forms a shell of charge.

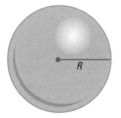

a sphere of mass of uniform density

a spherical shell of mass is similar to a charged conductor

a charged insulator has its charge distributed uniformly throughout the volume

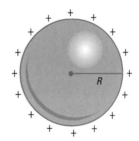

charged conductor has its charge on a thin layer at the surface

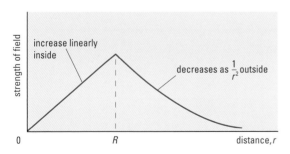

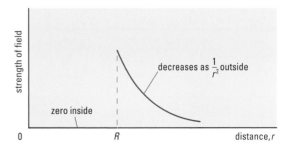

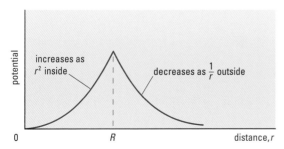

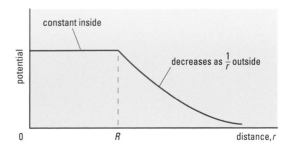

Fields associated with spherical objects.

- force on a moving charge, Bqv

- definition of magnetic field strength

- force on a current-carrying wire, BIl

MAGNETIC FORCES ON CHARGES

If you move a fridge magnet close to the metal door of a fridge then, as it gets right up to the surface, it will seem to want to jump out of your hand and lock onto the door. If you take two fridge magnets and hold them close to each other, you can feel that they are trying to push each other apart. Magnets hold a fascination that has attracted people since the discovery of the magnetic ore lodestone, which was first studied by the Ionian philosopher-scientist Thales (640–546 BC).

Simple experiments like the ones mentioned above hide the fact that permanent magnets are very complicated objects. At a fundamental level the magnetic effects are due to the magnetic fields produced by small currents within a material, which exert forces on other currents.

Magnetic force on a point charge

A magnetic field is a region of space in which a moving electric charge experiences a magnetic force. **Magnetic field strength** needs to be defined in a different way to electric field strength because magnetic forces are only exerted on *moving charges*. The size of the force acting is proportional to the size of the charge and to the speed at which it is moving:

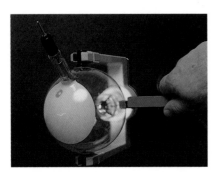

The electron beam's path is bent by the magnetic force exerted on the moving electrons within the beam. Notice that the magnet is being held at 90° to the direction of the beam.

Definition: magnetic force

magnetic force (N) = magnetic field strength (T) × charge (C) × speed (m s^{-1})

$$F_{\text{mag}} = Bqv \tag{1}$$

The unit of magnetic field strength is the **tesla**. In terms of base units this is

1 tesla, T, $= 1 \, \text{kg} \, \text{A}^{-1} \, \text{s}^{-2}$.

The table below will give a feel for the size of the tesla.

Typical magnetic field strengths.

	Earth's field	Typical bar magnet	Fridge magnet
Strength of field	~50 μT	0.1 T	~0.05 T

The simplest application of equation (1) is when the charge's motion cuts across the direction of the magnetic field lines at 90°. If the path of the charge is different from this, then the angle between the path and the field lines must also be taken into account. If the charge's velocity is at an angle θ to the field lines, then the force exerted is given by

$$F_{\text{mag}} = Bqv \sin \theta$$

When $\theta = 90°$, $\sin \theta = 1$ so the formula reduces to Bqv.

When $\theta = 0°$, $\sin \theta = 0$, so the force is zero.

Magnetic field

The magnetic field strength B is sometimes called the magnetic flux density.

B can be defined by:

$$B = \frac{F}{qv}$$

which is somewhat analogous to:

$$g = \frac{F}{m}$$

$$E = \frac{F}{Q}$$

Worked example

What is the force acting on an electron moving through a magnetic field of 1.6×10^{-5} T in the direction shown?

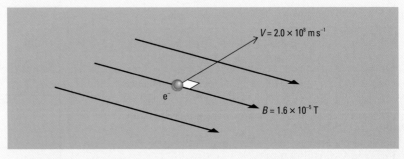

$$F_{\text{mag}} = Bqv = (1.6 \times 10^{-5} \, \text{T}) \times (1.6 \times 10^{-19} \, \text{C}) \times (2.0 \times 10^8 \, \text{m s}^{-1})$$

$$= 5.1 \times 10^{-16} \, \text{N}$$

Fleming's left-hand rule

The magnetic field strength, B, the velocity of the charge, v, and the magnetic force, F_{mag}, are all vector quantities. The equation $F_{mag} = Bqv \sin \theta$ relates the sizes of these vectors to each other. In order to find the direction in which the force acts, given the direction of the field and the velocity, Fleming devised a simple rule that uses two fingers and a thumb on the left hand.

NB Fleming's left-hand rule works for the direction in which a *positively charged* particle is moving. In other words it is defined for conventional current. In this case the particle is an electron, so when the rule is used the second finger must point in the *opposite direction* to that in which the electron is moving. In the example the force on the electron is *downwards*. Also note that the direction of the magnetic force is *always* perpendicular to the plane in which the magnetic field and the charge's velocity lie.

The force on a current-carrying wire

The magnetic force acting on an individual charged particle is too small to easily measure in the lab. The effect of the force can be seen in the deflection of an electron beam, but for measurement purposes the force on a current-carrying wire is more convenient. Given a wire of length l and cross-sectional area A, and carrying a current I then

$$I = nAev$$

where v is the drift velocity of the electrons and n is the number of free electrons per unit volume.

If the wire is crossed by a magnetic field of strength B, each electron drifting along the wire will experience a magnetic force equal to Bev. The total force acting on the wire is the sum of forces acting on each electron drifting through it. If the total number of drifting electrons in the wire is N, then

$$F_{wire} = N \times Bev = (n \times \text{volume of wire}) \qquad Bev = (nAl) \times (Bev)$$

Now

$$v = \frac{I}{nAe}$$

$$\therefore F_{wire} = (nAl) \times Be \times \frac{I}{nAe}$$

$$= BIl$$

This enables us to define one tesla as being the magnetic field strength required to exert a force of 1 N on a wire of length 1 m carrying a current of 1 A.

The magnetic force is acting on the electrons within the wire. However, the electrons form part of the structure of the metal and so the force is passed on to the wire as a whole. This indirect force on current-carrying conductors is very important technologically. Without this force, devices such as electric motors, loudspeakers, and electric bells would not exist.

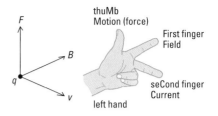

current is in the direction of a moving *positive* charge

Fleming's left-hand rule for the force on a moving positive charge. The importance of the rule is to show that the direction of the force is at 90° to the plane containing the charge's motion and the field lines.

Monopoles

The definitions of gravitational and electric field strengths are given in terms of the 'charge' involved with the force ie $E = F/q$, $g = F/m$. If magnetic fields followed the same pattern then the definition of magnetic field strength would be of the form $M = F/p$ with p being a single magnetic pole or **monopole**. As far as we are aware, magnetic poles never exist as single poles. North and south poles *always* come together, and so this sort of definition of the field strength is not possible.

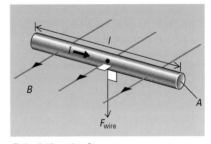

Calculating the force on a current-carrying wire.

Helpful hint

If the current-carrying wire is at angle θ to the field then:

$$F = BIl \sin \theta$$

PRACTICE

1 Use Fleming's left-hand rule to find the direction of the force exerted on the electron in the worked example.

2 A proton moving at $1 \times 10^8 \, m \, s^{-1}$ enters a region in which there is a magnetic field of strength 0.5 T. Calculate the force acting on the proton if the angle between the proton's initial path and the field lines is: **a** 0°, **b** 90°, **c** 45°, and **d** 30°.

3 The diagram shows a simple current balance. The straight part of the wire is being crossed by a magnetic field of 2.0×10^{-3} T. The current in the wire is 1.0 A. What mass m must be hung from the other side in order for the system to balance?

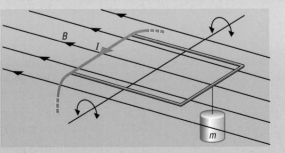

The unification of forces

Oersted's discovery brought together the separate subjects of electricity and magnetism. The unification of forces is a theme that modern physicists are pursuing, and they hope eventually to unify the four fundamental forces of nature.

The permeability of free space

μ_0 determines the strength of magnetic fields in a vacuum. Unification of electricity and magnetism, finally established by the work of the Scottish physicist James Clerk Maxwell (1831–79), showed that the three fundamental constants μ_0, ε_0, and c (the speed of light) are related to each other by

$$c^2 = 1/\mu_0\varepsilon_0$$

Fixing the definition of the ampere as shown on the next page has the effect of defining the value of μ_0 to be exactly $4\pi \times 10^{-7}$ kg m s^{-2} A^{-2}. The speed of light has also been fixed by definition at 299 792 458 m s^{-1}, so it follows that the value of ε_0 is also fixed at 8.854×10^{-12} s^4 kg^{-1} m^{-3} A^2.

The units usually quoted for μ_0 and ε_0 are

μ_0 H m^{-1}

ε_0 F m^{-1}

MAGNETIC FORCES BETWEEN CURRENTS

Many important advances in science are made by accident. A scientist performing an experiment notices a result that they had not expected, or something goes wrong with the equipment, giving rise to a new effect. In 1820, Hans Christian Oersted, a Danish lecturer in physics, noticed an interesting phenomenon while teaching a class on basic electricity. (In those days cells and currents in wires were undergraduate physics!) There happened to be a demonstration compass on the same bench as Oersted's electrical experiment. Whenever he turned on a current, he saw the compass responded by pointing in a different direction.

Oersted saw the significance of his accidental discovery. He had previously suspected that there might be a connection between magnetism and electricity – a connection denied by Coulomb. If electric currents generate magnetic fields, then perhaps electric currents within permanent magnets also generate magnetic fields.

Magnetic field of a current-carrying wire

The shape of the magnetic field produced by a long straight wire is revealed by placing a ring of small compasses on a card with the current-carrying wire passing through a hole in the centre. The compasses turn to point in a closed circle round the wire, showing that the magnetic field lines form circular loops centred on the wire.

If the card is moved along the length of the wire, the compasses stay pointing in circular loops. This shows that the compasses are drawing out a slice through a magnetic field pattern that is actually cylindrical and centred on the wire.

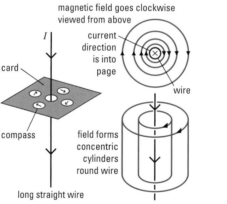

The magnetic field round a straight wire.

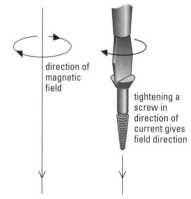

*The **corkscrew rule** can be used to find the direction of the magnetic field round a current.*

The strength of the magnetic field increases with the size of the current I, and decreases with distance from the wire, r:

$$B \propto \frac{I}{r} = \frac{\mu_0 I}{2\pi r}$$

μ_0 is the **permeability of free space** – a constant that plays a similar role in determining the basic strength of the magnetic field as $1/4\pi\varepsilon_0$ does for the electric field and G for gravity.

As magnetic fields are vector quantities, a convention is required to specify their direction at any point in space.

• The direction (or sense) of a magnetic field is specified by the way in which a small geographical compass would point if placed in the magnetic field at that point.

• This is also the direction in which the magnetic force would act on a 'free north pole' if monopoles existed.

• In the case of the field round the long wire, the sense is determined by the direction of the current in the wire.

Worked example

A long straight wire is positioned perpendicular to a uniform magnetic field of strength 2.0×10^{-5} T. If there is a current of 1 A in the wire, what will be the total magnetic field at points A and B in the diagram?

$$\text{Field due to current in wire} = \frac{\mu_0 I}{2\pi r}$$

$$= \frac{(4\pi \times 10^{-7}) \times (1.0\,\text{A})}{2\pi \times (0.01\,\text{m})} = 2.0 \times 10^{-5}\,\text{T} \quad (2\text{ sig. figs})$$

$$\therefore \text{ total field at } A = 2.0 \times 10^{-5}\,\text{T} + 2.0 \times 10^{-5}\,\text{T} = 4.0 \times 10^{-5}\,\text{T}$$

$$\text{total field at } B = 2.0 \times 10^{-5}\,\text{T} - 2.0 \times 10^{-5}\,\text{T} = 0.0$$

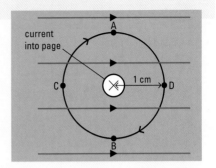

Force between two current-carrying wires

The diagram on the right shows two long parallel 'wires' carrying currents I_1 and I_2 in the same direction. Thin aluminium strips are used, as they are lightweight and will move easily under the magnetic force. Wire 1 is in a magnetic field and so experiences a force equal to $BI_1 l$. The magnetic field is generated by wire 2, and so

$$B = \frac{\mu_0 I_2}{2\pi r}$$

Hence the force acting on a 1 m length of wire 1 is

$$F = \frac{\mu_0 I_1 I_2}{2\pi r}$$

The argument, repeated from the point of view of wire 2, would produce the same result.

NB If the currents are in the *same* direction, the force is *attractive*. If the currents are in *opposite* directions, the force is *repulsive*. This can cause problems if wires carrying large currents need to be run next to one another, for example in overhead power lines.

> ### Definition: the ampere
>
> The force between two current-carrying wires is used to define the value of the **ampere**. If the current in both wires is 1 A and the distance between the wires is 1 m, then the force on each metre length of the wires will be
>
> $$F = \frac{\mu_0 I_1 I_2}{2\pi r} = \frac{(4\pi \times 10^{-7}) \times (1.0\,\text{A}) \times (1.0\,\text{A})}{2\pi \times (1.0\,\text{m})} = 2 \times 10^{-7}\,\text{N}$$
>
> So, 1 A is defined as that current in two long wires separated by 1m in a vacuum which produces a force of 2×10^{-7} N per metre between the wires.

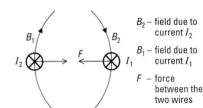

Two strips of aluminium foil can be used to demonstrate the force between current-carrying wires.

PRACTICE

1 Use the corkscrew rule and Fleming's left-hand rule to show that two current-carrying wires repel each other when the currents are in opposite directions.

2 Calculate the size and direction of the magnetic field at points C and D in the diagram at the top of this page.

3 Two parallel wires are separated by a distance of 0.6 m in a vacuum. One wire carries a current of 2 A and the other a current of 3 A. Both currents are in the same direction. Calculate the size of the force on a metre length of each wire.

4 Draw a diagram showing the arrangement of question 3 as seen from one end (current flowing into the page). Using the diagram, find where the total magnetic field will be zero. Calculate the position of this point.

5 The magnetic field strength at a certain point on the Earth's surface has a horizontal component of 18 µT due north, and a vertical component of 55 µT downwards. Calculate the force on a 1 m length of straight wire carrying a current of 4 A when:

a the wire is vertical with the current downwards;

b the wire is horizontal with the current in the direction from east to west.

MAGNETIC FIELDS GENERATED BY CURRENTS

Magnetic fields are invariably generated by electric currents. The shape of the magnetic field produced depends on the shape of the conductor carrying the current. The simplest example is the single wire loop. Many wire loops next to each other make up a coil or **solenoid**, and its magnetic field looks similar to the one produced by a bar magnet.

In fact, the field of a bar magnet can be modelled by considering small currents that run round the surface of the bar. Using this model the way in which bar magnets attract or repel one another can be explained in terms of the magnetic force between two currents.

The Earth's magnetic field is produced by electric currents. In this case the currents are generated by the complicated churning of molten iron within the Earth's core. The fact that a permanent magnet will always turn so that one end is pointing towards the geographical north pole of the Earth is again due to the magnetic force between the current in the Earth's core and the current in the magnet.

The field of a single loop of wire

A single loop of wire carrying a current sets up a magnetic field round it, as shown in the diagram on the left.

The sense of the magnetic field can be found by applying the corkscrew rule along the length of the wire.

Solenoid field

A solenoid is a tightly wound coil of wire.

• The coils should touch each other for the most uniform magnetic field.

• Therefore, the wire must be insulated to prevent the current shorting between the wires.

If one could wind an infinitely long solenoid from wire with a diameter that is much smaller than the diameter of the solenoid itself, the field inside would be uniform. There would be no field outside the solenoid. In practice, of course, no solenoid can be infinitely long, but the field outside is always very much weaker than the internal field.

$$B_{\text{inside}} = \mu_0 n I$$

where n is the number of turns per metre length of the solenoid (m^{-1}) and I is the current in the solenoid (A).

The direction of the field can be found by matching the direction of the current round the coil to arrows placed on the letters N and S. Magnetic field lines pass from N to S outside the solenoid and from S to N inside the solenoid. The use of the letters N and S refers to the fact that if the solenoid was hung from a thread and allowed to rotate freely, the N end would point in a geographically northern direction.

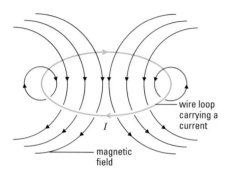

The magnetic field round a single loop of wire.

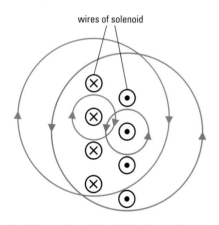

Inside a solenoid the fields from wires on opposite sides add up. Outside, the fields from opposite wires cancel.

Ferrous core

The magnetic field of a current is rather weak. To make a practical electromagnet, a solenoid can be wrapped around an iron core. The field of the current in the solenoid can then magnetize the iron, producing a much stronger overall field.

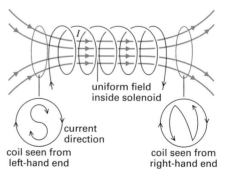

The magnetic field of a solenoid is very similar to that of a bar magnet. The end of the solenoid from which the field lines emerge is the north-seeking pole (N or north pole). The other end, into which the field lines pass, is the south-seeking pole (S or south pole).

Magnetic poles

It is often very difficult to work out the direction in which circulating currents would exert magnetic forces on each other. The task can be simplified by the idea of **magnetic poles**. The **north-seeking pole** (N) of the solenoid is the end from which the magnetic field emerges. The other end is the **south-seeking pole** (S).

A simple experiment with two solenoids shows that like poles *repel* (N to N, or S to S) and unlike poles *attract* each other (N to S). When like poles come together, the currents in the solenoids are circulating in a different direction, and so will be repelling each other.

Permanent bar magnets also have poles. Bringing two bar magnets together will produce forces between the currents within them that can lead to attraction or repulsion, depending on how they are aligned.

It is not always possible to identify suitable poles in a magnetic field. The field of a long, straight wire, for example, does not have poles. In such cases the forces must be found using Fleming's left-hand rule and the corkscrew rule.

Poles are not the magnetic equivalent of charges. There is no experimental evidence to suggest that magnetic charges (monopoles) exist.

The Earth's field

The magnetic field of the Earth is generated by powerful convection currents in the molten iron core. The churning of this liquid metal generates electric currents that produce the magnetic field. The effect is not well understood, but it is known that the field periodically reverses direction! (The reversal is spontaneous with no fixed pattern, but seems to happen with a rough regularity every million years or so.)

As long ago as the third century AD, the Chinese were using magnetized needles floating in dishes of water to point towards magnetic north. The end of the needle that points in a northerly direction is labelled N (for north-seeking). The other end of the needle is labelled S (for south-seeking). Modern electronic compasses use a semiconducting detector called a **Hall probe** to measure the strength and direction of the field.

The direction of the Earth's field can be confusing: as unlike poles attract, the north-seeking end of the magnetized needle must be attracted to a south-seeking pole when it points to geographical north. Near the Earth's *geographical* North Pole there is a south-seeking *magnetic* pole.

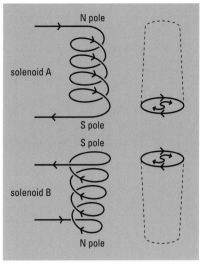

Bringing like poles together ensures that the currents in the solenoids are circulating in a different direction, so that the force between the currents is one of repulsion. The same thing happens when two bar magnets repel each other.

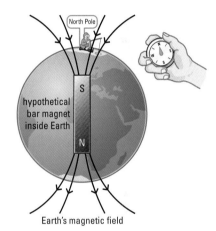

Earth's magnetic field

The north-seeking pole of the compass points towards the geographical North Pole of the Earth. This means that the south-seeking magnetic pole of the Earth is situated near the geographical North Pole, which can be rather confusing!

PRACTICE

1 Draw a diagram showing that when two solenoids are brought together with opposite poles facing, the currents are circulating in the same direction. Use Fleming's left-hand rule to show that the solenoids would attract each other.

2 At a particular point on the Earth, the magnetic field has a vertical component of 18 μT due north and 55 μT downwards. Calculate the magnitude and direction of the total field. Draw a diagram showing how a solenoid could be used to cancel the magnetic field of the Earth at this point. How large a current would be needed to do this if the solenoid is 60 cm long and has 100 turns?

3 Find the force per unit length between two successive turns of wire on a solenoid. The turns are separated by 5×10^{-4} m and carry a current of 0.46 A. In what direction do the forces act? Why do coils of wire carrying a.c. current often hum or buzz?

ELECTROMAGNETIC DEVICES –
1. THE MOTOR

The loudspeaker, relay, and electric motor are all dependent on the magnetic force. There are so many applications of the electric motor around the home, and in general life, that it is hard to list them all.

In a car electric motors operate the windows, sunroof, seats, door mirrors, and throttle. In the home, electric motors can be found in tape recorders, CD players, vacuum cleaners, hair dryers, washing machines, dish washers, microwave ovens, and central heating systems.

Cars powered by electric motors are becoming more common, in an effort to reduce the emissions from internal combustion engines. As electric motors become smaller and lighter, their usefulness increases. In October 1960 the physicist Richard Feynman offered a $1000 prize to the first person to construct an electric motor that occupied a volume less than 1/64 inch cubed. A month later, the prize was claimed by William McLellan.

The turning force on a motor coil

An electric motor is basically a coil of wire carrying a current in a magnetic field. The forces acting on the coil will, if properly arranged, tend to make the coil rotate. The top diagram shows a rectangular coil suspended in a uniform magnetic field.

The forces along edges AB and CD will rotate the coil, whereas the forces on BC and DA will tend to stretch it. Viewed from above, as in the bottom diagram, the pair of forces rotating the coil form a couple.

The torque exerted by the couple is

$$T = F \times w \cos \alpha$$

now

$$F = B \times h \times I$$
$$\therefore \quad T = B \times hw \times I \times \cos \alpha$$

hw is the area A of the coil, so the turning moment is

$$T = BAI \cos \alpha$$

This would be the turning moment on a single turn of wire.

If a multi-turn coil is used, with n turns, the whole coil would have a turning moment of

$$T = BAIn \cos \alpha$$

NB As the coil rotates the size of the torque changes, so the rotation is not smooth.

The size of the torque on the coil can be increased by:

- increasing the area of the coil;
- increasing the current in the coil;
- using more turns of wire in the coil;
- using a stronger magnetic field;
- making the magnetic field radial rather than uniform (so $\alpha \approx 0$ most of the time).

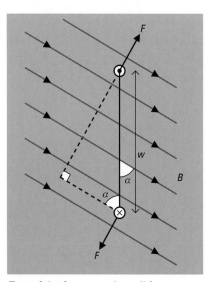

The forces acting on a rectangular coil in a magnetic field.

Two of the forces on the coil form a couple.

The electric motor

The diagram below shows a practical d.c. electric motor. Several features of the design have been developed to make the torque and rotation rate smooth and constant.

- The curved end pieces to the permanent magnet help to make the field radial.
- The cylindrical soft iron **armature** (the rotating part) increases field strength and helps to make it radial.
- More than one coil is used on the same axis – one is always nearly parallel to the magnetic field.
- A **split-ring commutator** is used, which swaps current direction every half turn. It also stops the wires being tangled!

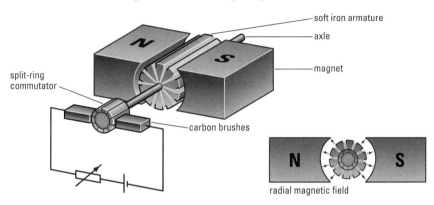

A practical design for an electric motor.

In most electric motors the magnetic field is generated by wrapping a solenoid round a piece of iron with curved end pieces. This is known as the **field winding**, to distinguish it from the **armature winding** (the rotating coil). The two windings can be connected in series or in parallel.

With a **series-wound** motor the same current flows in both coils.

- When the motor is stationary, a large current flows, because there is little resistance in the coils, and there is a large torque on the armature to get it started.
- When the motor is rotating, electromagnetic induction produces an e.m.f. that opposes the current which is consequently reduced in both coils.

Motors using a.c. are also series-wound. With a.c. field winding, the magnetic field is continually reversing. However, the torque in the armature does not reverse, as the current in that coil is reversed as well. Motors using a.c. are useful in situations where it is inconvenient to use d.c., such as in a mains-powered electric drill.

In a **shunt-wound** motor, the field coil is connected in parallel to the armature coil. The current in the coils is different, so there is no strong initial push, but the rotation rate is more constant under varying load conditions.

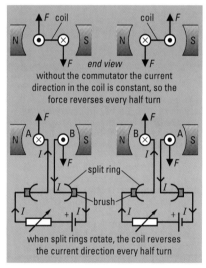

without the commutator the current direction in the coil is constant, so the force reverses every half turn

when split rings rotate, the coil reverses the current direction every half turn

The split-ring commutator rotates with the coil. Small blocks of carbon, called **brushes**, *are held against the split rings by springs, maintaining the electrical contact.*

PRACTICE

1 A coil of area 10 cm² is placed in a uniform magnetic field of strength 0.6 T. The coil has 100 turns and a resistance of 5 Ω. It is supplied with current from a 9 V cell. Calculate the maximum torque exerted on the coil.

2 Sketch a graph showing how the torque on the coil in question 1 varies with angle between the plane of the coil and the magnetic field direction.

3 A 25 turn rectangular coil 6 cm by 4 cm is placed with its long sides parallel to a long, straight wire. The wire carries a current of 6.8 A. The coil and the wire are in the same vertical plane, and the side of the coil nearest to the wire is 4 cm from the wire. If the current in the coil is 2.0 A, what are the forces on each of the long sides of the coil? Will the combination of forces acting on the coil cause it to rotate?

OBJECTIVES

- the moving-coil loudspeaker
- the relay
- the solenoid lock

ELECTROMAGNETIC DEVICES – 2. THE LOUDSPEAKER AND THE RELAY

The moving-coil loudspeaker

Although there have been many alternative loudspeakers developed, the moving-coil loudspeaker remains the cheapest way of reproducing sound.

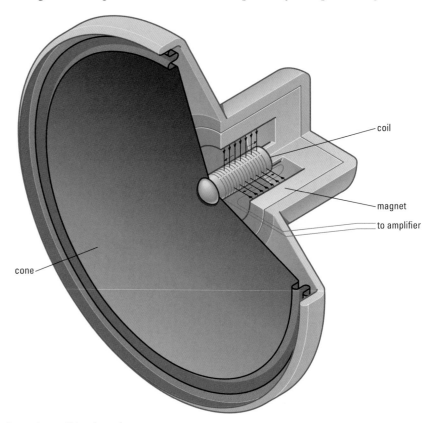

A moving-coil loudspeaker.

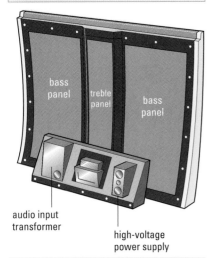

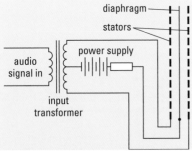

The electrostatic loudspeaker. In speakers of this type, thin diaphragms are stretched between stator panels and charged to several thousand volts above the stator panels. There is no current flow, however, because the diaphragm is insulated from the stators. The audio signal drives a step-up transformer that increases the voltage of the signal by about 50 times. The output of the transformer is connected to the stator panels. Potential difference across the stators causes a linear electric field between them, and this causes the diaphragm to be pushed and pulled towards one or other of the stators by electrostatic force. Advantages are that the diaphragm is very light and therefore capable of reproducing the entire audio spectrum. It is also uniformly driven so that it moves in a very controlled linear fashion. Its spring constant is high, so it couples very well to the air and is therefore well damped. Disadvantages are that the audio signal passes through a transformer and cancellation effects that plague all dipoles mean that the panels must be large to produce any bass at all – hence the separate bass panels.

A typical hi-fi system loudspeaker using two moving-coil 'drive units'.

The cut-away drawing of the loudspeaker shows that the magnet used consists of a central cylindrical pole piece surrounded by an outer ring, on which there are other poles. This produces a radial field that passes across the plane of the coil, and the forces acting on the coil's turns are at 90° to the plane of the coil. When the current in the coil is circulating in one direction it will be pushed out, and when the current's direction is reversed, it is pulled in. A continually reversing current will set the cone vibrating, so that a sound wave is generated.

Modern moving-coil loudspeakers have been designed very carefully to achieve the best sound quality.

- Thin plastic cones are used to achieve rigidity (to stop the cone resonating) and for lightness (so it can be started and stopped easily).

- Moulded cones are used with gently changing cross-sections, to control resonance.

- Cones are bonded to the frame with strong glue.

- Non-magnetic frames are used, to avoid interference with the coil.

A loudspeaker unit consists of two moving-coil loudspeakers (or **drivers**). The larger driver (the **woofer**) produces low-frequency notes, while high-frequency notes are produced by the smaller unit (the **tweeter**). Smaller loudspeakers are used for the tweeter, to try to make the unit nearly the same size as the wavelength that it is producing. If the cone's diameter was much bigger than the wavelength, the sound would only be beamed out directly in front of the unit, making it very hard to hear either side of the speaker.

The relay

Relays are used when large electric currents have to be switched indirectly. A small current in the coil magnetizes the soft iron core, and this attracts the armature, which pivots and presses contacts A and B together. At the same time contacts B and C are pushed apart. The contacts can be designed to carry currents far in excess of those required to magnetize the coil.

Relays have many applications in cars. For example, the starter motor on a car will draw up to 100 A from the battery. To survive this current the switch would have to be robust for reasons of safety, and would not look very attractive mounted on a dashboard.

- The current to the starter motor is controlled by the contacts of a relay mounted in the engine compartment.
- The current in the relay coil is very much smaller, and only requires an ordinary switch to operate it.

Other parts, such as the heated rear windscreen, fog lamps, and windscreen wipers, are all operated by relays.

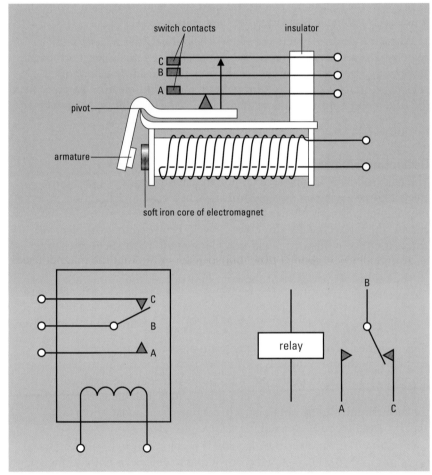

An electromagnetic relay.

The solenoid lock

In this device the current in the solenoid magnetizes the soft iron bolt. As the direction of magnetization in the soft iron is the same as that in the solenoid field, the pole at the end of the bolt will be opposite to that at the end of the solenoid. Consequently the bolt is drawn into the solenoid against the force of the spring. Once the current is removed, the spring forces the bolt out of the solenoid.

Such devices are used in secure buildings where visitors are 'buzzed' in when the electrical signal opens the bolt on the lock. They are also found in the central locking systems of cars, where the bolt actuates the door locks via a lever.

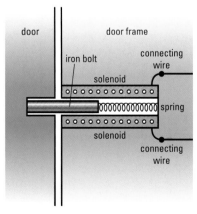

A solenoid lock.

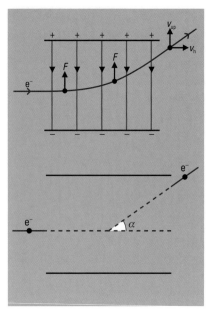

As an electron beam moves between two plates, the direction of the force acting does not change.

Maths box 1: derivation of electron's path in an electric field

If the p.d. between the plates is V and their separation is d, then the force acting on the electron is

$$F = Eq = e\frac{V}{d} \quad as \; E = \frac{V}{d}$$

The electron's vertical acceleration is

$$a = \frac{eV}{md}$$

where m is the mass of the electron.

At a particular point between the plates, the electron has travelled a horizontal distance x in a time $t = x/v_h$. The electron has also been accelerating upwards for this period of time, and so has travelled a vertical distance y given by

$$y = \frac{1}{2}at^2 = \frac{1}{2} \times \left(\frac{eV}{md}\right) \times \left(\frac{x}{v_h}\right)^2$$

$$= \left(\frac{eV}{2mdv_h{}^2}\right)x^2$$

The items in the brackets are constant, so the vertical distance y is proportional to the horizontal distance x^2. Hence the path is parabolic.

DEFLECTING ELECTRON BEAMS

Modern displays for televisions, computers, and laptops are virtually all LED or plasma based. However, the older technology based on cathode ray tubes (CRTs) still has some interest. The basic principle on which the CRT operates is to illuminate a small patch of the screen with a beam of electrons, which can be steered to different parts of the display at different intensities to draw a picture. This demands fine control over the path followed by the electron beam, something that can be done by deflection using either magnetic or electrical forces. TV tubes use magnetic deflection, while oscilloscopes use electrostatic deflection.

Deflecting electrons in electric fields

In the diagram opposite an electron beam is passing through a uniform electric field at 90° (generated by parallel, charged plates). The direction of the force acting on the electron is constant and towards the positive plate (upwards on the diagram). The electron's path curves upwards towards the positive plate and continues along a tangent line to the curve as it leaves the region of the electric field. The shape of the curved path followed is *parabolic* (provided the field is uniform).

When the electron enters the electric field it only has a horizontal velocity. The acceleration produced by the electric force gives it a vertical velocity, v_{up}, by the time it reaches the end of the field. The horizontal velocity, v_h, remains constant throughout, and so the electron emerges from the field along a straight line at an angle α to the original path given by

$$\tan\alpha = \frac{v_{up}}{v_h}$$

Deflecting electrons in magnetic fields

The path followed by an electron in a magnetic field is not parabolic. The electric force acting on the electron stays in a constant direction, whereas the magnetic force is always at 90° to the direction of motion. As the path curves, the force changes direction as well. Provided that the electron enters a uniform magnetic field in a plane at 90° to the field lines, the electron will move along a path that is an arc of a circle.

The radius of curvature of the arc r depends on the momentum p of the electron:

$$r = \frac{mv}{Be} = \frac{p}{Be}$$

This provides a convenient and accurate method for measuring a particle's momentum directly. If the strength of the magnetic field is well mapped and the path of the particle can be followed electronically (or via a photograph) the radius of curvature can be calculated from which the momentum can be obtained. This is one of the most important experimental techniques used in particle physics.

In most circumstances the electron is moving through some material (even air) and so will be slowing down due to collisions with atoms. As the electron slows down, r will decrease, and so the path will spiral inwards until the electron comes to rest. This produces some of the most interesting and pleasing experimental pictures.

Velocity selection

The combined action of an electric and magnetic field on a moving charge can be used to act as a velocity filter. If the fields are arranged at 90° to each other (see spread 8.3) then the forces can act in opposite directions along the same line. A beam of particles will be undeflected if:

$$Eq = Bqv \quad so \quad v = \frac{E}{B}$$

If the beam is aimed across a space containing the fields at a metal plate with a hole in it, then only the undeflected particles will get through. So, by tuning E and B, a specific velocity can be selected.

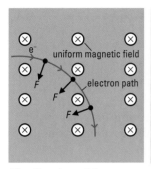

The direction of the magnetic force acting on a moving electron changes as the electron's direction of motion changes.

As a charged particle loses energy in a bubble chamber the curvature of its path tightens, producing a spiral.

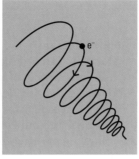

In three dimensions, the spiral is actually helical, because the particle's motion is not exactly at 90° to the magnetic field direction.

Maths box 2: derivation of electron's path in a magnetic field

The size of the force acting on the electron will be Bev, and its direction is always at 90° to the instantaneous direction of motion. The curve of the path will be determined by the amount of centripetal force that the magnetic field can provide. That is,

$$\frac{mv^2}{r} = Bev \quad \therefore \quad r = \frac{mv}{Be}$$

If the electron's initial path is not in a plane at 90° to the magnetic field lines, the resulting curve will move out of the same plane. The result is a helical path as shown in the diagram above left.

Mass spectrometers

The action of electric and magnetic fields on charged particles can be used to separate ions in a mass spectrometer.

The **mass spectrometer** separates a chemical sample into its components and records their relative abundances. The sample is vaporized and passed through an electric field, decomposing it into separate ions. Another electric field then accelerates the ions. Clearly the velocity achieved by the ions depends on their mass, but the mass spectrometer relies on ions of equal velocity passing through a magnetic field in order to separate them. A velocity selector is used in concert with the accelerating voltage to ensure that samples of each ion enter the device in turn, with equal velocity. The magnetic field is arranged at 90° to the ion's direction of flight, consequently:

$$\text{radius of curvature of ion's path} = B\frac{Q}{M}v$$

Hence as the velocity is fixed, the curvature depends on Q/M. The output of the spectrometer runs from a detector (a photographic plate or electronic detector) at the end of a path of very specific curvature. The B field is adjusted so that each ion sample in turn strikes the detector, which records the number of hits in a given period. The strength of the B field gives Q/M for that ion, and the hits record the relative abundance.

PRACTICE

1 A beam of electrons moving at $2 \times 10^7\,\text{m s}^{-1}$ is directed between a pair of parallel plates. The plates are 10 cm long and separated by 5 cm, and the p.d. between them is 100 V. The electrons enter the region of the plates at 90° to the field. Calculate the vertical velocity of the electrons as they leave the plates. What is the total velocity of the electrons after leaving the plates? The electrons continue at constant speed until they strike a screen 30 cm from the centre of the plates. If the electric field is turned off, the electrons hit the middle of the screen. With the field turned on, how far from the centre of the screen do the electrons now strike?

2 In which direction will the magnetic field be acting in the photograph above of a particle losing energy in a bubble chamber if the particle is negative?

3 An electron starts from rest and is accelerated through a p.d. of 200 V. What is the electron's kinetic energy at the end of the acceleration? At what speed will it be moving after the acceleration? The electron now enters a region in which the magnetic field strength is 0.2 T and at 90° to the electron's path. What is the force on the electron? What is the radius of curvature of the path followed? How long would the electron have to remain in the magnetic field for it to end up travelling at 90° to its initial direction?

O B J E C T I V E S

- the electron gun
- the cathode-ray tube
- the time base
- TV pictures

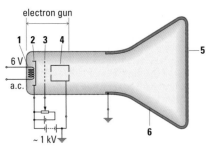

A cross-sectional diagram of a typical cathode-ray tube.

THE CATHODE-RAY TUBE

The electron gun

An **electron gun** produces and accelerates a beam of electrons. They are used in a variety of applications from TV monitors to electron microscopes, but the design is basically always the same. Modern colour TVs have three electron guns producing separate beams for each of the three primary colours produced on the screen. An electron gun contains the following parts (see figure).

1 **The heater** This is a small heating coil that raises the temperature of the cathode.

2 **The cathode** This is a metal plate warmed by the heater and held at a negative potential compared with the anode. (The anode is normally set to 0 V and the cathode to about –1000 V.) In a vacuum the heater provides electrons with enough energy to escape from the cathode (aided by repulsion), and replacement electrons are drawn from the power supply.

3 **The grid** This is held at a slightly negative potential with respect to the cathode and repels some electrons back. In this way the number of electrons in the beam can be controlled (only the fastest get through), and so the brightness can be adjusted.

4 **The anode** Shaped like a can with a small hole at either end, the anode accelerates the electrons by attracting them. It also helps to focus the beam as tightly as possible.

The screen

The screen of a **cathode-ray tube** (**CRT**) is coated with a phosphorescent material (**5**). When the electron beam strikes the screen, the kinetic energy of the electrons is transferred to the chemicals in the screen. This causes light to be emitted. If the beam is tightly focused, a small glowing spot will mark the impact point on the screen.

As more electrons arrive, the screen would slowly become negatively charged, and start to repel the beam. To avoid this the inner surface of the tube is coated with a graphite layer that conducts the electrons to earth (**6**).

The time base

To build up a picture the spot must be moved both across the screen and up and down. In an oscilloscope the spot simply needs to move across the screen at a regular rate while it is deflected up and down by the signal that is being measured. The horizontal deflection is controlled by a circuit known as the **time base**, and this produces a varying p.d. across the X plates.

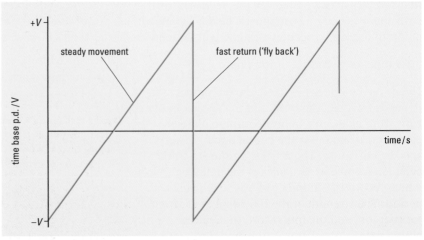

The time-base circuit produces this voltage variation, which drives the electron beam across the screen at a uniform rate.

As the p.d. increases uniformly, the spot moves across the screen from left to right at a steady speed. The rapid drop in p.d. returns the spot to the far left too rapidly for the eye to follow. This gives the illusion of a continual supply of spots marching from left to right across the screen. If the spot speed is fast enough, the eye cannot separate the individual spots, and a continuous line is apparent.

The signal to be measured is connected to the Y plates which deflect the spot vertically.

Using an oscilloscope to measure a.c. signals

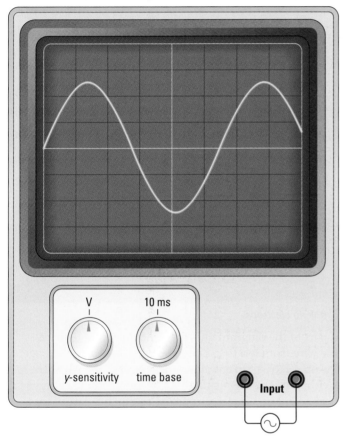

An oscilloscope can be used to measure the amplitude and period of an a.c. current.

An oscilloscope can be used to measure the amplitude and period (frequency) of an a.c. signal. Effectively the screen displays a graph of p.d. (y-axis) vs time (x-axis) for the signal.

The time-base control adjusts the time taken for the spot to traverse one square horizontally across the screen (the x-axis scale). In the diagram, one period corresponds to 5.5 squares along the x-axis. At $10\,\text{ms}\,\text{sq}^{-1}$ the period is measured to be $55\,\text{ms}$, and the frequency $1/T$ is $18.2\,\text{Hz}$.

The y-sensitivity sets the y-axis scale on the display. This a.c. p.d. rises to a maximum of 2.5 squares and given that the scale is $1\,\text{V}\,\text{sq}^{-1}$ the amplitude of the varying p.d. is measured to be $2.5\,\text{V}$.

D.c. voltages can also be measured using the same technique. The screen will show a horizontal line (as the p.d. is not changing with time) a certain number of squares above the x-axis (for a positive p.d.) or below it (for a negative p.d.). Counting the number of squares between the line and the x-axis, and multiplying by the y-axis scale, determines the p.d.

The oscilloscope can also be used to measure current indirectly, if it is set to display the p.d. across a known resistance. Determining the p.d. and applying Ohm's law provides the current.

- producing currents in moving conductors

- e.m.f. across a moving wire

MAGNETICALLY INDUCED ELECTRIC CURRENTS

The magnetic force on a moving charge can be used as a means of generating an electric current. Any piece of conducting material contains charges, and simply moving the material through a magnetic field will result in forces being exerted on those charges. We normally think of magnetic forces as acting on currents, but really they act on any moving charge – even the charges within a wire that is moving.

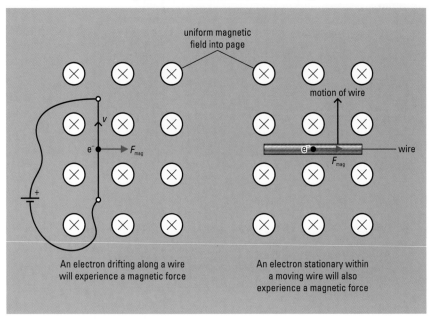

Magnetic forces on moving electrons.

In the right-hand part of the diagram above, the magnetic force drives electrons along the wire from left to right as the wire moves up the page. Effectively, there is an electric current along the length of the wire. The electrons collide with atoms as they drift along, so their kinetic energy is converted into thermal energy. The electrons are as much a part of the wire as the atoms within it, so the wire is also losing kinetic energy. This effect is known as **current braking** and can be used in practical situations.

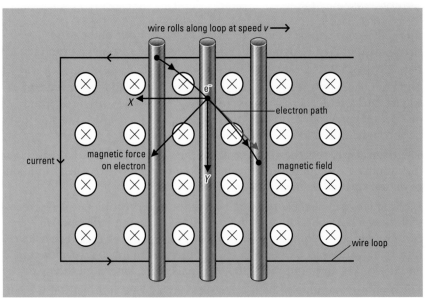

The magnetic force on the electrons in the wire has two components. X is part of the BII force on the wire; Y drives the electrons down the wire, establishing the current.

If there is some way of removing charge at one end of the wire and replacing it at the other, a continuous electric current can be maintained in the wire. This effect was investigated by NASA as a means of generating power for satellites in orbit.

As the satellite orbits with an electrical tether unwound, the force due to the Earth's magnetic field drives electrons along the wire. The positively charged end of the tether will attract negatively charged particles from space (many are trapped in the Earth's magnetic field). The other end of the wire will be negative and will attract positive particles. These particles discharge the ends of the wire, effectively closing the circuit.

In this situation the kinetic energy of the satellite is being converted into thermal energy in the wire (and in any load placed on the circuit).

Electromotive force (e.m.f.) across a moving wire

In the slightly simpler situation of a straight wire moving through a magnetic field, the e.m.f. across the wire can be calculated. A constant current can be maintained by having it roll along a wire loop as in the diagram on the previous page. The magnetic force acting on the electrons has a component perpendicular to the wire which is the BIl force that acts on any current-carrying wire.

magnetic force on current = BIl

This force acts in the opposite direction to the motion of the wire, and so an external force, F, is required to keep a constant speed.

∴ force required to keep a constant speed = BIl

∴ work done on wire per second = force × $\dfrac{\text{distance moved in one second}}{}$

= $BIlv$

This energy must replace that converted into thermal energy in the wire.

∴ $BIlv = I^2R$

(R is the resistance of the wire), so

$Blv = IR$

The left-hand term in this equation represents the conversion of kinetic energy into electrical energy and so is classed as an e.m.f. The right-hand term is the voltage drop across the wire. Hence we can say that the e.m.f. induced in the moving wire is given by

$E = Blv$

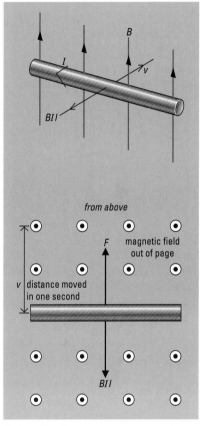

Calculating the e.m.f across a moving wire.

> **Broken tether**
>
> Unfortunately the first experiment carried out aboard the Space Shuttle failed. The tether broke as it was being unreeled – a shard of metal accidentally cut through the wire.

PRACTICE

1 The length of a train axle is 3 m. The train is moving at 40 km h⁻¹ at 90° to a magnetic field of 50 µT. What is the e.m.f. induced across the axle?

2 A conducting wire, 0.5 m long, is moving at 20 m s⁻¹ through a magnetic field of 25 µT so that the wire is at 90° to the field lines. The induced current is 250 µA. What is the force required to maintain the constant speed? What is the work done per second by this force? What is the resistance of the wire?

3 An orbiting satellite is travelling at 3100 m s⁻¹. It extends a 20 km long conducting tether so that the line of the tether is perpendicular to the Earth's surface. A current of 5 A is measured in the tether and the power generated is 15 kW. Estimate the average magnetic field strength along the length of the tether.

4 By making reasonable estimations of the required numbers, estimate the e.m.f. induced between the tips of a passenger aeroplane's wings while flying through the Earth's magnetic field (50 µT perpendicular to the wing's surface). Is this likely to be dangerous?

ELECTRICALLY INDUCED CURRENTS

In the previous spread a plan for producing an electric current in a tether extending from a satellite was described. Viewed from Earth, this effect is due to the magnetic force on the charges within the tether deflecting their motion along its length. Provided the ends of the wire are discharged (by attracting charged particles from space), a constant current can be maintained.

However, from the astronaut's point of view this explanation will not do. They see a stationary wire and so no magnetic forces can be acting on the charges within it. Yet they can measure a current in the tether.

The astronauts are observing the tether in a **frame of reference** in which it is stationary and the magnetic field is moving. People on Earth are observing from another frame of reference in which the tether is moving through a magnetic field. Both frames of reference must observe the same physical effect – an e.m.f. is acting. From one frame of reference that e.m.f. is due to a magnetic force; from the other frame of reference it is due to an **induced electric force**.

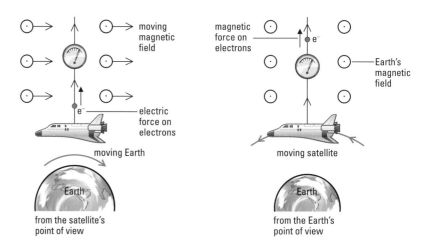

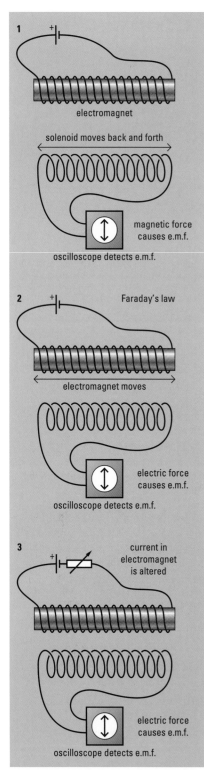

Three experiments that demonstrate induction effects. An experimenter who could only see the oscilloscope would not be able to tell which of the three experiments was taking place.

Observing the satellite tether experiment from two equivalent frames of reference.

Induced electric fields

This effect can be demonstrated in a laboratory. There are three possibilities:

1 Imagine holding an electromagnet still and moving a solenoid back and forth through the field. The magnetic force causes a current in the solenoid, and the e.m.f. can be detected by connecting the leads from the solenoid to an oscilloscope. The spot on the oscilloscope is deflected up and down as the electromagnet moves.

2 Now imagine keeping the solenoid still and moving the electromagnet back and forth. There can be no magnetic force acting, but the oscilloscope spot still moves. The two situations are equivalent to each other. Someone moving back and forth with the electromagnet would see it as being still and the solenoid moving – exactly the same as the previous experiment.

3 A final variation is a stationary solenoid in a magnetic field that is changing in strength. Nothing is moving, yet the oscilloscope will detect an e.m.f. in the solenoid. From the solenoid's point of view, all that is happening is that the magnetic field passing through it is changing. The change is not because the magnet producing the field is getting *nearer* (as before); this time the magnet is staying in the same place but is getting *stronger*.

In **1** the e.m.f. is produced by a magnetic field. In **2** and **3** the e.m.f. is due to an *induced electric field*. The strength of the induced electric field depends on the rate at which the magnetic field is changing. In order to quantify this it is helpful to introduce the **flux** of a magnetic field.

Magnetic flux

Definition: magnetic flux

$$\left(\begin{array}{c} \textbf{magnetic flux} \text{ in area } A \\ \text{of a magnetic field (Wb)} \end{array} \right) = \left(\begin{array}{c} \text{average component of} \\ B \text{ at } 90° \text{ to area (T)} \end{array} \right) \times \text{area } A \text{ (m}^2)$$

$$\phi = B_\perp \times A$$

The unit of flux is the weber (Wb), and $1\,\text{Wb} = 1\,\text{T m}^2$.

- The **flux linked** to a coil ($N\phi$) with multiple turns is the flux through one turn (ϕ) multiplied by the number of turns (N).
- The **flux cut** by a moving wire is the magnetic field strength multiplied by the area swept out by the wire.

Worked example 1

A 100 cm² wire loop is placed in a uniform magnetic field that passes through the plane of the loop at 90°. The magnetic field strength is 0.03 T. What is the flux through to the loop?

flux, $\phi = (0.03\,\text{T}) \times (100 \times 10^{-4}\,\text{m}^2) = 3 \times 10^{-4}\,\text{Wb}$

Worked example 2

What is the flux linked to the coil shown in the diagram on the top right?

$B_\perp = B \cos\alpha$ ∴ flux linked $N\phi = BAN \cos\alpha$

Worked example 3

A wire of length l is moving horizontally at constant speed v in a direction at 90° to its length. The wire is passing through a vertical magnetic field of uniform strength B. What is the flux cut by the wire per second?

In one second, the wire moves a distance v, and the area swept out $= lv$.

∴ flux cut by the wire per second $= Blv$

The e.m.f. induced in a circuit can be calculated from **Faraday's law of induction**.

Faraday's law of induction

e.m.f. = – rate of change of flux linkage $= -\dfrac{\Delta(N\phi)}{\Delta t} = -N\dfrac{\Delta\phi}{\Delta t}$ when N is constant

Or in calculus notation for the instantaneous rate of change of the flux:

$$\varepsilon = -\dfrac{\text{d}(N\phi)}{\text{d}t} = -N\dfrac{\text{d}\phi}{\text{d}t}$$

The minus sign indicates the direction in which the e.m.f. is induced (see above right).

Worked example 4

What is the e.m.f. across the moving wire from worked example 3?

e.m.f. = – rate of change of flux linkage = – flux cut per second $= -Blv$

Note: this is the same answer that we obtained previously using the magnetic force acting on the charges.

Worked example 5

The magnetic field passing through the 100 cm² coil in worked example 1 reduces from 0.03 T to 0.01 T in 20 s. What is the e.m.f. induced in the coil?

e.m.f. = – rate of change of flux $= -\dfrac{(0.01\,\text{T}) - (0.03\,\text{T})}{20\,\text{s}} = 1\,\text{mV}$

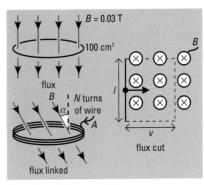

Flux, flux linked, and flux cut are essentially the same physical ideas in slightly different contexts.

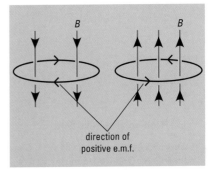

The convention for the direction of positive e.m.f.: if the magnetic field lines are pointing down into the page when viewed from above, a positive e.m.f. will be in a clockwise direction round a wire loop. In the case of a moving wire, the direction is taken round the perimeter of the area swept out.

PRACTICE

1 A flat loop of wire of area 50 cm² has a uniform magnetic field of 0.05 T passing through the plane of the loop at 90°. The loop is now rotated slowly until the loop lies parallel to the field lines. Sketch the situation before and after the rotation of the loop. If the loop takes 30 s to get to its new position, what is the e.m.f. induced in the loop while it is moving?

2 A 50 turn coil of wire of area 4 cm² is placed inside an electromagnet. The magnetic field strength is 0.03 T and passes through the coil at 90°. What is the flux linked to the coil? The electromagnet is turned off and the field drops uniformly to zero in 0.12 s. What is the e.m.f. induced in the coil?

ELECTROMAGNETIC INDUCTION

Faraday's law

Faraday's law is a remarkable piece of physics because it links three physically different effects with one equation:

$$\text{e.m.f.} = -\text{ rate of change of flux linkage} = -N\frac{d\phi}{dt}$$

1 $\dfrac{d\phi}{dt}$ can be the flux cut by a moving wire.

2 $\dfrac{d\phi}{dt}$ can be the change in flux due to a moving magnet.

3 $\dfrac{d\phi}{dt}$ can be the change in flux due to a stationary magnet which is changing in strength.

Faraday's contribution was to experimentally demonstrate the equivalence of **2** and **3**. The law of induction does not identify what physical effect is behind the e.m.f. (in **1** it is a magnetic force; in **2** and **3** it is an induced electric field), but it guarantees that however one looks at the situation the same physical result will always be seen.

Induction and relativity

It seems magical that an induced electric field can spring up in one frame of reference when there is a moving or changing magnetic field. Unlike any other electric field an induced field does not have electric charges acting as its source. The field lines of induced fields form closed loops – like magnetic field lines always do. Einstein's powerful physical intuition worked on the idea that a magnetic field in one frame of reference could appear as an electric field in another, and the result was the **special theory of relativity**.

The link between electric and magnetic fields suggests that there is only one *electromagnetic field* which interacts with stationary and moving charges. Any physical effect can be explained by an electric force, a magnetic force, or a mixture of both depending on the frame of reference from which the experiment is being viewed.

Think of this as being rather like the components of a vector.

- Any vector can be resolved into components, and there are many ways of doing this depending on the axes chosen.
- The electromagnetic field can be split into electric and magnetic 'components' in different combinations, depending on the frame of reference in which the field is viewed.

Lenz's law

The induced current is always in a direction that will help to counteract the change in flux that is producing it.

Non-conservative fields

Induced electric fields are not conservative. The work done in moving a charge round a loop in a conservative field is zero. An induced field can drive a current round a loop of wire. To do this requires energy as the current heats the wire, and so induced electric fields cannot be conservative. This is a direct consequence of them not being produced by charges.

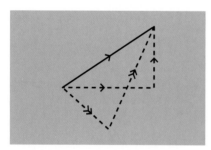

A vector can be resolved into components in many different ways.

Lenz's law

The minus sign in Faraday's law shows the direction in which the e.m.f. is produced. It is expressing a law of physics discovered in 1834 by the Russian physicist Emil Lenz.

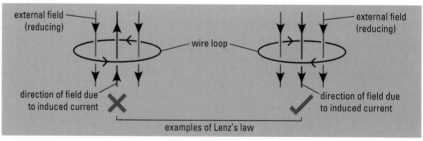

Lenz's law implies that the field due to an induced current must cancel the change in the field that is producing it.

In the diagram above, the magnetic field strength through the loop is being steadily *reduced*. The total magnetic field in the loop is made up of two parts – the external magnetic field (produced in some way not shown

advanced **PHYSICS**

on the diagram) and the magnetic field due to the induced current. The induced current could circulate either clockwise or anticlockwise, and Lenz's law helps to decide which is correct.

Anticlockwise The field of the induced current is in the *opposite* direction to the external field. The two fields tend to *cancel* inside the loop. This would *reduce* the flux linked, further inducing a greater current; but a greater current has a greater magnetic field, which improves the cancellation, and the problem gets worse. This situation would violate conservation of energy, as the current would increase almost without limit.

Clockwise The field of the induced current *adds* to that of the external field, partly compensating for its reduction, and this tends to *reduce* the size of the induced current, rather than increasing it. The magnetic field of the induced current can never completely replace the lost external field, so there is some flux loss, but the situation does *not* violate conservation of energy.

These examples show that Lenz's law is effectively a restatement of conservation of energy.

Current braking

A metal ring falling towards a solenoid will experience a braking force that slows its fall. As the ring is falling towards the solenoid, the magnetic field strength and so the flux linked to the ring will be increasing, and so a current is induced in the ring. Lenz's law shows that the current will circulate in the ring in the direction that allows the magnetic field of the current to compensate for the change in flux. The ring acts like a small coil, magnetized with the opposite pole facing downwards, and so is repelled by the solenoid.

The induced currents will not be large enough to cause the ring to hover, because the size of the current is limited by the induced e.m.f. and the resistance of the ring. If the same experiment is carried out using a superconducting ring, then it can be made to float. Alternatively, a small magnet will float above a superconducting dish. The currents induced in superconductors are always large enough to completely cancel the flux crossing their surface. (This is known as the **Meissner effect**.)

Currents can be induced in ordinary pieces of metal. Although the e.m.f.s are generally not very large, the **eddy currents** follow low resistance paths through the metallic structure and can be quite large. Lenz's law implies that they act to oppose the motion that produces them. They can act as a very efficient braking mechanism as the kinetic energy is dissipated as heat. Their main drawback is that as the motion slows down so the eddy currents become smaller, and the braking effect is reduced.

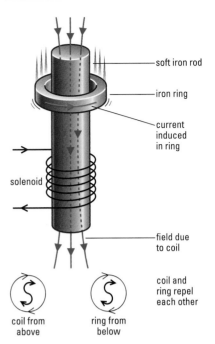

soft iron rod

iron ring

current induced in ring

solenoid

field due to coil

coil and ring repel each other

coil from above   ring from below

Current braking.

The eddy currents in a superconductor are so strong that the repulsion between them and the currents in a magnet can cause the magnet to float above the surface of the superconductor.

Whirlpools and eddies

The currents induced in metals follow paths that are small loops inside the body of the metal. As these remind people of the small whirlpools formed in streams (eddies) they are referred to as eddy currents.

PRACTICE

1 A straight wire is moving horizontally through a vertical magnetic field at a constant speed v. Sketch a diagram of the position of the wire at two moments A and B which are 1s apart. Consider the rectangle swept out by the wire as it moves from A to B. What happens to the flux linked to this rectangle as the wire moves from A to B? Use Lenz's law to work out the direction of the current induced in the wire. (Consider the magnetic field of the induced current.) Does this agree with the direction suggested by Faraday's law?

2 A soft iron rod is placed through the axis of a solenoid and an aluminium ring is threaded down the rod to rest on the solenoid. A constant d.c. current is now turned on in the solenoid. What happens to the ring? Explain your answer. The current is now turned off. State and explain what now happens to the ring. A ring is now cut so that it forms a 'C' shape when viewed from above, and the experiment is repeated. This time nothing happens. Explain.

3 Describe a simple experiment that you could use to demonstrate Lenz's law to a non-scientific audience.

Helpful hint

In the next few sections:

N is the total number of turns in a coil

n is the number of turns per unit length

Helpful hint

Differentiating sin and cos functions is covered in the Mathematics toolbox. Remember, in the equation for e.m.f. the angle ωt is in radians, and your calculator should be set accordingly.

The electrical tether developed by NASA is not a practical method for generating electricity on Earth – the speeds involved are too great, for one thing. However, the same basic effect – the magnetic induction of a current – can be used by rotating a coil in a magnetic field.

A rotating coil in a magnetic field

The diagram below shows a flat rectangular coil of N turns being rotated at a constant angular velocity ω in a magnetic field. At the instant shown, the coil is at an angle θ to the magnetic field direction. The flux linked to the coil is given by the component of the field strength passing through the coil at 90°, that is

$$N\phi = \text{area of coil} \times \text{number of turns} \times B \sin \theta$$

Assuming that the coil was horizontal when we started timing the rotation:

$$\theta = \text{angular velocity} \times \text{time} = \omega t$$

$$\therefore \quad N\phi = BAN \sin \omega t$$

The flux linked to the coil alters as the coil rotates, and Faraday's law can be used to calculate the e.m.f. induced:

$$\varepsilon = -\frac{\mathrm{d}(N\phi)}{\mathrm{d}t} = -\frac{\mathrm{d}(BAN \sin \omega t)}{\mathrm{d}t}$$

$$\therefore \quad \varepsilon = -BAN\omega \cos \omega t$$

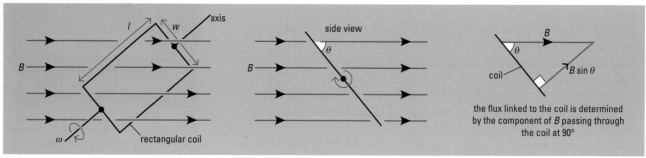

A rotating coil in a magnetic field.

the flux linked to the coil is determined by the component of B passing through the coil at 90°

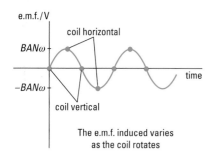

The e.m.f. in a rotating coil as a function of time.

The e.m.f. induced varies as the coil rotates

There are several important points relating to this result:

• When $t = 0$ or π/ω, the coil is *horizontal* and the *flux linked* is therefore zero.

• When the coil is *vertical* the *flux linked* is a *maximum*, $\phi = BNA$.

• When the coil is *horizontal*, the *induced e.m.f.* is a *maximum*, $\varepsilon = -BNA\omega$.

• When the coil is *vertical* the *induced e.m.f.* is *zero*.

This seems contradictory, but remember that the e.m.f. is determined by the rate at which the flux *changes*. As the coil passes through the horizontal position the flux is changing rapidly; as the coil is moving through the vertical position the flux is hardly changing.

• Every time the coil passes through the vertical the current reverses direction.

• The size of the e.m.f. is proportional to the rate at which the coil is rotating.

The diagram on the left shows an oscilloscope trace recording the e.m.f. induced in a rotating coil. In the second trace the coil speed has been doubled. Notice that the frequency at which the e.m.f. changes direction has increased as well as the size of the e.m.f.

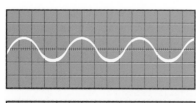

Oscilloscope traces of the e.m.f. in a rotating coil.

This technique can be used for generating electric current. In practice, turbines work by keeping a fixed coil and rotating the magnetic field generators.

Magnetically linked coils

The diagram on the right shows two coils wound round the same piece of soft iron. This arrangement ensures that the coils are magnetically linked to each other. Passing an alternating current through one of the coils induces an alternating e.m.f. in the other. This effect, known as **mutual induction**, is used in the design of transformers that can step voltages up or down.

The alternating current in coil 1 produces a varying magnetic field that magnetizes the soft iron in the same varying pattern. This ensures that there is a changing magnetic field passing through the centre of coil 2. A high-resistance voltmeter connected to this coil records the changing e.m.f. that is electrically induced. This e.m.f. is proportional to the rate at which the current is varying in coil 1.

$$\varepsilon_2 \propto -\frac{\mathrm{d}(I_1)}{\mathrm{d}t} = -M\frac{\mathrm{d}(I_1)}{\mathrm{d}t}$$

where I_1 is the size of the current in coil 1 at any time t and the constant of proportionality, M, is the **mutual inductance** of the two coils measured in the henry (H).

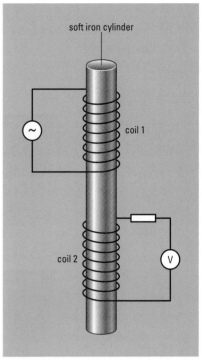

Mutual induction in two coils wound round the same piece of soft iron.

Maths box: deriving the equation for mutual inductance of two coils

If I_1 is the size of the current in coil 1 at any time t then the magnetic field in coil 1 is

$B_1 = \mu_0 n_1 I_1$ (n_1 is the number of turns per metre in coil 1)

The field passing through coil 2 is

$B_2 = \mu_r \mu_0 n_1 I_1$

where μ_r is the **relative magnetic permeability** of the soft iron. Consequently, the flux linked to the second coil is

$\phi_2 = A_2 N_2 B_2$ (N_2 is the number of turns in coil 2)

$= A_2 N_2 \mu_r \mu_0 n_1 I_1$

The e.m.f. induced by this changing flux is given by Faraday's law:

$\varepsilon = -\dfrac{\mathrm{d}(A_2 N_2 \mu_r \mu_0 n_1 I_1)}{\mathrm{d}t}$

Everything in the bracket is constant apart from I_1, so

$\varepsilon = -A_2 N_2 \mu_r \mu_0 n_1 \dfrac{\mathrm{d}(I_1)}{\mathrm{d}t} = -M\dfrac{\mathrm{d}(I_1)}{\mathrm{d}t}$ $\therefore M = A_2 N_2 \mu_r \mu_0 n_1$

Mutual inductance, M, is a constant and depends on the number of turns, area of the coils, etc. NB This equation applies only to the configuration shown in the diagram.

PRACTICE

1 What is the angular velocity equivalent to 30 r.p.m. (revolutions per minute)? What is the angular velocity of the minute hand on a clock?

2 What is the maximum e.m.f. that can be obtained from a rotating coil of 100 turns in a magnetic field of 0.2 T, if the area of the coil is 2.5 cm² and it is rotating at 20 r.p.m.?

3 Sketch a graph showing how the flux cut by the rotating coil in the diagram on the previous page varies with time for 2 complete rotations. On the same set of axes, sketch a graph showing how the induced e.m.f. varies with time. Be sure to include relevant information on the vertical axes of both graphs.

4 In the situation shown in the diagram above, the current in coil 1 is changing at 0.5 A s⁻¹. The e.m.f. in coil 2 is 0.1 V. What is the mutual inductance of the two coils? The situation is now reversed – the changing current is in coil 2 and coil 1 is connected to the ammeter. What rate of change of current in coil 2 is required to produce an e.m.f. of 0.3 V in coil 1? (Assume that the two coils are of the same length and cross-sectional area.)

5 What is the henry in base units?

- self-induction

- back e.m.f.

- back e.m.f. in motors

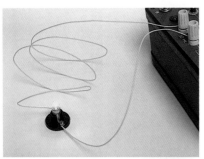

Top: a lamp connected to an a.c. power supply by resistance wire. Bottom: it is self-induction that causes the bulb to shine less brightly in the bottom photograph. Coiling the resistance wire around a nail has provided considerable a.c. resistance due to inductive effects in the coil.

SELF-INDUCTION

People who are determined to get the best out of their hi-fi systems insist that the leads to the loudspeakers should not be coiled up. This is to stop the current in one part of the wire from affecting another by induction. Coiling up wires of any sort makes them very much like solenoids, and that can have a big effect on a circuit.

In the top photograph a bulb is connected to a low-voltage a.c. supply via a length of resistance wire. The bulb glows at nearly full brightness. In the bottom photograph, the same piece of resistance wire has been coiled round a nail and the bulb is now much dimmer.

Coiling up the wire cannot have affected its resistance, but it has turned it into a solenoid. The alternating current in the solenoid is producing a changing magnetic field (aided by the nail) which is causing a flux change *within the coil itself*. This effect is known as **self-induction**. The result is an e.m.f. in the coil that opposes that of the current source (Lenz's law).

Self-inductance

Self-induction takes place when the changing magnetic field produced in a coil induces an e.m.f. in the same coil. The **self-inductance** is defined by

$$\varepsilon = -L\frac{\mathrm{d}(I)}{\mathrm{d}t}$$

where ε is the e.m.f. induced in the coil, L is the self-inductance (H), and I is the current in the coil.

- If I in the coil is increasing, ε will be in the *opposite direction* to the e.m.f. producing the current, and will take energy away from it.

- If I is decreasing, then ε will be in the *same direction* as the e.m.f. producing the current, and will provide the current with energy.

Self-induced e.m.f. is often termed **back e.m.f.** because of its tendency to oppose the external e.m.f.

In a simple d.c. circuit, such as in the diagram below left, the coil opposes the increase in current as the switch is closed, and the current does not instantly reach its maximum value.

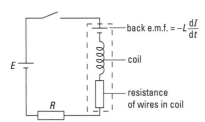

The back e.m.f. in a coil acts like a small cell in series with the coil, but in the opposite direction to the main cell that is driving the circuit.

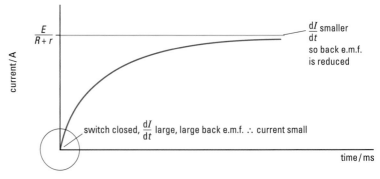

Variations in the rate of change of current affect the back e.m.f.

Back e.m.f. in d.c. motors

When an electric drill is running freely (no load) the motor is spinning at a high speed. As soon as the drill bit is pushed into a material the load on the motor increases, and the rate at which it is spinning is considerably reduced. At first glance, it seems strange that the current drawn by the motor increases when it is under load and not spinning very quickly.

As it is the magnetic force on the current that supplies the torque to drive the motor, this increase in current is exactly what is needed to enable the drill to cut into the material. The current drawn is regulated by the back e.m.f. in the motor.

If the motor is allowed to run freely, it will increase in speed until the back e.m.f. is nearly equal to the supply e.m.f. and so there will not be much current drawn, and little heat will be produced in the wires of the coil.

Once a load is placed on the motor, the speed decreases, and so does the e.m.f. This allows a greater current to be drawn from the supply. This means that more heat is produced and the efficiency of the motor is reduced. The speed does not decrease to zero; it settles at a value set by the balance between the back e.m.f. and the load on the motor.

Applying Kirchhoff's second law to the motor gives

$$E - IR - \varepsilon = 0$$
$$\therefore E = \varepsilon + IR \tag{1}$$

Multiplying through by I gives us an equation in terms of the various power losses and consumption in the motor:

$$IE = I\varepsilon + I^2R$$

where IE = electric power transferred to the motor, $I\varepsilon$ = power transformed in overcoming the back e.m.f. in the coil, and I^2R = power lost as heat due to the resistance of the coil.

As ε is determined by the load on the motor, $I\varepsilon$ can also be interpreted as the rate at which the motor is supplying mechanical work.

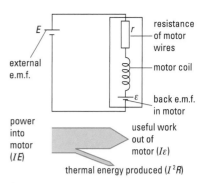

Calculating power transferred in a motor.

Maths box: deriving the equation for self-inductance of a coil

The magnetic field inside a coil is given by:

$$B = \mu_0 nI$$

where n is the number of turns per unit length of the coil.

So, the flux linked to the coil's own turns is

ϕ = field × area × number of coils

  = field × area × nl

  = $(\mu_0 nI) \times (Anl)$

where A is the area and l the length of the coil.

$$\therefore \quad \phi = \mu_0 n^2 VI$$

(V is the volume of the coil)

If I changes with time, the induced e.m.f. will be:

$$\varepsilon = \frac{\mathrm{d}(\mu_0 n^2 VI)}{\mathrm{d}t} = \frac{\mu_0 n^2 V \mathrm{d}I}{\mathrm{d}t} = \frac{L\mathrm{d}I}{\mathrm{d}t}$$

$$\therefore L = \mu_0 n^2 V$$

PRACTICE

1 A coil of resistance $1\,\Omega$ and self-inductance $0.2\,\text{mH}$ was connected to a cell via a $10\,\Omega$ resistor, a digital microammeter, and a switch. The digital ammeter was linked to a datalogging device which recorded the following currents in the short time period after the switch was closed:

$t/\mu s$	2	5	10	20	30	40
I/A	0.06	0.13	0.23	0.36	0.44	0.49

$t/\mu s$	50	60	70	80	90	100
I/A	0.51	0.53	0.54	0.54	0.54	0.54

Plot a graph of current against time. From your graph deduce:

a the gradient at the origin;

b the initial back e.m.f. in the coil;

c the e.m.f. of the cell.

2 A 12 V d.c. motor draws a current of 0.9 A when it is under load. If the resistance of the coils is $10\,\Omega$, what is the back e.m.f. in the motor? What is the useful power output of the motor? If the load is removed from the motor, would the current drawn increase or decrease? Explain.

3 The back e.m.f. in a d.c. motor can be written (approximately) as $NBA\omega$ where ω is the angular rate of rotation, B the magnetic field, N the number of turns, and A the area of the coil. Using equation (1) above, sketch a graph of the current drawn against ω. Given that the useful output power is $I\varepsilon$, sketch a graph of how the output power varies with ω. What is the maximum power output of the motor and at what value of ω will the motor deliver it?

- inductances in a.c. circuits
- reactance of an inductor
- capacitors in a.c. circuits
- reactance of a capacitor

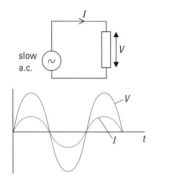

In a non-reactive circuit the current and voltage are in phase.

Helpful hint

Remember the phase relationships between current and p.d. for capacitors and inductors by using CIVIL. At the start of the word, C stands for capacitance and then I (current) is ahead of voltage (V). At the end of the word is L (inductance), and V (voltage) is ahead of I (current).

Maths box: back e.m.f.

The back e.m.f. produced in an inductor $\varepsilon = L\frac{dI}{dt}$, so with a sinusoidal a.c. current: $I = I_0 \sin(\omega t)$

the back e.m.f. is:

$$\varepsilon = L\frac{d(I_0 \sin(\omega t))}{dt} = L\omega I_0 \cos(\omega t)$$

with the peak value being $L\omega I_0$, and the reactance:

$$X_L = \frac{V_L}{I_0} = L\omega$$

For a capacitor, $V = \frac{Q}{C}$ and as:

$$Q = \int I dt = \int I_0 \sin(\omega t)dt = -\frac{I_0}{\omega}\cos(\omega t)$$

we have: $V = -\frac{I_0}{\omega C}\cos(\omega t)$,

and the peak value is $\frac{I_0}{\omega C}$.

It follows that the reactance of the capacitor:

$$X_C = \frac{V_C}{I_0} = \frac{1}{\omega C}$$

REACTIVE CIRCUITS

If a simple circuit containing a resistor is driven by a low-frequency a.c. generator (0.5–1.0 Hz), the current through the resistor will reach a peak value at the moment the e.m.f. reaches its maximum value. The current and e.m.f. are said to be **in phase** with each other. However, if the circuit includes a coil or a capacitor, or if the frequency is increased to a high value, then the current need not reach its peak value at the same moment as the e.m.f. A circuit in which the current and e.m.f. are out of phase is called a **reactive circuit**.

Inductive components

Many components have some self-inductance, the most obvious being a coil of wire. In a d.c. circuit the back e.m.f. in a coil prevents the current from instantaneously reaching a maximum. With a.c., the back e.m.f. acts against the changing supply e.m.f. In the diagram below left the coil has a self-inductance of 13 mH and the wire from which it is made has a resistance of 30 Ω. The a.c. generator is working at a frequency of 500 Hz. The graphs below show how the voltage across the wire's resistance, V_R, the back e.m.f., ε, and the current are related to each other.

At maximum, the rate at which the current changes becomes zero for a moment. At this moment, the back e.m.f. in the coil is also zero. The only p.d. in the circuit is V_R, so V_R is equal to the supply e.m.f., E.

When the current is zero, the rate of change is a maximum (given by the gradient of the line), so the back e.m.f. is at a maximum. The current is zero as the back e.m.f. is equal to the supply e.m.f.

The current through the coil is not synchronized with the back e.m.f. They are out of phase, and the back e.m.f. comes to a maximum *before* the current. In phase-angle terms ε is *90° ahead of the current*.

The supply p.d. is always equal to the sum of V_R and ε. It is also out of phase with the current, but the phase difference depends on the frequency of the a.c. supply.

The **reactance** of the coil is defined as:

$$X_L = \frac{V_0}{I_0} = \frac{V_{r.m.s}}{I_{r.m.s.}} = 2\pi fL = \omega L \qquad \omega = 2\pi f$$

L is the self-inductance (H) and f is the frequency of the supply (Hz).

Although reactance looks similar to resistance it does not determine the current at any moment – the current and back e.m.f. are out of phase.

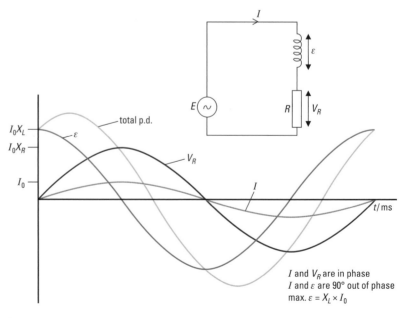

Graphs of p.d. and current for the inductive circuit.

Capacitive circuits

In modern electronics, capacitors are more often used in a.c. circuits than they are as charge-storing devices. Their great virtue as a.c. components is that their reactance (a.c. resistance) changes with frequency, so they can be used to remove certain frequencies from a signal. The diagram below shows a simple case of a capacitor in series with a resistance. The accompanying graphs have been calculated for a 25 µF capacitor in series with a 30 Ω resistance at 500 Hz.

As the a.c. changes direction the capacitor charges and discharges at a rate determined by its capacitance. Although no charge passes from one plate to the other, from outside the capacitor it appears as if it is conducting current. The p.d. between the plates depends on the size of the charge on the plates, and as they near full charge, the current drops (the charge on the plates is repelling any more charge trying to arrive), but the p.d. rises to a maximum. When there is little charge on the plates, the p.d. is very small, but the current in the circuit is high (no repulsion). The graph below shows that the p.d. between the plates and the current are out of phase, but this time the *current* is 90° *ahead* of the p.d.

The reactance of the capacitor is defined in the same way as that of an inductance:

$$X_C = \frac{V_0}{I_0} = \frac{V_{r.m.s}}{I_{r.m.s.}} = \frac{1}{2\pi f C} = \frac{1}{\omega C}$$

where C is the capacitance (F).

As with inductance, the reactance of a capacitor relates the r.m.s. currents and voltages in an a.c. circuit, but one must be careful of instantaneous values because they are not in phase with each other.

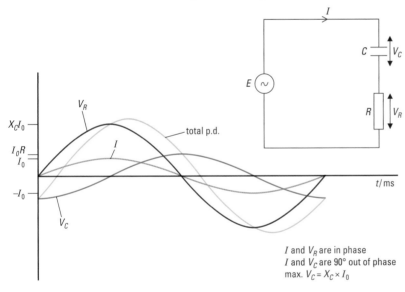

Graphs of p.d. and current for the capacitive circuit shown.

I and V_R are in phase
I and V_C are 90° out of phase
max. $V_C = X_C \times I_0$

Resistance, reactance, and impendence

For a resistive component such as a resistor the current and p.d. are always in phase. For a reactive component such as a capacitor or inductor there is a phase difference between the current and p.d.

Impedance is the collective term for reactance and resistance. In a circuit comprised of an inductor, capacitor, and resistor, the overall circuit has an impedance, the resistor a resistance, and the inductor and capacitor have reactance.

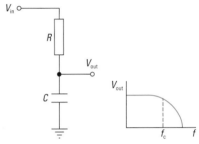

This circuit is a low-pass filter. If V_{in} contains a range of frequencies, the capacitor will filter out the higher frequencies so that V_{out} contains only the lower frequencies. Essentially at high frequencies the capacitor's reactance is low, so the signal is conducted to ground. The 'rollover' takes place at a frequency equal to $1/2\pi f C$.

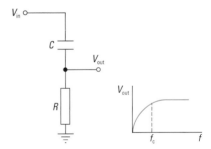

This circuit is a high-pass filter, passing the higher frequencies from V_{in} to V_{out} but blocking the lower frequencies. The circuit acts as a potential divider with the reactance X_C of the capacitor being large at low frequencies, but small at high frequencies. The 'rollover' takes place at a frequency equal to $1/2\pi f C$.

PRACTICE

1 Show that reactance is measured in ohms for both inductors and capacitors.

2 What is the reactance of the inductor in the bottom circuit shown on the previous page at: **a** 50 Hz, **b** 500 Hz, and **c** 5000 Hz? Comment on these values with respect to the resistance of the wires.

3 The graphs on the previous page have been plotted for a frequency of 500 Hz. Explain how you think each curve will change if the frequency is **a** reduced to 50 Hz, and **b** increased to 5000 Hz.

4 Sketch a graph showing how the reactance of a capacitor varies with frequency. Explain in terms of the charging and discharging of the capacitor why the reactance is very high at low frequencies and very small at high frequencies.

5 The graphs on this page have been plotted for a frequency of 500 Hz. Explain how you think each curve will change if the frequency is:

a reduced to 50 Hz;

b increased to 5000 Hz.

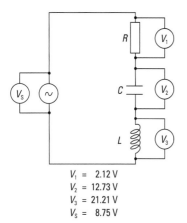

$$V_1 = 2.12\ \text{V}$$
$$V_2 = 12.73\ \text{V}$$
$$V_3 = 21.21\ \text{V}$$
$$V_s = 8.75\ \text{V}$$

In a reactive circuit with an a.c. supply, voltmeter readings do not follow Kirchhoff's second law.

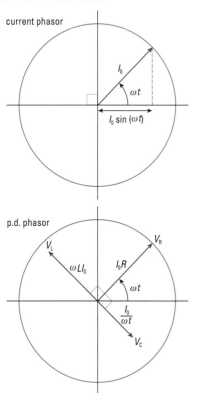

A phasor representation of an alternating current and an alternating p.d.

Helpful hint

On the phasor diagram, the angle θ of the current vector, which is the reference clock, is ωt at time t. Hence if it were animated, we would see the vector moving like the hand of a clock with angular velocity ω. A complete rotation of the current vector would take $2\pi/\omega$ seconds, which is the period T. An angle of 90°, which is $\pi/2$ radians, then corresponds to a time t given by $\pi/2 = \omega t$ or $t = \pi/2\omega$ seconds.

PHASORS AND REACTIVE CIRCUITS

In the diagram on the left, the sum of the *instantaneous* p.d.s across the components is equal to the *instantaneous* source e.m.f.. The voltmeters are not capable of reacting to moment-by-moment changes: they display the r.m.s. values, which is why the readings do not follow Kirchhoff's second law. The peak values, given by $V_p = \sqrt{2} \times V_{\text{r.m.s.}}$, will not sum properly either, as the reactive nature of the components ensures that the peaks are not reached at the same time.

One thing is certain: the current through each component is the same, as it is a series circuit. If:

$$I = I_0 \sin(\omega t)$$

then the p.d. across the resistor must be, from Ohm's law:

$$V_R(t) = I_0 R \sin(\omega t) = V_R \sin(\omega t)$$

where $V_R(t)$ is the instantaneous value and V_R the peak value of the resistor's p.d.

Calculating the instantaneous p.d.s across the inductor and capacitor is trickier. For the inductor:

$$V_L(t) = L\omega I_0 \cos(\omega t) = V_L \cos(\omega t)$$

where $V_L(t)$ is the instantaneous value and V_L the peak value.

For the capacitor:

$$V_C(t) = -\frac{I_0}{\omega C} \cos(\omega t) = -V_C \cos(\omega t)$$

So the e.m.f. of the source must be:

$$V_S(t) = I_0 \left[R \sin(\omega t) + L\omega \cos(\omega t) = -\frac{1}{\omega C} \cos(\omega t) \right] =$$

$$I_0 \left[R \sin(\omega t) + \left(L\omega - \frac{1}{\omega C} \right) \cos(\omega t) \right] \qquad (1)$$

which is the application of Kirchhoff's law at any time, t.

Phasors

The study of circular motion and simple harmonic oscillations (spread 3.33) shows how a sinusoidal variation can be related to a circular motion. In a **phasor diagram**, the same technique allows the sinusoidally varying currents and p.d.s in an a.c. circuit to be represented by vectors (phasors) rotating clockwise with time, which makes handling the phase differences much easier.

In the mapping of SHM to circular motion, the radius of the circle is the amplitude of the SHM. In a phasor diagram, the length of the phasor is the peak value of the current or p.d. concerned. The instantaneous value is given by the horizontal projection of the phasor.

Using the current phasor I_0 as a reference 'clock' the p.d. across the resistor is a phasor with length $I_0 R$ parallel to I_0 (the current and p.d. are in phase). As it is helpful to keep currents and p.ds on separate diagrams, the phasor for the resistance's p.d. becomes the reference clock on the p.d. phasor diagram.

It is easy to add phasors to represent the p.d.s across the inductor and the capacitor. The first step is to determine the length of these phasors (the peak values of the respective p.d.s), which are obtained from the reactances of the components (spread 5.26):

$$\text{peak p.d. across inductor} = V_L = X_L \times I_0 = \omega L I_0$$

$$\text{peak p.d. across capacitor} = V_C = X_C \times I_0 = \frac{I_0}{\omega C}$$

Relative to the V_R phasor, the p.d. across the inductor leads the current by 90° on the phasor diagram, which is equivalent to saying that the p.d. across the inductor reaches its peak $\dfrac{\pi}{2\omega}$ seconds *before* the current peaks.

The phasor representing the p.d. across the capacitor *lags* that of the current, and hence of resistor p.d., by 90°, meaning that the p.d. comes to a peak $2\omega/\pi$ seconds *after* the current does.

Projecting each of these phasors onto the horizontal axis would allow the instantaneous values of p.d. to be obtained and summed, producing equation (1) above.

Reactance and resonance

The phasor diagram can be simplified by combining V_L and V_C into one phasor, which is possible because they are 180° out of phase. Physically this means that the sense of the p.d. across the inductor is always opposite to that of the capacitor. Combining the remaining two p.d. phasors is not as straightforward due to the phase difference, but we can use the vector addition formula to produce:

$$V_S^2 = V_R^2 + (V_L - V_C)^2$$

which shows how the peak values combine.

A phasor representing the source e.m.f. would have length V_S and a phase angle given by:

$$\tan \theta = \frac{V_L - V_C}{V_R} \text{ measured from } V_R.$$

Substituting in the reactances gives:

$$V_S^2 = I_0^2 R^2 + I_0^2 \left[\omega L - \frac{1}{\omega C}\right]^2 \text{ and } \tan \theta = \frac{I_0(\omega L - 1/\omega C)}{I_0 R} = \frac{\omega L - 1/\omega C}{R}$$

from which we can define the overall impedance of the circuit:

$$X = \frac{V_{r.m.s.}}{I_{r.m.s.}} = \frac{V_S}{I_0} = \frac{I_0 \sqrt{R^2 + (\omega L - 1/\omega C)^2}}{I_0} = \sqrt{R^2 + (\omega L - 1/\omega C)^2}$$

Clearly the impedance of the circuit depends not only on the values of the components, but also on the frequency of the a.c. source ($\omega = 2\pi f$).

The impedance of the circuit will be a minimum (and hence the current will be a maximum) if:

$$\omega L - \frac{1}{\omega C} = 0 \quad \text{or} \quad \omega L = \frac{1}{\omega C}$$

A simple rearrangement gives $\omega^2 = \frac{1}{LC}$ and $\omega = \frac{1}{\sqrt{LC}}$ giving the **resonant frequency** as

$$f = \frac{1}{2\pi\sqrt{LC}}$$

Plotting current against frequency reveals a **resonance curve** for the circuit which peaks at the resonant frequency. A detailed analysis shows that the width of the resonant curve is given by the ratio $\omega L/R$ which is known as the **Q factor** of the circuit.

Tuned circuits

In simple application of an LCR circuit, the source e.m.f. is provided by a radio aerial. Adjusting the capacitance in the circuit allows it to resonate at the frequency of the desired station. Then the current passed to an amplifier will be from that station – this is how early radio receivers were tuned. Here it is advantageous to have circuits with a high Q value (in which case the peak of the resonance curve is narrow) so that there is minimal interference from stations with a similar frequency.

PRACTICE

1 A coil of inductance 40 mH and internal resistance 5 Ω is connected to a 10 uF capacitor. When an a.c. supply is connected to the circuit with a frequency of 250 Hz and a peak e.m.f. of 10 V, the r.m.s current in the circuit is 1.40 A. What are the peak p.d.s across the inductor and the capacitor? Determine the impedance and phase angle of the circuit. What is the resonant frequency of the circuit?

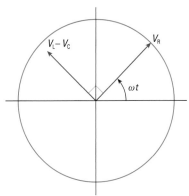

Phasors can be used to sum p.d.s in a reactive circuit.

> **Helpful hint**
>
> The same phasor technique can be applied to RC and LC circuits.

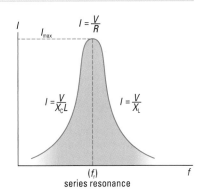

(f₁)
series resonance

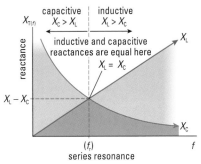

(f₁)
series resonance

Below the resonant frequency the reactance of the capacitor (which varies with 1/f) dominates the circuit's behaviour. Above the resonant frequency the reactance of the inductor (which varies with f) takes over. At the resonant frequency the circuit is purely resistive.

> **Helpful hint**
>
> The average power dissipated in perfect inductors and capacitors is zero. In real components their wires will always dissipate some power.

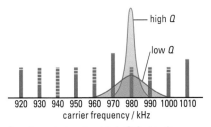

In a tuned circuit a high Q factor is desirable to prevent interference from other channels.

TRANSFORMERS AND POWER SUPPLIES

A **transformer** is a device for changing an alternating voltage from one value into another, and it plays a significant part in the distribution and use of electric power. Transformers step up the voltage produced by the generators to a very high value so that it can be transmitted efficiently to the user. Substations near to the user step down the voltage drop to 230 V r.m.s. In the home, small transformers that plug directly into a socket step down the voltage drop further to values that are used by electronic devices (often 9 V and **rectified** or turned into direct voltage).

The transformer

Transformers are constructed by wrapping two coils round a common piece of soft iron. In order to ensure the efficient magnetic linkage between the coils, the core of the transformer is usually a closed loop (a **magnetic circuit**). This helps the core to become uniformly magnetized.

Transformer cores

A transformer's core is often made of **laminated iron**, that is, layers of iron sandwiched by an insulator. This prevents the build-up of eddy currents in the core whose magnetic fields disturb the flux linkage, and whose heating effects reduce the efficiency of the conversion. The greater the a.c. frequency, the thinner the laminations have to be. At very high frequencies it may be necessary to use fine wires bundled together or iron dust packed into the core. Very high-frequency transformers (above 100 kHz) use cores made from non-conducting but magnetic materials, such as ferrites. These materials are similar to ceramics and so can be quite brittle, and are far less easily magnetized than soft iron.

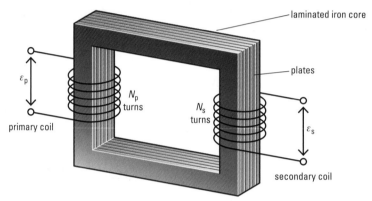

Transformers consist of two coils linked by a magnetic core.

If the primary coil is connected to an a.c. supply, an alternating magnetic field will be set up in the coil. This will cause an alternating flux linkage in the secondary coil producing an e.m.f. by mutual induction. The e.m.f. across the primary coil, ε_p, is related to the e.m.f. across the secondary coil, ε_s, by

$$\frac{\varepsilon_s}{\varepsilon_p} = \frac{N_s}{N_p}$$

where N_p and N_s are the numbers of turns on the primary and secondary coils respectively.

- If $N_s > N_p$ the transformer will **step up** the p.d. to a *higher* value.
- If $N_s < N_p$ the transformer will **step down** the p.d. to a *lower* value.

Helpful hint

ε_p, ε_s, I_p, I_r should be r.m.s. values.

Helpful hint

A *step-down* transformer will also step up the current. This can be seen from conservation of energy. The power in the primary coil is $I_p\varepsilon_p$, and this must equal the power in the secondary, $I_s\varepsilon_s$ (provided there are no heating losses). If the p.d. is decreased the current must increase, and vice versa.

Hence:

$$\frac{\varepsilon_s}{\varepsilon_p} = \frac{N_s}{N_p} = \frac{I_p}{I_s}$$

Maths box: deriving the transformer equation

The e.m.f. induced in the secondary coil is
$$\varepsilon_s = -M\frac{dI}{dt}$$

The back e.m.f. in the primary coil is
$$\varepsilon_b = -L\frac{dI}{dt}$$

As the primary coil has little resistance, the back e.m.f. will be approximately equal to the supply e.m.f. to which it is connected (the current that is drawn will be small).

$$\varepsilon_p \approx \varepsilon_b$$

$$\therefore \quad \frac{\varepsilon_p}{\varepsilon_s} = \frac{\varepsilon_b}{\varepsilon_s} = \frac{-LdI/dt}{-MdI/dt} = \frac{L}{M}$$

Remember that n stands for number of turns per unit length, and N for the total number of turns, so that $n = \frac{N}{l}$

So we have

$$\frac{\varepsilon_p}{\varepsilon_s} = \frac{\mu_0\mu_r n_p^2 A l_p}{\mu_0\mu_r N_s n_p A} = \frac{n_p l_p}{N_s} = \frac{N_p}{N_s}$$

Efficiency in a transformer

The efficiency of a transformer can be defined by:

$$E = \frac{I_s V_s}{I_p V_p} < 100\%$$

Energy dissipation in the transformer will reduce efficiency, and is caused by a combination of:

- the resistance of the wires in the primary and secondary coils – minimized by using thicker wires;
- eddy currents in the core – minimized by using a laminated core;
- energy losses in the core due to heating as it magnetizes and demagnetizes – minimized by using an appropriate magnetically soft alloy for the core.

Power supplies

To provide power for an electronic device such as a Blu-ray player a transformer is used to step down the p.d. from 230 V to about 9 V. The alternating voltage is then rectified, or turned into direct voltage.

A single diode will rectify a.c. to a certain extent because it will only conduct when the current is flowing in one direction. This is called **half-wave rectification**, and it works by cutting out half of the full a.c. wave. This is d.c. in the sense that the current is always in one direction, but it is still changing in time.

A better solution is to use a **diode bridge rectifier** to produce a **full-wave rectified** output.

The diagram above right shows how four diodes can be connected so that the current in the output is always in the same direction independent of the input.

The final step in producing d.c. is to *smooth* the full-wave output of the diode bridge. This can be achieved by connecting a large-value capacitor across the output of the bridge.

- As the current increases, the capacitor charges from the bridge.
- Then, as the output of the bridge drops, the capacitor will discharge through the load, maintaining the current until the bridge output has risen again.

Quite smooth d.c. can be generated in this way, although precision power supplies use **integrated-circuit voltage stabilizers** to provide d.c. with virtually no variations.

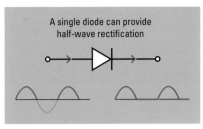

A single diode can provide half-wave rectification

Half-wave rectification.

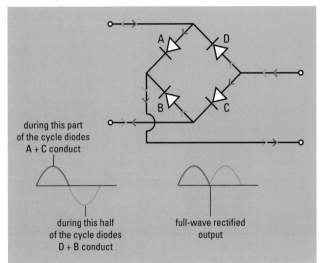

A diode bridge circuit.

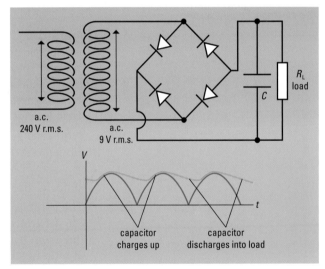

A simple power supply. As the capacitor discharges into the load, the time constant CR_L must be large compared with the period of the mains supply.

PRACTICE

1 What is the turns ratio required if a transformer is to step down a voltage from 100 V to 20 V?

2 A substation converts the 400 kV r.m.s. a.c. from a power transmission pylon into 230 V r.m.s. for domestic use. If a user turns on a light bulb of 100 W power, what will be the current drawn from the pylon? What assumption have you had to make in this calculation?

3 Transformer cores can be made from a variety of materials. What are the main features that you would require of a material to make a good transformer core? Suggest how well each of the following materials would perform:

air; solid soft iron; laminated soft iron; aluminium.

4 How would you explain to a GCSE-level physics student why transformers do not work with d.c.?

5 Why does half-wave rectification waste more power than full-wave rectification?

POWER GENERATION AND TRANSMISSION

The generation and transmission of electric power has become a central industry in both developed and developing countries. Much of the technology involved is simple, but must be efficient and reliable. This calls for a great deal of skill in the design, construction, and maintenance of the equipment. Below is a block diagram of the basic elements of the electrical supply system.

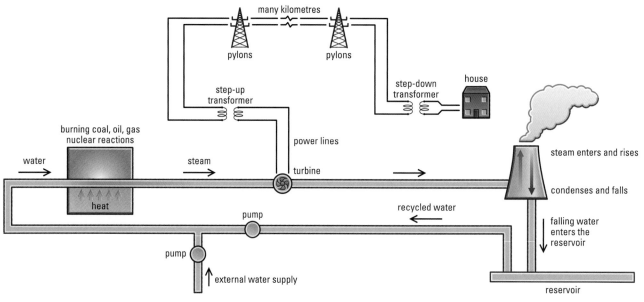

The electrical supply system.

Generators

Most electric power stations generate thermal energy that is used to turn water into high-pressure steam which drives turbines. The steam pressure acts on fins inside the turbine rotating the main shaft, and electromagnetic induction is used to transform this rotational energy into electric energy.

Industrial generators work by rotating a magnet (the **rotor**) inside a stationary set of coils (the **stator**).

- The rotor is an electromagnet wound round a soft iron core, and the stator is also made of soft iron.

- The combination of rotor and stator produces a uniform field inside the generator coils, with only a small air gap preventing the magnetic flux from being entirely inside the iron. This also means that it is greater in strength.

- The stator contains six generator coils which are wired in opposite pairs. The three outputs provide sinusoidally varying e.m.f.s which come to a maximum at different phases of the rotor's rotation. This is called a **three-phase supply**. It has the advantage of ensuring that the rotor runs smoothly, and that kinetic energy from the steam is used efficiently.

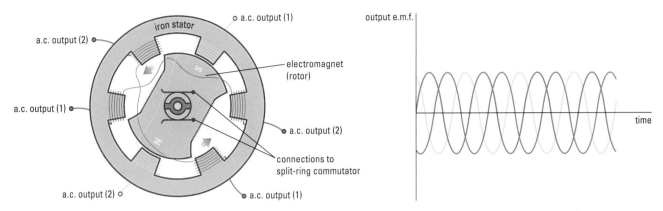

A three-phase a.c. generator. The current for the rotating electromagnet (the rotor) is supplied through the split-ring commutator (and is assumed here to be single-phase a.c.) to produce an electromagnetic field with constant poles, as shown. The three out-of-phase outputs from the three pairs of stator coils are shown in the graph.

Transformers

Alternating current is used in the generation and transmission of electric power for three reasons.

1 It is more easily generated than d.c. Generators naturally produce a.c. which is then converted into d.c.

2 It can be switched on and off more easily than d.c. Switching large d.c. can be a problem because of induction. With a.c. the operation can be timed to an instant when the current is small.

3 For efficiency, high power is transmitted at high voltage, so transformers (which only work with a.c.) are required to step up and step down the voltages.

When electric power is carried through cables, some energy is wasted as heat. To minimize this the resistance of the cables must be as small as possible. Power cables have to be many kilometres long, and the best technology can reduce their resistance to $0.06\,\Omega\,\text{km}^{-1}$. Power stations typically produce 1 GW of power, which will be shared between several towns. If we were to try and transmit 0.1 GW at 230 V the current would be 435 kA and the heat production along the wires would be 10 MW for every metre!

However, if we transmit at 400 kV then the current is

$$I_{\text{r.m.s}} = \frac{P}{V_{\text{r.m.s}}} = \frac{0.1 \times 10^9\,\text{W}}{400 \times 10^3\,\text{V}} = 250\,\text{A}$$

The power lost in the wires is

$$P_{\text{lost}} = I^2R = (250\,\text{A})^2 \times (0.06\,\Omega\,\text{km}^{-1}) = 3.75\,\text{kW per wire}$$

Over 50 km, this amounts to 375 kW (since there are two wires), which is 0.4% of the transmitted power.

NB Some care has to be taken in applying the equivalent power formula $P = V^2/R$ – the voltage has to be the value measured across the wires, not the 400 kV!

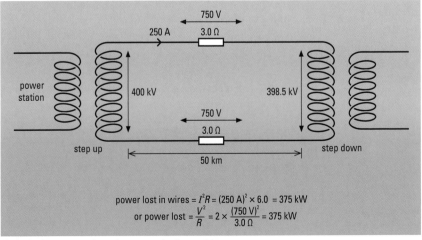

Calculating power loss in transmission cables.

PRACTICE EXAM QUESTIONS

1 A 600 W projector bulb is designed to operate from a mains supply of 230 V r.m.s., 50 Hz. A capacitor and series resistor are connected in parallel across the switch, as shown in Figure 5.1, to suppress radio-frequency interference.

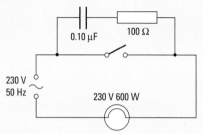

Figure 5.1

a Calculate

  i the r.m.s. current in the circuit when the switch is closed and the bulb is at its normal brightness,

  ii the resistance of the bulb under these conditions,

  iii the maximum value of the instantaneous current flowing under these conditions.

b The capacitor has a value of 0.10 µF and the resistor a value of 100 Ω.

  i Calculate the reactance of the capacitor at a frequency of 50 Hz.

  ii Estimate the current flowing in the circuit when the switch is open.

c The projector has a fan which is turned on and off by the same switch as the bulb.

  i Copy the diagram and show how the fan motor should be connected into the circuit.

  ii The inductance of the fan motor is 0.80 H. Show that the capacitor and inductor do *not* form a resonant circuit at the mains frequency.

2 A parallel plate capacitor (air-filled) has plates each of area A separated by a small distance d. The capacitor is charged from a battery of e.m.f. 12 V.

If $A = 0.24\,\mathrm{m}^2$ and $d = 0.50\,\mathrm{mm}$, calculate for this capacitor

a the capacitance,

b the energy stored.

3 A student is at a party. A balloon is rubbed against her hair. She notices that the hair is now attracted to the balloon and sticks to it.

a Explain the student's observations.

b She now places the balloon on the ceiling. It stays there for about half an hour and then falls towards the floor. Suggest explanations for these observations.

c Describe how the principles of the phenomena observed by the student can be used to improve paint-spraying processes.

4 a i Define electric potential.

  ii The electric potential V at a distance r from a point charge Q in vacuo (free space) is given by

  $$V = \frac{Q}{4\pi\varepsilon_o r}$$

Write down an expression for the electric potential energy of a point charge of q when at a distance r from a point charge of Q in vacuo (free space).

b An α-particle of charge $+3.2 \times 10^{-19}$ C and mass 6.8×10^{-27} kg is travelling at a speed of $1.2 \times 10^7\,\mathrm{m\,s^{-1}}$ directly towards a fixed nitrogen nucleus of charge $+11.2 \times 10^{-19}$ C.

Assuming that initially they are far apart calculate the closest distance of approach.

[Take $\dfrac{1}{4\pi\varepsilon_o} = 9.0 \times 10^9\,\mathrm{F^{-1}\,m}$]

5 a i Write down the equation giving the electrostatic force between two isolated point charges in a vacuum.

  ii When the equation in **i** is used to calculate the force between two given charges, the result is negative. What is the significance of the minus sign?

b Two small conducting spheres carry equal charges and are 5.0 mm apart.

Calculate

  i the charge on one of them for the force of repulsion to be 2.5×10^{-3} N,

  ii the number of electrons required to produce this charge.

c A small sphere of mass 6.0×10^{-6} kg remains stationary when placed in an electric field of intensity $5.0 \times 10^4\,\mathrm{V\,m^{-1}}$ acting vertically downwards. Calculate the sign and magnitude of the charge on the sphere.

6 Electrons projected along a horizontal path in a cathode ray tube may be deflected on their way to the screen by a horizontal magnetic field. Draw a diagram of this arrangement and show on your diagram the directions of the beam, the magnetic field and the deflection.

For an electron in a vacuum tube, $e\Delta V = \frac{1}{2}mv^2$. Explain what the term $e\Delta V$ corresponds to.

In television sets, electrons are typically accelerated through 25 kV. Calculate the speed that they reach. (The electronic charge = -1.6×10^{-19} C, electronic mass = 9.1×10^{-31} kg.)

7 Figures 5.2 and 5.3 both show a charged particle moving in an electric field. The direction of motion of the particle is shown by an arrow and, for each situation, the charged particle is labelled and also named underneath each diagram.

For each case

  i copy the diagram and show on it the direction of the electric force,

  ii calculate the magnitude of the electric force on the particle.

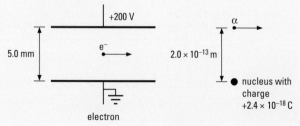

Figure 5.2   *Figure 5.3*

8

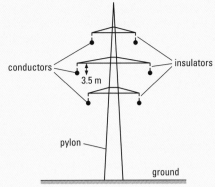

Figure 5.4

Figure 5.4 shows how electric power is transmitted from generating stations to consumers on high voltage overhead power lines supported by metal pylons, which are earthed. Each conductor is suspended vertically below the pylon structure by an insulator which is 3.5 m long. The distribution voltage is 380 kV r.m.s. It may be assumed that the voltage waveform applied to each conductor is sinusoidal.

a **i** Calculate the peak voltage applied to each of the conductors.

    ii Estimate the magnitude of the maximum electric field strength between a conductor and the pylon structure, if the field could be assumed to be uniform.

b Under certain atmospheric conditions it is possible to hear sharp, crackling noises coming from the region around the conductors.

    i State the atmospheric conditions under which the effect would become more pronounced and explain your answer.

    ii Suggest an explanation for this effect.

9 **a** A capacitor consists of two square parallel plates, made from thick metal. Each plate is of side length 0.25 m, and the plates are separated by an air gap of 1.0 mm.

A 100 V d.c. supply is connected across the plates.

    i Calculate the capacitance of the capacitor.

    ii Calculate the energy which is stored by the charged capacitor.

    iii The distance between the plates of the capacitor is then increased to 10.0 mm, with the 100 V supply still connected.

Which electrical quantity remains constant as the separation increases?

Calculate the charge stored after the separation has been increased.

b You are asked to investigate by experiment the relation between the capacitance of a parallel plate capacitor and the separation of its plates, using the metal plates described in **a**. You are provided with a charge-measuring device of appropriate range (or other alternative equipment of your choice) and you may assume that all the usual laboratory equipment is available.

    i State the result you would expect to find.

    ii Draw a diagram of the apparatus you would use and describe the experimental procedure you would follow.

10 **a** Define electric field strength.

    b State two pieces of information that can be deduced from drawings of electric field lines.

    c Figure 5.5 illustrates some of the electric charges in a thundercloud and on the surface of the Earth beneath it.

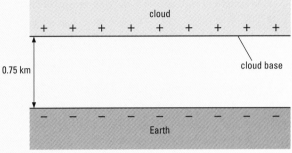

Figure 5.5

The base of the cloud and the surface of the Earth can be considered horizontal.

    i Copy Figure 5.5 and sketch on it the electric field between the cloud and the Earth.

    ii The cloud base is 0.75 km above the Earth. A lightning flash occurs in air containing raindrops when the electric field strength exceeds $5.0 \times 10^4 \, \text{N C}^{-1}$. Calculate the minimum electric potential difference between the cloud base and the Earth's surface for a lighting flash to occur.

11 This question is about the design of a new propulsion system for driving a boat at sea.

a In one proposal, shown in Figure 5.6, a copper bar is fixed under the boat and a large current is passed through it. A large electromagnet produces a vertical magnetic field.

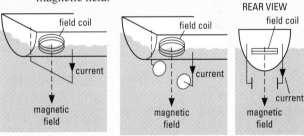

Figure 5.6   **Figure 5.7**

    i State the direction of the force on the bar when the current is flowing as shown in the diagram.

    ii Give *two* ways of increasing the force on the bar.

    iii Explain briefly why this system will not work.

b A second proposal suggests passing a current through the water instead of through the copper bar using an electrode system as shown in Figure 5.7. Each electrode face has a surface area of 0.75 m² and they are 1.5 m apart. The drag force on the boat, which has a mass of 150 000 kg, is 12 000 N when it moves at 8 m s⁻¹. The current used is to be 1000 A.

    i Explain what is meant by ionisation and how this enables a current to flow in sea water.

    ii Explain why the boat moves when a current is passed through the water.

    iii What is the strength of the magnetic field required so that the thrust is sufficient to maintain a constant speed of 8 m s⁻¹?

iv What will be the maximum possible initial acceleration of the boat when it starts from rest?

12 Outside the sphere, a charged conducting sphere behaves as if the charge were concentrated at its centre. The electric field strength inside the sphere is zero.

One sphere of radius 5.0 cm carries a positive charge of 6.7 nC.

a i Show that the potential at the surface of the sphere is about 1200 V.

ii Calculate the capacitance of the sphere.

Permittivity of free space, $\varepsilon_o = 8.9 \times 10^{-12}\,\text{F m}^{-1}$.

b i Calculate the electric field strength just outside the surface of the sphere.

ii Sketch a labelled graph to show how the electric field strength varies with distance from the centre of the sphere to a distance of 20 cm from the centre.

Include a suitable scale on your graph.

13 A long solenoid P, carrying a current, is shown in section in Figure 5.8. A small coil Q, which is connected to a sensitive voltmeter (not shown), is placed in the centre of the solenoid.

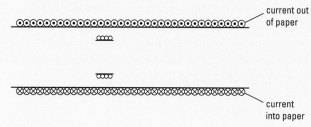

Figure 5.8

a Copy Figure 5.8 and sketch on it the flux pattern inside and outside the solenoid P due to the current in it.

b Explain why an induced e.m.f. is recorded on the voltmeter when the current in the solenoid changes.

c In further experiments, the current in solenoid P is changed at the same rate as in **b**.

Compare qualitatively the induced e.m.f. with that obtained in **b** when, separately,

i the coil Q is placed at the end of P,

ii a ferrous core is placed in P.

14 a State the laws of electromagnetic induction.

b Describe *one* simple demonstration to illustrate each law in a qualitative manner to a newcomer on an A-level Physics course.

c Give *two* examples of devices of technological importance which operate on the basis of these laws.

d In one system of braking an electric train, the motors are disconnected from the supply and placed in series with a bank of resistors of total resistance R when braking is needed.

i How is the braking action achieved?

ii By considering the situation at low speeds, say why some additional system of braking is required.

15 a A coil having both inductance and resistance, a d.c. source and a switch are connected in series. Sketch a graph showing how the current flowing through the coil varies with time from the instant when the switch is closed until the current reaches a steady value.

b A coil of inductance 200 mH and resistance 45 Ω is connected to the terminals of a 9.0 V battery of negligible internal resistance.

i What is the initial back e.m.f. across the inductor? Explain your answer.

ii Hence find the initial rate of increase of current.

iii Calculate the final steady current.

16

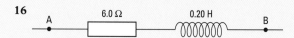

Figure 5.9

Figure 5.9 shows part of a circuit through which a current is to flow from A to B.

After switching on, the current increases uniformly from zero to 15 A in 5 s and then remains steady at 15 A.

What is the p.d. across AB

a after 4 s,

b after 6 s?

* 17 Figure 5.10 shows schematically one type of mass spectrometer that is used to investigate a sample of nickel ions each of which carry a charge of -1.6×10^{-19} C.

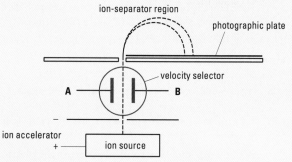

Figure 5.10

The ions from the source are accelerated and then pass into the velocity selector in which the magnetic flux density is 0.15 T directed out of the plane of the diagram. The ions that are selected have a speed of $1.3 \times 10^5\,\text{m s}^{-1}$. The ions are then separated by a magnetic field in the ion-separator region.

Assume that protons and neutrons have a mass of $1.7 \times 10^{27}\,\text{kg}$.

a Calculate the p.d. required to accelerate nickel-58 ions to a speed of $1.3 \times 10^5\,\text{m s}^{-1}$.

b Calculate the magnitude and direction of the electric field in the velocity selector.

c Ions of nickel-58 land on the photographic plate 300 mm from the slit S.

i Calculate the magnetic flux density in the ion separator.

ii Calculate the separation of nickel-58 and nickel-60 ions that fall on the photographic plate.

d Discuss ways in which the variables in the system may be changed to increase the separation of the ions.

*Pre-U-style question

e Data from a mass spectrometer shows the ratio of nickel ions to be:

68% nickel-58; 26% nickel-60; 1% nickel-61; 4% nickel-62; 1% nickel-64.

For nickel-58 the relative atomic mass (RAM) is 58. For nickel-60 the RAM is 60, etc. Calculate the RAM for the sample of nickel.

18 A transformer, to be used in a low voltage power supply, is connected to the 240 V r.m.s., 50 Hz, sinusoidal a.c. mains supply and gives an r.m.s. output voltage of 12 V. There are 1800 turns on the primary coil.

a i Calculate the number of turns on the secondary coil.

ii Assuming that there are no energy losses, calculate the current in the primary coil when the current in the secondary coil is 9.0 A.

iii A 2.5 Ω resistor is connected across the secondary coil. Calculate the rate at which electrical energy is transformed into internal energy in the resistor when the current in the secondary coil is 9.0 A.

b i Calculate the peak value of the output voltage of the transformer.

ii The output from the transformer is rectified by the use of a diode in series with the load. Sketch the output waveform, showing clearly how the value of the voltage across the load changes with time. Your sketch should include suitable voltage and time scales.

iii State and explain briefly how the output can be smoothed.

c State two causes of energy loss in the core of a transformer.

---

* 19 So that communications satellites can remain in the same place relative to a receiver on Earth they are placed in geostationary (or geosynchronous) orbits around the equator.

mass of the Earth = 6.0×10^{24} kg

universal gravitational constant = 6.7×10^{-11} N m^2 kg^{-2}

radius of the Earth = 6400 km

a i Calculate the radius of the orbit for such a satellite.

ii Assuming the Earth to be a sphere show that the greatest latitude at which the satellite is visible is 81°.

iii Ignoring delays introduced by electronic processing, calculate the delay between sending a signal from a person 81°N to someone receiving it at 81°S.

b i Calculate the kinetic energy of a 750 kg satellite in a geostationary orbit.

ii Calculate the difference between the total energy of the satellite when in a geostationary orbit and the energy it has when it is on the launch pad on the Earth's surface. Assume that it has no KE before launch and ignore frictional losses due to the atmosphere during take-off.

iii Explain why the energy can be reduced by launching the satellite from a site on the equator and calculate how much less energy is required.

Pre-U-style-question

---

20 Figure 5.11 shows a circuit for investigating the action of a transformer. No meters have been included.

Table 5.1 below shows some observations of primary and secondary r.m.s. currents and voltages, I_p, I_s, V_p, V_s, made during an experiment.

Figure 5.11

S	I_p/A	V_p/V	I_s/A	V_s/V
Open	0.20	12	0	3.0
Closed	1.5	12	4.0	3.0

Table 5.1

a Use electromagnetic principles to explain why there is an alternating voltage across the coil in the secondary circuit when an alternating voltage is applied to the primary coil.

b Sketch the circuit and include meters suitably positioned to obtain the data in the table above.

c Neglecting the resistance of the primary coil, use the data for S in the *open* position to determine:

i the reactance of the primary coil;

ii the inductance of the primary coil.

d Using the data for S in the *closed* position, estimate:

i the input power to the transformer;

ii the output power from the transformer;

iii the efficiency of the transformer;

iv the resistance of the load resistor.

---

21 a Sketch the magnetic field lines produced by a straight wire carrying an electric current. On your sketch show the directions of the field and current.

b A long straight conductor *P* carrying a current of 2 A is placed parallel to a short straight conductor *Q* of length 0.12 m carrying a current of 3 A, the directions of the currents being the same. The conductors are 0.20 m apart in air.

i What is the value of the force experienced by *Q*?

ii What is the direction of this force?

c State the magnitude and direction of the force experienced by **P** and briefly justify your statement.

$$B = \sqrt{\frac{\mu_0 I}{2\pi a}}$$

22 a Describe, with the aid of a labelled diagram, the basic structure of a cathode-ray tube in a cathode-ray oscilloscope (c.r.o.).

b In one type of c.r.o., the electrostatic deflection system consists of two parallel metal plates, each of length 2.0 cm, with a separation of 0.50 cm, as shown in Figure 5.12.

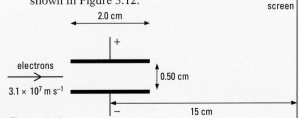

Figure 5.12

The centre of the plates is situated 15 cm from a screen. A potential difference of 80 V between the plates provides a uniform electric field in the region between the plates. Electrons of speed 3.1×10^7 m s^{-1} enter this region at right angles to the field. Calculate

   i the time taken for an electron to pass between the plates,

   ii the electric field strength between the plates,

   iii the force on an electron due to the electric field,

   iv the acceleration of the electron along the direction of the electric field,

   v the speed of the electron at right angles to its original direction of motion as it leaves the region between the plates.

c Hence, by considering your answer to **b v** and the original speed of the electron, estimate the deflection on the electron beam on the screen.

d i Figure 5.13 represents the front of the screen of the c.r.o.

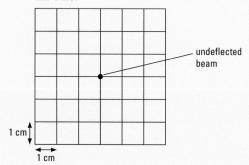

Figure 5.13

       Copy Figure 5.13 and mark on your diagram the position of the deflected beam of electrons.

   ii Draw similar sketch diagrams to show the trace on the screen if the p.d. across the plates is

       (1) varying sinusoidally with r.m.s. value 80 V,

       (2) a half-wave rectified sinusoidal voltage of r.m.s. value 80 V.

23 This question is about the force on the coil of a moving-coil loudspeaker.

In Figure 5.14 the moving-coil is suspended by the speaker cone, enabling it to move freely within the cylindrical space between the poles of the permanent magnet. Figure 5.15 shows a cross-section at the vertical plane marked XX on Figure 5.14, viewed from the left.

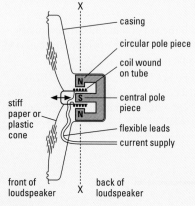

Figure 5.14

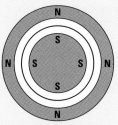

FRONT VIEW

Figure 5.15

a Copy Figure 5.15 and draw on it

   i lines with arrows to represent the magnetic field between the poles, and

   ii an arrow to represent the direction of the current in the coil which will cause the coil to move to the left.

b i Write down an expression for the force F on a wire of length l carrying a current I at right angles to a magnetic field B.

   ii Use the data below to calculate the force on the coil when there is a current of 50 mA in it.

The coil consists of 100 turns of wire of mean radius 15 mm, total length 9.4 m.

The radial field between the poles is 0.40 T.

c The coil, of mass 6.0 g, is made to oscillate in simple harmonic motion of amplitude 0.1 mm at f = 300 Hz. The cone's movement obeys the simple harmonic equation of motion:

$acceleration = -\omega^2 \cdot displacement$

where ω ($= 2\pi f$) is the angular frequency.

   i Show that the maximum acceleration of the coil is 360 m s^{-2}.

   ii Use parts **b i** and **c i** above to calculate the maximum amplitude of the current required to cause this motion.

   iii Suggest a reason why the cone of a high frequency loudspeaker, a tweeter, should be small.

24 a State, in words, Newton's law of gravitation (i.e. the force law between a pair of masses).

b i Calculate the gravitational potential at the surface of the Earth (assume the Earth to be a sphere of uniform density).

       Mass of Earth = 6.0×10^{24} kg

       Radius of Earth = 6400 km

       $G = 6.7 \times 10^{-11}$ N m^2 kg^{-2}

   ii Hence calculate the speed with which a body should be projected so as to escape completely from the Earth (the escape velocity). Neglect air resistance.

   iii *Briefly* comment on the fact that the Earth's atmosphere is largely composed of heavy gases.

25 a What is meant by a field of force?

b A particle has mass m and charge $+q$.

   i State the magnitude and the direction of the force on this particle when it is at rest in

       1. a gravitational field,

       2. an electric field,

       3. a magnetic field.

ii State the magnitude and the direction of the force on this particle when it is moving with velocity v in a direction normal to

1. a gravitational field,
2. an electric field,
3. a magnetic field.

c The Earth may be considered to be a uniform sphere of radius 6370 km, spinning on its axis with a period of 24.0 hours. The gravitational field at the Earth's surface is identical with that of a point mass of 5.98×10^{24} kg at the Earth's centre. For a 1.00 kg mass situated at the Equator,

i calculate, using Newton's law of Gravitation, the gravitational force on the mass,

ii determine the force required to maintain the circular path of the mass,

iii deduce the reading on an accurate newton-meter (spring balance) supporting the mass.

d Using your answers to **c**, state what would be the acceleration of the mass at the Earth's surface due to

i the gravitational force alone,

ii the force as measured on the newton-meter.

e A student, situated at the Equator, releases a ball from rest in a vacuum and measures its acceleration towards the Earth's surface. He then states that this acceleration is 'the acceleration due to gravity'. Comment on his statement.

26 a For a point at the surface of the Earth, the *acceleration due to gravity* is connected with the *gravitational constant* by the equation

$$g = \frac{-GM}{r^2}$$

i Explain the meanings of the terms in italics.

ii What do M and r represent?

b At the surface of a planet P of radius r_P and mean density ρ_P the acceleration due to gravity is g_P. If planet Q has a radius of r_Q and mean density of ρ_Q, show that the acceleration due to gravity g_Q at its surface is given by

$$g_Q = \frac{r_Q \rho_Q}{r_P \rho_P} g_P$$

c A catapult can fire a pellet to a vertical height of 10.0 m on Earth. It is taken to the Moon which has a radius 0.27 of that of the Earth and which is composed of material of mean density 0.61 of that of the Earth. Calculate

i the acceleration due to gravity on the Moon,

ii the height to which the pellet could be fired on the Moon.

27 The data in Table 5.2 below gives values for the absolute gravitational potential energy of a mass of 1500 kg in the Moon's gravitational field and the corresponding distance from the centre of the Moon. Neglect the effect of the Earth.

Table 5.2

Gravitational PE of a 1500 kg mass /10^9 J	Distance from the centre of the Moon /10^6 m
−4.20	1.74
−2.92	2.50
−2.44	3.00

Gravitational constant, $G = 6.7 \times 10^{-11}$ N m^2 kg^{-2}.

a State why the gravitational potential energy values are all negative.

b It is expected that the gravitational potential energy is inversely proportional to the distance from the centre of the Moon. Without drawing a graph show that the data is consistent with such a law.

c Use the data provided to determine the mass of the Moon.

28 The gravitational field strength at the surface of a star of radius 1.0×10^{10} m and mass 5.6×10^{33} kg is 4.0×10^3 N kg^{-1}.

a i Sketch a graph showing the variation of the magnitude of gravitational field strength with distance d from the centre of the star, for values of d which are greater than the radius of the star. The scale of the d-axis should be 0 to 5×10^{10} m.

ii Calculate the gravitational field strength of the star at a distance of 4.0×10^{17} m from its centre.

The universal gravitational constant, $G = 6.7 \times 10^{-11}$ N m^2 kg^{-2}.

b A second star, of mass 2.0×10^{30} kg is 4.0×10^{17} m from the first star. Calculate the force acting on the second star due to the gravitational field of the first.

29 a The gravitational field strength of the Earth at its surface is 9.81 N kg^{-1}.

Show that

i the acceleration of free fall at the surface of the Earth is 9.81 m s^{-2},

ii N kg^{-1} is equivalent to m s^{-2} in base units.

b Use the value of the gravitational field strength of the Earth quoted in **a**, together with the value of G, the gravitational constant, and of the radius of the Earth (6.38×10^6 m), to calculate the mass of the Earth.

c Calculate the Earth's gravitational field strength at a height of 0.12×10^6 m above the Earth's surface.

d Explain briefly why an astronaut in a satellite orbiting the Earth at this altitude may be described as weightless.

e The value of the gravitational potential ϕ at a point in the Earth's field is given by the equation

$$\phi = -GM/r,$$

where M is the mass of the Earth and r is the distance of the point from the centre of the Earth. (r is greater than the radius of the Earth.)

Explain

i what is meant by the term gravitational potential,

ii why the potential has a negative value.

f Use the expression given in **e** to calculate the gain in the potential energy of a satellite of mass 3000 kg between its launch and when it is at a height of 0.12×10^6 m above the Earth's surface.

6

WAVES

Waves on the sea are driven by the wind, but the wind energy itself originates as electromagnetic waves from the Sun. When a wave breaks on a beach, the crash of the surf imprints itself upon the air as a series of periodic variations in pressure that can be heard as sound. Most of what we know about the world comes from information carried by waves. Human beings are particularly sensitive to a narrow range of frequencies of sound. Other creatures use 'invisible' light and 'inaudible' sound. Now we are developing technologies that extend our own range of perception. To do this it is essential to understand wave types and properties.

In this chapter you will learn about electromagnetic and mechanical waves and their properties of reflection, refraction, diffraction, interference, and polarization. You will discover how we are able to measure the speed of light and sound and how to use seismic waves to locate an earthquake. Also discussed is the close link between the physics of waves and the sound of music.

Looking out into the immensity of the universe, we stand, in Carl Sagan's words, at the shores of the cosmic ocean. Yet our earthly seas are no less beautiful, and the crashing surf reminds us that most of what we know about our world comes from information carried by waves.

- transverse waves
- longitudinal waves
- wave speed
- waves and energy

Travelling or progressive waves

A wave that transfers energy from one place to another is called a travelling or 'progressive' wave. This is because points farther from the source vibrate with a progressively greater phase delay compared with a point at the source itself.

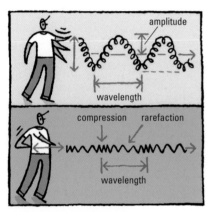

Transverse waves (top) and longitudinal waves (bottom).

WAVES

The 1970 Mexico World Cup was remarkable for something other than soccer – the crowds started what is now known as the **Mexican wave**, a communal movement that sweeps around a stadium and was soon being copied at major sporting events around the world. Joining in a Mexican wave could hardly be easier – you wait until the person to your left has sat down and then you stand, raise your arms, lower your arms, and sit, at which point the person on your right does the same thing. This simple repeated motion sends a pulse around the stadium and illustrates many of the properties of all waves.

- The Mexican wave is started by one group of people moving up and down.

  Wave sources are usually oscillations (mechanical or electromagnetic).

- You do not leave your seat, but the wave travels round the stadium.

  Mechanical waves are transmitted by vibrating particles, but the particles do not move from their mean positions.

- Your motion is vertical but the wave moves horizontally.

  This is a **transverse wave** *– the vibration direction is perpendicular to the wave direction.*

- Your up-and-down motion is out of step with that of your immediate neighbours.

  The vibrations of particles in a travelling wave are out of phase with those of their neighbours.

- If part of the stadium is empty the wave cannot pass through it.

  Mechanical waves need a material medium.

Longitudinal waves

Mexican waves are transverse. However, many waves, like sound, are **longitudinal**: their particles vibrate parallel to the direction in which the wave is moving. Imagine bumping into the end of a bus queue. The effect of the collision is to compress the queue at the end you hit, but as the people rebound they hit others further along the queue and a compression pulse travels along it. If you bump hard enough, someone at the far end may even be knocked over. Longitudinal waves consist of repeated **compressions** and **rarefactions** (regions where the particle spacing increases) that move through the medium. Both longitudinal and transverse waves can be demonstrated using a 'slinky' toy.

Properties of waves.

Property	Symbol	Unit	Definition/explanation
mechanical	–	–	waves transmitted through material medium by particle vibrations
electromagnetic	–	–	waves transmitted through electromagnetic fields by oscillations of electric and magnetic fields
transverse	–	–	oscillates perpendicular to wave direction (e.g. light)
longitudinal	–	–	oscillates parallel to wave direction (e.g. sound)
wavelength	λ	m	least distance between points on the wave that are in phase (e.g. from one ripple peak to the next in water waves)
frequency	f	Hz	number of waves per second passing a point $\left.\begin{array}{l}\\\\\end{array}\right\} f = \frac{1}{T}$
period	T	s	time for one wave to leave source or pass a point
phase	ϕ	rad	the position within its cycle of oscillation of a point on the wave. Two points one wavelength apart oscillate in phase (i.e. they change in the same way at the same time as the wave passes)
wavefront	–	–	a line joining points of the same phase in the wave, perpendicular to rays (e.g. the line of a ripple crest)
wave speed	v	m s^{-1}	the speed at which a wavefront moves (e.g. the speed of a ripple crest on a water wave)
amplitude	A	*	maximum disturbance measured from equilibrium
intensity	I	W m^{-2}	wave power per unit area

* Depends on type of wave. Vertical distance in metres would be appropriate for an ocean wave, but electric field strength would be appropriate for an electromagnetic wave, and pressure difference for a sound wave.

Wave speed

If a source of waves of wavelength λ emits f waves per second, then after t seconds ft waves will leave the source and the wavefront (leading edge of the first wave) will be a distance $ft\lambda$ from it. The wave speed can then be calculated from:

$$\text{wave speed (m s}^{-1}) = \frac{\text{distance travelled by wavefront (m)}}{\text{time taken (s)}}$$

$$v = \frac{ft\lambda}{t} \qquad \text{or} \qquad v = f\lambda$$

Worked example 1

(a) Ripple crests in a ripple tank are 4.0 cm apart and two complete waves leave the source every second. What is their speed?

$v = f\lambda = 2\,\text{Hz} \times 0.04\,\text{m} = 0.080\,\text{m s}^{-1} = 8.0\,\text{cm s}^{-1}$

(b) Atlantic 252 broadcasts radio waves of wavelength 252 m. To what frequency must a receiver be tuned to receive this station?

$v = f\lambda$ so $f = \dfrac{v}{\lambda} = \dfrac{3.0 \times 10^8\,\text{m s}^{-1}}{252\,\text{m}} = 1.2 \times 10^6\,\text{Hz} = 1.2\,\text{MHz}$

Waves and energy

Intensity All waves transmit energy from their source to whatever eventually absorbs them. The **intensity** or **radiation flux** of a wave is a measure of the power transmitted per unit area perpendicular to the wave. For example, the intensity of solar radiation reaching the edge of the Earth's atmosphere is approximately $1.4\,\text{kW m}^{-2}$.

Worked example 2

Calculate an approximate value for the total solar energy reaching the Earth in one 24 hour period. The radius of the Earth is 6400 km.

$E = Pt = IAt$

The area to use is a circle of radius equal to the Earth's radius, because this is the effective area that 'catches' the Sun's radiation.

$E = 1400\,\text{W} \times \pi \times (6.4 \times 10^6\,\text{m})^2 \times 24 \times 3600\,\text{s} = 1.6 \times 10^{22}\,\text{J}$

$$\text{intensity (W m}^{-2}) = \frac{\text{total power transmitted (W)}}{\text{total area through which the waves pass (m}^2)}$$

$$I = \frac{P}{A}$$

The inverse-square law For a point-like source of waves that radiate out evenly in all directions and are not absorbed in the medium around the source, the intensity falls off with distance as an inverse-square law. At a distance r from the source, the power P of the source passes through an area $4\pi r^2$ (surface area of a sphere of radius r centred on the source).

$$I = \frac{P}{A} = \frac{P}{4\pi r^2} \qquad \text{so} \qquad I \propto \frac{1}{r^2}$$

This means that going 10 times further away from a source reduces the intensity or radiation flux by a factor of 10^2.

Phase

Phase represents the particular point in a cycle of vibration or wave motion and is measured in radians (or degrees) with respect to some reference (e.g. the vibration at the wave source). One complete cycle of vibration corresponds to a phase change of 2π radians (or 360°). Points separated by an integer number of wavelengths along a wave will vibrate in phase.

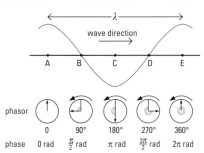

*A rotating vector, or **phasor**, is used to visualize phase. It completes one rotation when it is moved one wavelength along the wave. All phases are measured as delays with respect to particle A, which is why the phasor rotates anti-clockwise.*

Phase difference

The graph below shows two waves with a phase difference of $\pi/2$ (90°). This is equivalent to a path difference of $\lambda/4$ or a time delay of $T/4$ between the waves.

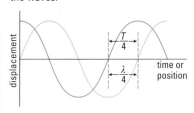

Amplitude and intensity

The energy of an oscillator is proportional to the square of its amplitude, and oscillators are wave sources, so it is not surprising that the intensity of a wave is also proportional to the square of its amplitude:

$I \propto A^2$

PRACTICE

1 Visible light has a range of wavelengths from about 400 nm (violet) to 700 nm (red). What is its range of frequencies?

2 A wave of frequency 2 Hz and wavelength 1.6 m travels along a stretched string.

a What is the wave velocity?

b What is the phase difference between the oscillations of points on the string if they are separated by **i** 1.6 m **ii** 0.8 m **iii** 0.4 m **iv** 2.4 m? (Give your answers in radians, degrees, and fractions of a cycle.)

3 Estimate the power entering your eye when you look directly at a 100 W lamp from a distance of 5 m. (Assume that 100 W is the radiated power.)

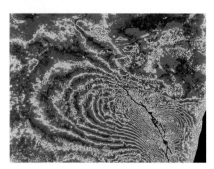

False-colour radar map taken from space, showing ground displacement contours following an earthquake.

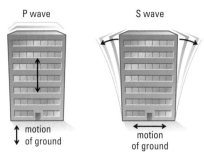

The effect of P waves and S waves on a building. Here the waves are arriving perpendicular to the ground. In practice they will arrive from various directions and then both S and P waves can excite the potentially devastating swaying mode.

Water waves

Water waves are sometimes used as examples of transverse waves because it is obvious that the surface moves up and down while the waves move horizontally, but this is misleading. In a swimming pool, if you tread water as waves pass you will notice that you move in a circular or elliptical motion. Surface water waves are actually a combination of transverse and longitudinal waves.

MECHANICAL WAVES

Earthquakes

Earthquakes occur when strain energy that has built up in the crust is released violently, usually along a fault line. The result is a violent shaking at the source of the earthquake that causes mechanical waves, called **seismic waves**, to travel outwards, carrying the energy of the earthquake with them. The initial shock gives rise to two distinct types of seismic wave:

- **P waves** These are **primary** or **pressure** waves, in which the direction of shaking is parallel to the direction of wave travel. They consist of regions where material is being compressed (compressions) and regions where it is being stretched (rarefactions). P waves are *longitudinal*.

- **S waves** These are **secondary** or **shear** waves, in which the direction of shaking is perpendicular to the direction of wave travel. They are an example of *transverse waves*.

The Earth responds differently to these two types of disturbance: P waves travel faster than S waves and reach distant points earlier. A comparison of the intensity and time of arrival of S and P waves at several seismic monitoring stations can pinpoint the position and strength of the earthquake. Also, P waves can travel through the Earth's liquid core whereas S waves cannot (because the liquid does not oppose shear forces). The actual speed of both types will depend on the nature of the rock through which they travel, but is usually around 2–3 km s^{-1}.

The magnitude of an earthquake is measured on the **Richter scale**. This is a logarithmic scale – an increase in magnitude of one unit corresponds to a release of ten times as much energy. A magnitude 6 earthquake (which is a major earthquake) releases 1000 (10^3) times more energy than one of magnitude 3 (which would barely be felt). The damage done by an earthquake at a particular site will depend on the distance of the site from the source (**epicentre**) and the geology of the rocks at the site.

If the epicentre occurs under the ocean it can raise the level of water over a large area and create **tsunami** (tidal waves). In deep water these travel fast, though they may be only a metre or so in height. However, when such a wave reaches shallow water the front of the wave slows up and the height can build to many tens of metres. The effect on a coastal region can be devastating.

Some devastating 20th-century earthquakes.

Year	Place	Magnitude	Approx. no. of deaths
1920	China	8.6	180 000
1923	Japan	8.3	100 000
1927	China	8.3	200 000
1935	Pakistan	7.5	40 000
1970	Peru	7.7	60 000
1976	China	7.8	240 000
1988	Armenia	7.0	25 000
2003	Iran	6.6	29 000
2004	Indonesia	9.1	250 000
2005	Pakistan	7.6	95 000
2008	China	7.9	80 000
2010	Haiti	7.0	250 000
2011	Japan	9.0	16 000

The Richter scale

2.5 not usually felt, but picked up by instruments (seismometers)

3.5 felt by many people as tremors

4.5 some local damage occurs and poorly built structures may collapse

6.0 a destructive earthquake: considerable damage to buildings

7.0 a major earthquake: many structures destroyed, ground badly cracked

8.0 great earthquakes

9.5 limit to Richter scale

Sound

Take the grille off the front of a hi-fi speaker and look at the cone. Watch it carefully as you play something with a strong bass beat: you should see it vibrating rapidly in and out. This is a good demonstration of the nature of sound – when moving forwards the speaker compresses the air in front of it; when moving back it rarefies it.

Sound waves are *longitudinal mechanical waves* and anything that produces sound must compress and rarefy the air like this. Look up the specifications of your hi-fi and you are likely to find that the speaker can handle signals from 20 Hz to 20 kHz. This more than covers the range of sounds audible to humans and corresponds to a wavelength range of 16.5 m to 1.65 cm (remember, $\lambda = v/f$).

The human ear responds to sounds over such a wide range of intensities that a logarithmic scale, the **decibel scale**, is used to compare them. This also corresponds to the individual's impression of how loudness varies with intensity. If you increase the sound intensity of a note and ask someone to say when it sounds 'twice as loud' you will find that the intensity of the sound has not doubled, but increased by a factor of 10. The weakest sound a person can normally detect is about 1 picowatt per square metre (defined as 0 dB) whereas a painfully loud sound (e.g. a jet taking off) is above 1 W m^{-2}, a thousand billion times (10^{12} times) more intense.

> ### The decibel scale
>
> Intensity levels (IL) are measured in bels (B)
>
> $$IL = \log_{10} \frac{\text{intensity of sound}}{\text{intensity at threshold of hearing (1 pW m}^{-2})} = \log_{10} \frac{I}{I_0}$$
>
> 1 decibel (dB) = 10 bels (B)
>
> $$IL\,(dB) = 10 \log_{10} \frac{I}{I_0}$$
>
> On this scale a jet taking off at 1 W m^{-2} has an intensity level
>
> $$IL\,(dB) = 10 \log_{10} \left(\frac{1\,W\,m^{-2}}{10^{-12}\,W\,m^{-2}} \right) = 120\,dB$$

Sound must have a mechanical medium. 'In space no one can hear you scream' because there are no material particles present to compress or rarefy. This is easily verified in a school laboratory by hanging an electric bell inside a container and pumping the air out. If the vacuum is good, and the bell supports do not transmit sound to the container, the bell will no longer be heard. However, sound is not confined to air; it can be transmitted through any solid, liquid, or gas and its velocity in each will depend on the density and elasticity of the medium (see spread 6.14).

Ultrasonic waves are sound waves above the frequency range of human hearing, so they start at about 20 kHz. These have many industrial and medical applications. For example, ultrasound at 40 kHz can be used for cleaning because it makes dirt particles resonate and effectively shakes them off things. At higher frequencies still (MHz) an ultrasonic beam can be used to make a prenatal scan of a fetus. This is possible because the waves are partially reflected at each tissue boundary and the timed reflections can be detected and used to determine the depth of the boundary. In the natural world, bats use ultrasonic sonar to locate their prey and to avoid crashing into objects as they fly around in the dark.

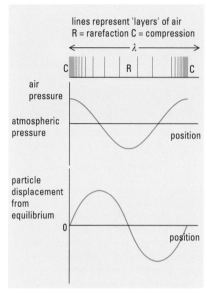

The disturbance of air due to sound can be represented graphically, but you have to be careful because waves can be represented in several ways. One approach is to plot the variation of pressure above and below atmospheric pressure along the wave. Another is to imagine a 'snap-shot' of the wave and plot the displacement of particles, in this case imaginary layers of air, from their mean positions. Note that the graphs look like transverse waves, even though the particle displacements are all parallel to the wave direction (i.e. longitudinal).

When sound reaches the ear the varying pressure differences across the ear-drum transmit information to the listener. It is sometimes necessary to plot a graph of pressure versus time at a fixed point (i.e. the ear-drum) to show this. The graph looks very much like the ones shown, but is now pressure versus time at one position rather than pressure versus position at one time!

PRACTICE

1 Explain why S waves cannot pass through the core of the Earth.

2 The speed of seismic waves is reduced when they enter denser rocks. Why does this happen? How might this complicate the process of locating the epicentre of an earthquake?

3 If the average speed through surface rocks is about 3 km s^{-1} for P waves and 2 km s^{-1} for S waves:

   a calculate the distance to the epicentre of an earthquake if the two types of wave arrive 5 minutes apart at a particular seismometer.

   b What assumption has been made in this calculation?

4 How much louder would your hi-fi sound if you doubled its power output?

- the electromagnetic spectrum

- wavelength

- frequency

- photon energy

Wear sunscreen. It protects against ultraviolet electromagnetic radiation.

Graphs of transverse waves

We can plot wave displacement against either position or time. This gives two different but related ways to represent the wave. Plotting displacement against position is like taking a snapshot of the wave at one moment; plotting displacement against time focuses on the vibration at one point in space.

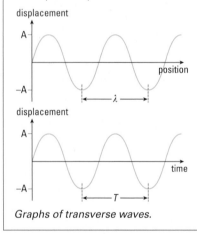

Graphs of transverse waves.

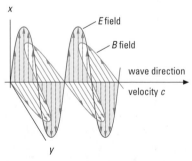

An electromagnetic wave. Electromagnetic waves involve periodic oscillations of the electric (E) and magnetic (B) fields in space. This wave is plane-polarized parallel to the x axis. The E field is varing parallel to x. The B field is varying parallel to y. NB The arrow lengths represent field strengths, not displacements.

ELECTROMAGNETIC WAVES

Mechanical waves are started by shaking massive particles; electromagnetic waves are emitted whenever charged particles are accelerated. This may be the result of a.c. currents in an antenna, which is how radio and TV waves are produced, or the quantum jumps made by electrons in atoms as they lose energy. In circular **particle accelerators (synchrotrons)** the centripetal acceleration of the electrons is responsible for synchrotron radiation, which is a very serious energy loss in high-energy experiments. Accelerated charges disturb the electric and magnetic fields and these disturbances spread out at the speed of light. No material medium is needed, because the wave does not consist of vibrating particles; but if the wave passes other charged particles it will force them to oscillate at the wave frequency and can transfer energy to them. They in turn may absorb or re-radiate this energy.

If electromagnetic waves are produced by a.c. electrical signals the wave frequency is fixed by the oscillator frequency. In a microwave oven a magnetron is used to generate electrical oscillations at 2.45 GHz. These have a wavelength of $\lambda = c/f = 0.12$ m. This frequency corresponds to one of the natural frequencies at which water molecules vibrate, so water molecules in the food resonate, absorb energy strongly from the radiation, and cook the food. Since the microwaves penetrate right through the food it will be cooked much more rapidly than by conventional methods that require the heat to conduct to the centre from the outside.

Waves and photons

Atomic and nuclear changes can also emit electromagnetic waves. These tend to be discrete events, quantum jumps involving fixed amounts of energy, and are best described by treating the electromagnetic 'wave' more like a particle, called a **photon**. The minimum energy that can be emitted or absorbed at a particular frequency is one photon. The frequency of the photon emitted is related to the energy of the quantum jump by:

$$E = hf$$

where h is the **Planck constant** and equals 6.6×10^{-34} J s.

Visible light is caused by electrons making quantum jumps in atoms involving photon energies of about 2–3 eV. Nuclear energy changes (e.g. following radioactive decay) involve much larger energy changes (MeV) and generate photons of much higher frequency and shorter wavelength (gamma rays). All electromagnetic radiation beyond the visible range (ultraviolet and higher frequencies) is potentially dangerous to humans because the photons are more energetic and can cause cell damage. Sunscreen can prevent some harmful ultraviolet radiation from being absorbed by the skin.

The wave model works well for long-wavelength, low-frequency waves because their photon energy is so low that we are usually dealing with enormous numbers of them and then they behave like a continuous wave. The discrete nature of the photon model becomes more obvious with higher-frequency radiation and on an atomic or nuclear scale where individual photons are exchanged. To help remember these two complementary descriptions, compare the uniform illumination produced by a filament lamp with the intermittent clicks of a Geiger counter detecting gamma rays from a radioactive source.

Gamma rays (γ rays)

These are emitted when excited nuclei make quantum jumps to lower energy states, usually after radioactive decay. γ-ray spectra give information about nuclear energy structure. γ rays are highly penetrating and dangerous to humans because they cause cell damage. γ rays are used to induce mutations in genetic experiments, and can be used to sterilize medical equipment. γ-ray astronomy has also come of age. The Compton γ-Ray Observatory was placed in orbit in 1991 and has identified hundreds of previously unknown sources of γ-rays and thousands of γ-ray bursts.

X-rays

These were discovered by Wilhelm Konrad Röntgen in 1895, and are produced when high energy (e.g. 50 keV) electrons collide with a heavy metal target (usually tungsten) and transfer some or all of their kinetic energy to X-ray photons (99% of the incident energy goes to thermal energy). They are used in medical imaging and security systems at airports, etc. If you break a bone an X-ray shadow photograph is taken because bones absorb X-rays more strongly than flesh. Dentists use the same technique. **Computed tomography** (CT) scans use an array of detectors around the patient to reconstruct a detailed image of one slice through the patient's body. X-ray telescopes such as Rosat (Röntgen satellite) and Chandra can detect X-rays from compact objects such as neutron stars and black holes formed after supernovae explosions.

Ultraviolet

Ultraviolet (UV) radiation stimulates the production of vitamin D in the skin and causes tanning. UV-A (320–400 nm) penetrates the skin most deeply and causes it to sag and wrinkle, but it is over-exposure to UV-B (290–320 nm) that is mainly responsible for skin cancers. The ozone layer in the Earth's atmosphere protects us from most of the Sun's UV radiation (absorbing almost all of the UV-C at 200–290 nm), so the developing hole in this layer is worrying. It is thought to have been caused by CFC pollutants. UV spectroscopy is used in the pharmaceutical industry to analyse drugs and in dentistry to polymerize materials used in some fillings. It is also used to detect forged banknotes by fluorescence. Bees can detect UV light and flowers attract them with beautiful patterns that are invisible to humans.

Visible

Most sources are hot bodies and light is detected by cells on the retina. It can also be detected chemically using photographic film, and electronically in **charge-coupled devices** (**CCDs**) using semiconductors in which charge carriers are freed to conduct by the absorbed photons. Spectra of galaxies, stars, and atoms provide an enormous amount of information about matter on the cosmic and atomic scales. Chlorophyll in green plants absorbs most of the violet, blue, orange, and red light that falls on it and uses the energy for photosynthesis. Fluorescent lights actually generate UV when electrons collide with mercury atoms inside the tube, but the inside surface is coated with a phosphor that absorbs the UV photons and emits visible light.

Infrared

Infrared (IR) radiation was discovered in 1800 by William Herschel. All warm bodies radiate in the infrared. Rattlesnakes and other pit vipers have IR detectors on their snouts that help them to detect warm-blooded prey. IR penetrates smoke and dust better than visible light so IR detectors are used by the rescue services to detect people, by the armed forces to make attacks under cover of a smoke screen, and on rifles for night vision. IR is also impeded less than light by interstellar dust and the Infrared Astronomy Satellite (IRAS) makes use of this in surveying the sky.

Microwaves

In 1946 Percy Le Baron Spencer noticed that the sweets he had in his pocket melted while he was working with microwaves. Within a year the company had developed and marketed the first microwave oven. Microwaves are used to transmit TV and telephone signals via small metal dish aerials on towers (like the BT tower in London) or on orbiting spacecraft (for satellite TV). Microwave background radiation is a remnant of the Big Bang and comes from all directions in space.

Radio waves

In 1901 Guglielmo Marconi transmitted 1 MHz radio signals across the Atlantic. The waves followed the curvature of the Earth by reflecting from a conducting layer in the atmosphere called the ionosphere. Nowadays radio frequencies are congested with signals from radio and TV stations. In 1932 Karl Guthe Jansky concluded that the source of a persistent radio interference came from the centre of our galaxy, and radio astronomy was born. The large radio telescopes, e.g. Jodrell Bank, operate on the same principle as a satellite dish, but are much larger because of the longer wavelength of radio waves.

The speed of light
All electromagnetic waves travel at the speed of light in a vacuum.

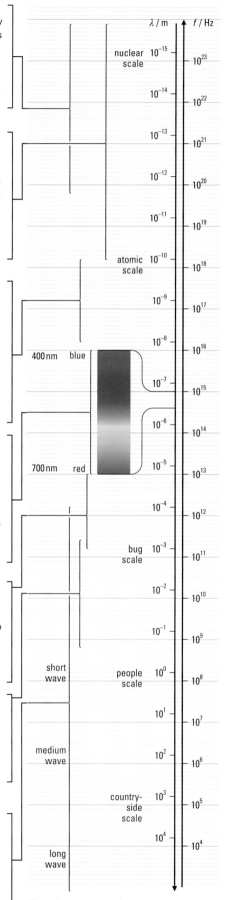

The electromagnetic spectrum. NB (1) the scale is logarithmic, (2) there are no clear-cut boundaries between wave types.

REFLECTION

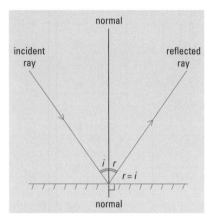

The **normal** is a line drawn perpendicular to the surface and is used in preference to the surface itself because it remains well defined even when the surface is curved.

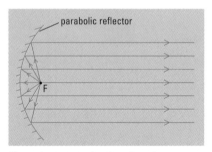

If a point source is placed at the focus, F, of a parabolic reflector, the reflected rays form a parallel beam. Conversely incident parallel rays are focused to a point.

Is anyone out there? The Arecibo radio telescope is being used as part of the SETI experiment (search for extraterrestrial intelligence), scanning the sky for radio signals from alien civilizations. It is capable of communicating with a similar dish anywhere in our galaxy.

Radar

Radar works by reflecting microwave pulses from solid objects and measuring the time it takes for the echo to return. **Sonar** does a similar thing with sound or ultrasound. Both depend on the **law of reflection** and the constant velocity of waves in a particular medium.

> **Law of reflection**
>
> The incident and reflected rays make equal angles to a normal to the surface at the reflection point.

> **The radar equation**
>
> $$s = \frac{ct}{2}$$
>
> where s is distance to 'target', t is time between emission of pulse and its reflection returning, and c is the speed of the waves used ($3 \times 10^8 \, \text{m s}^{-1}$ for radar and about $330 \, \text{m s}^{-1}$ for sonar).

Uses of radar

The first radar systems were used to give early warning of bombing raids during the Second World War, but they were soon being pointed into space. Ground-based radar is now used to make accurate measurements of distances and survey surface features on the Moon and inner planets. It can resolve details about 1 km across on Venus and 100 km across on Mercury. Return pulses take between 5 and 12 minutes on their journey to Mercury (depending on relative orbital positions). Radar has an advantage over visible light in that it easily penetrates cloud cover. Detailed radar surveys of several planets have been carried out by orbiting spacecraft.

The European Space Agency Satellite ERS-1 surveyed Mount Etna in May 1992 and October 1993 and showed it had 'deflated' by 11 cm. Inflation is expected to occur prior to an eruption as pressure builds up in the magma beneath it. The coloured bands indicate a drop of about 3 cm each.

To make an accurate measurement of direction and distance it is important to confine the emitted waves to a narrow beam. This is also done by reflection, using a parabolic reflector, like those in car headlamps. Nonetheless, the returning signal is extremely feeble and for astronomical measurements a very large collecting 'dish' is required. One of the first to be used was the world's largest radio telescope, the Arecibo Observatory in Puerto Rico. This is a 305 m parabolic reflector built into a natural depression in the Earth's surface, but this has a disadvantage in that the telescope points in a fixed direction relative to the Earth. Another way to make astronomical measurements using radar is to collect the signal in a linked array of radio telescopes; this effectively simulates the receiving characteristics of a much larger telescope.

The speed of a moving object can be measured using 'Doppler radar' (see spread 6.18). If waves are reflected from something that is moving towards the source their frequency is increased; if the reflector is moving away from the source the frequency falls and wavelength increases. This fractional change in frequency or wavelength is directly proportional to the speed. This technique is used by police to trap speeding motorists.

Reflection, absorption, and transmission

Whenever waves encounter a boundary they can be reflected, absorbed, or transmitted. An ideal transparent material would transmit all the light that falls onto it, but with real materials like glass some of the incident light is absorbed and some reflected. Inside the material impurities and cracks may also scatter the light. Extremely pure forms of glass are manufactured for use in optical fibres so that data can be transmitted over many kilometres before it becomes necessary to 'clean it up' and amplify it.

The amount of light reflected, transmitted, and absorbed is often frequency-dependent, and this explains why objects have distinctive colours. A red table, for example, reflects red light while absorbing most of the shorter wavelengths in the spectrum. If it was viewed in a blue light it would appear black since no light would reflect off it. A red stained-glass window appears red because it absorbs all other wavelengths and transmits only red. (Of course, there are many different 'reds' because what is actually absorbed or transmitted is a range of wavelengths.)

Change of phase

Get someone to hold the other end of a slinky still while you send a transverse pulse along it. The pulse will invert when it reflects at the fixed point. On the other hand, if the slinky is suspended vertically so the lower end is free, a pulse sent down it will not be inverted on reflection and the end of the slinky will wave about with a large amplitude. A similar thing is true for light. If light rays in air reflect from an air–glass boundary they have a π phase change, whereas if they reflect from a glass–air boundary they do not.

In general a wave will be inverted when it reflects from a boundary with a denser or less responsive medium. In the extreme case where the medium does not respond at all to the incident wave (i.e. is not 'waved' by it) the disturbance at the surface is constantly zero and this can be considered as the sum (**superposition**) of the incident wave and an inverted reflection.

These phase change effects are particularly important in experiments involving interference, since an inversion of phase in one beam will exchange the positions of maxima and minima in the resulting interference pattern.

Speed traps

Speed measurements using radar reflection must take account of a double Doppler effect:

$$\frac{\delta\lambda}{\lambda} = \frac{2v}{c}$$

For example, if the source is approaching, the rays are Doppler-shifted once on striking the reflector. Then these Doppler-shifted waves suffer a second similar shift because their point of reflection is moving.

Echo location

Dolphins and whales use sounds for both echo location and communication. Different densities, salinities, and temperatures of sea water create 'channels' that allow these sounds to be 'heard' tens or even hundreds of kilometres away. For communication these animals use a variety of pure tone whistles, pulsed squeals, screams, or barks, usually in the range 500 Hz to 20 kHz. For echo location they use short pulses of ultrasound with a broad range of frequencies up to 220 kHz. There are concerns that human noise pollution of the oceans (by shipping, etc.) could traumatize or disorient these creatures.

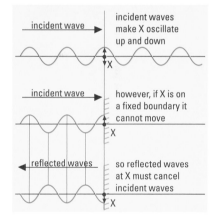

If the medium does not respond at all to the incident wave, then incident and reflected waves cancel each other out.

PRACTICE

1 If a radar reflection takes 14 minutes to complete a return trip to Mercury, how far is Mercury from the Earth at the time of the experiment?

2 In a police speed trap the small difference (δf) in frequency between the emitted waves (f_0) and the reflected waves (f) is a measure of the vehicle's speed.

   a How will δf change for a vehicle moving at constant speed as it approaches, passes, and moves away from a roadside speed trap?

b What implications does this have for the use of radar speed guns?

3 Prove that if two plane mirrors are placed perpendicular to one another, rays entering from any direction (in the plane perpendicular to both mirrors) will return along the same path.

4 Use a ray diagram to prove that the image is as far behind a plane mirror as the object is in front of it and lies on the same normal.

Waves at a beach.

Refraction

When a wave moves from one medium to another, or through a region in which the medium changes continuously, the velocity and wavelength change and these changes may cause the wave direction to change.

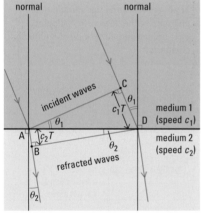

Light travelling from one medium to another.

Refracted sound waves

Sound travels faster in warmer air. At night air near the ground is often cooler than air higher up. Refraction of sound causes the waves to change direction so they travel more nearly parallel to the ground. This is why sounds carry further at night.

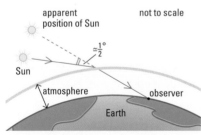

When the Sun is near the horizon, refraction of light on entering the atmosphere shifts its apparent position by about $\frac{1}{2}°$. This is about the same as the angular size of the Sun or Moon in the sky.

REFRACTION

Have you ever wondered why waves always seem to come up a beach even though they travel in all directions out at sea? A bird's-eye view of waves approaching a bay shows three things.

- The waves slow down.

- The wave peaks tend to bunch together.

- The waves turn towards the beach.

All three effects are caused by the reduction in depth of water in the bay. The velocity of shallow water waves is less than the velocity of deep water waves, so later waves catch up with earlier ones. The wave equation $v = f\lambda$ confirms this: the frequency of waves cannot change so when v drops λ must also drop. If the waves are approaching the beach obliquely, the side of the wave closest to the beach will be slowed down most and the whole wave will swing in towards the beach. This is similar to what happens to a car if the wheels on one side run off the road into mud – that side of the car is slowed and the car veers off the road. This is an example of **refraction**.

The law of refraction (Snell's law)

The diagram shows light travelling from one transparent medium (in which its velocity is c_1) across a boundary to another (in which its velocity is c_2).

The wavefronts are all one cycle apart, so the distance moved during that time will be c_1T in medium 1 and c_2T in medium 2, where T is the period of the wave. The part of the wave that has just hit the medium at A will move to B while the part at C will move to D. Since the wavefronts (AC and BD) are always perpendicular to the rays (AB and CD) two simple right-angled triangles are formed and these can be used to relate the incident and refracted angles (θ_1 and θ_2 respectively).

$$\sin\theta_1 = \frac{c_1T}{AD} \text{ (in triangle ACD)}$$

$$\sin\theta_2 = \frac{c_2T}{AD} \text{ (in triangle ABD)}$$

Eliminating AD gives

$$\frac{\sin\theta_1}{\sin\theta_2} = \frac{c_1}{c_2}$$

The ratio of the sines of incident to refracted angles is equal to the ratio of the velocities of the waves in the two media. This ratio is a constant for a ray moving between a particular pair of media and is called the **refractive index**, $_1n_2$. A ray travelling from medium 2 into medium 1 across this boundary would exactly retrace this path but in the opposite direction, with the incident and refracted rays swapped and a refractive index $_2n_1$, i.e. the reciprocal of $_1n_2$:

$$\frac{\sin\theta_1}{\sin\theta_2} = \frac{c_1}{c_2} = {_1n_2} \qquad {_1n_2} = \frac{1}{_2n_1}$$

If $c_1 > c_2$ then $_1n_2 > 1$ and the wave velocity reduces as the wave crosses the boundary, as with light travelling from air to glass or water waves approaching a beach. The waves then refract toward the normal. The velocity of light in a medium depends on electron density – the greater the density the slower the light – so a useful summary is:

- Light rays bend toward/away from the normal when entering an optically more dense/less dense medium.

Snell's law is sometimes written in terms of incident angle (i) and refracted angle (r) (both measured from the normal):

$$\frac{\sin i}{\sin r} = {_1n_2}$$

Absolute refractive index

For a ray travelling from medium 1 into medium 2, $_1n_2$ is a relative refractive index. If medium 1 is a vacuum (in which all electromagnetic waves travel at $3.0 \times 10^8\,\mathrm{m\,s^{-1}}$) then the refractive index for a ray travelling from the vacuum into a medium is a property of the medium called its **absolute refractive index**, n (e.g. $n_{glass} \approx 1.5$).

$$n = {_{vac}}n_{med} = \frac{c}{c_{med}}$$

The absolute refractive indices of a pair of media determine their relative refractive index:

$$_1n_2 = \frac{c_1}{c_2} = \frac{c_1}{c} \times \frac{c}{c_2} = \frac{n_2}{n_1} \quad so \quad \frac{n_2}{n_1} = \frac{\sin\theta_1}{\sin\theta_2}$$

Worked example 1

A ray of light travels from water into a glass block at an angle of 15° to the normal to the boundary between the two media. In what direction does it travel inside the glass? (Take $n_{water} = 1.33$ and $n_{glass} = 1.53$.)

$$n_w \sin\theta_w = n_g \sin\theta_g$$

$$\sin\theta_g = \frac{n_w \sin\theta_w}{n_g} = \frac{1.33 \sin 15°}{1.53} = 0.225 \quad so \quad \theta_g = 13.0°$$

so the ray travels at 13° to the normal inside the glass.

Worked example 2

What is the speed of light in diamond? (Take $n_{diamond} = 2.42$.)

$$n_d = \frac{c}{c_d} \quad so \quad c_d = \frac{c}{n_d} = \frac{3.0 \times 10^8\,\mathrm{m\,s^{-1}}}{2.42} = 1.2 \times 10^8\,\mathrm{m\,s^{-1}}$$

Deviation and dispersion

Newton showed that white light contains all colours of the spectrum. He passed a narrow beam of white light through a glass prism and it bent toward the normal on entering and away from the normal on leaving. These two refractions caused the light to deviate from its original path. However, the refractive index for light in glass increases with frequency so higher frequencies (e.g. blue) were deviated more than lower frequencies (e.g. red). This separation of frequencies is called **dispersion**.

Absolute refractive indices.

Material	Absolute refractive index at 589 nm wavelength*
flint glass	1.58–1.80 depending on density
crown glass	1.53
window glass	1.51
water	1.33
ice	1.31
diamond	2.42
air	1.0003

*Refractive index varies with wavelength. This wavelength is the yellow light emitted by a sodium lamp.

The law of refraction

This is the easiest form of the law of refraction to remember:

$$n_1 \sin\theta_1 = n_2 \sin\theta_2$$

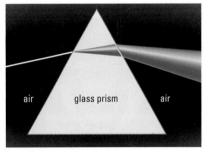

White light refracted through a prism.

Helpful hint

Higher frequencies deviate more than lower frequencies: remember 'blue bends best'.

PRACTICE

1 When Supernova 1987A was detected on Earth, neutrinos emitted in the explosion arrived about 20 hours earlier than the light. Suggest two possible explanations that are consistent with the idea that nothing can travel faster than the speed of light.

2 Prove that a light ray travelling through a rectangular glass block emerges parallel to its incident direction but displaced sideways. What determines the amount of displacement?

3 a A ray of light in air strikes the plane surface of a crown glass block at an angle of 50° to the normal. In what direction does the ray travel inside the block?

  b What is the maximum angle of incidence for a ray travelling from crown glass into air if the ray is to refract out of the block (this is called the **critical angle**)?

c Suggest what must happen if the ray in **b** strikes the glass–air boundary at an angle greater than the one calculated.

4 Large wax lenses are used to focus 3 cm microwaves in school experiments. What do you conclude about the velocity and wavelength of these microwaves in wax?

5 White light enters normally through one face of an equilateral glass prism and emerges from another. What is the approximate angular separation of red and blue light in the emerging beam?

6 Explain why you can see much better underwater if you wear a face mask.

7 Draw a labelled ray diagram to show why a swimming pool appears shallower than it actually is to someone looking down into the water.

6.6

O B J E C T I V E S

- critical angle

- total internal reflection

- optical fibres

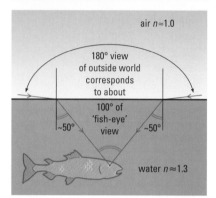

Outside this 100° range, the fish sees reflections of its underwater world owing to the phenomenon of total internal reflection.

A 'fish-eye lens' is used to compress a wide angle of view into a narrower angle.

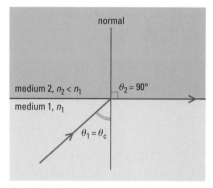

The critical angle.

An optical fibre.

TOTAL INTERNAL REFLECTION

You will be familiar with the effects of refraction when you look down into a pool, but what does the world look like to a fish looking out? Light entering the water from the outside world will be refracted at the surface and bent toward the normal. This means that a 180° view of the surface world is compressed into a smaller angle for the fish (see diagram).

Outside this range the fish cannot receive light from the outside world, only reflections of its own world from the underside of the surface. These reflections are examples of **total internal reflection**, an effect that can only occur for rays travelling out of a 'slower' medium.

Total internal reflection

When rays travel from a medium of higher refractive index to one of lower refractive index the angle of refraction is larger than the angle of incidence. However, the angle of refraction cannot exceed 90°, so there is a limiting angle of incidence above which no refracted ray can be formed. This is called the **critical angle**, θ_c. For incident angles greater than θ_c the rays reflect back into the first medium (total internal reflection).

> ### Critical angle θ_c
>
> The critical angle, θ_c, is the angle of incidence for a ray crossing the boundary from a medium of higher refractive index into one of lower refractive index at which the law of refraction predicts a refracted angle of 90°. No refracted ray can form so the incident ray undergoes total internal reflection.

The critical angle is related to the refractive index:

$$n_1 \sin \theta_1 = n_2 \sin \theta_2$$

But $\theta_1 = \theta_c$ and $\theta_2 = 90°$ at the critical condition, so

$$n_1 \sin \theta_c = n_2 \sin 90° = n_2$$

$$\sin \theta_c = \frac{n_1}{n_2}$$

If the ray is travelling out into a vacuum (or air), $n_2 = 1$, so:

$$\sin \theta_c = \frac{1}{n_1} \quad \text{where } n_1 \text{ is the absolute refractive index of the material.}$$

> ### Worked example
>
> What is the critical angle for light travelling from glass ($n_g = 1.5$) into air?
>
> $$\sin \theta_c = \frac{1}{n_g} = \frac{1}{1.5} = 0.67$$
>
> $$\theta_c = 42°$$

Fibre optics

Total internal reflection can be used to guide light along transparent fibres. Light rays moving along the fibre reflect from the inner surface because their incident angles are well above the critical angle at the boundary. This remains true even when the fibre is bent, as long as the radius into which it is bent is much greater than the radius of the fibre itself. In practice, fibres used for communications have diameters between about 0.1 mm and 0.5 mm, so this is not likely to be a problem. Light would be lost from a fibre, however, if it made contact with another fibre or something else of higher refractive index. To prevent this leakage, fibres are usually clad in another transparent material of lower refractive index than the core itself. Total internal reflection then occurs at the boundary between the two materials.

The cladding material typically has a refractive index only about 1% lower than that of the core, so the critical angle is about 82°, underlining the fact that rays travel close to the axis of the fibre even when it is bent.

The major use of optical fibres is for communications. The high frequency of light ($>10^{14}$ Hz) makes it ideal for transmitting information. Conventional telephone signals are carried by a.c. electric currents in metal wires, but these operate at much lower frequencies and so have much lower data transfer rates. The essential components of an optical fibre communications system are shown in the diagram. Such systems can carry telephone conversations, cable TV, internet links, videoconferencing signals, etc. Bundles of optical fibres are also used to carry images back from previously inaccessible places, especially inside the human body using an endoscope (see spread 13.11).

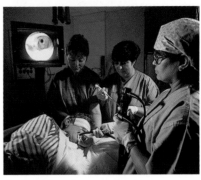

Optical fibres have enabled less invasive 'keyhole' surgical techniques to be developed, with greatly reduced recovery times.

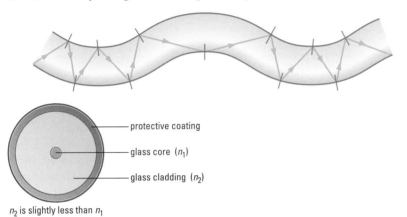

— protective coating

— glass core (n_1)

— glass cladding (n_2)

n_2 is slightly less than n_1

Transmission of light along the core of the optical fibre.

Light is absorbed and scattered to some extent in all transparent materials, so a great deal of research has gone into developing the plastics and glasses that are used in fibre optics. If a constant proportion of the signal is lost every kilometre the transmitted intensity will fall exponentially. For example, a loss of 2% per kilometre means that after 1 km 0.98 of the original intensity remains and after n kilometres 0.98^n remains. The incident intensity halves every n kilometres where:

$$0.98^n = 0.5$$
$$n \log 0.98 = \log 0.5$$
$$n = \frac{\log 0.5}{\log 0.98} = 34\,\text{km}$$

In addition to the loss of intensity, the signal will disperse because different wavelengths travel at different speeds (so that the shape of the transmitted signal begins to break up) and different rays travel along slightly different paths inside the fibre. These problems can be resolved using the methods in the box on the right. This allows signals to be transmitted many hundreds of kilometres without the need for **repeaters** (amplifiers).

Time division multiplexing

Optical fibres carry large numbers of telephone signals simultaneously. They do this by sampling each conversation about 10^4 times per second and encoding each sample into 8 light pulses lasting a total of about 10^{-7} s. This effectively compresses every conversation by a factor of about 1000, so that the encoded samples from 1000 conversations can be transmitted in the real time of just one of them. Of course, the equipment at the receiving end must recombine the correct pulses to decode the signal from each separate conversation.

Improving signals

1 Wavelengths are chosen for minimum absorption in the material (usually infrared).

2 Signals are carried by almost monochromatic light (avoiding dispersion due to refraction).

3 **Monomode** fibres are used for communications – these have a very narrow core, so the light is effectively confined to a single axial path.

4 Information is transmitted in digital form, since minor degradation can be corrected without loss of data.

PRACTICE

1 a What is the critical angle for light travelling from diamond to air? ($n_{\text{diamond}} = 2.42$, $n_{\text{air}} = 1.0$.)

  b What happens to the critical angle for rays leaving the diamond if it is submerged in water ($n_{\text{water}} = 1.3$)?

2 Light is travelling along an optical fibre made of glass of refractive index 1.52. What is the refractive index of the cladding material if the critical angle is 82°?

3 a What is the most direct route for light along the fibre-optic cable of question 2?

b The longest route is for light rays travelling at the critical angle. Calculate the path difference between the two rays after travelling 10 km of cable.

c What time delay does this correspond to?

d Would it be possible to transmit information at **i** 1 MHz and **ii** 100 MHz along this cable? Explain your answer.

e Why is it desirable that the refractive index of the cladding is only slightly less than that of the core?

- the principle of superposition

- two-source interference patterns

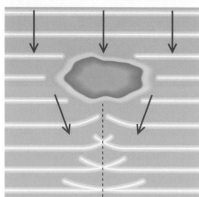

line of maximum amplitude

Waves diffracting around an obstacle.

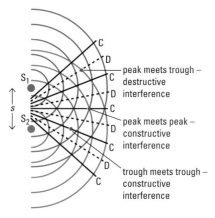

peak meets trough –
destructive
interference

peak meets peak –
constructive
interference

trough meets trough –
constructive
interference

Formation of a two-source interference pattern. A line drawn parallel to the source separation would cut through a regular pattern of maxima and minima.

Maxima

Maxima are not regions with a permanent large disturbance – they oscillate like any other part of the wave, passing through zero to negative values every cycle. They are positions where this oscillation has maximum amplitude.

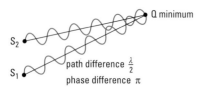

Q minimum

path difference $\frac{\lambda}{2}$
phase difference π

S_1 and S_2 are coherent sources in phase

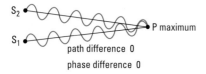

P maximum

path difference 0
phase difference 0

Path difference.

SUPERPOSITION

I once made a mistake no physicist should make. I had waded out to a small island carrying my young son, had a picnic, and then decided to wade back. In the meantime the tide had started to come in and rather gentle waves were diffracting around both sides of the island to cross over in the region between the island and the shore. The mean depth of water was no more than a metre, so I started to wade back across the 50 m or so that separated us from the beach. The waves from each side caused a swell of no more than 25 cm but where they crossed the water rose 50 cm and I had to hold my son above my head to keep him safe. I had forgotten the principle of superposition.

Principle of superposition

The resultant disturbance at a point where similar waves from several different sources cross is the vector sum of the individual disturbances.

I was wading along the most direct route back to the beach, and along this line the waves coming around both sides of the island had approximately equal amplitudes. Whenever two wavecrests crossed, the surface was lifted by the sum of their amplitudes. In between these 'super peaks', troughs would cross causing 'super troughs' and the whole choppy nature of the sea made progress very difficult. However, by moving to one side I was able to reach much calmer water where the peaks from one side of the island were out of phase with those from the other and so to some extent they cancelled one another. Back on the beach I realized what had happened – I had been wading through a large-scale **interference pattern**. My original path had been through a line of interference **maxima**, and my final path had crossed into a line of interference **minima**. Since the wave intensity is proportional to the square of the wave amplitude, the sea was approximately four times rougher in the maxima than it would be if only one of the waves were present.

Interference patterns

Stable interference patterns will only occur if overlapping waves satisfy the following conditions.

- They must be a similar type of wave (e.g. both light or both sound).

- The wave sources must maintain a constant phase relation to one another – they must be **coherent sources**. This is only possible if they have the same frequency and wavelength.

- They must have comparable (but not necessarily equal) amplitudes.

- If they are transverse waves they must vibrate in the same plane (have the same polarization).

Interference from two point sources

If two coherent, monochromatic point sources are set up then an interference pattern is formed in the region where their waves overlap. Assume that the two sources are in phase with one another. Maxima will be formed where the waves from both sources arrive in phase, and minima where they arrive exactly half a cycle (π) out of phase. The phase differences are caused by the different distances travelled by the waves from the source to the point concerned. An extra path of one half-wavelength from one source will introduce a phase difference of π radians resulting in cancellation (a minimum), whereas a path difference of any whole number of wavelengths results in the waves arriving in phase and adding (a maximum).

- **Path difference**, x = difference in distance from each source to a particular point.

- **Phase difference**, ϕ = difference in phase of the waves at a point. In general

$$\phi = \frac{2\pi x}{\lambda}$$

It is easy to set up an interference pattern with sound waves using two speakers driven from the same oscillator and held a short distance apart. If a suitable frequency is used you can walk through the interference pattern and hear maxima and minima at different places. Or you could move a microphone along the 'screen' line and detect maxima and minima on an oscilloscope. It is also easy to produce this pattern of maxima and minima using two dippers side by side in a ripple tank. It is rather harder to do this with light (see spread 6.8).

Two-source interference in a ripple tank.

Beats

If a musical instrument is slightly out of tune a note played on it will '**beat**' with the same note played on a tuned instrument. The beating is heard as a low-frequency variation in the volume of the sound. The maxima and minima which produce the beats are caused by constructive and destructive interference. A maximum occurs when sound from both sources arrives at your ear in phase. However, the higher note then gets progressively further ahead of the lower note until they are π out of phase at your ear and a minimum is formed. The higher note continues to get further ahead and soon another maximum forms, and so on. In the time between two successive maxima, the beat period T_b, the higher-frequency note completes exactly one oscillation more than the lower note:

$$f_1 T_b - f_2 T_b = 1$$
$$f_1 - f_2 = \frac{1}{T_b} = f_b$$

The beat frequency equals the difference between the source frequencies.

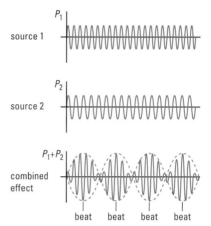

Beats.

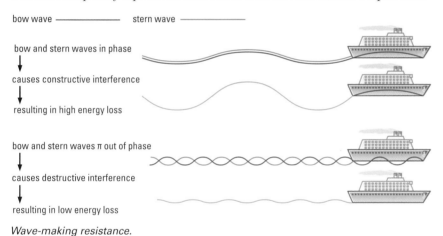

Wave-making resistance.

Wave-making resistance

When ships move across a flat water surface the bow and stern create waves that carry energy away from the ship. A drag force acts on the ship, making it do work to 'pay' for these waves as it moves forwards. The bow wave starts with a crest and the stern wave with a trough. Ships are designed so that the waves interfere destructively at the ship's cruising speed and minimize the energy loss and drag force.

PRACTICE

1 Explain why sound from two loudspeakers connected to the same audio signal forms an audible interference pattern, whereas overlapping headlamps from the same car do not form a visible interference pattern.

2 A light source emits light of a particular frequency uniformly in all directions. How do each of the following quantities vary with distance from the source?

a intensity b amplitude c wavelength d phase with respect to the source e speed f frequency.

3 Two loudspeakers emit 330 Hz sounds of equal amplitude A and intensity I in phase. The speakers are 2 m apart and face each other. (The speed of sound in air is 330 m s^{-1}.)

a What is the wavelength of the sound?

b Ignoring the reduction of amplitude with distance from each speaker, state the amplitude and intensity of sound i halfway between the speakers, ii 0.25 m from either speaker, iii 0.5 m from either speaker.

c What would happen to your answers to b if the connections to one of the speakers were reversed?

4 Two equal-amplitude sound sources are heard together. Their frequencies are 500 Hz and 502 Hz.

a Sketch a graph of pressure versus time at the observer's ear for a period of 2 seconds.

b What would be the effect of i halving the amplitude of one source, ii increasing the frequency of the higher frequency source to 503 Hz, and iii reversing the connections to one of the speakers?

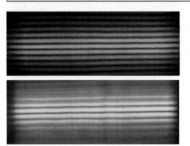

Young's fringes formed by monochromatic yellow (top) and white light (bottom).

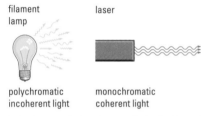

filament lamp   laser

polychromatic incoherent light   monochromatic coherent light

Polychromatic incoherent light (left) and monochromatic coherent light (right).

Conservation of energy

The amplitude at a maximum is double the amplitude that would be produced there by a single slit and four times the intensity. This increase in energy at the maxima balances the lack of energy at the minima and ensures that the energy in the interference pattern is the sum of the energies emitted by the two slits.

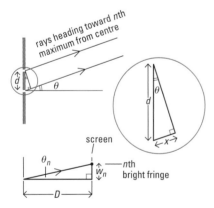

rays heading toward nth maximum from centre

screen

θ_n

w_n — nth bright fringe

D

The double-slit experiment.

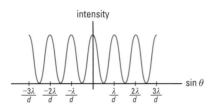

intensity

$-\frac{3\lambda}{d}$ $-\frac{2\lambda}{d}$ $-\frac{\lambda}{d}$ $\frac{\lambda}{d}$ $\frac{2\lambda}{d}$ $\frac{3\lambda}{d}$ $\sin\theta$

Intensity distribution for Young's fringes.

THE DOUBLE-SLIT EXPERIMENT

Young's experiment

In 1801 Thomas Young (a real polymath responsible for both the Young modulus *and* the key to interpreting Egyptian hieroglyphics!) devised his famous and highly significant 'double-slits experiment'. This was crucially important for two reasons:

• Nobody had measured the wavelength of light before.

• It showed that light *has* a wavelength, and seemed to settle the debate over whether light consisted of waves or particles in favour of the former.

The double-slit experiment relies on superposition and interference to create a regular pattern of interference fringes. Since particles do not interfere this pattern cannot be accounted for using a particle model (but this is not the end of the story – see spread 8.7 for a discussion of wave–particle duality).

Coherent sources

Waves emitted from hot bodies come from intermittent quantum jumps in individual atoms. These emissions are not correlated with one another so the light produced is incoherent. Therefore Young's main problem was to produce coherent sources. He achieved this by using a narrow single slit to diffract light from a small region (which is coherent) of a bright monochromatic source, and then a double slit placed in this diffracted light to act as two point (or line) sources. Nowadays we have a readily available source of coherent light, the laser. Laser light is produced by **stimulated emission**, where new photons are emitted *in phase* with those that are already present, resulting in a very intense coherent beam. If this is directed at double slits an interference pattern is formed on a distant screen. The pattern consists of equally-spaced bright maxima (fringes) separated by darker minima. This pattern is sometimes called **Young's fringes**.

Young's fringes

With light the separation of interference maxima, w, and the slit separation, d, are both very small compared to the distance to the fringes D, so all the angles are small. This leads to a very convenient simplification. Light rays travelling from the two slits to any particular point on the screen are more or less parallel. Their path difference can then be calculated from the angle at which they leave the slits, as shown on the left.

$$\text{path difference at an angle } \theta, \ x = d \sin\theta \approx d\theta$$

with θ in radians and using the small-angle approximation. For a maximum to occur, $x = n\lambda$. If the nth maximum (fringe) is formed at distance w_n from the centre of the pattern, then

$$\theta_n \approx \frac{w_n}{D} = \frac{n\lambda}{d} \quad \text{so} \quad w_n = \frac{n\lambda D}{d}$$

and the fringe separation is

$$w = w_{n+1} - w_n = \frac{\lambda D}{d} \quad \text{giving} \quad \lambda = \frac{wd}{D}$$

Notice that the fringes are evenly spaced as long as the small-angle approximations hold. This formula holds for other types of wave, but you must beware that this condition is satisfied.

The wavelength of light

Young's fringes can be used to measure the wavelength of light if slit separation, fringe separation, and screen distance are all measured. In practice the slit separation is measured with a travelling microscope, the fringe separation is measured as an average over many fringes using either a travelling microscope or a ruler, and the screen distance is measured using either a calibrated optical bench or a ruler. (The slit separation *can* be measured with a ruler if light is projected through the slits and the magnified shadow is measured from a screen.)

Worked example

A laser is used to produce Young's fringes with slits separated by 0.50 mm. The screen is 1.0 m from the slits and ten fringe separations occupy 12.5 mm. What is the wavelength of the laser light?

$$\lambda = \frac{wd}{D} = \frac{1.25 \times 10^{-3}\,\text{m} \times 0.5 \times 10^{-3}\,\text{m}}{1.0\,\text{m}} = 6.25 \times 10^{-7}\,\text{m} = 625\,\text{nm}$$

- If the sources have a phase difference but remain coherent the interference pattern is shifted along the screen, but this doesn't affect the fringe separation.

- The optical path of one ray can be changed by making it pass through a transparent medium on its way to the screen. The wavelength in a medium of greater refractive index is less, so the wave completes more oscillations in the medium and its path length measured in terms of wavelengths becomes longer. This results in an extra phase delay for that beam and again shifts the fringes.

- If sheets of polaroid are placed over each slit so that the light is plane-polarized (see spread 6.19) the interference pattern is only seen when both polarization directions are parallel. If they are perpendicular to each other the pattern disappears.

- If white light is used the pattern is a superposition of individual fringes for each wavelength. All have a central maximum, so this is white. The first fringe for blue light will appear closer to the centre than that for red so there is a spectrum from violet through to red on either side of the central maximum. After the first fringe the pattern gets very complicated because different orders for different colours overlap (see photograph opposite).

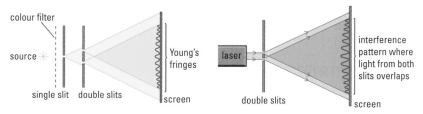

Young's fringes produced with a colour filter (left) and a laser (right).

Young's experiment can also be used to measure the wavelength of microwaves. Two slits are created using aluminium plates placed in front of a microwave source and the maxima are located using a microwave detector. The actual paths from slits to maxima can be measured using a ruler. The path difference from the slits to the first maximum on either side is equal to the microwave wavelength.

Lloyd's mirror

Two coherent sources can be simulated using a reflector as above. Young's formula still gives fringe separation but the 'central' fringe is dark because of a phase reversed in the reflected ray. An optical experiment of this kind was carried out by Lloyd in 1834.

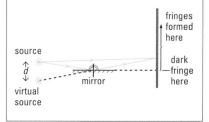

Interferometers

If a beam of light is split in two and the two beams are sent along different paths before being recombined, the resulting interference pattern can be used to compare the two paths to an accuracy comparable to the wavelength of light. This is the principle of an interferometer. The diagram shows the principle of the Michelson interferometer. A phase difference can arise from a path difference or by introducing a sample of material of different refractive index. Interferometers are used to make accurate distance measurements, to measure refractive indices, and to look for gravity waves (the passing waves would distort the interferometer and so shift the fringes).

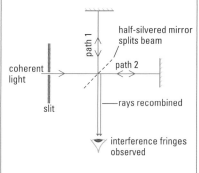

PRACTICE

1 An experiment is carried out using identical sound sources placed a short distance apart. A listener hears loud and quiet sounds as she walks along a line parallel to the source separation but some distance away from them.

  a She stops at a position of maximum sound intensity. Then one of the speakers is switched off. What happens to the amplitude and intensity of the sound at that point?

  b Both speakers are again switched on and she moves to a position of minimum intensity. What happens now when one speaker is switched off? Explain your answer.

  c She returns to a position of maximum intensity and the amplitude of one speaker is doubled. What does she hear?

2 Young's fringes are formed using slits separated by 0.60 mm and projected onto a screen at a distance of 1.4 m. Twelve fringe separations occupy a distance of 16 mm.

  a What is the wavelength of the light?

  b What colour is this?

  c If the whole apparatus was submerged in water of refractive index 1.3, what would be the new fringe separation?

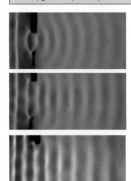

Diffraction through apertures of various widths.

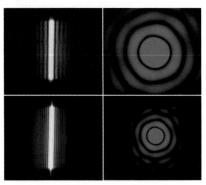

Diffraction patterns from a single slit and a circular hole. The upper patterns are produced by narrower apertures. The variation of intensity with distance from the centre of the pattern shows a broad central maximum and a series of diminishing secondary maxima on either side. This diffraction pattern is caused by interference of light from different parts of the aperture.

Diffraction patterns

On first sight the diffraction pattern produced by an aperture looks rather like Young's fringes. There is a very important difference – Young's fringes are all of the same width whereas the central maximum of the diffraction pattern is twice the width of any of the secondary maxima.

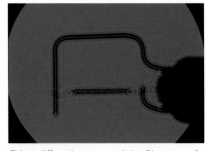

Edge diffraction around the filament of an electric light.

DIFFRACTION

To listen to someone you do not need to point your ears at them, and if you speak to a friend in front of you your words are also heard by someone standing beside you. This is because of **diffraction**, the spreading of waves as they pass through an aperture (i.e. a hole, like your mouth). The sound waves will also diffract around objects and edges in the room. All waves diffract, but the amount of diffraction depends on the ratio of the wavelength λ to the size a of the diffracting object. This shows up clearly in the sequence of ripple tank photographs.

$\dfrac{\lambda}{a} \approx 1$ (or more)   very significant diffraction and almost circular diffracted waves

$\dfrac{\lambda}{a} < 1$ less significant diffraction, centre part of wave more or less unaffected

$\dfrac{\lambda}{a} \ll 1$ no significant diffraction

TV and VHF radio use short waves (a few metres), which do not diffract significantly around natural barriers like hills and cliffs. This means that receiving aerials must be aligned to the transmitters – this is also true for the much shorter wavelengths used in microwave links. On the other hand, long-wave radio signals (wavelengths around 1 km) diffract around most objects and so can be received even in places that lie in a short-wave 'shadow'.

To see what someone is doing you *do* have to look at them. Our everyday experience of light is not obviously affected by diffraction. In fact, ray diagrams and all of geometrical optics are based on the idea of **rectilinear propagation** – light travelling in straight lines. Light *does* diffract, but its wavelength is typically 500 nm (about a million times smaller than the shortest wavelength of audible sound), so light is only significantly diffracted by objects of comparably small size.

The diffraction pattern from a single slit

To explain **diffraction patterns** we use a mathematical idea suggested by Christiaan Huygens, a Dutch physicist who was a contemporary of Newton and who argued for a wave model of light.

Huygens's principle

Each point on a wavefront is a point source of wavelets. These wavelets superpose and interfere to form future wavefronts.

This idea is not as strange as it seems. When a water wave passes, every point on the water surface is made to oscillate up and down. If any point were made to oscillate in this way by itself it would create circular ripples.

When a plane wave passes through a narrow gap the range of point sources is restricted (to the width of the gap) and interference between wavelets causes cancellation in certain directions. The amplitude of light arriving at any point on the screen is found by summing all wavelets moving in that direction. Unless the wavelets move straight ahead there will be a phase difference between wavelets from different points that may lead to constructive or destructive interference. When light passes through slits or a circular hole there is complete cancellation in certain directions. This creates the dark fringes between diffraction maxima.

Diffraction patterns are also created when waves pass the edge of an object. Wavelets near the edge spread into the region of geometric shadow creating a series of maxima and minima where they superpose (as in the photograph on the left).

Analysing the pattern

In practice we observe optical diffraction patterns a long way from the diffracting objects, and rays travelling to any particular point on the screen are more or less parallel to one another. This is used to simplify the analysis that follows.

For the first minimum, a ray near one edge of the slit cancels with one at the centre. The path difference between the centre and the edge is then $\lambda/2$. This occurs at an angle θ when

$$\sin \theta = \frac{\lambda}{a}$$

where a is the slit width. By similar arguments the nth minimum occurs at:

$$\frac{a}{2} \sin \theta = \frac{\lambda}{2}; \quad \sin \theta_n = \frac{n\lambda}{a}$$

The same analysis can be applied to a circular slit, but is complicated because of the geometry. The result is that minima form at angles such that

$$\sin \theta_n = \frac{1.22\, n\lambda}{D}$$

where D is the diameter. In most cases the angles are small enough to use the following results: the nth minimum occurs at an angle

$$\theta_n \approx \frac{n\lambda}{a} \quad \text{for a single slit of width } a,$$

$$\theta_n \approx \frac{1.22\, n\lambda}{D} \quad \text{for a circular hole of diameter } D.$$

Slit width

You might wonder why diffraction at the edge of a large gap is less significant than at the edge of a small gap. This is easily explained using Huygens's principle. If the gap is large the secondary wavelets remain in phase in the straight-ahead direction, but any rays moving away from this direction can be cancelled by waves out of phase with them from some distance along the wavefront. If the gap is narrow these waves are not available and there is greater spreading at the edge of the gap.

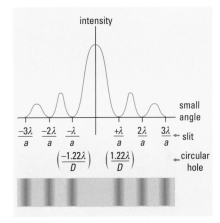

Diffraction patterns.

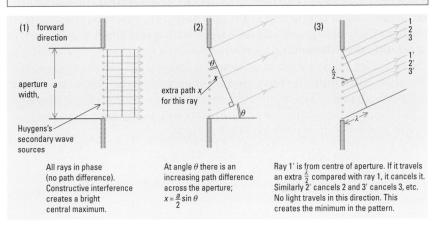

Huygens's secondary wave sources.

(1) forward direction

aperture width, a

Huygens's secondary wave sources

All rays in phase (no path difference). Constructive interference creates a bright central maximum.

(2)

extra path x for this ray

At angle θ there is an increasing path difference across the aperture; $x = \frac{a}{2} \sin \theta$

(3)

Ray 1' is from centre of aperture. If it travels an extra $\frac{\lambda}{2}$ compared with ray 1, it cancels it. Similarly 2' cancels 2 and 3' cancels 3, etc. No light travels in this direction. This creates the minimum in the pattern.

Young's fringes revisited

For Young's fringes it was assumed that each of the two slits was narrow enough to diffract light equally in all directions. This is not what happens. Each slit produces a (overlapping) single-slit diffraction pattern. Since the slit width is only a fraction of the slit separation, the central maximum is broader than the fringe separation. The combined effect of diffraction at each slit and interference between slits is that the intensities of the fringes vary in the shape of a single-slit diffraction pattern.

The large-scale structure of diffraction and interference patterns is determined by the fine detail in the object causing the pattern. The fine detail of the pattern is determined by the large-scale structure of the object. The slit size determines minima on the **envelope curve** whereas slit separation (which is larger) determines the separation of the fringes themselves.

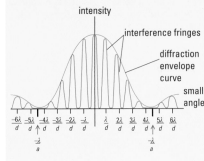

a is the slit width; *d* is the slit separation; $d > a$

Young's fringes intensity pattern.

Diffraction by small objects

The diffraction patterns formed by linear objects (e.g. a hair or wire) or small circular objects have the same shape and size as those formed by a slit or circular hole of the same dimensions.

PRACTICE

1 Sound diffracts through the aperture of a loudspeaker of radius 10 cm. How would you expect this to affect the distribution of sound intensity in the room for:

  a the high frequencies (around 8 kHz);

  b the low frequencies (around 200 Hz)? Take the speed of sound to be 330 m s⁻¹.

2 Laser light of wavelength 625 nm falls on a slit of width 0.30 mm and forms a diffraction pattern on a screen 1.6 m from the slit.

  a Sketch the pattern of intensity versus distance from the centre of the screen showing positions of all minima.

  b Why do successive secondary maxima become less and less intense?

6.10

O B J E C T I V E S

• diffraction limit to resolution

• the Rayleigh criterion

• polar diagrams

Jodrell Bank radio telescope. Incoming radio waves are diffracted by the circular dish, so objects with small angular separation may form overlapping images.

The headlamps of more distant cars subtend a smaller angle at the camera and remain unresolved (factors other than diffraction are most important in limiting resolving power here – e.g. scattering, graininess of film, etc.).

Resolving power

To see fine detail an instrument must form distinguishable images of objects that have a small angular separation. The smaller the angular separation for which it can do this, the greater the instrument's resolving power.

Matter microscopes

X-rays and electron beams (electrons have wave-like properties) are used to probe the structure of matter on a very small scale. Their resolving power is also limited by their wavelength – to measure atomic spacings requires wavelengths of 10^{-10} m or less, while to probe the internal quark structure of protons and neutrons needs much smaller wavelengths of 10^{-16} m and less. Electron wavelength at high energy is inversely proportional to energy, so enormous accelerators are required to generate the electron beams needed to investigate the fine structure of matter.

RESOLVING DETAIL

The Jodrell Bank radio telescope has a diameter of 76 m and is used to form radio images of astronomical objects. However, its ability to form distinguishable images of objects that appear close together in the sky is not as good as that of the unaided human eye. The reason for this surprising fact is that image resolution is limited by the diffraction that occurs at the collecting aperture (dish or pupil). Radio waves are so much longer than light waves that they are diffracted more significantly by the dish than light waves are when they enter the eye – the ratio of wavelength to diameter is greater for Jodrell Bank than for the eye.

Jodrell Bank: $\lambda = 0.15$ m $\qquad D = 76$ m $\qquad \dfrac{\lambda}{D} \approx 0.002$

Pupil of eye: $\lambda = 500$ nm $\qquad D = 0.003$ mm $\qquad \dfrac{\lambda}{D} \approx 0.0002$

Of course, the information from radio telescopes is still of vital importance because it is in a different part of the electromagnetic spectrum and so supplements what we learn from visible images.

The Rayleigh criterion and the limit of resolution

An optical instrument can resolve two point objects if their images can be distinguished. Diffraction at the aperture of the instrument will spread the light from each point, and for objects with a very small angular separation the diffraction patterns may overlap to such an extent that they produce a single blob of light on the image (they are not resolved). The theoretical **limit of resolution** is roughly when the maximum of one diffraction pattern falls on the first minimum of the other. This occurs when the angular separation of the objects equals the angle from the centre to the first minimum of either pattern.

$$\text{limit of resolution, } \alpha \approx \frac{1.22\lambda}{D} \approx \frac{\lambda}{D} \quad \text{(radians)}$$

This is called the **Rayleigh criterion**, after Lord Rayleigh, who first proposed it.

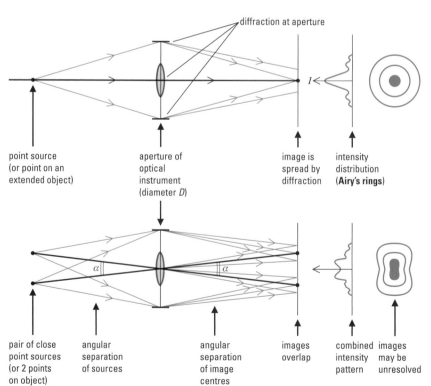

Forming images of a single point source (top) and two close point sources (bottom).

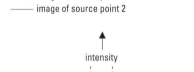
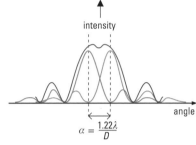

$$\alpha = \frac{1.22\lambda}{D}$$

This diagram shows the intensity distribution from two sources at the theoretical limit of resolution. Their angular separation is $\alpha = 1.22\lambda/D$ and the first maximum of one pattern coincides with the central maximum of the other (according to the Rayleigh criterion).

Worked example

Estimate the maximum distance at which you could resolve two fine black lines 1 mm apart on a white sheet.

Pupil diameter $\approx 3\,\text{mm}$

Wavelength of light $\approx 500\,\text{nm}$

$$\frac{1.22\lambda}{D} \approx 2 \times 10^{-4}\,\text{rad}$$

This means the lines will just be resolved when they subtend this angle at the eye.

$$\text{angle subtended at eye} \approx \frac{10^{-3}\,\text{m}}{\text{distance}} \approx \alpha$$

$$\text{distance} = \frac{10^{-3}\,\text{m}}{2 \times 10^{-4}\,\text{rad}} = 5\,\text{m}$$

In practice, other factors also limit resolving power (such as light intensity, sensitivity of receptor cells, and density of cells on the retina) so you are likely to come up with a value lower than this if you try it experimentally. The Earth's atmosphere is transparent to radio waves, so the resolving power of a radio telescope is 'diffraction-limited'. However, light interacts significantly with the atmosphere so optical telescopes do not come close to their theoretical resolving powers.

Polar response

Sound emitted by a loudspeaker passes out through the speaker aperture and diffracts. The sound intensity will vary in the room because of the diffraction pattern. This can be plotted as an intensity-versus-angle graph in the usual way, but it is more useful to produce a **polar response** diagram. This corresponds to the spatial distribution of sound and so is easier for most consumers to interpret.

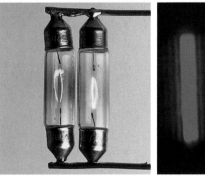
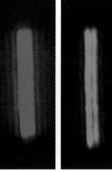
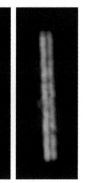

The photograph on the left shows a twin-filament light source. Each of the photographs on the right is taken through a narrow slit parallel to the source filament. The slit width has been adjusted so that the green images are just resolved. Red is then unresolved and blue quite clearly resolved.

Making the speaker smaller has the advantage that the central 'lobe' on the polar response diagram now fills most of the room. Of course, diffraction depends on wavelength as well as aperture size, so high frequencies (short wavelengths) will have more directional characteristics than the longer wavelengths (or bass sounds).

Any receiver or transmitter will have a polar response determined by diffraction, so diagrams like those shown for a loudspeaker could be drawn for a radio telescope or a satellite receiving dish. This is why it is important to align receivers to point towards the transmitter, but it also explains why a slight misalignment does not result in a complete loss of signal.

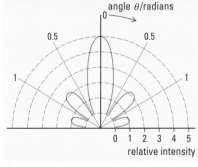

A polar response curve for a circular loudspeaker operating at a particular frequency.

PRACTICE

1 a Explain why the twin-filament light source in the photograph on this page is resolved in blue light but not in red.

  b How would the situation change if **i** the source was moved closer to the observer, **ii** the separation of the filaments was doubled? Explain your answers.

2 The Hubble Space Telescope has a mirror of diameter 2.4 m and works with cameras that operate with wavelengths from about 110–1100 nm. It can be pointed at a distant star with an accuracy of about 2×10^{-6} degrees.

  a In which parts of the electromagnetic spectrum are wavelengths 110 nm and 1100 nm?

  b Express the pointing accuracy in radians.

  c What is the theoretical limit of resolution using each of the wavelengths quoted?

  d What is the point of collecting information at a longer wavelength if the resolution of the images is poorer?

  e Compare the pointing accuracy with the limits of resolution and comment.

- the grating formula

- spectrometers

- uses of spectroscopy

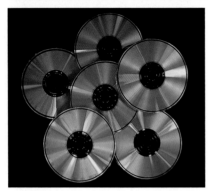

CDs act as reflection diffraction gratings.

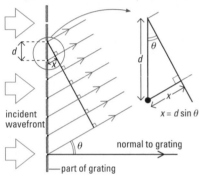

A diffraction grating. The incident wavefront is parallel to the grating so each slit source is in phase with all the others.

Diffraction grating formula

For the nth order: $n\lambda = d \sin \theta_n$

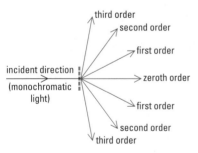

Orders of diffraction.

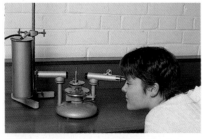

Using a spectrometer and diffraction grating to analyse the spectrum of light emitted by a gas discharge tube.

DIFFRACTION GRATINGS

If you look at white light reflected from a CD you will see bands of colour across its surface.

The CD is acting as a **diffraction grating** – each of the tens of thousands of equally-spaced lines of information on the surface of the CD (made up of microscopic pits in the aluminium substrate) diffracts the light that falls on it. Superposition from all these individual sources results in constructive and destructive interference that enhances certain wavelengths in certain directions, creating a **diffraction pattern**.

Diffraction maxima

Gratings can work by reflection (like the CD) or by transmission through a large number of parallel slits (like Young's double-slit experiment but with thousands of parallel slits). Both types of grating can be analysed in the same way – the diffraction pattern is caused by superposition of wavelets from a large number of point-like sources.

The analysis is identical to that for Young's slits: maxima will form in directions in which light from all sources is in phase. The path difference between adjacent slits for rays at angle θ is

$$x = d \sin \theta$$

For a maximum this must be an integral number of wavelengths: $x = n\lambda$. The nth maximum from the central maximum ($n = 0$) is at θ_n, where

$$\sin \theta_n = \frac{n\lambda}{d}$$

The diffraction pattern consists of a bright central maximum (**zeroth** order) and higher orders of diffraction (corresponding to $n = 1, 2$, etc.). Since the angle of diffraction cannot exceed 90°, there is a limit to the number of orders of diffraction:

$$\sin \theta_n \leq 1 \quad \text{so} \quad \frac{n\lambda}{d} \leq 1 \quad \text{giving} \quad n_{\max} \leq \frac{d}{\lambda}$$

Worked example

How many orders of diffraction will be visible if light of wavelength 500 nm falls on a grating with 600 000 lines per metre?

$$d = \frac{1\,\text{m}}{600\,000} = 1.7 \times 10^{-6}\,\text{m}$$

$$\frac{d}{\lambda} = \frac{1.7 \times 10^{-6}\,\text{m}}{500 \times 10^{-9}\,\text{m}} = 3.3$$

The 3rd order will be visible, but the 4th will not. So just 3 orders will be visible.

Analysing light

Diffraction gratings are used to analyse light into its constituent wavelengths. This is usually done by directing the incident light onto a diffraction grating and making accurate measurements of the positions of maxima for different colours (wavelengths), using a device called a **spectrometer**. The analysis of light is called **spectroscopy** and has played a major role in many discoveries about the nature of atoms and molecules, which are identified by their characteristic emission or absorption spectra, and about the structure and motion of stars and galaxies in which spectra are Doppler-shifted by motion and gravitational fields. Even source temperature can be calculated from the shape of the spectrum (see spread 12.12).

In principle, Young's slits (a two-slit diffraction grating) could be used for spectroscopy, but a diffraction grating with tens of thousands of parallel slits has two major advantages over the double-slit arrangement.

- **The maxima are much brighter**. If the amplitude and intensity from each slit are A and I, the amplitude at a maximum is NA (using N slits) and the intensity is N^2I (since $I \propto A^2$). This means having 10^4 slits rather than 2 gives an intensity 10^8I rather than $4I$.

- **The maxima are much sharper**. Maxima in Young's slits experiments reduce in intensity very gradually down to the next minimum. With a diffraction grating the maxima are incredibly sharply defined, dropping to zero almost immediately on either side. This is because wavelets from all slits across the grating combine at the maximum, so even a small angle away from the maximum will result in rays cancelling in pairs across the grating. This small-scale structure of the pattern is determined by the large-scale structure of the grating (its complete width) so that the maxima will fall to zero at angles of about λ/Nd on either side (where Nd is the width of the grating).

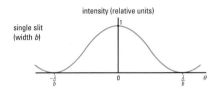

Using a diffraction grating to make a simple measurement of the wavelength of light from an LED: $\lambda = d \sin \theta_1$ where d is the slit separation in the grating.

Comments

- The diffraction grating has three increasing levels of structure – the slit width s, the slit separation d, and the grating width Nd. These have an inverse relation to the three levels of structure in the diffraction pattern – the overall intensity variation (minima at $\theta = n\lambda/s$); the orders of diffraction (maxima at $\sin \theta = n\lambda/d$); and the sharpness of the maxima (falling to zero at $\delta\theta = \pm \lambda/Nd$ either side of an order).

- If an order of diffraction is expected at the position of a minimum in the single-slit diffraction pattern, the maximum will not be formed and there will be a **missing order**. This occurs whenever $n\lambda/d = m\lambda/s$ where n and m are integers).

- If a grating is used to analyse a polychromatic source it is quite likely that the shorter wavelengths from higher orders may overlap the longer wavelengths from lower orders.

- Do not confuse the grating formula, which is a formula for *maxima*, with the single-slit diffraction formula, which has the same form but predicts *minima*!

- Planes of atoms in crystals act like slits in an optical diffraction pattern when X-rays are passed through the crystal. The crystalline structure can then be worked out from the size and symmetry of the diffraction pattern.

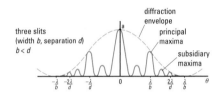

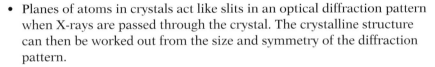

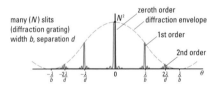

Angular positions of diffraction orders are as in Young's slit experiment, however many slits there are.

PRACTICE

1 a Explain why the intensity of the zeroth order using a diffraction grating with N slits is N^2 times that at the centre of the diffraction pattern formed by one of the slits.

b If N people each shone a torch at the same place on a white screen, would the centre of the bright spot be N or N^2 times the brightness when a single torch is shone there? Explain why.

2 Sodium lamps emit light strongly at two very close wavelengths. These are the 'sodium-D lines' at 589.0 nm and 589.6 nm. A student is observing the spectrum of sodium light through a spectrometer using a diffraction grating with 500 lines per millimetre.

a At what angles do the lines appear in the first and second order spectra?

b In how many orders will the lines be visible?

c In which order will their angular separation be greatest?

d Comment on which order you think will give the student the best chance of resolving the two lines.

3 A particular grating has slits of width 600 nm and a slit separation of 1800 nm. Will there be any missing orders if this is used to observe a line spectrum consisting of 450 nm, 600 nm, and 650 nm?

4 For Young's double-slits the first minimum is formed at an angle λ/d from the central maximum.

a By treating the whole grating like a single slit and looking for cancellation by pairs across the grating show that the zeroth-order beam falls to zero intensity at an angle λ/Nd from the centre.

b Explain why the maxima produced using a diffraction grating are sharper and more intense than when a double-slit source is used.

6.12

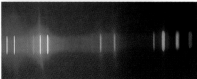

An electrical discharge lamp (top) emits a line spectrum (bottom).

SPECTRA

Spectroscopy is the analysis of light into its constituent wavelengths. This may sound rather unrewarding, but spectroscopy may be the single most important experimental technique to emerge in the last two centuries!

• Spectra emitted by different elements reveal their electronic structure.

• Spectroscopy of hydrogen atoms provided the main early tests of quantum theory.

• The solar spectrum tells us the Sun's temperature and composition.

• Stellar spectra tell us about the life-cycle of stars and the origin of the elements.

• Galactic spectra reveal the expansion of the universe.

• Spectra are used to identify tiny traces of elements in chemical processes on Earth and to search for the molecules of life in space.

Impressed? You should be, and this is just the tip of the iceberg!

Atoms and light

Isolated atoms have sharply-defined energy levels. Electrons within atoms can make quantum jumps from one energy level to another (see spreads 8.9 and 8.10). They do this when the atoms gain or lose discrete amounts of energy, which may take place when they absorb or emit photons. The energy of the photons equals the energy difference ΔE between the initial and final states of the atom, and this detemines the photon wavelength:

$$\Delta E = hf = \frac{hc}{\lambda} \text{ or } \lambda = \frac{hc}{\Delta E}$$

• Large energy jumps involve short-wavelength photons.

• Small energy jumps involve long-wavelength photons.

The range of energy jumps for a particular type of atom is unique to that element. This gives it a distinct spectrum (set of wavelengths) that can be used to identify it.

Emission spectra If many atoms are in an excited state (e.g. in a hot gas or a discharge tube) electrons will be making quantum jumps to lower energy states, emitting photons as they do so. This results in an **emission spectrum**.

Absorption spectra If many atoms are in low-energy states they will not emit much light. However, if light with a continuous spectrum is shone past them they can absorb any photons that have energies equal to the allowed quantum jumps. These will be photons with exactly the same wavelengths as those in the emission spectrum. The absorbed photons are subtracted from the light shining through the sample, so they form dark lines or bands in the otherwise continuous spectrum. This is an **absorption spectrum**.

Line spectra When the atoms that absorb or emit photons act independently of one another (e.g. in a low-pressure vapour or gas) their energy levels remain well defined and the spectral wavelengths are sharp. This is called a **line spectrum**.

Splitting of spectral lines If the atoms are in molecules or are affected by a magnetic or electric field their energy levels may split into a number of closely related but separate levels. This also splits the spectral lines into sets of close but separate wavelengths.

The dark absorption lines crossing the solar spectrum were first discovered by Fraunhofer, and are named after him.

advanced **PHYSICS**

Band spectra When atoms interact strongly with one another, as they do in a solid material, the energy levels spread into allowed energy bands and this results in a continuous range of photon energies that can be absorbed or emitted, resulting in a **band spectrum**.

Doppler broadening Atoms also have thermal motions. If you analyse the spectrum from a hot gas (e.g. a star) then, at any moment, some of the atoms emitting light are moving toward you and some are moving away. This results in a Doppler effect that changes the received wavelength, making it shorter if the atoms are approaching and longer if they are moving away. The effect is a broadening of the spectral line so that it covers a small range of wavelengths (rather than a sharp 'line'). The hotter the source the greater this **Doppler** (or **thermal**) **broadening**, so the width of the spectral line can be used to measure temperature.

| **Absorption spectra** |

When photons are absorbed by atoms the atoms jump to an excited state. A short time later they will re-emit one or more photons. The direction of re-emission is random. This is why the emitted photons do not 'fill in the gaps' in an absorption spectrum.

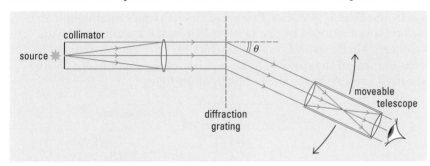

Analysing a spectrum. Each wavelength present in the emitted light is diffracted to a different position. The moveable telescope is used to measure these angular positions by focusing on images of the collimator slit. The wavelength can then be calculated from the diffraction grating formula ($n\lambda = d \sin\theta$).

Orders of magnitude

Visible light has wavelengths in the range (approximately) of 400 nm to 700 nm. Substituting these values into the equation above gives photon energies of a few electronvolts. This is the same order of magnitude as the quantum jumps of outer electrons in atoms. That is why the spectra of atoms are usually in the infrared, visible, or ultraviolet regions of the electromagnetic spectrum.

If violent collisions dislodge the inner electrons of atoms then the subsequent cascade of quantum jumps (as the electrons 'tumble down' into lower energy levels) results in the emission of photons of much higher energy, typically carrying tens of keV. These are X-rays.

The nucleus can also be excited and can absorb and emit photons. However, the strong forces between nucleons involves quantum jumps measured in hundreds of keV rather than eV, so these jumps involve much higher energies and much shorter wavelengths. The gamma rays emitted following a nuclear decay are created in this way, and some nuclei emit a spectrum of gamma rays just as atoms emit a spectrum of visible light.

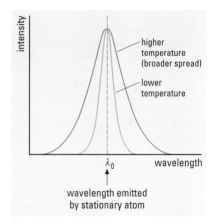

Doppler broadening.

PRACTICE

1 Sodium street lamps have a characteristic yellow colour. Why?

2 The Sun emits a continuous spectrum of radiation crossed by dark bands. What conclusions can you draw from this?

3 The energy levels of hydrogen are given by $E = -13.6\,\text{eV}/n^2$ where n is called the **principal quantum number**.

  a Why are the energies negative?

  b What is the energy difference between the $n = 5$ level and the $n = 2$ level?

  c What energy photon could excite an electron to jump from $n = 2$ to $n = 5$?

What is the wavelength of this photon? To what part of the electromagnetic spectrum does it belong?

  d What is the longest-wavelength photon that could ionize a hydrogen atom?
Assume the atom starts off in its ground state ($n = 1$).

4 Work out the wavelengths corresponding to photons with the following energies and in each case state what part of the electromagnetic spectrum is involved:

  a 0.1 eV **b** 2 eV **c** 1 keV **d** 50 keV **e** 1 MeV

Defining c

The velocity of light plays such a crucial role in physics that it now has a defined value that cannot be changed by future experiments. In fact if we discover that we have been underestimating the speed of light its value will not change. The metre will simply 'have to get longer' because the metre is now defined in terms of the distance travelled by light in a certain time. The speed of light is defined as:

$$c = 299\,792\,458 \text{ m s}^{-1}$$

and the metre is the distance travelled by light in

$$t = \frac{1}{299\,792\,458} \text{ s}$$

Looking back in time

If the Sun suddenly switched off it would be eight minutes before we noticed it. This is the time it takes light to travel the 1.5×10^{11} m from the Sun to the Earth. When you look up at the night sky the images of stars and galaxies you see show them as they were when the light now falling on your retina was emitted. Much of what is seen lies within our own galaxy which is 100 000 light years across, so you are seeing stars in our galaxy as they were from a few years to hundreds of thousands of years ago. Some stars may no longer exist; they may have turned into supernovae since emitting the light you are seeing (see spread 12.14). The Hubble Space Telescope can form images of galaxies as they were over 10 billion years ago, more than halfway back to the Big Bang. This is very useful because it allows cosmologists to see the universe as it was in the past and test theories about how it evolved.

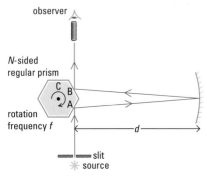

A simplified version of Michelson's rotating-mirror method for the speed of light (not to scale!).

THE SPEED OF LIGHT

A history of measurement of the speed of light

Before the 17th century most people who thought about it assumed light travelled instantaneously. Either that, or it was so incredibly fast that its time of travel was unmeasurable. Then in 1676 Rømer, a Danish astronomer, noticed that the (presumably) regular motion of Jupiter's moons seemed to vary as the relative positions of Jupiter and the Earth changed. As the distance to Jupiter increased the satellites emerged later and later from behind the planet. Rømer suggested that they only *seemed* to be late because the light was taking longer to reach us, and used the delay to make a surprisingly accurate estimate of the velocity of light (c).

The first measurement of the speed of light using terrestrial apparatus was by the French physicist, Fizeau, in 1849. Fizeau's rotating-mirror method was developed by one of his colleagues, Foucault, and by the American, Michelson, who spent much of his scientific life refining his measurement of the speed of light. In all cases the rotation of a mirror was used to measure the time of flight of the light.

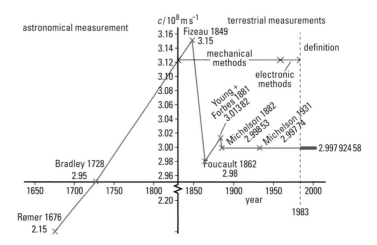

Variation in the measured value of the speed of light in recent history.

Michelson's method

The basic principles of Michelson's method are shown in the diagram. If the prism is stationary, light from the source reflects off one face to a distant mirror where it is reflected back to reflect off another face of the prism into the objective of an observer's telescope. If the mirror begins to rotate it is in a different position when the light returns and in most cases the image will be lost. However, at certain rotation frequencies the returning light will strike a different face of the prism in exactly the right position to reflect light to the observer. The lowest frequency f at which this occurs is when face C has moved to the position of B while the light travels to the distant mirror and back. If the prism has N faces this will take a time $t = 1/Nf$. During this time the light has travelled a distance $2d$ (typically several kilometres), so the speed of light is $c = 2d/t = 2dNf$. d can be measured by accurate surveying and f can be measured stroboscopically.

The importance of c

In the mid-nineteenth century James Clerk Maxwell showed that all electromagnetic waves must travel through a vacuum at the same speed c:

$$c = \sqrt{\frac{1}{\mu_0 \varepsilon_0}}$$

where μ_0 is the **permeability of free space** and ε_0 is the **permittivity of free space**.

(The permittivity and permeability measure the ability of space to support electric and magnetic fields.)

Maxwell calculated the value of c to be approximately $3.0 \times 10^8 \, \text{m s}^{-1}$, which is suggestively similar to the velocity of light. This confirmed what Faraday and others had suspected – *light is an electromagnetic wave*. However, the speed of light has an importance that goes far beyond electromagnetism and light. In 1905 Einstein showed that it is a kind of universal speed limit – it is impossible to accelerate anything up to the speed of light (although you can, with enough energy, make it move as close to c as you like) and *nothing at all* can travel faster than the speed of light.

Measuring c in a school laboratory

Modern electronic circuits can react very quickly to changing signals and are quite capable of recording the time delay for a pulse travelling over a path of just a few metres. This makes it possible to time a light pulse as it travels along an optical fibre.

Short electrical pulses are converted to light by an LED transmitter, and the pulses are transmitted to a **photodiode receiver** (this converts them back to an electrical signal). The oscilloscope records the transmitted and received electrical pulses and the time between them can be measured from the screen. Errors could arise because of electrical delays in the system, but these can be eliminated if the time delay for a long piece of fibre is compared with that for a short piece. The *additional* time taken for the signal to travel along the longer fibre must be due to the extra length of fibre. The electrical delays (if any) should be the same in both cases and the speed of light in the fibre can then be calculated using:

short fibre length = l_1 long fibre length = l_2

signal delay = t_1 signal delay = t_2

$$\text{speed of light} = \frac{\text{extra length of fibre}}{\text{extra time delay}}$$

$$c' = \frac{l_2 - l_1}{t_2 - t_1}$$

Typically this gives $c' \approx 2 \times 10^8 \, \text{m s}^{-1}$. The velocity in a medium is always less than the velocity of light in a vacuum because of the interaction of photons with electrons. The speed in a medium of absolute refractive index n is given by $c' = c/n$ (see spread 6.5).

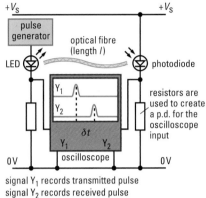

Measuring the speed of light in glass using optical fibres in a school laboratory.

signal Y_1 records transmitted pulse
signal Y_2 records received pulse
speed = $\frac{l}{\delta t}$

The speed of electricity

A similar experimental technique can be used to measure the speed of electrical pulses along a **coaxial cable** (this is like the lead that brings your TV signal from your TV aerial). A short pulse is fed into one end and timed to reach the other end. This shows that electrical signals also travel at the speed of light in a material.

PRACTICE

1 The eclipse of a moon of Jupiter by the planet varies by ± 8.25 minutes as the Earth moves around its orbit. (The Earth orbits the Sun at a distance of 1.5×10^{11} m.)

 a Estimate the velocity of light.

 b Explain why there is a 'delay'.

2 a Explain why vertical rain appears to be falling at an angle if you run through it.

 b In 1728 an English astronomer, James Bradley, noticed that the apparent positions of stars change during the year as the Earth completes its orbit – this is called **stellar aberration**. A telescope has to be constantly tilted in the direction of the Earth's motion in order to keep the star in the centre of the

field of view. Explain stellar aberration by analogy with the rainfall in **a**.

 c The Earth's orbit is about 1.5×10^{11} m across and stellar aberration for the pole star is about ± 20 seconds of arc. Use this to estimate the velocity of light. (Hint: think of vectors and relative velocities.)

3 In 1931 Michelson used a rotating prism with 32 faces to measure the time for light to travel through a 1.6 km evacuated tube to a distant mirror and back. From his results he calculated that the speed of light is $299\,774\,000 \, \text{m s}^{-1}$. Assuming the prism rotated by only one face during the time of flight of light calculate the rotation frequency of the prism.

THE SPEED OF MECHANICAL WAVES

Mechanical waves travel through material media by causing particles to vibrate. If one part of a medium is disturbed, the speed at which this disturbance is passed on depends on two properties of the medium.

- **Density** A denser medium has more mass per unit volume to set in motion and so responds more sluggishly to a disturbance. When one particle vibrates it exerts a force on its neighbours along the bonds that join it to them and they respond accordingly. If there is more mass to move, it gets a smaller acceleration.

- **Stiffness (or elasticity)** If the bonds between particles are stiff then a small disturbance of one particle will result in a larger force on neighbouring particles, giving them a greater acceleration and transferring the disturbance more rapidly.

Wave speed will be slow in a dense, weakly bonded medium and fast in a low-density, stiffly bonded one. A more detailed analysis shows that in many cases:

$$\text{speed of mechanical waves} \propto \sqrt{\frac{\text{stiffness factor}}{\text{density factor}}}$$

The **stiffness factor** and **density factor** take different forms in different circumstances, but the form of the wave-speed equation is always similar. The examples below are for a wide range of mechanical waves. In many cases the stiffness and density factors are fairly easy to see, but you may have to think hard to interpret the equations for water waves.

Transverse waves

Transverse waves tend to shear their medium rather than compress it, so the relevant stiffness must measure resistance to shear. The body of a liquid offers no resistance to shear so transverse waves cannot pass through the liquid. This is why P waves can pass through the core of the Earth but S waves cannot.

Waves in different media		**Terms used in the equations**	
Longitudinal waves (e.g. sound) in a solid rod:	$v = \sqrt{\dfrac{E}{\rho}}$	E	is the Young modulus
		ρ	is density
Transverse waves on a string:	$v = \sqrt{\dfrac{T}{\mu}}$	T	is tension (but temperature below)
Longitudinal waves on a spring:	$v = \sqrt{\dfrac{kl}{\mu}}$	μ	is mass per unit length
		k	is spring constant
		l	is spring length
Sound waves in a gas:	$v = \sqrt{\dfrac{\gamma p}{\rho}} = \sqrt{\dfrac{\gamma RT}{M}}$	γ	is a dimensionless constant (equal to 1.4 for air)
		p	is pressure
Ripples:	$v = \sqrt{\dfrac{2\pi\sigma}{\rho\lambda}}$	R	is the molar gas constant
		T	is the temperature (in kelvin)
Surface waves in shallow water:	$v = \sqrt{gh}$	M	is the molar mass
		σ	is surface tension
Surface waves in deep water:	$v = \sqrt{\dfrac{g\lambda}{2\pi}}$	λ	is wavelength
		g	is gravitational field strength
		h	is depth of water

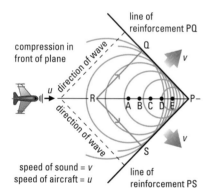

speed of sound = v
speed of aircraft = u
line of reinforcement PQ
line of reinforcement PS
compression in front of plane

*Breaking the sound barrier. When an aircraft flies faster than the speed of sound, the air in front of it is strongly compressed and creates a compression pulse that travels outwards. The diagram shows the progress of compressions from points A to E as the plane moves. Compressions reinforce along PQ and PS forming a shock wave (**sonic boom**) that travels outwards at an angle to the plane's path. If it takes the plane t seconds to move from R to P then RP = ut. In the same time compressions from R will have reached Q and S, a distance vt away. Simple trigonometry shows that the shock wave (sonic boom) travels out along a cone of half-angle θ and that cos θ = u/v.*

Even electromagnetic waves have a similar-looking equation:

$$c = \sqrt{\frac{1}{\varepsilon_0\mu_0}}$$

and mathematically the electromagnetic field can often be treated like an elastic medium (you may have used the idea of a '**catapult field**' to explain the motor effect).

Measuring the speed of sound

1 In air There are many simple ways to measure the speed of sound in air – a reasonable estimate can be made by timing echoes. A more accurate method measures the time sound takes to travel a measured distance using an oscilloscope and microphone. Two methods are illustrated.

In the first the apparatus is set up so that the signals from two separate microphones are in phase when they are quite close together (mic 2 at A). Mic 2 is then moved away from the other so that the sound travels further to reach it. The two signals go out of phase and only come back in phase whenever the microphone has been moved an integer number of wavelengths from its original (in phase) position. If this happens n times in a distance x then the wavelength is x/n. The period T of the sound can be measured from the oscilloscope screen in the usual way and then the frequency f can be calculated from $f = 1/T$. The velocity is then found from $v = f\lambda$.

In the second method, short pulses of sound are received at two microphones a distance x apart. The time t taken for sound to cover the extra distance x can be measured directly from the oscilloscope screen and the velocity is then simply x/t.

(A third method, using standing waves in a resonance tube, is described in spread 6.16.)

2 In a metal rod If a metal rod is dropped vertically onto another metal plate it will bounce. However, the two metals will stay in contact for the time it takes a compression pulse to travel along the rod, reflect from the far end and return. The returning pulse pushes the rod away from the plate and contact is lost. The contact time is equal to the time it takes a longitudinal pulse (like sound) to travel twice the length ($2l$) of the rod. If this time t can be measured it can be used to calculate the speed of sound v in the rod:

$$v = \frac{2l}{t}$$

One way to do this is to make the rod and plate part of a circuit so that making and breaking contact switches a high-frequency electrical signal to a storage oscilloscope, as shown. The oscilloscope is triggered (started) by the first contact and then a 25 kHz electrical signal is displayed on the oscilloscope until contact is broken. The signal is stored and can be used to estimate the contact time from the number of complete cycles displayed.

The velocity of sound in mild steel measured by this method is about 5000 m s^{-1}. This is much greater than in air (about 330 m s^{-1}) because of the stiff bonds between particles (despite the much greater density of mild steel).

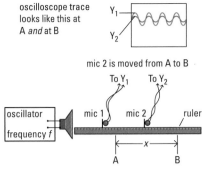

Measuring the speed of sound in air – method 1.

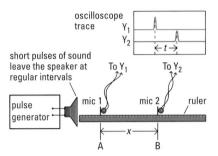

Measuring the speed of sound in air – method 2. Note that the oscilloscope must be triggered to the pulse frequency for a stable trace.

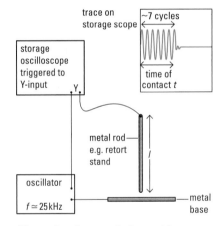

Measuring the speed of sound in a steel rod. In this experiment the rod remained in contact with the base plate for about 7 cycles of the 25 kHz signal. This is about 2.8 × 10^{-4} s during which a longitudinal pulse travelled the length of the rod and back (2l).

PRACTICE

1 Predict the speed of sound in aluminium and steel and give a physical explanation of why the values you calculate are so close despite the material densities and elasticities being different.

$E_{steel} = 2.1 \times 10^{11}\,\text{N m}^{-2}$ $\rho_{steel} = 7800\,\text{kg m}^{-3}$

$E_{Al} = 7.0 \times 10^{10}\,\text{N m}^{-2}$ $\rho_{Al} = 2700\,\text{kg m}^{-3}$

2 Is it ever, always, or never true to say that the speed of sound at sea level will be greater when the atmospheric pressure is higher? Explain your answer.

3 Explain why water waves near a gently-shelving beach turn towards the beach.

4 Before a concert, players in an orchestra tune their instruments to the note from an organ pipe whose fundamental frequency is 440 Hz. The temperature in the concert hall at this time is 18 °C. By the end of the concert the temperature has risen to 23 °C. Will this be a problem? (Take $\gamma = 1.4$ for air.)

5 What provides resistance to shear when a transverse wave is sent along a stretched string?

- nodes and antinodes

- fundamental and harmonics

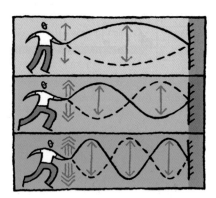

Resonance at different frequencies.

Comparing standing and progressive waves

A standing (or stationary) wave is characterized by nodes and antinodes in fixed positions. In a standing wave all points between two nodes oscillate in phase but their amplitude depends on position. In a **progressive wave** every point oscillates with the same amplitude but their phase depends on position, points further from the source being more delayed.

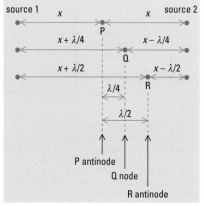

A standing wave.

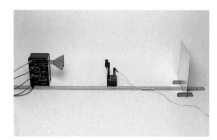

Measuring the wavelength of microwaves using standing waves. If the frequency of the microwave source is known this method can be used to measure the velocity of microwaves.

STANDING WAVES

Tie a rope to a rigid support, hold it fairly taut, and wave the free end up and down. You will discover that it resonates at certain frequencies. The lowest of these creates an extended oscillation with a large amplitude at the centre and zero amplitude at the fixed end. Wave it faster and the resonance dies away, but there are more resonances at higher frequencies and in each case the pattern of vibration has positions of large amplitude (**antinodes**) and positions of zero amplitude (**nodes**). These resonances are examples of **standing** or **stationary waves**.

> **Definition: standing waves**
>
> Standing waves are extended oscillations that store energy, and are formed when waves of the same amplitude and frequency moving in opposite directions superpose. In this case transverse waves travelling away from your hand superpose on their reflections from the fixed point at the wall.

Analysing a standing wave

Imagine two sources a distance $2x$ apart that emit monochromatic waves and remain in phase with one another. A standing wave pattern is set up along the line joining them. The essential characteristics of a standing wave can be understood by thinking about how the waves superpose at different points along this line.

The central point P is equidistant from both sources so waves always arrive there in phase and interfere constructively. They create a maximum amplitude oscillation (twice the amplitude from one source), called an antinode.

Move $\lambda/4$ closer to either source, e.g. at Q, and one wave is delayed by an extra quarter cycle ($\pi/2$ radians) and the other is advanced by a quarter cycle. The path difference from the sources to this point is now $\lambda/2$ and the phase difference is π, so that they interfere destructively and cancel out. This creates a node.

Move $\lambda/2$ from the centre in either direction, e.g. to R, and the path difference becomes λ. Now waves from the two sources combine in phase to create another antinode.

This pattern repeats so that nodes and antinodes alternate along the standing wave.

- Antinodes are maxima at regions of constructive interference where the waves always combine in phase.

- Nodes are minima at regions of destructive interference where the waves always combine π out of phase.

- The internodal distance is always $\lambda/2$.

If a microwave transmitter is set up in front of a metal reflector as shown and a small microwave detector is moved backwards and forwards in the region between the source and the reflector, it records a regular variation in intensity. This pattern of maxima and minima is a standing wave formed by the superposition of the incident and reflected microwaves. This set-up can be used to measure the wavelength of the microwaves (twice the average separation of nodes or antinodes) and calculate their speed (from $v = f\lambda$). The wavelength and speed of sound and ultrasound can be determined by a similar method.

Standing waves on strings

If an elastic string has one end fixed and the other attached to a vibrator, waves will travel down the string from the vibrator and reflect backwards and forwards along the string. However, this will not necessarily create a standing wave. The problem is that after several reflections there will be many waves superposing at each point and unless you are lucky they will have no fixed phase relation to one another. The net effect of adding waves of arbitrary phase is cancellation, since chance alone will cause as many positive disturbances as negative ones.

On the other hand, if you gradually increase the frequency of the oscillator the string resonates at certain frequencies, and standing waves are formed. The lowest-frequency resonance is called the **fundamental** mode or **first harmonic** of the string. It consists of an oscillation of the whole string, with the centre forming an antinode and the ends nodes (the point attached to the oscillator is not quite a node, but almost one since the vibrator amplitude is very small). Higher harmonics form a sequence with more and more antinodes and nodes along the string.

Standing waves will only form when the wavelength is related to the length of the string. The crucial condition is that nodes occur at either end of the string (since these points are effectively fixed). The first harmonic (fundamental) is the longest wave that fits these **boundary conditions** – one half wavelength is equal to the string length. The nth harmonic has n half wavelengths along the string. (Higher harmonics ($n > 1$) are sometimes called **overtones**, so the first overtone is the second harmonic.)

The velocity of waves along the string depends only on tension and mass per unit length of the string ($v = \sqrt{T/\mu}$), so the harmonic frequencies are related by simple ratios.

1st harmonic: $\quad l = \dfrac{\lambda_1}{2} \qquad \lambda_1 = 2l \qquad f_1 = \dfrac{v}{\lambda_1} = \dfrac{v}{2l}$

2nd harmonic: $\quad l = 2\dfrac{\lambda_2}{2} \qquad \lambda_2 = \dfrac{2l}{2} \qquad f_2 = \dfrac{v}{\lambda_2} = \dfrac{v}{l} = 2f_1$

3rd harmonic: $\quad l = 3\dfrac{\lambda_3}{2} \qquad \lambda_3 = \dfrac{2l}{3} \qquad f_3 = \dfrac{v}{\lambda_3} = \dfrac{3v}{2l} = 3f_1$

nth harmonic: $\quad l = n\dfrac{\lambda_n}{2} \qquad \lambda_n = \dfrac{2l}{n} \qquad f_n = \dfrac{v}{\lambda_n} = \dfrac{nv}{2l} = nf_1$

- Harmonics on a fixed string are at integer multiples of the fundamental frequency.

The strings on this instrument are fixed at both ends. When the string is plucked, the ends form nodes. The fundamental vibration has a single antinode at its centre. The musician can play higher notes by pressing the string down on the fingerboard. This shortens the wavelengths of all the harmonics, raising the frequency of the sound.

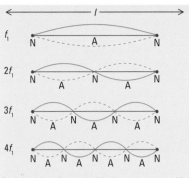

Each sketch is really a graph of displacement versus position along the string, showing extreme positions half a cycle apart represented by solid and broken lines.

The ends of a string may be fixed or free. These boundary conditions affect the shape of the standing waves that can be formed on it: a fixed end becomes a node, a free end an antinode.

Boundary conditions

The ends of a string may be fixed or free. These boundary conditions affect the shape of the standing waves that can be formed on it. A fixed end becomes a node, a free end an antinode.

PRACTICE

1 Draw a sketch showing how you would use a speaker, oscillator, microphone, oscilloscope, and metal sheet to measure the speed of sound using standing waves.

2 A car is moving at $20\,\mathrm{m\,s^{-1}}$ along a long straight road between two radio transmitters which are sending out the same programme using a frequency of $2.0\,\mathrm{MHz}$. When the car is roughly equidistant from the two transmitters the strength of the received signal fluctuates periodically.

  a Explain why the signal strength fluctuates.

  b How far apart are the positions at which maximum signal strength is received?

  c What is the frequency at which the signal fluctuates?

  d This is less of a problem when the car is closer to either transmitter; why?

3 A violin string has a mass per unit length of $0.35\,\mathrm{g\,m^{-1}}$ and is held under a tension of $25\,\mathrm{N}$.

  a What is the speed of transverse waves on the string?

  b What is the wavelength of the fundamental on this string if its length is restricted to $30\,\mathrm{cm}$?

  c What are the frequencies of the fundamental and second harmonic when the string is this length?

4 Middle C has a frequency of $256\,\mathrm{Hz}$.

  a What are the second, third, and fourth harmonics on this string in a piano?

  b Sketch diagrams to show how displacement varies with position along the string for the fundamental and each of these harmonics.

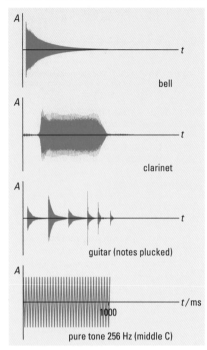

Oscilloscope traces of a bell, a guitar, a clarinet, and a pure tone.

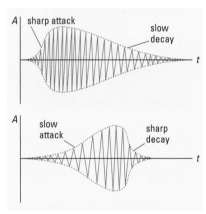

How a sound cuts in and dies away.

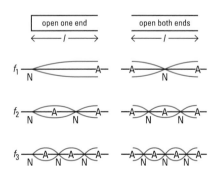

Standing waves in half-open and fully-open air columns. Note that the diagrams represent maximum longitudinal displacements of particles.

STANDING WAVES AND MUSIC

A double bass is much bigger than a violin, and the sounds it emits have longer wavelengths and lower frequencies. A similar comparison could be made between a church bell and a cow bell, or an organ pipe and a whistle. This is because musical instruments use standing waves to produce particular notes, and larger objects allow the **fundamental frequency** to have a longer wavelength and lower frequency.

Pure tones, such as those emitted when a sine-wave generator is connected directly to a loudspeaker, are not interesting to listen to, but the sound of a musical instrument is. The reason for this difference can be seen in the waveforms of the sounds – musical notes are more complicated than the regular sinusoidal variation of a pure tone. When you play a note on a musical instrument you 'excite' the instrument in some way: by plucking a string, hitting something with a stick or hammer, or blowing into or across it. In all these cases what you are doing is generating lots of waves of different frequencies that travel backwards and forwards along the string or air column or whatever, superposing on one another.

Out of all these waves certain frequencies will form standing waves (because their wavelengths in some way fit the instrument) of various amplitudes. The overall sound will be the combined effect of the fundamental frequency (which is the lowest-frequency standing wave, and determines the pitch or frequency of the note) and all the higher harmonics with significant amplitudes. The effect will also be changed by how the sound cuts in (**attack**) and dies away (**decay**); this is one reason why a stringed instrument sounds different when it is bowed to when it is plucked.

The exact mix of harmonics depends on lots of things, including how the note is played and how the instrument is constructed. For example, the vibrations of a guitar string will also excite resonances in the air inside the sounding box and in the sounding board of the guitar, so the sound of a particular instrument will depend on the detail of its construction. The difference heard when the same note is played on different instruments is a difference of **quality** or **timbre**.

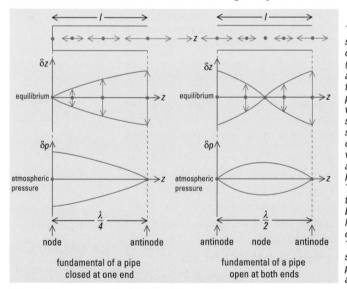

fundamental of a pipe closed at one end

fundamental of a pipe open at both ends

The top diagram shows a pipe closed at one end (on the left) and an open pipe (on the right) and particle vibrations within them. The second diagram shows particle displacement versus position at two times one half cycle apart. This looks like a transverse wave, but represents longitudinal displacements. The third diagram shows how the pressure varies along each pipe.

Air columns and wind instruments

If you blow across the top of a milk bottle you can make it 'sing'. Your blowing agitates the air near the opening, and sound waves of all frequencies travel down through the bottle and reflect back up. The sound you hear is 'picked out' of all the frequencies present because its wavelength fits the bottle and forms standing waves – the air column in the bottle resonates. If there is some water in the bottle the emitted sound has a higher frequency because the standing waves have shorter wavelength.

Air columns, like strings, have boundary conditions that determine what waves will form standing wave patterns. In the case of the bottle the top is open to the atmosphere and the bottom is closed. This means the top is constantly at atmospheric pressure (a **pressure node**), whereas the pressure at the bottom will vary as the longitudinal waves reflect there (a **pressure antinode**). Standing waves that fit these boundary conditions account for the fundamental and harmonics in the tube. The harmonics of an air column closed at one end form a series of frequencies in the ratio 1:3:5... and the nth harmonic has frequency $(2n - 1)f_1$ (where f_1 is the fundamental frequency).

There is an alternative way to describe standing waves in air: in terms of **particle displacements**. The waves are longitudinal and are free to move in and out of the open end, so this is a **displacement antinode**. The layer closest to the bottom of the tube cannot move, so this is a **displacement node**. Other nodes and antinodes may form between these two boundaries. The displacement description is the inverse of the pressure description – pressure antinodes occur at displacement nodes and vice versa.

Using a resonance tube

The velocity of sound in air can be measured using a transparent **resonance tube** containing a layer of fine dry dust (sometimes called a **Kundt's tube**). One end of the tube is closed with a moveable piston. At the other end the air is forced to vibrate by a loudspeaker connected to a signal generator. As the frequency of the oscillator is increased, the intensity of the sound varies. At certain frequencies there are strong maxima. These form when the wavelength of the sound emitted by the speaker is able to form standing waves between the vibrating cone and the reflecting surface at the end of the piston (which creates a displacement node). Inside the tube the dust collects in heaps that are regularly spaced. These are at displacement nodes in the standing wave (positions where the air vibration is a minimum). The average distance between adjacent heaps is equal to half the wavelength of the sound that is resonating in the tube. The frequency of the sound can be measured directly from the signal generator or by using a microphone and oscilloscope. Either way, velocity of sound is given by $v = f\lambda$. You can also look for resonances by fixing the frequency of the sound and moving the piston in the tube.

Musical intervals depend on the ratio of frequencies and not on the frequencies themselves. Notes an octave apart have frequencies in the ratio 2:1.

> **Standing waves and resonance tubes**
>
> In practice standing waves do not 'fit' the resonance tubes exactly. They disturb the air beyond an open end and the displacement antinode forms a distance e (the **end correction**) beyond the tube. e is about 0.6 times the radius of the tube. In accurate calculations this will affect estimates of wavelength based on resonance in air columns.

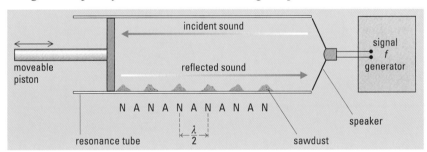

PRACTICE

1 Show that the harmonics of an air column closed at one end form a frequency series with ratios 1:3:5... etc.

2 A discarded hollow pipe is 1.65 m long and open at each end. Sometimes when the wind blows it resonates.

   a What is the wavelength of the fundamental frequency of sound produced?

   b What is the frequency of this sound? (Speed of sound = 330 m s^{-1}.)

   c Would there be any change in the fundamental frequency of this pipe if a hole were drilled in it 0.825 m from one end? Explain.

3 Why do displacement nodes correspond to pressure antinodes in a standing wave in an air column?

4 A clarinet acts like an air column open at one end and closed at the other. If it has a fundamental frequency of 147 Hz, how long is it?

5 A sound of frequency 2400 Hz is played into the end of a closed resonance tube containing a layer of sawdust. Small piles of sawdust form along the bottom of the tube. The separation between the 1st and 6th piles is 35 cm. Use this to calculate the speed of sound in air. Explain your method.

• standing waves in 1, 2, and 3 dimensions

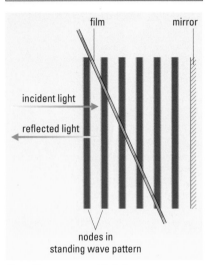

Standing waves with light.

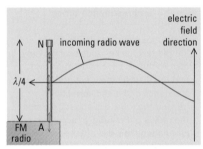

FM radio waves. Radio aerials respond most strongly to incoming radio waves if the electron oscillations excited by the wave's varying electric field set up a standing wave pattern in the aerial. Electrons near the end cannot leave the aerial, so this is a node; electrons at the base of the aerial can move in and out forming an antinode. The distance between a node and antinode in a standing wave is about a quarter wavelength, so an aerial of about this length resonates with the carrier wave and receives a strong signal. For FM reception $\lambda/4$ is about 0.75 m.

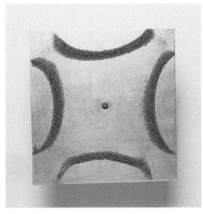

Chladni figures.

STANDING WAVE PATTERNS

Visible light has a wavelength of about 500 nm, so we are unlikely to see standing light waves directly (adjacent antinodes would form bright regions about 250 nm apart). However, they can and have been formed and detected using a very thin photographic film placed at a small angle to a mirror. Light reflected from the mirror superposes on incoming light. The film itself must be much thinner than the wavelength of light (otherwise the same bit of film would intersect nodes and antinodes) and is inclined so that it passes obliquely through the nodes and antinodes in the stationary wave pattern. This creates a regular pattern of maxima and minima on the film which appears as bright and dark bands. Optical standing waves have been used as diffraction gratings for atomic beams (atoms, like photons and electrons, have both wave and particle properties) – these experiments allow very accurate tests of quantum theory.

Catching a radio wave

Radio aerials respond most strongly to incoming radio waves if the electron oscillations excited by the wave's varying electric field set up a standing wave pattern in the aerial. Electrons near the end cannot leave the aerial, so this is a node; electrons at the base of the aerial can move in and out, forming an antinode. The distance between a node and antinode in a standing wave is about a quarter of a wavelength, so an aerial of about this length resonates with the carrier wave and receives a strong signal. For FM reception (frequencies of about 100 MHz) $\lambda/4$ is about 0.75 m. If you look at a portable radio you will see that this is roughly the length of its aerial.

AM transmissions on the medium- and long-wave bands have much longer wavelengths, so it is not practical to use a simple rod aerial; it would be too long. These transmissions are detected using a coil wound around a ferrite rod. The alternating magnetic field induces an alternating voltage in the coil. (Ferrite is used instead of iron because it is a non-conductor. At radio frequencies the large eddy currents induced in the iron would prevent the coil receiving the waves.)

Two-dimensional standing waves

If you tap a coffee cup with a spoon you will sometimes see a temporary but fairly stable pattern of concentric rings form on the surface of the liquid. This is a two-dimensional standing wave formed by waves reflecting from the edges of the cup. The elastic skin stretched across a drum is a similar two-dimensional surface, and has many possible modes of oscillation that satisfy the boundary conditions for standing waves to form. In this case the boundary condition is that the circumference of the skin is fixed, so it must form a displacement node.

The simplest mode has a single antinode at the centre so that the middle of the drum skin moves up and down with a large amplitude whilst the edges stay still. This mode determines the fundamental frequency of the drum (explaining why bass drums are big). In two-dimensional standing waves there are many more possible modes than in one-dimensional, so the sound from a drum is more complex than from a stringed instrument or air column (and sounds even less like a pure note). Drum sounds also have a very sharp attack because of the nature of the impact that produces the sound, and the relative loudness of different modes can be changed by striking the drum in different positions – a central impact will emphasize the fundamental frequency.

The complexity of two-dimensional standing waves can be seen in a simple experiment using a square steel plate connected centrally to a vibrator. Different modes are excited at different frequencies. If grains of salt or sand are sprinkled on the plate the patterns of vibration become visible as the grains are shaken off the antinodes and collect along the nodes. These patterns were first observed by a German physicist, Ernst Chladni, in the nineteenth century, and are known as **Chladni figures**.

A large-scale example of two-dimensional standing waves was seen after the Mexico earthquake in 1985. Aerial photographs revealed districts in which some buildings survived virtually intact whilst similar buildings in adjacent blocks on either side collapsed. The earthquake waves had forced the ground to vibrate like a drum skin and set up standing waves. Buildings unfortunate enough to be located on or near the antinodes were destroyed, while those near the nodes suffered significantly less damage.

Three-dimensional standing waves: atomic orbitals

According to quantum theory all 'particles' (e.g. electrons) also have wavelike properties (see spread 8.6). The distinct energy levels in atoms can then be explained by standing waves formed by the electron waves trapped in the attractive field of the nucleus. The simplest model (worked out by Niels Bohr) assumes that electrons move in two-dimensional circular orbits whose circumference must be an integer number of electron wavelengths. This is similar to the formation of standing waves on a guitar string, except that the electron waves must join up with themselves (rather than having nodes at the fixed ends). In excited states the electrons have more energy and momentum and a shorter de Broglie wavelength (see spread 8.6). These high-energy electrons complete more waves per orbit. In fact the number of waves per orbit is equal to the principal quantum number n.

Bohr's model of the atom was superceded by that of Erwin Schrödinger. Schrödinger worked out how the electron waves would form three-dimensional standing waves around the nucleus. Max Born suggested that the places where the amplitude of the standing wave is high represent regions in which there is a high probability of finding the electron. This leads to the idea of **electron clouds**, because the actual position and momentum of the electron at any instant is indeterminate. The shape of the electron clouds determines the crystalline structures that particular atoms can form.

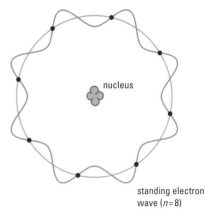
nucleus

standing electron wave ($n = 8$)

The existence of discrete atomic energy levels can be explained by the formation of electron standing waves.

Photograph of standing electron waves on the surface of copper. The waves are trapped inside a ring of specially positioned iron atoms. This is an image from an electron microscope.

PRACTICE

1 Under what conditions might you observe standing waves in the sea?

2 Why does your voice sound much more powerful if you sing in a shower cubicle (or a small room) than if you sing in a large hall?

3 Why is it difficult to carry a shallow rectangular tray of water across a room? (If you try to do this it is very difficult to stop it slopping out of the tray.)

4 The wavelength of an electron is linked to its linear momentum by the de Broglie relation $\lambda = h/mv$. Hydrogen atoms are about 10^{-10} m in diameter and their nuclei are about 10^{-15} m across. Suggest why it is possible for the nucleus to hold an electron in orbit around the atom, but impossible for an electron to remain trapped in the nucleus itself. (Think about the

kinetic energy associated with an electron if it has to form a standing wave in a small region of space, and the energy condition for a particle to be trapped in a bound state. Your answer also explains why electrons are expelled from the nucleus at high speed when they are created in beta decay.)

5 The frequencies at which a drum skin or Chladni plate forms standing waves do not form a simple sequence, whereas those formed on a guitar string or resonance tube do. Account for this difference.

6 Find out the frequency of one FM and one AM radio station. Work out the wavelengths of their carrier waves. How are each of these waves detected by your radio? Explain.

• Doppler shift formulae for sound and light

• relativistic effects

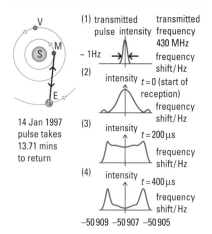

14 Jan 1997
pulse takes
13.71 mins
to return

(1) transmitted
pulse intensity

transmitted
frequency
430 MHz

~ 1Hz → ← frequency
shift/Hz

(2) intensity
$t = 0$ (start of reception)
frequency
shift/Hz

(3) intensity
$t = 200 \mu s$
frequency
shift/Hz

(4) intensity
$t = 400 \mu s$
frequency
shift/Hz

−50 909 −50 907 −50 905

Shortwave reflections from Mercury. The mean shift in the frequency of the reflected pulse is −50 907 Hz. This is caused by the relative motion of the Earth and Mercury. The sequence of reflected signals changes because of the curvature of the planet (reflections from the centre are received before those from the edges). Peaks at +2 Hz either side of centre correspond to rotational motion (Doppler effect).

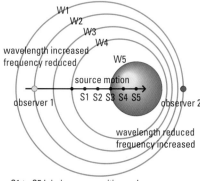

wavelength increased
frequency reduced

source motion

observer 1   observer 2

wavelength reduced
frequency increased

S1 to S5 label source positions when wave fronts W1 to W5 were emitted

Changes of wavelength from a moving source.

The expanding universe

Hubble's discovery that the light from distant galaxies is almost always **redshifted** (increased in wavelength) led to the conclusion that the galaxies are all receding from us and the universe is expanding. However, the modern view is that space itself is expanding and the light gets 'stretched' as part of this expansion, rather than as a result of the motion of galaxies *through space* – so the Hubble redshift is not a Doppler effect.

THE DOPPLER EFFECT

The colour of the Sun's disc looks fairly uniform, but spectral analysis shows that characteristic spectral lines from one side have slightly longer wavelengths than the same ones emitted from the other side. This is an example of the **Doppler effect** – *waves emitted from an approaching source are shortened while those emitted from a receding source are stretched*. The difference can be used to calculate the period of the Sun's rotation, about 27 days. The Doppler effect is also used to measure the rotation speed of stars about the nuclei of spiral galaxies. A similar effect with sound is familiar in the 'eeeeouwwwww' of a speeding car – the engine note seems to drop from a high to a low pitch as it passes you (even though the car's speed is constant). The details of the Doppler effect for mechanical and electromagnetic waves are different, so they are dealt with separately in the following sections.

Doppler effect with sound

The size of the Doppler effect will depend on the velocities of the source and listener relative to one another and to the air. To simplify the argument, assume that the air is still. The speed of sound relative to air is v_s and the source emits sounds of frequency f_0. In still air these waves would have a wavelength $\lambda_0 = v_s/f_0$.

1 Moving source If the source is moving towards the observer at velocity v and emits a wave at $t = 0$, the next wave will be released one period later, at $t = 1/f_0$, during which time the source has moved forward a distance $s = v/f_0$ and the wavelength is shortened by this distance. If the source moves away from the listener the wavelength will be increased by the same amount. In the equations below the transmitted frequency is f_0 and the received frequency is f'. The wavelength emitted by the source when it is at rest in the medium is λ_0 and λ' is the wavelength relative to the observer.

$$\lambda' = \frac{v_s}{f_0} \mp \frac{v}{f_0} = \left(1 \mp \frac{v}{v_s}\right)\lambda_0$$

The Doppler shift is

$$\delta\lambda = \mp \frac{v}{c}\lambda_0$$

The received frequency also changes:

$$f' = \frac{v_s}{\lambda'} = \frac{v_s}{\left(1 \mp \dfrac{v}{v_s}\right)\lambda_0} = \frac{1}{\left(1 \mp \dfrac{v}{v_s}\right)}f_0$$

If the velocity v is small compared to c this can be approximated by:

$$f' \approx \left(1 \mp \frac{v}{v_s}\right)f_0 \quad \text{so that} \quad \delta f \approx \mp \frac{v}{v_s}f_0$$

Note that in all cases above, where a plus/minus sign appears the top sign corresponds to an approaching source.

Worked example 1

A train approaches and passes a platform at a steady speed of 30 m s⁻¹ whilst sounding its whistle at a frequency of 500 Hz. How does the frequency of the whistle change as heard by someone standing on the platform? (Speed of sound = 330 m s⁻¹.)

$$f' = \frac{1}{\left(1 \mp \dfrac{v}{v_s}\right)} \qquad f_0 = \frac{1}{\left(1 \mp \dfrac{30 \, \text{m s}^{-1}}{330 \, \text{m s}^{-1}}\right)} \times 500 \, \text{Hz}$$

So $f'_{app} = 550$ Hz and $f'_{dep} = 458$ Hz where f'_{app} is the sound heard as the train approaches and f'_{dep} the sound as the train departs.

---

2 Moving observer and other cases If the observer moves through the waves emitted by a stationary source, the wavelength is unchanged but the wavespeed and frequency relative to the observer both change. A similar analysis to the one above gives:

$$f' = \left(1 \mp \frac{u}{v_s}\right)f_0$$

where u is the observer's velocity. In general the Doppler-shifted frequency is always given by

$$\text{shifted frequency} = \frac{\text{wave velocity relative to observer}}{\text{wavelength relative to observer}}$$

$$f' = \frac{v_s{}'}{\lambda'}$$

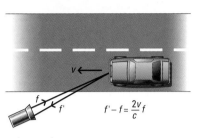

radar speed gun

A police speed trap. Note that the shift is due to a double Doppler effect. Frequency is increased as incoming waves hit the moving car and again as reflected waves from the moving car travel to the speed gun.

Doppler effect with electromagnetic waves

Light and other electromagnetic waves share a rather remarkable property. Their velocity is constant relative to any uniformly-moving observer. This means that if you fly away from a star at 90% of the speed of light, light from the star will still pass you with a relative velocity equal to 100% of the speed of light. Also, there is no material medium to consider, so all that matters is how much the waves get stretched or compressed as a result of relative motion between source and observer. From the point of view of the person receiving the light from a distant source, all that matters is the motion of the source, so (as long as the velocity v is much less than the speed of light) the Doppler shift for light gives exactly the same formula as for sound with a stationary observer and moving source:

$$\lambda' = \left(1 \mp \frac{v}{c}\right)\lambda_0 \qquad \delta\lambda = \mp \frac{v}{c}\lambda_0$$

$$f' = \frac{1}{\left(1 \mp \dfrac{v}{c}\right)}f_0 \qquad \delta f \approx \mp \frac{v}{c}f_0 \qquad \text{valid if } v \ll c$$

Once again the top signs in the equations represent source and observer approaching one another.

Worked example 2

Estimate the frequency shift of the sodium-D line at 589 nm as a result of the Sun's rotation. The period of rotation is 27 days and the radius of the Sun is 7.0×10^8 m.

At the edge of the Sun the surface is moving towards us with a velocity

$$v = r\omega = \frac{2\pi r}{T} = \frac{2\pi \times 7.0 \times 10^8 \,\text{m}}{27 \times 24 \times 3600 \,\text{s}} = 1900 \,\text{m s}^{-1}$$

$$\delta\lambda \approx \pm \frac{v\lambda}{c} = \pm \frac{1900 \,\text{m s}^{-1}}{3.0 \times 10^8 \,\text{m s}^{-1}} \times 589 \,\text{nm} = \pm 3.7 \times 10^{-3} \,\text{nm}$$

Relativistic effects

According to Einstein's special theory of relativity, moving clocks run slow. This means that the oscillating atoms that emit light from a moving source will appear to oscillate slowly compared to their counterparts on Earth. This gives an additional frequency shift on top of the usual Doppler effect which must be taken into account if v is a significant fraction of the speed of light.

If the source moves at velocity v relative to the observer then its oscillators slow by a factor

$$\sqrt{1 - \frac{v^2}{c^2}}$$

relative to a 'stationary' observer. The received frequency is now:

$$f' = \frac{\sqrt{1 - \dfrac{v^2}{c^2}}}{\left(1 + \dfrac{v}{c}\right)}f_0 = \frac{\left(1 - \dfrac{v}{c}\right)\left(1 + \dfrac{v}{c}\right)}{\left(1 + \dfrac{v}{c}\right)}f_0$$

$$\sqrt{\frac{\left(1 - \dfrac{v}{c}\right)}{\left(1 + \dfrac{v}{c}\right)}}f_0 = \sqrt{\frac{(c - v)}{(c + v)}}f_0$$

Many distant stars are moving fast enough relative to the Earth for the relativistic formula to be necessary.

PRACTICE

1 In 1969 the unmanned spacecraft *Luna 9* made a soft landing on the Moon. It transmitted information back to Jodrell Bank radio telescope using a transmission frequency of 78 MHz. Before landing the spacecraft decelerated by firing its retro rockets. During deceleration the frequency of the signal received at Jodrell Bank increased by 680 Hz.

a Explain why the frequency increased.

b Calculate how fast the spacecraft was moving away from the Earth before it slowed down (state the assumption that must be made in order to do this).

2 A police speed gun works by measuring the Doppler shift of microwaves reflected off an approaching vehicle. A typical microwave frequency is 10^{10} Hz and the Doppler shift is measured by combining the transmitted and reflected signals and measuring the beat frequency.

a How is the beat frequency related to the Doppler shift?

b Explain why the Doppler shift of an approaching vehicle gives a beat frequency $f_b = (2v/c)f$ rather than v/c.

c What is the beat frequency for a car approaching at 20 m s^{-1}?

Light reflected from the surface of a transparent medium is partially plane-polarized (top). Suitably positioned polarized sunglasses can cut out most of this reflected 'glare' (bottom).

Polarizing directions.

Polarization of microwaves

School microwave transmitters emit plane-polarized microwaves. If the filter or transmitter is rotated the received intensity varies with $\cos^2 \theta$ (Malus's law). When the polarization direction is parallel to the conducting rods currents are induced in the metal and the wave is absorbed.

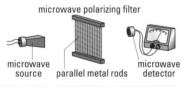

microwave polarizing filter

microwave source   parallel metal rods   microwave detector

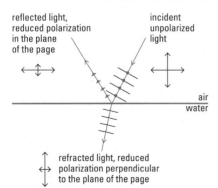

Polarizing by reflection. Dot size represents the strength of polarization perpendicular to the plane of the page. Line size represents the strength of polarization in the plane of the page.

POLARIZATION

Have you ever wondered whether that cheap pair of shades you bought is really **polarizing**? One way to check is to look through them at light reflecting off the windows of a building. Now rotate the glasses. If they are polarizing, the brightness varies, going bright and dark twice in a complete rotation. Light reflected from a transparent surface is partially polarized and this part of the light can only pass through polarized glasses when they are aligned with it.

Think about transverse waves on a horizontal string. These can vibrate in any direction perpendicular to the string itself. These alternatives are called **polarization directions**. If the end of the string is waved horizontally the waves lie in a horizontal plane and this is a **horizontally plane-polarized wave**. Vertical vibrations create a **vertically plane-polarized wave**. Longitudinal waves *cannot* be polarized because they have a unique vibration direction parallel to the wave direction.

If the string passes through a barrier with a long slit cut in it, waves polarized parallel to the slit can pass through, whilst those perpendicular to it are blocked. The slit is acting as a **polarizing filter**.

Polarized and unpolarized light

Electromagnetic waves are all transverse, so they can be polarized. By convention their polarization direction is defined as the direction in which their electric field varies. However, most sources of light, like filament lamps or the Sun, are *unpolarized*. This means rays coming from these sources contain all polarization directions superposed. Unpolarized light can be partly or completely polarized in several ways.

- **Pass it through a polarizing filter.** The transmitted light is plane-polarized parallel to the filter and has reduced intensity (an ideal filter would reduce the incident intensity by a factor of 2). Suitable filters for light usually consist of transparent polymers in which the long-chain molecules have all been aligned in some direction – e.g. **polaroid sheets**. (This action is similar to the slit in the example above, but with one major difference – the waves are *absorbed* if they are polarized parallel to the direction of molecular alignment).

- **Reflect it from a sheet of transparent material.** The reflected ray is partially plane-polarized parallel to the surface and the refracted ray is partially plane-polarized perpendicular to the surface. If the reflected and refracted rays are perpendicular to each other they are both completely plane-polarized. This occurs at an angle of incidence called the **Brewster angle**, θ_b. Polarization by reflection explains the variation in intensity you see if you try the simple experiment with sunglasses suggested above.

- **Let it scatter from a suspension of small particles.** The scattered light is partially plane-polarized. It is completely plane-polarized if the light is scattered at 90° to its incident direction. This is because the oscillations in the scattered light cannot have a component in the direction of the incident ray, and so there is only one vibration direction possible at this angle.

The brightness of the daytime sky is caused by scattered light, so there is a pattern of polarization in the sky. Many insects, including bees and ants, have receptor cells in their eyes that detect the polarization direction of incoming light. They use this extra sense to navigate. Many aquatic creatures can also detect polarized light – this may be because the light that penetrates water becomes polarized as it scatters off small suspended particles. It is not known how such creatures use this information.

Analysing polarized light

If plane-polarized light falls on a polarizing filter with its plane of polarization at an angle θ to the polarizing direction of the filter, then the component of light parallel to the filter direction is transmitted and the component perpendicular to it is absorbed (transferred to thermal energy in the filter). If the incident light has amplitude A_0 and intensity I_0, the transmitted light has amplitude $A_t = A_0 \cos \theta$ and intensity $I_t = I_0 \cos^2 \theta$ (Malus's law).

No light passes through two sheets of polaroid when they are placed with their polarizing directions at 90° – any light that makes it through the first will obviously be absorbed by the second. However, if a third polarizing filter is placed between the other two some light *does* pass through the whole arrangement. To understand how this happens, call the three filters X, Y, and Z. X and Z have their polarizing directions perpendicular to one another and Y is at some angle ϕ to X.

- The light emerging from X has amplitude A_0 and is polarized parallel to X, so it strikes Y polarized at an angle ϕ to Y's polarizing direction.

- Light emerging from Y has amplitude $A_Y = A_0 \cos\phi$ and is now polarized parallel to Y.

- This light from Y falls on Z at an angle of $(90° - \phi)$ to Z's polarizing direction so some of it will be transmitted. The amount transmitted is $A_Z = A_Y \cos(90° - \phi)$. The total light emerging from Z has amplitude $A_Z = A_0 \cos\phi \cos(90° - \phi)$. This is a maximum when $\phi = 45°$.

Optical activity and photoelastic analysis

Many complex molecules have a '**handedness**'. When polarized light passes through a solution containing such molecules its plane of polarization is rotated. This is called **optical activity**. The amount of rotation can be measured using a **polarimeter** and depends on the concentration of molecules in the sample. The sample is placed between crossed polarizing filters and the top filter is rotated until no light passes through the solution. The angle through which the filter must be rotated is equal to the angle through which the plane of polarization has been rotated by the sample (this is used industrially to measure and monitor the concentration of sugar solutions).

Long-chain molecules in transparent plastics like perspex can also rotate the plane of polarization. The amount of rotation depends on wavelength and on the amount of strain (which aligns the molecules) in the sample. If a strip of perspex is viewed through crossed polaroids as it is subjected to stress, a beautiful pattern of colours forms and can be used to investigate the stress distribution in the perspex. This is a useful aid in design where the stress distribution in a component can be determined by loading a perspex model.

filter polarizing direction

A_0 is amplitude of incident wave

ϕ is angle between filter's polarizing direction and polarization of incident wave

A_t is transmitted amplitude

A_a is absorbed amplitude

$A_t = A_0 \cos \phi$

Analysing polarized light.

Light from clouds is unpolarized, and light from the blue sky is partially plane polarized. A polarizing film in front of a camera lens can be used to enhance contrast by reducing the intensity of light from the sky.

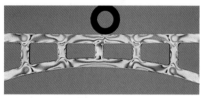

Stress patterns in transparent polymers show up as vivid coloured bands when they are placed between crossed polarizing filters.

Polarized TV transmission

In the UK most TV transmissions are carried by horizontally plane-polarized electromagnetic waves. However, in some regions of poor reception there are special booster transmitters. These are usually vertically polarized so that there is no interference between the signals from the main and booster transmitters. Aerial rods are oriented to lie parallel to the direction of the electric field vibrations in their area.

PRACTICE

1 Vertically plane-polarized light waves have an incident intensity I_0 and amplitude A_0. What is the transmitted amplitude and intensity in each of the following cases? Assume the polarizing filter is perfect.

  a The light falls on a polarizing filter at angle 30° to the vertical.

  b The light falls on a polarizing filter at angle 45° to the vertical.

  c Explain how energy is conserved when unpolarized light passes through a sheet of polaroid.

2 Prove that the transmitted intensity is half the incident intensity when unpolarized light falls on a polarizing filter.

3 Prove that the Brewster angle θ_b is given by the equation $n = \tan \theta_b$ where n is the refractive index for light travelling from air into a transparent medium.

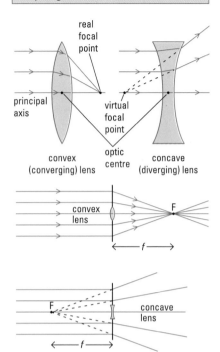

Real ('thick') lenses refract light at both surfaces. Often lenses are thin enough to treat this as a single refraction in the central plane of the lens. The lower diagrams show refraction by 'thin' lenses. These are represented by a vertical line. The type of lens is shown by a small symbol at the optic centre.

Helpful hint

The word 'power' is used in a very unusual way in connection with lenses: it has nothing whatever to do with rate of transfer of energy!

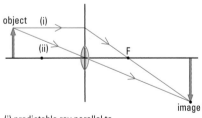

(i) predictable ray parallel to
    principal axis
(ii) predictable ray through
    optic centre

A convex lens forming a real image of a real object.

Convex (converging) and **concave** (diverging) lenses focus light by refraction. The curvature of the lens is such that rays further from the optic axis are bent more than those close to it. The action of a lens is like the action of a series of prisms of increasing refracting angle.

Convex lenses are manufactured so that rays parallel to the **principal axis** are focused to a *real* **focal point** beyond the lens. Similar rays passing through a concave lens diverge from the optic axis forming a *virtual* focal point on the same side of the lens as the source of the rays. If a lens is thin the two separate refractions at the surfaces of the lens are so close together that ray diagrams treat them as a single change of direction at the surface of a plane lens.

The simplest method of finding the focal length of a convex lens is to focus light from a distant object (near-parallel rays) onto a plane surface and then to measure the distance from the centre of the lens to the 'screen'.

The power of a lens

The shorter the focal length of a lens the more sharply it bends rays of light, so the power P of a lens is defined as:

$$P = \frac{1}{f}$$

f is measured in metres and gives power in dioptres (m^{-1}). A lens of focal length 0.20 m has a power of 5.0 dioptres.

Predictable rays

An image is a point-by-point representation of an object. The image position in a thin lens can be found using predictable rays.

- A ray parallel to the optic axis converges to (or diverges from) the focal point of the lens.
- Rays through the centre of the lens are not deviated.
- Rays directed through (or towards) the focal point are deviated so they become parallel to the optic axis.

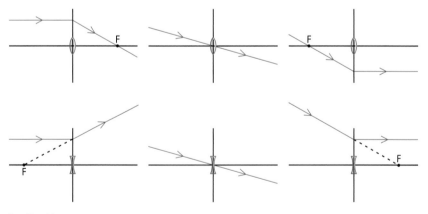

Predictable rays.

The diagram (left) shows how a convex lens forms a real image of a real object. Although every point of the object is reproduced in the image it is only necessary to locate one image point e.g. where two predictable rays from the corresponding point on the object cross. The object is drawn as an arrow and the obvious point to use is the tip of the arrow. The base point and all points in between will be focused in the same plane as the tip of the arrow.

The lens equation

Look at the ray diagram below.

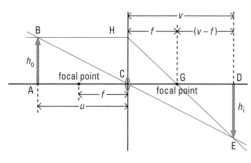

A convex lens forming a real image of a real object.

	Some definitions	
u =	object distance (from object position to centre of lens)	
v =	image distance (from image position to centre of lens)	
f =	focal length (from focal point to centre of lens)	
h_o =	object height	
h_i =	image height	
m =	$\dfrac{\text{image height}}{\text{object height}} = \dfrac{\text{linear}}{\text{magnification}}$	

Triangles ABC and DEC are similar so:

$$m = \text{linear magnification} = \frac{h_i}{h_o} = \frac{AB}{DE} = \frac{AC}{DC} = \frac{v}{u} \quad \text{giving} \quad m = \frac{v}{u}$$

Triangles GCH and GDE are also similar so:

$$\frac{DE}{CH} = \frac{h_i}{h_o} = \frac{v}{u} = \frac{v-f}{f}$$

which can be rearranged to give the **lens equation**:

$$\frac{1}{u} + \frac{1}{v} = \frac{1}{f}$$

This equation can be used to predict the image position, given the object position and focal length of the lens. It also works for concave lenses and for convex and concave mirrors, as long as a consistent **sign convention** is used: *real is positive*. All distances associated with real images and focal points are considered positive values, and those associated with virtual images or focal points are negative values.

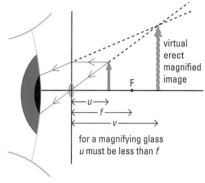

A magnifying glass creates a magnified virtual image of an object. The eyepieces of telescopes and microscopes are magnifying glasses. In both cases they magnify an intermediate image formed by the objective lens.

Worked example

Find the image, nature, position, and linear magnification when an object is placed 15 cm from (a) a convex lens of f = 10 cm and (b) a concave lens of f = −10 cm.

$$\frac{1}{u} + \frac{1}{v} = \frac{1}{f} \quad \text{giving} \quad \frac{1}{v} = \frac{1}{f} - \frac{1}{u}$$

(a) $$\frac{1}{v} = \frac{1}{10\,\text{cm}} - \frac{1}{15\,\text{cm}} = \frac{1}{30\,\text{cm}} \quad \text{giving} \quad v = \frac{30\,\text{cm}}{1} = 30\,\text{cm}$$

v is positive, indicating a real image.

$$m = \frac{v}{u} = \frac{30\,\text{cm}}{15\,\text{cm}} = 2.0 \quad \text{The image is 2.0 times the size of the object.}$$

(b) $$\frac{1}{v} = \frac{1}{-10\,\text{cm}} - \frac{1}{15\,\text{cm}} = \frac{-5}{30\,\text{cm}} \quad \text{giving} \quad v = -6\,\text{cm}$$

v is negative, indicating a virtual image.

$$m = \frac{v}{u} = \frac{-6\,\text{cm}}{15\,\text{cm}} = -0.4$$

The image is 0.4 times the size of the object.

PRACTICE

1 Draw simple ray diagrams to locate the image positions (in terms of f) and linear magnifications when a convex lens of focal length f is used to form images of an object placed at the following distances from the lens:

   a $4f$ **b** $2f$ **c** f **d** $f/2$.

2 Confirm your results from question 1 using the lens equation.

3 Repeat question 1 and question 2 for a concave lens of focal length f.

4 Find the image position, nature, and linear magnification when an object is placed 20 cm from:

   a a convex lens of focal length 15 cm;

   b a convex lens of power 4 dioptres;

   c a concave lens of focal length 10 cm.

5 **a** Use the lens equation to estimate the focal length and power of your eye when it is focusing objects at the near point (25 cm) and far point (infinity). (You will need to estimate the distance from the front to the back of your eye.)

   b Use your results for **a** to state the '**range of accommodation**' of your eye, i.e. the difference between its maximum and minimum power.

- visual angle

- magnifying power

- ray diagrams of instruments

The limitations of refracting telescopes mean that all large optical telescopes are reflectors. This is the 2.4 m diameter mirror of the Hubble Space Telescope.

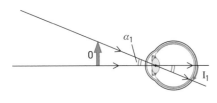

Apparent size diminishes with distance.

Small angles

In all the ray diagrams the rays are assumed to travel close to the principal axis and make small angles with it. This simplifies the analysis of magnifying power and allows you to use angles in radians in place of their sines and tangents (small-angle approximation).

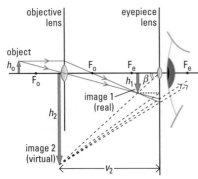

A compound microscope.

OPTICAL INSTRUMENTS

In the human eye most refraction occurs as light passes across the curved surface of the cornea. The eye lens has a fine adjustment that allows you to focus sharply on objects between the **near point** (about 25 cm) and the **far point** (infinity).

The apparent size of an object depends on how much of the retina is covered by its image. This is proportional to the angle subtended by the object at the eye – the visual angle α. If something moves away from you its apparent size diminishes because the visual angle reduces and its image on the retina is smaller. This effect is called **perspective**.

The purpose of optical instruments like telescopes and microscopes is to form an image that subtends a larger angle at the eye than would be formed by the object itself if observed directly. This increases the apparent size of the object.

$$\text{magnifying power} = \frac{\text{visual angle of object seen through the optical instrument}}{\text{visual angle of object viewed normally with the naked eye}} = \frac{\beta}{\alpha}$$

The magnifying power of an optical instrument depends on the position of the final image with respect to the eye (since this affects its visual angle) so magnifying power is usually quoted for '**normal adjustment**'.

- Normal adjustment for a telescope: final image at infinity.
- Normal adjustment for a microscope: final image at the near point (distance D).

Compound microscope

The object O is mounted just beyond the focal point of the objective lens. This lens forms a real image I_1 of the object, which acts as a secondary object for the eyepiece lens. The eyepiece itself is used like a magnifying glass (see spread 6.20) to form an enlarged virtual image I_2 that subtends a larger angle at the eye. To simplify calculations the visual angle at the eye is considered equal to the angle β subtended by I_2 at the centre of the eyepiece lens.

Assume the final image I_2 is formed a distance v_2 from the eyepiece lens (in normal adjustment this will be D). From the diagram

$$\tan \beta = \frac{h_2}{v_2}$$

but angles are small, so

$$\beta \approx \frac{h_2}{v_2}$$

For normal viewing (object at near point viewed by unaided eye)

$$\tan \alpha = \frac{h_o}{D} \quad \text{so } \alpha \approx \frac{h_o}{D}$$

$$\text{magnifying power, } M = \frac{\beta}{\alpha} \approx \frac{h_2}{v_2} \times \frac{D}{h_o} = \frac{h_2}{h_o} \times \frac{D}{v_2}$$

In normal adjustment

$$v_2 = D \quad \text{so} \quad M = \frac{h_2}{h_o} = m \text{ (linear } magnification\text{)}$$

and

$$\frac{h_2}{h_o} = \frac{h_2}{h_1} \frac{h_1}{h_o} = m_1 m_2 \quad \text{so} \quad M = m_1 m_2$$

m_1 and m_2 are linear magnifications of the eyepiece and objective lenses respectively.

Magnifying power of a compound microscope

The magnifying power of a compound microscope in normal adjustment is equal to the product of the magnifying powers of the eyepiece and objective lenses.

Astronomical refracting telescope

Rays from points on distant objects arrive near-parallel at the eye or telescope objective. If the distant object is extended (like the Sun) rays from either side of it will subtend a small angle α at the eye (or telescope objective). The job of the telescope is to create a final image that subtends a larger angle at the eye. The objective lens forms a real image I_1 inside the telescope. Since the object is at optical infinity, I_1 is formed in the focal plane of the objective. The eyepiece lens acts like a magnifying glass to form a larger virtual image I_2 at infinity (in normal adjustment). This happens when I_1 is formed on the focal plane of the eyepiece lens, so the separation of the two lenses (length of the telescope) is simply $f_o + f_e$, the sum of the focal lengths of the objective and eyepiece lenses. Rays from I_2 subtend an angle β at the eyepiece. From the diagram,

$$\beta \approx h_1/f_e \quad \text{and} \quad \alpha \approx h_1/f_o \quad \text{(using small-angle approximation)}$$

$$\text{magnifying power,} \quad M = \frac{\beta}{\alpha} = \frac{f_o}{f_e}$$

Magnifying power of a refracting telescope

The magnifying power of an astronomical refracting telescope in normal adjustment is equal to the ratio of focal lengths of its eyepiece and objective lenses.

Reflecting telescopes

Refracting telescopes are limited in several respects.

- **Chromatic aberration** The refractive index depends on the frequency of light, so different colours are focused at different distances and the image is not perfectly sharp.

- **Spherical aberration** Rays further from the optic axis are not focused to the same point as rays close to it, so the lens size must be limited. This reduces image intensity and resolving power (see spread 6.10).

Large optical telescopes use a curved mirror instead of a lens as an objective. This has several advantages.

- **No chromatic aberration** The mirror is silvered on its front surface, so no refraction takes place.

- **No spherical aberration** The mirror is parabolic rather than spherical.

- **Higher image intensity and resolving power** This is because the mirrors can be larger than the objective lenses used in refracting telescopes.

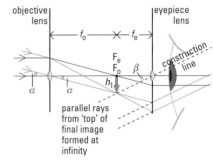

— rays from top of distant object
— rays from bottom of distant object

An astronomical refracting telescope.

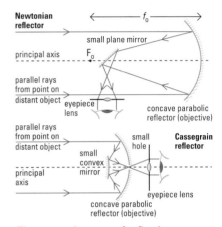

The two main types of reflecting telescope are shown in the diagrams. The magnifying power is again given by $M = f_o/f_e$ but the effective focal length of the Cassegrain reflector depends on both the concave objective and the small convex mirror in front of it. The main disadvantage of this design is that the small mirror blocks some of the incident light. This becomes more of a problem for small telescopes so these are usually refractors. The Hubble Space Telescope is a 2.4 m reflecting telescope and can resolve point sources with a minimum angular separation of just 14 millionths of a degree.

PRACTICE

1 a A man 1.8 m tall is standing 10 m away from you. Approximately what angle does he subtend at your eye?

b Later you see him in the distance, 150 m away. What angle does he subtend now?

c Is the small-angle approximation valid in either **a** or **b** or both?

2 A compound microscope has eyepiece and objective lenses of focal lengths 10 mm and 50 mm. It is used to view a small object on a microscope slide 12 mm from the objective lens.

a Find the position and linear magnification of the intermediate image formed by the objective lens.

b What must be the distance from the intermediate image to the eyepiece lens if the final image is to be formed at the near point (25 cm from the eyepiece)?

c What is the magnifying power of the microscope in normal adjustment?

3 a Draw a ray diagram showing how a magnifying glass (sometimes called a **'simple microscope'**) can form a magnified image of a small object a distance u from the lens.

b If this image is formed at the near point (distance D from the lens) show that the magnifying power of the lens is $M = D/u$.

4 The Moon's disc subtends an angle of about 0.5° when viewed with the unaided eye.

a What angle will its image subtend when viewed through an astronomical telescope with objective focal length 75 cm and eyepiece focal length 5 cm?

b How would it appear if the telescope were reversed and you looked through the objective?

PRACTICE EXAM QUESTIONS

1

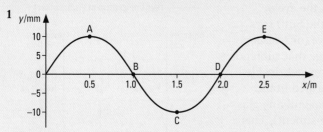

Figure 6.1

The graph in Figure 6.1 shows the displacement of particles in a transverse progressive wave against the distance from the source at a particular instant with points labelled A, B, C, D and E.

a Write down the letters of

  i all the points at which the speed of the particle is a maximum,

  ii all the points at which the magnitude of the acceleration of the particle is a maximum,

  iii two points which are in phase,

  iv two points which are 90° out of phase.

b State

  i the amplitude of the wave,

  ii the wavelength of the wave.

2 a Explain what is meant by the term displacement in a mechanical wave.

b Distinguish between transverse and longitudinal waves.

c With the aid of sketch graphs of a sinusoidal wave explain what is meant by

  i a progressive wave,

  ii wave speed.

3 Light, microwaves and VHF radio waves are all said to be electromagnetic waves.

a State the approximate range of wavelengths of each.

b For each of light, microwaves and VHF radio waves, describe a suitable experiment to measure the wavelength. Give typical dimensions for the apparatus used in each experiment. Show how to calculate the wavelength from the measurements.

c State two pieces of evidence which suggest that all three belong to the same family of waves.

4 Figure 6.2 shows a ray of light incident on the surface of a glass block.

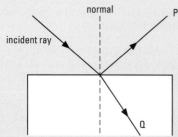

Figure 6.2

According to the laws of reflection and refraction, the incident ray, ray P, ray Q and the normal are coplanar.

a State one other law relating to

  i ray P,

  ii ray Q.

b A student discovers that the light in ray P is polarised. What may be deduced from this observation relating to the nature of light?

c The frequency of yellow light is approximately 5.0×10^{14} Hz. Calculate the wavelength of yellow light.

5 Figure 6.3 shows a glass pentaprism as used in the viewfinder of some cameras. Light enters face AB and leaves face BC. The faces AE, ED and DC are silvered and the refractive index of the glass is 1.52.

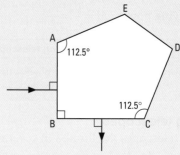

Figure 6.3

a On a copy of the diagram above draw the path of the incident ray from face AB to face CD.

b State why you have drawn the ray in this direction.

c Explain, with the aid of a calculation, why the face CD needs to be silvered if the ray shown is not to be refracted at face CD.

d Copy the diagram and on your copy continue the ray until it leaves the prism.

6

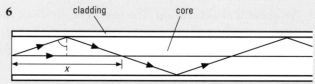

Figure 6.4

Figure 6.4 shows in section light travelling along an optical fibre. The refractive index of the core is 1.520.

a Calculate the speed of light in the fibre if the speed of light in vacuo is $3.00 \times 10^8 \, \mathrm{m \, s^{-1}}$.

b Light is shown travelling in a zig zag path along the fibre. If the smallest angle of incidence for which this can occur is 87.0° calculate the refractive index of the cladding.

c Give an expression for the distance travelled by light along the zig zag path corresponding to a distance x along the axis.

d The difference between the time taken for light to travel along the axis and the time for the reflected light to travel through a fibre is not to exceed 10^{-9} s. What is the length of the fibre?

7 An optical fibre consists of a core made from glass Y surrounded by cladding made from a different glass. Table 6.1 gives the refractive index for light travelling from air to glass for three glasses.

advanced **PHYSICS**

Table 6.1

Glass	X	Y	Z
Refractive index	1.45	1.47	1.49

Figure 6.5

Figure 6.5 shows how light incident at angle i on one end is refracted so that it is incident on the boundary with the cladding at the critical angle, θ_c, and reflected on through the fibre.

a i   Explain what is meant by critical angle.

   ii   State which of the glasses X or Z is suitable for the cladding, giving a reason for your answer.

b   For the arrangement in part **a ii**, calculate

   i   the critical angle, θ_c,

   ii   angle i.

8   In surveying the Earth for oil, longitudinal waves are generated at the air/rock interface, and travel into the rock where they may be refracted and reflected.

Table 6.2

Medium	Speed of waves/m s^{-1}
air	300
A	7000
B	5000
C	11 000

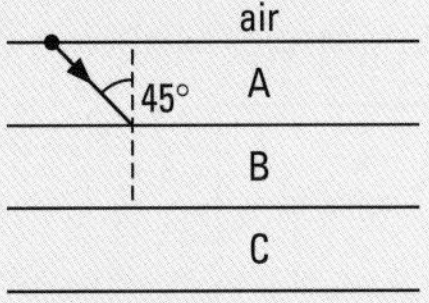

Figure 6.6

Figure 6.6 shows a narrow beam of waves directed into three horizontal, parallel layers of rock, A, B and C. The beam is incident on the AB boundary at 45° to the vertical, as shown. The speed of the waves in each medium is given in Table 6.2.

a   Calculate

   i   the refractive index for waves travelling from A to B,

   ii   the angle of incidence on the BC boundary.

b i   With the help of a calculation, show that total internal reflection will occur at the BC boundary.

   ii   Redraw the diagram, showing the complete path of the beam in the rock, and the magnitude of the angles.

State, with a reason, whether the waves will emerge into the air.

9   Figure 6.7 illustrates one section of the fringe pattern produced on a screen when coherent red light passes through a double-slit arrangement.

Figure 6.7

State, with a reason, the effect on the appearance and spacing of the fringes observed when, independently, the following changes are made:

a   the source of red light is replaced by one of coherent blue light,

b   the separation of the slits in the double-slit arrangement is increased,

c   the width of each slit in the double-slit arrangement is gradually increased.

10 Why is it *not* possible to see interference where the beams of light from the headlamps of a car overlap?

A filament lamp and a yellow filter are available to provide a light source of wavelength 600 nm. Describe how you would set up a double-slit apparatus to produce well-defined fringes 1 mm apart on a viewing screen. Suggest suitable values for the dimensions and spacings of the components, giving reasons for the values you choose.

In the experiment referred to above, what is the effect of making the following changes, *one* at a time? In your answer make clear what will and what will not change in each case.

a   The width of one of the slits is made twice as big as the other.

b   The length of the slits is doubled.

11 Red light of wavelength 7.00×10^{-7} m, incident normally on a diffraction grating, gave a first order maximum at an angle of 75°.

a   Calculate the spacing of the diffraction grating.

b   Calculate the angle at which the first order maximum for violet light of wavelength 4.50×10^{-7} m would be observed.

c   At what angle or angles would a detector receive radiation which is of wavelength 7.5×10^{-7} m transmitted by the grating? Explain your answer.

12 A laser emitting light of wavelength 6.0×10^{-7} m is used to illuminate two parallel slits, giving two coherent sources. Interference fringes are to be produced on a screen 2.0 m from the slits. The separation of the fringes required is 5.0 mm.

a   Calculate the distance between the centres of the two slits.

b   Interference takes place where light beams from the two slits overlap. With the aid of a diagram, explain how this overlap is produced.

c   State and explain what *two* changes you would expect in the fringe system if each of the slits was made narrower, but their separation was kept the same.

13 a   Distinguish clearly between diffraction and interference.

b   Monochromatic light of wavelength 5.0×10^{-7} m falls normally onto a diffraction grating which has 425 lines per millimetre.

   i   What is the highest order of diffraction?

   ii   What is the total number of lines produced by the grating?

   iii   Find the angular separation between the second and third order lines.

c   If the monochromatic light source used in **b** was replaced by a white light source, describe what would be seen in the zero and first order spectral positions.

14 A stretched string fixed at each end and of length 1.5 m is vibrating with its fundamental frequency. If, at the same time, a tuning fork of frequency 256 Hz is sounded, 5 beats per second are heard. The tuning fork now has its prongs lightly loaded with wax. When the fork is again sounded, 10 beats per second are heard.

Calculate:

a the fundamental frequency of the string;

b the speed of the transverse waves along the string.

Clearly show your reasoning.

15 a By considering the motion of a stretched wire fixed at both ends explain what is meant by the terms fundamental frequency and overtones.

b If the fundamental frequency of a stretched wire is f then calculate the frequencies of the overtones in terms of f.

c In the case of a musical note, what determines

   i the pitch,

   ii the quality,

   iii the loudness?

16 A ship at X is equidistant from two shore-based radio transmitters P and Q. Both transmitters operate on a wavelength of 300 m and radiate signals of equal amplitude.

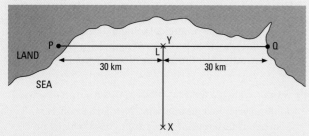

Figure 6.8

a On Figure 6.8, the ship at X detects zero signal amplitude. What information does this give about the signals from P and Q?

b The ship moves in a straight line from X to Y. Throughout the journey the amplitude of the signal detected by the ship is zero. Explain this.

c The ship moves in the direction YQ until the signal detected has an amplitude twice that from either transmitter alone. How far has the ship moved? Explain your answer.

d When the ship sails from Y to the harbour alongside transmitter Q the detected signal rises and falls in amplitude. Calculate how many dips in intensity will be passed.

17 This question is about a radar speed measuring device. In the device, shown schematically in Figure 6.9, a radio transmitter and receiver are placed side by side. Some of the waves from the transmitter are reflected from the approaching vehicle and are detected by the receiver. The signal detected by the receiver is added to the transmitter output in an electronic device called a signal adder.

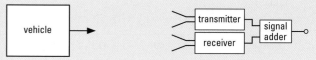

Figure 6.9

a With the vehicle in the position shown in Figure 6.9, the output signal from the adder is a maximum. When the vehicle moves 15 mm towards the device, this signal falls to a minimum and returns to a maximum after a further 15 mm. The transmitter emits radio waves of wavelength 60 mm.

   i Explain why the output signal from the adder decreases during the vehicle's movement.

   ii Why is the minimum not necessarily zero?

b The vehicle moves directly towards the device at a steady speed. The output signal from the adder fluctuates in intensity with a frequency of 500 Hz.

   i Calculate the time taken between successive signal maxima.

   ii How far does the vehicle travel in this time?

   iii Calculate the speed of the vehicle.

c A second vehicle, travelling at the same speed as the one above, approaches the device at a different angle, as shown in Figure 6.10.

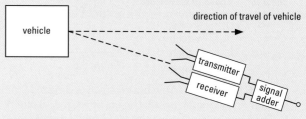

Figure 6.10

How would the output signal from the adder change? Justify your answer.

18 This question is about a recent application of two-slit interference using a beam of helium atoms all travelling with the same velocity.

Figure 6.11 shows the arrangement of the apparatus which uses a pair of parallel slits of spacing $s = 8.0 \times 10^{-6}$ m.

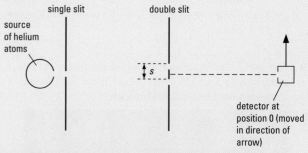

Figure 6.11

The variation in the intensity of the beam measured by the detector is shown in Figure 6.12.

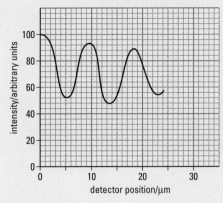

Figure 6.12

a What does this experiment tell you about moving helium atoms?

b **i** Use data from the graph (Figure 6.12) to show that the wavelength required to produce this pattern is approximately 1×10^{-10} m.

ii Use this result to calculate the velocity of the helium atoms.
(Mass of helium atom = 6.6×10^{-27} kg the Planck constant, $h = 6.6 \times 10^{-34}$ J s)

c When the velocity of the helium atoms is halved, what change in the interference pattern (Figure 6.12) takes place?

Give your reasoning.

19 This question is about one of the effects of the pantograph of an electric power unit on the overhead conductor of an electrified railway system (see Figure 6.13).

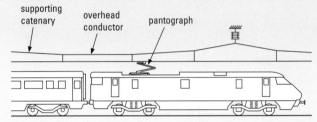

Figure 6.13

As the point of contact of the pantograph moves along the conductor, pressing upwards, it generates a transverse wave-pulse in the conductor. To avoid excessive forces on the pantograph the wave speed must be well in excess of the maximum design speed of the train.

The speed, u, of this pulse along the conductor is given by

$$u = \sqrt{\frac{T}{m}}$$

where T is the tension in the conductor and m is its mass *per unit length*.

a The overhead conductor has a cross-sectional area of 1.2×10^{-4} m^2 and a density of 8.8×10^3 kg m^{-3}. The tension in the wire is 20 kN. Calculate

i m, the mass per unit length of the conductor, and

ii u, the speed of a wave-pulse along the wire.

b Show that if ice were to form on the conductor, the magnitude of u would decrease.

c Show that both sides of the equation given above have the same units.

20 a The phenomenon of Fraunhofer diffraction may be demonstrated by illuminating a single slit with a coherent beam of monochromatic light.

i What do you understand by the term, a coherent beam of monochromatic light?

ii Determine the maximum width of a single slit if the angle subtended by the central maximum of the interference pattern produced in red light is not to be less than 2.0°.

b Explain how diffraction limits the sharpness of the image of a distant star produced by a telescope.

c When viewed using a spectrometer, a line in the spectrum of light emitted by a gas is broader at high temperatures than it is at low temperatures.

i Explain why this is so.

ii Determine the contribution to the width of the 4.7999×10^{-7} m line in the spectrum of cadmium when the temperature of the source is 800 °C.

The relative atomic mass of cadmium is 112.

* 21 In Bohr's model for the hydrogen atom the electron orbits the proton. The centripetal force is the force between the charges on the proton $(+e)$ and electron $(-e)$.

a Derive an expression for the angular momentum L of an electron of mass m, moving in an orbit of radius r with speed v.

b **i** Write down an expression for the magnitude of the force between two particles, each with charge e, separated by a distance r.

Hence deduce an expression for the speed v of an electron in an orbit of radius r in terms of r, e, m and ε_0.

ii Show that it is reasonable to neglect the gravitational force between the proton and electron.

c The angular momentum of the electron orbits is quantized so that the only possible momenta are given by

$$L = \frac{nh}{2\pi}$$

where n is an integer and h is the Planck constant.

Show that the only possible radii for the electron orbits is given by:

$$r = n^2 \frac{4\pi\varepsilon_0}{me^2} \left(\frac{h}{2\pi} \right)^2$$

d **i** Calculate the radius of the orbit when $n = 1$.

ii Calculate the kinetic energy of the electron in this orbit.

iii Calculate the total energy of the electron in this orbit.

iv Explain whether the answer to part **iii** is consistent with the observed ionization energy for hydrogen of 13.6 eV.

Pre-U-style question

7

HEAT

Few things conjure up the idea of heat better than this image of the molten rock inside a volcano's crater. Thermodynamics started off as the study of heat but has since become far more than that. Statistical thermodynamics studies gases by analysing the motion of the particles within them. This leads to an understanding of entropy and the flow of energy. The laws of thermodynamics apply to any system; in fact they make profound statements about the direction of time and the future of the universe as a whole.

In this chapter you will learn about the laws, the relationship between temperature and the kinetics of gas particles, the behaviour of gases, heat capacities, and conduction, convection, and radiation of energy.

The lava and cinder cone of the volcano Kilauea in the Hawaii Volcanoes National Park. Kilauea, Hawaii's youngest volcano, rises to some 1243 m above sea level and covers about one-seventh of the island of Hawaii. It is one of the most active volcanoes on Earth.

Tomorrow and tomorrow and tomorrow
Creeps in this petty pace from day to day
To the last syllable of recorded time...
William Shakespeare, *Macbeth*

VERY LARGE VERSUS VERY SMALL

Many aspects of life fascinated Shakespeare and time is certainly an aspect of life. (You see, even physicists read Shakespeare.) Time passes without interruption at a steady rate, although sometimes when you are very bored it can seem to pass very slowly. Newton thought that time and space are the stage on which the processes of the world took place. Einstein showed that space and time are intermixed and that measurements of them cannot be taken for granted. Space and time are part of the drama, not just the stage.

Despite the progress made in the understanding of how space and time behave, there is still the deep mystery about what time is. Physicists believe that all complex objects in the universe can be explained in terms of the behaviour of the very small objects of which they are composed (e.g. the kinetic theory of gases – see spread 7.6). This idea is known as **reductionism** – everything can be reduced to the actions of its parts. The reductionist method of doing science has been very successful, but it has its problems and one of them is to do with the nature of time.

Difficulties with reductionism

Studying large-scale (macroscopic) objects in terms of their very small (microscopic) components is difficult to do. A cup of tea contains about 1×10^{24} molecules. To understand the motions of all these molecules would mean keeping track of 6×10^{24} different variables – three coordinates for position and three components of velocity for each molecule! No computer could conceivably do this at present, so how can we ever hope to make progress? There are two ways of dealing with this problem.

- **Study macroscopic objects in macroscopic terms** i.e. do not worry about *why* the objects behave as they do, simply study *how* they behave. This was the attitude forced upon the scientists who first studied the properties of gases (see spreads 7.4 and 7.5). The science of **thermodynamics** has developed powerful theoretical tools for studying systems in this way. Ultimately this is unsatisfying. Scientists always want to take the thing apart to see how it works.

- **Study the motions of the microscopic components statistically** The mathematical techniques of statistics become more reliable with large numbers, and there are certainly large numbers of atoms and molecules. This method is known as **statistical thermodynamics** and it has been highly successful. The key insight is that large-scale objects, when isolated from the rest of the universe (e.g. by putting a gas in a flask), will over time settle into a state of **equilibrium** in which they are easier to study at the microscopic level.

Equilibrium and steady state

An insulated bar with a temperature difference between its ends will conduct energy along its length. This will heat up the cold end of the bar and cool down the hot end. If the situation continues, eventually all points along the bar will reach the same temperature and there will no longer be any net flow of energy along the bar. When this happens the bar is in **thermal equilibrium** – the state in which there is *no net transfer of energy between parts of the system* and as a result *all parts of the system are at a constant, equal temperature*. This will always happen provided no external interference with the system takes place – i.e. the system is isolated.

If some external action continually removes energy from the cold end of the bar and supplies energy to the warm end, then thermal equilibrium can never be reached. Energy will continue to flow down the bar in an effort to reach equilibrium, but to no avail. Points along the bar will reach a constant temperature, but they will not all be at the *same* temperature. In this circumstance the bar has reached a **steady state** characterized by a constant rate of energy flow and constant temperatures, but it is *not* in equilibrium.

All processes tend towards equilibrium (case 1). If the equilibrium is continually disturbed by an outside influence (case 2) then a steady state can be reached instead.

The arrow of time

An isolated system *always* approaches equilibrium. Equilibrium is a state in which nothing is going on, *on average*. Physically speaking, living systems are always out of equilibrium. When a leaf dies, falls to the ground, and decays it is approaching equilibrium with its environment (the forest). This natural movement towards equilibrium provides a definite direction for the flow of time. Some scientists have speculated that our perception of time moving forwards is due to our brains approaching equilibrium as we age. This single direction of time, from the past to the future, is very puzzling. At the microscopic level all the laws of physics are time symmetric – they do not say in which direction time should move. If an electron collides with another electron then the equations governing the situation do not indicate which way time is moving. The collision could be filmed and played backwards, 'reversing the flow of time'; it would not look odd, but a film of a cricket match shown backwards would look very odd. Yet every time the bat hits the ball electron collisions are taking place. The challenge to reductionism is to explain the macroscopic experience of time in terms of the time-neutral microscopic physics.

Entropy

The equilibrium state of a system is the one in which the particles from which it is made are in the most random arrangement that is possible for that situation. In order to understand this consider a box that contains gas. The box has a partition. On one side there is a single molecule, on the other there is the rest of the gas. The partition is then removed. If you come back and observe the gas after some time, you would be very surprised to find it with all the molecules except one crowded at one end of the box. It is far more likely that the molecules will be evenly spread throughout the box. But why? As far as an individual molecule is concerned one place is as good as another. Every possible arrangement of molecules within the box is equally likely, but there are far more possible arrangements with the molecules evenly spread out than there are with one molecule out on its own. Hence the probability of finding an arrangement with an even distribution of molecules is far greater than the other possibilities.

The number of possible arrangements of a system is related to the **entropy**, S, of the system:

$$S = k \ln (W)$$

where k is Boltzmann's constant $= 1.380 \times 10^{-23} \, \text{J K}^{-1}$ and W is the number of arrangements.

Scientists use entropy to measure a system's progress towards equilibrium. With an isolated system the entropy can only stay the same or increase with time. If the system is not isolated, its entropy can decrease, but only at the expense of a greater increase in entropy in the environment it is linked to (you can always expand your view to include the environment in your study and to regard the whole thing as isolated). This is the **second law of thermodynamics**. On a microscopic level increasing entropy means increasing randomness. Time moves in the direction of increasing entropy.

A collision between elementary particles has no obvious time sense. It is not possible to tell what the sequence of events was by looking at this photograph.

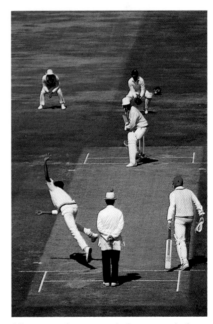

Macroscopic events do have an obvious time sense: you can tell what the sequence of events was by looking at the picture.

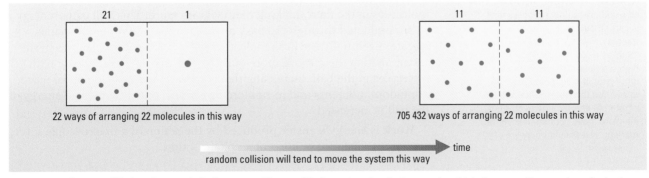

21   1

22 ways of arranging 22 molecules in this way

11   11

705 432 ways of arranging 22 molecules in this way

random collision will tend to move the system this way   time

The approach to equilibrium is a statistical process. The equilibrium situation is the one in which there are the most equivalent arrangements of the molecules within the system.

OBJECTIVES

- internal energy and total energy
- the first law of thermodynamics
- isothermal and adiabatic changes

A dictionary of thermodynamics

macroscopic variable a quantity that can be measured on a scale very much larger than atomic dimensions e.g. temperature, pressure, volume, mass, etc.

system any part of the universe that has been singled out for study

isolated system part of the universe that is prevented from exchanging energy with the rest of the universe (e.g. by putting it in a box that does not allow energy in or out)

state a specific situation in which a system's macroscopic variables have certain values (e.g. $P = 10$ Pa, $V = 100$ cm^3, and $T = 300$ K would be a state of a gas)

Random KE

The molecules in a cup of tea have KE due to their random thermal motion. If the cup of tea is on an aeroplane, then the molecules will also have KE due to their bulk motion with the plane. If the temperature does not change when the tea is moving it must be related to the random KE, hence the definition of the system at rest.

Helpful hint

Note that the word 'heat' is used as a verb, meaning a process that is carried out, not the name of a quantity. People often use the word 'heat' as a noun – e.g. water has a high temperature, so it contains a lot of heat. It would be more correct to say that the water contains a lot of internal energy. However, since the word 'heat' is often used in this manner, you should be aware of it.

THE FIRST LAW OF THERMODYNAMICS

You might not believe that there is enough similarity between a black hole and a bicycle pump to allow them both to be studied in the same way. There are many obvious differences. However, they both have definable temperatures and can both be investigated by the science of thermodynamics.

Thermodynamics can at first sight seem to be a rather abstract subject full of technical language, but it is one of the most fundamental in physics. It can apply to anything that we try to study, provided that it can be described in terms of a few macroscopic variables.

Internal energy

Consider a cup of tea. From the macroscopic point of view the system has volume, temperature, and pressure. From the microscopic point of view there are molecules vibrating, rotating, and drifting about the volume of the liquid. All these motions have energy. In addition, the molecules are weakly bonded to each other which gives them potential energy. The cup of tea is standing on a table and so is some height above the ground. This gives each of the molecules some gravitational potential energy. Finally the table could be on an aeroplane. The tea is then moving at considerable speed and all the molecules within it share in this motion. Totalling all these forms of energy gives the **total energy**, E, of the system:

$$E = \text{KE of molecules} + \text{PE (bonds)} + \text{PE (external forces e.g. gravity)}$$

Thermodynamics studies the properties of a system that are independent of its position and its motion. You would expect a cup of tea to have the same temperature when the plane is moving as when it is standing on the runway. For this reason the total energy in this system is often not as useful as the **internal energy**, U, which is defined as the sum of the kinetic energy and bond potential energies of the molecules of a system as measured when the system as a whole is not moving.

$$U = \text{KE of random molecular motion} + \text{PE (bonds)}$$

The internal energy is related to the total energy, but it is not the same:

$$E = U + \text{KE of collective molecular motion} + \text{PE (external forces)}$$

Heat and work

The first great advance made towards a modern understanding of thermodynamics was the realization that the temperature of a system could be altered either by heating or by doing work on the system. Heat was regarded as a form of fluid that flowed from hot objects (which held a lot of this 'fluid') to cold objects (which did not hold much 'fluid'). The modern understanding is that heat and work are ways of transferring energy from one object to another.

Heat is energy transfer without the action of a macroscopic force. Heating can only take place between objects at different temperatures.

An example is placing a ball in a flask of water which is then heated by a Bunsen burner flame. Fast-moving ions in the flame collide with molecules in the flask, making them vibrate more. This will cause energy to be conducted through the flask as more and more of the molecules vibrate. Eventually the flask molecules will collide with molecules in the water, making them move more quickly. These will in turn collide with molecules in the ball, increasing their speed. As the collisions take place at random moments and in random directions, only the internal energy of the ball is increased.

Work is energy transfer produced by the action of a macroscopic force, e.g. force × distance or current × voltage × time.

Acting on a the ball with a force can also increase its temperature. Squash players are familiar with having to hit the ball hard in order to warm it up. The molecules in a racket exert a force on the molecules of the ball. In this case the molecules are more coordinated in their motion – a macroscopic force is acting in a particular direction. The ball's molecules are squashed together as a result of the impact so their bond energy is increased. When the ball flies off the racket, it springs back into shape and the bond energy is converted into kinetic energy of the molecules. When this happens internal collisions soon randomize the motion of the molecules. The temperature increases. If the ball is simply thrown instead of being hit then the force acting increases the motion of all the molecules in the ball in a given direction, so the total energy increases but not the internal energy.

The first law of thermodynamics

The internal energy of a system can only change by exchanging energy with its surroundings, either by doing work or by heating.

change in internal energy = work done on system + heat

$$\Delta U = Q + W$$

If work is done on the system ($W > 0$), the internal energy *increases*. If the system is heated by the surroundings ($Q > 0$), U also *increases*. The first law expresses the equivalence of work and heat, as the change in internal energy (temperature rise) *is the same in both cases*.

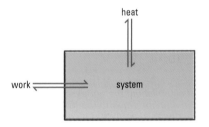

An *isothermal* change happens when the system changes its state, but remains at constant temperature.

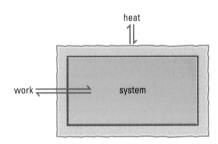

If heating is prevented then the system can only undergo *adiabatic* changes – which may *not* be at constant temperature.

Isothermal and adiabatic changes.

Isothermal change

An example is heating a gas and allowing it to expand; the energy supplied by heating is taken up in doing work against the surroundings when the gas expands i.e. $Q = -W$ and so $\Delta U = 0$, which for a gas implies no temperature change. The best way of ensuring that a change is isothermal is to do it slowly enough to allow the system to keep at the same temperature as its surroundings.

Adiabatic change

An example is squashing a gas rapidly e.g. in a bicycle pump with a finger over the end; the work done on the gas increases the temperature, and if this happens quickly then the gas has no time to lose this energy by heating the surroundings.

PRACTICE

1 Classify the following processes as either heating or doing work and say if the processes will increase the internal energy of the system:

kicking a football; stirring a cup of tea; warming a flask of water with an electrical element; causing an electrical heater to warm up by passing a current through it; striking a match.

2 A gas is compressed by a force that does 20 J of work on the gas. It is then allowed to cool down when 15 J of energy is conducted away. What is the internal energy change of the gas?

3 A gas is allowed to expand at constant temperature. As a result the gas does 30 J of work against the surroundings. What is the change in internal energy of the gas? How much energy enters or leaves the gas as heat?

4 A stream of water flows towards a waterfall. Compare the internal energy and the total energy of a volume of water at the following places:

a in the stream some way upstream from the waterfall;

b as the water is falling;

c when the water has hit the pool at the bottom of the waterfall.

TEMPERATURE SCALES

Heat is the transfer of energy without the action of a macroscopic force. This always involves a temperature difference. Our senses are not always very good at telling us which objects have different temperatures. For example, an iron gate on a frosty morning will seem colder than a wooden gate in the same circumstances. In fact the two gates are probably at exactly the same temperature. The difference is that the iron gate conducts energy more effectively than the wooden gate, so when you touch the iron gate it conducts energy away from your hand at a faster rate than the wooden gate. This is why it seems colder. You are reacting to the sensation of energy loss, not to the temperature of the gate.

Temperature is a macroscopic quantity that can easily be measured using a thermometer. There is a wide variety of different types of thermometer available, but they all rely on the same basic working principle.

Reminder

Energy flows from a region of high temperature to a region of lower temperature. If the temperatures are equal then the two regions will be in thermal equilibrium.

The zeroth law of thermodynamics

If A is in equilibrium with B, and B is in equilibrium with C, then A and C must be in equilibrium.

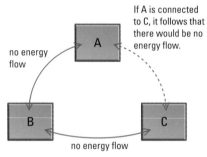

If A is connected to C, it follows that there would be no energy flow.

The zeroth law relates systems in equilibrium. If connecting B to A does not produce an energy flow, and neither does connecting B to C, then we can deduce that A and C would also be in equilibrium.

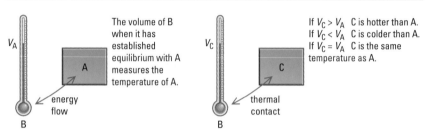

The volume of B when it has established equilibrium with A measures the temperature of A.

If $V_C > V_A$ C is hotter than A.
If $V_C < V_A$ C is colder than A.
If $V_C = V_A$ C is the same temperature as A.

Without the zeroth law there would be no way of establishing the temperature of a system by using a reference – objects that were in equilibrium with the thermometer (they both have same reading) would not necessarily be in equilibrium with each other (i.e. they could have different temperatures).

A thermometer is an object which has a property that depends on temperature (for example, volume). In the figure above, B represents a thermometer. When B is linked to A, energy will flow until they come into equilibrium, at which point the volume of B will have settled to a certain value. This value can be used as an indication of the temperature of A. If B is now put in contact with another object, C, and you observe how its volume changes, you can compare the temperatures of A and C. If the volume of B decreases, then C is at a lower temperature than A. If the volume of B increases, then C is at a higher temperature than A. If the volume stays the same then A and C are at the same temperature. The zeroth law ensures that you always get the same comparative values for A and C, no matter what you choose to use as the thermometer.

Centigrade, Kelvin, and Celsius temperature scales

The centigrade temperature scale is based on dividing the interval between the **ice point** and the **steam point** (the temperatures at atmospheric pressure of freezing pure water and boiling pure water, respectively) into 100 equal parts. This is a temperature scale based on human choices. However, there is an absolute temperature scale that is partly independent of human choice. It is based on the fact that all gases have zero volume at the same temperature –273.15 °C (see spread 7.4). It is impossible to have a temperature less than this as the gases would then have to have negative volume! This provides one fixed reference point called the **absolute zero of temperature**. To complete such a scale either a second fixed point, or alternatively some way of fixing the size of the scale intervals, is needed.

Helpful hint

For this investigation to work you must make sure that system B – the thermometer – does not exchange much energy with the systems it is testing. The chosen system B would not be much use as a thermometer if it cooled down or warmed up the object it was testing. You must choose a system that has a macroscopic variable that changes a great deal with a small temperature change.

In 1967 the **triple point of water** (the temperature at which water, ice, and water vapour are in equilibrium), 0.01°C, was adopted as a reference temperature. Using this, the definition of the absolute, or Kelvin, temperature scale can be completed.

Absolute temperature scale

The triple point of water is defined as 273.16 kelvin (K).

1 degree on absolute temperature scale = 1/273.16 of the temperature of the triple point of water

The Celsius temperature scale is defined relative to the absolute scale:
temperature on Celsius scale = temperature on Kelvin scale −273.15

The meaning of absolute temperature

For an ideal gas (one in which there are no forces between molecules) the temperature of the gas is related to the average kinetic energy of the molecules in the gas. This would mean that at absolute zero (0 kelvin) the molecules had no kinetic energy – they would be stationary. Quantum mechanics shows that this is an impossible situation. There would always be some fluctuating kinetic energy. For a solid, liquid, or gas in which there are evident forces between the molecules (i.e. a gas that is at a low enough temperature to be liquefied) it is more difficult to define what is meant by temperature. The idea of entropy must be employed.

The entropy of a system is related to the number of ways in which molecules can be arranged within the system. This counting of arrangements must include the ways in which the system's energy can be divided amongst the molecules. At high temperatures there are a large number of possible arrangements, at low temperatures the number of arrangements open to the system is lower. It is possible to define the temperature of a system from:

$$\frac{1}{kT} = \frac{\Delta \log W}{\Delta U}$$

where W is the number of arrangements, U is the internal energy, and k is the Boltzmann constant (see spread 7.7). In most cases, this reduces to the simple situation where temperature is related to the mean internal energy of the molecules. Note that two different systems with the same internal energy need not be at the same temperature – one system may have more arrangements open to it than the other. If this all makes temperature sound as if it is not a 'thing', remember that our sensations of 'hot' and 'cold' are actually more to do with the conduction of energy – they are not a direct sensing of temperature!

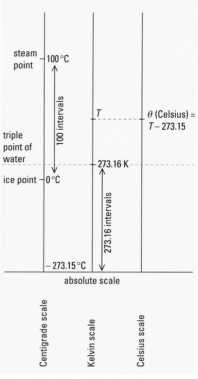

Comparing scales.

Absolute zero

At absolute zero any substance would have minimal internal energy (i.e. only quantum fluctuations would remain). The laws of physics prevent us ever achieving absolute zero, but the absolute temperature scale does provide a scale independent of the properties of any substance.

PRACTICE

1 A mercury thermometer has no markings on its scale. When placed in a flask of iced water the length of the mercury column is 3.0 cm. When placed in a flask of boiling water, the length of the column is 30 cm. What will be the length of the column when placed in a flask of water at 40 °C? What is the temperature of a flask of water when the column length is 15 cm?

2 A thermocouple is a type of thermometer that produces an e.m.f., E, that is related to temperature, θ, by

$$E = a + b\theta$$

where a and b are constants depending on the metals involved. Use the following data to obtain values of a and b from a suitable graph:

Temp. (°C)	10	20	30	40	50	60	70	80
e.m.f. (mV)	0.29	0.69	1.09	1.49	1.89	2.29	2.69	3.09

3 A platinum resistance thermometer uses the change in resistance of a platinum wire to measure temperature. The resistance of the wire is given by

$$R = R_0 (1 + a\theta + b\theta^2)$$

where R_0, a, and b are constants and θ is the temperature in °C.

Given that $R = 10\,\Omega$ when $\theta = 0\,°C$, $R = 14\,\Omega$ when $\theta = 100\,°C$, and $R = 24\,\Omega$ when $\theta = 420\,°C$,

a calculate R_0, a, and b;

b sketch the graph of R against θ for $0\,°C < \theta < 500\,°C$.

- gas laws

- definition of an ideal gas

- thermodynamic temperature scale

IDEAL GAS BEHAVIOUR

Individual gases behave in very different ways. It would be a great advantage to physicists if the physical properties of gases were totally independent of their chemical properties. Chlorine (Cl_2) is a highly reactive gas and argon (Ar) is very unreactive, but under certain conditions their physical properties – pressure, temperature, volume, and mass – relate to each other in exactly the same manner. When gases start to lose their individuality they are said to be behaving in an **ideal manner**. All gases will behave ideally under the right conditions, making life considerably easier for the physicist who can then talk about the physics of gases without having to specify which gas is being referred to. On the microscopic level this shows that the physical properties of a gas are not related to the nature of the particles involved – particles of chlorine gas behave physically just like those of argon.

Boyle's law

In 1662 Robert Boyle discovered that the pressure and volume of a gas are related to each other in a manner that does not depend on the nature of the gas. If the temperature of the gas is kept constant then

$$\text{pressure (Pa)} \propto \frac{1}{\text{volume (m}^3)}$$

or

$$pV = \text{constant at constant temperature}$$

Boyle's law, as it is now known, works well for all gases provided that: (a) they are well above the temperature at which they start to condense, (b) the pressure is not too high, (c) the volume is not too small (the gas molecules must have room to move about and collide), and (d) no gas escapes.

Charles's law

In 1787 the French physicist Jacques Charles experimented with different gases, keeping their pressure constant and altering their temperature. He discovered that as the temperature increased, every gas expanded in proportion to the temperature increase. Using the volume of the gas at $0\,°C$ as a reference, he showed that for every degree rise in temperature each gas increased its volume by $1/273.15$ of the volume at $0\,°C$:

$$V = V_0 \left(1 + \frac{t}{273.15} \right)$$

This rule, which is now known as **Charles's law**, holds true as long as the gas is at a temperature well above that at which it starts to condense.

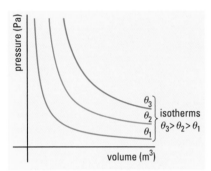

For an ideal gas, plotting pressure against volume at fixed temperatures produces curved isothermal lines as $p \propto 1/V$.

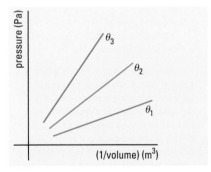

Plotting p against $1/V$ will give a straight line, showing that Boyle's law is being obeyed.

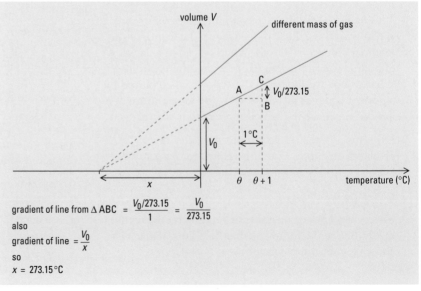

$$\text{gradient of line from } \triangle ABC = \frac{V_0/273.15}{1} = \frac{V_0}{273.15}$$

also

$$\text{gradient of line} = \frac{V_0}{x}$$

so

$$x = 273.15\,°C$$

Using Charles's law.

If every gas expands by the same proportion per degree temperature rise, then every gas line must intercept the temperature axis at exactly the same point: –273.15 °C. This fact suggests that there is an absolute temperature scale independent of the mass or nature of gas used in the experiment and with an absolute zero value of –273.15 °C.

Pressure law and thermodynamic temperature

The relationship between pressure and temperature, at constant volume, mirrors Charles's law: for every degree rise in temperature the pressure of the gas increases by 1/273.15 of its pressure at 0 °C. Between them these two laws relate easily measurable properties of a gas to its temperature. Consequently, they can be used as the basis of a thermometer.

Kelvin suggested that the product of the pressure and volume of an ideal gas could be used as a property to measure temperature. This system would be a way of defining a temperature scale. An ideal gas is placed in thermal contact with a system and allowed to come into equilibrium. The pressure and volume of the gas is then measured. After this, the gas is placed in thermal contact with water at its triple-point temperature and the pressure and volume measured again. From these measurements the thermodynamic temperature is defined by:

$$T(K) = \frac{(pV)_T}{(pV)_{tr}} \times 273.16$$

Effectively this is saying that the temperature is *defined* so that equal increases in pV produce equal increases in T. Interestingly, the temperature scale defined in this way turns out to be indentical to the absolute (Kelvin) scale defined earlier (unfortunately the proof of this is too advanced for this treatment).

What is an ideal gas?

In macroscopic terms, an ideal gas is best defined as one that obeys Boyle's law at all temperatures. In practice, real gases tend to deviate from ideal behaviour if the volume is very small or the temperature is too low. In microscopic terms an ideal gas is one for which the forces between the molecules (or atoms) in the gas can be neglected. This happens for all gases at high temperatures – the molecules are moving so quickly that the forces between them hardly deflect their motion and so can be neglected. At small volumes the molecules are more crowded together and so the forces, which change strength with distance, are greater. Of course there is no such thing as a genuinely ideal gas – such a gas could never be turned into a liquid without some forces between its molecules. Ideal gas behaviour is an approximation that can be successfully applied in the right conditions.

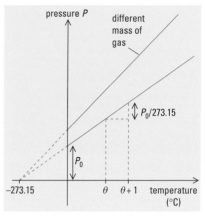

The pressure law.

PRACTICE

1 In an experiment the pressure p and volume V of a gas were measured at constant temperature. The following readings were obtained:

p/kPa	102	143	178	200	233
V/cm³	40.5	28.7	23.4	20.7	17.8

Plot a graph of **a** p against V and **b** p against $1/V$. Does the second graph show that Boyle's law is correct for this gas?

2 The pressure of air in a car tyre is 1.90×10^5 Pa at 18 °C. Assuming that the volume of the tyre is constant:

a What will the pressure of the tyre be at 25 °C?

b By how much does the pressure of the tyre increase for every °C rise in temperature?

3 A gas occupies a volume of 6.0×10^{-3} m³ and exerts a pressure of 80 kPa at a temperature of 20 °C. What pressure does it exert if, separately

i the temperature is raised to 40 °C;

ii the volume is halved;

iii the volume is changed to 7.7×10^{-3} m³ and the temperature becomes 57 °C.

THE PHYSICS OF IDEAL GASES

Avogadro's hypothesis

The table below shows the volumes of 1 mole of different gases at s.t.p.

Gas	Volume/cm³
hydrogen	22 432
helium	22 396
oxygen	22 392
carbon dioxide	22 263

These figures suggest that Avogadro's hypothesis works well for temperatures and pressures close to s.t.p.

Moles

The SI unit of number is the mole (mol) and is defined as being the number of $^{12}_{6}C$ atoms in 0.012 kg of pure carbon-12. This corresponds to 6.022×10^{23} molecules (known as the **Avogadro constant**, N_A). The ideal gas equation uses the molar gas constant R and the number of moles of gas.

The molar mass M is related to the relative molecular mass M_r:

$$\frac{M}{kg} = \frac{M_r}{1000}$$

$$\frac{\text{number of}}{\text{moles } n} = \frac{\text{total mass } m}{\text{molar mass } M}$$

Helpful hint

The gas equation can also be written in the form:

$$pV = NkT$$

where N is the number of molecules:

$$N = nN_A$$

and

$$k = \text{Boltzmann's constant} = \frac{R}{N_A}$$

γ

A more advanced treatment can establish that pV^γ is constant for an ideal gas undergoing an adiabatic change and will also show that γ is the ratio between the specific heat capacities of the gas. **Specific heat capacity** records the amount of energy required to increase the temperature of 1 kg of a substance by 1 K. For a gas the temperature increase can be achieved while allowing the gas to increase its volume or by constraining the gas to remain at a constant volume (in which case the pressure must change). This gives rise to two values of specific heat capacity, c_V (constant volume) and c_p (constant pressure). $c_p/c_V = \gamma$ and $c_p - c_V = R$.

The ideal gas equation

The definition of the thermodynamic temperature scale is constructed in such a way as to make the product of pressure and volume for an ideal gas proportional to its temperature. This is *not* an experimentally deduced relationship. It follows from the definition quoted earlier as

$$T = \frac{(pV)_T}{(pV)_{tr}} = 273.16 \quad \text{(definition of thermodynamic temperature)}$$

so

$$pV = T \times (pV)_{tr} = \text{constant} \times T$$

since the value of pV evaluated at the triple point of water is a fixed quantity.

In 1811 Amadeo Avogadro published a paper in which he suggested what is now known as **Avogadro's hypothesis** – that equal volumes of gas under the same conditions of temperature and pressure should contain the same number of molecules. The **molar mass** of a gas is the mass of one mole of that gas (which is by definition 6.022×10^{23} molecules) so that 1 mole of hydrogen (H_2) has a mass of 2 grams and 1 mole of oxygen (O_2) has a mass of 16 grams. If this is combined with Avogadro's hypothesis, 1 mole of any gas should always occupy the same volume at a given temperature and pressure. Generally the comparison is made at 0 °C and atmospheric pressure (a combination known as **standard temperature and pressure – s.t.p.**). If 1 mole of any gas at s.t.p. has the same volume then the constant in the equation $pV = \text{constant} \times T$ must have the same value for all gases at s.t.p. This is known as the **molar gas constant**, R:

$$pV = RT$$

where R has the value $8.31 \text{ J K}^{-1} \text{mol}^{-1}$. As the thermodynamic temperature has been defined by equal intervals of pV then a 1 K temperature rise is an increase of R in the value of pV. Hence the value of the constant in the relationship is R for all gases and all combinations of pressure, volume, and temperature. If the gas contains more than one mole of molecules then the relationship becomes

$$pV = nRT$$

with n being the number of moles. Sometimes a different form of the relationship is more convenient:

$$\frac{p_1 V_1}{T_1} = \frac{p_2 V_2}{T_2}$$

in which p_1, V_1, and T_1 represent the quantities before some change has taken place and p_2, V_2, and T_2 are the values after the change.

Isothermal and adiabatic changes

The ideal gas equation allows one to calculate any one of the temperature, volume, or pressure of a gas given the values of the other two. Furthermore, it relates the combination of these properties before the gas is subjected to some process to those after the process is complete. However, under certain circumstances more restrictive relationships can be applied. For example, if the temperature of the gas remains the same then Boyle's law applies and $pV = \text{constant}$. It is important to realize that this will still hold true if the temperature changes during the process, provided that by the end of the process the temperature is restored to the original value. This will be an isothermal change. If this is the case, then Boyle's law will apply.

If the gas is encased in a container that prevents thermal contact with its surroundings, then no heat can flow. Any process that is subsequently carried out on the gas can only produce an adiabatic change and invariably a change in temperature. Under these conditions Boyle's law will not apply, but interestingly a variation on the law does hold true:

$$pV^\gamma = \text{constant}$$

γ is different for different gases but is generally between 1.3 and 1.7.

Worked example 1

A fixed mass of an ideal gas has a volume of $200 \, \text{cm}^3$, a temperature of $31 \, °C$, and a pressure of 1 atmosphere. The gas has a γ value of 1.67. It is allowed to expand adiabatically until the volume is $300 \, \text{cm}^3$, after which it is kept at constant pressure until the temperature returns to $31 \, °C$. What is the final pressure and temperature after the adiabatic expansion and the final volume once the temperature is restored to $31 \, °C$?

During the adiabatic expansion pV^γ is a constant, so

$$pV^\gamma = (1 \, \text{atm}) \times (200 \, \text{cm}^3)^{1.67} = p \times (300 \, \text{cm}^3)^{1.67}$$

$$\therefore p = \frac{200^{1.67}}{300^{1.67}} = \left(\frac{200}{300}\right)^{1.67} = 0.508 \, \text{atm}$$

We can now apply the ideal gas equation (pV/T = constant) to calculate the temperature:

$$\frac{pV}{T} = \frac{1 \times 200}{(273 + 31)} = \frac{0.508 \times 300}{T_2}$$

$$\therefore T_2 = \frac{0.508 \times 300 \times (273 + 31)}{200} = 231.7 \, \text{K} = -41.3 \, °C$$

The next stage is performed at constant pressure, so Charles's law applies:

$$\frac{V}{T} = \frac{300}{231.7} = \frac{V}{(273 + 31)}$$

$$\therefore V = \frac{300 \times (273 + 31)}{231.7} = 394 \, \text{cm}^3$$

Work done on an ideal gas

Whenever a gas is compressed energy will be transferred to the gas. This is because the gas exerts a pressure on its container and the process of compressing the gas must involve reducing the size of the container against the outward force exerted by this pressure. Consider a gas in a cylindrical container that has one end wall capable of sliding back and forth (without any friction). A force slightly greater than that exerted by the gas pressure must be applied to the sliding wall in order to push it inwards. If the gas pressure is p and the area of the moving wall A then the force applied from outside must be

$$(p + \Delta p) \times A$$

where Δp represents the slight excess in the external pressure. Assuming that the sliding wall moves inwards a small amount Δx, the work done by the external force will be

work done = force × distance = $(p + \Delta p) \times A\Delta x \approx pA\Delta x = p\Delta V$

where ΔV is the change in volume of the cylinder – and hence of the gas.

This equation can be applied provided the pressure of the gas does not change during the compression. An expanding gas will do work on its surroundings, which can be calculated in the same manner.

If the pressure does not remain constant during the volume change then the work done can be calculated from either the area under a graph of p against V for the gas, or from an integral, provided there is a mathematical expression for how the pressure changes.

Worked example 2

Calculate the work done by an ideal gas when it expands by 2% at constant temperature (initial volume of gas = $0.6 \, \text{m}^3$; pressure = $10^5 \, \text{Pa}$).

work done = $p \, \Delta V$

$$= (10^5 \, \text{Pa}) \times (0.6 \times 0.02 \, \text{m}^3)$$

$$= 1.2 \, \text{kJ}$$

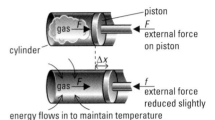

When a gas expands and pushes on a frictionless piston the gas does work.

Large volume changes

If the gas expands over a large volume change, then we can no longer assume that the pressure will remain constant even if the temperature does. Under these circumstances the work done must be found by using:

$$\text{work done} = \int_{V_1}^{V_2} p \, dV$$

$$= \int_{V_1}^{V_2} \frac{nRT}{V} \, dV = nRT \ln\left(\frac{V_2}{V_1}\right)$$

The situation is even more complicated if the temperature does not remain constant. (See also spread 7.10.)

Helpful hint

In a solid or liquid ΔV is small unless the pressure is enormous. In consequence W is small and so from $\Delta U = Q + W$, ΔU is generally provided by heating, $\Delta U \approx Q$.

PRACTICE

1 Calculate the volume of 1 mole of an ideal gas when the temperature is $0 \, °C$ and the pressure is $10^5 \, \text{Pa}$.

2 Oxygen gas has a volume of $1.0 \, l$ and a pressure of $10^5 \, \text{Pa}$ at $40 \, °C$. It expands to a volume of $1.5 \, l$ and then its pressure is $1.06 \times 10^5 \, \text{Pa}$.

   a How many moles of gas are present?

   b What is the final temperature of the gas?

3 A very good vacuum produced in a laboratory

corresponds to a pressure of about $10^{-9} \, \text{Pa}$. Estimate the number of molecules per metre cubed in such a vacuum at room temperature.

4 $100 \, \text{cm}^3$ of air at $0 \, °C$ and $10^5 \, \text{Pa}$ are compressed adiabatically to $20 \, \text{cm}^3$.

   a What is the new pressure?

   b What is the new temperature?

   (γ is 1.41 for air.)

• assumptions of kinetic theory

• collisions

• pressure of a gas in molecular terms

⟨ ⟩ means 'mean'

In this spread, angular brackets, ⟨ ⟩, stand for the average (mean) of a quantity. For example $\langle v^2 \rangle$ is the mean of v^2.

Outline kinetic theory

The pressure in a gas is determined by how hard a molecule hits the container walls, and how often. If you reduce the volume then it will hit the walls more often. Hence $p \propto 1/V$ at constant T. If you increase the temperature it will hit harder and more often, hence $p \propto T$. However, if you allow the volume to change the excess pressure will cause the volume to increase, reducing the number of collisions per second until the pressure goes back to the initial value (the external pressure). Hence $V \propto T$ at constant pressure.

THE KINETIC THEORY OF GASES

What is kinetic theory?

Kinetic theory explains the properties of gases in terms of the motions of the particles that make up the gas. It is an attempt to explain macroscopic behaviour in terms of microscopic motion. The number of particles involved is so large that keeping track of each particle's individual motion is impossible. The theory relies on the average behaviour of all the molecules being similar.

In order for the theory to work the properties of gases have to be related to aspects of the motion of the particles:

Pressure of the gas: caused by the particles of the gas colliding with the walls of the container.

Volume of the gas: total volume of space in which the particles are free to move about.

Mass of the gas: total mass of the particles.

Assumptions

The kinetic theory of gases is the simplest example of a complete theory of physics that can be studied at this level. It depends on the following set of assumptions.

• Gases are composed of spherical molecules that cannot be broken up.

• A very large number of such molecules exist in a small volume of the gas.

• The volume of the molecules is very much smaller than the volume of the gas.

• Collisions between molecules or between molecules and the walls of the container are elastic.

• No long-range forces exist between molecules (there must be short-range forces when they collide), so the molecules travel in straight lines between collisions.

• The time the molecules spend in contact with each other, or in contact with the wall, during a collision is very small compared to the time between collisions.

The justification for these assumptions is that their consequences can be fully worked out and found to agree with experiments.

Collisions with walls

Using the above assumptions as a starting point it should be possible to calculate the average effect of having a large number of molecules colliding with the walls of the container.

The diagram to the left is a force–time graph for one of the walls. The sharp spikes are the individual molecules rebounding off the wall. The molecules are much less massive than the wall so it is safe to assume that the wall 'soaks up' a negligibly small amount of the molecules' energy – i.e. the wall does not recoil from the collision.

The area under each spike is the impulse exerted by the molecule on the wall. Molecules remain in contact with the wall for a very short period of time – much shorter than the time delay between collisions (as assumed). Any measurement of the gas pressure will be working on a time scale very much longer than either the duration of a collision or the time between collisions. Consequently, it is reasonable to assume that the pressure measured will be the result of a great number of molecular collisions. The average impulse exerted on the wall over a time period can be found by taking the total area under all the spikes and spreading that area uniformly over the graph.

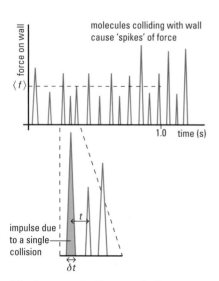

The force exerted by a gas is the average of the forces due to individual molecular collisions.

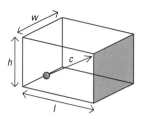

Calculating pressure

Consider a rectangular box as shown in the diagram to the right.

Initially, the box contains a single molecule of mass m and velocity c. That velocity can be split into three components parallel to the sides of the box:

$$c = v_x + v_y + v_z$$

or in terms of magnitudes $\quad c^2 = v^2{}_x + v^2{}_y + v^2{}_z$

When this molecule strikes the shaded wall it rebounds elastically (the wall has negligible recoil). The x and z components of its momentum are unaffected as they are parallel to the wall. Only the y component is reversed.

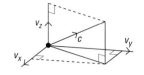

$$\text{initial } y \text{ component of momentum} = mv_y$$
$$\text{final } y \text{ component of momentum} = -mv_y$$
$$\therefore \text{ change in momentum} = -2mv_y$$

This is the impulse of the wall on the molecule. By Newton's third law of motion, the impulse of the molecules on the wall is the same size but positive. These impulses are the spikes in the previous graph.

If the duration of the collision, δt, could be measured, the force exerted, f, could be obtained from

$$f\delta t = 2mv_y$$

As δt is very small the individual forces are quite large.

After colliding with the shaded wall the molecule proceeds to rebound off the other walls until it strikes the shaded wall again. The time taken for it to complete this path back to the shaded wall is

$$t = \frac{2l}{v_y}$$

In the time T over which the pressure is measured, the number of times this molecule will collide with the wall is given by T/t, so

$$\text{number of collisions} = \frac{Tv_y}{2l}$$

$$\therefore \text{ total impulse on the wall in this time} = \left(\frac{Tv_y}{2l}\right) \times 2mv_y$$

$$= \frac{Tmv^2{}_y}{l}$$

This is the impulse due to an average force $\langle f \rangle$ acting over the whole period T:

$$\frac{Tmv^2{}_y}{l} = \langle f \rangle T$$

$$\therefore \text{ average force on the wall, } \langle f \rangle = \frac{mv^2{}_y}{l}$$

If the box is now considered to contain N molecules, each of these will be colliding with the wall and contributing to the total average force

$$\langle F \rangle = \frac{Nm\langle v^2{}_y \rangle}{l}$$

Note that you are now obliged to consider the average $v^2{}_y$ as the molecules will all be moving at different speeds. However, this average $v^2{}_y$ can be related to the average c^2 of the molecules:

$$c^2 = v^2{}_x + v^2{}_y + v^2{}_z$$

$$\text{so} \quad \langle c^2 \rangle = \langle v^2{}_x \rangle + \langle v^2{}_y \rangle + \langle v^2{}_z \rangle$$

$$= 3\langle v^2{}_y \rangle$$

as $\langle v^2{}_y \rangle = \langle v^2{}_x \rangle = \langle v^2{}_z \rangle$ (otherwise the molecules within the box would all be drifting in one direction on average).

Using this gives

$$\langle F \rangle = \frac{Nm\langle c^2 \rangle}{3l}$$

The pressure on the wall is the force divided by the area, $w \times h$.

$$\therefore \quad p = \frac{Nm\langle c^2 \rangle}{3lhw} \quad \text{and} \quad pV = \frac{1}{3}Nm\langle c^2 \rangle$$

as the volume of the box V is lhw.

In its final form the relationship

$$pV = \frac{1}{3}Nm\langle c^2 \rangle$$

is true for all walls of the box, and for any shape of box.

The expression can also be written as

$$p = \frac{1}{3}\rho\langle c^2 \rangle$$

where ρ is the density of the gas.

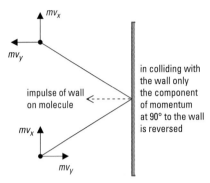

Box from above

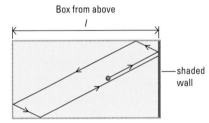

As the molecule collides with the other walls the y component of the velocity is unchanged. The only distance that matters is 2l, so the time taken between hits on the shaded wall is $2l/v_y$.

PRACTICE

1 Show that the pressure of a gas is related to its density, ρ, by

$$p = \frac{1}{3}\rho\langle c^2 \rangle$$

2 The root mean square (r.m.s.) speed of molecules in a gas is defined as $\sqrt{\langle c^2 \rangle}$.

Atmospheric pressure is 1.0×10^5 Pa and its density is 1.2 kg m^{-3}. Estimate the root mean square speed of the molecules in air.

3 A box is cube-shaped with a side length of 20 cm. Inside is air at 1.0×10^5 Pa and 1.2 kg m^{-3}. Given that the average mass of air molecules is 4.8×10^{-26} kg, estimate the number of molecules in the box.

LINKING TEMPERATURE TO KINETICS

O B J E C T I V E S

• temperature of an ideal gas

• molecular meaning of temperature

• internal energy of an ideal gas

• equipartition theorem

Models

Every model has a range of applicability outside which it does not work. The kinetic theory of gases is a model based on the assumptions set out in spread 7.6. One of the key assumptions is the lack of long range forces between molecules. If this were really the case a gas would never liquefy. Hence the model works well at high temperatures, far from the condensation point of a gas. The model also falls down if the container volume is very small so that the volume of the molecules becomes significant.

Evidence for kinetic theory

Kinetic theory took a long time to become accepted, as the molecues of which gases are supposedly made are far too small to be seen. The turning point came surprisingly late. In 1905 Einstein published a paper that explained how Brownian motion could be analysed using the kinetic theory. Few people realize that it was this work (along with the explanation of the photoelectric effect) that won him the Nobel Prize – not his work on relativity!

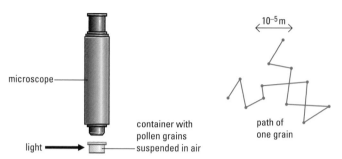

Brownian motion was first observed by Robert Brown in 1827. The pollen grains moved about with a random jerky motion as if being pushed about by other objects that were too small to be seen in the microscope.

Explaining Brownian motion in molecular terms was a problem before Einstein. At first glance you would think that, as the pollen grains are being struck on all sides by molecules moving with the same speed, the grain should stay still. Einstein showed that as the molecules had the same *average speed*, their effect would only cancel *on average*. The pollen grain should jitter about staying roughly in the same place. He was able to calculate how the pollen grain's motion should look, a result that paved the way for the general acceptance of the existence of atoms and, with that, the kinetic theory.

Kinetic theory of temperature

The microscopic theory of ideal gases leads to the equation first established by R. Clausius in 1857:

$$pV = \frac{1}{3}Nm\langle c^2 \rangle \tag{1}$$

The macroscopic study of the behaviour of ideal gases in the laboratory leads to the ideal gas equation in the form

$$pV = nRT \tag{2}$$

If both pieces of physics are correct then it must be possible to establish a link between them. This was partially done in spread 7.6 where molecular interpretations of pressure, volume, and mass were suggested. The pressure relationship has been quantified by Clausius's equation, but the molecular interpretation of temperature is still rather vague. However, combining equations 1 and 2 leads to

$$nRT = \frac{1}{3}Nm\langle c^2 \rangle$$

Now, the number of molecules, N, is equal to the number of moles of gas, n, times the number of molecules in one mole (the Avogadro number, N_A).

$$\therefore \quad nRT = \frac{1}{3}nN_Am\langle c^2 \rangle$$

$$\therefore \quad RT = \frac{2}{3}N_A(\frac{1}{2}m\langle c^2 \rangle)$$

$$= \frac{2}{3}N_A\langle KE \rangle$$

Helpful hint

The ideal gas equation is only correct for a gas in thermal equilibrium, so the link that is being made is only strictly applicable in that situation.

Temperature and energy

Note the important consequence that at absolute zero of temperature the kinetic energy of molecules must be zero, i.e. the molecules are not moving.

So the average kinetic energy of molecules in the gas is given by

$$\langle \text{KE} \rangle = \frac{3RT}{2N_A} = \frac{3}{2}kT$$

where k is Boltzmann's constant, $k = R/N_A = 1.380 \times 10^{-23}\,\text{J}\,\text{K}^{-1}$.

Temperature in other systems

Linking the molecular theory to the observed properties at the macroscopic level has led to a connection between the temperature of the gas (a property measured on a thermometer) and the average kinetic energy of the molecules within the gas. This connection is only strictly true for an ideal gas, but it can be extended to all macroscopic objects:

absolute temperature $\propto$ average internal energy per molecule

Internal energy of an ideal gas

Calculating the internal energy of an ideal gas is quite straightforward. The only energy to worry about is the kinetic energy of the molecules, as there are no long-range forces acting (hence no potential energy):

internal energy (U) = total KE of molecules in the gas

= N × average KE

$$= \frac{3}{2}NkT = \frac{3}{2}nRT$$

This is an important result: the internal energy of an ideal gas depends only on the absolute temperature of the gas.

Any change in an ideal gas that is carried out isothermally (i.e. at constant temperature) must also be at constant internal energy. If work is done isothermally on an ideal gas, then the gas must lose the same amount of energy by heating the surroundings.

Equipartition of energy

The relationship between the average kinetic energy of molecules in an ideal gas and the temperature, $\langle \text{KE} \rangle = \frac{3}{2}kT$, is an example of a much more general and powerful theorem called the **equipartition of energy**.

This theorem states that, with a system in thermal equilibrium, any form of energy that depends on some quantity *squared* will have an average value of $\frac{1}{2}kT$.

For example, kinetic energy depends on velocity squared so the theorem applies. However, there are three components of velocity to worry about:

$$\left\langle \tfrac{1}{2}mv^2 \right\rangle = \left\langle \tfrac{1}{2}mv_x^2 \right\rangle + \left\langle \tfrac{1}{2}mv_y^2 \right\rangle + \left\langle \tfrac{1}{2}mv_z^2 \right\rangle = \tfrac{1}{2}kT + \tfrac{1}{2}kT + \tfrac{1}{2}kT = \tfrac{3}{2}kT$$

As another example, the potential energy of bonds between the atoms in a molecule depends on the bond length squared (roughly), so in thermal equilibrium we expect to find the atoms vibrating about their bonds with average KE = $\frac{1}{2}kT$.

Temperature and molecules

People often say things like 'when a gas is heated the molecules within it are moving more quickly as they are hotter'. The molecular interpretation of temperature should make it quite clear that temperature is a property that can only be understood for a *collection of molecules* – the temperature of a *single molecule* has no meaning.

Internal energy of a real gas

In the case of a real gas the internal energy depends on the absolute temperature and the volume of the gas. There are forces acting between the molecules and so potential energy needs to be included in the internal energy. If the volume of the gas is reduced then the molecules get nearer to each other and the potential energy changes. Hence U depends on T and V.

vibrational motion = $\frac{1}{2}kT$

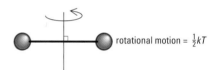
rotational motion = $\frac{1}{2}kT$

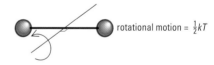
rotational motion = $\frac{1}{2}kT$

A dumb-bell shaped molecule can have three types of motion in addition to its transitional motion through the gas. In thermal equilibrium each of them has the same average kinetic energy. Rotational motion about the axis of the bond contributes a negligible amount of energy as the moment of inertia is small about this axis.

PRACTICE

1 What is the internal energy of 1 mole of gas at room temperature? What is the average kinetic energy of the molecules in this gas?

2 How much energy does it require to increase the temperature of 1 mole of gas by 10 °C?

3 The r.m.s. speed of the molecules in a gas is 600 m s⁻¹ and the mass of the molecules is 4.6×10^{-26} kg. What is the temperature of the gas in degrees Celsius?

4 Two moles of gas are cooled from 40 °C to 10 °C. How much energy is extracted from the gas?

5 A certain gas molecule has six degrees of freedom (three translational motion, two rotational motion, and a bond vibration).

   a What is the internal energy of 1 mole of this gas at 30 °C?

   b How much energy is needed to raise the temperature of 1 mole of this gas by 1 °C?

   c How could a simple experiment distinguish between such a gas and one that has only three degrees of freedom?

- molar heat capacity
- specific heat capacity
- measuring specific heat capacity

SPECIFIC HEAT CAPACITY

Adding the same amount of energy to different objects does not necessarily bring about the same temperature rise. Adding 1000 J of energy to 1 kg of water will increase its temperature by 0.26 °C. Adding 1000 J to 1 kg of iron will increase its temperature by 10 °C. The materials have different **heat capacities**, the physical property that determines the amount of energy required to bring about a temperature change.

Molar heat capacity

Conduction and convection (if it is a fluid) within the object will ensure that any energy added is evenly shared between the molecules. As temperature measures the average energy per molecule,

temperature change, $\Delta T \propto \dfrac{\text{internal energy change}}{\text{number of molecules}}$

or

internal energy change = constant × no. of molecules × temperature change

$$E = k \times N \times \Delta T$$

where E is the amount of energy supplied or removed

Now, for 1 mole of a substance the number of molecules is defined to be the Avogadro number, N_A, or 6.02×10^{23}. In order to achieve a certain temperature change in n moles of a substance an energy change must be produced given by

$$E = k \times n \times N_A \times \Delta T$$
$$= n \times c_M \times \Delta T$$

where c_M is the **molar heat capacity** – the amount of energy needed to raise the temperature of one mole of a substance by 1 K – and n the number of moles.

The Dulong and Petit law

Interestingly, experimental measurements of the molar heat capacity of a variety of different substances show them to be remarkably similar. This is known as the **Dulong and Petit law**. Specifically, the molar heat capacity of all substances at high enough temperature is $25 \, \text{J K}^{-1} \, \text{mol}^{-1}$. For most solids the law is true at room temperature. The equipartion of energy theorem (see spread 7.7) goes some way towards explaining this rule. Each molecule in the lattice of a solid has three degrees of freedom for vibrational kinetic energy as well as three degrees of freedom for bond potential energies. This is a total of 6 degrees of freedom per molecule.

$$\therefore \text{ mean energy per molecule} = 6 \times \tfrac{1}{2}kT = 3kT$$

$$\therefore \text{ total energy in 1 mole of molecules} = N_A \times 3kT$$

So, to achieve a temperature change ΔT in 1 mole, the size of the energy change is $3N_A k \Delta T$ – making the molar heat capacity equal to $3N_A k$. As $k = R/N_A$ the molar heat capacity is also equal to $3R = 3 \times 8.31 \, \text{J K}^{-1} \, \text{mol}^{-1} = 24.93 \, \text{J K}^{-1} \, \text{mol}^{-1}$, in remarkable agreement for such a simple theory.

Specific heat capacity

In some circumstances it is more convenient to deal with quantities specified by mass rather than number of moles. The **specific heat capacity** is the amount of energy required to raise the temperature of 1 kg of a substance by 1 K. This can be obtained from the molar heat capacity. The number of moles in a sample of mass M is M/mN_A where m is the mass of one molecule:

$$E = nc_M \Delta T = Mc_M \Delta T / mN_A$$

so

$$E = Mc_S \Delta T$$

with the specific heat capacity ($\text{J K}^{-1} \, \text{mol}^{-1}$), c_S, being c_M/mN_A and so different for different samples of materials of the same *mass*.

Heat or work

In the case of a solid or liquid, it is difficult to exchange energy as work. Solids and liquids are resistant to volume changes. Hence for them it is most likely that:

$$Q = nc_M \Delta T$$

Helpful hint

The specific heat capacity is mis-named as it tends to suggest that the object has a capacity for absorbing heat – as if heat were a thing.

Also, the units of c_S suggest that the temperature change should be put in kelvin. However, temperature *changes* are the same in degrees Celsius or kelvin. It is a common mistake to calculate the temperature change in degrees Celsius and then to add 273 to get the change in kelvin!

Specific heat capacity

The specific heat capacity of a substance can be measured at constant volume or at constant pressure. This gives two different answers! When measuring at constant pressure you are allowing the substance to expand – which requires it to do work on the surroundings. Consequently a greater amount of energy must be provided to achieve a temperature change.

The heat capacity of a substance, defined as $M \times c_S$ ($J\,K^{-1}$), is sometimes used instead of the specific heat capacity.

Worked example 1

How much energy must be added to a 1.6 kg aluminium wok in order to heat it to the temperature required to cook a good stir fry, say about 80 °C? (Specific heat capacity of aluminium is $510\,J\,kg^{-1}\,K^{-1}$.)

Assuming that the wok starts at 20 °C, the energy required is

$Q = mc_S\Delta T$

$Q = (1.6\,kg) \times (510\,J\,kg^{-1}\,K^{-1}) \times (80\,°C - 20\,°C) = 49\,kJ$ (2 sig. figs)

Worked example 2

The heater for my morning shower has a power of 8 kW. It raises the temperature of the water from 18 °C to 34 °C. What mass of water passes through the shower per second? (Specific heat capacity of water is $4200\,J\,kg^{-1}\,K^{-1}$.)

The shower can provide 8000 J of energy per second

$Q = mc_S\Delta T$

$\therefore m = \dfrac{Q}{c_S\Delta T} = \dfrac{8000\,J}{(4200\,J\,kg^{-1}\,K^{-1}) \times (34\,°C - 18\,°C)} = 119\,g$ per second

Measuring specific heat capacity

For metals with good thermal conductivities (such as copper, aluminium, brass, etc.) a simple electrical method can be used to measure the specific heat capacity. A similar method can be used for a liquid, but then it is important to continually stir the liquid to ensure a uniform temperature throughout. The principle is illustrated on the right. The ammeter and voltmeter enable the power of the heater to be measured and the thermometer allows periodic measurements of the temperature rise. A graph of temperature against time will have a linear section, the gradient of which gives the specific heat capacity of the material. In most cases the gradient will decrease at higher temperatures as the rate of heat loss to the surroundings (which depends on the temperature difference between the sample and the surroundings) increases. Consequently it is important to minimize heat losses by using a lid (in the case of a liquid), and providing thermal insulation (such as expanded polystyrene). Some liquid or oil in the hole for the thermometer ensures a satisfactory thermal constant with a metal block.

The energy in from the heater is $Q = I \times V \times t$, which produces a temperature rise given by $Q = mc_S\Delta T$, assuming no energy loss.

$\therefore IVt = mc_S\Delta T$

If the block starts at a temperature T_{room} and is at a temperature T some t seconds later,

$(T - T_{room}) = \dfrac{IV}{mc_S}t$

$\therefore T = \dfrac{IV}{mc_S}t + T_{room}$

The graph will have a gradient of IV/mc_S.

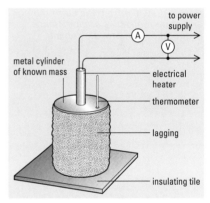

Measuring the specific heat capacity of a metal block. The lagging and insulating tile are to minimize energy loss to the surroundings.

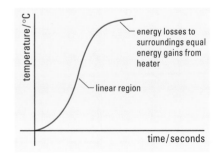

The gradient of the linear section gives the specific heat capacity of the material.

PRACTICE

1 A metal cylinder of mass 200 g and specific heat capacity $500\,J\,kg^{-1}\,K^{-1}$ is being warmed at a constant rate by an electrical heater of power 40 W. The block is well-lagged to prevent any energy loss to the surroundings. Calculate the rate at which the temperature of the block is increasing.

2 The lagging is removed from the block in question 1 and the heater turned off. The temperature of the block starts to fall at an initial rate of $3.0\,K\,min^{-1}$. Calculate the rate of energy loss to the surroundings.

PHASE CHANGES

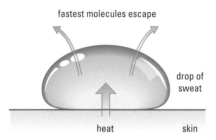

Cooling by evaporation: as the fastest molecules escape, the liquid cools. Heat is drawn in from the surroundings and more molecules escape. Hence evaporation can provide cooling.

Energy losses when objects are heated

There are various reasons why the energy calculated from the specific heat capacity might bring about a different temperature change to that calculated.

• Energy is lost to the surroundings by conduction, convection, and radiation. The rate at which the object loses energy to the surroundings is determined by the temperature difference between the object (θ_O) and the surroundings (θ_S).

conduction	energy loss rate	$= kA(\theta_S - \theta_O)/l$
radiation	energy loss rate	$= \sigma A(\theta_O^4 - \theta_S^4)$
convection	energy loss rate depends on whether the air is forced to flow (by a fan for example) or moves by natural convection	

The total effect of these three processes is to give a rate of energy loss per second that is proportional to the temperature difference between the object and its surroundings. This is known as **Newton's law of cooling**.

• The object may be a liquid contained in a flask (or similar), in which case some of the energy will be absorbed by the container, causing its temperature to rise.

• If the object being heated is a fluid, then there will be some evaporation of the fluid.

• The object might go through a phase change – i.e. a solid might melt to a liquid, or a liquid might boil into a gas. Melting involves breaking the strong bonds between particles in a solid, allowing them freer but localized motion in the liquid. Boiling breaks the weaker bonds between particles in the liquid, resulting in much greater freedom of motion in the gas.

The phases of matter

There are six distinct forms, or **phases**, in which matter can be found:

solid	strong molecular bonds and quite rigid shape
liquid	weaker molecular bonds and capable of flow
gas	virtually no molecular bonds and free flow
plasma	a gas that has been fully ionized
nuclear matter	as found in a nucleus, strong interparticle bonds and very high density ($10^{18}\,\text{kg m}^{-3}$)
quark matter	hypothetical matter composed of subatomic particles (quarks), extraordinarily high density ($10^{21}\,\text{kg m}^{-3}$)

The last three are far from ordinary states and will not be experienced in everyday life!

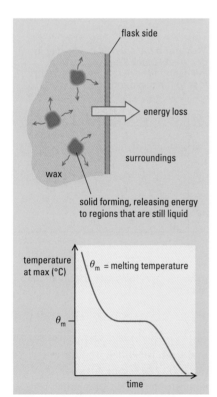

At the melting point of an object the temperature remains constant while bonds form. If the object is being heated, then the process happens in reverse. It is characteristic of phase changes that they take place at a specific temperature.

Latent heat

Whenever matter changes from one phase to another some energy is either released or absorbed. For example, if a test tube of molten wax is allowed to cool and a graph of its temperature against time plotted, then during the time when the wax is solidifying the temperature will remain largely constant.

This makes it appear as if the wax has stopped losing energy to its surroundings while it is solidifying. In fact the energy loss continues at the same rate as just before solidification started. The energy is being replaced in the solidification process. What happens is that at various points throughout the liquid wax the temperature drops sufficiently to allow the molecules to start to bond together. As bonds are formed energy is released which warms up the liquid surrounding the solid as it is forming.

For the period of time that the solid is forming out of the liquid the two phases remain in equilibrium. Eventually not enough energy is being released to compensate for the loss to the surroundings and the temperature starts to drop again.

The amount of energy released (absorbed) when a phase change takes place is the **latent heat** of the sample of material:

energy released or absorbed = latent heat (J)

$$= \text{mass (kg)} \times \text{specific latent heat (J kg}^{-1})$$

or

$$E = mL_S$$

Worked example

What is the energy released when 2 g of steam at 100 °C condenses into water at the same temperature?

$E = mLS$

$= (0.02 \, \text{kg}) \times (2.3 \times 10^6 \, \text{J kg}^{-1}) = 46 \, \text{kJ}$

Measuring latent heat

Any experiment to measure the energy required to perform a phase change will be made considerably easier if energy losses to the surroundings can be minimized. This can be done with some insulation, but a better method is to use a water bath. For example, measuring the latent heat of melting of wax can be achieved by placing some molten wax in a container with an electrical heater and a thermometer. A voltmeter and ammeter should be connected to the heater so that its power can be accurately determined. The container is then placed in a water bath, which is thermostatically controlled to the melting temperature, θ_m, of wax. Once the heater is switched on, the energy flow into the wax (coupled with some conduction in from the water bath) will start to raise the temperature of the wax. The temperature can be recorded as a function of time using the thermometer. A better, and more convenient, method is to use a temperature probe connected to a datalogging device. Once the temperature has reached θ_m, conduction from the water bath will stop and the temperature will remain constant for a period of time until all the wax has melted. This length of time coupled with the power of the heater enables the energy delivered during melting to be calculated, from which the latent heat is obtained.

> **Steam scalds**
>
> Steam scalds are much more serious than water scalds at the same temperature. Water droplets will cool down towards body temperature and the energy is conducted through the skin. Steam in contact with the skin will first condense into water releasing energy that is conducted through the skin as well.

> **Fusion and vaporization**
>
> The specific latent heat of vaporization (phase change from liquid to gas) is greater than the specific latent heat of fusion (solid to liquid) for a given substance. In changing state from a liquid to a gas, energy is needed to break bonds and also to allow the particles to spread out much more – doing work against an external pressure. Hence vaporization takes more energy than fusion.

> **Vapours and gases**
>
> The distinction between a vapour and a gas lies in the idea of critical temperature. Above the critical temperature a gas cannot be converted into a liquid no matter how much pressure is applied. Below the critical temperature applying pressure can cause liquid to form. A gas below its critical temperature is called a vapour.

PRACTICE

1 a Calculate the amount of energy required to melt 20 g of ice at 0 °C.

 b Calculate the amount of energy released when steam at 100 °C condenses into water at 100 °C.

 c Calculate the amount of energy released when 1 g of water is cooled by 1 °C.

 d A flask contains 20 g of ice floating in 100 g of water at 0 °C. Steam at 100 °C is passed into the flask until all the ice has just melted.
 What is the mass of water in the container after all the ice has been melted in this way?

 Specific latent heat of steam = $2.3 \times 10^6 \, \text{J kg}^{-1}$

 Specific latent heat of ice = $3.4 \times 10^5 \, \text{J kg}^{-1}$

 Specific heat capacity of water = $4.2 \times 10^3 \, \text{J kg}^{-1} \text{K}^{-1}$

O B J E C T I V E S

- efficiency of a heat engine
- expansion and compression cycles
- work done by an ideal gas

HEAT ENGINES

The internal combustion engine, the turbines that generate electricity in power stations, and simple steam engines all operate in an essentially similar way. All these devices are examples of **heat engines**.

The principle of a heat engine is to extract useful mechanical work from a machine and to replace the energy lost by heating the machine. In this way a certain amount of heat can be converted into work.

---

pV diagrams

It is often convenient to represent a process on a graph of pressure against volume (a *pV* or **indicator diagram**).

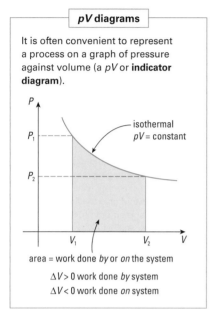

area = work done *by* or *on* the system

$\Delta V > 0$ work done *by* system
$\Delta V < 0$ work done *on* system

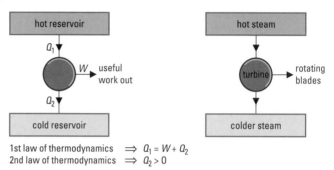

1st law of thermodynamics $\Rightarrow Q_1 = W + Q_2$
2nd law of thermodynamics $\Rightarrow Q_2 > 0$

The principle of a heat engine.

Energy is extracted as heat from a high-temperature reservoir – a source of energy with a very high thermal capacity so that a great deal of energy can be extracted without changing the temperature noticeably – and some of it is converted into useful work. The remaining energy is dumped into a low-temperature reservoir (sometimes called a sink).

The first law of thermodynamics implies that the amount of work done by the engine cannot be greater than the heat drawn in from the hot reservoir (see spread 7.2). The second law requires that the engine cannot be 100% efficient – which is the same as saying that some energy must be dumped into the low-temperature reservoir (see spread 7.13).

Efficiency of a heat engine

The first person to study the operation of heat engines was Sadi Carnot in 1824. He made the remarkable deduction that the **efficiency** of a heat engine is determined by the temperatures of the hot and cold reservoirs. The design and operation of the device is irrelevant: the efficiency is fixed by the laws of thermodynamics.

$$\text{efficiency of heat engine} < \frac{\text{useful work out of engine}}{\text{heat absorbed from hot reservoir}} = \frac{W}{Q_1}$$

$$< 1 - \frac{T_{\text{cold reservoir}}}{T_{\text{hot reservoir}}}$$

Worked example 1

The turbine in a power station extracts kinetic energy from steam at a temperature of 800 K. The steam emerging from the turbine has a temperature of about 370 K. What is the maximum efficiency of the turbine?

$$\text{maximum efficiency} = 1 - \frac{T_{\text{cold reservoir}}}{T_{\text{hot reservoir}}} = 1 - \frac{370\,\text{K}}{800\,\text{K}} = 0.54 = 54\%$$

This will be the maximum efficiency of the power station as a whole.

The Carnot cycle

Carnot derived his formula for the efficiency of a heat engine by considering an engine based on the expansion and compression of an ideal gas. The specific sequence of events is now known as the **Carnot cycle** and represents the most efficient way of extracting useful work. The cycle consists of several individual steps assumed to be operating on an ideal gas in a frictionless piston.

Imperfect efficiency

This unfortunate consequence is due to the increase in entropy that is required of any process (because macroscopic processes must move forward in time). Extracting energy from the hot reservoir actually reduces its entropy. If this energy could be completely converted into work then the overall entropy of the universe would decrease. But according to the second law this cannot happen. Hence some energy must be dumped into the colder reservoir – a process that increases that reservoir's entropy by more than the decrease in the hot reservoir.

---

1. The system starts in state A on the pV diagram – i.e. temperature T_{HOT}, pressure p_A, and volume V_A. The gas is then allowed to expand to a new volume V_B while remaining in thermal contact with a hot reservoir of temperature T_{HOT}. As the gas can draw in energy as heat (ΔQ_H), the expansion is isothermal. The gas does work on the surroundings during this expansion losing an amount of energy equal to the heat it gained – i.e. $\Delta U = 0$ as $\Delta W = -\Delta Q_H$.

2. The gas is now thermally isolated from the hot reservoir, but its expansion continues. This is now an adiabatic expansion and so $\Delta U < 0$ as the energy required to do work on the surroundings has to be provided by the internal energy of the gas. Once the temperature has dropped to T_{COLD} the expansion is halted at volume V_C and pressure p_C.

3. Next the gas is placed in thermal contact with a cold reservoir of temperature T_{COLD} and compressed (isothermally again) until the volume and pressure are now V_D and p_D respectively. During this part of the cycle work is done on the gas, but the energy gained is dumped to the cold reservoir (ΔQ_C) so $\Delta U = 0$.

4. The final stage is the adiabatic compression from D to A, restoring the gas to its initial conditions so that it can be taken round the cycle again. The work done on the gas during this compression is equal to the work that it did in moving from B to C.

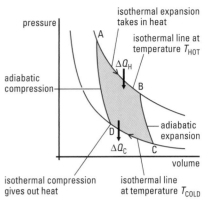

Heat pumps

Energy can be moved from a cold reservoir to a hot reservoir, against the natural flow, by using a heat pump (e.g. a refrigerator). To achieve this, work must be done on the device by an external energy source (usually electrical). The coefficient of performance (COP) for a heat pump is defined as:

$$COP_{heat\ pump} = \frac{\text{energy delivered to hot reservoir}}{\text{work done on pump}} = \frac{Q_2}{W} = \frac{Q_2}{Q_2 - Q_1}$$

The Cardiff Senedd (National Assembly for Wales) building uses a heat pump to extract geothermal energy from 100 m below ground.

Refrigerators use a refrigerant chemical, which is compressed and pumped by a motor. The warm liquid is passed through a heat exchanger, normally at the back of the appliance, where it transfers energy to the surroundings by radiation and convection. The liquid is then allowed to expand, drawing in energy, as it passes through the ice box. For a refrigerator, a more useful COP compares the energy extracted from the cold reservoir with the work required:

$$COP_{refrigerator} = \frac{\text{energy extracted from cold reservoir}}{\text{work done on pump}} = \frac{Q_1}{W} = \frac{Q_1}{Q_2 - Q_1}$$

The Carnot cycle for an ideal gas on a pressure–volume graph. The work is done on and by the gas during the adiabatic parts of the cycle. Heat is taken in or given out during the isothermal parts of the cycle. As the gas is taken round a closed cycle that restores it to its initial conditions, the net internal energy change must be zero. However, the net work done by the gas on the surroundings is not. The work done by the gas from A to B is the area under the line connecting A to B and the work done on the gas from C to D is the area under that line. Consequently the net work done is the area between the lines – i.e. the area of the closed loop of the cycle (shaded).

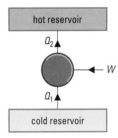

Heat can be pumped from a cold reservoir to a hot reservoir, if work is done on the pump.

PRACTICE

1. What is the maximum efficiency of a heat engine with a hot reservoir of 100 °C and a cold reservoir of 20 °C? The heat engine works at a rate of 1 kW. What is the rate at which energy is absorbed from the hot reservoir?

2. Outline how a heat engine could be used in reverse to act as a refrigerator.

3. A piston of volume 1000 cm³ contains a gas at a pressure of 1.1×10^5 Pa. The gas expands at constant temperature increasing the volume of the piston by 10%.

   a. Calculate the work done by the gas.

   b. How much energy must be absorbed by the gas from the surroundings in order to maintain the constant temperature?

4. A piston contains 0.05 mol of an ideal gas at 27 °C. The pressure of the gas is 10^5 Pa.

   a. What is the volume of the cylinder?

   b. What is the internal energy of the gas?

   The gas is now warmed to 77 °C, causing the piston's volume to increase. The pressure remains constant.

   c. What is the change in internal energy of the gas?

   d. What is the work done by the gas?

   e. How much energy is absorbed by the gas from the surroundings?

OBJECTIVES

- internal and external combustion engines
- the Otto cycle
- the Diesel cycle

Torque and power

An internal combustion engine such as this uses the motion of the pistons to turn a crankshaft (via rotational joints) which provides the mechanical power to the wheels (in the case of a car). The engine consequently produces torque, T, at a given engine revolution rate, ω. From this the brake power can be calculated as $P = T\omega$. It is important to realize that engines do not produce their maximum power at the same engine revs as their maximum torque. Torque produces acceleration; power produces energy at a rate to overcome friction at the given speed.

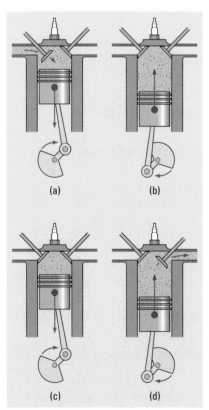

The four-stroke cycle in a gasoline engine: (a) intake stroke; (b) compression stroke; (c) ignition followed by the power stroke; (d) exhaust stroke.

REAL ENGINES

An **external combustion engine** (for example, a steam engine) is powered by heat conducted through the walls of the cylinder containing the working substance. Although such engines can generate a great deal of power, they have several disadvantages, including a limit to the maximum temperature (and hence efficiency), a tendency for the cylinder walls to soften, and the slow diffusion of energy through the working substance. These disadvantages have led to them being superseded by **internal combustion engines** in most applications.

Internal combustion engines use air to which has been added a small amount of a combustible gas as their working substance. The ignition of this combustible substance provides the heat necessary to power the cycle.

The various cycles that can be employed in internal combustion engines can be categorized on the basis of whether the combustion takes place isothermally, at constant volume, or adiabatically. The Carnot cycle is an example of an isothermal cycle – adapted for an internal combustion engine, the cycle would involve the adiabatic compression of a combustible mixture until the temperature reaches that at which the mixture ignites. The ignited mixture is then effectively the hot reservoir for the rest of the working substance and the cold reservoir is the surroundings into which the exhaust is dumped when a fresh mixture is supplied each cycle. The Carnot cycle is the most efficient cycle between any two working temperatures, but in practice considerable difficulties are encountered making an engine based on this cycle. Other, less efficient cycles are more realizable.

The four-stroke internal combustion engine

A practical internal combustion engine working on a petrol–air mixture can be produced using a cylinder containing a piston and a pair of valves. The process involves four '**strokes**', which are the up-and-down movements of the piston that turn the crankshaft to provide the required drive. Although the cycle involves these four strokes, there are six identifiable stages to the cycle, which in a real engine is carried out by a set of cylinders that will each be at different stages of the cycle.

1 **Intake stroke** The drive provided by another cylinder in the engine rotates the crankshaft which pulls the piston in this cylinder down. This reduces the pressure in the cylinder by an amount sufficient to cause the fuel–air mixture to be forced into the cylinder by the external atmospheric pressure when the intake valve opens.

2 **Compression stroke** The rotation of the crankshaft now drives the piston back up, compressing the fuel–air mixture to a much higher pressure and temperature. Friction and conduction of energy through the cylinder walls restrict the efficiency of this process.

3 **Explosion** An electrical spark provided by a spark plug ignites the fuel–air mixture, causing a rapid rise in temperature and pressure. This is timed to take place just before the piston reaches the top of its stroke, so that by the time combustion has been completely established the piston has come to rest ready for the power stroke.

4 **Power stroke** The hot gases expand rapidly, forcing the piston down, and in the process their temperature and pressure drop. Friction and conduction again limit efficiency. This is the part of the cycle that produces useful work.

5 **Exhaust** The exhaust gases in the cylinder are still at a higher temperature and pressure than the surroundings, so that they start to flow out of the cylinder when the exhaust valve opens.

6 **Exhaust stroke** Crankshaft rotation forces the piston up, expelling any remaining exhaust gases into the surroundings.

A full analysis of the processes going on in this cycle is extremely complex and impossible to carry out in mathematical detail. A simplified idea can be gained by considering the idealized cycle, which was first put into practice by N. Otto in 1876. To do this it is necessary to assume that the working substance is air behaving as an ideal gas and that there is no friction in the system. The pV (indicator) diagram for the **Otto cycle** is shown opposite. As combustion take place at constant volume the Otto cycle is an example of the second type of cycle listed above.

Relationship between the Otto cycle and the four-stroke cycle

1 OA – intake stroke.

2 AB – compression stroke: compression is rapid so can be considered to be adiabatic.

3 BC – explosion: a constant-volume increase in pressure and temperature.

4 CD – power stroke: a fast, hence essentially adiabatic, expansion.

5 DA – exhaust: drop in pressure (as gas is escaping) at constant volume.

6 AO – exhaust stroke.

Heat is drawn in during BC, ΔQ_{in}, and emitted during DA, ΔQ_{out}. In both cases the heat flow is taking place over a range of temperatures, rather than the fixed temperatures of the Carnot cycle, which is a significant reason why the Otto cycle is less efficient. The work produced by the cylinder is represented by the area bounded by the closed part of the cycle. Consequently, the theoretical output power of the whole engine can be written as

indicated power = area of pV loop × no. of cylinders × no. of cycles per second

The input power to the engine is derived from the '**calorific value**' of the fuel–air mixture – i.e. the amount of energy released by burning unit mass of fuel.

Consequently the input power can be written as

input power = calorific value (J kg^{-1}) × fuel flow rate (kg s^{-1})

In practice the actual power derived from the engine (the **brake power** – as measured on an engine test bed or **brake**) is less than the indicated power, as there are power losses due to friction:

brake power, $T\omega$ = indicated power – power lost overcoming friction

The Diesel engine

This cycle was first proposed by Rudolph Diesel in 1900 and is an example of the third type of cycle listed earlier. The basic structure of the cycle is the same as that of the Otto cycle. The main difference is that the fuel employed does not require a spark to ignite it.

1 During the intake stroke the cylinder is filled with pure air at atmospheric pressure.

2 The air is now compressed adiabatically during the compression stroke, with a corresponding rise in temperature. The fuel is now injected at a rate that ensures that as the combustion forces the piston down, allowing the contents of the cylinder to expand, the expansion takes place at constant pressure. At C the fuel intake is cut off and the rest of the working stroke produces an adiabatic expansion of the working substance. The temperature falls in the process. At D an exhaust valve opens allowing the exhaust to escape and the pressure immediately falls to that at point A on the curve.

3 The exhaust stroke removes the remaining exhaust gases and the cylinder is ready to be charged with air for the next cycle.

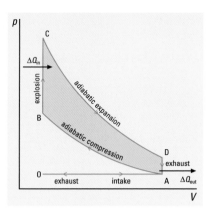

The idealized Otto cycle.

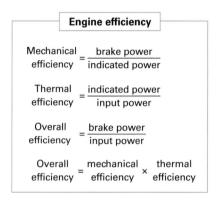

Engine efficiency

$$\text{Mechanical efficiency} = \frac{\text{brake power}}{\text{indicated power}}$$

$$\text{Thermal efficiency} = \frac{\text{indicated power}}{\text{input power}}$$

$$\text{Overall efficiency} = \frac{\text{brake power}}{\text{input power}}$$

$$\text{Overall efficiency} = \text{mechanical efficiency} \times \text{thermal efficiency}$$

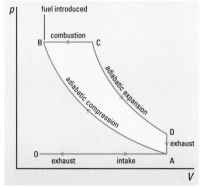

The Diesel cycle.

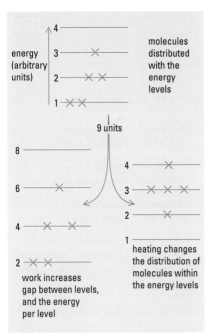

Work and heat have different effects on the arrangements of molecules within energy levels.

Heat, work, and energy levels

In a simple quantum theory of atoms the electrons in orbit about the nucleus are limited to certain set values of energy called **energy levels** (see chapter 8). The idea can be extended to large systems of many molecules if the energy levels apply to the atoms and molecules, not just the electrons within them. The diagram on the left shows a qualitative representation of this idea. In reality the pattern of energy levels is much more complicated than this.

If work is done on or by the system without any heating, then the macroscopic properties are bound to change (volume, temperature, pressure, etc.). This has the effect of altering the energy difference between the levels and moving the whole pattern of levels up or down in energy, but not altering the way that the molecules are distributed within the levels.

If the system heats the surroundings (or the surroundings heat the system), then the energy levels stay where they are, but the distribution of molecules within them changes. In many cases working and heating are both taking place so that a combination of both changes would be seen.

This provides a nice way of *visualizing* the difference between work and heat at the molecular level.

Entropy and energy

In spread 7.1 the nature of entropy and the arrow of time were discussed.

- Entropy measures the number of different ways the molecules in a system can be arranged without altering its macroscopic properties.

- Entropy continually increases. Random collisions within the system will tend to move the system towards the state with the largest number of microscopic arrangements. This may well be the reason for the direction of time.

The example used to illustrate this was that of a gas confined to half of a box and then released. The state with the greatest number of microscopic arrangements had the gas evenly distributed in the box.

However, this cannot be the whole story. In the example given, the energy of the molecules does not depend on where they are although in many systems the energy does depend on position; in such situations the entropy is calculated from the different ways of *arranging the energy between the molecules*.

A simple system of 3 molecules with a total energy of 6 units can have its energy levels arranged in several different ways. However, one way (the arrangement with the three molecules in separate levels) has many equivalent possibilities. This makes it the most likely arrangement to be found. As the number of molecules increases one arrangement dominates over all the others. With macroscopic numbers of molecules, once the system is in equilibrium only one arrangement need be considered. Boltzmann showed that the equilibrium arrangement of molecules leads to the probability of finding an individual molecule with energy E being:

$$\text{probability} \propto e^{-E/kT}$$

known as the **Boltzmann factor**. E is the energy of an individual molecule; the system as a whole will have an energy E which is the sum of that of the individual molecules.

As molecules cannot be distinguished, these arrangements are all exactly equivalent.

The number of different ways that three molecules can be placed in energy levels so that the total energy is 6 units.

Entropy and thermal contact

If two objects of different temperature are put into thermal contact then energy will flow from the one with the higher temperature to the one with the lower temperature. This is such a well confirmed fact experimentally that it is given the status of a law. It is one of the many equivalent ways of expressing the **second law of thermodynamics**.

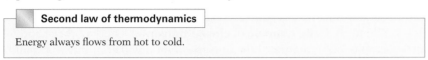

Second law of thermodynamics

Energy always flows from hot to cold.

However, this does not provide any physical reason *why* energy should flow in this manner. A rough idea of the physics involved can be gained by considering a simple system consisting of only three molecules. The diagram below shows some of the various ways in which three molecules can be placed into energy levels.

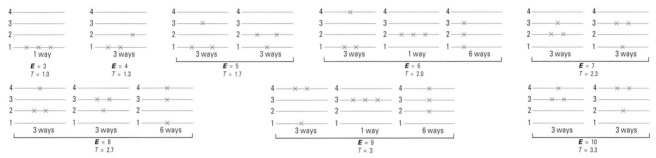

*All the possible arrangements for three molecules with total energy up to 10 units. Also shown is the number of equivalent diagrams with the same arrangement of molecules (remember: molecules cannot be distinguished, and can be re-arranged within the diagram). **E** = total amount of energy in the system for each arrangement; T = temperature (average energy per molecule).*

Now, think about what would happen if two such systems were connected together and energy allowed to flow from one to the other via heating. Suppose that at the start one of the systems has energy 10 and the other energy 3, then the temperatures of the two will be 3.3 and 1.0 respectively (consult the diagram above to check). What will happen as energy flows from one to the other? One thing is certain, the total amount of energy in the two systems must always total to 13.

The key to understanding what happens is to realize that the total number of ways of arranging the molecules in the *joint system* formed when they are linked together is the *product of the number of arrangements for each system*. At the start, one has energy 3 which can be done in 1 way only, and the other has energy 10 which can be achieved by any one of 6 arrangements. So, the number of ways of arranging the whole system with energy 3 and 10 in the separate subsystems is 6 × 1 = 6 ways.

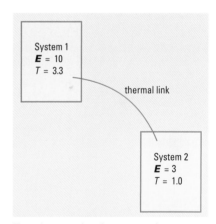

Two three-molecule systems become a joint system when thermally linked. However, as only energy can flow via heat the two systems can still be distinguished from each other.

All the ways that the molecules can be arranged with total energy 13.

Energy 1	Energy 2	Temperature 1	Temperature 2	Arrangements
10	3	3.3	1.0	6 × 1 = 6
9	4	3.0	1.3	10 × 3 = 30
8	5	2.7	1.7	12 × 6 = 36
7	6	2.3	2.0	6 × 10 = 60

Notice that the situation with the largest number of arrangements has the temperature of the two parts of the joined system most nearly equal. Furthermore, to get to this situation the system with the highest temperature to start with has lost the most energy.

In a more complicated example with more than just three molecules it would be possible to get the temperatures almost exactly equal, but this does not detract from the important message: *when systems are joined energy flows from hot to cold until the temperatures are equal, as that is the way in which the entropy (number of arrangements) of the joined system increases.*

OBJECTIVES

- different ways of expressing the second law
- Maxwell's demon
- the heat death of the universe
- the third law of thermodynamics
- approaching absolute zero

THE SECOND AND THIRD LAWS OF THERMODYNAMICS

Which ever way you look at it, physics does not hold out a very promising future for the universe. On the one hand, if there is enough matter in the universe then its expansion after the Big Bang will at some stage in the future reverse, and all the galaxies will collapse in on one another and eventually form a '**Big Crunch**' (a **closed universe**). On the other hand, if there is not enough matter in the universe then it will continue to expand indefinitely (an **open universe**). However, this is not a comfortable situation either. With an indefinite amount of time to operate in, the second law of thermodynamics can have a significant effect.

The second law of thermodynamics

The second law of thermodynamics can be stated in many different but exactly equivalent ways.

- The total entropy in the universe can never decrease.
- Heat can only flow from hot objects to cold objects.
- Heat engines cannot be 100% efficient.

Of these, the first statement is the most fundamental and may be equivalent to saying that time can only run forwards. The second follows from the first, as the entropy of the universe increases when energy flows from hot objects to cold ones (the entropy of the hot object decreases, but that of the cold one increases by a larger amount). Similarly the third statement also follows. If energy is extracted from a hot object its entropy decreases: if this energy were to be completely converted into work then there would be no corresponding increase in entropy to balance the books.

Does energy only flow from hot to cold?

If the most fundamental way of expressing the second law is through the idea of entropy, and as entropy is a measure of the most probable arrangement of molecules, then the second law must also be rooted in probabilities. This is not so obvious when it comes to the direction of energy flow.

The statistical nature of the second law is rather neatly illustrated by the fable of **Maxwell's demon**. He suggested that a 'small finite creature' (a 'demon') could arrange for energy to flow from a cold object to a hot object. Consider a box with a partition dividing it in half. One half of the box contains a high-temperature gas, the other a low-temperature gas. The demon is given control of a door between the two halves. The demon opens the door whenever a slow molecule in the hot gas approaches and when a fast molecule in the cold gas comes near. At other times the door remains closed. By doing this the demon can cheat the normal probabilities and arrange for the molecules in the cold gas to get slower and those in the hot gas to be faster (on average) – so energy has passed from cold to hot.

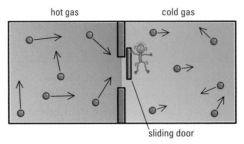

Maxwell's demon at work.

As there is no conceivable way of making such a device or creature, the importance of the argument is simply to demonstrate that the second law is an expression of what is highly likely to happen.

The heat death of the universe

According to the second law of thermodynamics the efficiency of heat engines is restricted so that energy taken in via heating cannot be fully converted into work. Although, so far, heat engines have been considered as pieces of technology, such as turbines and steam engines, the law also applies to any process that might take place in nature. Energy taken in from the hot reservoir via heat decreases the amount of internal energy in the reservoir. The second law implies that internal energy cannot be converted into any other form with complete efficiency, but the law does not apply to the conversion of other forms of energy into internal energy. Internal energy is *less available* – it cannot be converted easily into other forms.

Over the time scales that would be available in an open universe, all forms of energy would be steadily converted into internal energy. Once this had happened and the universe had come into thermal equilibrium (all places at the same temperature) then processes would stop. This is rather grandly and depressingly called the **heat death of the universe**. Stars would be cold burnt-out embers; galaxies would be so far apart as to be unobservable from each other. However, this need not be a bleak situation. Quantum mechanics suggests that slow, gentle processes might continue to take place. The English-born physicist Freeman Dyson has studied this situation and believes that life might not come to an end, but rather would flourish in the much more gentle and less violent conditions in such a universe.

The third law of thermodynamics

The third law does not really fit into the pattern established by the others.

0 The zeroth law – establishes the existence of thermal equilibrium and allows temperature to be defined.

1 The first law (conservation of energy) – establishes that work and heat are both perfectly good ways of increasing the internal energy of an object.

2 The second law – the entropy of the universe can never decrease.

3 The third law – objects can never be cooled down to absolute zero.

In order to cool something down energy must be drained from it into a colder object. As an object's temperature is reduced towards absolute zero this becomes an increasingly difficult task. Nevertheless scientists have devised increasingly clever ways of extracting energy, and temperatures have been reduced to a few thousandths of a kelvin.

Freeman Dyson won the 2000 Templeton Prize for contributions to science and religion. The prize's cash value is guaranteed greater than that of the Nobel Prize.

Entropy of the universe

The second law of thermodynamics implies that the universe must have started in a very low entropy state. Consequently it must also have been a very low probability state. This is a good argument against a cyclic universe (big bang → crunch → big bang → . . .) as it is hard to imagine any physics that would dramatically reduce entropy at the end of a universe's lifetime.

Superconductivity and superfluidity

The quest for lower temperatures has largely been a case of pure research, but on the way important scientific discoveries have been made:

Superconductivity The resistivity of some materials drops to zero at temperatures of a few kelvin. In 1972, John Bardeen, Leon N. Cooper, and J. Robert Schrieffer were awarded the Nobel Prize for jointly developed the theory of superconductivity.

Superfluidity Helium-3 cools to a liquid but never solidifies: instead, at a distinct temperature it becomes a fluid that flows with no viscosity. For this discovery David M. Lee, Douglas D. Osheroff, and C. Richardson were awarded the Nobel Prize in 1996.

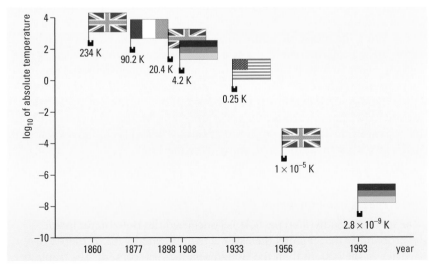

Graph showing the progress towards achieving absolute zero.

- activation processes

- the Boltzmann factor

- Maxwell–Boltzmann distribution

Ludwig Boltzmann was an Austrian physicist whose theoretical research developed the atomic theory of matter into statistical thermodynamics, leading to the iconic equation for entropy, $S = k \ln W$. This equation is inscribed on his tombstone. He committed suicide in 1906, depressed by both his failing eyesight and strong opposition to his theories from influential scientists and philosophers such as Ernst Mach. Ironically, in 1905 a young physicist Albert Einstein had just published a paper on the statistical interpretation of Brownian motion. This ultimately led to experimental confirmation of the statistical atomic model and wide acceptance of Boltzmann's statistical mechanics.

Maths box: simple derivation of the Boltzmann factor

W = original number of ways

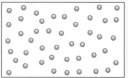

W' = number of ways when ΔE removed

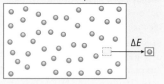

1 Imagine a large thermal reservoir is split into two systems by removing one particle.

2 Allow a reversible transfer of energy between them.

3 The entropy change for the reservoir will be: $\Delta S = -\Delta E/T$

$$S = k \ln W \text{ therefore } W = e^{S/k}$$

$$W' = e^{S'/k} = e^{(S - \Delta S)/k} = e^{(S/k - \Delta E/kT)}$$

$$W' = e^{S/k} e^{-\Delta E/kT} = W e^{-\Delta E/kT}$$

Probability = $W'/W = e^{-\Delta E/kT}$
i.e. the Boltzmann factor!

THE BOLTZMANN FACTOR

Activation processes

If you leave water in a glass for long enough it will evaporate, even though the surrounding temperature is well below the boiling point of water. Water molecules break free of the bonds holding them to each other and escape from the surface, even though the average thermal energy of a molecule in the water is far less than the energy needed to escape. They are able to do this because of the random thermal motion of the molecules – there is a chance that a series of collisions will all increase the energy of a particular molecule near the surface and eject it.

This is an example of an activation process. There is a minimum extra energy needed for a molecule to escape. The greater this 'activation energy' is, the smaller the chance that a molecule will gain enough extra energy to escape. The higher the temperature, the more violent the molecular motions, and the higher the probability that an individual molecule will gain enough energy to escape. This is why water evaporates faster on a hot day than on a cold day.

Many physical and chemical processes are activation processes, for example:

- evaporation

- many chemical reactions

- viscous flow in a liquid

- creep in a polymer

- nuclear fusion

- conduction of electricity in a semiconductor.

In all cases the rate at which the process or reaction proceeds depends on the ratio of the activation energy ΔE to the typical thermal energy kT. The larger this ratio, the smaller the probability that a molecule will gain this extra energy, and the slower the rate of the process. If the temperature is increased the ratio falls and the rate increases. This is why the rate of chemical reactions increases with temperature, the viscosity of oil falls with temperature (increased flow rate), and the conductivity of a thermistor increases with temperature.

Activation and probability

Imagine a thermal reservoir containing a large number of molecules and a certain amount of energy at a temperature T. What is the probability that one of these molecules gains an extra amount of energy ΔE? One way to approach this question is by splitting the system into two parts by removing the molecule in question and asking what effect removing energy ΔE has on the rest of the system. If energy is removed there will be fewer ways in which the remaining energy can be distributed amongst the molecules. Therefore the number of arrangements W for the system will fall and so will its entropy ($S = k \ln W$). If the number of ways before and after the energy is removed are W and W' then the chance of finding the system in the lower-energy state is simply W'/W. It can be shown (see maths box) that this ratio is given by the equation:

$$\frac{W'}{W} = e^{-\frac{\Delta E}{kT}}$$

This is the 'Boltzmann factor'.

As far as the individual particle is concerned the Boltzmann factor is the probability that it gains the extra activation energy ΔE. If there are N particles in the system then the Boltzmann factor represents the fraction of these that have gained this extra energy at any moment. Since the rate of an activation process is directly proportional to the fraction of particles with this energy, the rate R is directly proportional to the Boltzmann factor. This leads to the following very general result for such activation processes:

advanced **PHYSICS**

$$R = R_0 e^{-\frac{\Delta E}{kT}} \quad \ln R = \ln R_0 - \frac{\Delta E}{kT}$$

where R_0 is a rate constant.

It is clear that a graph of $\ln R$ against $1/T$ will be a straight line with negative gradient equal to $\Delta E/k$. Such a graph could be used to find either k or ΔE.

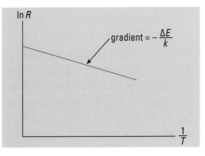

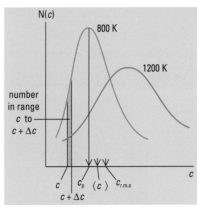

*In 1860 Maxwell derived the speed distribution for molecules of a gas in thermal equilibrium. He did this by assuming they had random motion and exchanged energy in collisions. The area under the graph represents numbers of molecules. As temperature increases the probability of a particular molecule gaining higher energy increases, so the curve flattens out – the peak (most likely speed, c_p) moves to a higher value, but the area under the entire curve remains constant (same number of particles). Maxwell's speed distribution can be rigorously derived from Boltzmann's ideas and the distribution of speeds or kinetic energies in a gas is usually referred to as the **Maxwell–Boltzmann distribution**.*

Case study – semiconductor conductivity

Semiconductors have relatively low conductivity because very few charge carriers are free to move through the material; most are trapped in the bonds between atoms. However if an electron can gain an activation energy ΔE from random thermal collisions it will break free and become available as a charge carrier (as will a hole in the space left behind). The fraction of electron-hole pairs in the material at any moment is given by the Boltzmann factor. The conductance G ($= 1/R$) of the semiconductor is directly proportional to the number of charge carriers so we should expect:

$$G = G_0 e^{-\frac{\Delta E}{kT}} \quad \ln G = \ln G_0 - \frac{\Delta E}{kT}$$

This can be tested in a very simple experiment using an ohmmeter to measure the resistance of a thermistor as its temperature is changed in a water bath. Temperature T (kelvin) and resistance R (ohm) are recorded, and conductance G (seimen) is calculated. If $\ln G$ is plotted against $1/T$ then the activation energy can be found from the gradient of the straight-line graph produced.

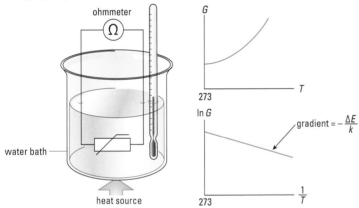

Measuring the resistance R and conductance G = 1/R of a thermistor as a function of temperature.

Evaporation revisited

The experiment above allows us to find the activation energy if we already know the Boltzmann constant. However, if we could measure activation energy by a different method then we could use it to find the value of the Boltzmann constant. In fact the activation energy for water molecules to leave the liquid surface can be found experimentally (and approximately) by measuring the energy required to boil off a measured mass of water and then calculating the energy per molecule. It turns out to be about $40\,\text{kJ}\,\text{mol}^{-1}$ or $6.64 \times 10^{-20}\,\text{J}$ or $0.42\,\text{eV}$ per molecule. This is about 17 times the average thermal energy per molecule ($0.025\,\text{eV}$) at room temperature (around $300\,\text{K}$).

The Boltzmann factor for evaporation at room temperature is therefore about e^{-17}, a tiny number. This means that only about 1 molecule in e^{17} (1 in about $20\,000\,000$!) has enough energy to evaporate at any moment. However, the number of molecular interactions per second is enormous so this does result in a measurable rate of evaporation. As a very rough rule of thumb, for activation processes, a process will take place at an observable rate if the ratio $\Delta E/kT < 20$. If this ratio is much larger, say $50+$, then don't expect any appreciable activity. If it is much smaller, say 1–5, then the process will run to completion very quickly.

FLOW

One of the most important features of physics is the comparison that can be made between different physical effects. This sometimes enables us to apply lessons learned in one area to the problems of another. A good example of this is in the study of flow. Water flows along rivers, charge flows along wires, and energy flows between regions at different temperatures. Although the details are different, the same sort of results are seen in each case. We can apply the methods of electrical physics to the study of energy flow and vice versa.

Comparison of electrical flow to energy flow

Energy will always flow from a region of high temperature to a region of low temperature. A lagged bar will have an energy flow that continues as long as energy is removed from one end and supplied to the other. The bar reaches steady state, but not equilibrium. *To maintain the energy flow a temperature difference must be maintained.*

If an electrically conducting bar is charged at one end but uncharged at the other, there will be a p.d. across the bar. Charge will flow along the bar, giving rise to an electric current. The current continues until the charge distribution in the bar is such that there is no p.d. between any two points on the bar. The bar has formed an **equipotential** (see spread 5.9) and it is in equilibrium (note that electrical heating will have ensured that it is at a constant temperature along its length).

If some external agency such as a battery continually adds charge to one end of the bar and removes it from the other, then the bar never reaches equilibrium. It does achieve steady state with a constant current. The p.d. between one end of the bar and any point along its length will change uniformly along the bar.

To maintain the charge flow the p.d. must be maintained.

Charge and energy flow equations

The previous section suggested that the charge flow in a circuit and the energy flow in a bar are quite similar effects. A comparison can also be made between a temperature difference and an e.m.f. This is borne out by an inspection of the equations that govern the two processes:

1 **Energy flow**

$$\text{rate of energy flow } (H) = \frac{\Delta Q}{\Delta t} = -kA \frac{\Delta T}{\Delta x}$$

2 **Charge flow**

Combining the equations gives

$$I = \frac{\Delta Q}{\Delta t} = -\frac{\Delta V}{R} \quad \text{and} \quad R = \frac{\rho \Delta x}{A}$$

in which Q is charge, ΔV the potential difference, and Δx the length. Hence

$$I = \frac{\Delta Q}{\Delta t} = -\frac{A}{\rho} \frac{\Delta V}{\Delta x} = -\sigma A \frac{\Delta V}{\Delta x}$$

This equation has some interesting points of comparison with the energy flow equation.

- The charge flow (I) is the equivalent of the energy flow, (H).
- Area and length of sample appear in both equations in the same fashion.
- The potential difference ΔV is fulfilling the same role as ΔT, the temperature difference.
- The thermal conductivity (k) is a material property similar to the **electrical conductivity** (σ) which is 1/resistivity ($1/\rho$).

Comparisons of this sort between two different physical phenomena can be useful in solving problems.

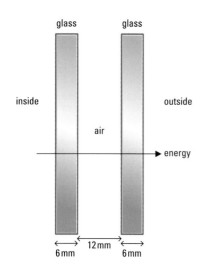

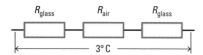

The energy flow through three materials can be treated like the flow of charge through three resistances in series.

Worked example

A sheet of double glazing consists of two pieces of glass of thickness 6 mm, between which is trapped a layer of air that is 12 mm thick. The area of the glass sheets is 2.0 m². Calculate the rate of energy flow through the double glazing if there is a temperature difference of 3.0 °C between the inside and the outside.

One way of solving this is to define the thermal resistance of a material:

$$R_T = \frac{1}{kA}$$

The thermal resistance of the glass sheet is

$$R_g = \frac{6 \times 10^{-3}\,\text{m}}{(1.0\,\text{W m}^{-1}\,\text{K}^{-1}) \times (2.0\,\text{m}^2)} = 3 \times 10^{-3}\,\text{K W}^{-1}$$

and for the air layer

$$R_{air} = \frac{12 \times 10^{-3}\,\text{m}}{(0.02\,\text{W m}^{-1}\,\text{K}^{-1}) \times (2.0\,\text{m}^2)} = 3 \times 10^{-1}\,\text{K W}^{-1}$$

The problem can be regarded as being like that of three resistors in series:

$$\text{total resistance} = 2R_g + R_{air}$$
$$= 2 \times 3 \times 10^{-3}\,\text{K W}^{-1} + 3 \times 10^{-1}\,\text{K W}^{-1}$$
$$= 0.306\,\text{K W}^{-1}$$

$$\therefore \quad \text{rate of energy flow} = \frac{\text{temperature difference}}{\text{thermal resistance}}$$
$$= \frac{3\,\text{K}}{0.306\,\text{K W}^{-1}}$$
$$= 9.8\,\text{W (2 sig. figs)}$$

U values

In the building industry the ability of a material to conduct energy is measured by its **U value**, which is related to the thermal resistance:

$$U = \frac{1}{R} = \frac{k}{l}$$

so that

$$\frac{\Delta Q}{\Delta t} = UA\Delta T$$

Regulations for the conservation of energy in housing specify the maximum U values that materials used in construction can have. For example, in the United Kingdom the U value for a house wall must not exceed $0.60\,\text{W m}^{-2}\,\text{K}^{-1}$.

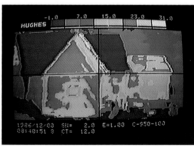

*Thermogram showing energy flow through the roof and walls of a house. The energy loss through walls can be cut down by placing some insulating material in the cavity between the double row of bricks. This is called **cavity wall insulation**.*

PRACTICE

1 Ice is forming on top of a pond. When the layer of ice is 4.6 cm thick, the temperature of the ice in contact with the water is 273 K and the temperature of the ice in contact with the air is 260 K. Calculate the rate of energy loss per square metre through the ice.

(Thermal conductivity of ice = 2.3 W m⁻¹ K⁻¹.)

2 a A brick wall is 6 m by 3 m. The wall is broken by a window that is made from five panes of glass each 0.5 m by 1 m. Calculate the average U value for the wall.

b Given that this is the only outside wall to the room, what power of heater would be needed to keep the room at a constant temperature of 18 °C if the outside temperature was 5 °C?

(U (cavity brick wall) = 1.8 W m⁻² K⁻¹, U (glass) = 5.5 W m⁻² K⁻¹.)

- black-body spectra
- Wien's law
- Stefan's law

Radiation terminology

The term 'radiation' is used in physics to apply to many different situations in which something is emitted:

- Nuclear radiation – the emission of particles by unstable nuclei (α, β, γ).
- Visible radiation – the emission of visible wavelengths of electromagnetic (EM) radiation.
- Thermal radiation – the emission of infrared (IR) wavelengths of electromagnetic radiation by warm objects.

Although these are all forms of radiation, the term 'radioactive' refers only to nuclear processes.

 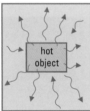

Radiation and absorption.

Black-body radiation

Black objects look black because they absorb most of the visible light falling on them. At room temperature objects mostly produce infrared radiation. There is no reason why an object that looks black must be a good absorber or emitter of infrared – it may not be 'black' at these wavelengths. The colour of an object in visible light need not be a good guide to its behaviour at other wavelengths.

THERMAL RADIATION

Electromagnetic radiation

Electromagnetic waves are emitted whenever electrons lose energy. In single atoms this happens when electrons in orbit drop into lower atomic energy levels, an effect that is used to generate light in discharge lamps such as sodium street lights. In solid objects the electrons are involved with more than one atom and the energy levels are much more complex. This is why solids do not produce the characteristic line spectra of the individual atoms. Electrons can still drop into lower levels and emit electromagnetic radiation in the process. The spectrum is more 'broad band', covering a range of wavelengths rather than individual lines. However, there are some general rules that apply to the rate and wavelength of emission.

Thermal radiation

All objects emit electromagnetic radiation at a rate that depends on their temperature. The hotter the object the greater the rate of energy emission. The temperature also helps to determine the range of wavelengths (the spectrum) of the emitted radiation, along with the nature of the material.

When electromagnetic radiation falls on an object it will be partially reflected and partially absorbed. This will happen at any temperature and the amount of reflection versus absorption depends on the nature of the object. However, if the object is at a higher temperature than its surroundings the amount of energy that it is emitting is much greater than that which it is absorbing. The reverse is true for a cold object in warm surroundings.

There are some rules of thumb that can be applied to get a feel for how the nature of the object affects its ability to radiate.

- An object will radiate more energy than it absorbs if its temperature is *greater than* the temperature of its surroundings. Good radiators and good absorbers tend to be matt black with rough surfaces.

- An object will absorb more energy than it emits if its temperature is *less than* the temperature of its surroundings. Bad radiators and bad absorbers tend to be light-coloured (white), shiny, and with smooth surfaces.

Perfect thermal sources

If an object is capable of emitting and absorbing radiation of every wavelength then it can come into thermal equilibrium with its radiation. When this happens at every wavelength the amount of energy being emitted and absorbed per second is the same. Under these conditions the spectrum of radiation produced only depends on the temperature. Objects behaving in this manner are known as **perfect thermal sources (black bodies)** and the spectrum of radiation is called **black-body radiation**. Although it sounds rather unlikely, there are many situations in which this can come about, especially with astronomical objects. The Sun, for example, has a spectrum that is approximately black-body over most wavelengths produced. The most perfect black-body spectrum that has been measured is that of the **cosmic microwave background radiation** (see spread 12.23).

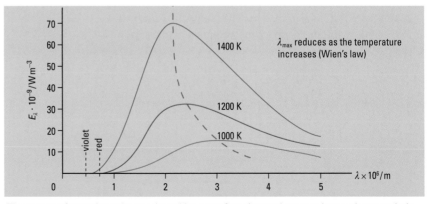

The range of wavelengths produced by a perfect thermal source has a characteristic spectrum.

Wien's law for black bodies

$$\text{peak wavelength (m)} \propto \frac{1}{\text{absolute temperature (K)}}$$

$$\lambda_{\text{peak}} = \frac{W}{T}$$

W is Wien's constant $= 2.898 \times 10^{-3}\,\text{m K}$

Stefan's law for black bodies

$$\text{total radiated power (W)} \propto \text{area of object (m}^2) \times (\text{absolute temp (K)})^4$$

$$P = \sigma A T^4$$

σ is the Stefan constant $= 5.6696 \times 10^{-8}\,\text{W m}^{-2}\,\text{K}^{-4}$

Helpful hint

A black body will absorb the radiation falling on it. The rate of radiation produced by the surroundings is proportional to the temperature of the surroundings, T_s, to the fourth power. So, the net energy production by the object is proportional to $T_o^4 - T_s^4$.

Worked example

A human being has a surface area of $1.9\,\text{m}^2$ and a body temperature of $37\,°\text{C}$. What is the total power radiated and the peak wavelength, assuming the human to be a perfect thermal source? (Assume the surroundings to be at a temperature of $20\,°\text{C}$.)

$$\text{energy radiated by body} = \sigma A T_{\text{body}}^4$$

$$\text{energy radiated by surroundings falling on body} = \sigma A T_{\text{surroundings}}^4$$

Hence,

$$\text{net power radiated} = \sigma A (T_{\text{body}}^4 - T_{\text{surroundings}}^4)$$

$$= (5.669 \times 10^{-8}\,\text{W m}^{-2}\,\text{K}^{-4}) \times (1.9\,\text{m}^2)$$
$$\times ((310\,\text{K})^4 - (293\,\text{K})^4)$$

$$= 200\,\text{W}$$

$$\text{peak wavelength } (\lambda) = \frac{\text{Wien's constant}}{\text{temperature}} = \frac{2.898 \times 10^{-3}\,\text{m K}}{310\,\text{K}} = 9.35\,\mu\text{m}$$

Ball lightning

It is possible that ball lightning – a rare and controversial phenomenon in which a glowing hot ball can be seen floating through the sky – is a ball of plasma created by the high temperatures in a lightning flash.

Are black bodies black?

Although temperature is the determining factor in the spectrum of a black body, the radiation produced need not be infrared. As the temperature increases, so the spectrum moves up the wavelength range. The spectrum has a broad maximum so a black body will start to emit visible light and glow a dull red colour before the peak wavelength reaches the visible region. At higher temperature most of the peak will be in the visible region so the object is emitting the full range of visible wavelengths and the object glows white hot. Blue–white indicates even higher temperatures with the peak starting to move into the ultraviolet region and the rest of the range moving into the bluer part of the visible spectrum.

PRACTICE

1 a The radius of the Sun is approximately $7 \times 10^8\,\text{m}$ and its surface temperature is $5800\,\text{K}$. Estimate the total emitted power of the Sun.

b The solar constant is the average power incident on the Earth per square metre and has the value $1.4\,\text{kW m}^{-2}$. How far away from the Sun is the Earth on average?

2 A water pipe is made of thin copper and has a diameter of $3.0\,\text{cm}$. It carries water at a temperature of $30\,°\text{C}$ above the surrounding air. Estimate the amount of energy lost by radiation per second along a $1\,\text{m}$ length of pipe if the surrounding air has a temperature of $18\,°\text{C}$.

3 The solar radiation striking the Earth has an intensity of $1.4\,\text{kW m}^{-2}$. A black metal plate is placed so that the sunlight strikes its surface at $90°$. The temperature of the surrounding air is $20\,°\text{C}$.

a Explain why the temperature of the piece of metal rises to a constant value.

b Calculate the value of this constant temperature.

PRACTICE EXAM QUESTIONS

1 A research organisation has been commissioned to investigate the variation of temperature inside a new model of domestic refrigerator. It is necessary to monitor the temperature both inside and outside the refrigerator.

   a State *three* important specifications which must be considered in selecting the type of thermometer to be used. (You are not required to give details of particular thermometer structures.)

   b Name *one* type of thermometer which would be suitable for use in both the locations, and explain why it is suitable in terms of the specifications selected.

2 This question is about energy changes in an ideal gas.

   a State, in words, the first law of thermodynamics applied to the energy changes for a fixed mass of ideal gas.

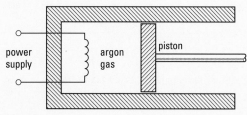

Figure 7.1

   b 6.0 g of argon gas, molar mass 0.018 kg mol^{-1}, are contained at atmospheric pressure, 1.0×10^5 Pa, within a large thermally insulated cylinder by a perfectly fitting frictionless piston, as shown in Figure 7.1. A 5.0 W electrical heater is used to heat the gas for 25 s. The experiment is carried out twice under different conditions.

   i When the piston is fixed in position, the temperature rise is 30 K. Calculate the molar heat capacity of argon which these data give.

   ii When the piston is free to move, the temperature rise is found to be 18 K. Calculate the molar heat capacity of argon which these data give.

   iii Suggest a reason why the two values calculated in **i** and **ii** above are different.

   iv By how much does the volume of the gas change when the piston is free to move?

3 This question is concerned with aspects of the operation of a shower system designed to work from the mains supply.

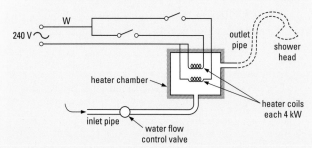

Figure 7.2

The system shown in Figure 7.2 can be operated with one or two 4 kW heaters connected.

   a Explain why the water achieves a constant temperature when the shower is operating.

   b On a particular day the inlet temperature is 10 °C and the water flow rate is adjusted to 3.5×10^{-3} m^3 min^{-1}.

   The specific heat capacity of water is 4200 J kg^{-1} K^{-1}.

   The density of water is 100 kg m^{-3}.

   i Calculate the temperature of the water leaving the heater chamber when *one* of the heaters is operating.

   ii Calculate the temperature when *both* heaters are used.

   c i Calculate the resistance of each heater.

   ii What current must the wire W be capable of carrying?

   iii A heater is manufactured from wire with a resistivity of 4.5×10^{-7} Ωm at its operating temperature. The wire in the element has a diameter of 0.30 mm.

   Calculate the length of wire in each element.

   iv An engineer makes a mistake and instead of connecting the heaters in parallel connects them in series. Calculate the output power of the heaters in this case.

4 The ground floor of a house has an area of 50 m^2. The floor is fitted with a carpet 15 mm thick which completely covers the floor. The carpet rests on a layer of concrete 200 mm thick. The top surface of the carpet has a temperature of 15 °C and the lower surface of the concrete has a temperature of 10 °C.

   Thermal conductivity of concrete = 0.75 W m^{-1} K^{-1}.

   Thermal conductivity of the carpet material = 0.06 W m^{-1} K^{-1}.

   a Calculate the rate at which energy transfer would occur *without* the carpet, assuming the temperatures of the top and bottom surfaces of the concrete are 15 °C and 10 °C respectively.

   b Calculate the temperature at the carpet/concrete boundary.

   c Calculate the energy transfer rate with the carpet in place.

5 a i State the first law of thermodynamics.

   ii Distinguish between internal energy and heat.

   b 2.30 mol of the monatomic gas helium occupies a volume 5.49×10^{-2} m^3 at a temperature of 290 K and pressure of 1.01×10^5 Pa.

   i The gas undergoes a change of volume by a factor of 1.2, doing work against a constant external pressure of 1.01×10^5 Pa. Calculate the work done.

   ii During this process, the temperature of the gas rises from 290 K to 348 K. Calculate the increase in the internal energy.

   iii Use these results, together with the fact that the total heat transfer to the gas in the above process is measured to be 2.77×10^3 J to show that the gas is behaving ideally.

6 a i State *four* assumptions of the kinetic theory of an ideal gas.

ii A gas molecule in a cubical box travels with speed *c* at right angles to one wall of the box. Show that the average force the molecule exerts on that wall is proportional to c^2.

b Table 7.1 gives measured values of pressure and density for a fixed mass of gas at a constant temperature of 27 °C.

Table 7.1

Pressure / 10^5 Pa	0.60	0.80	1.00	1.20	1.40
Density / kg m^{-3}	0.68	0.91	1.14	1.37	1.60

i Plot a graph of pressure against density. Does your graph indicate that the gas behaves as an ideal gas under these conditions? Justify your answer.

ii Use your graph to calculate the root mean square speed of the molecules of the gas.

iii The temperature of the gas is raised to 57 °C. Calculate the pressure when the density is 1.00 kg m^{-3}, and hence draw the corresponding graph of pressure against density at 57 °C, using the same axes as before.

c A container holds a mixture of helium and argon, relative atomic masses 4 and 40 respectively.

i For the gases in the container, calculate the ratio

$$\frac{\text{root mean square speed of helium atoms}}{\text{root mean square speed of argon atoms}}$$

ii There are approximately equal numbers of helium and argon atoms in the container, when gas starts to leak slowly out of a small hole in the side. After a short time, will the number of argon atoms remaining in the container be greater or less than the number of helium atoms?

Give a reason for your answer.

7 a State Boyle's law. Give the assumptions upon which the kinetic theory of gases is based and describe qualitatively how the theory gives an explanation of Boyle's experimental law.

b A car tyre of volume 1.6×10^{-2} m^3 contains air at a temperature of 17 °C and a pressure of 2.6×10^5 Pa. A foot pump of volume 3.0×10^{-4} m^3 is used to pump air, at a temperature of 17 °C and an initial pressure of 1.0×10^5 Pa, into the tyre to increase the pressure to 3.1×10^5 Pa. Assume that pumping does not change the temperature or the volume of the air in the tyre.

i What amount of air in mol does the tyre contain initially?

ii Calculate the minimum number of strokes of the pump needed to raise the pressure to 3.1×10^5 Pa.

iii Is the assumption of constant temperature justified? Explain your answer.

iv Calculate the root-mean-square speed of the air molecules in the tyre. (Assume an average value of the molar mass of air of 0.029 kg mol^{-1}.)

8 a State the relationship between the energy of the molecules of a perfect gas and

i the temperature of the gas,

ii the internal energy of the gas.

b A sample of hot liquid is allowed to cool naturally in a test tube. The temperature of the liquid is recorded at various times as the liquid cools and the results are shown in Figure 7.3.

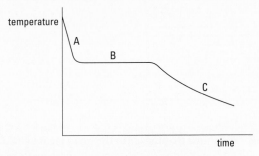

Figure 7.3

i Describe and explain what is happening in the test tube during each of the time intervals corresponding to the regions labelled A, B and C.

ii By reference to A, B and C explain the shape of each portion of the graph.

iii State whether the material in the test tube has a higher specific heat capacity in region A or in region C. Explain your answer.

9 Two cylinders A and B contain ideal gas under the conditions shown in Figure 7.4. The cylinders are connected by a tube of negligible volume which incorporates a closed tap.

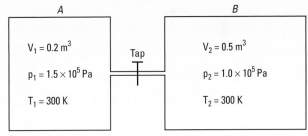

Figure 7.4

a Find the number of moles of gas in each cylinder.

b The tap is opened and steady conditions attained.

i What is the pressure in the gas?

ii Why is there no change in temperature?

10 a An ideal gas, with initial state properties p_1, V_1 and T_1 expands adiabatically to a state with properties p_2, V_2 and T_2.

i Write down an equation relating p_1, V_1 and T_1 to p_2, V_2 and T_2.

ii Write down an equation relating p_1, V_1, p_2, and V_2.

b The above equations can be combined in the form

$$\frac{T_1}{T_2} = \left(\frac{P_1}{P_2}\right)^{\left(\frac{\gamma-1}{\gamma}\right)}$$

where γ is the ratio of the principal specific heat capacities of the gas. Figure 7.5 shows a pump that raises the pressure of air for pneumatic road construction equipment from atmospheric pressure of 1×10^5 Pa to an absolute pressure of 4×10^5 Pa. The pump acts almost adiabatically. External air temperature is 280 K. Estimate the air temperature immediately after compression.

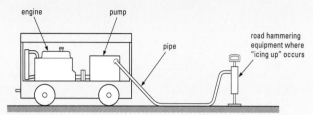

Figure 7.5

    c Assuming that the air in the pipes feeding the pneumatic road construction equipment falls to ambient temperature before reaching the equipment, explain why the operators sometimes experience 'icing-up' of the equipment when the air is released.

        Ratio of the principal specific heat capacities of air $(\gamma) = 1.4$

11 a List the assumptions of the kinetic theory of gases.

    b Prove that, with the usual symbols, the pressure p of a gas is given by

$$p = \tfrac{1}{3}\rho\langle c^2 \rangle$$

    c i By comparing this equation with the equation of state for 1 mole of an ideal gas, find the translational kinetic energy for 1 mole of the monatomic gas.

    ii Hence show that the average translational kinetic energy per molecule of a monatomic gas is given by $\tfrac{3}{2}kT$ where k (called the Boltzmann constant) is

$$\frac{R}{N_A}$$

    d 5 moles of helium at 27 °C are allowed to mix with 3 moles of neon at 127 °C. Calculate the temperature of the resulting mixture of gases.

        Assume no heat losses during the mixing process.

12 Figure 7.6 shows the theoretical pressure-volume diagram of an engine in which a fixed mass of air is heated $(A \rightarrow B)$, allowed to expand adiabatically $(B \rightarrow C)$ and finally returned to its initial state $(C \rightarrow A)$ before repeating the same cycle.

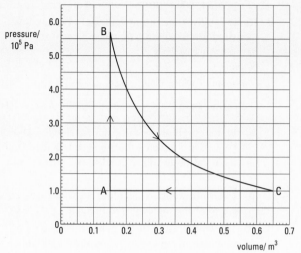

Figure 7.6

    a During the process $A \rightarrow B$ a heat source supplies 2.0×10^5 J of energy to the air, which is at an initial temperature of 20 °C.

    i Calculate the number of moles of air which are taken through this process.

        The molar gas constant $R = 8.3$ J mol^{-1} K^{-1}.

    ii Show that the temperature of the air at B is approximately 1400 °C.

    iii Calculate the maximum possible efficiency of any engine working between these temperature limits.

    b During the process $B \rightarrow C$ the expanding gas does work on an external load.

    i Using values taken from the diagram, show that the *net* output work done by the engine during one cycle is approximately 58 kJ.

    ii Determine the output power of the engine if it completes 180 cycles per minute.

    iii Estimate the overall thermal efficiency of the engine.

13 A fixed mass of gas in a heat pump undergoes a cycle of changes of pressure, volume and temperature as illustrated in the graph, Figure 7.7. The gas is assumed to be ideal.

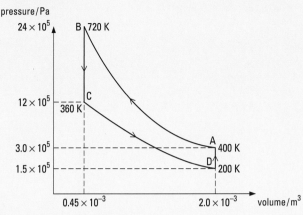

Figure 7.7

Table 7.2 shows the increase in internal energy which takes place during each of the changes A to B, B to C and C to D. It also shows that in both of sections A to B and C to D, no heat is supplied to the gas.

Table 7.2

	Increase in internal energy /J	Heat supplied to gas /J	Work done on gas /J
A to B	1200	0	
B to C	−1350		
C to D	−600	0	
D to A			

a Using the first law of thermodynamics and necessary data from the graph, complete a copy of the table. You will find it helpful to proceed in the following order.

  i work done on gas for A to B and C to D

  ii work done on gas for B to C and D to A

  iii heat supplied to gas for B to C

  iv increase in internal energy for D to A

  v heat supplied to gas for D to A

b Calculate *P*, the coefficient of performance of the heat pump, given that

  P = Heat delivered by gas (during change B to C).

  Net work done on gas

14 The indicator diagram in Figure 7.8 below is for one cylinder of a four-stroke petrol engine.

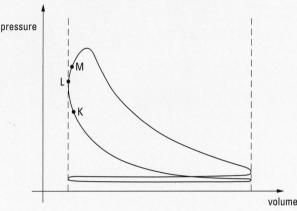

Figure 7.8

a The spark which ignites the mixture can be made to occur at one of the points on the diagram marked K, L or M. State at which of these points the spark is usually timed to occur, giving two reasons for this choice.

b i Explain what is meant by the indicated power of an engine.

  ii Explain fully how you would use the indicator diagram shown, if the pressure and volume axes were calibrated, to estimate the indicated power of this engine. State what extra information you would need.

c In order to calibrate the pressure axis, a small sensor is fitted inside the cylinder to measure the pressure of the gas at any time when the engine is running. State *one* factor, apart from the need to work at high pressures, which you consider to be important when choosing a sensor for this application.

* 15 When a semiconductor material such as germanium is connected to a constant p.d. the current flowing through it varies according to the equation $I = Ae^{-\Delta E/kT}$, where ΔE is the energy required for the release of each charge carrier (electrons and holes).

Table 7.3 shows how the current *I* through the smallest area of a slice of germanium of dimensions 10 mm × 10 mm × 1 mm varied with temperature *T* when a constant p.d. of 3.0 V was connected across it.

Table 7.3

T / K	300	340	380	420	460
I / mA	6	30	100	266	592

a Calculate the resistivity of the germanium slice at 340 K.

b For this experiment the Boltzmann factor is $e^{-\Delta E/kT}$. Explain how this factor arises in the mathematical model for the variation of current with temperature.

c Describe one other situation where the Boltzmann factor describes the effect of temperature on the process.

d By plotting a suitable graph determine the values of the constant *A* for the slice and the value of ΔE for germanium. Give your answer in eV.

e Discuss the significance of the constant *A*.

*Pre-U-style question

8

Modern PHYSICS

Three incredible years at the end of the nineteenth century saw the birth of modern physics. In 1895 Wilhelm Röntgen (1845–1923) discovered X-rays. In 1896 Henri Becquerel (1852–1908) discovered radioactivity. And in 1897 J. J. (Joseph John) Thomson (1856–1940) discovered the electron.

The major new disciplines of atomic, nuclear, and particle physics, relativity, and quantum theory developed rapidly during the next few decades. By 1925 the world of physics had been turned on its head. We are still trying to come to terms with the startling new ideas and the amazing technologies that rely on them. X-rays, which were initially used only for shadow pictures of broken bones, can now be used to generate sophisticated and detailed three-dimensional images of the human body. Radioactivity has many uses in industry, archaeology, and medicine, and while it can damage tissue it may actually have been the driving force behind evolution. Understanding the electron revolutionized the second half of the twentieth century through electronics. Today we can use electron microscopes to locate individual atoms and move them from one place to another, holding out the real prospect of nanotechnology – engineering on the molecular scale. Albert Einstein's (1879–1955) realization that mass and energy are intimately linked offers the prospect of unlimited and relatively safe energy, once we are able to control nuclear fusion reactions on Earth. The technology behind the Internet and global communications derives from the physics that enables integrated circuits to be made ever more complex and smaller. Underlying all of this is quantum theory. It will shatter your common-sense view of the world, but then even quantum physicists find it rather puzzling.

In this chapter you will start by looking at ideas about the atom and the electron. This leads on to an introductory discussion of one of the great theories of modern physics, quantum mechanics, in which we consider wave–particle duality and other fundamental quantum ideas that influence the atomic model. Once these ideas have been established, you will then study atomic spectra, radioactivity, and radiation. The other important theory of the twentieth century, relativity, is covered next, and the concept of the equivalence of mass and energy prefaces a discussion of the nucleus and nuclear energy.

A scanning electron micrograph of an ant, clasping an integrated circuit in its serrated jaws: modern physics at its most ingenious. (Magnification: 503.)

About models

Atomic models are used to help understand the properties of matter: if they do that successfully they are good models. As scientists have learnt more about the physical and chemical properties of matter they have had to adapt their current model to explain the new observations. A really good model allows the scientist to make predictions and then test them in experiments. It is wrong to think that all the old models are wrong and only the most up-to-date quantum model is right. In one sense none of them is right, and in another sense they all are. For example, the simplest way to explain the hexagonal close-packing of atoms in a metal lattice is to treat the atoms as featureless solid spheres, but this model is useless when trying to explain **thermionic emission** from a heated source (thermal radiation) or radioactive decay (see spread 8.11). Each model is most appropriate in a *particular situation* – so the most important thing to understand about a model is its limitations.

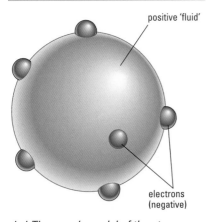

positive 'fluid'

electrons (negative)

J. J. Thomson's model of the atom became known as the plum-pudding model. Thomson and others used the model to explain many properties of elements including bonding, reactivity, and chemical periodicity.

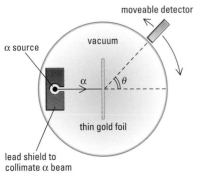

moveable detector

α source

vacuum

α

θ

thin gold foil

lead shield to collimate α beam

The Rutherford scattering experiment.

MODELS OF THE ATOM

Atomic theory goes back at least as far as the ancient Greeks. It is usually associated with the Greek philosopher Leucippus and his pupil, Democritus, in the 5th century BC. The name 'atom' comes from the Greek word *atomos* meaning 'uncuttable'. The Greeks' theories were described by the Roman poet Lucretius. The following short extract from his writings sounds remarkably modern, although it was written in the 1st century BC:

Things that seem hard to us and stiff must be composed of deeply indented and hooked atoms and held firm by their intertangling branches. In the front rank of this class stand diamonds…Liquids…must owe their fluid consistency to atoms that are smooth and round…A third class is constituted by things…such as smoke, cloud and flames. If their atoms are not all smooth and round…yet they cannot be jagged and intertangled.

This is a good mechanical atomic model. It is similar to the theory that grew up to explain chemical combinations and the kinetic theory of gases in the nineteenth century.

The electronic atom

Thomson's discovery of the electron in 1897 ended the idea of the 'indivisible atom'. He showed that electrons have a tiny mass (about 1/1840 of the mass of a hydrogen ion) and are present in atoms of all elements.

If atoms contain smaller particles, what are these particles and how do they fit together? Thomson soon turned his attention to atomic models and suggested the **plum-pudding** model, where the bulk of the atom consists of a positive material, in which electrons are embedded. In this model the negative charge of the electrons balances the positive charge in the rest of the atom, and the whole thing is held together by electrostatic forces. Ions form when atoms gain or lose electrons.

This model is simple and explains many of the same things as were explained by the early idea of indivisible atoms. It also offers some hope of explaining bonding in terms of electrical forces. However, the lack of detail about atomic structure makes it useless for tackling problems like atomic spectra (see spread 8.10) or radioactivity.

Rutherford scattering and the nuclear atom

Radioactivity was discovered in 1896, and Rutherford soon set to work using the **alpha particles** emitted by some radioactive sources as probes to investigate the atom. His most famous experiment was actually carried out under his guidance by two of his students, Geiger and Marsden. They fired collimated (parallel) beams of alpha particles at very thin pieces of gold foil, expecting the alpha particles to be deflected only slightly, if at all, because they knew how well alpha particles penetrated air and various thicknesses of absorbers, such as paper or card. They measured the rate of arrival of alpha particles at different scattering angles, using an arrangement similar to the one illustrated below left.

The results were startling.

- The vast majority of the alpha particles did penetrate the foil with little or no deflection.

- However, a few more were deflected through large angles.

- A very small proportion (about 1/8000 in the original experiment) bounced back off the foil.

Rutherford said it was as if, very occasionally, a fifteen-inch shell bounced back off a sheet of tissue paper. He knew this was telling him something about the internal structure of the atom, but what? It certainly meant that the atoms were not uniform as Thomson had suggested: if they were, then all the alpha particles would do more or less the same thing. The ones that bounced back must have 'hit' something with a large mass and probably a high charge too – he called this the **nucleus** of the atom. Since most of the alpha particles got through virtually undeflected, the *chance* of hitting the nucleus must be very small (remember the gold foil was very thin but still consisted of hundreds of layers of atoms), so the nucleus must be much smaller than the atom. He went on to derive a mathematical equation (the **Rutherford scattering equation**) to predict how the alpha particles would be scattered from the foil, based on the following assumptions.

- Most of the atom is virtually empty space.
- In its centre is a tiny nucleus, which has most of the mass of the atom and all its positive charge. See spread 8.26 to find out how Rutherford used the closest approach of an alpha particle to estimate the radius of the nucleus
- Electrons orbit the nucleus rather like planets orbit the Sun, bound to the nucleus by an electrostatic attraction.
- Alpha particles scatter from the nucleus because of the electrostatic repulsion between like charges.

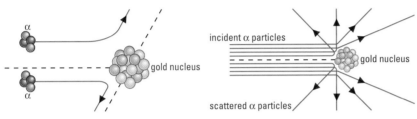

The closer an alpha particle comes to the nucleus, the larger the angle through which it is deflected. Since nuclear radii are about 10^{-5} of the atomic radius, most alpha particles are barely scattered at all. This diagram shows scattering for the tiny minority that pass relatively close to the nucleus. Nuclear radii vary in size from about 1 to 6 fm (1 fm = 10^{-15} m).

Alpha particles fired at foils made from different materials confirmed that the charge on the nucleus increases with the atomic number of the element in the periodic table; different elements must have different nuclear charges. Hydrogen has a single positive charge on the nucleus, helium has two, lithium has three, and so on. So the nucleus too seems to be made up of smaller particles, some of which are positively charged. However, although a helium nucleus has double the charge of a hydrogen nucleus, it has *four times* the mass (roughly), so helium nuclei are not just two hydrogen nuclei stuck together.

For a while the nucleus was thought to contain only protons and electrons, but this theory was abandoned for reasons discussed later. Rutherford suggested that the nucleus might contain yet another subatomic particle to make up the mass. This particle would have about the same mass as a proton and no electric charge. It was named the **neutron**, and was finally discovered by James Chadwick, one of Rutherford's students, in 1932. Protons and neutrons are collectively known as **nucleons** (because they are the constituents of all nuclei).

This model explained the differences between elements – they have different numbers of protons in their nucleus. It also explained why atoms of the same element may come in forms which have different masses – the number of neutrons can vary. Atoms of the same element that contain different numbers of neutrons in their nuclei are called **isotopes**.

Some useful terminology

Terminology	Explanation
Electron	tiny negatively charged particle orbiting the nucleus
Nucleus	central massive positive charge in an atom
Nucleon	a proton or a neutron
Proton	positively charged nuclear particle (also hydrogen nucleus)
Neutron	neutral nuclear particle, like an uncharged proton
Nuclide	particular type of nucleus, e.g. carbon-12
Isotopes	nuclei of the same element with different numbers of neutrons (same Z, different N and A)
Alpha particle	a helium nucleus ejected in radioactive decay
Ion	charged atom, i.e. one that has gained or lost one or more electrons

---

Symbolic representation

The structure of a particular nuclear species or nuclide can be written down symbolically as:

$$^A_Z X$$

- X is the chemical symbol for the element.
- A is the mass number (or nucleon number), equal to the number of neutrons plus protons in the nucleus.
- Z is the atomic number (or proton number), equal to the number of protons in the nucleus. It is also equal to the number of orbital electrons in the neutral atom.
- N is also sometimes used. This is the neutron number, and is equal to A – Z.

E.g. $^{12}_6 C$ and $^{14}_6 C$ are two isotopes of carbon. The former is the normal form and the latter is the radioactive isotope carbon-14 used in radiocarbon dating.

Mass number and atomic mass

A is a *whole number*, not a mass. It is equal to the number of nucleons in the nucleus. However, it is close, numerically, to the mass in grams of 1 mole of atoms. For example, iron-56 has a mass number 56 and an atomic mass 55.847 g. This is no coincidence – it occurs because of the way the mole is defined (as the number of atoms in 12 g of carbon-12) so there is a similar correspondence for all elements. It is useful too – if accuracy to three figures is not essential then A g can be used as the atomic mass.

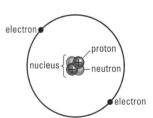

Rutherford's planetary atom. This is helium-4. If it were drawn to scale with the nucleus this size, the nearest electrons would be several hundred metres away.

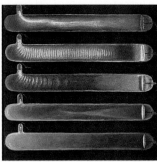

Crookes tubes. Accelerated electrons collide with gas molecules and ionize them. Light is emitted when the positive and negative ions re-combine to form neutral molecules.

Specific charge

Specific charge $= \dfrac{q}{m}$

q = particle charge

m = particle mass

For an electron: $\dfrac{e}{m} = 1.76 \times 10^{11}\,\text{C kg}^{-1}$

For a hydrogen
ion (proton): $\dfrac{q}{m} = 9.6 \times 10^{7}\,\text{C kg}^{-1}$

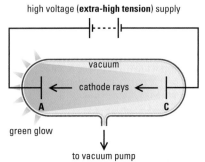

high voltage (**extra-high tension**) supply

vacuum

← cathode rays ←

A

C

green glow

to vacuum pump

A is anode (+)   **C** is cathode (−)

The region of glass behind the anode glows green where the cathode rays hit it.

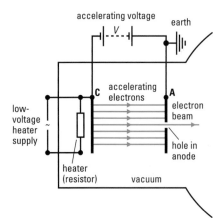

accelerating voltage

earth

V

C accelerating electrons

A

electron beam

low-voltage heater supply

heater (resistor)

vacuum

hole in anode

An electron gun, similar to those in a TV or an oscilloscope. In practice the electrodes are more complicated than this in order to control the beam intensity and focus.

THE DISCOVERY OF THE ELECTRON

The electron was 'discovered' by J. J. Thomson in 1897. It could not have been discovered much earlier, certainly not before William Crookes's experiments in the 1870s. Crookes worked out how to produce very low pressures in sealed glass tubes, and began to use these **Crookes tubes** to investigate the flow of electricity through gases at low pressure. Having demonstrated that low-pressure gases do indeed conduct electricity, he noticed some strange and interesting effects. Patterns of glowing gas appeared in the tubes, and their colours were characteristic of the gases used. Further investigations of these gaseous discharges by Crookes and others provided vital clues that Thomson eventually used to identify the electron.

It was not only the gas in the tube that glowed. The glass behind the anode glowed as well, and it was assumed this was because it was being hit by rays streaming from the cathode. These rays became known as **cathode rays**. Thomson wondered whether they might be streams of negatively charged particles. To carry out experiments using cathode rays, a hole was made in the anode so that some of the rays could stream through into the near-vacuum beyond. A range of experiments established that:

• Cathode rays transfer energy, momentum, and mass.

• Cathode rays transfer a negative charge.

• The charge-to-mass ratio (called **specific charge**) of cathode rays is much larger than that for hydrogen ions.

• Cathode rays have the same properties whatever gas is used in the tube, and whatever metal is used as the cathode.

If Thomson's particle hypothesis was correct these particles must have mass and be negatively charged, but the most radical conclusion was based on the specific charge, which for cathode rays is about 1840 times bigger than for hydrogen ions. It seemed highly unlikely that each particle in a cathode ray would carry greater charge than a hydrogen ion (since the existence of such particles would have shown up in other experiments, e.g. in electrolysis). So Thomson assumed each 'cathode-ray particle' carried a charge equal to that on a hydrogen ion.

If this was true each one must have a mass *many times (about 1840 times) smaller than that of the hydrogen ion.* Hydrogen is the lightest element, so this implied that the negative particles were much less massive than atoms. Cathode rays from all elements are identical, so these particles must be found in all atoms. Thomson concluded that the electron is a *subatomic* particle.

Maths box: calculating the speed of the electrons

The speed with which the electrons emerge from the hole in the anode depends on the potential difference V between the cathode and anode. Assume that electrons leaving the cathode by thermionic emission have negligible speeds.

Work done on an electron as it moves from the cathode to the anode:

$$W = eV$$

The electron is in a vacuum so this work creates kinetic energy:

$$\tfrac{1}{2}mv^2 = eV$$

The speed of the electron is therefore

$$v = \sqrt{\frac{2eV}{m}}$$

For an accelerating voltage of 1000 V the electron velocity is

$$v = 1.9 \times 10^7\,\text{ms}^{-1}$$

This is about 6% of the speed of light.

If the accelerating voltage is much higher than this, we must use relativistic equations (see spread 8.25).

Thermionic emission and the electron gun

The experiments that established the ideas described above were carried out using Crookes tubes. Nowadays cathode rays (or electron beams) are produced by heating the cathode. This increases the thermal energy of electrons in the cathode and they leave the surface more easily – a process called **thermionic emission**, which is rather like the evaporation of molecules from a liquid surface. Once free they are accelerated in the electric field between the cathode and anode. The whole device is mounted in a vacuum and called an **electron gun**. These are used in TVs and oscilloscopes to generate the electron beams that scan across the screen.

<table><tr><td>

Anode potential

You might be wondering why the electron beams are not slowed down after they have passed through the hole in the anode. Shouldn't they be attracted back to it if it is positive? Electrons accelerate or decelerate because a force is applied to them. This force is exerted by an electric field, and electric fields only exist between points of different potential, like the anode and cathode. If the anode is earthed it will be at the same potential as the surroundings, so there will be no electric field beyond the anode to affect the motion of the electrons in any way.

</td></tr></table>

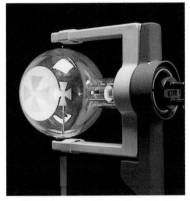
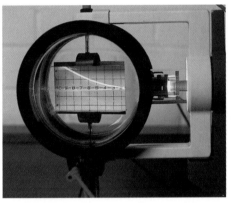

A Maltese cross tube. If the cross is earthed and connected to the anode, the electrons travel in straight lines to form a sharp 'shadow'. If the cross is unconnected it will gain a negative charge, and the shadow distorts as the electrons are deflected away from it.

A deflection tube. Electrode plates above and below the beam provide a vertical electrostatic deflection. A pair of Helmholtz coils either side of the beam provides vertical magnetic deflection (Fleming's left-hand rule).

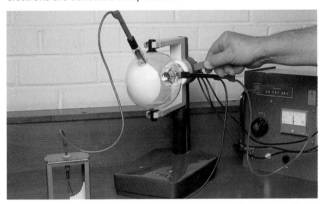

The Perrin tube. A bar magnet can be used to deflect the cathode rays into the collecting can at the end of the tube. The tube, which is connected to the electroscope, becomes negatively charged. This shows that cathode rays transfer negative charge.

The fine-beam tube. Low-pressure gas inside this tube fluoresces as electrons collide with it, making the beam path visible as a coloured line.

PRACTICE

1 Why do gases at low pressure conduct electricity more easily than gases at high pressure?

2 Descibe and explain the effect on the operation of an electron beam tube if:

   a The heater supply voltage is increased.

   b The accelerating voltage is increased.

   c A leak develops and the tube slowly fills with air.

   d The polarity of the accelerating voltage is reversed.

   e The accelerating voltage is switched to a.c.

   f The separation of anode and cathode is reduced.

3 Calculate the velocity of electrons emerging from the anode of an electron gun if the accelerating voltage is:

   a 100 V

   b 1000 V

   c 10 000 V.

In each case express your answer as a fraction of the speed of light, 3.0×10^8 m s^{-1}, and say whether you think it was valid to use non-relativistic equations.

4 Give evidence to support the ideas that:

   a Cathode rays transfer negative charge.

   b Cathode rays transfer energy and linear momentum.

OBJECTIVES

- the deflection tube
- the fine-beam tube
- Millikan's experiment

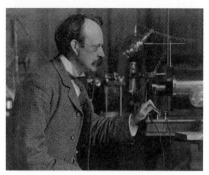

Sir J. J. Thomson won the 1906 Nobel Prize for Physics and was knighted in 1908.

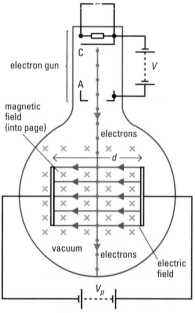

The experimental set-up to measure e/m using a deflection tube. Helmholtz coils are not shown. V_p is adjusted until the beam is undeflected. The magnetic and electrostatic forces on the electrons are then balanced.

Properties of the electron

- Charge magnitude*
  $e = 1.602 \times 10^{-19}$ C.

- Charge sign is negative.

- Mass $m = 9.110 \times 10^{-31}$ kg.

- Specific charge
  $e/m = 1.759 \times 10^{11}$ C kg^{-1}.

- The electron also has one half unit of spin which is responsible for its magnetic moment, and is thought to be a structureless point.

*Although the quarks inside hadrons carry fractional charge, they cannot become free particles, so e is the smallest unit of free electric charge – the **quantum** of electricity.

THE MASS AND CHARGE OF AN ELECTRON

The crucial piece of evidence that confirmed J. J. Thomson's idea that cathode rays consist of streams of subatomic, negatively charged particles (**electrons**) was the large value of their charge-to-mass ratio, e/m. This can be measured in various ways, by deflecting electron beams in electric and magnetic fields. Two methods that are often carried out in school laboratories are described below. The existence of the electron shows that electric charge is **quantized** (i.e. discrete not continuous), but it was over ten years after the discovery of the electron that this **quantum** of charge was measured, in an ingenious experiment carried out by Robert Millikan. He showed that the tiny charges on oil drops in a fine spray are always multiples of one value, the electronic charge e.

The deflection tube method for measuring e/m

This method balances the electric and magnetic forces on moving electrons as they pass through a region of crossed electric and magnetic fields.

The electrostatic force F_{es} on an electron in an electric field of strength E is

$$F_{es} = eE$$

This force is in the opposite direction to the lines of the electric field (because of the electron's negative charge). The magnetic force F_m on an electron moving at velocity v perpendicular to a magnetic field of strength B is

$$F_m = Bev$$

Free-body diagram.

The force direction is given by Fleming's left-hand rule. If these forces are equal and opposed to one another, the electron beam is undeflected, and

$$eE = Bev \qquad \text{giving} \quad v = \frac{E}{B} \tag{1}$$

However, the velocity is also related to the accelerating voltage V by

$$eV = \frac{1}{2}mv^2 \qquad \text{giving} \quad v = \sqrt{\frac{2eV}{m}} \tag{2}$$

Equations (1) and (2) lead to an equation for the specific charge, e/m:

$$v^2 = \frac{2eV}{m} = \frac{E^2}{B^2} \qquad \text{from which} \quad \frac{e}{m} = \frac{E^2}{2VB^2} \tag{2}$$

If the electric field is between parallel plates separated by a distance d with a potential difference V_p between them, then the magnitude of the field strength E is

$$E = \frac{V_p}{d}$$

Substitute this into equation (3) to get

$$\frac{e}{m} = \frac{V_p^2}{2VB^2d^2}$$

This equation is useful because everything on the right-hand side can be measured. Both voltages can be measured with voltmeters and the plate separation can be measured using a ruler. The magnetic field is usually set up using Helmholtz coils, one either side of the vacuum tube. The field strength (B) is then calculated from the current in the coils and their arrangement using a standard formula for Helmholtz coils:

$$B = \frac{0.72\mu_0 NI}{r}$$

(N is the number of turns, I the current, and r the radius of the coils.)

- The accepted value for the specific charge of the electron is
  $$e/m = 1.759 \times 10^{11}\,\text{C kg}^{-1}$$

The fine-beam tube method for measuring *e/m*

Unlike deflection tubes, which show the beam only where it hits a fluorescent screen, the fine-beam tube makes *all* of the electron beam visible. Inside the tube is some low-pressure helium. Helium atoms are ionized by electrons in the beam and emit photons of visible light as they re-combine with electrons to become helium atoms once again. If a magnetic field is applied perpendicular to the electron beam direction, the magnetic force, which is always perpendicular to the beam and the field, acts as a centripetal force and makes the electrons move in a circular path:

$$\frac{mv^2}{r} = Bev \quad \text{or} \quad v = \frac{Ber}{m}$$

where r is the radius of the electron path. From equation (2) opposite,

$$v = \sqrt{\frac{2eV}{m}}$$

giving

$$\frac{Ber}{m} = \sqrt{\frac{2eV}{m}} \quad \text{or} \quad \frac{e}{m} = \frac{2V}{B^2r^2}$$

when simplified. Once again, all the quantities on the right-hand side of the equation can be measured experimentally.

The Millikan oil drop experiment

The purpose of this experiment is to measure the charge on an electron. In a fine spray many oil droplets become charged by friction as they emerge from the nozzle of the spraying device. Millikan used the uniform electric field between a pair of parallel plates to exert an upward electrostatic force on individual drops, in order to balance their weight so they remained suspended between the plates. This allowed him to measure the charge on a suspended drop.

He repeated the experiment thousands of times, recording the charge on each oil drop. All charges turned out to be a multiple of a unique value, 1.6×10^{-19} C (the highest common factor). He concluded that this must be the charge on an individual electron, and that oil drops become charged by gaining or losing electrons.

electrostatic force, $F_{es} = qE$ (q is the charge on the drop.)

weight, $W = Mg$ (M is droplet mass)

so, when the drop is in equilibrium, $qE = Mg$ or $q = \dfrac{Mg}{E}$

(and $E = V_p/d$). Once the charge was known, the electron's mass was calculated by substituting e into e/m. Its mass is about 9.1×10^{-31} kg, a little over 1/2000 of the mass of a nucleon.

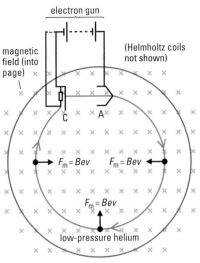

electron gun

magnetic field (into page)

(Helmholtz coils not shown)

$F_m = Bev$ $F_m = Bev$

$F_m = Bev$

low-pressure helium

The fine-beam tube method for measuring e/m.

Rotating the tube

If the fine-beam tube is rotated so that the electron beam is not perpendicular to the field, then the electrons' general path is a helix.

The component of electron velocity parallel to the field is not affected by the magnetic field, so this is the direction of the axis of the helix. The motion of the electrons in the plane perpendicular to the field is circular.

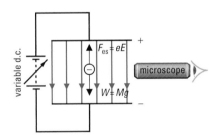

variable d.c.

$F_{es} = eE$

$W = Mg$

microscope

The experimental arrangement for Millikan's intricate oil drop experiment. The drops were illuminated from the side and viewed through a microscope.

PRACTICE

1 An electron beam is accelerated through a potential difference of 250 V. The electrons then enter a region of uniform magnetic field, perpendicular to the beam, of strength 0.02 T.

   a Calculate the speed and kinetic energy of the electrons when they leave the electron gun.

   b If the anode and cathode are separated by 1.6 cm, calculate the acceleration of an electron between these electrodes.

   c An electric field is now applied so that the electrostatic force opposes the magnetic force on the electrons and the beam goes straight. Calculate the size of the electric field.

2 Calculate the magnetic field strength required to deflect electrons into a circle of radius 8 cm if the accelerating voltage in the electron gun is 400 V.

3 Below are the magnitudes of some of the charges Millikan measured on his oil drops. In each case the charge is given as a multiple of 10^{-18} C. Use these values to suggest a value for the charge on an electron and use this and *e/m* to estimate the mass of an electron.

1.90, 2.24, 1.76, 0.803, 1.13, 1.29, 1.13, 0.966, 0.803, 0.643

Explain how you arrived at your result.

QUANTA

If you were a physicist in the late nineteenth century you would probably feel quite pleased with yourself. There was a particle model to explain the properties of matter and a wave model to explain light. Mechanics and electromagnetism together appeared to be able to explain everything. All that remained was to find a link between the two and the Theory of Everything would be complete. But life is never that easy! Very few late nineteenth-century physicists ever came to terms with the revolutionary new discoveries that came with the new century.

Black-body radiation

All hot bodies radiate. An ideal radiator is called a **black body** and the spectrum of radiation from a black body was well known to 19th-century physicists. The problem was to *derive* the spectrum from mechanics and electromagnetism. Until 1899, no one had managed to do this, and that was not for want of trying! The obstacle they had encountered became known as the **ultraviolet catastrophe** (see left).

In 1900 Max Planck, a German physicist, came up with a 'desperate remedy'. He showed that an accurate equation for the spectrum *could* be derived as long as one new assumption was added to those of classical physics. He assumed that the oscillators that emit radiation can only have **discrete energies**. Each oscillator can have zero energy or some multiple of a fixed amount (quantum) which depends on the frequency f of oscillation according to the formula

$$E = nhf$$

n is an integer, 0, 1, 2, ..., and h is a new constant, now known as the **Planck constant**:

$$h = 6.626 \times 10^{-34} \, \text{J s}$$

How does this fix the ultraviolet catastrophe? The shorter wavelengths correspond to higher frequencies, so the oscillators responsible for radiation in this part of the spectrum need a lot more energy to get into even the first vibration state than those emitting radiation at a longer wavelength (lower frequency). Thermal energy is randomly distributed, so the chance that high-frequency oscillators will get enough energy to start vibrating (at least hf) is much smaller than for the lower frequency oscillators. The result is that if energy is quantized in this way the high-frequency oscillators are 'switched off' and the intensity of the spectrum at high frequencies drops down rapidly to zero – exactly as observed. (In classical physics *all* oscillation frequencies would have been excited, and the cumulative effect was the ultraviolet catastrophe.)

Planck and other physicists were uneasy about this new idea, but there seemed to be no other way to explain the black-body spectrum. The inescapable conclusion was that

- electromagnetic radiation is emitted in discrete energy packets or **quanta**.

The photoelectric effect

Another problem that arose late in the nineteenth century concerned the way light falling on some metal surfaces could eject electrons from them. This is called the **photoelectric effect**. According to wave theory, light energy is spread evenly across the wavefront, so electrons should be emitted only if enough energy is delivered close to an electron on the surface. Also, the ejection should depend only on the *intensity* of the incident light, and not on its frequency. Neither of these expectations was borne out in practice. Experiments led to these 'laws' of photoelectricity.

- For any metal, electrons are only emitted if the frequency of the incident light is above some threshold value f_0. (So weak ultraviolet can emit electrons from zinc, whereas very intense infrared *cannot*, even

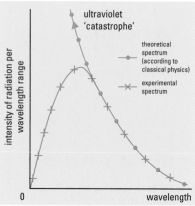

The spectrum of radiation emitted by a perfect radiator (black body) proved impossible to derive from the assumptions of classical physics. Classical physics assumed that the thermal energy of a hot body is shared out among the atoms, making them vibrate, resulting in the emission of electromagnetic waves. The equations that came out of this were a good fit to the black-body spectrum at long wavelengths but did not work for short wavelengths. In fact, they predicted that an infinite amount of energy would be radiated at short wavelengths. This was obviously wrong (among other things, it violated the law of conservation of energy), and it came to be known as the ultraviolet catastrophe (because ultraviolet light has a shorter wavelength than visible light).

Emission of radiation

In 1900 there was no clear atomic theory, and no one knew what 'oscillates' to emit radiation. A good way to think about the radiation is to imagine the electromagnetic field excited into various modes of oscillation at different frequencies. Each mode can have an integer number of quanta of energy. These quanta of the electromagnetic field are what later became known as photons. In fact, the black-body radiation in an enclosure can be treated like a gas of photons in thermal equilibrium with the walls of the enclosure.

though it is delivering far more energy per second to each unit area of the zinc surface.)

- The threshold frequency depends on the metal and is usually lower for more reactive elements (so electrons are emitted from potassium more readily than from zinc, and from zinc more readily than from copper).

- The maximum kinetic energy of the ejected electrons depends only on the *frequency* of the incident radiation and is proportional to the difference between the light frequency and the threshold frequency:

$$KE_{max} \propto (f - f_0).$$

Einstein, who was aware of Planck's work, tackled the photoelectric effect in 1905. He saw that all the experimental laws could be explained if it was assumed that atoms can only absorb light energy in discrete 'energy packets' or **quanta**, and that the size of one quantum is proportional to the frequency of the light and given by

$$E = hf$$

These quanta became known as **photons**, and Einstein won the 1921 Nobel Prize for Physics for this work. Photons solve all of the problems with which wave theory had difficulty.

- Photons are indivisible, so each photon gives all its energy to one electron. If there is a minimum or threshold energy required to eject electrons from a particular metal surface, then there will be a minimum photon energy that can do this. Photon energy is proportional to frequency, so electrons are only ejected with light above a certain threshold frequency. Increasing the intensity of light does not affect the energy of individual photons, only the number arriving per second.

- The minimum energy required to free an electron from the surface depends on the metal, so the threshold frequency changes from one metal to another. Reactive metals lose electrons easily, so less energy is required and their threshold frequency is lower.

- If the light frequency is only just above the threshold frequency, the photon energy is only just sufficient to eject electrons, so there is little left over for kinetic energy. The maximum kinetic energy of ejected electrons can be no greater than the difference between the photon energy and the threshold energy. This is directly proportional to the difference between light frequency and threshold frequency.

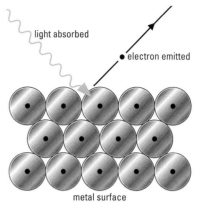

The photoelectric effect. If light of high enough frequency falls onto a metal surface, it can eject electrons.

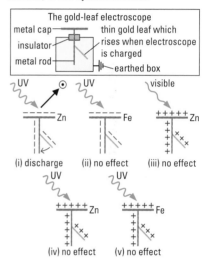

Each diagram is of a charged gold-leaf electroscope illuminated by ultraviolet or visible light. Neither ultraviolet nor visible light emits electrons from iron. Electrons are only emitted from the negatively charged electroscope with the zinc plate when it is illuminated with ultraviolet light.

PRACTICE

1 Explain in terms of photons why, when you turn on a filament lamp, the filament changes from red-hot to white-hot.

2 a Estimate the number of photons emitted per second from an electric lamp that emits 20 W of visible light. (You will need to estimate the average frequency of visible light.)

  b Explain why the 'lumpiness' of light is not obvious when you look at things.

3 Electrons can be emitted from the surface of zinc by ultraviolet light, but not by visible light. On the other hand, electrons can be emitted from potassium even by visible light.

  a Why is this?

  b How would electrons emitted from zinc compare with those emitted from potassium if both metals were illuminated with ultraviolet light of the same frequency?

4 Explain each of the following.

  a If the intensity of ultraviolet light directed at a piece of zinc is doubled the number of electrons leaving the surface per second also doubles, but their maximum kinetic energy is unchanged.

  b The maximum kinetic energy of photoelectrons is directly proportional to the difference between the frequency of light falling on the surface and the threshold frequency for that metal.

  c Gamma-ray photons are more harmful to people than infrared photons.

5 Compare the frequencies and photon energies (in J and eV) for microwaves (wavelength 3 cm), light (wavelength 500 nm), and X-rays (wavelength 10^{-10} m). In each case compare your values with a typical ionization energy of about 30 eV.

6 Explain why atomic spectra are usually made up of visible or near-visible radiation and nuclear spectra are usually made up of gamma rays.

- measuring h
- photocells
- optoelectronics

The quantum eye

There is some evidence that suggests the human eye can detect *individual* photons in very dark conditions.

Work functions

caesium	1.4 eV
potassium	2.2 eV
sodium	2.3 eV
zinc	4.2 eV
iron	4.5 eV
silver	4.7 eV

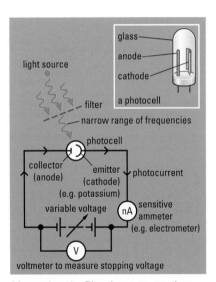

Measuring the Planck constant using a photocell.

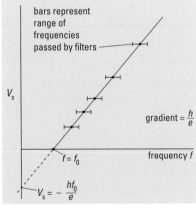

Typical experimental results.

PHOTON PHYSICS

Planck's constant fixes the scale of energy on which quantum effects will be obvious. Its tiny value means that most events on a human scale involve so many quanta that the intermittent nature of energy transfer cannot be distinguished from the continuous energy transfer assumed in classical physics. For example, if a lamp emits 10 W of visible light this corresponds to about 2.5×10^{19} photons per second streaming away from the source! On the other hand, gamma-ray photons carry about a million times more energy than visible photons, and the intermittent clicking of a geiger counter near a gamma-ray source does seem to suggest that the energy is given off in discrete chunks. The quantum nature of light has a direct effect on our lives through **optoelectronics**, a branch of modern electronics that relies on the interaction of photons with atoms and electrons inside semiconductors.

Measuring the Planck constant

The Planck constant can be measured in the school laboratory by finding out how the kinetic energy of the ejected electrons varies with the frequency of the incident radiation. If light falls onto a metal plate and ejects electrons, the photon energy is transferred directly to individual electrons. The minimum energy required to eject an electron from the surface is called the **work function** for the metal. The minimum frequency at which electrons will be emitted is found when the photon energy is equal to the work function, ϕ:

$$hf_0 = \phi$$

If the frequency is above this threshold ($f > f_0$), each photon transfers more than enough energy to eject an electron, so the electron gains kinetic energy as well. The maximum kinetic energy (E_k) of any electron is the difference between the energy of the incoming photon and the work function of the metal $(hf - \phi)$.

photon energy, $\quad E = hf$
work function, $\quad \phi = hf_0$
kinetic energy, $\quad E_k \leq E - \phi$

or

$$E_k \leq hf - hf_0$$

If E_k and f are measured, h is the gradient of a graph of E_k versus f. In practice E_k is found indirectly by measuring the potential difference required to stop electrons ejected from the surface. This potential difference is called a **stopping voltage**. The first experiment of this kind was carried out by Millikan, and the value found for h agreed with the value obtained from Planck's own work on the black-body spectrum. A suitable experimental arrangement is shown in the diagram (left).

Filters are used to reduce the continuous band of frequencies from the white light source to a narrow range, but this means that the frequency is still not a unique value and a bar rather than a point must be plotted on the graph. The voltage supply V makes the collecting electrode negative with respect to the emitting electrode. This means the electrons (which are also negative) must transfer kinetic energy to electrical potential energy as they approach the collector. The work they do is eV. For each filter the p.d. is increased until the current in the circuit is reduced to zero. The voltage which is needed to do this is called the stopping voltage, V_s.

At this voltage the work needed to move between the electrodes is equal to the maximum kinetic energy of the electrons.

$$eV_s = E_{k\,max} = hf - hf_0 \qquad \text{or} \qquad V_s = \left(\frac{h}{e}\right)f - \left(\frac{hf_0}{e}\right)$$

A graph of V_s versus f has a gradient h/e and intercepts $-hf_0/e$ on the voltage axis and f_0 on the frequency axis. If e is known, the Planck constant and the work function can be calculated.

Light-emitting diodes (LEDs) and *h*

Semiconductor diodes consist of two types of semiconductor in contact with one another. One type (**n type**) has an abundance of negative charge carriers, electrons, while the other type (**p type**) has an abundance of positive charge carriers, called **holes** because each arises from the absence of an electron. When they conduct, electrons at the junction can make a **quantum jump** into the holes and release a photon of energy. LEDs work because the material is transparent and the quantum jump has an energy that corresponds to a photon in the visible part of the spectrum. Some simple measurements on LEDs give a surprisingly accurate value for the Planck constant.

Assume electrons lose energy only to emitted photons as they pass through the LED (i.e. ignore the resistance of the material). They will emerge at a lower electrical potential energy because of the energy they have lost to the photon, so there must be a forward voltage drop across a conducting LED. Call this voltage *V* and measure it with a voltmeter. Now measure or estimate the wavelength of the light emitted by the LED, and call it λ.

$$\text{photon energy,} \quad E = hf = \frac{hc}{\lambda}$$

electron energy loss, $\quad E = eV$

So by the conservation of energy

$$eV = \frac{hc}{\lambda} \quad \text{or} \quad h = \frac{\lambda eV}{c}$$

When you try this experiment, you can make the result more accurate by plotting a graph of *V* against the current through the LED and extrapolating back to zero current to find the voltage drop caused only by photon emission (ohmic heating is zero when $I = 0$).

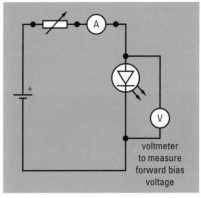

Measuring the Planck constant using an LED. The wavelengths emitted by the LED can be measured using a diffraction grating and spectrometer.

Reversed LEDs

Photon theory arose when it was realized that light is absorbed and emitted in quanta. LEDs emit quanta when electrons drop down a potential difference in a forward-biased semiconductor diode. Can this too be reversed? In other words, if you shine light of the right frequency onto a reverse-biased LED, will the photons give electrons a large enough 'kick' to get them up the potential gradient and cause a current to flow? The diagram shows a circuit that could be used to test this idea. If green LEDs are used then photons from blue light should be energetic enough to get the electrons over the potential barrier, while those from red light will not be. Colour filters can be used to restrict the range of photon energies. Try this experiment. (You could also investigate how the intensity of the incident light affects the current.)

An array of solar cells.

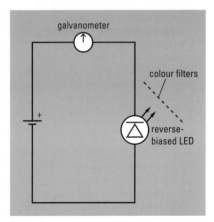

Investigating reversed LEDs in photon theory.

PRACTICE

1 Use the table of work functions opposite to answer the following questions.

   a Which of the metals in the table will emit electrons in visible light?

   b Which of these metals has the greatest threshold frequency?

   c Will any of the metals emit electrons when exposed to infrared radiation ($\lambda > 700\,\text{nm}$)?

   d What is the maximum kinetic energy of electrons emitted from the surface of zinc when it is illuminated by electromagnetic radiation of wavelength 300 nm?

   e What stopping voltage would be needed to prevent electrons emitted from a sodium electrode from reaching a copper electrode if the sodium is illuminated with light of wavelength 500 nm?

2 What stopping voltage must be applied to a photocell with a potassium emitter if it is illuminated by light of wavelength 350 nm?

3 What is the maximum velocity of electrons emitted from the surface of caesium when it is illuminated by ultraviolet light of wavelength 380 nm? Will all the emitted electrons have this velocity? Explain your answer.

• de Broglie's hypothesis

• electron diffraction

Prince Louis de Broglie was the first to suggest that matter, like light, has both wave-like and particle-like properties.

Wave–particle duality

De Broglie's idea was radical, but the wave-like aspect of electrons was soon demonstrated experimentally by Davisson and Germer, and independently by G. P. Thomson, the son of J. J. Thomson (discoverer of the electron). It is rather ironic that J. J. Thomson got the Nobel Prize for showing that the electron 'is' a particle and G. P. Thomson got one for showing it 'is' a wave. No wonder **wave–particle duality**, as it came to be called, was so controversial!

Relativistic momentum

At very high energies the Newtonian equations for kinetic energy and momentum are no longer valid. In fact the relation between total energy and momentum for electrons (or any other matter particle) gets closer and closer to the relation for photons: $p = E/c$. This equation must therefore be used to calculate the de Broglie wavelength, $\lambda = h/p = hc/E$. This will be necessary whenever v is close to c or $E \gg E_0$.

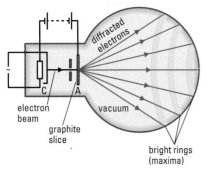

The principle of an electron diffraction tube.

MATTER WAVES

Prince Louis Victor Pierre Raymond de Broglie studied history at the Sorbonne in Paris, but shifted his interest to physics when he was posted to a signals unit based at the Eiffel Tower during the First World War. In 1924, as part of his doctoral thesis, de Broglie proposed that electrons, previously thought of as particles, might also have a wave-like character. This would make the physical description of matter more like that of light, which also has wave and particle-like properties.

The de Broglie relation

De Broglie's argument ran roughly as follows.

• Wave-like properties of light are fixed by a wavelength, λ.
• Particle-like properties are fixed by a linear momentum, $p = mv$.
• These can be linked for a photon by using Einstein's mass–energy relation and his formula for photon energy:

$$\text{photon energy}, E = hf = mc^2$$

$$\text{photon momentum}, p = mc$$

so

$$p = \frac{hf}{c} = \frac{h}{\lambda} \qquad \text{or} \quad \lambda = \frac{h}{p}$$

De Broglie proposed that the same relation would hold for an electron – its wavelength would be inversely proportional to its linear momentum with a value fixed by the Planck constant.

Worked example

What is the 'de Broglie wavelength' of an electron which has been accelerated through a potential difference of 1000 V? (Assume non-relativistic formulae are suitable.)

The de Broglie relation links wavelength to momentum, so we need the momentum, p. We get p from kinetic energy.

$$\text{loss of electrical potential energy} = eV$$

This is transferred to kinetic energy:

$$E_k = \tfrac{1}{2}mv^2 = \frac{p^2}{2m} = eV$$

(This is a very useful expression for kinetic energy since it links it directly to linear momentum.)

$$\frac{p^2}{2m} = eV \qquad \text{or} \qquad p = \sqrt{2meV}$$

Then

$$\lambda = \frac{h}{p} = \frac{h}{\sqrt{2meV}} = \frac{6.6 \times 10^{-34}\,\text{J s}}{\sqrt{2 \times 9.1 \times 10^{-31}\,\text{kg} \times 1.6 \times 10^{-19}\,\text{C} \times 1000\,\text{V}}} = 3.9 \times 10^{-11}\,\text{m}$$

This is comparable to the wavelength of hard X-rays, so interference effects might occur when electron waves pass through structures on this scale, as they do for X-rays. (Of course, the de Broglie relation applies to all kinds of matter, not just electrons.)

Electron diffraction

Light has a wavelength of the order of 10^{-7} m, and its wave-like properties only become obvious when light interacts with regular structures on a similar scale (as in Young's double-slit experiment). The predicted wavelengths for electrons produced by electron guns in the early part of this century were about a thousand times smaller, 10^{-10} m or less, a scale of distance comparable to the spacing of atoms in a crystal lattice. So electron diffraction would be expected from regular structures on an atomic scale, crystals for example. Sure enough, electron diffraction patterns from crystals were observed and measured in experiments carried out on both sides of the Atlantic in 1925.

The Americans Clinton Davisson and Edmund Germer directed electron beams at single crystals of nickel. They found that electrons scattered by the crystal formed a regular pattern, with many electrons detected in certain directions and very few in others. This is like the maxima and minima of X-ray diffraction patterns (see spread 8.34). Davisson and Germer were able to show that the wavelength of X-rays which would create the same pattern was equal to the de Broglie wavelength for the electrons they were using in the experiment. Their conclusion was that the de Broglie waves had scattered from atoms in the crystal and interfered to produce maxima and minima in the reflected beam. The electrons were then detected close to interference maxima, and not near the minima.

G. P. Thomson used a slightly different approach. He fired electron beams at thin sheets of metal in a vacuum tube. Metals have a crystal lattice, but all consist of many small grains, in each of which the atomic layers are in different planes. This means that incoming electrons can diffract from any layers that are in the right orientation to form a maximum, so if a particular maximum occurs for a glancing angle θ, then electrons emerge anywhere on the surface of a cone of half-angle θ and form a ring on a screen or photographic plate. This is exactly the same as the way X-ray powder photographs are formed in X-ray crystallography (see spread 8.34). Once again, the X-ray wavelength that would be needed to create the same pattern of concentric rings was equal to the de Broglie wavelength for the electrons used in the experiment. De Broglie's hypothesis was confirmed.

De Broglie received the 1929 Nobel Prize for Physics, and Davisson and Thomson shared the 1937 Prize.

Seeing with electrons

The ability of an optical instrument to resolve detail is limited by the wavelength of light and the aperture diameter D of the instrument – the **Rayleigh criterion** (see spread 6.10) gives a theoretical resolving power of about λ/D. X-rays have a much shorter wavelength than visible light so they can be used to study crystalline structures, but they are not much good for forming images because they are very difficult to refract and focus. Electrons have two great advantages over light and X-rays.

- Their wavelength can be controlled by controlling their kinetic energy (i.e. adjusting the accelerating voltage).

- They can be focused electromagnetically, because they are charged.

For these reasons electron microscopes of various kinds have been used to create highly detailed images of tiny objects right down to the atomic scale.

An electron diffraction pattern. The ring diameters can be used to calculate the spacing of atomic layers in the crystal causing the diffraction.

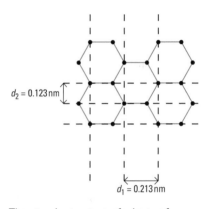

The atomic structure of a layer of graphite. Two prominent layers of atoms are indicated. These act rather like the slits in an optical diffraction grating and diffract the electron beams that strike the graphite target.

PRACTICE

1 Calculate the de Broglie wavelength of electrons accelerated through **a** 100 V, **b** 1 kV, **c** 100 kV, **d** 10 MV (the rest mass of an electron is 9.110^{-31} kg and the speed of light is 3.0×10^8 m s^{-1}).

2 Atomic spacings in a crystal are typically a few tenths of a nanometre. Through what voltage must an electron be accelerated if it is to have a de Broglie wavelength of 0.3 nm?

3 School electron diffraction tubes usually direct a beam of electrons at a thin layer of graphite. The transmitted electrons form a diffraction pattern on the end of the tube that is a series of concentric rings.

   a Why are rings formed?

   b How does the pattern change if the accelerating voltage is increased? Explain your answer.

4 The planes of atoms in graphite act like the slits in an optical diffraction grating as the electron waves pass through the graphite. The diagram shows that there are two distinct values for the 'slit separation' d. (Take the de Broglie wavelength as 6.15×10^{-11} m.)

   a What are the two values d_1 and d_2?

   b Quote the diffraction grating formula (see spread 6.11).

   c Use the formula to calculate the angular positions for the first and second orders of diffraction maxima for both sets of 'slits'.

   d The angles you calculated in **c** determine where the bright rings in the electron diffraction pattern will be formed. Draw a diagram to show this.

   e The crystal is 10 cm from the fluorescent screen. Calculate the radii of the first 4 bright rings.

   f Derive a formula linking the accelerating voltage V in the electron gun to the ring diameter D.

   g Explain why a graph of ring diameter D against $1/\sqrt{V}$ is a straight line through the origin. What would you need to know to be able to use this graph to measure the Planck constant?

- the Copenhagen Interpretation
- wave function
- measurement problem
- Schrödinger's cat

Niels Bohr is generally regarded as the architect of the Copenhagen interpretation of quantum theory. His arguments with Einstein about the new theory are physics legends. Recent experiments reinforce Bohr's point of view and the strangeness of quantum theory.

The Copenhagen interpretation

- The most complete description of matter or radiation is given by the wave function ψ.
- The wave function interacts with apparatus like a continuous classical wave.
- The probability of finding a photon (or electron, etc.) in a small volume of space is proportional to the square of the amplitude of the wave function at that point and the volume of the space concerned:

  probability that the photon is found in a volume $\delta V \propto |\psi|^2 \, \delta V$
- Photons, electrons, etc. are always observed as complete quanta.
- When an observation is made the wave function 'collapses' instantaneously everywhere.

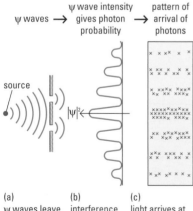

ψ wave intensity gives photon probability → pattern of arrival of photons

ψ waves →

source

$|\psi|^2$

(a) ψ waves leave the source and interact with the double slits

(b) interference leads to a pattern of maxima and minima of $|\psi|^2$ at the screen

(c) light arrives at the screen as photons. The distribution of photons is governed by $|\psi|^2$

Using wave–particle duality.

WAVE–PARTICLE DUALITY

Thomas Young used the double-slit interference pattern to measure the wavelength of light for the first time. However, the photoelectric effect could not be explained by wave theory, and needed Einstein's photon hypothesis, a particle theory. So what *is* light? It cannot be *just* a stream of particles and yet it cannot be a continuous wave either. This dilemma is sometimes described as **wave–particle duality**, and is not restricted to light. De Broglie predicted that matter also suffers from it. This was soon confirmed by electron diffraction and, more recently, by the diffraction and interference of whole atoms.

Wave–particle duality means that neither the particle nor the wave model can be used consistently to explain everything that light or matter does. The best model we have is that of quantum mechanics, an abstract mathematical theory which does not lend itself to any simple pictures of what is going on. This becomes obvious if you try to explain the double-slit interference pattern using photon theory. In the end, most physicists settled for the **Copenhagen interpretation**, which really abandons any hope of a 'physical picture'. It is essentially a way of relating the readings obtained on detectors (geiger counters, photographic plates, etc.) to the mathematical model of quantum theory. Not all physicists agree with this interpretation, so it's still the subject of many late-night arguments!

Some sort of interpretation

One of Einstein's most famous sayings is, 'God does not play dice'. He was referring to the so-called Copenhagen interpretation, with which he did not agree. Surprisingly though, the central idea of the Copenhagen interpretation, that the waves are linked to probability, was first suggested by Einstein himself in a letter to his friend Max Born. At the time Einstein was trying to understand how light could sometimes behave like a particle and sometimes like a wave. He wondered whether the strength of the waves at each point might be linked to the probability that photons are found there.

This idea was developed by Born into a probability model for wave–particle duality. Assume that the wave-like aspect of a photon (or electron) is represented by a **wave function**, ψ ('psi'). This behaves just like a classical wave but with one very important difference – it is not directly observable. Its *intensity* in any small volume of space is proportional to the chance that the photon is there.

This provides a sensible way to think about interference without losing the particle aspect of radiation. The wave functions interact with the apparatus, diffract through apertures, and form interference patterns, but the intensities of these patterns determine where the photons are likely to be found. So even if just one photon interacts with double slits in Young's experiment its wave function forms an interference pattern and the photon is more likely to hit the screen near a maximum (high intensity and therefore high probability) than near a minimum.

In the probability model the distribution of photons or electrons is fixed by the way the wave function interacts with the apparatus, but we only ever detect complete photons or electrons. It is a bit like tossing a coin – the probabilities of heads or tails are equal at 0.5 each, but the coin is always observed to land in one or the other state.

Collapse of the wave function

Something strange happens when a photon (or electron) is detected. Before detection the wave function represents the probability that the photon *might* be detected at a particular position. After detection it must change to represent the new state of affairs – either the photon is there (probability 1) or not (probability 0). This means the wave function changes *everywhere* whenever a measurement or observation is made.

This causes problems for physicists, because it implies that a part of the wave function must respond to something that happens in a particular position far away with *no time delay* for a signal of any kind to reach it. This problem is referred to as the **quantum measurement problem** or simply the **collapse of the wave function**.

Quantum theory and the future

The nature of probability involved in quantum theory is deeper than that involved in tossing a coin. The way the coin lands could (in principle) be predicted if you knew enough about the coin and the mechanics of the toss. This is *not* the case in quantum theory. If you knew everything there was to know about a photon approaching a double slit, you could predict what would happen to its wave function, but not where among the maxima and minima the photon would end up. In classical physics unique initial conditions lead to unique outcomes – this is called a **deterministic theory**. In quantum theory the same initial conditions can lead to many different outcomes, so the future is not uniquely determined by the past. This is another radical change of thinking brought about by the new theory.

Schrödinger's cat

Erwin Schrödinger was so shocked by the implications of quantum theory that it was claimed he wished he had never had anything to do with it! To show how strange it is he invented a very famous thought experiment. If quantum theory applies to atoms it must also apply to cats and people since we are made of atoms. He imagined a cat locked in a box with a capsule of poison and a radioactive atom. If the atom decays it triggers a detector that smashes the capsule and poisons the cat. Before the box is opened there is no way of telling what has actually happened.

In the classical world the cat is either dead or alive, but in the quantum world the unobserved atom is in a **superposition of states** represented by wave functions for 'decay' and 'no decay'. If the cat too is described by a wave function, it is in a superposition of dead and alive states. It only becomes actually dead or alive when we open the box and look in. It is as if two possible realities co-exist in our universe until an *act of observation* causes the wave function to collapse and one or other state to become real! Strange though this sounds, it represents the conventional interpretation of quantum theory used by most physicists.

Alternative interpretations

If you don't like the Copenhagen interpretation there are alternatives. Two of the most popular are outlined below.

- **The many worlds theory** Imagine an experiment has two alternative outcomes. According to this theory both occur, but in parallel universes. Surprisingly this idea was made into a consistent model of quantum theory by Hugh Everett III and is particularly popular among quantum cosmologists. Everett's interpretation solves the measurement problem. If all worlds are equally real then the wave function does not collapse!

- **The sum over histories** Another way to explain quantum theory is to assume that everything that can happen does, but in a world of potentialities. The actualities (that is, what actually does happen) are like weighted averages of all these possible histories. For example, in the double-slit experiment, the interference pattern can be derived by superposing all possible routes the photons could take from the source to the screen, including the crazy ones (like stopping halfway and taking a detour via New York). In the end the effects of the crazy ones cancel out, leaving just what can be expected by more conventional derivations.

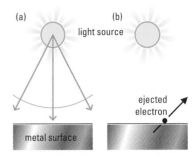

(a) shows a continuous ψ wave just before a photon hits a metal surface. In (b) the ψ wave has collapsed and all the photon energy has been transferred to a single electron at one point in the metal surface (the photoelectric effect).

Schrödinger's cat. Just before he opens the box the quantum mechanical description has a superposition of live and dead states. When an observation is made the wave function collapses into one of these states and the fate of the cat is sealed.

Indeterminacy and uncertainty

The more precisely we can measure the position of a particle, the shorter its corresponding de Broglie wavelength. For example, if we measure the x-position of an electron with an uncertainty of $\pm \Delta x$ then its de Broglie wavelength must be of order $\pm \Delta x$. But the de Broglie wavelength is linked to the electron's momentum in the x-direction so there is a large uncertainty in this momentum, of order: $\Delta p \sim \pm h/\Delta x$. In simple terms, the smaller the uncertainty in position, the larger the uncertainty in momentum and vice versa. This is often called Heisenberg's uncertainty principle and can be written: $\Delta x \, \Delta p \sim h$.

A similar principle holds between the energy of a process and the time over which it takes place (for example, a very short-lived particle has a large uncertainty in its rest energy). The energy–time uncertainty principle is: $\Delta E \, \Delta t \sim h$.

Indeterminacy is fundamental to quantum theory and this was one of the ideas that Einstein found hard to take. Its implication is that particles such as electrons *do not have* well defined positions or momenta. Einstein thought these were essential criteria for particles to be real – he could not accept that quantum theory was a complete description of nature.

So, naturalists observe, a flea
Hath smaller fleas that on him prey;
And these have smaller fleas to bite 'em,
And so proceed ad infinitum.
Thus every poet, in his kind,
Is bit by him that comes behind.

Jonathan Swift, *On Poetry: a Rhapsody*

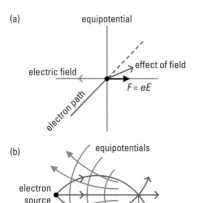

Electrostatic focusing. (a) Electrons
'refract' as they cross equipotential
surfaces. (b) A suitably shaped field will
focus a diverging beam of electrons.

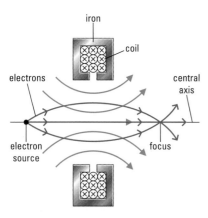

Cross-section through a short magnetic
lens. The force on the electrons (given
by Fleming's left-hand rule) causes them
to rotate around the central axis. All
magnetic lenses cause a rotation of the
image.

ELECTRON MICROSCOPES

There seems to be structure on every scale that can be observed, from the clusters of galaxies that can be seen using the Hubble Space Telescope to the regular layers of atoms in crystals, or the quarks in protons and neutrons. For a long time it was only possible to look at structures using a very narrow range of frequencies in the electromagnetic spectrum, and whilst the view was inspiring, it was limited.

In the last hundred years the limitations of the visible spectrum have gradually been exceeded. Astronomers now have radio, infrared, ultraviolet, X-ray and gamma-ray telescopes. The microscopist, too has moved beyond the visible, enhancing resolving power by using shorter wavelength ultraviolet sources, and the X-ray crystallographer can produce diffraction patterns from crystalline structures (see spread 8.34). But the most remarkable images of all have come by using electron microscopes rather than electromagnetic waves. Electrons have two major advantages over visible light:

- They are charged, so they are reasonably easy to accelerate and focus using electric and magnetic fields.

- Their wavelength is inversely proportional to their momentum, so the more they are accelerated the smaller their wavelength and the finer the detail they can resolve.

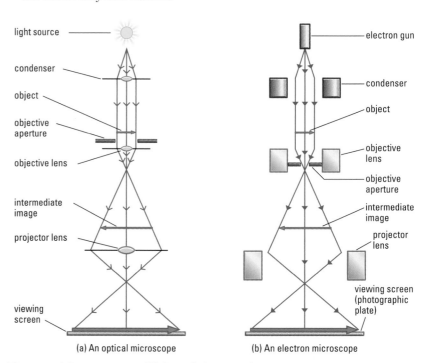

The essential components of light and electron microscopes.

Electron optics

The electron wavelength for non-relativistic electrons is

$$\lambda = \frac{h}{p} = \frac{h}{\sqrt{2mK}}$$

where K is the kinetic energy of the electron. The electron wavelength for relativistic electrons is:

$$p \approx \frac{E}{c} \qquad \lambda \approx \frac{hc}{E}$$

where E is the total energy.

As a rule of thumb, the relativistic equations should be used if the electron is travelling faster than about 10% of the speed of light. This corresponds to an accelerator voltage of about 2500 V and a de Broglie wavelength of about 2.5×10^{-11} m.

Electrostatic focusing

Electron beams can be focused as they pass through a pair of cylindrical anodes held at different potentials. The electron paths 'refract' towards the field lines (which are perpendicular to the equipotential surfaces).

Magnetic focusing

The magnetic force on moving charges can also be used to focus the electron beam as it passes through a suitably shaped region of magnetic field.

Electron beams do have some *disadvantages* compared to conventional optics. Although much shorter wavelengths can be used, the aperture size of the magnetic lenses cannot be very large because it is very difficult to eliminate spherical aberration (see spread 12.4). In addition to this, bombarding an object with electrons can severely damage it, and so object preparation is involved (e.g. freeze-drying); this may alter the very features being investigated. For example, live samples cannot be used, and many surfaces can only be imaged indirectly by making a thin replica (using metallic deposition) and using that as the object.

Types of electron microscope

Scanning electron microscope (SEM) This is particularly useful for forming three-dimensional images of the surface of a sample. A tightly focused electron beam is scanned across the surface and scattered electrons, along with secondary electrons ejected from the surface, are collected and counted. This enables an image to be built up pixel by pixel rather like a TV picture. SEMs can create images with a magnification greater than 100 000 times. An advantage of the SEM is that the sample does not need elaborate preparation.

Transmission electron microscope (TEM) Unlike the SEM, this microscope does not scan across the specimen. An electron beam is transmitted through the specimen and electrons hit a screen on the far side. The pattern of electrons hitting the screen is measured and used to create an image whose magnification can be in excess of 1 million times. One of the main drawbacks of the TEM is the need for very thin samples to be used – perhaps only a hundred atoms thick.

Scanning tunnelling electron microscope (STM) This microscope uses quantum tunnelling (see spread 8.30) and can be used to resolve individual atoms on the surface of a material. An extremely fine needle, whose tip may be just one atom wide, is held a fraction of a nanometre above the surface to be imaged and is scanned across it. A small voltage is applied between the tip of the probe and the surface, and electrons are then able to tunnel across the gap. The electron current gets larger if the tip comes closer, and smaller if it is moved away. By monitoring the changing electron current as the tip moves, a three-dimensional image of the surface can be constructed.

Atomic force microscope (AFM) This works in a similar way to the STM, but now the force of repulsion between the electrons in the surface and the probe tip is monitored as the probe scans across the surface. Again, a three-dimensional image of the surface can be built up, and the positions of individual atoms in the surface can be measured.

The STM and AFM can even be used to pick up and move individual atoms. The first time this was done was in April 1990 when Don Eigler used the tip of an STM to write IBM in letters 5 nm high. A group of 'atomic artists' used these electron microscopes to create a wide range of images made up of individually arranged atoms. The technology that has been developed to allow this is being used to build tiny machines on a molecular scale, and to improve data storage capacity by writing information ever more densely.

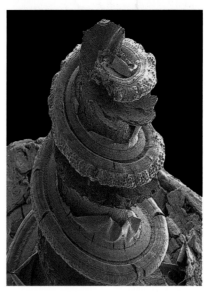

A scanning electron micrograph of the human cochlea in the inner ear.

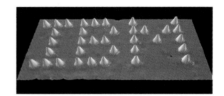

These amazing images show individual atoms cleverly arranged by the company that produced the image. They are computer enhanced to give a three-dimensional effect and show charge distribution across the surface.

Light, X-rays, and electrons

The wavelength of visible light is about 0.5 μm, more than a thousand times greater than atomic dimensions, so light is useless for forming images of atoms. X-rays have a short enough wavelength, but they cannot be focused well enough, so are used more indirectly to form diffraction patterns from crystalline structures. Electron beams can be formed with any wavelength and can still be focused. They have been used to investigate atoms and nuclei, and even to probe inside nucleons themselves.

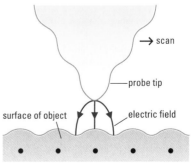

An STM.

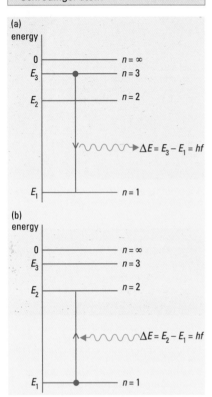

(a) Emission of a photon from an excited atom. (b) Absorption of a photon causes excitation.

Maths box: Bohr's quantum condition

Bohr quantized the angular momentum (mvr) of electrons, setting it equal to an integer multiple of $h/2\pi$. This condition, added to the classical equations for electrostatic attraction and centripetal force in the Rutherford model, led to an equation for the allowed energy levels in the hydrogen atom:

$$E_n = -\frac{me^4}{8\varepsilon_0^2 h^2} \frac{1}{n^2} \quad \text{where } n = 1, 2, 3, \text{ etc.}$$

This equation accurately predicts the ground state and many of the excited states. Differences between levels can be used to calculate the wavelengths of hydrogen's spectral lines. These agree with observed values – e.g. in the Balmer series (see spread 8.10).

It also gives the ionization energy of hydrogen. The electron will leave the atom if its total energy is greater than or equal to zero (all bound states have negative total energy). This is equivalent to promoting an electron from the $n = 1$ to $n = \infty$ levels.

$$E_{\text{ionization}} = E_\infty - E_1 = \frac{me^4}{8\varepsilon_0^2 h^2} \approx 13.6 \, \text{eV}$$

in agreement with measured values.

THE QUANTUM ATOM

Rutherford's model of the atom was useful, but had serious drawbacks. For example, Maxwell's theory of electromagnetism predicts that all accelerating charges should radiate. Electrons orbiting nuclei have a centripetal acceleration, so they should radiate. Atoms ought to collapse in a fraction of a second as the orbiting electrons radiate their energy and fall into the nucleus. On a larger scale this **synchrotron radiation** is a major power loss from circular accelerators, so it should not be dismissed lightly. How then can atoms exist?

Niels Bohr was aware of this problem. He was also aware that quantization of energy had solved other problems in classical physics, so he quantized the atom. In Rutherford's theory an electron can orbit with any energy and any radius, so that the atom has a continuous range of allowed energies. Bohr added a quantum condition (see below left) which meant the electrons could only orbit at certain radii and have certain energies. Atomic collapse was prevented because there were no allowed states between these **energy levels** and because the lowest, or **ground state**, was not at zero energy. Bohr's original theory involved circular orbits; Sommerfeld developed the theory to include the more general elliptical orbits.

Quantum jumps

If an atom is in its ground state it cannot lose energy, because there are no available states of lower energy. However, if something collides with the atom or if it absorbs a photon of sufficient energy, an electron can make a **quantum jump** to a higher allowed energy level. The atom is now in an **excited state**. Although it may remain in an excited state for some time (typically 10^{-8} s), there are now lower allowed energy levels and the electron will eventually make a quantum jump back down into one of these. As it does so, it loses an amount of energy equal to the difference between the energies of the original and final states, and radiates this as a photon. The photon frequency and wavelength are therefore fixed by the size of the energy jump.

Energy levels are labelled by the principal quantum number n which can take any integer value from 1 upwards. If an electron jumps from the $n = 3$ state to the $n = 1$ state it will emit a photon of frequency f. For this quantum jump

$$\Delta E = E_3 - E_1$$

$$\text{photon energy, } hf = \frac{hc}{\lambda} = \Delta E = E_3 - E_1$$

so

$$f = \frac{\Delta E}{h} = \frac{E_3 - E_1}{h} \quad \text{and} \quad \lambda = \frac{c}{f} = \frac{hc}{E_3 - E_1}$$

- Notice that the larger the energy of the quantum jump, the shorter the wavelength of the radiation emitted.

Atoms of a particular element have a characteristic structure of allowed energy levels, so there are only certain energy jumps the electrons can make. This means the photons emitted from atoms should have certain well-defined wavelengths. They do – if a gas of atoms is excited by heating it or passing an electric current through it, it emits a characteristic **line spectrum**.

Electron waves in atoms

Rutherford treated electrons in atoms like tiny charged particles. De Broglie's idea of matter waves changed all this and led to a simple pictorial way of understanding why the orbits are quantized. If the electrons in atoms act like waves, they might behave like the standing waves that are formed on a string of a musical instrument when it is plucked. Only certain wavelengths 'fit' the string and reinforce, others cancel out. This leads to a set of discrete frequencies giving the fundamental and harmonics of the instrument. It turns out that the 'fundamental' and 'harmonics' of electron waves form the ground state and excited states of the atom.

For Bohr's circular orbits the condition for electron waves to reinforce as standing waves is simply that their wavelength fits the circumference of the orbit a whole number of times. The principal quantum number n is equal to the number of complete waves that fit the circumference of the orbit, and Bohr's quantum condition becomes

$$2\pi r = n\lambda$$

The de Broglie relation gives

$$\lambda = \frac{h}{mv}$$

so

$$2\pi r = \frac{nh}{mv} \quad \text{or} \quad mvr = \frac{nh}{2\pi}$$

which is Bohr's original condition for the quantization of electron angular momentum.

The Schrödinger atom and beyond

The piecemeal quantization of physics by Planck, Einstein, Bohr, and others was very successful, but there was no real quantum theory that could be applied to all phenomena. This was supplied by Schrödinger and Heisenberg in 1925. Their versions of quantum theory used different mathematics but were soon shown to be equivalent.

Schrödinger's version is easiest to think about. He set about finding an equation for electron wave functions like Maxwell's electromagnetic equations for light, and came up with what is now called the Schrödinger equation. He applied it to the hydrogen atom. The result was a series of three-dimensional equations for the way the electron wave function varies with position around the nucleus. The intensity of the wave function at any point (given by its magnitude squared) is proportional to the probability that the electron is in that region of space. The solutions are a discrete set of equations for the electron wave function, each one uniquely defined by three separate quantum numbers, n (corresponding to the Bohr theory), l (to do with orbital angular momentum), and m (to do with the magnetic effect of the orbit).

The 'electron clouds' that are sometimes mentioned in connection with the atom are simply a poor description of the probability distribution surrounding the nucleus. In the Schrödinger atom the electron has lost its orbit; it could be anywhere. However, the peak of the probability still corresponds to the radius of the Bohr orbit.

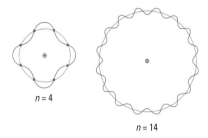

Bohr's atom. In the n = 4 state the circumference of the electron orbit equals 4 de Broglie wavelengths. On the right is a highly excited state (n = 14).

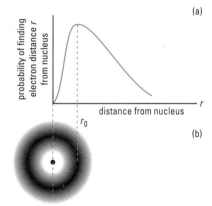

Schrödinger's atom. (a) Radial probability density for the 1s state of a hydrogen atom derived from the Schrödinger equation. (b) A two-dimensional representation of the 1s electron orbital. Darker shading represents greater probability and maximum probability is at $r = r_0$, equal to the Bohr radius. Higher energy orbitals are not spherical.

Chemical notation

If you have come across 1s and 2p states in chemistry, these correspond to $n = 1$, $l = 0$ and $n = 2$, $l = 1$ respectively.

PRACTICE

1 The table gives some of the allowed energy levels for a hydrogen atom.

Principal quantum number	Energy/10^{-18} J
1	−2.180
2	−0.545
3	−0.242
4	−0.136
5	−0.087
6	−0.061
7	−0.044
8	−0.034
9	−0.027
10	−0.022

  a Draw a vertical axis to represent energy in electron volts from −15 eV (at the bottom) to 0 eV at the top.

Now add a series of horizontal rungs to represent allowed energy levels for hydrogen. Label each rung with the principal quantum number and energy in electron volts.

b Show that the ionization energy for hydrogen is about 13.6 eV.

c In what part of the electromagnetic spectrum would the photon be that was emitted by a quantum jump from $n = 5$ to $n = 3$?

d What is the minimum frequency of light needed to eject an electron from a hydrogen atom in its $n = 4$ state?

e Show that the energy of these levels is proportional to $1/n^2$.

8.10

OBJECTIVES

- the hydrogen spectrum
- spectral series
- ionization
- band and continuous spectra

ATOMIC SPECTRA

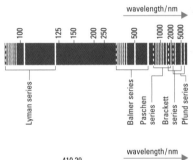

An atomic line spectrum, here the Balmer lines (see below) of the hydrogen atom.

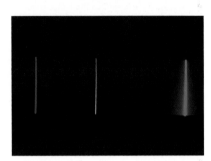

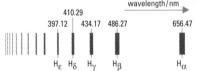

The hydrogen spectrum. Top: the five spectral series of lines. Bottom: the Balmer series of lines in more detail.

Ionization

The **first ionization energy** for hydrogen is the minimum energy that must be supplied to remove an electron from the $n = 1$ level (ground state). The ground-state energy is $-13.6\,\text{eV}$ and free electrons have positive energy, so the first ionization energy for hydrogen is $+13.6\,\text{eV}$. Another way of looking at ionization is to say that energy must be supplied to 'lift' the electron from $n = 1$ to $n = \infty$. The energy needed is then $(E_\infty - E_n)$. Since the energy approaches zero as n approaches infinity this gives the same result as the argument above.

Atomic line spectra arise from a process similar to what happens when a child jumps down stairs. When the child stands on a particular stair she has a well-defined energy which is greater than that at floor level – there are a set of discrete energy levels, one for each stair (and the child cannot stand halfway between two stairs). If she jumps from the 5th stair to the 4th she loses a fixed amount of energy that depends on the height of one step. Some of the energy she loses is given out as sound as she thumps down on the lower stair. If she jumps two or three stairs in one go she makes a bigger thump as she lands. To get back up the stairs into higher energy levels she needs an input of energy.

Electrons in atoms also have a discrete set of allowed energy levels and cannot remain in intermediate states. When an electron drops to a lower level (a quantum jump) the energy lost is released as a photon of electromagnetic radiation. The bigger the quantum jump the higher the energy (and frequency) of the emitted photon. To send it back up into a higher energy level an energy input is required – the atom must absorb a photon or gain energy in a collision. The main difference between energy levels in an atom and stairs is that the atomic energy levels are not equally spaced.

The hydrogen spectrum

The energy levels in atoms are calculated using quantum theory. The photons emitted from atoms have wavelengths fixed by the energy difference between atomic energy levels, so the first question to be asked of quantum theory was whether the energy levels it predicted could explain atomic spectra. The first atomic spectrum to be tackled was that of hydrogen, since hydrogen is the simplest atom, having just one orbital electron. (In fact it is very difficult to apply quantum theory precisely to any other atom because in all other cases the outer electrons interact with inner electrons as well as the nucleus. For this reason, most atoms are modelled using mathematical approximations.)

For hydrogen, quantum theory allows energy levels:

$$E_n = \frac{13.6}{n^2}\,\text{eV}$$

where n is the principal quantum number and can be any positive integer.

Worked example 1

Calculate the energy of the ground state and 1st excited state of the hydrogen atom.

$$E_1 = -\frac{13.6}{1^2}\,\text{eV} = -13.6\,\text{eV} \quad \text{(ground state at } n = 1\text{)}$$

$$E_2 = -\frac{13.6}{2^2}\,\text{eV} = -3.4\,\text{eV} \quad \text{(1st excited state at } n = 2\text{)}$$

Worked example 2

Calculate the frequency and wavelength of the photon emitted when an electron makes a quantum jump from the $n = 3$ state to the ground state of the hydrogen atom.

$$E_1 = -13.6\,\text{eV} = -2.18 \times 10^{-18}\,\text{J} \qquad E_3 = -\frac{13.6}{3^2}\,\text{eV} = -1.51\,\text{eV} = -2.42 \times 10^{-19}\,\text{J}$$

$$\Delta E = E_3 - E_1 = hf$$

$$f = \frac{E_3 - E_1}{h} = \frac{-2.42 \times 10^{-19}\,\text{J} - -2.18 \times 10^{-18}\,\text{J}}{6.6 \times 10^{-34}\,\text{J s}} = 2.9 \times 10^{15}\,\text{Hz}$$

$$\lambda = \frac{c}{f} = \frac{3.0 \times 10^8\,\text{m s}^{-1}}{2.9 \times 10^{15}\,\text{Hz}} = 1.0 \times 10^{-7}\,\text{m}$$

This is in the ultraviolet part of the electromagnetic spectrum.

The spectral series

In 1885 Johann Jakob Balmer, a Swiss school teacher, noticed a simple mathematical link between the frequencies of lines in the visible part of the hydrogen spectrum.

$$f_n = k\left(\frac{1}{2^2} - \frac{1}{n^2}\right)$$

where k is a constant equal to 3.29×10^{15} Hz and n has values 3, 4, 5 etc.

This formula was derived directly from experimental values and neither Balmer nor anyone else at the time knew *why* it worked.

Quantum theory now shows that this series is formed by quantum jumps from excited states above $n = 2$ as the electrons fall back into the $n = 2$ level. These energy jumps correspond to photons in the visible part of the spectrum, which is why the Balmer series was the first series to be discovered. Transitions to $n = 1$ involve larger energies and result in ultraviolet photons (the Lyman series), those to $n = 3$ or above involve smaller energies and result in infrared photons (Paschen series to $n = 3$, Brackett to $n = 4$, and Pfund to $n = 5$).

A general formula for the frequencies of all hydrogen lines can be derived from the quantum formula for the energy levels by calculating the energy difference for a quantum jump from the mth to the nth level ($m > n$).

$$\Delta E = E_m - E_n = hf_{mn}$$

$$f_{mn} = \frac{E_m - E_n}{h}$$

In general the nth energy level has a value $1/n^2$ times the ground state energy.

$$E_n = \frac{E_1}{n^2} \quad \text{and} \quad E_m = \frac{E_1}{m^2}$$

$$f_{mn} = \frac{E_1}{h}\left(\frac{1}{m^2} - \frac{1}{n^2}\right)$$

$$\frac{E_1}{h} = -\frac{13.6 \times 1.60 \times 10^{-19}\,\text{J}}{6.63 \times 10^{-34}\,\text{J s}} = -3.28 \times 10^{15}\,\text{Hz}$$

$$f_{mn} = 3.28 \times 10^{15}\left(\frac{1}{n^2} - \frac{1}{m^2}\right)$$

(The order inside the bracket changed because of the minus sign.) The Balmer series corresponds to $n = 2$.

The yellow light from sodium street lamps comes from two very similar quantum jumps in the sodium atom and consists of two very close spectral lines with wavelengths 589 nm and 589.6 nm.

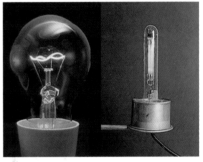

A filament lamp (left), emits the characteristic continuous spectrum of a hot solid and a significant amount of its energy is in the infrared part of the spectrum. In a strip light (right), mercury vapour is excited by passing a current through it. It then emits visible and ultraviolet radiation. Sometimes the inside of the tube is coated with fluorescent material that converts the ultraviolet to visible radiation, providing an efficient source of white light.

PRACTICE

1 Explain why energy levels are all given negative values.

2 Calculate the wavelengths of the first four spectral lines in the Lyman series. In what part of the electromagnetic spectrum would these be found?

3 Radiation from a particular quantum jump in the caesium-133 atom is used as a frequency standard to define the second. The energy levels concerned are separated by 3.8×10^{-5} eV.

   a Calculate the frequency of this radiation.

   b How many oscillations of this radiation occur in 1 second?

4 Fluorescent paints can absorb light at one colour and re-emit it as light of another colour.

   a Suggest an explanation for this in terms of energy levels inside the paint atoms.

   b Would you expect a red luminous paint to fluoresce when illuminated with green light?

   c Would you expect green luminous paint to fluoresce when illuminated with red light?

5 Why would you normally expect a continuous spectrum from a hot solid and a line spectrum from a hot gas?

6 What is the shortest wavelength likely to be emitted from an excited hydrogen atom?

Marie Curie won the 1903 Nobel Prize for Physics (for her work on radioactivity) and the 1911 Nobel Prize for Chemistry (for her discovery of the new elements radium and polonium). In the early 1900s it was very difficult for women to find support for a career in fundamental scientific research, and it is interesting that in 1904 Pierre Curie was given a chair at the Sorbonne in Paris whilst Marie was offered a part-time post as a physics teacher in a girls' school in Sèvres.

Lord Rutherford of Nelson.

Rutherford and his followers

No one did more to explain the mysteries of radioactivity than Ernest Rutherford, a New Zealand physicist who did most of his important work in the UK. He and his pupils discovered the nature of radioactive decay, the structure of the atom, and the constituents of the nucleus, and set the agenda for the developments that led to high-energy physics research in the rest of the twentieth century.

RADIOACTIVITY

Radioactivity was discovered by accident in 1896. Antoine Henri Becquerel was trying to detect X-ray emission from fluorescent salts when they absorb ultraviolet light. He was inspired to do this by the discovery of X-rays a year earlier by Wilhelm Röntgen. Becquerel's method was simplicity itself. The salts (uranyl sulphate) were placed in sunlight next to a photographic plate wrapped in thick black paper. If X-rays were emitted they should penetrate the paper and darken the film. The films were darkened, and this seemed to support his hypothesis, but then he noticed something mysterious. Wrapped films kept in a drawer next to similar salts were also darkened, even though these crystals had not been exposed to sunlight. So whatever darkened the film was not caused by fluorescence – it must be an emission from something in the crystal. But what? He carried out further tests and showed that the new radiation came from uranium in the salt.

Becquerel's discovery was confirmed by the Polish physicist Marie Curie, who showed that the activity of uranium salts depended only on the amount of uranium they contained and not at all on the physical state or chemical composition of the salt. We now understand why this is – radioactive emissions come from the *nuclei* of particular atoms so they are not affected by bonding (which concerns the outer electrons) or physical conditions (like temperature, pressure, etc.).

Marie Curie made several other important discoveries in this field. She showed that thorium is also radioactive and noticed that some ores of uranium, pitchblende and chalcolite, are more active than uranium itself. She thought this must be due to new radioactive elements inside them and soon discovered radium and polonium, which are highly radioactive. Because Curie was unaware that these radiations were dangerous, she took no safety precautions as she worked with them, and her notebooks are still too radioactive to handle even today! She shared the 1903 Nobel Prize for Physics with her husband Pierre, and with Becquerel.

Some important early discoveries

- There are three types of radioactive emission, called alpha, beta, and gamma radiation. These can be separated by electric or magnetic fields.

- Radioactive emissions cause ionization, alpha radiation being most strongly ionizing and gamma radiation least strongly ionizing. Ionization is the property used to detect and measure radioactivity.

- The more strongly ionizing the radioactivity is, the more rapidly it dissipates its energy when it passes through materials. The ranges for similar energy emissions are in the order

    gamma > beta > alpha

- The activity of a source is independent of physical conditions and chemical bonding, depending only on the type of atom involved and the number of these atoms present.

- Each atom has a nucleus and radioactive decays involve nuclear transformations.

- Alpha particles are helium nuclei emitted from the nuclei of some radioactive atoms. This was shown by stopping the alpha particles inside a sealed container where they captured two electrons and became helium atoms. The presence of helium was shown by finding its spectral lines in the light emitted when an electrical discharge was passed through the tube.

- Beta particles are electrons emitted from the nuclei of some radioactive atoms (they are created in the decay and are not related to the orbital electrons in any way). This was shown by measuring their deflection in a magnetic field and showing they had the same charge-to-mass ratio (e/m) as an electron.

- Gamma rays are high-energy electromagnetic photons emitted from the nuclei of some radioactive atoms following an alpha or beta decay. These are not deflected in electric or magnetic fields. Their properties are identical to those of hard X-rays.

Nuclear transformation

Rutherford and Soddy worked out the rules for radioactive decay. All decays conserve two fundamental properties:

- charge
- number of nucleons.

Alpha decay is the emission of a helium nucleus from the nucleus of a heavy radioactive element.

A general equation for alpha decay: $\qquad {}^{A}_{Z}X \rightarrow {}^{A-4}_{Z-2}Y + {}^{4}_{2}\alpha$

An example of alpha decay: $\qquad {}^{238}_{92}U \rightarrow {}^{234}_{90}Th + {}^{4}_{2}\alpha$

The nucleus created in alpha decay is that of a new element two places below the original one in the periodic table. In the example above thorium is created from uranium.

Beta decay is the emission of a fast electron from the nucleus of a radioactive element.

A general equation for beta decay: $\qquad {}^{A}_{Z}X \rightarrow {}^{A}_{Z+1}Y + {}^{0}_{-1}\beta\,(+ {}^{0}_{0}\overline{\nu}_e)$

An example of beta decay: $\qquad {}^{14}_{6}C \rightarrow {}^{14}_{7}N + {}^{0}_{-1}\beta\,(+ {}^{0}_{0}\overline{\nu}_e)$

In order to conserve energy and momentum during beta-minus decay another particle, the **anti-neutrino**, must be emitted. This is explained on spread 8.20.

Notice that beta decay is more complex than alpha decay. In alpha decay some of the nucleons are hurled out of the nucleus. In beta decay a neutron inside the nucleus changes to a proton and the electron (and an anti-neutrino) is created in the process:

$${}^{1}_{0}n \rightarrow {}^{1}_{1}p + {}^{0}_{-1}e$$

This electron is the beta particle emitted. The nucleus created in beta decay is of a new element one place above the original in the periodic table.

Gamma rays are emitted when excited nuclei make quantum jumps to lower energy levels (similar to the photons emitted by quantum jumps of electrons in atoms, but of much higher energy). Often an alpha or beta decay leaves the new nucleus in an excited state, so gamma-ray emission often accompanies other radioactive decays. Gamma-ray emission does not affect the type or number of particles present in the nucleus. Sometimes excited states are represented using an asterisk.

A general equation for gamma emission: $\quad {}^{A}_{Z}X^* \rightarrow {}^{A}_{Z}X + {}^{0}_{0}\gamma$

${}^{0}_{0}\gamma$ is a photon of gamma radiation.

> ### Definition: activity
>
> The **activity, A,** of a radioactive source is the number of disintegrations per second in the source. The SI unit for activity is the becquerel (Bq):
>
> $\qquad$ 1 becquerel = 1 disintegration per second
>
> $\qquad$ $1\,Bq = 1\,s^{-1}$

> ### Balancing equations
>
> The A, Z notation for nuclides has been extended to beta particles in order to emphasize the conservation laws. To be consistent the mass number A has now become **baryon number** (protons and neutrons both have B = 1 and electrons have B = 0), and the atomic number Z has now become charge Q, so that electrons (to which atomic numbers do not apply) can be recorded as Z = –1 because of their single quantum of negative charge.
>
> $\qquad {}^{1}_{0}n \rightarrow {}^{1}_{1}p + {}^{0}_{-1}e$
>
> $\qquad B \;\; 1 = 1 + 0$
>
> $\qquad Q \;\; 0 = 1 - 1$

Smoke detectors contain a small amount of americium-241, an alpha emitter. When smoke (or other invisible products of combustion) enters the detector it reduces the number of alpha particles reaching an electronic detector. This triggers an alarm.

PRACTICE

1 Use your understanding of energy levels and quantum jumps in atoms to explain why you would not expect X-rays to be emitted from fluorescent salts when they have been irradiated by ultraviolet light.

2 Potassium-40 (${}^{40}_{19}K$) decays by beta emission to calcium.

  a Write down a transformation equation for this decay.

  b What quantities are conserved in the decay? Show how they are conserved.

  c How is it possible for one element to change into another in a radioactive decay?

  d Explain where the electron emitted in the decay came from.

  e Suggest a reason why some nuclei are prone to beta decay.

3 Plutonium-239 (${}^{239}_{94}Pu$) decays by alpha emission to an isotope of uranium.

  a How many protons and neutrons are present in the uranium isotope?

  b Why are alpha particles so strongly ionizing?

  c Write down a transformation equation for this decay.

RANDOM DECAY

We are resigned to the fact that we grow old. The alchemists searched for an 'elixir of life' that would make a person eternally young, never ageing, always the same – but their search was in vain. Radioactive nuclei, however, have solved the problem of eternal youth. They don't grow old; but there is a price to pay. They have to live with a constant probability that they will explode during each successive moment of their existence. Humans approach death gradually, accumulating injuries and illnesses so that the chance of death increases with time. Not so for a nucleus: if it survives a million years it is as fresh as at the moment it came into being.

There is no way, not even in principle, that anyone can predict when a *particular* nucleus will decay: it is a totally random quantum mechanical process. However, if every nucleus in a particular sample has the same chance of decaying per unit time it is possible to predict (statistically) the fraction of nuclei that are likely to decay in any given time – even though it is not possible to say *which* nuclei these will be. This is often compared to predicting what happens when you roll one or many dice. If one is rolled there is an equal chance that we get any outcome from 1 to 6. If 600 are rolled together you can predict fairly confidently that about 100 of them will end up showing a 6 (even though you cannot predict which dice these will be).

A model for random decay

Imagine a population of aliens whose lives are governed by the same rules as unstable atomic nuclei. They have a 1 in 5 chance of dying each year (probability of dying equals 0.2 per year). A particular group of 100 (celibate) aliens have become separated from the rest: how does the population of this group change in future years? This can be modelled using a spreadsheet and considering steps of one year.

About probability

The dice analogy is helpful in some ways but misleading in others. The outcome of rolling a dice can be predicted if you know the initial conditions in great detail, so the 1 in 6 chance of each outcome is really a result of your *ignorance* of the set-up. With radioactive decay, complete knowledge of the initial state of the nuclei is still insufficient to predict which will decay – even the nuclei themselves 'don't know'!

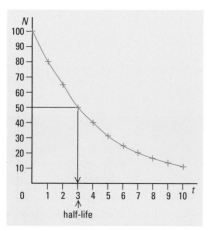

Random decay produces an exponential law!

Approaching zero

It may have occurred to you that an exponential curve never quite reaches zero: it is **asymptotic** to the time axis. At first sight this seems to be saying that populations (of aliens or nuclei) will never quite die out. This is incorrect. The exponential model does not take into account the discrete nature of the things that are decaying. When only a few remain the step-like nature of decay becomes very important and the statistical fluctuations make the exponential model invalid. In the extreme case of a single remaining nucleus this will decay at some moment, but this moment is unpredictable. Once it has decayed no nuclei remain.

Population change calculated statistically.

Year	Number surviving (start of year) N	Population change during year $\Delta N = -0.2N$	Number surviving (end of year) $N' = N + \Delta N = N - 0.2N$
0	100	−20	80
1	80	−16	64
2	64	−13	51
3	51	−10	41
4	41	−8	33
5	33	−7	26
6	26	−5	21
7	21	−4	17
8	17	−3	14
9	14	−3	11
10	11	−2	9

A graph of N versus t shows a characteristic *exponential* decay curve. The defining characteristic of an exponential curve is that its rate of change (gradient) is directly proportional to the number remaining: this is true here since ΔN is equal to $-0.2N$. The negative sign ensures decay, rather than exponential growth.

Here it is assumed that exactly one in five of all surviving aliens die each year. With statistical predictions there are always fluctuations, and these become more significant if the prediction is applied to a small number of things – so in reality the population curve would not be so smooth.

Looking at the table or graph, you can see that it takes about three years for the population to halve from 100 to 50. It takes another three years to halve again to about 25. Similarly for 40 to 20 or 20 to 10, the 'half-life' is about three years.

- All exponential decay curves have this constant proportion property.
- The time taken for any initial number to halve is constant. This is called the **half-life**, $t_{\frac{1}{2}}$.
- If the probability of death had been greater than 0.2 per year, the population would fall off more rapidly and the half-life would be shorter.

Because of statistical fluctuations, half-life is only useful when applied to a large initial number of things.

Radioactive decay

The decay of a population of unstable nuclei follows exactly the same pattern as the decline of the population of aliens in the example above.

- Each nucleus has a fixed probability of decay per unit time. This is called the decay constant, λ.
- The number of nuclei remaining falls exponentially.
- The decay curve has a constant proportion property, leading to a constant half-life.
- The greater the decay constant, the shorter the half-life.
- Statistical predictions based on the exponential model will be accurate if they are applied to a large population of nuclei, but will have significant statistical uncertainties if applied to a small population.

A couple of nuclides illustrate these ideas:

- Iodine-135 has a decay constant $2.9 \times 10^{-5}\,\text{s}^{-1}$ and a half-life of 6.7 h.
- Copper-64 has a decay constant $1.5 \times 10^{-5}\,\text{s}^{-1}$ and a half-life of 12.8 h.

The probability of decay per unit time for the iodine isotope is about double that of the copper isotope, so the iodine half-life is about half that of the copper. This also means that samples of iodine-135 and copper-64 that contain the same number of atoms will have different activities – the iodine will be about twice as active.

PRACTICE

1 a A particular (unbiased) coin is tossed repeatedly until heads comes up four times in a row. What is the probability that the next toss of the coin results in tails?

b Give one similarity and one difference between the probability of a coin landing heads or tails and the probability that a particular unstable nucleus decays in the next second.

2 Look at the curve of N versus t for the alien population in the example.

a How long does it take for the initial population to drop to 0.75 times its original value?

b Now choose three different values of N on the graph and see how long each of these takes to drop to 0.75 times its value. Comment on your answer.

3 The two most common isotopes of uranium are U-238 and U-235, which have half-lives of $4.6 \times 10^{9}\,\text{y}$ and $7.1 \times 10^{8}\,\text{y}$ respectively.

a If you had one atom of each, which would be the most likely to decay in the next second?

b If you had 1 gram of each which would be most active?

c Natural uranium is about 99.3% U-238 and 0.7% U-235 at the present day. The Earth is roughly $4.6 \times 10^{9}\,\text{y}$ old. Estimate the percentages of these two isotopes in natural uranium on Earth just after the planet formed. Explain how you arrive at your answers.

d Sketch a graph to show how the amount of U-235 has changed since the Earth formed.

Comparing radioactive decay and capacitor discharge

The graphs of the radioactive decay of a nuclide and of the decay of charge as a capacitor decays through a resistor are both exponential. The reason for this is because in both cases the rate of decay is directly proportional to the 'amount remaining' (N or Q). This can be written mathematically:

- $\dfrac{dN}{dt} \propto -N$ (radioactive decay)
- $\dfrac{dQ}{dt} \propto -Q$ (capacitor discharge)

The capacitor equation follows because dQ/dt is current, which is directly proportional to the voltage across the capacitor ($Q = CV$).

Both graphs have a constant proportion property – i.e. the graph reduces by the same proportion in the same time. Both decays have a constant half-life. However, in practice we use different proportions when we discuss the decays:

- For radioactive decay $t_{\frac{1}{2}}$ is the time for N to fall to half of any initial value.
- For capacitor discharge the 'time constant' $T = RC$ is the time for Q to fall to 1/e (about 0.37) of any initial value.

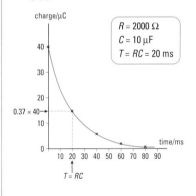

8.13

OBJECTIVES

- the decay equation
- the decay constant
- half-life
- activity

Decay constant and probability

The equation

$$\frac{\delta N}{\delta t} = -\lambda N$$

can be rearranged to give an expression for the decay constant λ:

$$\lambda = \frac{1}{\delta t}\left(\frac{-\delta N}{N}\right)$$

The bracketed term is the fractional change in N during a time δt. This is equal to the chance of decay in that time, so λ is equal to the probability of decay per unit time in the limit of short time intervals.

THE DECAY EQUATION

The mathematical analysis of radioactive decay is based on two simple assumptions:

- Decay is completely random.
- The rate of decay is directly proportional to the number of unstable nuclei present.

Maths box: the decay equation

The second assumption above can be written down as an equation by considering the change in number of nuclei δN in a short time δt:

rate of change of number of nuclei = $\delta N/\delta t$

This is proportional to the number of nuclei present:

$$\frac{\delta N}{\delta t} \propto -N$$

The constant of proportionality is called the decay constant λ.
The negative sign shows that N is decreasing.

$$\frac{\delta N}{\delta t} = -\lambda N$$

If the time interval δt is made smaller and smaller ($\delta t \to 0$) the equation becomes:

$$\frac{dN}{dt} = -\lambda N$$

This final equation shows that the rate of change of N is proportional to N itself, a recipe for exponential decay. It is a first-order differential equation whose solution can be written as:

$$N = N_0 e^{-\lambda t}$$

where N_0 is the original number of unstable nuclei present at $t = 0$.

The decay equation

The number of nuclei N that remain at time t from an initial population of N_0 unstable nuclei is given by:

$$N = N_0 e^{-\lambda t}$$

where λ is the decay constant for the nuclide concerned.

The term $e^{-\lambda t}$ can be interpreted as the fraction remaining after time t: this fraction decays exponentially with a constant half-life, $t_{\frac{1}{2}}$.

It is worth pointing out that this equation was derived by assuming that radioactive decays occur completely at random. The fact that the exponential decay equation works confirms the idea that there is no law of nature that determines when a particular nucleus decays.

The activity of a source

Activity is defined as the number of disintegrations per second taking place inside the source, and is measured in becquerels (Bq).

1 Bq = 1 disintegration per second

There is an older unit which is still often used for activity: the curie (Ci).

1 Ci = 3.7×10^{10} disintegrations per second = 3.7×10^{10} Bq

School sources have a total activity which is typically $5.0 \,\mu\text{Ci} = 1.9 \times 10^5 \,\text{Bq}$.

Activity A (a positive number) is equal to the number of decays per second, so it is simply minus the rate of change of N:

$$A = -\frac{dN}{dt} \quad \text{or} \quad A = \lambda N$$

In most experiments in radioactivity you will measure activity, not N. The fact that the activity is proportional to N means that it will fall off exponentially with exactly the same half-life as N does. So measuring the half-life of the activity is equivalent to measuring the half-life of the source.

$$A = A_0 e^{-\lambda t} \quad \text{where } A_0 = \text{initial activity (Bq)}$$

Half-life and decay constant

The half-life of a nuclide is the time taken for half of an initially large number of unstable nuclei to decay.

The larger the probability of decay per unit time, the shorter the half-life of the nuclide. When $t = t_{\frac{1}{2}}$ half the nuclei have decayed, so

$$N = \frac{N_0}{2}$$

But $\quad N = N_0 e^{-\lambda t} \quad$ so $\quad e^{-\lambda t_{\frac{1}{2}}} = \frac{1}{2}$

Taking logarithms of both sides,

$$-\lambda t_{\frac{1}{2}} = \ln\frac{1}{2} = -\ln 2$$

gives

$$t_{\frac{1}{2}} = \frac{\ln 2}{\lambda}$$

This is a very useful link.

The number of nuclei remaining after any particular time can be calculated using the decay equation. It can also be calculated using the half-life: if the time is a multiple of the half-life, e.g. n half-lives, then the number remaining is simply $1/2^n$ of the original number.

Calculations using half-lives

The number of nuclei remaining and the activity of a radioactive source both fall off with the same half-life. We can use this to calculate values of N or A after any number of half-lives, even if this is not a whole number. The method is simple:

Step 1: Work out how many half-lives have passed: $x = \dfrac{t}{t_{\frac{1}{2}}}$

e.g. if the half-life is 400 years and you want the activity after 1000 years then $x = 1000/400 = 2.4$.

Step 2: Find the initial value and divide it by 2^x

e.g. the activity after time t is given by: $A = \dfrac{A_0}{2^x}$

e.g. the number of nuclei remaining is given by: $N = \dfrac{N_0}{2^x}$

Worked example

Americium-141 has a half-life of 433 y. (a) What is its decay constant? (b) What is the total activity of 1 g of this nuclide? (c) How much of this 1 g sample remains after 1000 y?

(a) $\lambda = \dfrac{\ln 2}{t_{\frac{1}{2}}} = \dfrac{\ln 2}{433 \times 3.16 \times 10^7\,\text{s}} = 5.1 \times 10^{-11}\,\text{s}^{-1}$

(b) In 1 g there are approximately $(1\,\text{g}/241\,\text{g}) \times 6.0 \times 10^{23}$ atoms and the same number of nuclei, hence activity,

$A = \lambda N = 5.1 \times 10^{-11}\,\text{s}^{-1} \times \dfrac{1\,\text{g}}{241\,\text{g}} \times 6.0 \times 10^{23} = 1.3 \times 10^{11}\,\text{Bq}$

(c) Fraction remaining $= e^{-\lambda t} = e^{-(5.1 \times 10^{-11}\,\text{s}^{-1} \times 1000 \times 3.2 \times 10^7\,\text{s})} = 0.20$

Summary of useful equations

Rate of decay	$\dfrac{dN}{dt} = -\lambda N$
Activity	$A = -\dfrac{dN}{dt} = \lambda N$
Decay equation	$N = N_0 e^{-\lambda t}$
Half-life	$t_{\frac{1}{2}} = \dfrac{\ln 2}{\lambda}$

PRACTICE

1 The activity of 0.32 mol of element X is four times the activity of 0.16 mol of element Y. Element Y has a half-life of 20 000 y.

  a What is the half-life of X?

  b How much of each will remain after

    i 40 000 y and **ii** 5000 y?

  c How long will it be before equal amounts of X and Y remain?

2 Iodine-131 has a half-life of 8 days.

  a What is the decay constant for iodine-131?

  b What is the activity of 1 milligram of iodine-131?

  c How long will it take for the activity of a 1 milligram sample of iodine-131 to fall to 10% of its original value?

3 Iodine-131 (discussed in question 2) is used as a tracer to check whether the thyroid gland is taking up iodine from the blood as it should. It emits both beta particles and gamma rays, and the latter are monitored from outside the body close to the thyroid gland. A dose of iodine of activity 1×10^8 Bq is injected into a patient's blood, and 20% of this is absorbed by the thyroid gland. The detector records 0.4% of the gamma rays emitted.

  a Why are the beta particles not monitored?

  b Suggest reasons why the detector only records 0.4% of the gamma rays emitted from iodine-131 nuclei inside the thyroid.

  c What activity would you expect the detector to record **i** soon after absorption, **ii** 16 days later, and **iii** three weeks after injection? State any assumptions.

4 Schools sometimes use 5 microcurie strontium-90 sources as beta sources. Strontium-90 has a half-life of 28 years. What is the total activity of a strontium-90 source 10 years after purchase?

5 The activity of a particular source falls from 5×10^7 Bq to 2×10^7 Bq in 20 minutes.

  a What is the half-life and decay constant for this nuclide?

  b How many atoms were there in the original source?

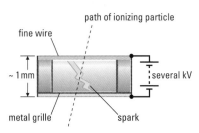

A spark counter.

A large rectangular sheet of plastic scintillator as used in an experiment at the European Laboratory for Particle Physics (CERN – for Conseil Européen pour la Recherche Nucléaire).

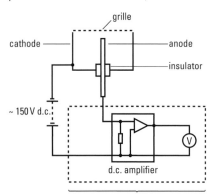

this circuit is effectively a nano-ammeter to measure ionization current

An ionization chamber.

The d.c. amplifier

The currents generated in ionization chambers are too small to measure with a conventional ammeter. The d.c. amplifier converts this tiny current into a significant voltage by making it flow through a very large resistance (e.g. $10^{10}\,\Omega$). It then measures the voltage. If the resistance is known the current can be calculated.

RADIATION DETECTORS

Alpha, beta, and gamma radiation are all detected by their electromagnetic effects on matter. Alpha and beta particles are charged and cause a lot of ionization as they collide with atoms and molecules. Gamma radiation is part of the electromagnetic spectrum and consists of oscillating electric and magnetic fields that are also capable of ejecting electrons from atoms. Most detectors are made sensitive by putting them into a state that is on the verge of a dramatic physical change (e.g. electrical breakdown, change of state, etc.) and the incoming radiation triggers this change.

Early detectors

The spark counter   This consists of a fine wire running close to a metal grille. A high voltage is applied between the wire and grille. As the voltage is gradually increased a point is reached where the air breaks down near the wire (where the electric field is most intense) and sparks jump between the two electrodes. To detect radiation the voltage is reduced from this value until the sparking just stops. If an alpha source (strongly ionizing) is then held close to the gap between the electrodes, ions created by alpha particles are accelerated in the field and crash into other air molecules. This causes further ionization and sets off an electrical 'avalanche', resulting in more sparks.

- The rate of sparking gives some idea of the activity of the source.

The scintillator (spinthariscope)   When Geiger and Marsden carried out the famous alpha-scattering experiments leading to the discovery of the atomic nucleus, they had to count individual alpha particles scattered in various directions. They used an eyepiece focused onto a screen coated in zinc sulphide. When an alpha particle hits such a screen it transfers energy to atoms which then lose this energy by radiating photons of visible light. The tiny 'scintillations' are just about visible to the unaided eye when it has become accustomed to the dark. Geiger and Marsden spent many hours in darkened rooms counting scintillations!

- Many modern detectors rely on the rapid response of (more sophisticated) scintillation counters to act as triggers for slower detectors, to ensure they record significant events.

The ionization chamber

The ionization chamber works in a similar way to a spark counter. The difference is that the detector itself is a conducting box and the source to be studied is usually placed inside it. The box is connected to one side of a voltage supply, and the other electrode is a central metal rod inside the box (but insulated from it). Ions created by the source drift toward the two electrodes, forming a small ionization current (typically nA) that can be measured using a sensitive ammeter. In the school laboratory this is likely to be a d.c. amplifier.

- Ionization chambers are useful because ionization current is proportional to activity.

Current is rate of flow of charge, so the ionization current is a measure of the rate at which the source creates charged particles (ions) in the chamber. If the rate of particle emission from the source (the **effective activity**) is known, the ionization current can be used to calculate the number of ions created per particle. If the ionization energy is also known this gives a measure of the average particle energy.

The Geiger–Müller (GM) tube

The GM tube or **Geiger counter** is very similar to an ionization chamber, but differs in several important ways:

- It can register the arrival of *individual* alpha or beta particles or gamma rays (rather than an average ionization current).

- It can be connected directly to a counter to record the number of 'events' in a set time (e.g. 10 s or a minute), which gives a 'count rate'.
- It is much easier to set up and use.

The GM tube consists of a central fine-wire electrode and an outer cylindrical electrode with an electric field between them. Ionizing particles cause an avalanche of charge in the tube, which contains air or argon at low pressure (gases conduct electricity more readily at low pressure because the ions travel further, on average, between collisions). The pulse of current caused by this avalanche flows through an external resistor connecting the two electrodes, and creates a voltage pulse that is counted electronically.

However, all this takes time, and if the pulse of current lasts too long another particle might enter the tube before it has finished and so give a continuous output rather than a series of pulses. This problem is limited, but not removed, by introducing a small quantity of bromine into the tube. This absorbs energy from the ions, reducing their speed sufficiently for them to re-combine rapidly and switch off the current pulse. The bromine is called a **quenching agent**.

In practice each pulse still lasts about 200 μs, so if ionizing particles follow one another more frequently than this they will not be recorded separately. The GM tube is said to have a **dead time** of about 200 μs. This gives a maximum practical count rate of less than about 5000 counts per second.

There are other points to consider when using a GM tube:

- The radiation to be detected must first enter the tube. This is a particular problem for alpha particles, so GM tubes are fitted with a very thin end window (made of a slice of mica).
- The efficiency of a GM tube is near 100% for alpha and beta radiation, but much lower for gamma (because it is less strongly ionizing).
- Whilst alpha and beta particles must enter via the end window, gamma rays pass easily through the sides of the tube (from which they eject electrons), so it is worth using the tube side-on for gamma rays, because this presents a larger catching area toward the source.

The cloud chamber

The air inside a cloud chamber is saturated with alcohol vapour. When a saturated vapour is cooled it condenses out on any available surface or particles, in much the same way that water vapour condenses from warm air in a bathroom onto the colder surface of a glass mirror, or as clouds form when warm damp air rises into a cooler region of the atmosphere.

The air in the cloud chamber is cooled suddenly, either by expanding the chamber (**expansion chamber**) or by placing some dry ice under the base of the chamber (**diffusion chamber**). In either case a region that is **super-saturated** in alcohol vapour is formed. It temporarily holds more alcohol than it should at that temperature and any slight disturbance causes condensation and droplet formation. The trail of ions left in the wake of an ionizing particle are an ideal place for condensation to occur, so droplets form along the line and the track becomes visible.

Solid-state detectors

Semiconductors owe their electrical properties to the fact that their charge carriers are neither free nor tightly bound to particular atoms. This means they can be 'freed' for conduction by various atomic processes that provide the small amount of energy required. A solid-state detector is essentially a reverse-biased semiconductor diode. When an ionizing particle hits it, it supplies enough energy to release many charge carriers inside the semiconductor (holes and electrons near the **pn junction**, that is, the junction between p-type and n-type). The pulse of current can be turned into a voltage pulse through a resistor and counted electronically. This kind of detector is particularly useful for alpha particles.

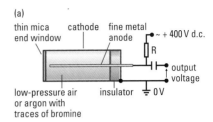

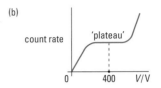

(a) The Geiger–Müller tube. (b) The operating voltage is set in the plateau region.

Operating voltage

The responses of the spark chamber, ionization chamber, and GM tube all change with the voltage between their electrodes. If this is too low they will not respond to all the particles that enter them. If the voltage is too high they will respond continuously as the gas inside breaks down. Between these two extremes is a region in which they respond to pretty well all the ions inside them. This region is the voltage **plateau**.

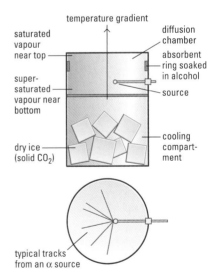

A diffusion cloud chamber.

OBJECTIVES

- background radiation
- biological effects
- radiation dose
- radon
- safety precautions

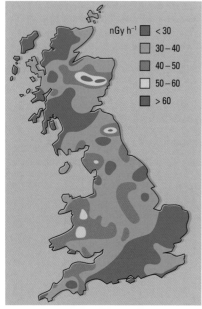

12% internal from eating, drinking, and breathing

14% terrestrial gamma rays from rocks and soil

10% cosmic rays from outer space

47% radon and 4% thoron inside our homes

0.1% waste from the nuclear industry

0.2% occupational from medical and industrial uses

0.4% fallout from weapons tests and Chernobyl

0.4% miscellaneous, mainly from air travel and luminous watches

12% medical mainly from X-rays

Sources of radiation exposure for the UK population.

nGy h⁻¹
- < 30
- 30 – 40
- 40 – 50
- 50 – 60
- > 60

Radiation levels in Great Britain from rocks and soils.

Safety limits

Public: 1000 μSv per year

Radiation workers: 50 000 μSv per year

It is estimated that a dose of 10 000 μSv is needed to give a reduced lymphocyte (white blood cell) count. Lymphocytes fight infection, so a reduced count depresses the immune system, leaving you more likely to develop other diseases like pneumonia. In this respect the after-effects of a significant radiation dose are similar to diseases like AIDS which also reduce the body's defences.

RADIATION HAZARDS

Over 10% of your annual radiation dose is likely to come from nuclides inside your own body (as a result of eating, drinking, and breathing). About 15% will come from gamma rays emitted by nuclides in rocks and soil, and 10% from outer space in the form of high-energy cosmic rays. But the greatest radiation hazard in our natural environment comes from radon, an alpha-emitting gas (the only naturally occurring radioactive gas) which is created in the uranium decay series and accumulates in buildings. Your exposure to radon will vary depending on where you live, because of the variation in amount and depth of radon-producing rocks – the high-risk counties in the UK are Cornwall, Devon, Somerset, Northamptonshire, and Derbyshire. On average about 50% of every British person's annual radiation dose comes from radon. An independent study for Public Health England (2009) estimated that radon causes over 1100 deaths from lung cancer in the UK each year.

Biological effects

The typical energy of a radioactive emission is around 1 MeV. This is about a hundred thousand times greater than the energy of a chemical bond, so if radiation is absorbed by human flesh it can have devastating results on a molecular scale. In some cases it can damage the genetic material inside cells, causing them to mutate or reproduce uncontrollably, leading to cancer.

Of course, we have evolved in a radioactive environment, so we do have some resistance to radiation damage: our cells have repair mechanisms that can rectify copying errors in our genes. However, as the absorbed radiation dose increases, the repair mechanisms may be overwhelmed. In cases where a large dose has been absorbed in a short time the cells break down completely and severe radiation burns occur. It is more difficult to assess the risks involved if you are exposed to a lower level of radiation for a long time, so a useful rule of thumb is to ensure your total dose is small compared to the annual average dose you would get simply by living on Earth.

Measuring radiation exposure

Radiation dose is measured as energy absorbed per kilogram of body tissue.

- 1 gray = 1 joule absorbed per kilogram

  $1 \, \text{Gy} = 1 \, \text{J kg}^{-1}$

However, the effects of alpha, beta, and gamma radiations are different. The damage done by highly ionizing alpha particles is far more severe than by beta or gamma, so a more useful measure is the **dose equivalent** which is calculated by multiplying the actual dose absorbed (in Gy) by a **quality factor** Q which depends on the type of radiation absorbed. Q is 20 for alpha particles and 1 for beta and gamma ($Q = 10$ for irradiation by protons or neutrons). The dose equivalent is measured in sieverts (Sv):

- $$\text{dose equivalent (Sv)} = \frac{\text{quality factor} \times \text{energy absorbed (J)}}{\text{mass of absorbing tissue (kg)}}$$

  $1 \, \text{Sv} = 1 \, \text{J kg}^{-1}$

People are often confused by the high quality factor for alpha particles. Alpha sources *outside* the body are the least hazardous because the alpha particles are stopped by a few centimetres of air or the outer (dead) layers of skin. The real hazard from alpha emitters is if they get inside the body. This is why radon is so dangerous: it can be breathed in. Also, alpha-emitting powders (e.g. the chemicals used in school radon generators) are extremely dangerous if they become airborne, which is why they are kept in sealed containers (if they are kept at all).

Worked example

A dose calculation

Before you carry out an extended investigation using radioactive sources you must do a risk assessment. This is a calculation of the maximum radiation dose you could receive in carrying out the experiment.

Imagine you are using a school gamma source (cobalt-60, for example) for 2 hours and the experimental set-up means your minimum distance from the source will be 10 cm.

Data
gamma-ray energy, $E = 1.33\,\text{MeV} = 2.1 \times 10^{-13}\,\text{J}$,
source activity, $A = 5\,\mu\text{Ci} = 185\,\text{kBq}$,
distance, $r = 10\,\text{cm}$, $Q = 1$, time $t = 2\,\text{h} = 7200\,\text{s}$,
area of body exposed, $a \approx 0.05\,\text{m}^2$
mass of absorbing tissue $\approx 2.0\,\text{kg}$
total energy radiated in 2 h is EAt; this is radiated through an area of $4\pi r^2$ at a distance r. If your estimated 'catching area' is a, the fraction you will absorb is $a/4\pi r^2$, so

$$\text{your total dose} = \frac{QEAta}{4\pi r^2 m}$$

$$= 1 \times 2.1 \times 10^{-13}\,\text{J} \times 185 \times 10^3\,\text{s}^{-1} \times 7200\,\text{s} \times \frac{0.05\,\text{m}^2}{4\pi \times (0.1\,\text{m})^2 \times 2.0\,\text{kg}}$$

$$= 50\,\mu\text{Sv}$$

This is, of course, a very rough calculation. The effective activity of the source is likely to be considerably lower than its total activity, and the 'catching area' is unlikely to be constant. Nonetheless a dose estimate of this kind gives an order of magnitude which can be compared against known risks to decide whether to proceed or to modify your plans.

Changing attitudes

Nowadays people are far more careful about the risks of radiation than in the past. Rutherford's notebooks are now considered too radioactive to handle directly, almost a century after his original experiments! Rutherford himself lived to the age of 68, but others involved in the early use of radioactivity were not so lucky. In 1915 Dr von Hoffman set up the Radium Luminous Materials Company in New Jersey. He employed hundreds of girls, some as young as 14, to paint watch dials and other artefacts with luminous paint, which was a mixture of radium and zinc sulphide. The girls used to roll the ends of the brushes between their lips to make a fine point for an accurate result. Some of the girls even painted their teeth to make them glow when they went out to parties! By the end of 1926 four of the girls had died and many others were developing cancer.

Corrected count rates

When you carry out an experiment in which you need to measure and record the activity of a radioactive source it is important to eliminate the effects of background radiation. This can be done by setting up the apparatus without the radioactive source and then measuring the average count over a suitable time period, such as 5 minutes. From this the average background count can be calculated (in counts per minute, c.p.m.). Once this has been done the source can be placed in position and the experimental measurements taken. The average background count must then be subtracted from each experimental value; this gives the corrected count rate.

$$\begin{array}{ccc} \text{corrected count} & = & \text{experimental count} & - & \text{average background} \\ \text{rate (c.p.m.)} & & \text{rate (c.p.m.)} & & \text{count rate (c.p.m.)} \end{array}$$

It is important to recalculate the average background count rate for each experiment since this will depend on location, shielding (e.g. by the apparatus), and the sensitivity of the detector.

Handling radioactive sources in a school laboratory

School sources are weak and should not pose any significant danger if they are handled carefully and sensible precautions are taken. The point of these precautions is to reduce the dose you receive to a minimum and to avoid contaminating yourself or the laboratory with radioactive materials.

For **sealed** sources:

- Only remove sources from their shielded containers when they are required for experiments.
- Never handle sources directly with your fingers; always use tongs or special holders.
- Always point sources away from you and other people.
- Do not bring food or drink into a laboratory where radioactive experiments will take place, and do not eat or drink in the laboratory.
- Always wash your hands immediately after carrying out experiments involving radioactive sources.
- If you drop a source or suspect there may be a leak or some damage to the source inform your teacher immediately.
- Cover any cuts or scratches with a plaster while carrying out experiments.

It is unlikely that you will use any **open** sources, but if you do, remember that these pose a greater risk of contamination. Check all safety precautions with your teacher before proceeding.

OBJECTIVES

• measuring half-life of radon-220

• decay series

• radioactive equilibrium

• decay and recovery

Warning!

The descriptions of experiments given here are not intended as complete practical instructions. Experiments involving radioactive sources must be carried out following strict safety guidelines provided by your local **radiation protection supervisor**. If this is done the dose you are likely to receive from school sources is negligible.

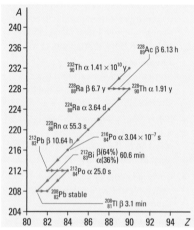

The radioactive decay series for thorium-232.

Radon in the home

Radon is generated in the Earth's crust, particularly where there is a lot of granite. It can cause a radiation risk if it accumulates in houses. Nowadays it is possible to buy radon ventilators which pass air through the rooms to prevent the concentration of radon from reaching undesirable levels.

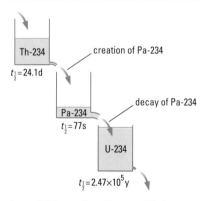

A model for radioactive equilibrium. 'Water' depth represents amount of nuclide. Size of exit hole represents the decay constant for that nuclide. Equilibrium is reached when the depth is such that the inflow rate equals the outflow rate.

Radon-220 is a radioactive gas that decays to polonium-216 by alpha emission. Its half-life can be measured in the school laboratory. Radon is created naturally in the thorium decay series and in this experiment it is generated from a compound of thorium such as thorium carbonate or hydroxide.

It is because radon is a gas that it can leave the rocks and accumulate in buildings. The same principle is used to separate it from other substances in a radon generator (which contains many other members of the thorium decay series as well as radon) so that its decay can be monitored in isolation. A sample of radon is pumped into an ionization chamber, and the ionization current is measured as it falls back to zero. Ionization is proportional to the activity of the source and so decays with the same half-life. This can be calculated by measuring the time taken for the current to halve, or better, the average half-life can be measured by looking at several halvings (e.g. the half-life $t_{\frac{1}{2}}$ is 1/3 of the time taken for the ionization current to fall to $1/8 = 1/2^3$ of its original value). However, the best method is to plot the logarithm of the ionization current against time.

$$I \propto A \text{ so } I = I_0 e^{-\lambda t} \quad (A \text{ is activity})$$

$$\ln I = \ln I_0 - \lambda t$$

So a graph of $\ln I$ versus t is a straight line with a gradient of $-\lambda$ and intercept on the $\ln I$ axis of $\ln I_0$. The half-life is then found from the equation

$$t_{\frac{1}{2}} = \frac{\ln 2}{\lambda}$$

The accepted value for the half-life of radon-220 is 55.5 s.

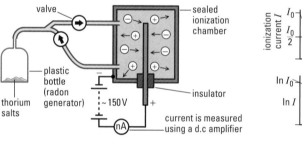

Using a radon generator and an ionization chamber to measure the half-life of radon-220.

Radioactive equilibrium

Radioactive minerals contain many different nuclides. The amount of each one is fixed by a balance between its rate of creation and its rate of decay. For example, protactinium is a member of the **uranium decay series**. It is generated by the beta decay of thorium-234 and decays by another beta decay to another isotope of uranium. When it is in radioactive equilibrium with its parent (thorium-234) and daughter (uranium-234) the rate at which thorium decays to protactinium must be equal to the rate at which the protactinium decays to uranium:

$$A_{\text{Th-234}} = \lambda_{\text{Th-234}} N_{\text{Th-234}} \quad A_{\text{Pa-234}} = \lambda_{\text{Pa-234}} N_{\text{Pa-234}} \quad (\text{activities})$$

giving

$$\lambda_{\text{Th-234}} N_{\text{Th-234}} = \lambda_{\text{Pa-234}} N_{\text{Pa-234}}$$

so

$$\frac{N_{\text{Pa-234}}}{N_{\text{Th-234}}} = \frac{\lambda_{\text{Th-234}}}{\lambda_{\text{Pa-234}}} = \frac{t_{\frac{1}{2}}(\text{Pa-234})}{t_{\frac{1}{2}}(\text{Th-234})}$$

• In equilibrium, the amount of each nuclide is proportional to its half-life.

In this case the half-life of protactinium-234 is only 1.17 minutes compared to thorium-234's 24.1 days, so only a trace amount of protactinium will be present.

(Actually, radioactive equilibrium is never quite achieved, because the thorium is being generated by the alpha decays of uranium-238 and this is not being replaced. Ultimately all nuclides in the series will decay, leaving the stable nuclide lead-206.)

Decay and recovery

If protactinium is removed from the series it will decay and its half-life can be measured by monitoring its activity. This is very similar to the approach used for radon described above. Removing protactinium from the rest of the nuclides in the decay series disturbs radioactive equilibrium and for a while the protactinium is being generated more rapidly than it is decaying. This continues until the equilibrium state is recovered. A neat experiment can be carried out in the school laboratory to show this process of decay and recovery.

An aqueous solution of uranyl nitrate generates the decay series. Above it is a less dense layer of organic solvent in which the protactinium compounds will dissolve. The two solutions are held in a sealed plastic bottle. If the bottle is shaken the protactinium is removed from the aqueous layer, and its decay can be monitored by placing a Geiger counter close to the top of the bottle once the two solutions have again settled out. The experiment can then be repeated with the Geiger counter next to the lower aqueous layer to monitor the recovery (the count rate rises as the proportion of protactinium in that layer rises).

Note that only the beta particles from the protactinium are recorded because they are higher energy (1.3 MeV) than those from the thorium (0.19 MeV), and the uranium-234 that is formed is an alpha-emitter – the alpha particles cannot penetrate the walls of the bottle.

Both of the experiments described above are ideal to interface with a computer so that count rates are taken automatically and data is stored in a spreadsheet.

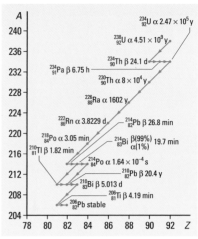

The uranium-238 radioactive decay series.

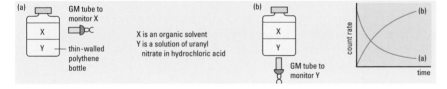

The decay and recovery of protactinium. After the bottle has been shaken protactinium chloride dissolves in X. Once the layers have settled out its decay in X (a) and recovery in Y (b) can be separately monitored as shown. The graph shows typical results.

PRACTICE

1 Radon-220 has a half-life of approximately 1 minute.

a What fraction of a sample of radon-220 would remain after 4 minutes?

b What fraction of a sample of radon-220 has decayed after 2 minutes?

c What has happened to the nuclei that have decayed?

2 In the experiment described above to measure the half-life of radon-220, it is actually the ionization current that is measured rather than the activity of the source. Explain why the half-life for the decay of the current is also equal to the half-life of the source.

3 Look at the uranium decay series above.

a What is the half-life for thorium-90?

b How would you expect the amount of thorium-90 in rocks containing uranium to compare with the amounts of uranium-234 and radium-226? Explain how you arrive at your answer.

4 A sample containing radioactive potassium is known to contain two isotopes, potassium-42 and potassium-44. Both of these are beta emitters but potassium-42 has the longer half-life. The data below shows the count rate from the sample (corrected for background).

Time/h	0.0	0.5	1.0	1.5	2.0	2.5	3.0	4.0	5.0	6.0	7.0	8.0	9.0	10.0
Counts per min	1000	398	213	126	95.5	89.0	83.2	78.8	75.2	71.0	66.8	63.0	60.5	57.5

Use this data to estimate:

a the half-life of each isotope;

b the ratio of potassium-42 to potassium-44 in the original sample.

5 There are only four different radioactive decay series. Find out why this is. (Hint: the only decays involved in the series are alpha and beta; think about the effect of these on *A* and *Z*).

LONGER HALF-LIVES

When a nuclide has a half-life measured in seconds, minutes, hours, or even days it is quite reasonable to measure its half-life by monitoring its decay. However, half-lives have an extremely wide range of values and very long half-lives cannot be measured in this way, because no significant change in activity would occur during a human lifetime. Uranium-238, for example, has a half-life of 4.51 billion years! How do we know that?

These long half-lives can be calculated using the decay constant λ:

$$A = \lambda N$$

If the activity is measured and the number of nuclei in the sample is calculated (from its mass) the decay constant can be calculated. Half-life is once again calculated from:

$$t_{\frac{1}{2}} = \frac{\ln 2}{\lambda}$$

Naturally occurring isotopes with very long half-lives have been used to work out the age of the Earth and solar system.

Radiocarbon dating

When cosmic rays (which are often heavy nuclei) strike atoms in the upper atmosphere, one of their effects is to knock neutrons out of the atomic nuclei. Some neutrons go on to collide with nitrogen nuclei (the atmosphere contains about 80% nitrogen) and create carbon-14, an unstable, beta-emitting isotope of carbon:

$$^1_0 n + {}^{14}_7 N \rightarrow {}^{14}_6 C + {}^1_1 p$$

then

$$^{14}_6 C \rightarrow {}^{14}_7 N + {}^0_{-1}\beta + {}^0_0 \overline{\nu}$$

with a half-life of 5730 y. (the final particle is an anti-neutrino – see spread 8.20.)

The creation and decay of carbon-14 lets this isotope reach an equilibrium level, so that a constant fraction of all carbon atoms in the atmosphere and in all living things (which continually exchange their carbon compounds with the atmosphere) is carbon-14. However, when a living organism dies it stops exchanging carbon with the atmosphere and the remaining carbon-14 within it decays exponentially. This decay is used to date archaeological finds that contain dead organic matter. There are two main ways in which the dating process is carried out.

- The activity of a small sample of the material is compared with the activity of the same mass of living material. The sample activity will be lower. If, for example, the sample activity is 1/4 of the value from a similar living specimen, then two half-lives have passed and the sample is about 11 400 years old. In general:

$$\frac{\text{sample activity}}{\text{expected activity if sample were living}} = e^{-\lambda t}$$

where

$$\lambda = \frac{\ln 2}{t_{\frac{1}{2}}}$$

- The ratio of carbon-14 atoms to carbon-12 atoms in a very small part of the sample can be measured directly using ionic separation (as in a mass spectrometer). Ion beams are passed through a vacuum at ninety degrees to a magnetic field. The more massive ions are deflected less than the less massive ones and the rate of arrival of each type can be measured separately. This has the advantage that little damage is done to the sample, which may well be of great historical importance.

Experimental measurements show that the ratio of carbon-14 to carbon-12 atoms in living material is about $1:10^{12}$.

Radiocarbon dating. Samples of carbon dioxide are being prepared before being loaded into a linear accelerator, which analyses the amount of carbon-14 in the gas, produced by the combustion of the sample in oxygen.

The Turin shroud and the Dead Sea scrolls

Carbon dating has been particularly important in trying to assess the authenticity and significance of certain religious artifacts. Ionic separation was used to date the Turin shroud. Three independent laboratories were sent samples to analyse: some were from the shroud, others were of known age to check the accuracy of the laboratory measurements. The results dated the shroud to the middle ages, somewhere between 1260 and 1390AD. It is interesting that there is no historical mention of the shroud before the 14th century either.

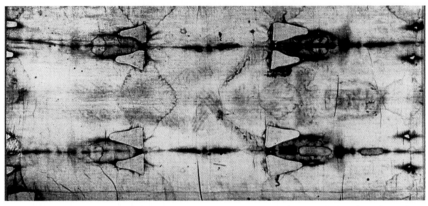

The Turin shroud is a cloth that bears the ghostly image of a man injured at the wrists, feet, chest, and head, and has been claimed by many to be the cloth used to cover Christ when he was brought down from the cross.

However, this may not be enough to convince everybody that the shroud is a fake. When the radiocarbon dating results were officially published in February 1989, a Harvard University physicist named Thomas Phillips suggested that they might be unreliable because the shroud may have been subjected to an intense burst of radiation during the resurrection. This could have altered the chemical and isotopic balance in the material.

In 1947 an Arab shepherd made one of the most remarkable finds in biblical archaeology. He found seven ancient scrolls hidden in a cave at Qumran near the Dead Sea. Some of these contained copies of the Book of Isaiah that turned out to be over a thousand years older than any Hebrew versions of the Old Testament then known. They also contained accounts of the life and beliefs of a Jewish religious community known as the Essenes who may have come into contact with Christ, particularly during his early adulthood, a time which is not dealt with in the Bible.

The crucial question about these manuscripts is whether or not they describe a time before Christ (in which case he could have been influenced by their ideas) or if they were written after Christ, in which case they may have been influenced by him. Carbon dating at the University of Arizona placed crucial texts between 150BC and 5BC with 95% confidence.

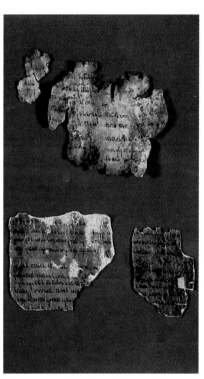

Fragments of the Dead Sea scrolls.

PRACTICE

1 The total activity of a 1 g sample of uranium-234 is 7.2×10^6 Bq. What is the half-life of uranium-234?

2 The ratio of carbon-14 to carbon-12 has not always remained constant.

   a Suggest why this ratio is not constant.

   b In practice the carbon dating techniques can be calibrated against tree rings. How?

3 In a carbon dating experiment a sample of wood from an object was burnt and the carbon dioxide produced was collected. The activity of the carbon dioxide was equivalent to 5.4 counts per minute per gram of carbon, after correction for background. When the same experiment was repeated using wood from a modern source, the corrected count rate was 18 counts per minute per gram of carbon. What is the likely age of the find?

4 Carbon dating techniques make Stonehenge about 4000 years old. Comment on the use of this technique to measure the age of a *stone* monument.

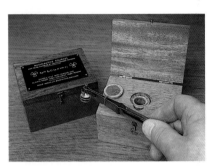

A sealed radioactive source, in this case radium-226, for use in a school laboratory.

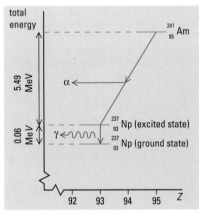

Radioactive decay scheme for americium-241 (not to scale).

AN ALPHA SOURCE

Most radioactive sources used in schools are **sealed sources**. The radioactive element itself is surrounded by a thin layer of foil and held in a steel cup, as shown in the photograph. This reduces the risk of contamination by parts of the source material, but allows most of the radiation to escape through the foil and through an opening in the front of the steel holder.

Americium-241 is unstable and decays by emitting an alpha particle. It is often used as a source of alpha radiation in school laboratories.

$$^{241}_{95}\text{Am} \rightarrow \, ^{237}_{93}\text{Np} + \, ^{4}_{2}\alpha$$

Most of the energy released in the decay goes to the alpha particle; in this case the alpha particle's energy is 5.49 MeV. (The decay of americium-241 also results in a low-energy gamma emission, 0.06 MeV, but the low ionizing power of gamma radiation means this is unlikely to have a significant effect on count rates in experiments designed to detect the much more strongly ionizing alpha particles).

Alpha particles from a particular source have a discrete set of energy values. Most sources, like americium-241, emit alpha particles with a unique energy. Others release alpha particles with two or more different but distinct energies. When this happens the alpha particles with lower energy leave the nucleus formed in the decay in an excited state. The nucleus then decays by emitting one or more gamma rays. The gamma rays too have a discrete energy spectrum.

Alpha decays occur in heavy nuclei with too many protons. By emitting an alpha particle these heavy nuclei reduce their proton-to-neutron ratio. The overall effect is to reduce the mass number (number of nucleons in the nucleus) by 4 and the atomic number by 2. This takes them 2 places lower in the periodic table.

Maths box: alpha decay and conservation of momentum

Alpha decays can be treated like simple explosions. The nuclear energy released in the decay is shared between the two particles in such a way that they conserve linear momentum. If the original nucleus is at rest the new nucleus and alpha particle will move off in opposite directions.

For any (non-relativistic) particle,

momentum, $p = mv$ and kinetic energy, $K = \frac{1}{2}mv^2$

These are linked by

$$K = \frac{p^2}{2m} \quad \text{or } p = \sqrt{2mK}$$

In the equations below, subscripts α and n label quantities associated with the alpha particle or new nucleus respectively. It is assumed that the original nucleus is stationary when it decays.

initial linear momentum of americium-241 nucleus = 0

final linear momentum of alpha particle, $p_\alpha = \sqrt{2m_\alpha K_\alpha}$ to the right.

final linear momentum of neptunium-237 nucleus, $p_n = \sqrt{2m_n K_n}$ to the left.

Momentum conservation gives

$$p_\alpha = p_n \quad \text{(magnitudes)}$$
$$\sqrt{2m_\alpha K_\alpha} = \sqrt{2m_n K_n}$$
$$\frac{K_\alpha}{K_n} = \frac{m_n}{m_\alpha}$$

Masses are approximately proportional to mass numbers, so $m_\alpha = 4$, $m_n = 237$. So

$$\frac{K_\alpha}{K_n} = \frac{237}{4}$$

The alpha particle gets $\frac{237}{244}$ parts of the total energy (i.e. almost all of it).

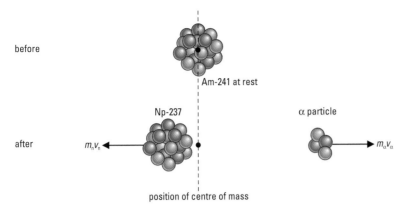

before — Am-241 at rest

after — Np-237 — $m_n v_n$ — α particle — $m_\alpha v_\alpha$

position of centre of mass

Momentum conservation in alpha decay. The much lower mass of the alpha particle means it gains a high velocity and takes away most of the kinetic energy.

Absorption

Alpha particles are helium nuclei emitted with energies of several MeV travelling at about 5% of the speed of light. It only takes about 35 eV to ionize an air molecule so the massive, highly charged alpha particles are very strongly ionizing. They use up their initial energy in a very short distance by forming large numbers of ions per millimetre as they travel through materials. The distance they travel before losing the ability to ionize is called their **range**. In air this is typically between about 3 and 8 cm depending on the energy of the alpha particles, which can be calculated from the lengths of their tracks in a cloud chamber.

When an alpha particle ionizes an air molecule it creates a massive positive ion and ejects an electron. These **ion pairs** can act as charge carriers, making the air temporarily conducting. If an alpha source is held close to a charged electroscope it will soon discharge (by attracting ions of opposite charge from the ionized air and repelling ions of the same charge).

The strong ionizing power from alpha sources makes them easy to detect in ionization chambers. It is more difficult and less efficient to detect alpha particles using a Geiger counter because most of them are stopped before they enter the counter. To reduce this problem Geiger counters with very thin end windows (made of mica sheets) are manufactured.

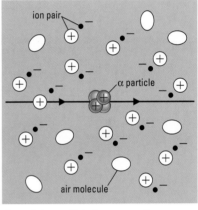

One alpha particle has enough energy to create tens of thousands of ion pairs in the air.

PRACTICE

1 Why are alpha particles so strongly ionizing?

2 Write down a transformation equation for the decay of:

   a $^{238}_{92}U$ to thorium;

   b $^{239}_{94}Pu$ to uranium.

3 Why do you think helium-4 nuclei (alpha particles) are often emitted from unstable heavy nuclei whereas bundles of neutrons or protons alone are not?

4 The alpha particles emitted by americium-241 have an energy of 5.49 MeV.

   a What is their range in air?

   b How many air molecules will they ionize? (Assume that about 30 eV is needed per ion pair formed.)

INVESTIGATING ALPHA PARTICLES

If a beam of alpha particles is fired at right angles to a magnetic field the particles are deflected only slightly, much less than beta particles. This suggests that they have a much larger mass than beta particles. The curvature of their path in the magnetic field can be used to measure their specific charge (charge-to-mass ratio). This is equal to that of a helium nucleus.

In 1909 Rutherford and Royds carried out an experiment in which radon, an alpha-emitting gas, was trapped in a thin-walled tube. Alpha particles emitted from radon nuclei travelled through the walls of the tube, and were stopped inside a surrounding vacuum tube. After several days a low-pressure gas was collected from the 'vacuum' tube and analysed by passing an electric spark through it and looking at the spectrum of the light emitted. It was the helium spectrum. The alpha particles had stopped, collected two electrons from the walls of the vacuum tube, and become helium atoms. This is the destiny of all alpha particles.

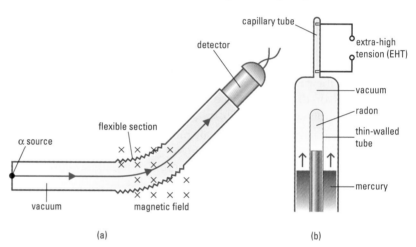

(a) Specific charge (Q/M) for an alpha particle was measured by deflecting a beam of alpha particles in a known magnetic field. (b) The Rutherford–Royds experiment. Alpha particles from the decay of radon escaped from a thin-walled tube and were discharged to helium in the outer enclosure. The helium produced was moved into the capillary tube by raising the level of mercury in the outer containers. An electric discharge through the gas caused an emission spectrum from which the gas was identified.

Ionizing power and range

Alpha particles are about 7000 times more massive than electrons, so they are hardly deflected by electrons in the atoms they hit and their tracks are very straight (a few go very close to a nucleus and are significantly deflected, but this is very rare because the nucleus is so small). They are very strongly ionizing, and the number of ions produced per unit length of an alpha-particle track is much larger than for a beta or gamma track. This means they transfer their initial kinetic energy very quickly and have a relatively short range. Alpha particles are stopped by thick paper or by the outer layers of our skin, and only travel a few centimetres in air at atmospheric pressure.

The ionizing power of the alpha particles depends on their velocity, being larger when they are travelling *more slowly* (when they move slowly they spend longer closer to any particular molecule). This means they produce most ions per millimetre of path just before they stop.

Their range can be measured by moving an alpha source away from a detector (this could be a solid-state detector, a spark counter or an ionization chamber) and measuring the activity. As the source distance increases, fewer ions are produced in the detector (the alpha particles have less energy when they reach it) and its reading decreases.

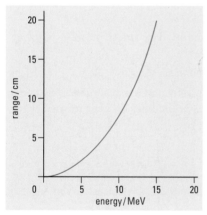

The range of alpha particles in air at atmospheric pressure and temperature. Naturally occurring alpha emitters emit alpha particles with energies up to about 10 MeV.

Measuring the ionizing power of an alpha source

If an alpha source is enclosed in an ionization chamber, all the ion pairs produced remain in the chamber until they re-combine. If the source is attached to the central electrode of the ionization chamber and the outside electrode is made positive, a small current flows as the ions are swept through the chamber onto the electrodes. The size of the current is limited by the rate of production of ions in the chamber and can be used to measure the number of ions created per second by the source.

If all the ions are singly charged then each ion pair is discharged when a single electron moves around the external circuit. So the number of electrons per second is equal to the number of ions pairs per second.

If the ionization chamber current is I, the number of electrons per second is I/e. If the effective activity of the source is A alpha particles per second then the average number of ion pairs per alpha particle is

$$\frac{I}{Ae}$$

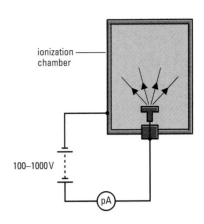

Measuring the alpha particle's ionizing power.

Effective activity

The **total activity** of a source is equal to the number of decays per second taking place inside it, A_0. However, many of the particles emitted will be stopped inside the source (especially if they are alpha particles), so the total activity (which can be calculated from the mass of the source material) is less useful than the **effective activity**, A, which can be measured experimentally.

One way to measure the effective activity is to place a sensitive solid-state detector a measured distance d from the source and record the mean number of particles per second that reach it, A_d. The source emits particles randomly in all directions and the detector only stops a fraction of these. This fraction is

$$\frac{a}{4\pi d^2}$$

where a is the area of the detector facing the source and $4\pi d^2$ is the surface area of a sphere of radius d (centred on the source).

The number of alpha particles per second leaving the source is called the effective activity A and is given by

$$A = \frac{4\pi d^2}{a}A_d$$

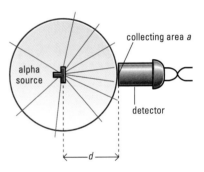

Measuring the effective activity of an alpha source.

PRACTICE

1 a Write down an equation for the force (F) on a moving charged particle in terms of its charge (q) and velocity (v), and the strength of the magnetic field (B). (Assume the field is perpendicular to the path of the particle.)

b Write down an equation for the magnitude of centripetal force in terms of mass (m), velocity (v), and radius of curvature (r).

c Use these two equations to derive an equation for the radius of curvature of a moving charge in a magnetic field.

d The specific charge of an electron (charge divided by mass) is about 4000 times larger than that of an alpha particle. Explain why alpha particles and beta particles of similar energy are deflected very differently in the same magnetic field.

e How does the direction of deflection in the field tell you the sign of charge of the particle?

2 True or false: 'If you absorb a large radiation dose from alpha particles you will become radioactive'. Explain your answer.

3 Explain why the useful activity of a sealed source is always considerably less than its total activity.

4 An experiment to measure the number of ion pairs per alpha particle produced the following results:

ionization current = 0.28 nA
useful activity = 12 000 Bq

a If the energy required to create one ion pair in air is about 35 eV, calculate the number of ion pairs per alpha particle and the energy of the alpha particles.

b What would be the range of these alpha particles in air?

8.20

O B J E C T I V E S

- beta decay

- neutron decay

- the neutrino

- beta-plus decay

Neutrinos they are very small.
They have no charge and have no mass
And do not interact at all.
The earth is just a silly ball
To them, through which they simply pass.
Like dustmaids down a drafty hall
Or photons through a sheet of glass.
They snub the most exquisite gas,
Ignore the most substantial wall,
Cold-shoulder steel and sounding brass,
Insult the stallion in his stall,
And, scorning barriers of class,
Infiltrate you and me! Like tall
And painless guillotines, they fall
Down through our heads into the grass.
At night they enter at Nepal
And pierce the lover and his lass
From underneath the bed – you call
It wonderful; I call it crass.

John Updike, *Cosmic Gall*

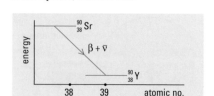

Beta decay energy diagram. The energy
released is shared between the beta
particle and an anti-neutrino (the yttrium
nucleus also gains a little kinetic energy).

Neutron decay

The underlying process in beta decay
is the conversion of a neutron in the
nucleus to a proton. This occurs in
neutron-rich nuclei (like those formed
in nuclear fission).

$$^1_0 n \rightarrow {}^1_1 p + {}^0_{-1} e + {}^0_0 \overline{\nu}$$

Free neutrons are also unstable and
decay to protons with a half-life of
about 11.7 minutes.

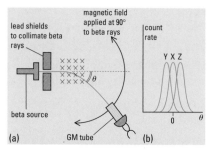

(a) Deflection of beta particles in a
magnetic field shows that they have a
negative charge. (b) Typical results: X is
with no field applied, Y is with the field
out of the page, and Z is with the field
into the page.

BETA PARTICLES

Strontium-90 is often used as a source of beta radiation in school
laboratories.

$$^{90}_{38} Sr \rightarrow {}^{90}_{39} Y + {}^0_{-1} \beta$$

The yttrium-90 nucleus formed in the decay is in its ground state, so no
gamma rays accompany this decay. This is an advantage because gamma
rays would make the results of experiments on beta radiation difficult to
interpret.

The energy released in the decay is 0.546 MeV. This is shared between
all the particles formed in the decay. If the electron and yttrium nucleus
were the only particles they would share this energy in a unique way
(because of conservation of linear momentum and total energy) so that all
electrons emitted from decaying strontium-90 nuclei should have exactly
the same energy. In fact, the beta particles from *any* beta decay have a
continuous range of energies from zero up to a maximum equal to the
value predicted using the laws of conservation of energy and momentum.

When this was first realized it baffled physicists (some even thought
that energy conservation might not apply to beta decays). In the end
Wolfgang Pauli proposed that the missing energy was carried off by a
new particle, which is now known to be an **anti-neutrino**. Neutrinos and
anti-neutrinos are neutral, probably massless, and almost impossible to
detect (Pauli predicted their existence in 1932, but they didn't appear in
an experiment until 1956!) The full decay then becomes:

$$^{90}_{38} Sr \rightarrow {}^{90}_{39} Y + {}^0_{-1} \beta + {}^0_0 \overline{\nu}$$

(The bar over the neutrino symbol indicates that it is an anti-neutrino.)

The energy expected to go to the electron is actually shared between
it and the anti-neutrino. It can be shared in any ratio, explaining the
continuous spectrum of beta particle energies. A scheme for the decay is
illustrated.

Deflection of beta particles in a magnetic field

An experiment to measure the deflection of beta particles uses a
collimated source (absorbers are arranged so that a narrow beam
emerges) and a uniform magnetic field perpendicular to the emerging
beta rays. A radiation detector (e.g. Geiger counter) is used to measure
the activity (counts per second) at various angles to the initial direction
of the beam when the magnetic field is (i) absent (i.e. a control),
(ii) vertically downward, (iii) vertically upward. After the background
count has been subtracted from all the results, count rate versus angle
can be plotted to show all the data on the same graph.

When the magnetic field is present the peak count rate is shifted from
its central position (where it appears in the absence of a magnetic field),
and the direction of shift is reversed when the field is reversed. If Fleming's
left-hand rule is applied to the deflection, it shows that the particles
carry negative charge (i.e. the current points towards the source, since
conventional current flow is opposite to the way electrons actually move).

The greater the energy and momentum of the electrons, the harder
they are to deflect from their path. Since beta particles (high-speed
electrons) from any source have a continuous spectrum of energies, the
most energetic ones are deflected less than those of lower energy. This can
be used to measure the energy spectrum.

The charge-to-mass ratio for beta particles can be calculated from
their radius of curvature in a magnetic field. This is equal to the value for
electrons of similar energy.

Penetrating power

The penetrating power of beta particles can be investigated by placing various absorbers between the source and the detector. Their penetrating power depends upon their energy, but they can usually penetrate paper and skin quite easily, and are stopped by a few millimetres of aluminium. The diagram shows a simple arrangement that could be used to investigate the absorption of beta radiation in a metal (e.g. aluminium).

The beta particles lose energy as they interact with atoms of the absorber. This may result in ionization of the absorber atoms, or the emission of energy in the form of X-ray photons. A 1 MeV beta particle has enough energy to ionize tens of thousands of atoms. The result of the various processes which dissipate energy is that the intensity of beta radiation in the material falls off with distance more or less exponentially. The denser the material the more interactions per unit distance in the medium, and the steeper the curve.

If the logarithm of the transmitted beta activity is plotted against the thickness of the absorber the graph shows a linear decrease at first and then levels out (as shown in graph (b) on the right). The 'range' of beta particles in a medium is taken to be the thickness at which the gradient of this graph changes. This is found by extending the linear parts of the graph to the point where they intersect, as shown. (The horizontal section of the graph shows that some radiation is continuing to emerge from the absorber even when the beta particles have all been stopped. This radiation is due to X-rays and secondary electrons ejected by the beta particles as they stop.)

Beta-plus (positron) emission

Some proton-rich nuclei can become more stable by converting a proton to a neutron and emitting an anti-electron or **positron**. One important example is the isotope oxygen-15:

$$^{15}_{8}O \rightarrow \ ^{15}_{7}N + \ ^{0}_{1}\beta^+ + \ ^{0}_{0}\nu$$

This is very similar to normal beta decay, but the positron emitted is an anti-particle and soon annihilates with an electron to form a pair of gamma rays:

$$^{0}_{1}e^+ + \ ^{0}_{-1}e \rightarrow 2\,^{0}_{0}\gamma$$

Gamma rays from these annihilations are used to form medical images in positron emission tomography (PET) scans (see spread 8.22).

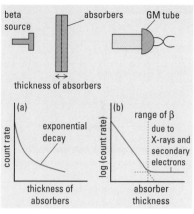

The penetrating power of beta radiation. (See below for the distinction between thickness of absorber and 'absorber thickness' or surface density).

Low-energy beta particles are deflected quite significantly by collisions with air molecules; their tracks appear tortuous and faint (because of their low ionizing power). High-energy tracks are straighter, but hard to see because of the low number of ion pairs per centimetre along them.

Absorber thickness

The ability of an absorber to block radiation depends on its thickness and density, so the absorbing power of a thin piece of dense material can match that of a thick piece of low density material. To compare absorbers a new quantity is invented – absorber thickness or surface density. This is the product of thickness and density and so has units $kg\,m^{-2}$.

For example, 1 mm of lead (density $9800\,kg\,m^{-3}$) has about the same stopping power as about 3.6 mm of aluminium (density $2700\,kg\,m^{-3}$):

surface density for the lead absorber
$= 10^{-3}\,m \times 9800\,kg\,m^{-3} = 9.8\,kg\,m^{-2}$

surface density of aluminium absorber
$= 3.6 \times 10^{-3}\,m \times 2700\,kg\,m^{-3} = 9.7\,kg\,m^{-2}$

Beta and alpha

Beta particles have a much greater range than alpha particles because they react less strongly with matter. This is because their smaller mass and charge and their higher speed make them less strongly ionizing than alpha particles. The tracks made by beta rays in a cloud chamber are less distinct than those of alpha particles because there are fewer ions per cm around which droplets can form.

PRACTICE

1 a What change occurs in the nucleus when it undergoes beta decay?

b Suggest a reason why carbon-14 is a beta emitter and yet carbon-12 is stable.

c Carbon-11 is also unstable. Can you suggest how it might decay?

2 Cobalt-60 ($Z = 27$) decays to nickel by emitting a beta particle. The maximum energy of the beta particle emitted is 0.31 MeV. The decay is followed by the emission of two gamma rays of energy 1.33 MeV and 1.17 MeV.

a Draw an energy diagram to show this decay.

b Explain why the energy of the beta particle is quoted as a *maximum* value.

c If a beta particle of 0.21 MeV is emitted, what happens to the other 0.10 MeV?

d Cobalt-60 is often used as a gamma source in school laboratories. Why is it particularly suitable?

- energy spectra
- the inverse square law
- half-thickness

Overlap

Low-energy gamma rays and hard X-rays overlap in the electromagnetic spectrum. There is no difference between the photons themselves in this region; the name simply refers to how they were created. Gamma rays are emitted by excited nuclei. X-rays are emitted by the deceleration of electrons when they hit a target, or by large quantum jumps inside atoms as electrons 'drop' into vacant energy levels in the inner electron shells close to the nucleus (usually following the ejection of an inner electron by another electron).

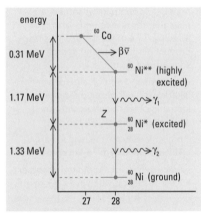

Energy diagram for the decay of cobalt-60.

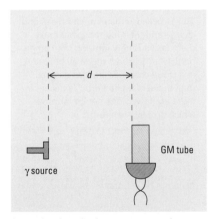

Investigating the inverse square law for gamma rays in air. Note that the GM tube has been turned sideways to increase its sensitivity to gamma rays and increase its catching area. In practice the actual separation of source and detector is (d+x) where x is not directly measurable. This means the count rate C is proportional to $1/(d+x)^2$ rather than $1/d^2$. From this relation you should be able to see that a graph of d versus $1/\sqrt{C}$ should give a straight line (from which x can be found).

GAMMA RAYS

Gamma rays are photons of high-frequency electromagnetic radiation. They have typical wavelengths of about 10^{-11} m, which is slightly smaller than the distance between atoms in a solid and shorter than the wavelength of hard X-rays. Their wave-like characteristics were first demonstrated by the diffraction patterns they form when they pass through crystals. The clicks recorded in a Geiger counter are a reminder of their particle-like properties.

A school gamma source

Gamma radiation is a by-product of other decays, and does not affect the number or type of nucleons that remain in the nucleus. If an alpha or beta decay leaves the new nucleus in an excited state it will soon lose this 'extra energy' by making one or more quantum jumps as it returns to its ground state. This means that the source of the gamma rays is not really the original nucleus but the one into which it decays.

Cobalt-60 is often used as an experimental source of gamma radiation in school experiments. It decays to nickel-60 by beta emission, but the beta particle (and accompanying anti-neutrino) carry away only about 1/6 of the total energy available. This leaves the nickel nucleus in a highly excited state, and it decays in two almost equal quantum jumps to its ground state. This means that every cobalt-60 nucleus that decays to nickel-60 results in the emission of a pair of gamma rays (in this case their energies are 1.17 MeV and 1.33 MeV respectively).

All gamma spectra are discrete, and studying them can give information about nuclear energy levels in much the same way that optical spectra tell us about atomic energy levels.

Ionization and penetrating power

Gamma rays have much lower ionizing power than alpha or beta particles. This is because they interact with matter more weakly and in a very different way. Alpha and beta particles are both charged and create thousands of ions as a result of collisions. Gamma rays are uncharged electromagnetic photons. They can ionize atoms but they do this by giving up all their energy in one go (as in the photoelectric effect), creating an energetic ion pair that can go on to create more ion pairs as it crashes into surrounding atoms and molecules. Even so, this is a relatively rare event, so gamma rays leave indistinct tracks in cloud chambers, and Geiger counters are inefficient at detecting them.

Gamma rays can also lose energy by scattering from atomic electrons or, if they have enough energy, by creating electron–positron pairs in the strong electric field near the nucleus. However, the weakness of their interaction with matter means they have a much longer range in all materials than alpha or beta particles.

Inverse square law for gamma rays

Gamma rays interact so weakly with air that if we are reasonably close (e.g. <1 m) to the source, absorption in the air can be completely ignored. This means that all the gamma rays emitted from a particular source will be radiated like light from a point source in space. The intensity of the gamma rays will follow an inverse square law – i.e. if you move twice as far from the source the intensity will drop to 1/4 of its previous value.

If A is the effective strength of the source then the intensity I of gamma radiation at a distance d from the source is:

$$I = \frac{a}{4\pi d^2}$$

where $4\pi d^2$ is the surface area of a sphere of radius d.

This means that by doubling your distance from the γ-ray source you will reduce your radiation dose by at least a factor of 4.

Half-thickness of lead

In solid materials the most useful model for gamma-ray absorption is to assume that the chance that a particular photon is absorbed in any short length δx of its path is a constant that depends only on the length and the material. To understand this, imagine 1000 gamma ray photons entering a sheet of lead, and assume each photon has a 1 in 10 chance of being absorbed in each millimetre of its path through the lead. It is simple to make a numerical model to predict the intensity of gamma rays emerging from any thickness of lead.

$$\text{initial number of photons, } N_0 = 1000$$

$$\text{probability of absorption in the first millimetre, } p_a = 0.1$$

$$\text{number absorbed in first millimetre, } p_a N_0 = 0.1 \times 1000 = 100$$

$$\text{number remaining after 1 mm, } N_1 = N_0 - p_a N_0$$
$$= N_0(1 - p_a)$$
$$= 1000 - 100 = 900$$

This process can be repeated to generate the number passing through 2 mm and so on. The results are tabulated below.

Absorption of gamma-ray photons.

Thickness passed through/mm	Number of gamma-ray photons remaining	Number absorbed in next millimetre
0	$N_0 = 1000$	$0.1 \times 1000 = 100$
1	$N_1 = 1000 - 100 = 900$	$0.1 \times 900 = 90$
2	$N_2 = 900 - 90 = 810$	$0.1 \times 810 = 81$
3	$N_3 = 810 - 81 = 729$	$0.1 \times 729 = 73$
4	$N_4 = 729 - 73 = 656$	$0.1 \times 656 = 66$
5	$N_5 = 656 - 66 = 590$	$0.1 \times 590 = 59$
6	$N_6 = 590 - 59 = 531$	$0.1 \times 531 = 53$
7	$N_7 = 531 - 53 = 478$	$0.1 \times 478 = 48$
8	$N_8 = 478 - 48 = 430$	$0.1 \times 430 = 43$
9	$N_9 = 430 - 43 = 387$	$0.1 \times 387 = 39$

The constant probability of absorption per unit length means that the number actually absorbed per millimetre is proportional to the number that make it that far. This makes the intensity of gamma radiation fall off exponentially with distance into the material. Exponential curves like this have a constant proportion property, and a particularly useful property is the 'half-thickness' of the material for gamma rays.

> **Definition: half-thickness**
>
> The **half-thickness** of a particular absorber is the thickness that absorbs exactly half of the incident gamma radiation, whatever that incident intensity may be.

In this case the half-thickness of lead for gamma rays of this energy is 6–7 mm.

Maths box: linear absorption

The numerical model can be developed into an equation for gamma-ray absorption.

The key idea is that the probability of absorption per unit length is constant.

If N is the incident gamma-ray intensity on a layer of thickness δx and the probability of absorption per unit length of path is μ, then the change in number of gamma ray photons is

$$\delta N = -\mu N x \delta x$$

This can be written as:

$$\frac{dN}{dx} = -\mu N_x$$

This is a first-order differential equation which can be solved by separation of variables to give

$$N = N_0 e^{-\mu x}$$

μ is called the **linear absorption coefficient**.

The half-thickness $x_{\frac{1}{2}}$ can also be derived from this:

$$e^{-\mu x_{\frac{1}{2}}} = \frac{1}{2} \quad \text{or} \quad x_{\frac{1}{2}} = \frac{\ln 2}{\mu}$$

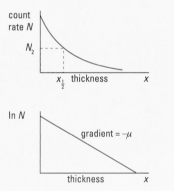

The penetrating power of gamma rays. The half-thickness can be measured (a) directly from a graph of count rate versus absorber thickness or (b) by plotting the logarithm of count rate versus thickness. The absorption coefficient μ is then the negative gradient and the half-thickness is ln 2/μ.

PRACTICE

1 Explain why gamma-ray intensity drops like an inverse-square law in air, whereas alpha intensity does not.

2 Rubidium-90 ($Z = 37$) decays by emitting a beta particle of maximum energy 6.6 MeV to form an isotope of strontium. A short time after the decay (typically about 10^{-8} s) the strontium nucleus emits a gamma-ray of energy 0.83 MeV.

  a Write down an equation for the beta decay and another for the gamma emission.

  b Draw an energy diagram for this decay.

3 What conclusions do you draw from the fact that gamma-ray spectra are discrete?

4 Gamma emitters are often used as radioactive tracers in medical diagnosis. What is the advantage of gamma rays over alpha or beta particles?

5 The half-thickness of lead for 0.5 MeV gamma rays is about 4.2 mm.

  a Why does half-thickness depend on
    i material density and ii particle energy?

  b Predict a value for the half-thickness of aluminium for 0.5 MeV gamma rays. Explain how you obtained your answer.

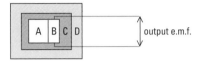

A radioactive battery. A: encapsulated radioactive source. B: thermoelectric transducer. C: thermal insulation. D: biological shield.

Radioactive batteries

The thermal energy generated when ionizing radiation stops in matter can be transferred (in part) to electrical energy using a **thermocouple**. The e.m.f. across the thermocouple can power various applications, such as heart pacemakers and satellite equipment. The advantage of such an energy source is that it is unaffected by external physical conditions and, if the half-life of the source is long, it will last a very long time. However, there are some drawbacks:

- The longer the half-life of the source used the lower the activity from a given mass of source: the shorter the half-life the more rapidly the battery 'runs down'.

- The radioactive battery must be shielded so that it emits negligible amounts of radiation. This is difficult if the source emits gamma rays as well as alpha or beta.

The power available from a particular source will depend on its activity A and on the energy E of the particles emitted. If all of this could be converted to electrical energy the maximum available power would be $P = AE$ watts.

Lord Kelvin. Kelvin matriculated at Glasgow University when he was just 10 years old! He went on to become professor of Natural Philosophy at Glasgow, a chair he held for 53 years. He made first-rank contributions to electromagnetism and thermodynamics and was much concerned about how to account for the apparent age of the Earth using physical principles.

RADIOACTIVITY IN ACTION

The age of the Sun and Earth

In 1654 Bishop Ussher added up the ages of the patriarchs as they are listed in the Bible and dated the creation to 4004 BC. This agreed pretty well with records of human history and was accepted as approximately correct by most people. However, two new ideas developed in the next two centuries both implied that the Earth must be much older than this.

- James Hutton proposed that the landscape had been shaped by gradual geological erosion.

- Charles Darwin proposed the theory of evolution of species by natural selection.

Both processes must have operated over millions of years if they are to explain the present state of the world.

Meanwhile William Thomson (later Lord Kelvin) was working on a related problem in physics – if the Sun generates its energy by chemical combustion he calculated that it could only last for about 10 000 years, nowhere near long enough for erosion or evolution. He tried another approach – if the Earth and planets formed from the same matter as the Sun, they must have been formed in a molten state and have been cooling ever since. He did a few calculations on heat flow and thermal radiation rates and showed that the Earth must have been cooling for between 25 and 400 million years. This seemed long enough to allow evolution and erosion to fashion the world, but how could the Sun shine for so long without running out of fuel?

These problems could not be resolved using the physics which was known at the time. We now know that:

- Kelvin's cooling calculation was reasonable, but missed one essential point. Thermal energy is continuously generated inside the Earth as radioactive minerals undergo radioactive decays. The Earth is much older than indicated by Kelvin's calculations.

- The Sun's energy comes from nuclear energy released when nuclei of hydrogen fuse together to form helium. This releases about a million times more energy per nucleus than chemical combustion, so the fuel lasts much longer.

Rutherford showed that if heating from radioactive decays inside the Earth is included in the calculation, its age might be measured in hundreds of millions of years. He then set about finding a more accurate way to measure the age by using these decays. One modern method is based on the nuclide potassium-40 which sometimes (89%) decays to calcium-40 and sometimes (11%) to argon-40. The latter decay, which is by beta-plus (positron) emission is interesting because argon is a gas. This means it is reasonable to suppose that no argon was present in the rocks when they solidified from a molten form. Any argon found in them now must have come from the decay of potassium-40. By measuring the ratio of argon-40 to potassium-40 in rocks and knowing the half-life for the decay (1.3 billion years), the age of the rock can be calculated.

The same technique has been applied using different nuclides and has been used to date meteorites (which presumably formed with the solar system) and Moon rocks. From these measurements it seems that the solar system formed about 4.6 billion years ago.

What about the Sun? Estimates based on the rate at which it is radiating energy and the rate at which it is consuming its nuclear fuel in fusion reactions show that it could have a lifetime of about 10 billion years. It is about halfway through this.

Tracers

Radioactive nuclides retain their activity whatever compound their atoms are attached to. This is very useful – they can be used to replace stable atoms in particular chemicals and act as 'labels' to follow the path of that compound through some process or system. Radioactive **tracers** added to oil can be used to look for leaks from underground pipelines. A gamma emitter is suitable since gamma rays are sufficiently penetrating to give a measurable count rate at ground level. If there is a leak there will be an unusually high count rate close to the crack.

Traces of radioactive elements or compounds can also be introduced into the body to investigate metabolic pathways or blood flow. One particular example is the use of small quantities of iodine-131, a beta and gamma emitter with a half-life of 8.1 days, to check kidney function. The iodine is injected into the patient's bloodstream and builds up in the kidneys. If the kidneys are functioning properly it should then be passed on to the bladder. The progress of the iodine can be monitored using a Geiger counter outside the body pointed at the kidney. The count rate should rise and then fall. If there is a blockage it will rise and then remain constant. Iodine-131 is also useful for investigating the function of the thyroid gland which is in the neck. This gland controls growth and causes problems if it takes up too much (i.e. is **overactive**) or too little (i.e. is **underactive**) of a person's iodine intake. The rate of uptake can be monitored following an injection of iodine-131.

An alternative approach to the kidney problem described above is to use a gamma camera. Technetium-99 is a gamma emitter with a half-life of 6 hours. The technetium is used to label a compound which is taken up by the tissues to be studied. The total gamma radiation can be monitored to show the amount and rate of uptake of the compound by the tissue, or the pattern of gamma ray emission can be recorded by an array of detectors whose outputs are analysed and combined electronically to produce an image.

Yet another imaging technique is **positron emission tomography** (**PET**). Positron emitters like carbon-11, nitrogen-13, and oxygen-15 are particularly useful because they are isotopes of common elements in the body. PET scans of the brain have played a very important role in diagnosis and research in recent years. Water is labelled with oxygen-15 and is used to monitor blood flow in the brain (blood flow increases in regions which become more active, so brain functions are made visible).

Artificial radioisotopes

Many of the radioactive isotopes used in medical imaging or diagnosis can be made by bombarding stable nuclei with neutrons. This is done by placing samples of stable nuclides inside a nuclear reactor. The intense neutron flux results in some of the stable nuclides absorbing a neutron and becoming unstable. For example, if copper is placed inside a reactor core this reaction sometimes takes place:

$$^{63}_{29}\text{Cu} + ^{1}_{0}\text{n} \rightarrow ^{64}_{29}\text{Cu}$$

This isotope of copper is unstable and decays to zinc:

$$^{64}_{29}\text{Cu} \rightarrow ^{64}_{30}\text{Zn} + ^{0}_{-1}\beta + ^{0}_{0}\bar{\text{v}} \quad \text{(half-life = 12.8 h)}$$

Cobalt-60 can also be created like this:

$$^{59}_{27}\text{Co} + ^{1}_{0}\text{n} \rightarrow ^{60}_{27}\text{Co}$$

This is followed by:

$$^{60}_{27}\text{Co} \rightarrow ^{60}_{28}\text{Ni} + ^{0}_{-1}\beta^{-} + ^{0}_{0}\bar{\text{v}} \quad \text{(half-life = 5.3 y)}$$

Research hospitals sometimes have a small reactor on site to produce radioisotopes. Some also have **cyclotrons** (particle accelerators) which are used to create useful radioisotopes when protons crash into fixed targets. Oxygen-15, a positron emitter used for PET scans, is produced in this way. In fact it has to be produced locally, because its half-life is only 2 minutes.

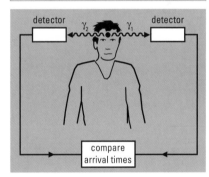

PET (positron emission tomography). Positrons emitted by the nuclide annihilate with atomic electrons very close to their point of creation and emit a pair of gamma rays moving in opposite directions. If the head is surrounded by gamma detectors, the delay between a gamma ray arriving on one side and its partner arriving opposite is used to calculate the position of the decay. In this way a map of decays taking place in the brain can be built up. If the subject is given different stimuli – music or speech for example – the pattern of blood flow in the brain changes.

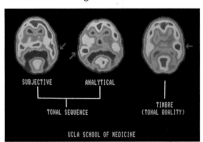

PET scans showing the varying rates of activity in different parts of the brains of three people listening to music. Red indicates the greatest activity. The technique shows how untrained listeners react to music more with the right, intuitive hemisphere of the brain, while trained musicians rely on the left, logical side of the brain.

- ether theory
- the Michelson–Morley experiment
- the principle of relativity

Einstein's special theory of relativity (1905) and his general theory (1915) revolutionized our ideas about space, time, matter, and gravity.

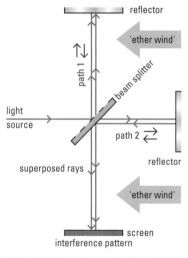

A modern version of the Michelson–Morley experiment. If the Earth moves through the ether it must carry this apparatus with it. This should result in an 'ether wind' through the laboratory that introduces different amounts of delay into the light beams following each of the two paths.

Absolute space?

The Michelson–Morley experiment can be regarded as an attempt to measure the velocity of the Earth relative to Newton's hypothetical 'absolute space'. Michelson and Morley assumed that light would travel at a constant velocity of c relative to this reference frame, so that the Earth's orbital motion through space ought to affect the measured speed of light. The null result can be interpreted as an indication that there is no such thing as absolute space.

SPECIAL RELATIVITY

When Einstein was about 16 he wondered what a light wave would look like if he could catch one up. He knew that light was thought to be an electromagnetic wave, so hitching a ride on one should be a bit like surfing on an ocean wave where the surfboard stays on the same part of the wave and is carried in to the beach. As far as the surfer is concerned he is *at rest* on the wave. The problem was that Einstein knew electromagnetic theory did not allow for that kind of stationary electromagnetic wave. This worried him.

He was not the only person to be worried about light. A lot of physicists had been asking the apparently reasonable question – if light is a wave, what is '*waving*'? Sound waves move the air, water waves move water, seismic waves move the crust of the Earth, but what do light waves move? Another way of asking this question is to ask what *medium* supports electromagnetic waves. This is difficult to answer because light and other electromagnetic waves reach Earth through the vacuum of space. What is in the space to support the electromagnetic vibrations that make light? To answer this question a new substance was invented and named the **ether**. The ether would have to fill all of space, support electromagnetic vibrations and energy, and yet not interfere in any obvious way with the motion of planets and stars through space. This idea led to new problems.

Does the Earth move?

By the early twentieth century few people doubted that the Earth orbits the Sun and is not fixed in any privileged place in space. This meant that it must be *moving through the ether*, so an 'ether wind' must blow past the Earth. This has a serious consequence for the measured speed of light. If light moves in the ether and we move through the ether then the speed of light relative to us will be different in different directions as it moves with or against the ether wind. This is a similar effect to the way an aircraft's ground speed is increased if it flies in the same direction as the wind, and reduced if it flies against the wind, even though its airspeed is the same in both directions.

This apparent change in the velocity of light could be used to measure the velocity of the Earth's motion through the ether and test the ether hypothesis, so a number of experiments were carried out to detect and measure the change. The most famous of these was the Michelson–Morley experiment, described below. But all of them had the same result – there is *no change* in the velocity of light as a result of the Earth's motion. This was a great surprise, and no one was able to give a completely consistent explanation in terms of known physics.

The Michelson–Morley experiment

Michelson and Morley's experiment raced two light beams over paths of equal length which were perpendicular to one another. The two beams were made by splitting a single beam so both started the race in phase. The two beams were expected to be affected differently by the ether wind (if it existed) – one would be delayed more than the other so they would return with a *phase difference*. The two returning beams were re-combined and made to form an interference pattern like that in a Young's slits experiment.

The whole apparatus was floated on a bath of mercury so that once an interference pattern was seen, it could be rotated through 90° and the two paths would exchange roles, the one that was delayed before now getting back first. This would reverse the phase difference between the returning beams and shift the interference pattern they would form when re-combined. The faster the ether wind, the greater the phase difference between the rays would be, and the larger the expected shift in position of the interference fringes when the apparatus was rotated. The idea was to measure this shift and from it calculate the velocity of the Earth's motion through the ether.

But there was no shift: the two rays took exactly the same time to travel back and forth along either route. It was as if the ether played no role in the experiment.

A clue

If a magnet is moved toward a closed coil of wire, the changing magnetic flux through the coil induces an e.m.f. in the coil and an induced current flows. This is electromagnetic induction, discovered by Faraday and incorporated into Maxwell's equations of electromagnetism. But what happens if the magnet is stationary and the coil moves towards it? In this case a static magnetic field is cut by a moving conductor and electrons in the wires are moved by the **motor effect**, also incorporated into electromagnetism. Einstein thought it was strange that the same effect was explained by two different methods depending on what moved. It seemed to him that the current in the coil resulted from *the same laws of physics* in both cases. All that mattered was the *relative* motion of coil and magnet, not the question of which was actually at rest and which was actually moving.

This started him thinking about light again. The speed of light is a natural consequence of the laws of electromagnetism, so if the laws of electromagnetism are the same for everyone, whether they are 'at rest' or 'moving' (as long as the motion is at constant velocity), then *the speed of light will also be the same for everyone.*

The principle of relativity

The **special theory of relativity** was published in 1905 and can be summarized by two simple statements, the first of which is known as the principle of relativity:

- The laws of physics are the same for everyone who is not accelerating.
- The speed of light is the same for everyone (it is an **invariant**). It does not depend on the speed of the source or the speed of the observer.

These do not sound particularly radical ideas, but they sparked a revolution in physics that changed everyone's ideas about the nature of space and time. They also solved all the problems with light and the ether, as demonstrated below.

Consequences of the special theory

According to Einstein, the laws of physics are the same in all **inertial reference frames**. This means *any experiment*, including an experiment to measure the speed of light, will give exactly the same results in all inertial reference frames. There is no experiment you could carry out to tell you whether you are moving or not. This is why the Michelson–Morley experiment gave a null result – any other result would distinguish rest from motion and identify the ether. If the laws of physics are the same whether the Earth moves through the ether or not, then the result of the experiment must be the same in either case, i.e. no shift in fringe positions.

Another way of looking at this is to say that the speed of light relative to the Earth is the same in all directions, so no delay occurs. This also challenges the idea of an ether – if there is no way to detect its presence what purpose does it serve? So the special theory of relativity effectively 'killed off' the ether theory.

Relativity can also explain why you can't catch a light beam. However fast you move the speed of light relative to you remains the same, so it is a hopeless task! This means that however much you accelerate something, it can never reach the speed of light. Nothing can travel faster than light and nothing can be accelerated quite to the speed of light.

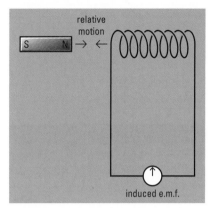

In electromagnetic induction the outcome (induced e.m.f.) depends only on relative motion. Einstein interpreted this symmetry to imply that there is nothing special about being 'at rest' – rest and uniform motion are equivalent physical states and the laws of physics should be the same when viewed from either reference frame.

Speed of light is a constant

Although Einstein treated the invariance of the speed of light as a fundamental principle, you could also regard it as a consequence of the principle of relativity. If the laws of physics are the same for all inertial observers and the speed of light comes from the laws of electromagnetism then it must be constant for all inertial observers.

Reference frames

If you are in a school laboratory you are likely to measure the positions and motions of things relative to the laboratory. The laboratory is your reference frame. If you are in a train or aircraft you refer all your measurements and observations to a reference frame that is moving with the aircraft. Reference frames that move at constant velocity with no acceleration are called inertial reference frames (since any free object moves like this because of its own inertia).

These players are happy to assume their playing area is at rest. However it is moving at about 300 m s⁻¹ relative to the centre of the Earth and at about 30 000 m s⁻¹ relative to the Sun! The Sun also moves relative to the distant stars.

The velocity of light is the same relative to all uniformly moving (inertial) observers, regardless of their relative velocities to one another or to the source of the light.

Maths box: deriving the time dilation formula

Assume the separation of mirrors in the light clocks is l and the speed of the rocket is v. Now calculate the time t on *your* clock for the light in your friend's clock to complete the extended journey and compare it with the time t_0 for your own clock to tick once.

For your own clock the path length is $2l$ so the time between ticks is:

$$t_0 = \frac{2l}{c} \qquad (1)$$

Using Pythagoras's theorem and distance $= vt$ we get

$$\frac{c^2 t^2}{4} = l^2 + \frac{v^2 t^2}{4}$$

Substituting for l from (1) gives

$$\frac{c^2 t^2}{4} = \frac{c^2 t_0^2}{4} + \frac{v^2 t^2}{4} \qquad (2)$$

Equation (2) can now be used to find how t is related to t_0:

$$t = \gamma t_0$$

where γ is a very useful shorthand for the time dilation factor:

$$\gamma = \frac{1}{\sqrt{1 - \dfrac{v^2}{c^2}}}$$

light clock moving past laboratory at velocity v

B — B' — B''

l

A — A' — A''

vt

light clock at rest in laboratory

l

Imagine a light beam passing you at $3.0 \times 10^8 \, \text{m s}^{-1}$ as a friend flies past in the same direction in a rocket moving at $2.0 \times 10^8 \, \text{m s}^{-1}$. How fast does the light gain on your friend? In each second of *your* time the light moves $3.0 \times 10^8 \, \text{m}$ and the rocket moves $2.0 \times 10^8 \, \text{m}$, so *by your reckoning* it gains at $1.0 \times 10^8 \, \text{m s}^{-1}$. But the principle of relativity says that if your friend measures the light ray's speed relative to *her* rocket she will find that it is *equal* to the velocity it has relative to you – the invariant speed of light $3.0 \times 10^8 \, \text{m s}^{-1}$. This is very hard to understand at first.

- *Question:* How can the same beam of light gain $1.0 \times 10^8 \, \text{m}$ *in each second as judged by you* and $3.0 \times 10^8 \, \text{m}$ *in each second relative to your friend?*

- *Answer:* It can if your friend's seconds last longer than yours!

If special relativity is correct then moving clocks must run slow (more of your time passes between two 'ticks' of your friend's clock than between two 'ticks' of your own clock)!

Time dilation

A simple way to understand time dilation is to imagine a 'light clock' made of two parallel mirrors with a beam of light bouncing back and forth marking time. A light clock is used because we know that all observers must agree on the speed of the light inside it. Now compare your own light clock with one carried by your moving friend (with the mirrors parallel to her direction of travel). The paths of light in the moving clock (relative to you) are longer than those on your clock (see diagram below left). But light in the moving clock travels at the same velocity relative to you as light in your own clock (principle of relativity), so the time between ticks on the moving clock is longer than on your own. *Moving clocks run slow*.

However, as far as your friend is concerned, she would be equally justified in thinking herself at rest and you in motion. She would see light rays in *your* clock stretched out and, by the same argument as you used, think that your clock was running slow. Time dilation is a reciprocal effect.

It is not just clocks that are affected. If you could tell that your own clock was running slow (say by comparing it with your pulse rate) there would be a real difference between motion and rest. The principle of relativity implies that there is no difference as far as physics is concerned. If this is correct (and all the experimental evidence suggests that it is) then your light clock must seem normal *to you*, and your friend's light clock must seem normal *to her*. The inescapable conclusion is that it is not just light clocks that are affected like this, it is the rate at which *time* passes in each reference frame. As your friend passes, you see her age more slowly than yourself. She in turn will see you age more slowly.

Testing time dilation

The slowing of time is well tested. Many subatomic particles have very short lifetimes so they cannot travel far from the point where they are created before they decay. **Muons** are a good example. They are created about 60 km above sea level when cosmic rays collide with molecules in the Earth's atmosphere, and then shower down on us at speeds close to the speed of light. However, their half-life is about $2 \, \mu\text{s}$, and it takes about $200 \, \mu\text{s}$ to travel 60 km, so we would expect only about 1 in 2^{100} of those created at the top of the atmosphere to survive at sea level. In fact about 1 in 8 do! This is possible because the muons are travelling at such high speed. Time dilation slows their time in our (the Earth's) reference frame, so that only three half-lives pass for them even though 100 times the muon half-life passes for us.

Length contraction

Imagine you were riding on one of the muons created in the upper atmosphere. You would pass through the entire atmosphere in a time (on your clock) of about 6 μs at a speed close to the speed of light. By your reckoning the distance travelled, which is equal to the depth of the atmosphere, is about $6 \times 10^{-6}\,\text{s} \times 3 \times 10^{8}\,\text{m s}^{-1} = 1800\,\text{m}$. If time dilation occurs so does length contraction. For us, at rest in the atmosphere, the depth of air is about 60 km. For the muon, the atmosphere rushes past at close to the speed of light and contracts to a little under 2 km! *Moving objects contract along their direction of motion.*

The contraction factor is equal to the time dilation factor γ, in this case about 33. The general formula for length contraction is

$$l = l_0 \left(1 - \frac{v^2}{c^2}\right)^{\frac{1}{2}} \quad \text{or} \quad l = \frac{l_0}{\gamma}$$

where l_0 is the length of an object measured with a ruler at rest beside it and l is the length of the object as it rushes past the measuring instruments at velocity v. The length l_0 is called the **proper length** of the object.

The twin paradox

A lot of physicists objected to the principle of relativity because it clashed so violently with common sense views of space and time – the views Newton had adopted in mechanics. To try to show how crazy the new ideas were, a 'thought experiment' was invented – **the twin paradox**. It goes like this.

Twins celebrate their 21st birthday together on Earth and then one of the two (the boy) leaps into a high-speed spacecraft for a round trip to the stars. The trip lasts 20 Earth years and the speed is such that γ has an average value of 10. According to the twin who stays on Earth her brother has aged at one tenth of the rate she has aged during their separation, so she is not at all surprised when they re-unite, to find that her brother is just two years older (a youthful 23-year-old) whilst she is now a 41-year-old.

No one has ever tried this with people, but it has been tried with atomic clocks, and, sure enough, less time passes according to the travelling clock than the one that stays behind.

However, the apparent paradox is this – the travelling twin argues that he is equally justified in viewing everything from his reference frame (the rocket) and so should expect that the stay-at-home twin ages less than he does. This is the opposite of the prediction above (and the opposite of what actually happens). Why?

In fact the travelling twin's argument is incorrect. He cannot claim his reference frame is equivalent to the stay-at-home twin's frame *for the whole journey*, because the traveller undergoes accelerations and decelerations. These periods of non-inertial motion change his frame of reference and it is during these changes that the real difference between reference frames comes about.

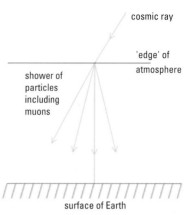

The high rate of arrival of muons at the Earth's surface is strong evidence to support the idea of time dilation.

Proper time

The measured time interval between events at one place will depend on the motion of the clock(s) used to calculate it. The proper time is the interval measured by a clock at rest with respect to these two events. The proper time for an observer is the time measured by a clock moving with the observer (i.e. at rest in the observer's reference frame).

Interstellar spaceflight

It might seem that distant stars are inaccessible simply because it would take someone more than one human lifetime to get to them. Length contraction changes this. Imagine setting out on a journey of 1000 light years. What is the minimum time it would take to complete the journey? Relative to Earth the minimum time is 1000 years if you travelled at just under the speed of light. Relative to you the journey time can be as short as you like. The faster you travel the more the length of the journey contracts, and the less of your time passes as you complete it. There is, however, a drawback. On your return to Earth 2000 years will have passed. Your friends and family will have perished. You have taken a one-way journey into the future.

PRACTICE

1 The significance of relativistic effects can be judged by the size of γ.

  a Copy and complete the following table.

Velocity v/c	0.01	0.05	0.1	0.3	0.7
Gamma γ					

  b Now sketch a graph of gamma (γ) versus velocity (v).

  c What happens to gamma as v approaches the speed of light, c?

  d What would happen to time in a moving rocket viewed from Earth if the rocket's speed relative to Earth was close to the speed of light?

2 The bright star Canopus is about 230 light years away from the Earth.

  a How long does it take light to reach us from Canopus?

  b Explain how it is possible to travel to Canopus in less time than it takes light to reach us from Canopus *without exceeding the speed of light*.

  c At what speed would you have to travel to reach Canopus in 23 of your years?

  d If you travelled to Canopus at the speed you calculated in **c** how far, by your reckoning, would you have travelled?

- variation of mass with velocity
- rest mass
- rest energy
- mass–energy relation

In high-energy collisions kinetic energy is transferred to the rest energy of new particles, and these may have rest energies much greater than those of the colliding particles.

Photons and mass

If we try to accelerate an electron its mass increases so much that we would need an infinite amount of energy to make it reach the speed of light. Photons transfer energy and mass so how can they travel at the speed of light? The difference is that electrons have a distinct rest mass whereas photons have **zero rest mass**. This means that the mass increase with velocity equation does not 'blow up' to infinity for a photon, so it can travel at the speed of light. In fact photons *must* travel at the speed of light, because if you slow down or stop a photon it ceases to exist.

Conservation laws

Some people think that Einstein's discovery that energy has mass means that we have to give up on separate laws of conservation of mass and energy. This is not true. The equation simply says that when 1 kg of mass is transferred from A to B then 9×10^{16} J of energy goes with it, so if energy is conserved, mass must be conserved. Both laws are valid.

MASS AND ENERGY

Imagine applying the same constant force to something as it goes faster and faster. As its velocity gets closer and closer to the speed of light it gets harder and harder to accelerate. Its inertia (reluctance to change its state of motion) increases. Inertia is a property associated with mass, so mass must increase with velocity. The equation that shows how its total mass m increases is:

$$m = \frac{m_0}{\sqrt{1 - \dfrac{v^2}{c^2}}} \quad \text{or} \quad m = \gamma m_0$$

m_0 is the mass the object has when it is at rest. This is called its **rest mass**.

Notice that the mass increases without limit as the velocity approaches the speed of light.

The increase in mass can be measured when high-speed charged particles are deflected in a magnetic field. The radius of the path they follow is given by $r = mv/Bq$ where B is the strength of the magnetic field, q is the charge, and r the radius of curvature of the particle path. If m increases, r will be larger than expected (i.e. larger than it would be if the mass remained equal to the particle's rest mass m_0) and the particles will not be deflected so easily. This has been measured many times and the results are all in agreement with Einstein's predictions. In fact, the variation of mass with velocity has to be allowed for when designing large circular particle accelerators (**synchrotrons**).

Worked example 1

What is the fractional increase in mass of a soccer ball of rest mass 0.50 kg when it is kicked at $10 \, \text{m s}^{-1}$?

$$m = = = \frac{m_0}{\sqrt{1 - \dfrac{v^2}{c^2}}} = m_0 \left(1 - \frac{100 \, \text{m}^2 \, \text{s}^{-2}}{9.0 \times 10^{16} \, \text{m}^2 \, \text{s}^{-2}}\right)^{-\frac{1}{2}} = 1.000\,000\,000\,000\,000\,55 \, m_0$$

This is an increase of only about 6 parts in 10^{16}, so it is hardly surprising that this does not cause problems with soccer. (On the other hand, if the speed of light was just $12 \, \text{m s}^{-1}$ this soccer ball would have almost doubled its mass – a very noticeable effect if you try to head it!). In practice relativistic effects like time dilation, length contraction, and mass increase are only really significant for things moving at velocities comparable to the speed of light. Most everyday objects move much less rapidly, so Newtonian mechanics is perfectly adequate to describe them.

Mass and energy

Applying a force to a particle that is already travelling near the speed of light has very little effect on its velocity, but increases its mass. Electrons in the Large Electron–Positron (LEP) collider at the European Laboratory for Particle Physics (CERN – for Conseil Européen pour la Recherche Nucléaire) are travelling so fast that their mass is increased more than 100 000 times. Where does the extra mass come from? All that has been supplied to the particle is energy by the work that has been done accelerating it. Perhaps energy itself has mass? Einstein showed that all forms of energy have a mass given by:

$$E = mc^2$$

This means a particle of rest mass m_0 has a rest energy $E_0 = m_0 c^2$. This may well be the most famous equation in physics, and it applies not just to photons but to all energy and mass transfers. Unfortunately its meaning is often misunderstood. It tells us that energy *has* mass, not that it can be converted to or from mass. *Every* energy transfer involves a mass transfer (this includes striking a match or standing up, not just nuclear explosions!).

Worked example

How much energy is stored in the mass of a glass of water? (Assume there is about 500 ml of water in the glass. This water has a mass of 0.50 kg.)

$$E = mc^2 = 0.50 \times (3.0 \times 10^8 \, \text{m s}^{-1})^2 = 4.5 \times 10^{16} \, \text{J}$$

This is enough to run a large power station for about a year and a half! Unfortunately the only way we know to release this energy is to annihilate matter with anti-matter, and anti-matter is in rather short supply.

In particle accelerators large amounts of kinetic energy are given to the particles. This causes their mass to increase dramatically. When they collide with something and stop moving this kinetic energy is transferred to other forms, much of it going to the rest energy of new particles. The more energetic the collision, the more energy is available to create new particles; this is why particle physicists want to build bigger and more energetic machines.

If $E = mc^2$ is combined with the equation for mass and velocity we get

$$E = \frac{m_0 c^2}{\sqrt{1 - \dfrac{v^2}{c^2}}} = \gamma m_0 c^2 = \gamma E_0$$

This shows how total particle energy depends on velocity.

Rest mass, rest energy, and various units

SI units for mass and energy are kg and J respectively, but these are rarely used for the masses and energies of subatomic particles.

Energies are usually measured in electronvolts (eV), megaelectronvolts (1 MeV = 10^6 eV) or gigaelectronvolts (1 GeV = 10^9 eV). The link between mass and energy leads to a convenient new mass unit, the MeV/c^2. A proton has rest energy $E_0 = 938$ MeV and rest mass $M_0 = 938$ MeV/c^2.

There is yet another commonly used mass unit in atomic and sub-atomic physics, the atomic mass unit, u. This is equal to 1/12 of the mass of a carbon-12 atom.

$$1 \, \text{u} = 1.661 \times 10^{-27} \, \text{kg} = 931.5 \, \text{MeV}/c^2$$

Important terms

Rest mass m_0
the mass of a particle measured in its own rest frame.

Rest energy $E_0 = m_0 c^2$
the energy associated with the rest mass.

Total energy E
the sum of rest energy and KE. $E = mc^2$.

Total mass $m = E/c^2$
the rest mass plus the mass due to the particle's KE.

Kinetic energy
the difference between the particle's total energy and its rest energy.

Units for mass and energy.

Unit/symbol	Used for	SI equivalent
a.m.u. or u	atomic and nuclear masses	1.66×10^{-27} kg
MeV/c^2	atomic, nuclear, and particle masses	1.78×10^{-30} kg
GeV/c^2	atomic, nuclear, and particle masses	1.78×10^{-27} kg
MeV	atomic, nuclear, and particle energies	1.60×10^{-13} J
GeV	atomic, nuclear, and particle energies	1.60×10^{-10} J

PRACTICE

1 a By what factor would the mass of a rocket increase if it were accelerated from rest to half the speed of light?

b Would passengers inside the rocket be able to measure this increase? Explain.

2 The rest energy of an electron is about 0.5 MeV.

a How much kinetic energy will it gain if it is accelerated through a potential difference of 500 kV? (Give your answer in both joules and MeV.)

b How does the kinetic energy of this electron compare with its rest energy?

c By what factor has its total energy increased as a result of acceleration?

d The rest mass of the electron is about 0.5MeV/c^2. What is its total mass after acceleration?

e By what factor has its total mass increased as a result of acceleration?

f Explain how the acceleration of an electron in an accelerator obeys the laws of conservation of mass and energy.

g Convert the rest mass and kinetic energy of the electron into kg and J and use these values in Newton's equation for kinetic energy ($KE = \frac{1}{2}mv^2$) to calculate the velocity of the electron. Do you think this method is valid? Explain.

h Write down a relativistic equation for kinetic energy by subtracting the rest energy of the electron from its total energy. Now use this to calculate the velocity of the electron and compare your answer with the one you got in **g**.

i When do you think the relativistic equation should be used?

3 Copy and complete the following table.

Particle	Rest mass/kg	Rest mass/ MeV/c^2	Rest mass/u	Rest energy/J	Rest energy/MeV
electron					0.511
proton		938.3			
neutron	1.675×10^{-27}				
^{12}C atom			12		
photon					0

- closest approach

- measuring the nucleus

- the strong nuclear force

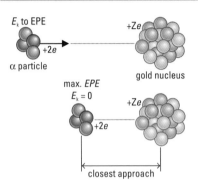

At closest approach in a head-on collision, the electrical potential energy stored in the field is equal to the incident energy of the alpha particle.

The strong nuclear force

Neutrons are unstable and protons repel one another, so the proton–neutron model of the nucleus needs some explaining! Protons and neutrons bind to one another by the strong nuclear force. This is a short-range force which is strong enough, on the nuclear scale, to overcome the mutual electrostatic repulsion of protons and inhibit the decay of neutrons. We now know that both protons and neutrons are made of quarks, and the strong force is related to the force that binds quarks themselves together. The variation of strong force with distance is shown in the diagram. The strong force only acts on particles that are made of quarks; these are called hadrons. It has no effect on electrons.

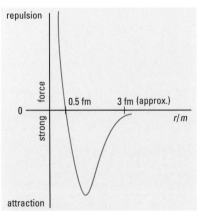

The strong nuclear force:
- *short-range attraction – range about 3 fm;*
- *very short-range repulsion – range below about about 0.5 fm;*
- *acts on nucleons (protons and neutrons).*

THE NUCLEUS

Rutherford was able to estimate the radius of a nucleus by working out how close an alpha particle gets to one when it hits it head on. At closest approach the alpha particle stops moving, so all its original kinetic energy is now stored in the field as electrical potential energy. It is as if an invisible spring is being compressed between the particles as they come closer and closer and eventually all the initial kinetic energy of the alpha particle (E_k) is stored in this 'spring'. The equation for electrical potential energy depends on the separation d of the two charges:

$$EPE = \frac{q_1 q_2}{4\pi\varepsilon_0 d}$$

At closest approach

$$\frac{q_1 q_2}{4\pi\varepsilon_0 d} = E_k$$

If E_k and the two charges are known, d can be calculated:

$$d = \frac{q_1 q_2}{4\pi\varepsilon_0 E_k} = \frac{2e \times Ze}{4\pi\varepsilon_0 E_k}$$

where Z is the atomic number of the target nucleus.

If the target nucleus is gold ($Z = 79$) and the incident alpha particles have a kinetic energy of about 4 MeV (which is typical for naturally produced alpha particles) the closest approach is

$$d = \frac{2 \times 79 \times (1.6 \times 10^{-19}\,\text{C})^2}{4\pi \times 8.9 \times 10^{-12}\,\text{F m}^{-1} \times 4.0 \times 10^6 \times 1.6 \times 10^{-19}\,\text{J}} \approx 10^{-13}\,\text{m}$$

This is the separation of the centres of the alpha particle and gold nucleus at closest approach, so it gives an upper limit for the sum of their radii. From this it seems reasonable to assume that the radius of a typical nucleus is of the order of 10^{-14} m. More accurate measurements (e.g. by scattering high-energy electrons) confirm this – the gold nucleus has a radius of about 6.9×10^{-15} m.

Rutherford's experiment was repeated some years later using accelerated alpha particles of considerably higher energy than those emitted from radioactive sources. At these energies the Rutherford scattering formula began to break down, that is, the distribution of scattered alpha particles no longer fit the prediction. This meant that one of the assumptions used to derive the equation was valid at low energy and not at high energy. At high energies the alpha particles get much closer to the target nucleus, so perhaps a new short-range force operates over these short distances and changes the interaction. This was the first evidence for the **strong nuclear force**, a short-range interaction that binds neutrons and protons together in the nucleus but has no long-range (i.e. greater than about 10^{-15} m) effects, and does not act on electrons.

Using electrons to measure the diameter of the nucleus

High-energy electrons have very short de Broglie wavelengths, so they form useful diffraction patterns from very tiny structures. If the energy is high enough, when they pass nuclei they form diffraction patterns that are similar to those formed by light passing through a small circular hole or around a small circular object of diameter D. The diffraction pattern formed by such an obstacle has its first minimum at an angle θ, where $\sin\theta \approx \lambda/D$. When electrons diffract from nuclei the first minimum of the diffraction pattern is located, and D can then be calculated from the formula if the de Broglie wavelength of the electrons is known. Electrons have one major advantage over alpha particles for these measurements; they are not affected by the strong force, so they respond only to the charge distribution on the nucleus.

At the high energies involved the electrons are moving relativistically, so their momentum and wavelength must be calculated from relativistic equations:

$$\lambda = \frac{hc}{E}$$

In practice, electron energies of about 420 MeV are used. This is much greater than the rest energy of the electrons (about 0.5 MeV), so it can be taken as their total energy E and used in the equation above:

$$\lambda = \frac{hc}{E} = \frac{6.6 \times 10^{-34}\,\text{J s} \times 3.0 \times 10^8\,\text{m s}^{-1}}{420 \times 10^6 \times 1.6 \times 10^{-19}\,\text{J}} = 2.9 \times 10^{-15}\,\text{m}$$

Nuclear radii and nuclear densities

Talking about the radius of the nucleus implies it is some kind of ball-like object in the centre of the atom. Once again this is a crude model, and the radius measured depends on the probes used to measure it – values obtained by scattering alpha particles are different from those obtained using electrons, and all the results are open to different interpretations. The table below right gives a set of values calculated from electron scattering experiments assuming a 'uniform charge density model' – this is a crude picture in which the nucleus is thought of as a sphere of positive charge with constant charge density dropping to zero at its surface.

The gradual increase in nuclear radius shows that nucleons take up space – the more there are the larger the nucleus. But there is a more interesting link here. The radii are approximately linked to the mass number (A) by the equation:

$$r = r_0 A^{1/3}$$

where r_0 is a constant (equal to about 1 fm). This means the volume of the nucleus (which depends on r^3) is proportional to A and so the number of nucleons per unit volume is roughly the same for all nuclei:

$$V = \frac{4}{3}\pi r^3 = \frac{4}{3}\pi r_0^3 A$$

So the number of nucleons per unit volume is

$$\frac{A}{V} = \frac{3}{4\pi r_0^3}$$

which does not depend on A. The nuclear density is

$$\rho = \frac{Am}{V} = \frac{3m}{4\pi r_0^3}$$

where m is the mass of a nucleon. Taking $m \approx 1.7 \times 10^{-27}$ kg and $r_0 = 10^{-15}$ m, the density of nuclear matter is about

$$\rho = \frac{3m}{4\pi r_0^3} = \frac{3 \times 1.7 \times 10^{-27}\,\text{kg}}{4 \times \pi \times (1.5 \times 10^{-15}\,\text{m})^3} = 1.2 \times 10^{17}\,\text{kg m}^{-3}$$

This is enormous. A teaspoonful of this material would have a mass of about 600 billion tonnes!

Calculating nuclear diameters

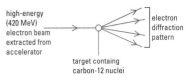

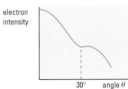

Elastic scattering of high-energy electrons results in a diffraction pattern from which nuclear diameters can be calculated. The graph shows the result of such an experiment using carbon-12 nuclei as targets and 420 MeV electrons as probes. There is a distinct first minimum (although it is not zero) at around 30°. For the first minimum,

$$\sin\theta \approx \frac{\lambda}{D}$$

so

$$D \approx \frac{\lambda}{\sin\theta} = \frac{2.9 \times 10^{-15}\,\text{m}}{\sin 30°}$$

$$= 6 \times 10^{-15}\,\text{m}$$

The radius of the carbon nucleus is therefore about 3×10^{-15} m.

Nuclear radii.

Nucleus	Mass number A	Radius/fm (1 fm = 10^{-15} m)
^{1}H	1	1.00
^{4}He	4	2.08
^{12}C	12	3.04
^{16}O	16	3.41
^{28}Si	28	3.92
^{32}S	32	4.12
^{40}Ca	40	4.54
^{51}V	51	4.63
^{59}Co	59	4.94
^{88}Sr	88	5.34
^{115}In	115	5.80
^{122}Sb	122	5.97
^{197}Au	197	6.87
^{209}Bi	209	7.13

PRACTICE

1 a What is the closest approach of a 5 MeV alpha particle to an aluminium nucleus ($Z = 13$)?

b Use your value from **a** to put an upper limit on the radius of the aluminium nucleus.

c To calculate closest approach for gold, it was assumed the gold nucleus was fixed in place. In fact it must recoil. Why? How will this affect the value for closest approach?

d Would the assumption that the target nucleus is at rest be equally good for aluminium?

2 a The strong nuclear force cannot fall off as an inverse-square law. Why not?

b Give a reason for thinking that the strong nuclear force must become a repulsion at very short range.

3 a Use the data above to show that nuclear radii are roughly proportional to $A^{1/3}$.

b Use the same data to determine a value for r_0.

4 Why are high-energy electrons particularly useful for investigating the size of nuclei?

5 a What is the de Broglie wavelength of a 350 MeV electron?

b Explain why electrons of energy 50 keV are not much use for investigating the nucleus.

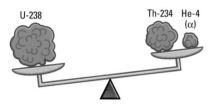

The rest mass of the U-238 nucleus is greater than the sum of rest masses of the Th-234 and He-4 nuclei into which it fissions. The mass defect is equal to the mass of the kinetic energy shared by the products of this fission reaction.

| The energy equivalent of 1 u |

$1\,u = \dfrac{1}{12}$ mass of a carbon-12 atom

$\qquad = 1.661 \times 10^{-27}\,\text{kg}$

$E = mc^2$

$\qquad = 1.661 \times 10^{-27}\,\text{kg} \times (2.998 \times 10^8\,\text{m\,s}^{-1})^2$

$\qquad = 1.493 \times 10^{-10}\,\text{J}$

$1\,\text{eV} = 1.602 \times 10^{-19}\,\text{J}$

$E = 1.493 \times 10^{-10}\,\text{J} \div 1.602 \times 10^{-19}\,\text{J/eV}$

$\qquad = 931\,\text{MeV (3 sig. fig.)}$

The value is 931.5 MeV to 4 sig. figs.

| Conserving mass and energy |

Of course, if the energy released in the alpha decay of uranium-238 goes simply to kinetic energy of the two particles created in the reaction their *total* masses (including the mass of their kinetic energy) will still equal the rest mass of the original uranium nucleus. What has actually happened is that there has been a reduction of *total rest mass* of the system, so there is a sense in which there has been a shift from rest mass to energy.

NUCLEAR ENERGY

People often think that Einstein's mass–energy equation only applies to exotic events like nuclear fusion or radioactive decay, but this is not true. It really shows that all mass has energy and all energy has mass, so every transfer of energy involves a transfer of mass. Chemical reactions are an example of this.

- Think of forming water by combining hydrogen and oxygen. This releases energy (liquid hydrogen and oxygen are combined explosively to propel some rockets), and energy has mass.

- This means *the water molecule must be slightly less massive than the three atoms from which it is formed*, even though it contains exactly the same number of protons, neutrons, and electrons as there are in the three separate atoms!

In practice the mass change associated with chemical reactions is so small that it is extremely difficult to measure, and chemists never need to include it in their calculations.

Nuclear reactions involve energy changes about a million times bigger per atom than in chemical reactions. This means that the associated mass changes are much more significant and nuclear physicists must take them into account in their calculations. This is done by comparing the atomic masses of the reactants and products of nuclear reactions and calculating the **mass defect** for the reaction.

- mass defect = (sum of rest masses of reactants) – (sum of rest masses of products)

Worked example 1

Calculate the energy released during the alpha decay of uranium-238.

$$^{238}_{92}\text{U} \rightarrow {}^{234}_{90}\text{Th} + {}^{4}_{2}\text{He}$$

From tables the atomic masses are

$\qquad ^{238}_{92}\text{U}\quad 238.050\,82\,u$

$\qquad ^{234}_{90}\text{Th}\ 234.043\,64\,u$

$\qquad ^{4}_{2}\text{He}\qquad 4.002\,60\,u$

Alpha decay is a nuclear reaction, so really we should use nuclear masses. These can be calculated from atomic masses by subtracting Z times the mass of an electron from each atomic mass. In practice this is unnecessary, since the numbers of electrons on either side of this reaction would balance.

$$\text{mass defect } \Delta m = (238.050\,82)\,u - (234.043\,64 + 4.002\,60)\,u$$

$$\Delta m = 0.004\,58\,u$$

The energy equivalent of 1 u is 931.5 MeV, so the energy equivalent of the mass defect is

$$E = 0.004\,58 \times 931.5\,\text{MeV} = 4.3\,\text{MeV}$$

This is the energy released in the decay, and would be shared between the alpha particle and the thorium nucleus in such a way that linear momentum is conserved. This results in the lighter particle (the alpha particle) getting almost all (234/238) of the available energy.

- The energy released in a nuclear reaction is the energy equivalent of the mass defect for the reaction.

Limits of the possible

Chemical reactions will only occur spontaneously if (i) there is a mechanism for them to occur and (ii) they are **exothermic**, that is, they result in energy release. The same is true for nuclear reactions. The best way to see if they result in a net energy output is to calculate the mass defect for the proposed reaction and look at its sign. If the defect is positive (products have lower total rest mass than the reactants) then the reaction can happen if there is some way to bring it about. If the mass defect is negative the reaction cannot occur spontaneously, but it may be able to occur if there is an input of energy (e.g. in a collision or by absorbing a photon).

- A nuclear reaction can only happen spontaneously if the products of the reaction have a smaller total rest mass than the reactants.

Worked example 2

Can oxygen-16 undergo spontaneous alpha decay to form carbon?

$$^{16}_{8}O \rightarrow {}^{12}_{6}C + {}^{4}_{2}He?$$

To find out, work out the mass defect for the reaction:

$^{16}_{8}O$	15.994 92 u
$^{12}_{6}C$	12.000 00 u (by definition)
$^{4}_{2}He$	4.002 60 u
Δm	= 15.994 92 u − (12.000 00 + 4.002 60) u = −0.007 68 u

In this case the mass defect is negative – the products of the reaction have greater rest mass than the reactants. This means that the reaction can only proceed if mass (and energy) are supplied. It cannot happen spontaneously, so oxygen-16 is not an alpha emitter.

Geothermal energy is becoming an important energy resource. The high temperatures inside the Earth are maintained by the decay of radioactive isotopes contained in the rocks. The thermal energy we convert to electricity derives from the mass defect of decaying nuclei.

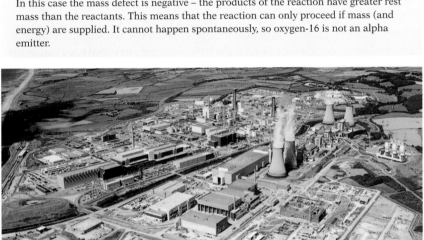

Sellafield, Cumbria, UK, is the largest such site in the country. The main activity here is the reprocessing of used fuel from nuclear power stations worldwide. Also on the site is Calder Hall, the world's first industrial-scale nuclear power station, which opened in 1956 and stopped producing electricity in 2003.

PRACTICE

1 A kettle containing 2 litres of water (2 kg mass of water) at 20 °C is switched on for a short time and then switched off. The final temperature of the water is 70 °C. The specific heat capacity of water is 4200 J kg⁻¹ K⁻¹.

   a Calculate the total energy supplied to the water.

   b Ignore evaporation and calculate the change in mass of the water. Comment on the size of your answer.

   c Where did this extra mass come from?

   d Explain how the laws of conservation of mass and energy apply to this example.

   e Do you think it was justifiable to ignore evaporation?

2 Is the following statement true or false? The mass of an ice cube is exactly equal to the mass of the water from which it formed. Explain your answer carefully.

3 Radium-224 (element 88) decays to radon by emitting an alpha particle.

   a Write down an equation for this decay.

   b Explain where the kinetic energy of the emitted alpha particle comes from.

   c Explain why the radon nucleus recoils.

   d Calculate the energy released in this decay.

   e Could radon-220 undergo an alpha decay to form polonium-216 (element 84)?

   Atomic masses: Ra-224, 224.020 20 u; Rn-220, 220.011 39 u; Po-216, 216.001 92 u; He-4, 4.002 60 u.

4 Can lithium-7 (element 3) undergo beta decay to form beryllium-7 (element 4)?

   Atomic masses: Li-7, 7.016 004 u; Be-7, 7.016 004 u.

5 Explain why a free neutron can spontaneously decay to a proton by emitting an electron, whereas a free proton cannot spontaneously decay to a neutron by emitting a positron.

   Masses: proton, 1.007 276 u; neutron, 1.008 665 u; electron or positron, 0.000 549 u.

8.28

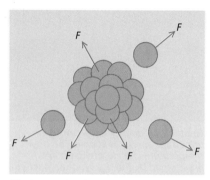

Work must be done to remove nucleons from a nucleus, so free neutrons have more energy and mass than those that are bound to a nucleus.

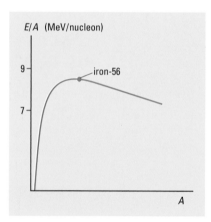

Binding energy per nucleon versus nucleon number.

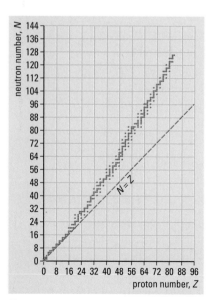

Neutron number versus proton number for stable nuclei.

NUCLEAR BINDING ENERGY AND STABILITY

It takes a lot of energy to send a rocket into space. Work has to be done to lift it against the attractive gravitational force of the Earth, and this work increases its gravitational potential energy (GPE). A similar argument applies to nucleons in the nucleus. They are bound in place by the strong nuclear force and a lot of work must be done to drag them out as free particles. The total work that must be done to free all the nucleons in a particular nucleus is called the **binding energy** of that nucleus. Notice that this is not an energy contained *within* the nucleus, but an amount of energy *that must be supplied* to break the nucleus apart. In this respect the binding energy is a measure of the stability of the nucleus; it is the energy 'price' we must pay to completely disintegrate it.

> **Definition: nuclear binding energy**
>
> **Nuclear binding energy** is the energy that must be supplied to break a nucleus into separate, non-interacting nucleons (i.e. to move them far away from one another).

If work must be done to separate nucleons from the nucleus they must be in higher energy states as free particles. Energy has mass, so nucleons in the nucleus have less mass than free nucleons, and the rest mass of a particular nucleus is less than the sum of the rest masses of all the nucleons it contains.

- The binding energy of a nucleus is the energy equivalent of its mass defect.

> **Worked example**
>
> Calculate the binding energy of a uranium-238 nucleus.
>
> nuclear mass of $^{238}_{92}\text{U}$ = atomic mass of $^{238}_{92}\text{U}$ − 92 × the electron mass
>
> $$= 238.050\,82\,\text{u} - 92 \times 0.000\,55\,\text{u} = 238.000\,22\,\text{u}$$
>
> mass defect = mass of (92 protons + 146 neutrons) − nuclear mass of $^{238}_{92}\text{U}$
>
> $$= 92\,(1.007\,28\,\text{u}) + 146\,(1.008\,67\,\text{u}) - 238.000\,22\,\text{u} = 1.94\,\text{u}$$
>
> The binding energy E is the energy equivalent of this mass defect:
>
> $$E = 1.94\,\text{u} \times 931.5\,\text{MeV} = 1803\,\text{MeV}$$

To compare the stability of different nuclei it is more helpful to use the **binding energy per nucleon** than the total binding energy:

binding energy per nucleon = binding energy ÷ number of nucleons

$$= \frac{E}{A}$$

For uranium-238 this gives

$$\frac{E}{A} = \frac{1803\,\text{MeV}}{238} = 7.6\,\text{MeV nucleon}^{-1}$$

The binding energies of all except the lightest nuclei lie between about 7 and 9 MeV per nucleon. Binding energy is not the only factor that determines nuclear stability, but if other things are equal a larger value of binding energy per nucleon usually implies a more stable nucleus.

If binding energy per nucleon is plotted against nucleon number an interesting pattern emerges.

- Hydrogen-1 has zero binding energy because it has only one nucleon.

- E/A increases rapidly for light nuclei and reaches a peak at iron-56.

- Beyond iron-56 E/A falls gradually as the nuclei get larger.

- There are a series of peaks for light nuclei which are multiples of the helium nucleus, in particular helium-4, carbon-12, and oxygen-16. These are nuclei that are particularly stable compared to their immediate neighbours.

Nucleon potential energies

Another way of thinking about nucleons in the nucleus is to say that they are trapped in a potential well created by the strong nuclear force. This is very like a golf ball stuck in the bottom of a hole; it needs to be given some energy to lift it out of the hole. If the surface of the ground was taken as the zero of GPE then the golf ball has negative GPE when it is in the hole, and the act of lifting it out increases its energy from a negative value up to zero. In the hole the golf ball is in a **bound state** and the binding energy is equal to the minimum work that must be done to lift it out.

The potential energy (PE) of each nucleon in a nucleus is also negative and equal in magnitude to the binding energy per nucleon for that nucleus. If nucleon potential energies are plotted against nucleon number we get a graph which is simply the inverse of the one above. Nucleons in iron-56 have the lowest potential energy of all and so are most difficult to prise out of their nucleus.

This graph helps explain the energy released in two very important nuclear processes.

- **Nuclear fusion** – the joining together of two light nuclei to make a heavier one. This decreases the average nucleon potential energy (and increases the binding energy per nucleon), and the PE lost is emitted in other forms such as gamma-ray photons or kinetic energy of particles.

- **Nuclear fission** – the splitting of a heavy nucleus to form two smaller nuclei. This also decreases nucleon potential energy and releases it in other forms.

Nuclear stability

One of the factors that determines nuclear stability is the ratio of protons to neutrons in the nucleus. If all the nuclides are plotted on a graph of N (neutron number) versus Z (proton number) there is a clear pattern.

- For light nuclei $N = Z$; for heavy nuclei $N > Z$. The extra neutrons in heavy nuclei help overcome the cumulative repulsion of all the positive charges on the protons.

- There is a band of stability that runs up the centre of the graph.

- Nuclei above the stability band have 'too many neutrons'. Heavy (Z greater than about 80) are likely to decay by emitting alpha particles (these reduce N and Z by 2 each) and their products, which also lie above the stability band, are likely to be beta emitters. Beta decay converts a neutron to a proton. This reduces N by 1 and increases Z by 1, moving the nucleus down and to the right at 45°. This brings it closer to stability.

- Below the stability band nuclei are proton-rich and tend to emit positrons (beta-plus decay).

- Some proton-rich nuclei decay by **electron capture**. An inner (K-shell) electron combines with a proton in the nucleus to create a neutron and emit a neutrino:

$$^1_1\text{p} + {}^{\ 0}_{-1}\text{e} \rightarrow {}^1_0\text{n} + {}^0_0\bar{\nu}$$

If we look more closely at the stable nuclei it is clear that there is yet another factor at work – nuclei with even numbers of protons and neutrons seem to be favoured over those with odd numbers (see table). This is strong evidence that nucleons tend to pair up inside the nucleus.

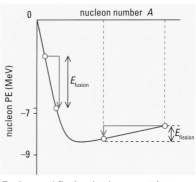

Fusion and fission both convert less stable nuclei to more stable nuclei. Fusion releases more energy per nucleon, but the nuclei involved contain fewer nucleons than those involved in fission. This means fusion releases more energy per kilogram, whereas fission releases more energy per atom. Taking rough values, fusion results in about 3–4 MeV per nucleon, whilst fission releases less than 1 MeV per nucleon.

N, Z, and stable nuclei.

N	Z	Number of stable nuclei
even	even	160
even	odd	56
odd	even	52
odd	odd	4

The shell model

There are also certain numbers of Z or N for which a large number of stable nuclei exist. These are called the **magic numbers**:

2, 8, 20, 28, 50, 82, 126

Something similar to this occurs with electron orbits in atoms. Whenever an electron orbit is filled the atom created is particularly stable – the inert gases, for example, have full outer electron shells. Evidence like this suggests nucleons may also form a shell structure inside the nucleus, with magic numbers corresponding to filled shells. It turns out that this **shell model** of the nucleus is useful for explaining some nuclear properties, but it is not so dominant in nuclear physics as the electron shell model is in atomic physics. There are other models (e.g. the charged liquid drop model) which seem to capture other aspects of the behaviour of nuclei.

PRACTICE

1 a Calculate the binding energy and binding energy per nucleon for each of the following nuclei.

$^{14}_{7}\text{N}$	atomic mass 14.003 07 u
$^{16}_{8}\text{O}$	atomic mass 15.994 92 u
$^{56}_{26}\text{Fe}$	atomic mass 55.934 93 u

$^{206}_{82}\text{Pb}$ atomic mass 205.974 47 u

$^{239}_{94}\text{Pu}$ atomic mass 239.052 18 u

$m_\text{p} = 1.007\,276\,\text{u}$ $m_\text{n} = 1.008\,665\,\text{u}$

b Which of these nuclei is 'magic'?

c Which of these nuclei is most stable? Explain your answer.

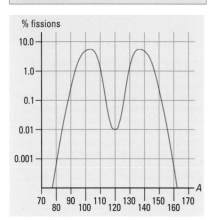

U-235 can split many ways in a fission reaction. This graph shows the percentage yield (on a logarithmic scale) versus nucleon number of daughter nuclei. Why is the graph symmetrical?

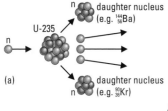

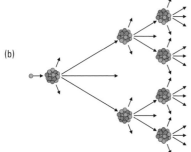

(a) Induced fission. (b) A chain reaction.

The development of nuclear weapons took place during World War 2 at Los Alamos, a top secret research site in the New Mexico desert. There Robert Oppenheimer (pictured here, centre, inspecting the Trinity test site) was in charge of a group of physicists and engineers drawn from the Allied countries and occupied Europe. At the time physicists thought that Germany, which had led the world in fission research before the war, was close to producing its own atomic bomb. After the war it was discovered that this had not been the case.

NUCLEAR FISSION

The first indication that it was possible to split the nucleus came in 1938, just before the outbreak of World War 2. Two German physicists, Otto Hahn and Fritz Strassmann, discovered traces of light elements present in a sample of uranium that had been irradiated by neutrons. This was explained by Otto Frisch and Lise Meitner, who suggested that the uranium nucleus had absorbed a neutron and deformed to such an extent that it had split in two. The lighter elements detected by Hahn and Strassmann were **daughter nuclides** formed by the fission (splitting) of uranium. The following year John Wheeler and Niels Bohr analysed fission using the **charged liquid drop model** (see spread 8.30) and concluded that:

- The isotope uranium-235 (which accounts for just 0.7% of all natural uranium) is readily fissionable, especially with **slow** neutrons (i.e. those with low kinetic energy).

- Uranium-238 (99.3% of natural uranium) is more likely to absorb neutrons than to undergo fission (see below).

- There is no single way that a uranium nucleus splits, so many different pairs of daughter nuclei are possible.

- Daughter nuclei are likely to be beta emitters because they retain the ratio of N to Z from uranium and this is bigger than the stable ratio for lighter nuclides.

- One or more fast neutrons will be emitted along with the daughter nuclei in most fissions.

If fission is brought about by the absorption of a neutron it is called **induced fission**. If a nucleus simply splits in two it is called **spontaneous fission**. Both processes are important in nuclear reactions. The equation below shows one possible mode of induced fission.

$$\,^{1}_{0}\text{n} + \,^{235}_{92}\text{U} \rightarrow \,^{144}_{56}\text{Ba} + \,^{90}_{36}\text{Kr} + 2\,^{1}_{0}\text{n}$$

Atomic masses:

$^{1}_{0}\text{n}$	1.009 u
$^{235}_{92}\text{U}$	235.044 u
$^{144}_{56}\text{Ba}$	143.923 u
$^{90}_{36}\text{Kr}$	89.920 u

Using these values, we find that $\Delta m = 0.192\,\text{u}$. The energy equivalent of this is 179 MeV.

This is an enormous amount of energy. The energy released per atom in a chemical reaction is measured in tens of electronvolts, not millions. Physicists soon realized that nuclear fission could release about a million times more energy per kilogram of fuel (or explosive) than conventional chemical resources. But how could they make sure that the neutrons hit the rare uranium-235 nuclei and did not get swallowed up by uranium-238?

A chain reaction

The key to getting energy from fission is to use the neutrons emitted in fission to induce more fission reactions. As long as more than one neutron per fission goes on to induce another fission, the reaction rate and power generated will rise. This is called a **chain reaction**. There are two major problems, both of which prevent neutrons from inducing fission.

- Uranium-238 absorbs neutrons.

- Neutrons are lost from the surface of the assembly.

On top of this, the neutrons emitted in fission are fast neutrons (high energy), but fission is more likely to occur if the neutrons have low energies. (In a reactor, slow neutrons are called **thermal neutrons** because they have kinetic energies comparable to the thermal kinetic energy of particles in the reactor core, approximately kT (see spread 7.7)).

There are three main ways in which these problems can be overcome:

- Use pure uranium-235, or else highly enriched uranium (that is, natural uranium in which the proportion of uranium-235 has been artificially increased).

- Mix the uranium with a material that does not absorb neutrons, but slows them down – this increases the ratio of neutrons that continue to induce fission compared to those that are absorbed by the uranium-238.

- Control the shape and size of the assembly (reactor core or explosive) to reduce the surface-area-to-volume ratio so that the proportion of neutrons lost from the surface is reduced.

The assembly of fuel must reach a certain **critical size** before a self-sustaining chain reaction is possible. With a subcritical assembly too many neutrons are lost from the surface. As the size of the assembly increases its surface-area-to-volume ratio falls, so the ratio of lost neutrons (which depends on the surface area) to total number of neutrons released (which depends on volume) also falls. When a fission bomb (often misleadingly called an 'atom bomb') is detonated it becomes a **supercritical** assembly – more than one neutron per fission goes on to induce more fissions. The trick with a nuclear reactor is to keep the assembly balanced in a critical state so that, in normal operation, an average of one neutron per fission induces another fission and the reaction rate and power output stay at a steady level.

Plutonium

Plutonium, element 94 in the periodic table, does not occur naturally in any significant quantities. It is created in nuclear reactors when uranium-238 absorbs neutrons:

$$^{238}_{92}U + ^{1}_{0}n \rightarrow ^{239}_{92}U$$

$$^{239}_{92}U \rightarrow ^{239}_{93}Np + ^{0}_{-1}\beta + ^{0}_{0}\overline{v}$$

$$^{239}_{93}Np \rightarrow ^{239}_{94}Pu + ^{0}_{-1}\beta + ^{0}_{0}\overline{v}$$

While the half-lives of U-239 and Np-239 are 23.5 minutes and 2.35 days respectively, the half-life of Pu-239 (which is an alpha emitter) is 24 360 years, so it accumulates in the core of the reactor. Plutonium is also fissionable, and most of the early nuclear reactors were built in order to produce plutonium for weapons rather than electricity for power, although this was a useful by-product. The advantage of plutonium over uranium-235 is that it is a different element so it can be separated chemically, whereas U-235 and U-238 are isotopes of the same element and so are chemically identical. Separation techniques must use processes (like **diffusion** or **centrifuging**) which depend on the difference of only 3 parts in 238 in their atomic masses. Much of the world's nuclear arsenal is based on the physics of plutonium, and the close link between plutonium production and peaceful nuclear power poses many political problems.

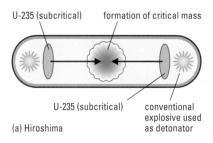

U-235 (subcritical) formation of critical mass

U-235 (subcritical) conventional explosive used as detonator

(a) Hiroshima

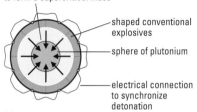

spherical shock waves compress the plutonium to form a supercritical mass

shaped conventional explosives

sphere of plutonium

electrical connection to synchronize detonation

(b) Nagasaki

The Hiroshima and Nagasaki weapons used two different methods to create a supercritical mass of fissile material.

A devastated city – Hiroshima, August 1945. The only nuclear weapons to have been used in warfare were the two fission weapons dropped on the Japanese cities of Hiroshima on 6 August 1945 and Nagasaki three days later, by the Americans at the end of World War 2. About 130 000 people were killed, injured, or missing following the bombing at Hiroshima, and 180 000 were made homeless. The blast also destroyed more than 10 km² or about 60% of the city. About 66 000 people were killed or injured at Nagasaki.

PRACTICE

1 a Explain why it is so difficult to separate the isotopes U-238 and U-235.

  b Why would anyone want to separate them?

  c One separation technique involves diffusion using the gaseous compound, UF_6. How can diffusion help to separate these isotopes?

2 A large enough lump of uranium-235 or plutonium will explode like an 'atom bomb', but natural uranium will not, however big the lump. Explain why this is.

3 Fission of uranium-235 releases about 200 MeV per fission.

a How many atoms are there in 1 kg of U-235?

b Convert 200 MeV to joules.

c How much energy would be released if all the atoms in 1 kg of U-235 underwent fission?

d A nuclear reactor generates 500 MW of electrical power. It is 35% efficient at transferring the energy released in fission to electricity. How much U-235 does it consume in one year?

e If the reactor above uses fuel containing 4% U-235 and 96% U-238 how much uranium (in tonnes) does it consume per year?

The forces that bind water molecules together have a similar mathematical form to those which bind nucleons in a nucleus. In some situations nuclei behave just like charged liquid drops.

Spherical nuclei?

The simple assumption that nuclei are basically spherical is often inappropriate. Recent experiments have shown that nuclei can become severely distorted, and rapid rotation can even make them stretch out like cylindrical rods!

MODELLING NUCLEAR PROCESSES

A water droplet can be any shape, but the most stable shape is the one with minimum energy. For water droplets in free fall this shape is a sphere. A sphere is the shape with the smallest ratio of surface area to volume, so this minimizes the number of surface molecules. The reason this shape also minimizes energy is explained by the way water molecules bind together. Water molecules bind most strongly to their nearest neighbours. Surface molecules have fewer nearest neighbours than molecules in the body of the liquid so they are not bound so tightly to the drop. These surface molecules therefore have higher than average potential energies. The fewer surface molecules there are, the more stable the drop.

Nucleons in nuclei are also bound to one another by short-range forces, so their behaviour is similar in some respects to that of a *charged liquid drop*. Niels Bohr was the first person to use this model to explain induced nuclear fission. If a heavy nucleus is hit by a neutron it absorbs the neutron, but begins to wobble about because of the extra energy. As it vibrates, it distorts so that it is like two smaller drops joined by a small section of nuclear material. Because the two smaller drops are both positively charged they repel one another. If the distortion is great enough they may fly apart and form two new lighter nuclei, and perhaps eject some nucleons as well.

The model can be made quantitative and gives reasonable agreement with experiments for the energy released in some processes. The sequence of diagrams below shows how the charged liquid drop model can be used to explain various nuclear processes.

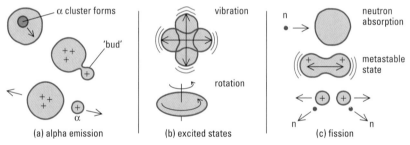

The charged liquid drop model.

Alpha clusters and quantum tunnelling

Think of a heavy nucleus like a 'bag of marbles' being shaken about, each marble representing a nucleon. Every now and again two protons and two neutrons will come together as an **alpha cluster**, or helium nucleus ($^4_2$He). Helium nuclei have a much larger binding energy per nucleon than other nucleon clusters of comparable size, so if they form by chance within a nucleus they will get a big 'kick' from the energy released, about 28 MeV per alpha cluster. However, this boost of kinetic energy is still not enough to throw them out of the nucleus. Something else is happening. To understand it in detail you need to consider how the potential energy of an alpha particle depends on its distance from the centre of the nucleus.

The potential energy of an alpha particle is affected by two interactions: the electrostatic repulsion between its positive charge and the rest of the protons in the nucleus, and the strong nuclear attraction to other nucleons. The combined effect is to trap the alpha cluster in a potential well centred on the nucleus. This is shown in the diagram. In stable nuclei the extra energy gained by an alpha cluster still leaves it in a negative potential energy state, so there is no chance it can escape from the nucleus. However, in some nuclei the 28 MeV 'boost' is enough to lift the alpha cluster to a positive energy state, but not enough to lift it over the potential energy 'hill' at the edge of the nucleus.

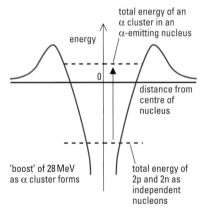

Alpha clusters get a 'boost' of about 28 MeV when they form. This may (in unstable nuclei) lift them to positive total energies.

This is like having a golf ball trapped in the bottom of a hole on top of a hill. The ball has more energy than free golf balls on the fairway below but not enough kinetic energy to climb out of the hole.

According to quantum theory a trapped particle, like the alpha particle in the nucleus, can also be described by a wave function bouncing backwards and forwards between the walls of the potential barrier. The intensity $|\psi|^2$ of the waves at any point is related to the probability of finding the alpha particle there. If the alpha particle has a net positive energy which is less than the 'height' of the barrier, this amplitude decays exponentially into the barrier.

However, the barrier itself is not infinitely wide, so the amplitude is not zero on the far side of it. This means there is a small but non-zero probability that the alpha particle will emerge on the far side of the barrier, even though, in classical mechanics, this takes it through a region inside the barrier where its kinetic energy would be negative!

It is as if the particle does not actually pass over the energy hill but **tunnels** through it. The emitted alpha particle then has an energy equal to that of the alpha cluster inside the nucleus.

This strange idea can be tested. If it is true, alpha particles with a high positive energy would have a greater chance of penetrating the barrier (because their wave amplitude starts higher and has less width of barrier in which to decay). A higher chance of barrier penetration means a larger probability of decay (bigger decay constant) and a shorter half-life. This is borne out by experimental data – high-energy alpha particles are associated with short half-life decays. Browse through a data book to convince yourself.

Nuclear shells

The periodic table shows that atomic structure has a periodicity. For example, the first ionization energies increase with atomic number Z and have peaks for the inert gases. It is now known that this periodicity corresponds to the filling of electron shells and that the peak value corresponds to the energy required to remove an electron from a full shell. Many nuclear properties also show a periodicity peaking at certain numbers of neutrons or protons, so maybe nucleons inside the nucleus have orbits and form shells.

The 'magic numbers' have already been mentioned – nucleon, neutron, or proton numbers that form particularly stable nuclei. These numbers may correspond to closed shells of nucleons. The strong interaction and close proximity of nucleons makes it much more difficult to model the nucleus than the atom, and the interaction of the orbital motions with effects of electron spin make it especially complex. However, the magic numbers *were* eventually derived from a shell model and some excited states of nuclei have been predicted quite accurately.

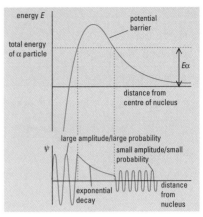

Quantum tunnelling.

Without the quantum tunnel effect this computer would not work.

The tunnel effect

The tunnel effect is very important in:

* semiconductor electronics, e.g. the tunnel diode

* electron microscopy, e.g. the scanning tunnelling electron microscope

* nuclear fusion reactors (two colliding nuclei, which classically should repel one another, tunnel through their coulomb potential barrier and fuse. This reduces the temperature of the plasma necessary to achieve fusion)

* fusion in stars

PRACTICE

1 In the charged liquid drop model, nuclear fission results in an increase of total nuclear surface area.

  a Surely this should prevent fission, since it increases total surface energy? Comment on why fission occurs for some heavy nuclei but does not occur for light nuclei.

  b How could the charged liquid drop model account for gamma-ray emission?

2 **a** Explain how the shell model would account for:

    i gamma-ray emission;

    ii gamma-ray absorption.

  b Explain why nuclear spectra involve gamma rays, whereas atomic spectra are often in the visible part of the spectrum.

3 Use a data book to collect values for the half-lives and alpha particle energies of about 20 nuclides (make sure you get a good range of values). Plot the energy against the half-life and use your results to discuss the theory of alpha tunnelling.

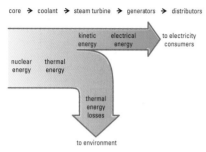

Energy transfers in a nuclear reactor. The only difference between this and a fossil-fuelled reactor is the origin of the thermal energy. The UK generates about one-sixth of its electricity in nuclear reactors.

NUCLEAR REACTORS

Power stations generate electricity in essentially the same way regardless of whether nuclear or chemical fuel is used. Chemical or nuclear potential energy in the fuel is transferred to thermal energy in the **core** of the reactor which is used to turn water to steam. Steam jets then spin turbogenerators which generate electricity by electromagnetic induction.

Commercial nuclear reactors must sustain a critical chain reaction in which an average of one neutron per fission goes on to induce another fission in the next generation of reactions. The fission fragments fly apart with very high kinetic energy and collide with other atoms inside the core, spreading this energy randomly among a very large number of particles. In normal operation the rate at which thermal energy is generated is balanced by its rate of removal by a **coolant** (often water) pumped through the core. This coolant is used to transfer thermal energy to the steam generators. The detailed design of nuclear reactors varies considerably, but the main features are similar.

- **Fuel** The early British Magnox reactors used natural uranium. However, uranium oxide has a higher melting point and this is preferred, because the efficiency of any thermal reactor increases if it operates at a higher temperature. Most modern reactors use fuel whose U-235 content has been enriched from 0.7% to, typically, 2–4%.

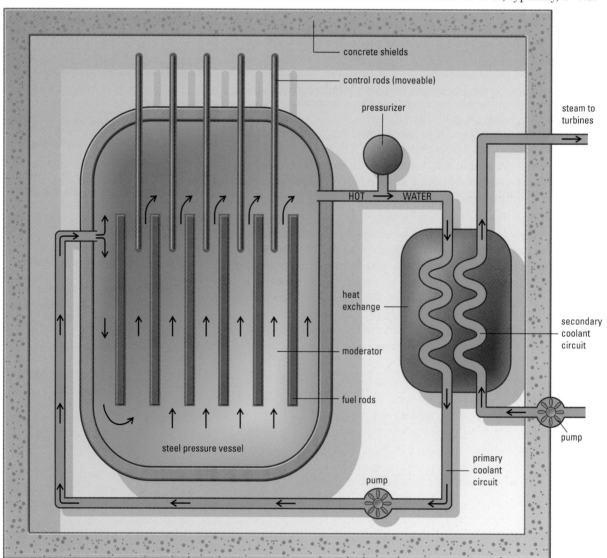

A nuclear reactor. The design illustrated is based on the pressurized-water reactor (PWR), the world's most common type. This type of reactor uses water both as a coolant and a moderator. If the coolant were to fail, the core temperature of a reactor would rise rapidly and might even cause the core to melt down. To avoid this, neutron-absorbing control rods (also shown here) can be inserted into the core to keep it subcritical.

This makes it easier to achieve a critical assembly. The fuel is assembled into pellets and these are stacked inside metal fuel rods which can be removed from the reactor and replaced when they are depleted in U-235.

- **Moderator** The balance between fission in U-235 and absorption by U-238 tips in favour of fission if the neutrons are slowed down. The core of the reactor is therefore filled with a material that does not absorb the neutrons but allows them to bounce off again and again, losing momentum and energy in each collision. Graphite was used in the early Magnox reactors, carbon dioxide in the British advanced gas cooled reactors, and water (doubling as coolant and moderator) in pressurized-water reactors. A low nuclear mass is desirable because this increases the fraction of neutron kinetic energy transferred in each collision.

- **Control rods** These are made of materials with a large probability of absorbing neutrons (they are said to have a large '**cross-section**' for neutron capture). Boron and cadmium are suitable elements.

- **Coolant** There is usually a primary coolant that circulates in the core, and this may be water or carbon dioxide. This transfers thermal energy to water in a heat exchanger; the water in the secondary circuit is used to raise steam for the turbogenerators.

- **Containment and shielding** Whatever goes wrong, a nuclear reactor will not explode like an atom bomb. However, if the core overheats, chemical explosions can occur, and if these breach the reactor containment they will result in massive contamination from the complex cocktail of radioisotopes that accumulate in the core. This is what happened in the Chernobyl accident in 1986. Reactor cores are surrounded by a strong steel pressure vessel and beyond that there is a thick layer of concrete shielding to absorb the neutrons and gamma rays emitted from reactions taking place in the core.

- **Waste** Nuclear reactors generate a very much smaller mass of waste than fossil-fuelled reactors of similar power. However, high-level nuclear waste (such as spent fuel rods and reactor materials) is toxic, highly radioactive, and must be kept isolated from the natural environment for tens of thousands of years. When it is first removed from the reactor (which must be done by remote handling techniques) it is stored under water for several weeks. This allows the more active, shorter half-life isotopes to decay, and the water acts as both a shield and a coolant. So far no permanent disposal method for this type of waste has been agreed, but the most likely process is vitrification – conversion into glassy slabs which can then be buried deep underground in stable geological formations from which they cannot contaminate water supplies. Waste of a lower level, such as the used clothing of workers at nuclear plants, is buried at specially designated sites e.g. Drigg in Cumbria.

- **Reprocessing** Spent fuel can be reprocessed to retrieve plutonium and depleted uranium (i.e. uranium from which most of the U-235 has been removed). Plutonium is used in the production of nuclear weapons, although the demand for and value of plutonium have both dropped dramatically since the end of the Cold War and the breakup of the old Soviet Union. Depleted uranium is also used in fast reactors, and has military uses such as armour-piercing shells and armour plating on military vehicles.

- **Decommissioning** is a costly and lengthy process. Ideally decommissioning should involve the removal of all radioactive parts, the dismantling of the reactor, and site reclamation to leave a **greenfield site** which could then be used for any other purpose.

The accident at Chernobyl

The world's worst nuclear accident occurred at Chernobyl, near Kiev, on 26 April 1986. Soviet engineers were carrying out an experiment to test the effectiveness of backup safety systems. This required some parts of the emergency systems to be disabled. The experiment went horribly wrong. Coolant circulation in the core of the reactor dropped off, the core temperature rose exponentially, and a series of chemical explosions blew the roof off the reactor. About 7 tonnes of radioactive material with a total activity of around 3×10^{18} Bq were ejected into the atmosphere. Fallout from the accident contaminated large areas of western Europe. The area immediately around the reactor was worst affected and over a hundred thousand people were evacuated from their homes.

Storing and handling nuclear fuel

There are three primary safety objectives in fuel fabrication and handling plants.

- Avoid criticality – particularly important when fabricating fuel rods using enriched uranium or when storing spent fuel. Strict controls on the amount of fuel that can be present at any time and the allowed arrangement of fissile materials must be maintained.

- Provide adequate cooling – radioactive decay and fission inside new and used fuel rods generates heat; this must be removed to prevent melting, often by pumping a coolant through the storage facility.

- Provide adequate radiation protection – materials must be handled remotely and isolated from workers and the environment by suitable shielding. Radioactive materials are usually handled in a sealed chamber kept below atmospheric pressure so that radioactive materials will not escape even if the shielding is breached.

Individual radiation doses for workers in nuclear facilities are monitored by using radiation badges so that a cumulative dose can be calculated. This must remain below officially allowed maximum values (20 mSv per year for nuclear industry workers in the UK).

The hydrogen bomb

Having shown the terrible power of fission weapons, some of the Los Alamos team turned their attention to the possibility of building a fusion weapon – the '**Super**'. Some were horrified by what they had created and returned to their university posts or even left physics altogether, but the development of the 'Super' or '**H-bomb**' (so-called because it uses an isotope of hydrogen) seemed inevitable. The Americans tested their first weapon ('Mike') in 1952, and the Soviet Union followed in 1953. The accumulation of nuclear arsenals capable of destroying the world several times over went under the enlightened policy acronym of '**MAD**' – security through the capability of Mutually Assured Destruction. In other words it was assumed that there could be no gain from a first-strike attack because the response would be devastating and both sides would be destroyed. It is not clear to what extent this apparently crazy policy is responsible for the fact that there have been no nuclear conflicts since 1945 and the world (at least at the time of writing) has not yet been destroyed by nuclear weapons.

*JET, the Joint European Torus, a prototype fusion reactor which uses the TOKAMAK design (TOKAMAK is a Russian acronym for **toroidal magnetic chamber**).*

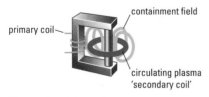

Plasma heating analogue.

NUCLEAR FUSION

The Big Bang seeded the universe with hydrogen and helium. All the other elements were made subsequently by fusing light nuclei to make heavy nuclei inside stars. The sunlight that sustains us on Earth is a result of energy released in fusion reactions in the core of the Sun (these are mainly reactions that convert hydrogen to helium). The elements in our bodies mostly originated in other stars whose remains became the gas and dust cloud from which our solar system formed some 5 billion years ago.

Fusion of light nuclei results in a very large release of energy; typically, several MeV per nucleon are involved in the reaction. This can be seen from the graph of nucleon potential energy (binding energy) versus nucleon number on spread 8.28 – the potential energy per nucleon drops rapidly for the light nuclei. However, although this process releases energy for all nuclei formed up to iron-56, it will not happen spontaneously at low temperatures. This is because all nuclei are positive and repel each other with the long-range electromagnetic force. The strong nuclear force responsible for binding them together is very short-range, and can only overcome electrostatic repulsion if the nuclei get to within a few times 10^{-15} m of each other.

This can only occur if they collide very violently, so there is an **activation energy** for fusion and the reactions will only take place at extremely high temperatures. These conditions arise naturally at the centres of stars and can be generated on Earth by using a fission weapon as a detonator for fusion reactions in a '**hydrogen bomb**'. In the long term it is hoped that fusion reactors will provide a safe, economical, and effectively limitless source of electrical energy.

Fusion reactions for power

Fusion may be the solution to future energy problems. What is required is a reactor in which self-sustaining fusion reactions can take place transferring thermal energy to steam to generate electricity. Several prototype reactors exist, but there are great technical problems involved in achieving and maintaining the extreme conditions in which fusion reactions take place, and it is likely to be several decades before a commercial reactor design exists.

Although there are several possible reactions, the one that is of most interest fuses two isotopes of hydrogen, deuterium and tritium, to form helium:

$$_1^2\text{H} + _1^3\text{H} \rightarrow _2^4\text{He} + _0^1\text{n}$$

This fusion reaction is chosen because it has a high reaction rate and can be achieved at a lower temperature than others. It releases about 17 MeV, most of which goes to the neutron.

Deuterium is plentiful, making up 0.015% of all hydrogen. Tritium is rare but can be created when lithium (another abundant element on Earth) is bombarded with neutrons. The plan is to surround the reactor core with a **lithium blanket** in which the neutrons will be absorbed. This generates tritium in the blanket and also dissipates the neutron energy. If water is passed through the blanket, it will heat up and can be used to raise steam for electrical generators. Lithium breeding reactions include

$$_3^7\text{Li} + _0^1\text{n} \rightarrow _2^4\text{He} + _1^3\text{H} + _0^1\text{n}$$

$$_3^6\text{Li} + _0^1\text{n} \rightarrow _2^4\text{He} + _1^3\text{H}$$

(Lithium-7 and lithium-6 are the two common isotopes of lithium.)

To achieve fusion reactions the deuterium–tritium mixture must be heated to around 100 million K. It is then in a **plasma state** – collisions between atoms are so violent that they break apart and form a gas of nuclei and electrons. The fact that a plasma contains charged rather than neutral particles has two advantages for reactor design.

- The hot plasma can be kept away from the walls of the reactor (where they would lose their energy and vaporize part of the wall) by applying suitable magnetic fields and using the magnetic forces on moving charges to confine the plasma. This is easier said than done, and **plasma instabilities** are one of the major problems in fusion reactor design. The plasma tends to writhe about, and cannot be confined for a long enough time to create an economic and efficient reactor.

- Plasma heating can be achieved by inducing very large electric currents in the conducting plasma. This is done using a changing magnetic field (electromagnetic induction) so that the torus-shaped plasma acts rather like the secondary coil of a very large transformer.

To be of any use for power generation a fusion reactor must generate more power than it consumes. For this to be the case the plasma must be confined for a minimum time T that depends on the particle density n in the plasma. If the density is higher, more reactions per second take place and the break-even time is less. This is summarized in the **Lawson criterion**:

For the deuterium–tritium reaction, $n \times T > 10^{20}\,\text{s m}^{-3}$

temperature, $T > 10^8\,\text{K}$

The advantages of fusion are:

- readily available and virtually inexhaustable supplies of fuels (lithium and deuterium);

- none of the toxic and highly radioactive wastes associated with fission;

- greater energy yield per kilogram of fuel consumed.

Of course, there *are* waste products – in particular irradiated materials that have absorbed neutrons – but these pose nothing like the magnitude of the problem of nuclear waste from fission. As with all kinds of power station there is also a large amount of thermal pollution – waste thermal energy from the reactor that cannot be converted into electricity and so gets dumped in the environment. But even an ideal reactor would generate thermal pollution. It is not possible to convert thermal energy to electricity with 100% efficiency – this is a consequence of the second law of thermodynamics.

Energy released in a fusion reaction

$^2_1\text{H} + ^3_1\text{H} \rightarrow ^4_2\text{He} + ^1_0\text{n}$

Atomic masses:

^1_0n	1.009 u
^2_1H	2.014 102 u
^3_1H	3.016 050 u
^4_2He	4.002 603 u

Using these values we find that $\Delta m = 0.0185$ u.

The energy equivalent of this is 17.3 MeV.

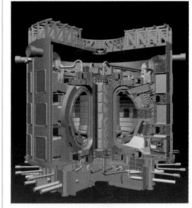

34 nations are collaborating to build ITER, the world's largest prototype nuclear fusion reactor in Cadarache, France. The target is for its first plasma to be created in 2020 and for the reactor to generate 500 MW output from 50 MW input. This would be a major step toward commercial nuclear fusion reactors.

PRACTICE

1 Explain why spontaneous fusion reactions cannot create nuclei heavier than iron-56.

2 The main series of fusion reactions in the Sun is the **proton cycle**. This consists of several fusion reactions and has an overall effect of combining four protons to form one helium-4 nucleus.

$^1_1\text{H} + ^1_1\text{H} \rightarrow ^2_1\text{H} + ^0_1\text{e}^+ + ^0_0\nu$ (twice)   (1)

$^1_1\text{H} + ^2_1\text{H} \rightarrow ^3_2\text{He} + ^0_0\gamma$ (twice)   (2)

$^3_2\text{He} + ^3_2\text{H} \rightarrow ^4_2\text{He} + ^1_1\text{H} + ^1_1\text{H}$ (3)

  a Where do the deuterons (deuterium nuclei) for equation (2) come from?

  b Where do the helium-3 nuclei come from for equation (3)?

  c What happens to the two protons in equation (3)?

  d What happens to the positron and neutrino in equation (1)?

  e What happens to the gamma rays?

  f Write down a single equation for the overall transformation brought about in the proton cycle.

  g Calculate the total energy released in the proton cycle. (Use the data in question 2 and below.)

  Atomic masses: ^1_1H, 1.007 825 u; ^4_2He, 4.002 603 u; ^1_1p, 1.007 276 u; $^0_{-1}\text{e}^-$ and $+^0_{+1}\text{e}^+$, 0.000 549 u.

  Neutrinos and photons have zero rest mass.

Tunnelling nuclei

On first analysis most of the fusion reactions that take place in the Sun should not occur, because the nuclei do not get close enough for the strong force to overcome electrostatic repulsion. However, nuclei must be described by quantum theory and this allows a small probability that they can *tunnel* through the barrier. The same effect makes the temperature that must be reached to fulfil the Lawson criterion a little easier to achieve.

OBJECTIVES

- the discovery of X-rays
- the X-ray tube
- X-ray spectra
- Moseley's law

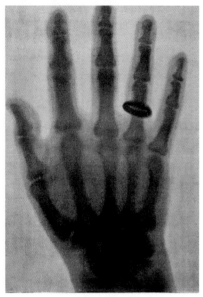

An early X-ray photograph.

The discovery of X-rays

X-rays were discovered by chance. In 1895 Wilhelm Röntgen noticed that some photographic plates, which had been kept in darkness in his laboratory, had become fogged (i.e. exposed) during the course of some experiments with a cathode-ray tube. He concluded that a new penetrating radiation was being emitted from the tube when the electron beam hit it. He went on to investigate the nature of these rays, and their well-publicized ability to form shadow pictures of bones inside the body helped make Röntgen internationally famous.

Keeping cool: rotating-anode X-ray tubes

Less than 1% of the energy in the electron beam is transferred to X-rays; the rest heats the target anode. To prevent it melting:

- It is continually rotated during use.
- A liquid coolant is pumped through it.
- It is made of a metal with a high melting point (e.g. tungsten).

The rotation allows heat generated in one place to conduct away before that part is struck again.

X-RAYS

When an electron beam crashes into something the electron kinetic energy must go somewhere. Most of it is given to a large number of atoms inside the target material as the electrons undergo a complex series of collisions. This eventually becomes thermal energy that raises the temperature of the target. Some of the energy goes to photons of electromagnetic radiation as the electrons decelerate. Very occasionally an electron will give all or most of its initial kinetic energy to a single photon. This results in the emission of a high-energy **X-ray photon** way beyond the high-frequency end of the visible spectrum.

- X-rays are generated by directing a high-energy electron beam at a heavy metal target.

The maximum energy of an emitted photon in such a collision cannot be greater than the energy the electron had just before the collision. This means there is a maximum frequency and minimum wavelength for photons given off in any particular X-ray tube. There will then be a continuous spectrum of X-rays given off with wavelengths having any value from this minimum up. X-rays produced in this way are called **Bremsstrahlung** or **braking radiation**.

Worked example

Calculate the minimum wavelength of X-rays emitted from an X-ray tube in which the accelerating voltage is 80 kV.

$$\text{electron kinetic energy} = eV$$

$$\text{photon energy} = hf = \frac{hc}{\lambda}$$

If all electron kinetic energy is given to a single photon,

$$\frac{hc}{\lambda_{min}} = eV$$

$$\lambda_{min} = \frac{hc}{eV} = \frac{6.6 \times 10^{-34}\,\text{J s} \times 3.0 \times 10^{8}\,\text{m s}^{-1}}{1.6 \times 10^{-19}\,\text{C} \times 80\,000\,\text{V}} = 1.5 \times 10^{-11}\,\text{m}$$

This is a little smaller than an atom and about 30 000 times shorter than the wavelength of visible light.

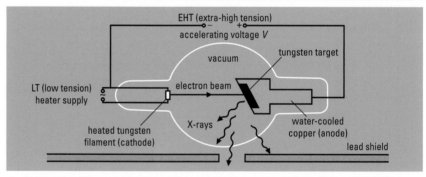

An X-ray tube. The accelerating voltage varies from tens to hundreds of kilovolts. The anode is shaped to make a narrow X-ray beam.

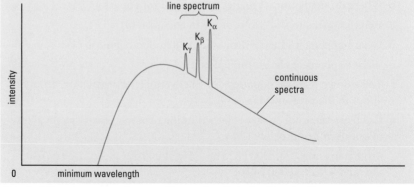

A typical X-ray spectrum from an X-ray tube. Only the K-series line spectrum is shown. L and M series may also appear (at longer wavelengths than the K series).

The line spectrum

In addition to the continuous spectrum described above, certain distinct X-ray frequencies are also emitted. These are characteristic of the metal used as a target. This suggests that they are related to the structure of the target atoms in some way. What happens is that some of the incident electrons penetrate the target atoms and eject electrons from inner electron shells close to the nucleus. X-ray photons are radiated when electrons in outer shells make quantum jumps into these now vacant states. The discrete-line spectrum corresponds to the structure of energy levels available deep within the target atoms.

The inner shells are labelled K, L, M, etc. as you move further from the nucleus, and X-ray lines are labelled according to which of these the electron jumps into. There are, of course, many higher levels from which electrons can jump, so there is a series of lines corresponding to each of the inner shells. The most likely jump is from the level immediately above the vacant level, so this is the strongest line in each case (K_α, L_α, M_α, etc.). This is also the smallest energy jump that can occur into that level, so it has the longest wavelength in that series.

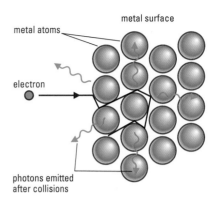

The continuous spectrum results from a complex series of collisions when electrons strike the metal surface. Photons of various energies are emitted when the electrons undergo violent accelerations during the collision process.

Moseley's law

In 1913 Henry Moseley discovered that the frequency of the K_α lines for a number of target elements was related to the atomic number of the target atoms in a simple mathematical way:

$$f = 2.5 \times 10^{15} (Z - 1)^2$$

The pattern was so convincing that he predicted the existence of several new elements from gaps in the graph, and also discovered that cobalt has a lower atomic number than nickel even though its atomic mass is greater. This gave a new and unambiguous way to arrange atoms in the periodic table and confirmed the importance of atomic number in defining the elements. Moseley died two years later in World War 1, shot by a sniper at Gallipoli. He was 27 years old.

Moseley's law can be explained by applying the Bohr theory of the atom to the inner electron shells and calculating the energy of quantum jumps from the L to the K shell. This is similar to the way the Balmer series is explained by electron jumps in the hydrogen atom, but is complicated by the presence of another electron in the inner shell and by all the outer electrons, both of which affect the inner energy levels. However, it is easy to see where the $(Z - 1)$ term comes from. When a L-shell electron drops into the K shell it is 'falling' towards the nucleus (charge $+Ze$) and the lone electron in the K shell (charge $-e$): the effective charge of the combination is $+(Z - 1)e$. It is this value that determines the size of the quantum jump and, here, the photon energy and frequency.

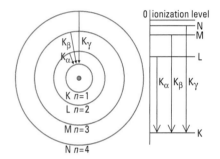

Production of the K-series X-ray line spectrum. The L series is formed in a similar way but by quantum jumps down to the L ($n = 2$) shell. This is like the formation of the Lyman and Balmer series in hydrogen, but at much higher energies and shorter wavelengths.

PRACTICE

1 Calculate the minimum wavelength of X-rays emitted by an X-ray tube operating with an accelerating voltage of 65 kV.

2 What accelerating voltage would be needed to generate an X-ray spectrum with a minimum wavelength 2.5×10^{-10} m (this is roughly the diameter of an atom)?

3 The electrons striking the inside surface of a TV tube have been accelerated through about 20 kV. How are you protected from these X-rays when you watch TV?

4 Explain why electrons striking a metal target emit both a continuous and line spectrum of X-rays.

5 a How can the short wavelength limit of the continuous X-ray spectrum be used to measure the Planck constant?

  b In a particular X-ray tube the accelerating p.d. is 50 kV and the short wavelength cutoff is at 2.50×10^{-11} m. What value does this give for the Planck constant?

Maths box: Bragg's law

d is the separation of atomic layers.

λ is the X-ray wavelength.

θ is the angle between incident (or reflected) X-rays and the atomic planes.

The path difference between rays reflected from the top plane and from the corresponding atom on the plane below is:

path difference = $2d \sin \theta$

(This is distance BC + CD on the diagram.)

This will result in a maximum if it is a whole number of wavelengths.

Maxima occur when:

$$n\lambda = 2d \sin \theta$$

(the **Bragg equation**)

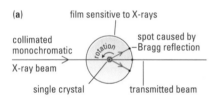

(a)

collimated monochromatic X-ray beam

film sensitive to X-rays

spot caused by Bragg reflection

rotation

single crystal

transmitted beam

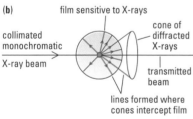

(b)

collimated monochromatic X-ray beam

film sensitive to X-rays

cone of diffracted X-rays

transmitted beam

lines formed where cones intercept film

X-ray cameras for crystallography: (a) using a single crystal and (b) using a powdered sample of crystalline material.

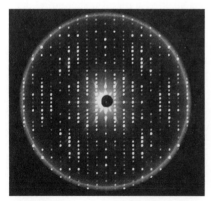

X-ray diffraction images. Above: using a single crystal. Right: using a powdered crystal.

USING X-RAYS

X-ray wavelengths are comparable to the spacing between atoms in matter. This means that the regularly spaced layers of atoms in a crystal lattice can act as a diffraction grating for X-rays and create a diffraction pattern. This idea can be exploited in various ways to investigate the structures of crystalline materials like metals. One of the most useful techniques involves **Bragg reflection**.

When monochromatic X-rays strike a crystal lattice, each atom diffracts the X-rays, and superposition of diffracted waves results in a pattern of maxima and minima from which the crystal structure can be worked out. Maxima are formed in directions in which X-rays scattered from particular layers are in phase with those scattered from the layers beneath. This leads to a condition rather like the law of reflection and gives the process its name.

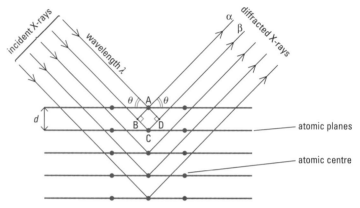

Bragg 'reflection'. Look at rays α and β. β travels an extra path equal to (BC + CD). If this is a whole number of wavelengths α and β will 'reflect' in phase. Rays parallel to these from lower layers will also be in phase, so there is a strong 'reflection' in this direction.

X-ray crystallography

Crystalline materials are analysed using X-ray cameras. Two different types of camera that use Bragg reflection are shown in the diagram on the left. The first directs a beam of monochromatic X-rays at a single crystal, the second directs one at a powdered sample of a crystalline material.

If the single crystal is used it is rotated about a vertical axis aligned with one of its own axes of symmetry. Each time a set of atomic planes moves into a position in which Bragg reflection can occur, a strong beam of X-rays bounces off the crystal and hits the surrounding film causing a spot to appear. The film is then removed from the camera and the positions of spots are measured. The angles at which Bragg reflections occurred can then be calculated and the set of crystal plane separations can be worked out using the Bragg equation. These uniquely identify the symmetry and size of the crystal structure.

If a powder sample is used it will contain atomic planes in *all* orientations, so it does not have to be rotated. Planes for which the Bragg reflection angle is ϕ will create a cone of scattered X-rays around the beam direction. This cone will have a half-angle ϕ and X-rays will hit the film in a slightly curved line. A series of these lines will be formed rather than spots. Once again the lines from a particular crystal are measured and used to calculate its symmetry and size.

X-ray images

Medical and dental X-ray images are usually 'shadow' photographs. They rely on the fact that bone and teeth absorb X-rays more strongly than soft tissues that surround them. To take a medical X-ray, the incident X-rays are passed through a metal filter to remove the low-energy photons which would only be absorbed by the patient, causing heating and possibly tissue damage. The photons with higher energy penetrate the filter and pass through the body. Those that emerge then hit a grid of lead plates designed to absorb scattered X-rays which would otherwise reduce the contrast of the X-ray photograph. The filter allows the rest of the X-rays, i.e. those that have passed through with little or no scattering, to hit the film.

X-rays are also used to look for weapons and bombs in sealed cases at airports, and X-rays are passed through structural materials to test them for cracks. Cracks scatter X-rays so they show up as darker regions on a shadow photograph.

If a moving X-ray image is required the X-rays are allowed to fall onto a fluorescent screen rather than photographic film. X-rays can also be used to investigate soft tissues if an X-ray absorbing material is added – barium sulphate (a **barium meal**) is often used to investigate problems in the intestines. More detailed images can be constructed by recording the *energy* of the X-rays as well as their positions of arrival, and slice-by-slice pictures of the body are made using a process of **X-ray tomography**.

Hard X-rays (with high frequency and large photon energy) are also used to kill cancer cells as an alternative to radium treatment. In both cases, the source is moved around the tumour to reduce the irradiation and damage caused to surrounding tissues.

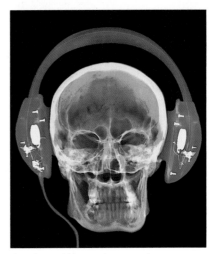

A coloured X-ray photograph.

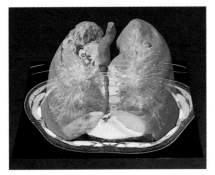

A coloured three-dimensional computed tomography (CT scan) of a patient with a cancerous tumour (yellow) in a lung.

PRACTICE

1 An X-ray tube has an electron beam current of 1.5 mA and an accelerating voltage of 50 kV.

   a How many electrons per second hit the target?

   b If 99% of the incident energy is transferred to thermal energy in the target, at what rate must it be cooled in order to stay at a constant temperature?

   c What is the minimum wavelength of X-rays emitted by this X-ray tube?

2 Suggest how Bragg reflection could be used to create a monochromatic X-ray beam from collimated X-rays emerging from an X-ray tube.

3 a At what angles would you expect Bragg reflections when X-rays of wavelength 4×10^{-11} m scatter from atomic planes separated by 0.50 nm?

   b Explain (using a diagram) why these are not the only Bragg reflections from a cubic crystal with atomic spacing 0.50 nm.

4 Derive an equation to show that there is a unique wavelength at which X-ray photons and electrons have the same energy. Calculate this wavelength and energy.

PRACTICE EXAM QUESTIONS

1 a For a radioactive material explain what is meant by activity and define the becquerel.

b Write down an equation which shows how the activity **A** of a radioactive sample varies with time. State clearly what each symbol in your equation represents and give the appropriate SI unit for each physical quantity.

c Use your equation in **b** to derive the relationship between half-life and decay constant.

2 a What is meant by

i the decay constant λ of radioactive material,

ii the half-life $t_{\frac{1}{2}}$?

b The decay constant and the half-life are related by the equation

$$\lambda = \frac{0.693}{t_{\frac{1}{2}}}$$

The half-life of $^{60}_{27}\text{Co}$ is 5.26 years.

i What do the numbers 27 and 60 represent?

ii Calculate the decay constant of $^{60}_{27}\text{Co}$.

iii Calculate the activity of 1.00 gram of $^{60}_{27}\text{Co}$.

(60 grams of $^{60}_{27}\text{Co}$ contain 6.02×10^{23} atoms.)

3 a Explain the meaning of atomic mass number.

b i If the mass of a deuterium atom, ^2_1H is 2.008 032 unified atomic mass units, calculate the mass defect.

ii Hence calculate the binding energy per nucleon for deuterium.

$$\begin{bmatrix} 1 \text{ Unified Atomic Mass Unit} = 1\,m_u = 931\,\text{MeV} \\ \text{Neutron mass} = 1.008\,987\,m_u \\ \text{mass of } ^1_1\text{H} = 1.008\,145\,m_u \end{bmatrix}$$

4 a State *two* precautions you would take when performing an experiment using radioactive materials.

b In one natural radioactivity decay series a nucleus of thorium-232 decays to thorium-228.

How many alpha and how many beta particles are emitted during this process?

c An alpha particle from thorium has a kinetic energy of 1.4×10^{-12} J and travels 86 mm in air at normal atmospheric pressure.

The alpha particle produces pairs of ions when it collides with molecules in the air. The average energy required to produce a pair of ions is 5.6×10^{-18} J.

i How many ion pairs are produced per mm when the alpha particle passes through air at normal atmospheric pressure?

ii Explain why reducing the pressure of the air increases the range of an alpha particle.

*5 a State the two postulates of the theory of special relativity.

b Michelson and Morley carried out an experiment that gave a '*null result*' and was stated to be the greatest null result in history by a scientist named Bernal.

State the purpose of Michelson and Morley's experiment and why the 'null result' was so important.

c Andy is at the middle front of a fictitious bullet train that is 1000 m long and is travelling at constant velocity of 1.5×10^8 m s^{-1} relative to Beth who is standing on a platform. The train has two signals at the front and rear which operate when a beam of light falls on them.

At the moment that Andy and Beth are directly opposite each other Andy switches on a lamp and they both start their stopwatches.

i Explain why Andy notices that both signals operate simultaneously but Beth says that one of the signals operates before the other.

ii Andy measures the length of the train as 1000 m. Calculate the length of the train as measured by Beth.

6 a When α particles are projected at a thin metal foil in a vacuum enclosure they are scattered at various angles.

i In which direction will the maximum number of α particles coming from the foil be detected?

ii Describe the angular distribution of the scattered α particles around the foil.

iii What do the results suggest about the structure of the metal atoms?

b In this arrangement explain why

i the foil should be thin,

ii the incident beam of α particles should be parallel and narrow.

7 Aluminium is bombarded by alpha-particles. One reaction which occurs is the production of a phosphorus isotope with the emission of a neutron. This reaction is described by the equation:

$$^{27}_{13}\text{Al} + ^4_2\text{He} \rightarrow ^{30}_{15}\text{P} + ^1_0\text{n}$$

a Explain how this equation demonstrates the conservation of both nucleon number and charge.

b The phosphorus isotope is radioactive and breaks down by emitting a positron to give a stable isotope of silicon, Si.

i Explain what is meant by the term radioactive.

ii Write the equation representing this reaction.

c The half-life of the phosphorus isotope is 200 s. Calculate the fraction of a sample of phosphorus which will remain 600 s after it was formed.

d The positron generated in the reaction in part **b** has a short life. It slows down and annihilates with an electron to produce two γ-ray photons of the same frequency.

i Explain how this reaction can be reconciled with the conservation of energy.

ii Hence calculate the energy of each of the photons produced.

*Pre-U-style question

8 **a** An isotope of americium decays by the emission of an alpha particle to form neptunium-237. The proton number of neptunium is 93.

    i Determine: the proton number of the americium isotope; the nucleon number of the americium isotope.

    ii Describe briefly the test you would carry out to demonstrate that the emission from americium is in fact alpha radiation.

    iii The energy of the alpha particle emitted is 8.8×10^{-13} J.

      The mass of a proton or neutron is 1.7×10^{-27} kg.

      Show that the speed of the alpha particle is $1.6 \times 10^{7}\, \text{m s}^{-1}$.

    iv Assuming that the americium nucleus is stationary when the alpha particle is emitted, determine the recoil speed of the neptunium nucleus.

b Figure 8.1 shows the results of an experiment to investigate the relationship between the de Broglie wavelength and the momentum of a particle.

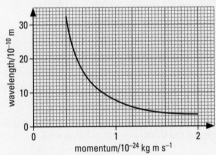

Figure 8.1

The de Broglie equation suggests that the wavelength λ is related to the momentum p by the equation:

$$\lambda = \frac{h}{p}$$

    i Use data from Figure 8.1 to show that the results support this relationship.

    ii What do the results of the experiment suggest is the value for the Planck constant, h ?

    iii Determine the de Broglie wavelength of the alpha particle in **a**.

    iv Explain briefly why it is difficult to demonstrate interference using alpha particles of this energy.

    v State what information can be gained from knowing the amplitude of the de Broglie wave associated with a particle, at a given point.

9 The isotope of strontium, $^{90}_{38}\text{Sr}$, decays by β^- emission and forms the radioactive isotope of yttrium, $^{90}_{39}\text{Y}$.

a Write down an equation which describes the decay process.

b The half-life of $^{90}_{38}\text{Sr}$ is 17.8 years and the half-life of $^{90}_{37}\text{Y}$ is 64.0 hours.
Calculate the mass of yttrium present in a stable specimen (where the rates of decay of $^{90}_{38}\text{Sr}$ and $^{90}_{39}\text{Y}$ are equal) which contains 1.12 µg of $^{90}_{38}\text{Sr}$.

c Figure 8.2 shows a beta source near a Geiger counter and with a sheet of aluminium of thickness x between the source and the G-M tube.

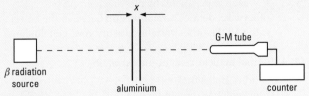

Figure 8.2

If I_0 and I are the number of beta particles counted per second with the aluminium sheet absent and in place respectively, then it is expected that
$$I = I_0 \cdot e^{-\mu x},$$
where μ is constant.

Given a supply of aluminium sheets of different known thicknesses, explain how you would test the above equation. How would you determine a value for μ?

10 $^{60}_{27}\text{Co}$ is produced by irradiation of the stable isotope $^{59}_{27}\text{Co}$ with neutrons, and decays by the emission of a beta particle and two gamma rays.

a Write down nuclear equations to represent these processes.

b $^{60}_{27}\text{Co}$ is used as a gamma ray source. Explain how the contributions of beta particles to the total radiation may be removed. Will this affect the gamma ray beam?

c Two experiments were carried out on the gamma radiation from $^{60}_{27}\text{Co}$. In both you should assume that the source S behaved as a point source radiating equally in all directions and that the recorded count is due solely to $^{60}_{27}\text{Co}$.

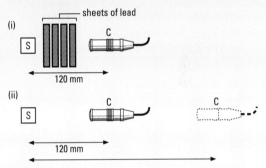

Figure 8.3

In experiment (i) sheets of lead were placed between the source S and the counter C which was at a fixed distance from the source. The number of counts in 10 minutes was measured for various total thicknesses of lead (Figure 8.3 (i)) and the readings were tabulated.

Experiment (i)

number of counts in 10 min	4250	2510	1500	850	500
total thickness of lead / mm	0	10	20	30	40

In experiment (ii) no absorber was used. The counter C was moved further away from the source S after each reading (Figure 8.3 (ii)). Readings of the number of counts in 10 min against distance from source to counter were recorded.

Experiment (ii)

number of counts in 10 min	4250	1060	480	250	170
distance from source to counter/mm	120	240	360	480	600

For each experiment, by plotting a graph or otherwise, deduce a relationship between the count rate and the other variable. Justify your conclusions.

11 A thermal nuclear reactor may contain

   (I)   uranium as <u>fuel</u>,
   (II)   graphite as <u>moderator</u>,
   (III) <u>control rods</u> of boron steel.

   a Why is the reactor described as thermal?

   b *Briefly* describe how the substances in (I), (II) and (III) fulfil the underlined functions.

12 This question is about the energies and energy changes within atoms and nuclei.

   a Explain the terms ground state, excitation energy and ionisation energy as applied to the hydrogen atom. Illustrate your answer with an energy level diagram.

   b **i** Explain what is meant by the term nuclear binding energy.

       ii Describe how the binding energy per nucleon varies with atomic mass number for stable nuclei. Illustrate your answer with a simple sketch graph.

      iii Explain why energy is released in nuclear fission.

   c Using your answers to **a** and **b** above, compare and contrast the emissions of a photon in the visible region of the spectrum and a gamma ray photon. What conditions are necessary for each to occur?

13 In a sodium discharge tube, light is emitted when accelerated electrons collide with atoms of sodium.

   a Describe briefly the process which leads to the emission of light from the tube.

   b Calculate the energy of a photon of light which has a wavelength of 590 nm:

      i in joules;

      ii in electron-volts.

         Planck constant, $h = 6.6 \times 10^{-34}$ J s.
         Speed of light in free space, $c = 3.0 \times 10^8$ m s^{-1}.
         Charge on an electron, $e = -1.6 \times 10^{-19}$ C.

   c The average distance travelled by an electron between collisions in a sodium lamp is 1.0×10^{-4} m. Estimate the electric field strength required to produce photons of wavelength 590 nm in a sodium discharge tube.

14 A monochromatic light source provides 5 W of light of wavelength 4.50×10^{-7} m. This light falls on a clean potassium surface and liberates 3.2×10^{11} photoelectrons per second. The photoelectrons are collected by an electrode just above the metal surface and the photoelectric current measured.

   charge of electron $= -1.60 \times 10^{-19}$ C.
   speed of light in vacuo $= 3.00 \times 10^8$ m s^{-1}

   a Calculate the photoelectric current given by this arrangement.

   b Estimate the photoelectric current given by a similar arrangement using a source which provides 10 W of light of wavelength 4.50×10^{-7} m.

   c Explain whether or not photoelectrons would be emitted if a 20 W source operating at a wavelength of 6.00×10^{-7} m were to be used. The threshold frequency for potassium is 5.46×10^{14} Hz.

15 **a** The spectrum of atomic hydrogen contains a prominent red line having a wavelength of 6.60×10^{-7} m. Calculate the energy of a photon with this wavelength.

   b The ionisation energy of hydrogen is 2.18×10^{-18} J. The next allowed energy level above the ground state in hydrogen has an energy -5.40×10^{-19} J. Show by calculation that the lowest energy level cannot be involved in the production of the prominent red line in **a**.

   c Calculate the longest wavelength line that would be possible involving the lowest energy level and state in which part of the electromagnetic spectrum that this wavelength lies.

16 The first example of an induced nuclear transformation detected in the laboratory resulted from the bombardment of nitrogen by alpha particles:

$$^{14}_{7}\text{N} + \alpha \rightarrow {}^{17}_{8}\text{O} + \text{p}$$

Use the following data to calculate the increase in rest mass which occurs.

   Mass of nitrogen-14 atom = 23.2530×10^{-27} kg
   Mass of oxygen-17 atom = 28.2282×10^{-27} kg
   Mass of alpha particle = 6.6442×10^{-27} kg
   Mass of proton = 1.6725×10^{-27} kg

State the source of this increased rest mass.

Calculate the minimum kinetic energy of the incident alpha particle for the transformation to be possible. You may neglect the recoil energy of the oxygen-17 atom and the proton.

(Speed of light in vacuum, $c = 3.0 \times 10^8$ m s^{-1}.)

17 **a** With reference to electron energy levels in atoms, explain what is meant by **i** excitation, **ii** ionisation.

   b Figure 8.4 represents the energy level diagram for atomic hydrogen.

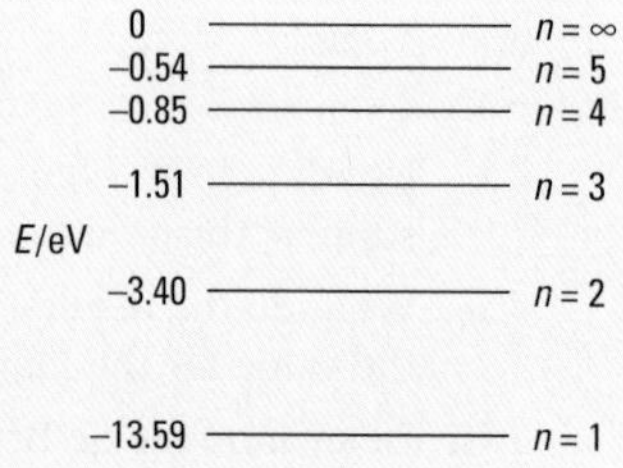

E/eV	
0	$n = \infty$
-0.54	$n = 5$
-0.85	$n = 4$
-1.51	$n = 3$
-3.40	$n = 2$
-13.59	$n = 1$

Figure 8.4

      i Calculate the ionisation energy in joule.

      ii Calculate the wavelength emitted for the transition $n = 3$ to $n = 2$.

      iii Calculate the energies of the photons which could be emitted after a hydrogen atom in its ground-state gains 12.08 eV.

18 This question is about a helium-neon laser which produces red light. Figure 8.5 shows some of the electron energy levels of helium and neon atoms.

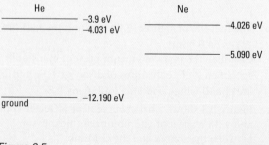

Figure 8.5

a Why are the energy levels in Figure 8.5 given negative values?

b **i** The red light emitted by this laser has a wavelength of 6.33×10^{-7} m. Show that this corresponds to a photon energy of 1.96 eV.

DATA: The Planck constant = 6.63×10^{-34} J s, speed of light, $c = 3.00 \times 10^8$ m s^{-1}, electronic charge = 1.60×10^{-19} C

ii Copy Figure 8.5 and mark on it an arrow between two of the energy levels, representing the transition responsible for the emission of these photons.

c The laser works because energy is transferred between colliding helium and neon atoms. When a helium atom in the first excited state (–4.031 eV) collides with a neon atom in the ground state, energy is transferred between electrons enabling the neon to move directly to its second excited state (–4.026 eV). The small amount of additional energy, 0.005 eV, required to allow the transition to occur, is thought to be provided from the kinetic energy of the colliding atoms.

i Show that the mean energy of a particle in a monatomic gas at 293 K is approximately 6×10^{-21} J. (The Boltzmann constant = 1.4×10^{-23} J K^{-1})

ii Show that this energy is sufficient for the above transition to occur.

19 a One of the two postulates of Einstein's theory of special relativity is that the speed of light in free space is invariant.

i What is meant by this statement?

ii What is the other postulate?

b K$^+$ mesons are sub-atomic particles of half-life 8.6 ns when at rest. Figure 8.6 shows a diagram of an accelerator experiment in which a beam of K$^+$ mesons travelling at a speed of $0.95c$ is created where c is the speed of light.

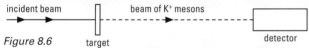

Figure 8.6

i Calculate the half-life of the K$^+$ mesons in the beam measured in the laboratory frame of reference.

ii What is the greatest distance that a detector could be sited from the point of production of the K$^+$ mesons to detect at least 25% of the K$^+$ mesons produced?

20 The Michelson–Morley experiment outlined in Figure 8.7 was designed to detect absolute motion of the Earth. A fringe pattern due to interference of the two light beams reflected at plane mirrors M$_1$ and M$_2$ respectively was observed using a telescope.

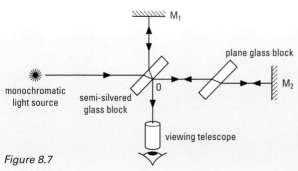

Figure 8.7

i Explain why it was thought that rotating the apparatus through 90° might cause a change of the fringe pattern.

ii No change of the fringe pattern could be detected when the apparatus was rotated through 90°. What was the significance of this null result in terms of attempts to detect absolute motion?

21 a Calculate the kinetic energy in MeV of an electron travelling at a speed of precisely $0.9900c$.

b Show that the kinetic energy of the electron is about six times its rest energy.

c Show that the linear momentum p of a particle is related to the radius r of its orbit when it is travelling in a plane perpendicular to a magnetic field B by
$$p = eBr$$
where e is the charge on the particle.

d Calculate the radius r of the orbit of the electron in a in a magnetic field B of 0.1 T.

22 a Explain carefully what is meant by the terms:

i atomic number;

ii atomic mass number;

iii isotope;

iv nuclear binding energy.

b **i** Sketch, giving a rough scale, a graph showing how the binding energy per nucleon varies with atomic mass number and indicate on it the region of maximum stability.

ii Use the graph to explain briefly why energy is released in the processes known as fission and fusion.

c **i** The fusion of deuterium nuclei can be represented by the equation
$$^2_1\text{H} + ^2_1\text{H} = ^4_2\text{He}.$$
Calculate the energy released by this reaction.

ii 2 kg of deuterium are caused to fuse and it is proposed that the energy released by the fusion is used to generate electricity in a power station. If the efficiency of the conversion process is 52% and the output of the station is to be 5.0 MW, for how long would the fuel (deuterium) last?

iii What are the difficulties associated with using fusion as a source of power?

[Relative molecular mass of deuterium = 2
Nuclear masses: ^2_1H = 2.014 19 u
^4_2He = 4.002 77 u
1 u is equivalent to 1.49×10^{-10} J]

* 23 a Explain what is meant by *time dilation* and *length contraction*.

b In an experiment, muons of charge 1.6×10^{-19} C are accelerated to a speed of 2.5×10^8 m s^{-1} as measured by a laboratory observer. The muon beam passes first through a detector **A** and then through a detector **B** that is 2000 m from detector **A**. The count rate at detector **A** is found to be 4 times that at detector **B**.

i Calculate the half-life of the muons as measured by the observer in the laboratory.

*Pre-U-style question

ii Calculate the half-life of the muons as measured in the frame of reference in which the muons are at rest.

iii How far apart are the detectors as measured in the muons' frame of reference?

c The muon beam enters a mass spectrometer where the mass of a muon is measured to be 3.5×10^{-28} kg.

i Calculate the rest mass of a muon in kg.

ii Calculate the potential difference used to accelerate the muons to 2.5×10^8 m s^{-1}.

Mass at speed v is given by: $m = \dfrac{m_0}{\sqrt{1 - \left(\dfrac{v^2}{c^2}\right)}}$

24 a State *three* advantages nuclear fusion would offer compared to nuclear fission as a potential source of power.

b One of the most productive fusion reactions occurs when a deuterium nucleus fuses with a tritium nucleus to form an alpha particle and a neutron.

i Write down the equation which represents this reaction.

ii Calculate the Q-value in MeV for this reaction. Your calculation should make it clear how you have allowed for the electron masses.

c i Calculate the radius of a deuterium nucleus and the radius of a tritium nucleus, using $R = r_0 A^{\frac{1}{3}}$ and taking r_0 to be 1.3 fm.

ii By considering these two nuclei as charged spheres in contact, calculate the minimum energy in MeV which must be supplied to them if they are to overcome their Coulomb barrier and fuse together.

d i Use the equation

$$\frac{1}{2}mc^2 = \frac{3}{2}kt$$

to estimate the temperature at which deuterium and tritium nuclei would have enough kinetic energy to undergo fusion.

ii How would you expect deuterium and tritium atoms to be changed by such a high temperature?

25 An α-source with an activity of 150 kBq is placed in a metal can as shown in Figure 8.8. A 100 V d.c. source and a $10^9 \, \Omega$ resistor are connected in series to the can and the source. This arrangement is sometimes called an ionisation chamber.

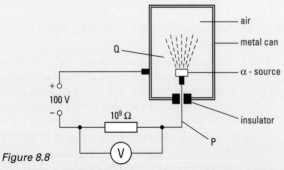

Figure 8.8

a What is meant in this case by an activity of 150 kBq?

b Describe how the nature of the electric current in the wire at P differs from that in the air at Q.

c A potential difference of 3.4 V is registered on the voltmeter.

i Calculate the current in the wire at P. State any assumption you make.

ii Calculate the corresponding number of ionisations occurring in the metal can every second. State any assumption you make.

d With the α-source removed from the metal can, the voltmeter still registers a potential difference of 0.2 V. Suggest two reasons why the current is not zero.

e The half-life of the α-source is known to be 1600 years. Calculate the decay constant and hence deduce the number of radioactive atoms in the source.

26 Three adjacent elements in a long radioactive series are actinium, francium and radium.

$\rightarrow$ actinium (Ac) $\xrightarrow{\alpha}$ francium (Fr) $\xrightarrow{\beta^-}$ radium (Ra) $\rightarrow$

a An atom of $^{223}_{87}$Fr is formed when an atom of actinium decays with the emission of an α-particle. The masses of the atoms in atomic mass units (u) are: actinium atom 227.0278 u, francium atom 223.0198 u, helium atom 4.0026 u.

i Write a nuclear reaction equation describing this change.

ii Compare the mass of the parent actinium atom with the sum of the masses of the decay products. Show how the values you obtain are consistent with the emission of the α-particle with considerable kinetic energy.

b i The mass of a proton is 1.0073 u, of an electron is 0.0005 u and that of a neutron is 1.0087 u. Calculate the difference in mass between a $^{223}_{87}$Fr nucleus and the sum of the masses of its nucleons.

ii How do you account for this difference?

c The $^{223}_{87}$Fr atom decays with the emission of a β$^-$-particle into an atom of radium. Write a nuclear reaction equation for this change.

d The half-life of the actinium isotope is 19 days and that of the francium isotope is 21 mins.

i Explain what is meant by the term half-life.

ii A sample of pure actinium is monitored for one day. Without any calculation, comment briefly on how the concentration of francium varies during this time.

($1 \, \text{u} \equiv 931 \, \text{MeV}$)

27 Figure 8.9 shows a narrow beam of electrons in an evacuated tube. The electrons are emitted by the hot-wire filament and are attracted to the anode, passing through the hole in the anode to form the beam.

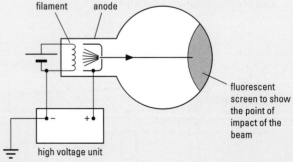

Figure 8.9

a i Why is it necessary for the filament wire to be hot?

ii Why must the tube be evacuated?

b i Calculate the speed of the electrons in the beam if the anode potential is 2000 V. Neglect relativistic effects.

ii Why is it reasonable to assume relativistic effects are negligible?

28 Figure 8.10 shows an arrangement of apparatus which can be used to determine the charge on an oil drop.

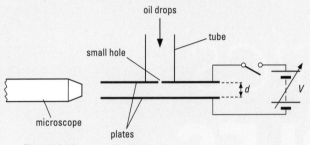

Figure 8.10

Charged oil drops are sprayed into the tube and are allowed to fall freely until they enter the region between two horizontal plates separated by a distance d. For the purposes of this question, it may be assumed that each of the oil drops has the same mass m.

a i Explain why some of the oil drops may become stationary when a potential difference V is applied across the plates.

ii Show that the charge q on a drop which becomes stationary when the potential difference is applied to the plates is given by

$$q = \frac{mgd}{V}$$

iii A set of data for one of the stationary oil drops is given below.

$d = 5.00 \text{ mm}$
$V = 904 \text{ V}$
$m = 1.18 \times 10^{-14} \text{ kg}$

Use the data to calculate a value for the charge on the oil drop.

b The potential difference across the plates is now varied, and the charges on several drops are measured. The results from these measurements are shown in Table 8.1.

Table 8.1

drop number	charge on drop/10^{-19} C
1	6.4
2	22.4
3	16.0
4	19.2
5	9.6

i Use these experimental results to calculate a value for the elementary charge.

ii Explain why it is necessary to take some further readings in order to obtain a reliable value for the elementary charge.

29 This question is about the use of the photoelectric effect in a lightmeter. The basis for the device is a photocell which consists of two electrodes in an evacuated tube. One of the electrodes has a surface which is photosensitive and has a suitable work function. The device is shown schematically in Figure 8.11.

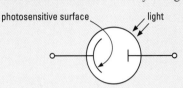

Figure 8.11

a State what is meant by the photoelectric effect?

b i The photocell is connected in series with a d.c. supply and a micro-ammeter. Using the symbol shown in Figure 8.12 sketch the circuit which is used. Indicate on it the positive and negative terminals of the supply and the meter.

ii Explain why the arrangement would not work if the supply were connected the wrong way round.

c It is suggested that potassium is a suitable material for use in the light meter. This has a work function of 2.9×10^{-19} J. The photocell should ideally respond to all visible wavelengths which lie in the range 400 nm to 800 nm.

Planck constant	$h = 6.6 \times 10^{-34}$ J s.
Speed of light in vacuo	$c = 3.0 \times 10^8$ m s^{-1}.
Mass of an electron	$m_e = 9.1 \times 10^{-31}$ kg.

i Show, by means of a suitable calculation, that when potassium is used, the photoelectric effect occurs for wavelengths in the visible range.

ii What will be the maximum kinetic energy of the electrons emitted from this surface when all visible frequencies are present?

iii Calculate the speed of these electrons.

iv What is the de Broglie wavelength of these electrons?

30 a Estimate the de Broglie wavelength of an electron that has been emitted thermionically in a vacuum from a filament and then accelerated through a p.d. of 30.0 kV.

b Outline the principle of operation of a transmission electron microscope (TEM).

c State *one* reason why a transmission electron microscope in operation is unable to achieve its theoretical resolving power.

9

The Physics of **PARTICLES**

How do we know that the particles of modern physics are actually there? Just as we deduce that birds have been in the garden when we see their tracks in the snow next morning, so particle physicists deduce that various particles have been present by the tracks that they leave. From the early cloud chambers to modern solid-state detectors, this is essentially what particle physicists have always done: the shape of the tracks and the reactions that can be seen enable them to draw conclusions about the properties of the particles themselves.

In this chapter you will be presented with the techniques used in detecting particles as well as the equipment used to accelerate beams of particles to give them the energy required to take part in various reactions. Simple relativity is revisited. The basic constituents of matter, their properties, and the fundamental forces that they experience are outlined.

The beginnings of an 'electron tree' produced in a block of plastic by a beam of electrons. Electrons bombarding the plastic penetrate a short distance into the block. If this trapped energy is released by a sharp blow to the block, the electrons shoot out as an electrical discharge, leaving a pattern of tracks.

9.1

OBJECTIVES

- financing physics
- identifying particles
- fundamental forces

CERN facts

Percentage contributions of the member states to the CERN annual budget:

Austria	2.02	Italy	10.27
Belgium	2.62	Netherlands	4.23
Bulgaria	0.26	Norway	2.29
Czech Republic	0.90	Poland	2.67
		Portugal	1.23
Denmark	1.69	Slovakia	0.44
Finland	1.28	Spain	7.47
France	14.46	Sweden	2.54
Germany	18.65	Switzerland	4.74
Greece	1.51	United	
Hungary	0.58	Kingdom	12.51

Total budget for 2012 was 1174.78 million Swiss francs

CERN staff:

2400 full time

1500 part time

10 000 visiting scientists and engineers of 113 nationalities and from 608 universities

Helpful hint

It is conventional in particle physics to fix the unit of charge as +1. Hence the charge of the up quark would be written as $+\frac{2}{3}$ not $+\frac{2}{3}e$. This convention will be followed from now on.

Mass of neutrinos

There is good evidence to suggest that neutrinos do have a very tiny mass (much smaller than that of electrons). However, syllabuses still regard them as being massless.

BIG SCIENCE

'Big science' is the current jargon for scientific research that costs a great deal of money, with **particle physics** and **astrophysics** at the top of the list. Collaborative teams of people have been forced to pool their financial and technical resources and bid for time to carry out experiments on large international telescopes and particle accelerators. The largest European centre for particle physics research is the European Laboratory for Particle Physics (CERN – for Conseil Européen pour la Recherche Nucléaire) in Geneva. The European countries contribute money to its operating budget in proportion to their national income, the annual CERN budget being some £800 million. The electricity budget alone runs to £40 million.

In addition, money has to be found by the university collaborations to design and build the experimental devices that detect new particles produced by the accelerators. This money is also funded by the tax payer via government. In the United States, Fermilab (near Chicago) and the Stanford Linear Accelerator Centre (SLAC) rival CERN in their facilities. It is clear that a great deal of money goes into particle physics. Why? What does particle physics do and how much does it contribute to civilization in order to justify such enormous expense?

What is particle physics?

Particle physics is the modern version of an age-old quest – to find the smallest pieces of matter that cannot be broken down. The Ancient Greeks were the first people to suggest that there might be such 'indivisible' pieces of matter from which everything else is made. Democritus (460–370 BC) believed that nature consisted of a wide variety of 'a-toms' ('un-cuttable') that combined together into the stuff that we see in the world. Modern science has discovered that Democritus was wrong on two counts: there is not a wide variety of different atoms (there are only 92 stable atoms) and they are not indivisible; they can be split into smaller objects.

The smallest pieces of matter that are now known divide into two groups, the **quarks** and the **leptons**, and there are six varieties of each.

Quarks and leptons.

The six quarks		
up (u)	**charm (c)**	**top (t)**
charge = $+\frac{2}{3}e$	charge = $+\frac{2}{3}e$	charge = $+\frac{2}{3}e$
mass = $\frac{1}{3}m_p$	mass = $1.7m_p$	mass = $186m_p$
down (d)	**strange (s)**	**bottom (b)**
charge = $-\frac{1}{3}e$	charge = $-\frac{1}{3}e$	charge = $-\frac{1}{3}e$
mass = $\frac{1}{3}m_p$	mass = $0.5m_p$	mass = $4.9m_p$

The six leptons		
electron (e⁻)	**muon (μ⁻)**	**tau (τ⁻)**
charge = $-e$	charge = $-e$	charge = $-e$
mass = $0.0005m_p$	mass = $0.1m_p$	mass = $1.9m_p$
electron-neutrino (ν_e)	**muon-neutrino (ν_μ)**	**tau-neutrino (ν_τ)**
charge = 0	charge = 0	charge = 0
mass = 0	mass = 0	mass = 0

NB $e = 1.6 \times 10^{-19}$ C, m_p = mass of proton = 0.938 GeV/c^2 = 1.67×10^{-27} kg.

Hadrons

One important difference between quarks and leptons must be emphasized. Quarks can *never* be found in isolation. They are only ever found inside other particles called **hadrons**. For example, the proton is a hadron as it has three quarks inside it (two up quarks and a down – uud). The neutron is also a hadron (two down quarks and an up – ddu). There are many other types of hadron made up from the other quarks.

The fundamental forces

A fundamental force is one that cannot be explained by the action of another force. (For example, friction is not a fundamental force as it is due to the electromagnetic forces between the atoms in one object and those in another.)

Fundamental forces.

Force	Felt by	Range*	Relative strength**
electromagnetism	any charged particle	infinite	1
gravity	any particle with mass	infinite	10^{-38}
strong force	quarks	10^{-15} m	100
weak force	any particle	10^{-17} m	10^{-5}

* The range of a force is the distance that two objects must be separated before the force becomes negligible.

** The strengths of the forces relative to the electromagnetic force have been estimated for two protons that are just touching.

Exchange particles

When an electron drops from a high atomic energy level to a lower one, the energy difference is emitted in a packet (or quantum) of electromagnetic energy called a **photon**. Photons play a very important role in the theory of electromagnetic forces at the level of elementary particles. The other fundamental forces are also associated with quanta of energy, which play similar roles. Collectively they are known as **exchange particles** (or **gauge bosons**). The quantum theories of the fundamental forces suggest that the continual bouncing back and forth of exchange particles between elementary particles gives rise to the forces between them.

Exchange particles.

Force	Particle	Charge	Mass
electromagnetic	photon (γ)	0	0
strong	gluon (g)	0	0
weak	W$^+$	$+e$	$89m_p$
	W$^-$	$-e$	$89m_p$
	Z^0	0	$99m_p$

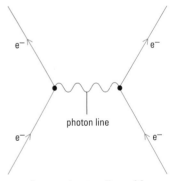

The modern understanding of forces is in terms of exchange particles that fire back and forth between elementary particles. In this diagram the electromagnetic repulsion between two electrons is visualized as the exchange of electromagnetic energy in the form of a photon.

PRACTICE

1 Show that a hadron composed of two up quarks and a down quark has the correct basic properties (i.e. charge, mass) to be a proton. Show that the neutron can be made up of two downs and an up.

2 Write down all the possible combinations of three quarks using just the u, d, and s quarks. For each one list the charge and approximate mass of the hadron formed from the combination.

3 Compare the gravitational attraction between two protons separated by 1 m with the electrostatic repulsion between them. What is the ratio of the forces in this case?

4 If the electromagnetic force is so much stronger than gravity, why is it that the large-scale structure of the universe is dominated by gravity?

OBJECTIVES

- momentum
- energy and mass
- massless particles
- using standard units

USING RELATIVITY IN PARTICLE PHYSICS

Any understanding of particle physics relies on Einstein's **special theory of relativity**. His general theory, which updates Newton's ideas about gravity, is not required, as elementary particles are not massive enough to produce strong gravitational fields. By its nature, particle physics is high-energy physics, and high energies often imply that speeds are close to that of light, so relativity is required.

The basic ideas of the special theory can be summed up as follows.

1 Any two observers in relative motion must agree on the laws of physics.

2 In order to make **1** work, it is necessary to compare measurements of length, time, mass, and energy between the two observers. To do this we can look over the shoulder of one observer (and so have that person stationary with respect to us) and compare the measurements of the other observer who is moving at a constant speed v from our point of view. The measurements compare as shown below.

Space, time, and mass in relativity

Our point of view

length of our measuring stick = l_0

time intervals on our clock = t_0

mass of our standard object = m_0

Other observer

length of observer's stick as measured by us = l

time taken for observer's clock to 'tick' one interval = t

mass of observer's standard as measured by us = m

$$l = l_0 \sqrt{1 - \frac{v^2}{c^2}} \qquad t = \frac{t_0}{\sqrt{1 - \frac{v^2}{c^2}}} \qquad m = \frac{m_0}{\sqrt{1 - \frac{v^2}{c^2}}}$$

These equations are often simplified by using the abbreviation $\gamma = \dfrac{1}{\sqrt{1 - \dfrac{v^2}{c^2}}}$

So

$$l = \frac{l_0}{\gamma} \qquad\qquad t = \gamma t_0 \qquad\qquad m = \gamma m_0$$

For a fuller explanation, see spread 8.24.

Representing rest mass

In this spread, m_0 will be used to represent the mass of a particle as measured at rest. For the rest of this section the standard convention in particle physics will be adopted, using just m.

Momentum

One significant realization brought about by the special theory of relativity is that the mass of a moving object is apparently not the same as the mass of an otherwise identical stationary object. This is *apparently* the case as the mass of a moving object cannot be measured directly; it must be calculated from a measurement of momentum and velocity. If the object has a mass m_0 when it is stationary (often called its **rest mass**), then when it is moving with a velocity of v relative to us, its momentum and energy will be

$$p = \frac{m_0 v}{\sqrt{1 - \frac{v^2}{c^2}}} = \gamma m_0 v \qquad\qquad E = \frac{m_0 c^2}{\sqrt{1 - \frac{v^2}{c^2}}} = \gamma m_0 c^2$$

which are the relativistic equations for momentum, p, and total energy, E.

Particle physicists find it very useful to identify a particle by its mass. It would be inconvenient to have to say what the speed is whenever a mass is mentioned. For this reason it has become conventional in particle physics just to refer to the mass of a particle (assumed always to be measured at rest) and only to talk about the momentum when it is moving. You will hardly ever see the term 'rest mass' in particle physics.

Energy

Another dramatic change in our thinking brought about by relativity is the connection between energy and mass as symbolized in the famous equation $E = m_0 c^2$. This is actually a simplified version of a full relationship:

$$E^2 - p^2c^2 = m_0^2c^4 \tag{1}$$

which reduces to $E = m_0c^2$ when the object is at rest ($p = 0$). In Newton's version of mechanics a particle on its own, away from any force fields like gravity or electromagnetism, could only have one form of energy – *kinetic energy*. If the particle was at rest, then it had no energy at all. This is not the case in relativity. A particle in such a situation has energy equal to m_0c^2. Newton could not have known this as at the time there was no way to convert this energy into any other form, so he did not realize that it was there. Now this energy has been seen being released in other forms in radioactive decay and nuclear fission.

The photon loophole

Photons are quanta of electromagnetic radiation. They are packets of energy travelling at the speed of light. The energy in a photon is given by $E = hf$ where f is the frequency of the equivalent electromagnetic wave.

If a photon is travelling at the speed of light, then its momentum should be

$$p = \gamma m_0 v \frac{m_0 c}{\sqrt{1 - \dfrac{c^2}{c^2}}} = \frac{m_0 c}{0}$$

If you divide by zero the answer is infinity, which is impossible. But if the mass of the photon m_0 is zero, then the momentum becomes

$$p = \frac{m_0 c}{\sqrt{1 - \dfrac{c^2}{c^2}}} = \frac{0}{0}$$

which is not defined mathematically. Consequently this equation for momentum *only applies for particles with non-zero masses*.

Another way of calculating momentum is to use

$$E^2 - p^2c^2 = m_0^2c^4$$

Hence

$$p = \sqrt{\frac{E^2 - m_0^2c^4}{c^2}} = \frac{E}{c}$$

if $m_0 = 0$. Inserting the equation for the energy of a photon $E = hf$ gives

$$p = \frac{E}{c} = \frac{hf}{c} = \frac{h}{\lambda}$$

where λ is the wavelength of the electromagnetic radiation.

This shows that the photon can have a perfectly well defined momentum, *even though its mass is zero*. Newton could not have imagined massless particles travelling at the speed of light that also have momentum.

Photons are not the only objects that travel at the speed of light and have zero masses. **Gluons** (which are the equivalent objects for the **strong force**) are also massless.

Standard units

The standard SI units of energy, mass, and momentum are inconveniently large to be used in particle physics. For this reason it is conventional to express the results of calculations in **standard units**.

Unit of energy

Energy is measured in electronvolts (eV). $1\,\text{eV} = 1.6 \times 10^{-19}\,\text{J}$, the energy collected by an electron when it passes through a p.d. of 1 volt. In practice GeV (gigaelectronvolts, $10^9\,\text{eV}$) and MeV (megaelectronvolts, $10^6\,\text{eV}$) are commonly used.

Unit of mass

Mass is measured in multiples of GeV/c^2. This rather odd unit comes from $E = m_0c^2$. If the energy on the left-hand side is expressed in units of GeV, then the mass must be in GeV/c^2 to balance the units. (However, people never actually bother to divide by c^2 – they just work in multiples of this quantity.)

Unit of momentum

Momentum is measured in GeV/c. This follows from $p = E/c$, although the unit applies to all particles.

PRACTICE

1 Compare the momentum of an electron as calculated from the Newtonian and relativistic equations when the electron has a speed of

 a $27\,000\,\text{m s}^{-1}$;

 b $2.7 \times 10^8\,\text{m s}^{-1}$.

 Use SI units.

2 Calculate the energy and momentum of a proton moving at 90% of the speed of light. Use SI units.

3 Use the equations $E = \gamma m_0 c^2$ and $p = \gamma m_0 v$ to derive equation (1).

4 In the Large Electron–Positron (LEP) accelerator at CERN electrons are accelerated to 45 GeV energy. What is their momentum at this energy? (The electron mass is 0.511 MeV/c^2.)

5 A particle of energy 0.5 GeV has a momentum of 0.48 GeV/c. What is its mass?

9.3

O B J E C T I V E S

- collisions
- conservation laws
- decay and lifetime

Helpful hint

Note that the definitions of elastic and inelastic collisions used in particle physics are different to those in Newtonian mechanics. A Newtonian elastic collision conserves kinetic energy. In particle physics the possibility of creating new particles must be considered as well.

A particle reaction in a bubble chamber.

COLLISIONS AND DECAYS

Collisions

When particles collide, the total energy in the reaction is their kinetic energy plus the energy that they have by virtue of their masses. After the collision the kinetic and mass energies can be different, but the total must remain the same (conservation of energy). This means that the particles coming out of the reaction can be different to those that went in.

$$A + B \rightarrow C + D$$

In a typical reaction equation, such as the one above, the particles listed on the left hand side (A and B) are the ones that collide and those on the right are the ones that emerge. If the same particles come out as went in it is an **elastic collision**:

$$p + p \rightarrow p + p \qquad \text{elastic collision}$$

If new particles are created then it is an **inelastic collision**:

$$p + p \rightarrow p + p + \pi^0 \qquad \text{inelastic collision}$$

In this case some of the energy in the reaction has gone into producing the mass of a new particle – a **pi zero** (the π^0 is a very common hadron – see spread 9.11). The reaction is also subject to conservation of momentum, which can mean that not all of the energy is available to create new particles (see spread 9.5).

Forces and collisions

Every collision is governed by one of the fundamental forces (except the force of gravity, which is so weak that it cannot influence particles with such tiny masses). The other forces come into play to different extents according to the circumstances. With 5 GeV electrons the relative likelihoods of reactions involving the forces are

electromagnetic	1
strong force	10
weak force	10^{-8}

These figures are only approximate and will change with energy.

- The **electromagnetic force** leads to simple collisions between charged particles,

  e.g. $p + p \rightarrow p + p$

- The **strong force** dominates reactions between hadrons (which contain quarks),

  e.g. $p + p \rightarrow p + n + \pi^+$

- The **weak force** is more likely to be seen in lepton reactions (leptons do not feel the strong force, which dominates hadron reactions), especially if one of the leptons is a neutrino so that there can be no electromagnetic reaction (neutrinos being neutral),

  e.g. $\nu_e + \mu^- \rightarrow e^- + \nu_\mu$ lepton reaction

  e.g. $\nu_e + n \rightarrow p + e^-$ lepton–hadron reaction

Conservation laws

The fundamental forces do not have a free hand in bringing about reactions. They are all subject to certain **conservation laws**. Energy, momentum, and electric charge are all familiar conserved quantities: in addition physicists have now discovered other properties of particles that are conserved by some forces but not by others (e.g. lepton number, baryon number, strangeness, etc. – see spreads 9.8, 9.9, 9.12, and 9.13).

Decays

Most particles are unstable, which means that they will decay into other particles. This process is very similar to radioactive decay. Any particle will decay if there are particles of lighter mass into which it can decay.

$$A \rightarrow B + C + ...$$

provided $M_A > M_B + M_C$ ...

Decays are subject to conservation laws as well and these often prevent decays that, in mass terms alone, seem to be possible. In general, the greater the mass difference (and so energy released) the faster the decay will take place.

Lifetime

A collection of unstable particles of the same type will not all decay at the same moment. This is true even if the particles were created together and so are of the same 'age'. Like radioactive decay, particle decay is a random process. There is no internal 'alarm clock' that 'rings' after a certain time to tell the particle to decay. However, if the number of decays per second is recorded then the familiar exponential curve can be plotted. Radioactive isotopes are characterized by their **half-lives** – the amount of time taken on average for half the number of isotope atoms to decay. Particles are characterized by their **lifetimes** – the average time taken for a particle to decay. Lifetime and half-life are not the same, although they are related. As

$$N = N_0 e^{-\lambda t}$$

$$\text{lifetime, } \tau = \frac{1}{\lambda} \text{ and half-life, } t_{\frac{1}{2}} = \frac{\ln 2}{\lambda} = \tau \ln 2$$

where λ is the probability of a single particle decaying per second.

Lifetimes and forces

The lifetime of a particle is determined partly by the fundamental force that is responsible for the reaction. The forces act with different strengths in decays, and so have different characteristic lifetimes.

Particle lifetimes.

Force	Typical lifetime	Example of decay	Energy release	Example of lifetime
strong force	10^{-23} seconds	$\Delta^{++} \rightarrow p + \pi^+$	154 MeV	10^{-23} seconds
weak force	10^{-6} seconds	$\mu^- \rightarrow e^- + \bar{v}_e + v_\mu$	105 MeV	2.2×10^{-6} seconds
electromagnetic	10^{-16} seconds	$\pi^0 \rightarrow \gamma + \gamma$	125 MeV	0.8×10^{-16} seconds

The examples of decay in the table above have been chosen to be representative of the forces involved, and also to have reasonably similar energy releases, so that a fair comparison of lifetimes can be made.

NB Other factors can make the lifetimes rather different from that expected due to the nature of the force. For example, the neutron is unstable when isolated and it has a lifetime of about 15 minutes. This is a weak-force decay, but it is much slower than normal as the energy released in this decay is very small.

Decay equations

Some care needs to be taken in interpreting the equations of particle decays. For example the β⁻ decay of a neutron takes place via

$$n \rightarrow p + e^- + \bar{v}_e$$

so it would be natural to assume that the proton, electron, and anti-neutrino are 'wrapped up' inside the neutron and the decay process enables them to escape. But the equivalent equation with β⁺ decay would be

$$p \rightarrow n + e^+ + v_e$$

which would suggest that the neutron is inside the proton! A better way to look at β⁻ decay is to consider that the neutron has changed into a proton and in the process the energy released has become the electron and anti-neutrino *that did not exist before the decay took place.*

PRACTICE

1 Two protons of kinetic energy 270 GeV collide head on. What is the total energy in the reaction?

2 Two protons of equal energy collide head on. The result is the reaction

$$p + p \rightarrow p + p + \pi^0$$

If all the particles are produced at rest (a highly unlikely situation), what must be the kinetic energy of the original protons? ($m_{\pi 0} = 135$ MeV/c^2).

3 The π^0 is an unstable hadron that decays into two photons. Given that the π^0 has a mass of 135 MeV/c^2, calculate the energy and momentum of the two photons.

4 Calculate the frequency and wavelength of the photons released in the π^0 decay. To what part of the electromagnetic spectrum do they belong?

ACCELERATING PARTICLES

The next time that you see someone running through a car park shouting 'I am going to be famous' do not dismiss the person as being totally crazy. He or she just might be a physicist running from a library after reading an obscure paper in German that provided the clue needed to invent the first particle accelerator. Such was the sequence of events that lead E. O. Lawrence to design the first machine to accelerate particles in 1928. His prototype machine was 13 cm in diameter and accelerated protons to 80 keV. The CERN Large Hadron Collider (LHC) is 8.6 km in diameter and accelerates protons to 4 TeV. However, despite this huge difference in scale, many of the features of Lawrence's design can be seen in modern accelerators.

Lawrence's cyclotron

Lawrence's first machine was built from two metal D shapes placed between the poles of a magnet. A proton source was placed at the centre of the machine between the two Ds. The magnetic field bent the paths of the protons so that they curved round and crossed from one D into the other. One of the D shapes was connected to the positive terminal of a power supply, and the other to the negative terminal. As a proton crossed from one to the other it was accelerated by the p.d. between the Ds. While inside the hollow metal shapes, the protons were not acted on by any electric field (see chapter 5) so they coasted at constant speed.

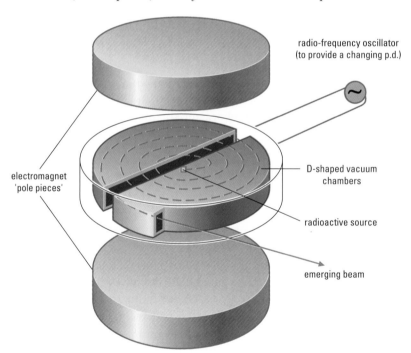

radio-frequency oscillator
(to provide a changing p.d.)

electromagnet
'pole pieces'

D-shaped vacuum
chambers

radioactive source

emerging beam

The first of Lawrence's cyclotrons (for which he received the Nobel Prize in 1939) was 13 cm in diameter and accelerated protons to 80 keV.

If the Ds remained connected to the power supply as in the diagram, then protons would be accelerated on one side of the machine and decelerated on the other. Lawrence's brilliant idea was to switch the polarity of the Ds repeatedly so that the protons were always accelerated. This was only possible because the radius of the protons' path increases as they speed up – so they always take the same time to complete an orbit. If this were not the case, then the switching over would have to be carefully timed to correspond to the moment when the proton crossed from one D to the other, and this would be at a different interval every orbit.

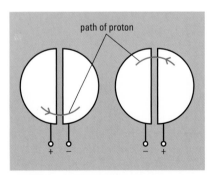

path of proton

In order to accelerate the proton every time it crosses the gap, the polarity must be reversed.

The cyclotron frequency

The magnetic force acting on a proton in the cyclotron is Bev, where v is the proton's velocity. This provides a centripetal force that causes the path to be curved into a circular arc. The centripetal force needed for a radius r is $m_p v^2/r$

$$\therefore \quad Bev = \frac{m_p v^2}{r}$$

so

$$v = \frac{Ber}{m_p}$$

The time taken for the proton to go round a complete circle is

$$T = \frac{2\pi r}{v}$$

so

$$T = \frac{2\pi r m_p}{Ber} = \frac{2\pi m_p}{Be}$$

which is independent of the proton's velocity. As the protons are accelerated they move into a path with a greater radius, and so take the same amount of time to complete the path. This allows the frequency at which the polarity of the accelerating voltage is switched to be fixed at twice the frequency of the proton's orbit.

$$\text{proton frequency} = \frac{1}{T} = \frac{Be}{2\pi m_p}$$

$$\therefore \text{ switching frequency} = \frac{Be}{\pi m_p}$$

which is typically similar to radio transmissions.

Developments

Over a period of 27 years Lawrence built a sequence of larger and larger machines, culminating in a 4.7 m machine that accelerated protons up to 30 MeV. At this energy relativistic effects start to become important and the cyclotron frequency is no longer constant.

The first measure developed to compensate for this was gradually to increase the frequency of the a.c. supply to match the increasing frequency with which the protons moved round the machine. Such machines are called **synchrocyclotrons**. The disadvantage of this technique is that you cannot accelerate a continuous stream of particles. Slow-moving ones just leaving the source require a different frequency to faster moving ones nearing the end of their acceleration cycle. The particles have to be accelerated in **bunches** that clear the machine before the next bunch can start. In all modern machines several bunches at a time can be accelerated round a large ring.

The development of the synchrocyclotron allowed energies to be pushed up to around 200 MeV. The limit then became the physical size of the magnet. This led to the development of the **synchrotron**, which uses an evacuated circular tube with many magnets placed round the circumference. As the particles are accelerated the frequency of the a.c. supply is increased, and the strengths of the magnetic fields are also increased to keep the particles moving in a ring of constant radius. This is the principle on which all modern accelerators are based.

Linear accelerators

One major disadvantage of using a ring accelerator is the amount of energy lost as electromagnetic radiation. Any charged particle that accelerates must radiate energy. In a circular path particles accelerate even if they move at constant speed. So a continual supply of energy is required just to keep the particles moving at a constant speed, to say nothing of accelerating them further. The intensity of this **synchrotron radiation** can be lethal near to the machine. Electron machines suffer more from this problem as the emitted power is $\propto 1/m^4$. However, this disadvantage can be useful in certain applications. Linear machines use a straight path and a sequence of accelerating voltages as the particles move down the line.

PRACTICE

1 In the LHC, 1232 dipole magnets each of length 14.3 m are used to bend the beam in each circuit. The magnetic flux density can be varied up to a maximum of 8.3 T.

a i Explain why the magnetic flux density needs to be changed.

ii Explain why magnets use superconductors to produce this field.

b i Calculate the angle turned through, in radian, as the beam passes through each dipole magnet so that it travels around the LHC.

ii Calculate the radius of the path of a proton when travelling through the field of one of the dipole magnets.

iii Calculate the mass of the proton when travelling at its maximum speed of $0.9999c$.

iv By what factor does the mass of a proton change as it travels from rest to its maximum speed?

v Calculate the maximum total energy of each proton in eV.

Mass at speed v:

$$m = \frac{m_0}{\sqrt{1 - \left(\frac{v^2}{c^2}\right)}}$$

MODERN ACCELERATORS

Imagine a big bag made out of netting with a large number of tennis balls in it. If you started to throw other tennis balls at the bag the chances are quite high that one of the balls that you threw would hit one in the bag. Now imagine standing 5 m away from a friend and throwing tennis balls towards each other. The object of this game is to try to get the balls to hit each other while in flight. This is obviously very much harder.

The situation is similar in particle physics experiments. **Fixed-target** experiments have a volume of stationary particles (usually the protons in a hydrogen bubble chamber or the atoms in a block of material) into which a beam of accelerated particles is fired. The chances of a collision are very high in such circumstances. **Collider experiments** take two beams of particles that have been separately accelerated and steer them together so that they are approaching from opposite directions. It is very much harder to achieve a collision in this case.

Despite their obvious disadvantages, almost all the major experiments taking place today, or planned for the near future, are collider experiments. This is because they are much more efficient users of energy.

Energy and momentum in fixed targets and colliders

When two particles of equal momentum collide head on, the total momentum in the collision is zero. After the reaction has taken place, the total momentum must also be zero. This being the case, it is possible that all the particles present after the reaction are not moving, in which case there is no kinetic energy. All the reaction energy must be locked up in the masses of the new particles. This is exactly what particle physicists want. It is a difficult and expensive process to accelerate particles, so it is vital to make sure that as much energy as possible is used to create new particles to study. This is the attraction of a collider experiment.

On the other hand, if a moving particle strikes another that is at rest, then there is some net momentum in the reaction. An equal net momentum must be present afterwards, so at least some of the particles produced must be moving. If they are moving they have kinetic energy so some of the reaction energy is not available to create new particles.

Luminosity

The rate at which an experiment can generate collisions is measured by its **luminosity** – the number of collisions per second per square centimetre of beam area. The Large Electron–Positron (LEP) collider at CERN achieved luminosities of $2.2 \times 10^{31} \, \text{s}^{-1} \, \text{cm}^{-2}$.

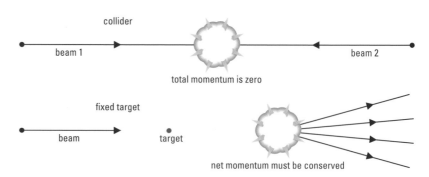

In a collider experiment all of the energy can go into creating new particles. If a fixed target is used, some of the energy must go into the kinetic energy of the moving particles, needed to conserve momentum.

In summary, the pros and cons are:

- with a fixed target, it is easy to get collisions but energy is wasted;
- with a collider, it is more difficult to get collisions, but energy is used more efficiently.

The first experiments were with fixed targets, as the technology did not exist to steer particles accurately. Now greater and greater energies are needed to test our theories, and this has forced the improvement in technology that has lead to the widespread use of colliders.

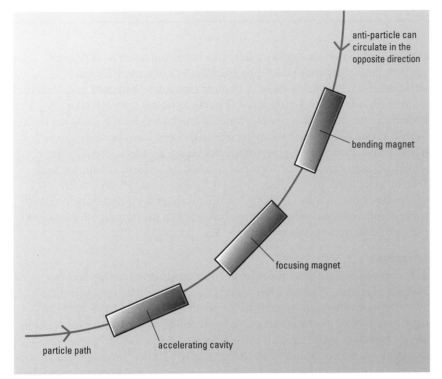

Part of an accelerator ring.

A bending magnet. This is a 10 m-long prototype for the LHC, which reached a field of 8.67 T.

Modern accelerators

Almost all modern accelerators are of the ringed collider type. The sole major exception is the Stanford Linear Collider (SLC), which uses a 3 km long linear accelerator to take bunches of electrons and positrons up to energy and then bring them into a loop and collide them at the end.

In order to accelerate and guide particles round a ring three components are needed:

1 **Radio-frequency accelerating cavities** Alternating voltages of radio frequency continually 'kick' the particles as they move from cavity to cavity.

2 **Bending magnets** These keep the particles in the same circular path as they move round the ring and are accelerated.

3 **Focusing magnets** An experimental collision requires a small bunch of particles; as the particles are deflected and accelerated mutual electrostatic repulsion tends to spread out the bunch. Focusing magnets bring them back together again.

A focusing magnet. This is a sextupole magnet in CERN's 27 km LEP electron–positron collider which operatated at above 90 GeV per beam. The LEP was used from 1989 till 2000.

The LEP accelerator at CERN used 816 focusing magnets and 1696 bending magnets round its 27 km circumference ring.

Current and proposed accelerators.

Year	Name	Laboratory	Feature	Energy/GeV	Particle	Type
1989	LEP1	CERN	Z^0 factory	45 + 45	e^+e^-	collider
1989	SLC[2]	Stanford	linear accelerator	50 + 50	e^+e^-	linear collider
1992	HERA[3]	DESY[4]	study quarks	820 + 30	p e$^-$	collider
1992	Tevatron	Fermilab	discovered top quark	900 + 900	p$\bar{\text{p}}$	collider
1996	LEP2	CERN	W^+W^- factory	80.5 + 80.5	e^+e^-	collider
2010	LHC[5]	CERN	using LEP tunnel	3500 + 3500	pp	collider
2014	LHC	CERN	upgrade to LHC	7000 + 7000	pp	collider

[1] Large Electron–Positron collider.

[2] Stanford Linear Collider.

[3] HERA, Hadron–Electron Ring Accelerator.

[4] DESY, German Electron Synchrotron (Deutsches Elektronen Synchrotron).

[5] Large Hadron Collider.

9.6

OBJECTIVES

- ionization
- electromagnetic showers
- tracking particles

DETECTING PARTICLES

It is no use having accelerators that can boost particles to very high energies, and steer them with precision into collisions with other particles, if the results of these collisions cannot be detected and recorded in some manner. In the early days of particle physics research the recording was done with photographs of tracks left by the particles in **bubble chambers**. Since then electronic devices have been developed that have revolutionized experiments by being able to record far more detail and many more reactions per second. As a result computer and electronic technology has been pushed forwards, with very useful applications in other areas. Many of the modern developments in data storage and processing have come about due to pressure from physicists involved in particle physics experiments.

Ionization

All forms of particle detector rely on the ionization of a material by charged particles. When a high-speed charged particle, for example a proton, passes through a material it will strip electrons from some of the atoms by electrostatic forces. At each collision the proton loses some energy, which is split between the electron and the ionized atom. The proton slows down. Sometimes the rate at which the particle loses energy is so great that it comes to rest inside the material and becomes absorbed. Whatever happens, there will be a trail of ionized atoms marking out the path followed by the particle.

The number of ions created per centimetre is known as the **ionizing power**, and depends on the charge of the particle and its speed through the material. Ionization obviously requires that the particle carry an electric charge – the direct detection of neutral particles is much more difficult.

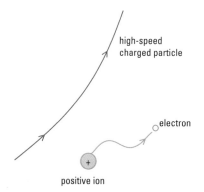

Ionization takes place whenever a charged particle passes close to an atom.

Showers

Another aspect of the manner in which electrons can interact with matter is important in the measurement of particle properties. When a fast-moving electron is sharply decelerated, or deflected by an atom, it radiates a great deal of its energy in the form of photons. This is known as **bremsstrahlung (braking radiation)** and is the main way in which X-rays are produced (see spread 8.33). In some types of particle detector, such as electromagnetic calorimeters, material is chosen specifically to make this process very efficient. As the photons produced travel through the material, each one has a good chance of converting into an electron and a positron (this process, known as **pair creation**, is discussed in more detail in spread 9.9). Both the electron and the positron produced go on to create more photons as they are deflected by atoms. Very rapidly an **electromagnetic shower** of photons, electrons, and positrons builds up. The energy of the electron that started the shower going is proportional to the distance the shower penetrates into the material.

Electromagnetic calorimeters are used to detect electrons, positrons, photons, and any neutral particle that can decay into these.

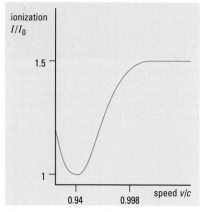

The number of ions per centimetre depends on the speed of the particle.

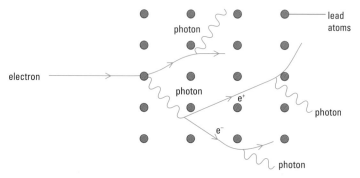

Electromagnetic showers are very useful in measuring the energy of fast-moving electrons and positrons. They can also be used to detect photons.

The bubble chamber

Bubble chambers were developed in the early 1950s based on the ideas of Donald Glasser. He realized that the ions left by a charged particle passing through a liquid could be converted into a trail of bubbles that could then be photographed. In a bubble chamber, the liquid, in a suitable vessel, is held at a temperature above its boiling point, but under pressure so that it does not boil (this is a **superheated liquid**). This is the **sensitive state** of the chamber, when it is able to record particle tracks. After a short while the pressure is released and the liquid starts to boil first in the regions where there are ions, producing bubbles that draw a trace of the ionization trail. These bubbles are allowed to grow for a time and then the whole chamber is photographed. The pressure is then restored, bursting the bubbles and returning the chamber to its sensitive state.

Most bubble chambers used liquid hydrogen. This had the advantage of providing a stock of protons within the chamber that could be used as target particles as well as detecting the reactions that were produced.

> ### 'Seeding' bubbles
>
> When a liquid boils the bubbles have to form round some 'seeds'. Normally dust or dirt in the liquid provides such seeds. Often a spot on the side of a glass containing a fizzy drink is seen to have a stream of bubbles rising from it – a rough spot in the glass is seeding bubble formation. As there is no dust in the bubble chamber and the sides are quite smooth, the seeds are the ions left by particles.

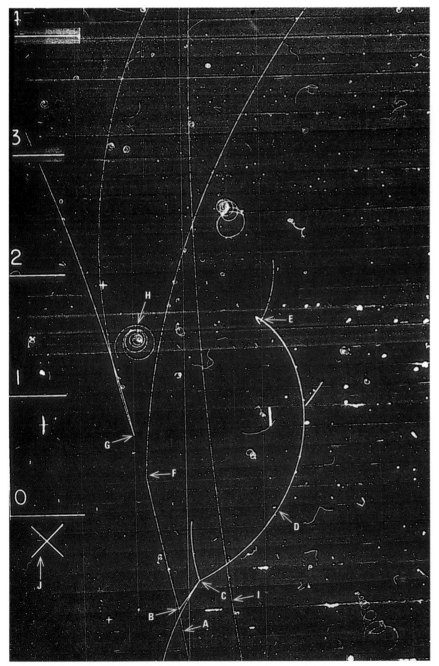

A bubble chamber photograph showing a small amount of activity.

A Track left by particle entering the bubble chamber.

B Reaction – the incoming particle has collided and reacted with a proton in the chamber's liquid hydrogen. Two new particles are produced.

C One of the new particles decays into three other charged particles.

D The chamber is in a magnetic field so the tracks are curved. Oppositely charged particles curve in opposite directions: the greater the curvature, the smaller the particle's momentum.

E This particle has in turn decayed. The sharp 'kink' in the track is due to a neutral particle being emitted that is not detected. Here there are two kinks, indicating that there have been two decays, one after the other.

F The other particle from the main interaction has decayed at this point. This time the decay has produced a charged particle, and a neutral particle that is not detected as it does not show a trace.

G At this point the neutral particle also decays into two oppositely charged particles.

H This is an electron spiral. A low-energy electron has been knocked out of one of the hydrogen atoms by a passing charged particle. As it has little momentum the track is very curved. The electron slows down in the liquid, so the track becomes even more curved, forming a characteristic spiral.

I This is another incoming particle that has passed through the chamber without reacting.

J This is a **fiducial mark** – a marking etched on the chamber window to allow positions to be determined accurately within the chamber.

THE UA1 DETECTOR – A CASE STUDY

UA1 and UA2

UA1 stands for Underground Area 1. The companion experiment UA2 also detected W and Z particles. It was constructed on the very sound reasoning that every experiment must be checked by performing another. The two experiments were designed to complement each other.

In January 1983, in a packed meeting hall at CERN, Carlo Rubbia announced the worst-kept secret in the particle physics community. After a six-year development, construction, and data-taking period the UA1 experiment had found evidence for the existence of the long suspected W exchange particles that are responsible for the action of the weak force. Such was the enormity of this discovery that Rubbia shared the Nobel Prize for physics with Simon Van der Meer (who had developed some of the technology required for the experiment) the next year – an unusually short period of time after the discovery for the awarding of a Nobel Prize.

At the time the UA1 detector was the largest single particle detector in the world. It was larger than a two-storey house and weighed as much as five jumbo jets. The gigantic machine cost £20 million and over a hundred physicists from all over the world were involved in its construction, as well as countless numbers of technicians, draughtsmen, and electronics experts.

The UA1 detector.

UA1 was designed to collect as much information as possible about all the particles produced by a collision taking place at the centre of the detector. The particles produced would stream away from the reaction point in all directions, so it was important that the detector had complete coverage. Also, the events that the team were looking for were known to be extremely rare compared to many other reactions that might take place, so the detector was designed to be 'intelligent' – very fast electronics looked at each event as it happened, to decide if it was potentially of any interest. If it was, the electronic information from the detector for that event was sent to a **computer buffer** where it was stored until it could be recorded on magnetic tape. This may seem a very complicated procedure, but consider the following information: 1200 collisions were taking place every second, of which one or maybe two were of any interest; the detector had to decide if the reaction was interesting within 4 µs or too many other reactions would be missed; writing out the information from the detector on to magnetic tape took 20 seconds for each event.

Eventually, after running the experiment 24 hours a day for six months, the UA1 team found just five events that they thought contained the W particles. By the end of the experiment's life this number had been increased to about 30, with 10 or so Z^0 events also having been observed. It is a measure of the progress made in this field that in 1992 the LEP accelerator was producing 50 000 Z^0 events per week.

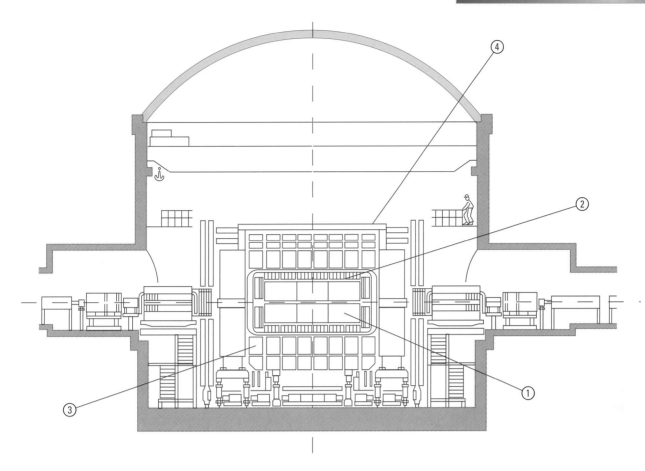

Cross-section though UA1.

1 The central part of the detector is a series of six **drift chambers**. These surround the collision point and are bathed in a magnetic field. Charged particles can be tracked through the drift chambers, and from the bending of their paths the momentum can be calculated.

2 The **electromagnetic calorimeter** is constructed from sandwiches of lead sheet and a plastic material called scintillator (see spread 8.14). Showers set up in the lead generate electrons and positrons that have to pass through the scintillator, producing brief flashes of light. The light is collected and guided along plastic tubes towards photosensitive detectors. About 100 tonnes of lead are involved in this part of the device.

3 The **hadron calorimeter** works in a similar manner to the electromagnetic calorimeter. However, it is not the photons produced by bremsstrahlung that make up the shower. Hadrons produced in the reaction collide with the atoms in sheets of iron, producing more hadrons that in turn go on to collide. This hadronic shower also produces light as it passes through the scintillator sheets between the iron layers. As with the electromagnetic shower, the distance the shower penetrates into the calorimeter indicates the energy of the initial hadron. The iron has a double purpose in that it forms part of the magnet generating the field in the central drift chamber.

4 Anything that can get through the electromagnetic and hadron calorimeters has to be either a muon or a neutrino. Being neutral, the neutrinos are not picked up at all. Muons are 200 times more massive than electrons, so for a given energy they do not have as much speed. For this reason they do not trigger electromagnetic showers and have to be detected by simple ionization in the **muon detectors**.

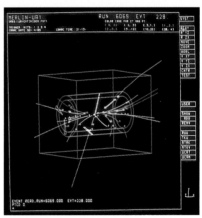

Event information was graphically displayed on computers – the successor to the old bubble chamber photograph.

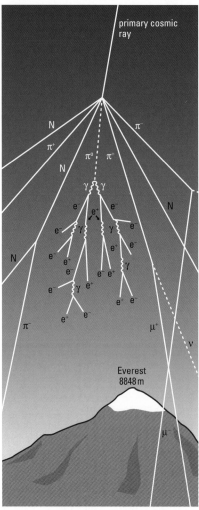

primary cosmic ray

N   π⁻

π⁺

π⁰   π⁺

N

γ   γ

N

e⁻   e⁻   e⁻

e⁻   γ   e⁺   γ   e⁻

γ   e⁻

N   e⁻   e⁺   γ   e⁻

e⁻   e⁺   e⁻   e⁻

e⁻   γ   e⁺

e⁺   e⁻

π⁻   e⁺   e⁻

μ⁺

γ   e⁺

ν

Everest
8848 m

μ⁻

Cosmic radiation in the atmosphere.

Cosmic rays in space

In space the *Apollo* astronauts occasionally used to report seeing flashes of light in their eyes while trying to sleep. The effect was caused by high-energy cosmic rays hitting their retinas.

Cloud chambers

Cloud chambers work in a similar manner to bubble chambers, except that they contain a gas that is close to condensing (a **supercooled gas**). Charged particles passing through leave a vapour trail that can be photographed. Cosmic ray experimenters took cloud chambers up in balloons and hauled them up mountains in order to get nearer to the energetic particles.

LEPTONS

Cosmic rays

In 1911–12 Victor Hess (1936 Nobel Prize winner) made a series of balloon flights to measure radiation in the atmosphere. He found that the intensity of radiation increased to five times the amount at sea level by about 5000 m altitude; hence the radiation must be coming from space. Robert Millikan coined the name **cosmic rays** for this penetrating and ionizing radiation, and believed it was electromagnetic in nature. However, by studying the intensity of radiation at different latitudes, Compton showed that the radiation was being deflected by the Earth's magnetic field and so must be charged.

Most of the cosmic rays striking the Earth are protons. Some are the nuclei of heavy elements (uranium nuclei have been detected) and there are also some very high-energy electrons. Typically they have energies between 1 and 10 GeV, but occasionally extraordinarily high-energy particles (up to 20 million GeV) are detected, smashing into protons and producing hundreds of particles. The nature and origin of these cosmic rays is still a mystery.

The muon

In the 1930s the study of cosmic rays was the 'best game in town'. Cloud chambers were used to photograph tracks left by cosmic rays. In 1936 Carl Anderson discovered cloud chamber tracks produced by a particle with the same charge as the electron, but with about 200 times the mass. One year before, Hideki Yukawa had predicted the existence of a particle of similar mass, which he needed to explain the forces between protons and neutrons in the nucleus. It did not take long to show that Anderson's particle (now known as the **muon**) could not be the particle that Yukawa was looking for, as it did not react strongly enough with nuclei. In 1947 another cosmic ray particle, the **pion**, was discovered with a mass 270 times that of the electron. This one *did* react strongly with nuclei and was Yukawa's particle. It is a nice irony that pions produced in the upper atmosphere, when protons from space hit nuclei, decay to produce the muons detected at sea level.

Since its discovery the muon has been extensively studied. It is known to be unstable, and decays with a lifetime of $2.197\,03 \times 10^{-6}$ seconds. This sort of time scale is characteristic of a weak-force decay (see spread 9.3). Muons appear to be unstable 'heavy electrons'. It is possible to create muonic atoms in which a muon replaces one of the electrons in orbit round the nucleus. The biggest mystery that remains concerning the muon is why nature needs such a particle to exist.

The tau

In 1974 Nature's generosity in giving us the electron *and* the muon overflowed when another heavy electron, the **tau**, was discovered (see spread 9.1). The discovery came from the SPEAR collider (Stanford Positron–Electron Asymmetric Ring) built on a car park at SLAC. The particle was totally unexpected. It did not take long to confirm that it was another unstable lepton. However, the tau's enormous mass ($1.78\,\text{GeV}/c^2$ – 3.5 million times the mass of the electron!) and very short lifetime ($3.3 \times 10^{-13}\,\text{s}$) mean that it is very unlikely to be seen in cosmic rays.

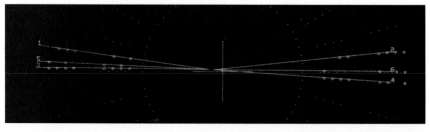

A typical tau decay. In this electronic track reconstruction a tau and an anti-tau have been produced, travelled a short distance and each decayed into three pions and a neutrino. The small gap at the centre shows the distance travelled by the taus.

Lepton number and generations

So now there are three particles, the electron, the muon, and the tau, that appear to be versions of each other with increasing mass. They all carry the same electric charge and they all react via the weak and electromagnetic forces, but not the strong force. In each case there appears to be a partner: a neutral, massless object called a **neutrino** (see spread 9.10).

In order to categorize these particles physicists invented a labelling system called **lepton number**. According to this scheme the particles are split into three **generations** of increasing charged particle mass. Each generation has its own property: electron number (L_e), muon number (L_μ), or tau number (L_τ), as shown in the table below.

Lepton generations.

1st generation ($L_e = 1$)	2nd generation ($L_\mu = 1$)	3rd generation ($L_\tau = 1$)
electron • found in atoms • involved in electrical and thermal conduction in wires • produced in β^- radioactivity	**muon** • produced in large numbers when cosmic rays collide with nuclei in the upper atmosphere – most cosmic rays that reach the surface are muons	**tau** • so far only seen in laboratories
electron-neutrino • produced in β^- radioactivity • produced in large numbers by atomic reactors • produced in huge numbers by the nuclear reactions in the Sun	**muon-neutrino** • produced by atomic reactors • produced in the upper atmosphere by cosmic rays • produced in the Sun by nuclear reactions	**tau-neutrino** • so far only seen in laboratories

Lepton number turns out to be more than simply a convenient way of tabulating the particles – it is also a conserved quantity.

Lepton number conservation

In any reaction involving leptons the total number in each generation must always remain the same (see spread 9.13).

> **Worked example**
>
> Which of the following reactions conserves lepton number?
>
>   (a) $\nu_e + \mu^- \rightarrow e^- + \nu_\mu$
>
> L_e 1 + 0 = 1 + 0
>
> L_μ 0 + 1 = 0 + 1
>
> This reaction conserves lepton number for both generations involved.
>
>   (b) $\nu_e + n \rightarrow p + \mu^-$
>
> L_e 1 + 0 $\neq$ 0 + 0
>
> L_μ 0 + 0 $\neq$ 0 + 1
>
> This reaction violates conservation of both lepton numbers, and so never happens.

PRACTICE

1 Cosmic-ray reactions that create muons take place between 15 000 m and 40 000 m above sea level.

  a How long would it take a particle moving at the speed of light to cover this distance?

  b Compare this number with the muon lifetime. How is it that so many muons can reach the surface of the Earth?

2 Which of the following reactions conserve lepton number?

  a $\nu_e + n \rightarrow p + e^-$

  b $\mu^- \rightarrow e^- + \nu_\mu + \nu_e$

  c $\nu_e + {}^{35}_{17}Cl \rightarrow e^- + {}^{35}_{18}Ar$

  d $e^- + u \rightarrow d + \nu_e$

  e $e^- + \nu_e \rightarrow \mu^- + \nu_\mu$

ANTI-MATTER

In 1932 Carl Anderson discovered a track on a cloud chamber photograph that had been left by a cosmic-ray particle with the same mass as an electron, but a positive electrical charge. He had discovered the **positron** – the anti-matter partner of the electron.

In this first cloud chamber photograph of a positron, the line across the centre is a 6 mm lead sheet. The chamber was in a magnetic field which caused the track to be curved. The positron was moving upwards on the photograph.

This may sound rather mundane compared with some of the amazing concepts proposed in science fiction such as 'negative mass' (whatever that might mean) but in many ways anti-matter is rather 'ordinary stuff'. Every particle has an anti-matter partner with the same mass, but some of its properties are opposite (electrical charge, for example). Particle physicists have been experimenting with anti-matter for many years – CERN sent a 'bottle' of anti-protons to the Expo 92 exhibition in Spain. The bottle was a magnetic device that kept the anti-protons away from the walls of a container. In 1995 CERN announced that a small experiment using LEAR (Low-Energy Anti-proton Ring) had managed to create atoms of anti-hydrogen (a negatively charged anti-proton with a positively charged positron in orbit). The overenthusiastic press predicted the advent of limitless energy based on anti-matter; in fact, nine atoms had been created in three weeks.

Anti-leptons

The **anti-muon** was discovered at virtually the same time as the muon. Like the positron it has the same mass as its matter partner but the opposite charge. All the anti-leptons carry a negative lepton number.

Matter	$L_e = 1$	e^-	$L_\mu = 1$	μ^-	$L_\tau = 1$	τ^-
		ν_e		ν_μ		ν_τ
Anti-matter	$L_e = -1$	e^+	$L_\mu = -1$	μ^+	$L_\tau = -1$	τ^+
		$\bar{\nu}_e$		$\bar{\nu}_\mu$		$\bar{\nu}_\tau$

In the table above the anti-neutrinos have a 'bar' over the top of the symbol. Since they are massless and neutral, one of the few ways to distinguish between a neutrino and an anti-neutrino is by lepton number. If a beam of neutrinos is allowed to strike a metal target then electrons are produced from the reaction:

$$\nu_e + n \quad \rightarrow \quad p + e^- \qquad (1)$$
$$L_e \quad 1 + 0 \quad = \quad 0 + 1$$

but never positrons. On the other hand, an anti-neutrino beam always produces positrons:

$$\bar{\nu}_e + p \quad \rightarrow \quad n + e^+ \qquad (2)$$
$$L_e \quad -1 + 0 \quad = \quad 0 + -1$$

Clearly other elementary particles can distinguish between a neutrino and an anti-neutrino!

β⁺ decay

Antimatter is not necessarily rare and exotic. Some nuclei adjust their n/p ratios to achieve greater stability by undergoing β⁺ decay:

$$p \rightarrow n + e^+ + \nu_e$$

The e^+ particle emitted is the positron. ν_e is the electron-neutrino (see spread 9.10).

Anti-quarks

Quarks also have their anti-quark partners. They have the opposite charge to the matter versions, and also negative **baryon number**. Baryon number plays a similar role for quarks to that of lepton number for the leptons. It is also a conserved quantity. However, in the case of the quarks it seems as if there is no difference between one generation and another – they all have the same baryon number, $B = \frac{1}{3}$ for quarks and $B = -\frac{1}{3}$ for anti-quarks.

Matter and anti-matter.

	Baryon number	Charge	Type of quark		
matter	$B = \frac{1}{3}$	charge $= +\frac{2}{3}e$	u	c	t
		charge $= -\frac{1}{3}e$	d	s	b
anti-matter	$B = -\frac{1}{3}$	charge $= -\frac{2}{3}e$	$\bar{u}$	$\bar{c}$	$\bar{t}$
		charge $= +\frac{1}{3}e$	$\bar{d}$	$\bar{s}$	$\bar{b}$

Annihilations

One thing that science fiction has been right about is the tendency for matter and anti-matter to annihilate each other. If a matter particle and an anti-matter particle of the same type come into contact, then they spontaneously convert into energy. For example, an electron and a positron can annihilate into a high-energy photon, which is extremely 'unstable' and will rapidly convert back into a pair of particles – one of matter and one of anti-matter. However, because the photon is a 'clean slate' (in other words it does not have any charge, baryon number, lepton number, etc.), any pair of particles can emerge:

$$e^- + e^+ \rightarrow \gamma \rightarrow e^- + e^+$$
$$e^- + e^+ \rightarrow \gamma \rightarrow \mu^+ + \mu^-$$
$$e^- + e^+ \rightarrow \gamma \rightarrow \tau^+ + \tau^-$$
$$e^- + e^+ \rightarrow \gamma \rightarrow q + \bar{q}$$

In the last reaction q can stand for any of the quarks.

Annihilations can take place between any particle of matter and its anti-matter partner. However, electromagnetic radiation is not always produced. Quarks interact with each other far more readily via the strong force, so if a quark and an anti-quark annihilate a burst of strong-force energy is produced in the form of the strong-force exchange particle:

$$q_a + \bar{q}_a \rightarrow \text{gluon} \rightarrow q_b + \bar{q}_b$$

The gluon will convert back into quarks shortly after it has been created, but they need not be the same quarks as before.

A beautiful photograph showing pair creation in a bubble chamber. Two photons have separately converted into $e^- e^+$. In the event towards the top some of the photon's energy has been absorbed by an electron in one of the hydrogen atoms causing it to form the long straight downward track. The $e^- e^+$ consequently have less energy so their tracks curve more, creating the spirals. The lower event is an ordinary pair creation.

PRACTICE

1 Study the photograph of a positron at the top of the previous page. How can you tell that the particle is travelling upwards in the photograph?

2 What is the quark composition of an anti-proton? Show that it has negative charge.

3 What is the quark composition of an anti-neutron? Will anti-neutrons be experimentally different to neutrons in the way they behave?

4 Write down reactions similar to those in equations 1 and 2 that would apply for muon neutrinos and anti-muon neutrinos reacting with matter. Show that the reactions conserve lepton number.

5 The muon decays into an electron and two other neutral particles.

   a Use lepton number conservation to decide what the other two particles are.

   b Write down a similar decay for an anti-muon.

- β decay and neutrinos

- reaction rate and decay

Wolfgang Pauli.

Explaining the electron energies

If the neutron decays into just a proton and an electron, then the amount of energy produced in the reaction is shared between just these two particles. Once momentum conservation is taken into account as well, it turns out that there is only ever one way to share the energy between the two particles. The electrons and the protons must always have the same energy in that reaction (of course the decays of different isotopes will give rise to different fixed energies). If there are three particles produced in the reaction, then conservation of energy and momentum cannot give rise to fixed values, and so a spread of electron energies is expected for each isotope.

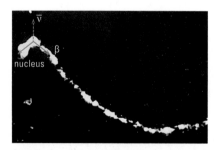

In this cloud chamber photograph an atom of He⁶ has entered from the top. One of the neutrons in the nucleus has decayed. The short thick track to the left is the recoiling nucleus, the lighter curved track to the right is the electron. The fact that they do not start in a line indicates that there is a third particle that does not leave a track (otherwise momentum would not be conserved).

NEUTRINOS

Neutrinos are probably the most bizarre of the fundamental particles. They have no electric charge, an extremely small mass (if any at all), and can only interact by the weak force. Experiments have to be very carefully designed to catch any sign of them at all. Despite their insubstantial nature there are vast numbers of neutrinos in the universe – up to $10^8\,\mathrm{m^{-3}}$. The majority are swimming about the universe, left over from the Big Bang. Some come in a continual torrent from the nuclear reactions in the Sun, although for some reason we do not detect as many on Earth as our theories suggest we should.

β decay and the discovery of the neutrino

The first person to suggest that neutrinos might exist was the Austrian physicist Wolfgang Pauli. In the late 1920s the study of nuclear β decay revealed a mystery. The nucleus involved decayed when one of its neutrons converted into a proton and an electron. The electron was produced moving at very high speed, so it left the nucleus and was detected by Geiger counters (for example). However, measurements of the electron's energy revealed that a sample of a radioactive isotope would produce a *range* of electron energies.

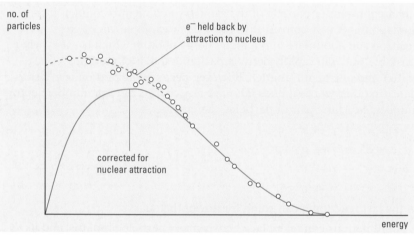

Graph of energy curve for electrons from the β decay of a typical isotope.

If the decay involved just the electron and the proton (a two-body process), then the electron's energy should be fixed at one value. Pauli took the bold step of suggesting that there might be a third particle involved. Every time an electron was emitted with only a little energy the third particle had taken a lot of energy, and vice versa. The third particle would have to be neutral (otherwise electrical charge would not be conserved) and also of very small mass (otherwise the energy curve of the electrons would be a different shape). In 1933 Enrico Fermi produced a more complete theory of β decay using quantum mechanics and incorporated Pauli's particle, naming it the **neutrino** ('little neutral one'). The full β-decay equation then becomes

$$n \rightarrow p + e^- + \bar{\nu}_e$$

Notice that it is the **anti-electron-neutrino** involved in this reaction. This is in order to conserve lepton number (see spread 9.13).

Discovery of the neutrino

Although it provides a neat solution to the problem of β-decay energies, the existence of the neutrino had to be proved experimentally before many physicists would take it seriously. This was not done until 1956. As neutrinos are so weakly interacting, it takes an experiment that is dealing with a very large number of them to stand any chance of detecting them. The physicists who devised the experiment, Clyde Cowan and Fred Reines, built a device to detect the flood of neutrinos produced by the Savannah River reactor in South Carolina. The idea was to look for evidence of the reverse reaction to β-decay:

$$p + \bar{v}_e \rightarrow n + e^+$$

They used a large tank of water (with some cadmium chloride dissolved in it) to provide the protons, and looked for evidence of gamma rays being produced within a very short time of each other. The first burst of gamma rays would come from the positron annihilating with an electron in the water, the second would be due to the neutron being absorbed by an atom of cadmium which then emitted the radiation. They sent a telegram to Pauli on 14 June 1956 to announce the discovery of gamma-ray bursts 5.5 μs apart. Pauli had waited 20 years to find out that he was right.

Two types of neutrino

In the late 1950s physicists started to suspect that there were two different types of neutrino. The first hint came from the study of muon decay.

By measuring the energy distribution of the electrons produced, physicists established that muon decay was at least a three-body process. If the decay only involved two particles then the electron's energy should be fixed. However it was clear that the electrons had a range of energies and so more than two particles must be involved.

At this stage, the idea of separate lepton numbers for the generations was not firmly established, although conservation of total lepton number was (i.e. the total number of leptons of all types was conserved). The easiest way of conserving total lepton number and electric charge was a decay of the form:

$$\mu^- \rightarrow e^- + v_1 + \bar{v}_2$$

but there was no evidence to decide if the neutrinos were of the same type.

The matter was resolved in a classic experiment at Brookhaven in the 1960s. A beam of neutrinos produced from muon decay (i.e. muon-neutrinos) was allowed to strike a dense target. Over a period of 25 days some 10^{14} neutrinos struck the target. During this time 51 muons were detected coming from the target. No electrons were seen. This implied that the neutrinos produced by muon decay were capable of reacting with matter to produce muons, but not electrons. In contrast electron-neutrinos readily produced electrons but not muons. This experiment established the difference between the muon-neutrino and electron-neutrino, and so put the existence of separate lepton numbers for each generation on a secure experimental footing.

Neutrino mass

There is now good evidence that neutrinos have a small mass. This is a matter of some interest to cosmologists – as neutrinos exist in such huge numbers, even a very small neutrino mass could be enough to provide the gravity to slow the expansion of the universe and bring it back to a '**Big Crunch**'. Astronomical and cosmological data places an upper limit on the total mass of all these neutrinos, which must be < 0.3 eV.

The experiment that first detected the neutrino was called Project Poltergeist, in reflection of the neutrino's elusive nature. Here Cowan and Reines check the main control panel.

Reaction rate

Of the 10^{20} anti-neutrinos produced per second by the Savannah River reactor, one reacted with the equipment every twenty minutes.

Muon decay

The muon decay is actually

$$\mu^- \rightarrow e^- + v_\mu + \bar{v}_e$$

Muon-neutrinos

The best measurements from CERN now suggest that the ratio of electrons to muons produced by muon-neutrinos is 0.017 ± 0.05 – completely consistent with being zero. Leon Lederman was awarded the Nobel Prize in 1988 for his work on this experiment.

PRACTICE

1 Given that the volume of the observable universe is about 10^{31} cubic light years, what would be the total mass of neutrinos in the universe (in kg) if each neutrino had a mass of 30 eV/c^2?

   (1 light year = 9.46×10^{15} m, mass of the Sun = 2×10^{30} kg.)

2 By comparing the mass of the neutron to that of the proton, calculate the maximum energy with which an electron can be emitted in β decay (you may assume that the anti-neutrino takes no energy).

Murray Gell-Mann (top) and George Zweig (bottom) invented the idea of quarks to explain the properties of hadrons. Zweig was convinced that there were real objects inside the quarks, but it is unclear if, at first, Gell-Mann thought of them as anything other than a helpful mathematical idea.

Protons as baryons

As protons are the lowest mass particle in the baryon family, all baryons must eventually decay into protons. In turn, protons may decay (see spread 9.17) which would violate baryon number conservation. However, the proton lifetime is thought to be at least 10^{32} years.

HADRONS AND QUARKS

Why would anyone give the small objects that might exist inside protons, and other hadrons, the name 'quarks'? The man who came up with the idea was one of the two physicists who first proposed that they might exist, Murray Gell-Mann. The name comes from a line in *Finnegan's Wake* – 'three quarks for Muster Mark' (you see, even physicists read James Joyce). According to Gell-Mann the name should be pronounced like 'quork', not 'quark'. He suggested this name because it does not mean anything. Previously, every time a fundamental particle had been given a name with some meaning – such as atom, meaning indivisible – it turned out to be wrong!

Hadrons

Any particle that contains quarks is called a **hadron**.

According to the best theories at the moment, it is impossible for a quark to exist on its own. Quarks must always be bound into particles by the strong force that exists between them. As it happens there are only three stable combinations that the strong force will allow.

Combinations of quarks and anti-quarks.

Combination	Symbol	Name
three quarks	qqq	baryon
three anti-quarks	$\bar{q}\bar{q}\bar{q}$	anti-baryon
a quark and an anti-quark	$q\bar{q}$	meson

Baryons, anti-baryons, and **mesons** were discovered long before quarks and their properties were well known. The success of the quark idea was to draw together and explain their properties in terms of the properties of the quarks inside them.

Worked example

Into what category of hadron can the following particles be placed?

1 uud 3 quarks ∴ a baryon with charge +1 (in fact the proton)

2 udd 3 quarks ∴ a baryon with charge 0 (in fact the neutron)

3 $u\bar{d}$ quark + anti-quark ∴ a meson with a charge of +1 (this is one of the lightest mesons and is called a π^+ (pi plus))

4 $\bar{u}d$ quark + anti-quark ∴ a meson with a charge of –1 (a π^-)

5 $u\bar{u}$ quark + anti-quark ∴ a meson with a charge of 0 (a π^0)

The three pi particles have very similar masses, and belong to a family known as the pions.

Baryon number

Baryon number is a property of hadrons that was discovered by studying various reactions taking place in particle accelerators. It is possible to give every hadron a value of baryon number (B) based on a simple scheme:

any baryon $B = 1$

any anti-baryon $B = -1$

any meson $B = 0$

e.g. protons and neutrons have $B = 1$, an anti-proton has $B = -1$, all pions have $B = 0$. In all reactions baryon number is found to be a conserved quantity (see spreads 9.12 and 9.13). The baryon number of the various hadrons can be explained if the quarks are given values of baryon number according to the scheme

all quarks $B = \dfrac{1}{3}$

all anti-quarks $B = -\dfrac{1}{3}$

A few moments of consideration should convince you that this will produce the required baryon numbers of the hadrons. The baryon number of a hadron is effectively the number of quarks minus the number of anti-quarks inside it. Baryon number is important as it is a conserved quantity, like lepton number.

Strangeness

In the early 1950s particles with large masses and long lifetimes started to emerge in cosmic-ray experiments. In 1954 Gell-Mann, Nishijima, and Nakone independently categorized these particles and gave them a new property – **strangeness**. The odd lifetimes of the strange particles could then be explained by assuming that strangeness was conserved in strong-force reactions, but not by the weak force (see spread 9.12). Faced with the need for an arbitrary choice, physicists chose the K^+ meson to have strangeness $S = +1$, and categorized the other particles relative to this. For example in the reaction

$$p + p \rightarrow p + K^+ + \Lambda$$

which happened often enough to indicate that it was a strong-force reaction, strangeness was conserved, so

$$p + p \rightarrow p + K^+ + \Lambda$$
$$S \quad 0 + 0 \quad = \quad 0 + 1 \quad + ?$$

showing the Λ to have $S = -1$. In this way all the strange particles could be given S values.

It is now understood that the strangeness of a baryon indicates the presence of strange quarks within it. An unfortunate consequence of the choice of the K^+ meson to have $S = +1$ is that the strange quark (s) has $S = -1$, but the choice is arbitrary, so not really important. The anti-strange quark has $S = +1$.

The eightfold way

Between 1960 and 1961 Gell-Mann and Yuval Ne'eman published a full classification of hadrons that is known as the **eightfold way** (after the Buddhist path to enlightenment). The cornerstone of this classification is to divide the hadrons into families of nearly equal mass and then to plot them on a graphical diagram which includes their strangeness, mass, and electrical charge.

The eightfold way patterns can now be explained by the various combinations of the three lightest quarks – u, d, and s (see practice questions below). Now, with more energy in accelerators, physicists are able to create the other more massive quarks and all their particles (which do not appear on this diagram).

Quarks and energy levels

The eightfold way pattern gives some particles the same quark content (e.g. the Δ^+ and the proton). However, these particles have different masses. The reason for this is that the hadrons have energy levels inside them (just as an atom does) and the quarks can be arranged in these levels differently. An arrangement with higher energy gives a hadron with more mass. Some arrangements, like uuu, ddd, and sss are not possible in the lower mass versions, because of the rules about how the quarks can fit into the energy levels.

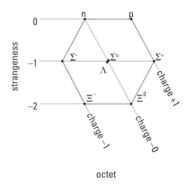

octet

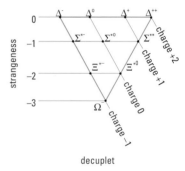

decuplet

The eightfold way classification of particles. Notice that some of the particles (e.g. the Δ^+ and the proton) have the same quark combinations; however, they have different masses.

PRACTICE

1 What are the values of charge, baryon number, and strangeness of the following particles: uus, $d\bar{s}$, uuu, dss, sss?

2 What is the quark content of each of the following particles?

   a K^+ meson, strangeness = +1, mass = 493 MeV/c^2

   b K– meson, strangeness = –1, mass = 493 MeV/c^2

   c Λ baryon, strangeness = –1, mass = 1.12 MeV/c^2

   d Ξ^+ baryon, strangeness = +2, mass = 1.32 GeV/c^2

   e Δ^+ baryon, strangeness = 0, mass = 1.23 GeV/c^2

3 Write down in a sequence along one line all the possible combinations of the u and d quarks that give rise to baryons. On the line beneath, write down all the baryon combinations of u, d, and s quarks *using one s quark only*. On the next line do the same thing including two s quarks. Now, for each combination write down the charge and strangeness of the particle. Compare the result with the baryon pattern in the eightfold way diagram above. What is the quark content of the Ω^-?

O B J E C T I V E S

- conservation of charge
- conservation of baryon number
- strangeness and conservation

Anything that is not expressly forbidden by Nature is compulsory.

Richard Feynman

CONSERVATION LAWS – 1

The often quoted little rule on the left is an example of Richard Feynman's ability to put his finger on very important truths. He was speaking in the context of particle physics, and drawing attention to the fact that as much can be learnt from the things that do *not* happen as from those that do. His point was that any reaction that you can think of that ought to be possible (i.e. does not violate known conservation laws) should happen in your experiments. If it does not, then there is some rule of physics that you were not aware of that is preventing the reaction taking place. The best way to find this rule is to study the reactions that *are* happening alongside those that are not. When this is done, the situation can normally be explained in terms of some conservation law.

What is a conservation law?

Conservation laws place limits on the events that can take place in Nature. The two most well known conservation laws, those of energy and momentum, state that whenever objects collide, explode, stick together, or repel each other, the total energy and momentum in the universe must remain the same. The essence of a conservation law is to compare the total amount of some quantity, such as energy, before an event has taken place with the total amount afterwards. If the total has not changed, then the quantity is conserved. Conservation of energy and momentum are important in particle physics (always remembering to use the relativistic equations), but there are many other rules that introduce some new physics. By studying the quantities that are conserved in reactions and decays, particle physicists can learn more about the fundamental forces driving these processes.

Conservation of charge

From its humble beginnings in relation to Kirchhoff's first law, conservation of charge has come a long way. It is now one of the most important conservation laws in particle physics. It is this law that prevents the following reactions from taking place:

$$
\begin{array}{lccc}
& p + p & \rightarrow & p + n \\
Q & 1 + 1 & \neq & 1 + 0
\end{array}
$$

$$
\begin{array}{lccc}
& p + n & \rightarrow & p + n + \pi^+ \\
Q & 1 + 0 & \neq & 1 + 0 + 1
\end{array}
$$

However, some surprising reactions can take place, provided there is enough energy involved:

$$
\begin{array}{lccc}
& p + p & \rightarrow & p + p + p + \bar{p} \\
Q & 1 + 1 & = & 1 + 1 + 1 + (-1)
\end{array}
$$

$$
\begin{array}{lcccccc}
& e^+ & + & e^- & \rightarrow & \mu^+ & + & \mu^- \\
Q & (+1) & + & (-1) & = & (+1) & + & (-1)
\end{array}
$$

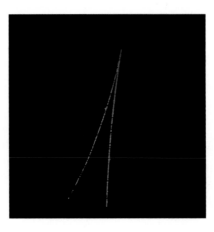

A beautiful example of conservation of charge is shown in this bubble chamber photograph. A photon (γ), which does not leave a track, converts into an electron–positron pair in the bubble chamber. The tracks leaving the conversion point must be of opposite charge as they are bending in opposite directions in the chamber's magnetic field.

Conservation of baryon number

Baryon number was discovered when physicists noticed that some reactions that conserved electrical charge were not taking place:

$$p + p \nrightarrow p + \pi^+ \tag{1}$$

$$p + n \nrightarrow p + \pi^0 \tag{2}$$

$$p + p \nrightarrow \pi^+ + \pi^+ \tag{3}$$

It seemed that the total number of 'proton-like' objects must remain the same. Pions, which can appear in any number given enough energy, are not proton-like. Neutrons are. This suggested a simple scheme in which the proton and neutron had baryon number $B = 1$, and the pions had $B = 0$. Given this scheme it is easy to show that reactions 1–3 do not conserve baryon number:

$$\begin{array}{cccc} & p + p & \nrightarrow & p + \pi^+ \\ B & 1 + 1 & \neq & 1 + 0 \end{array}$$

$$\begin{array}{cccc} & p + n & \nrightarrow & p + \pi^0 \\ B & 1 + 1 & \neq & 1 + 0 \end{array}$$

$$\begin{array}{cccc} & p + p & \nrightarrow & \pi^+ + \pi^+ \\ B & 1 + 1 & \neq & 0 + 0 \end{array}$$

Baryon-number conservation indicates that the fundamental forces must all keep the net number of quarks (i.e. quarks minus anti-quarks) constant.

Strangeness – sometimes conserved, sometimes not

Gell-Mann, Nishijima, and Nakone introduced strangeness to explain the odd behaviour of strange particles. Strange particles were first observed in events in which cosmic-ray protons hit protons inside a sheet of lead, e.g.

$$p + p \rightarrow p + K^+ + \Lambda$$

$$p + p \rightarrow p + p + K^+ + K^-$$

Baryon number conservation shows that the K^+ and K^- must be mesons, but the lambda (Λ) is a baryon. These reactions, and others like them, are quite easily produced in accelerators, suggesting that they are caused by the strong force. This is another indication that the strange particles must be hadrons (only quarks feel the strong force). What was strange about them was their mass and their lifetimes. The lambda has a mass of $1.115 \, \text{GeV}/c^2$, which is slightly less than the non-strange Δ^0 baryon ($1.23 \, \text{GeV}/c^2$). However, the lambda's lifetime is $2.6 \times 10^{-10} \, \text{s}$, which is enormous compared to $10^{-25} \, \text{s}$ for the Δ^0. The Δ^0 lifetime is typical for a strong decay (see spread 9.3), but lambda lifetime is more like that produced by a weak decay. Why should the strange baryons be produced by the strong force, but decay by the weak force? After all, the two decays look very similar:

$$\begin{array}{lll} \Delta^0 \rightarrow p + \pi^+ & \tau \sim 10^{-25} \, \text{s} & \text{strong decay} \\ \Lambda \rightarrow p + \pi^- & \tau = 2.6 \times 10^{-10} \, \text{s} & \text{weak decay} \end{array}$$

The answer is that strong force decays must conserve strangeness. There is no strange baryon with less mass than the Λ for it to decay into, so it cannot decay and conserve strangeness. Its only choice is to decay into non-strange particles, and that can only happen with the longer lifetime of the weak force.

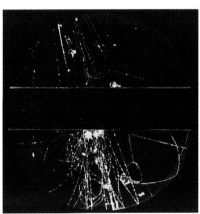

This is the picture of the first strange particle observed. The thick black band across the chamber is a sheet of lead. A cosmic-ray proton coming from above on the picture has interacted in the lead, producing a K^0 which, being neutral, has left no track. The K^0 has decayed into a π^+ and a π^- which have produced the characteristic 'V' shape.

OBJECTIVES

- heavy hadrons and quarks
- conservation laws at the quark level
- using conservation laws
- the Ω^- particle

CONSERVATION LAWS – 2

A balance in all things

By the late 1960s the quark model was becoming established. Combinations of the u, d, and s quarks could account for all the hadrons that were known. However, in 1974 a new meson was discovered and with it the **charm** (c) quark. For a short while the world of the elementary particles was very satisfying. There were four quarks (u, d, s, c) and four leptons (e^-, ν_e, μ^-, ν_μ) – a very appealing symmetry.

However, even as 1974 was drawing to a close the first hints that the pattern was about to be broken came. Then in 1975 a team of physicists working at the Stanford Positron–Electron Asymmetric Ring (SPEAR) electron–positron collider at the SLAC accelerator announced that they had seen evidence for a fifth type of lepton. The **tau** (τ), as it is now known, is more massive than the proton at 1.78 GeV/c^2, and with it goes the tau-neutrino (ν_τ), making six leptons in total.

As soon as the tau's discovery was confirmed in other experiments, physicists began to suspect that there were more quarks to be found to restore the balance between quarks and leptons. On the 30th June 1977 Leon Lederman's team at Fermilab announced the discovery of another new meson and with it the **bottom** (b) quark. There was then a long wait for the sixth quark. Finally in 1995 Fermilab announced the first reasonable experimental evidence for **top** (t) quark production. Once again the balance between the leptons and the quarks has been restored.

More massive hadrons

With the c, b, and t quarks come the properties **charm** (C), **bottomness** (B), and **topness** (T) into particle physics. Like strangeness, these properties influence how particles react with each other and decay. All are conserved by the strong and electromagnetic forces, but not by the weak force. A particle's properties can be used to work out its quark content according to the scheme shown in the table on the left.

Of course the anti-quarks have the opposite values, e.g. $\bar{c}$ has C = –1 etc.

The assignment of –1 strangeness to the strange quark is an historical accident which forces –1 bottomness on the b quark.

Quark properties. S, strangeness; C, charm; B, bottomness; T, topness.

Quark	S	C	B	T
u	0	0	0	0
d	0	0	0	0
s	–1	0	0	0
c	0	+1	0	0
b	0	0	–1	0
t	0	0	0	+1

Conservation laws at the quark level

It is instructive to examine some hadron reactions at the quark level:

$$p + p \rightarrow p + n + \pi+ \qquad (1)$$
$$uud \quad\quad uud \quad\quad uud \quad\quad udd \quad\quad u\bar{d}$$

Notice that the total number of u quarks has not changed, although they have moved about between particles. However d and $\bar{d}$ quarks have been created. This is the reverse of the annihilation reactions discussed in spread 9.9. The energy of the reaction has been used by the strong force to produce the mass of a d and $\bar{d}$. The strong force can only produce a quark in company with its anti-quark – so the net number of quarks must always remain the same. This is why baryon number is conserved.

$$p + p \rightarrow p + \Lambda + K+ \qquad (2)$$
$$uud \quad\quad uud \quad\quad uud \quad\quad uds \quad\quad u\bar{s}$$

This is a very similar reaction. In this case the energy has been used to materialize an $s\bar{s}$ quark combination. Notice how the need for the strong force to materialize both the s and the $\bar{s}$ automatically forces conservation of strangeness in strong-force reactions.

Using the conservation laws

Consider the following reaction in which X is a newly discovered particle:

$$K^- + p \rightarrow K^0 + K^+ + X$$

What can be deduced about the properties of X by applying the various conservation laws?

Firstly, conservation of electric charge shows that X must carry a charge of –1:

$$K^- + p \rightarrow K^0 + K^+ + X$$
$$Q \;\; (-1) + (+1) = 0 + (+1) + \;?$$

so ? must be (–1).

Secondly, conservation of baryon number shows that X must be a baryon:

$$K^- + p \rightarrow K^0 + K^+ + X$$
$$B \;\; 0 + (+1) = 0 + 0 + \;?$$

so ? must be 1, indicating that X is a baryon.

The next thing to decide is whether strangeness is conserved in this reaction. As the reaction comes about due to the interaction of a K meson and a proton, it is highly likely that the strong force is playing a significant part, in which case, strangeness is a conserved quantity:

$$K^- + p \rightarrow K^0 + K^+ + X$$
$$S \;\; (-1) + 0 = (+1) + (+1) + \;?$$

so that ? is –3. X is a very strange baryon indeed. In point of fact X is the Ω^- particle, and its existence was predicted by Murray Gell-Mann from his eightfold way of organizing particle properties.

The Ω^-

Gell-Mann suggested that the Ω^- would be massive ($1.67\,\text{GeV}/c^2$) and have a very long lifetime (0.8×10^{-10} seconds). It is instructive to see why the Ω^- has such a long lifetime, as on the face of it there are other strange particles with less mass, so it should decay via a rapid strong-force decay. A possible decay for the Ω^- would be

$$\Omega^- \rightarrow \Xi^- + \overline{K}^0$$
$$S \;\; -3 = (-2) + (-1)$$

conserving strangeness and so open to the strong force.

This decay is very similar to an equivalent one for the Δ^{++} baryon:

$$\Delta^{++} \rightarrow p + \pi^0$$

which has a lifetime of the order of 10^{-25} seconds. Compared to this the Ω^- takes an eternity to decay. The reason is that the strong-force decay is impossible here. However, this time it is not for any subtle reason to do with conservation laws. The reason is more obvious than that. If the masses of the various particles concerned are examined you find that

Ξ^-	mass = $1.197\,\text{GeV}/c^2$
$\overline{K}^0$	mass = $0.498\,\text{GeV}/c^2$
	total mass = $1.692\,\text{GeV}/c^2$
Ω^-	mass = $1.672\,\text{GeV}/c^2$

The particles into which you might expect the Ω^- to decay are more massive than it is – the decay cannot take place as there is no energy advantage involved. Gell-Mann was aware of this when he predicted the mass of the Ω^-, so he was able to suggest that the strong-force decay was not open to it, and that it would have to 'wait' for the much slower weak-force decay:

$$\Omega^- \rightarrow \Lambda + K^- \quad \text{or} \quad \Omega^- \rightarrow \Xi^0 + \pi^- \quad \text{or} \quad \Omega^- \rightarrow \Xi^- + \pi^0$$

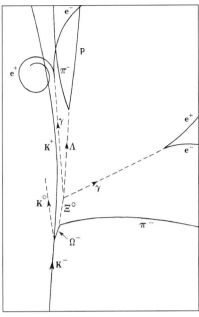

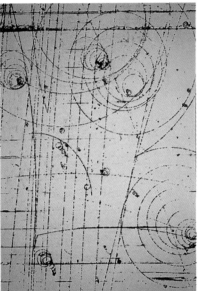

This is the first picture taken of the Ω^- particle. The experiment involved firing a beam of K^- particles into the 200cm hydrogen bubble chamber at Brookhaven. This picture caused a sensation when it was published in 1964, as the discovery of the Ω^- showed that the ideas behind the eightfold way were correct. Piecing together the tracks is a very complicated process because of the number of neutral particles involved. The diagram shows what the main tracks represent.

PRACTICE

1 Use the following information to deduce what you can about the particle X.

$$X \rightarrow \mu + \nu_\mu \quad \tau = 1.2 \times 10^{-8}\,\text{s}$$

2 Why are the proton and the electron stable particles?

EVIDENCE FOR QUARKS

The properties of hadrons

The first people to suggest that quarks might exist inside hadrons were Murray Gell-Mann and George Zweig in 1964. They independently came up with the idea that the properties of the hadrons known at the time could be explained as combinations of the properties of three smaller objects combined in various ways. These objects are now known as the u, d, and s quarks. While the idea was very elegant and of some interest to theoreticians, it did not gain very widespread acceptance as there was no experimental evidence to suggest that protons, for example, had any smaller objects inside them.

Deep inelastic scattering

The reason that no evidence had been found for objects inside hadrons, such as protons, was that nobody had looked at them with a powerful enough 'microscope' yet. The situation was starting to change even as Gell-Mann and Zweig published their ideas. In the late 1960s SLAC had been developing a machine that was capable of accelerating electrons up to 20 GeV energy. These electrons were then collided with a proton target. As a result of these collisions the electrons scattered off at some angle to the target, and the detectors used were able to measure this angle.

What the results showed was that at low energies the electrons collided 'softly' with the protons and scattered away at quite small angles; the protons simply recoiled from the collision. However, once the energy of the electrons passed a certain minimum value they started to see much 'harder' collisions – the electrons scattered off at greater angles and the protons shattered into a stream of hadrons rather than simply recoiling. This effect became known as **deep inelastic scattering**.

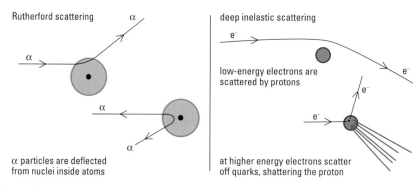

Deep inelastic scattering is a modern version of an old principle (Rutherford scattering) – using one particle to probe the structure of another.

A careful study of the results, notably by Bjorken and Feynman in 1968, suggested that the protons contained three very small charged objects. At high energies the de Broglie wavelengths of the electrons (see spread 8.6) were small enough to 'resolve' the very small objects inside the protons – in effect the electrons were starting to diffract off the quarks inside the protons. As a result, one of the quarks was kicked away from the other two and this shattered the proton into a stream of hadrons.

At lower energies the de Broglie wavelength was only the same as the size of the whole proton, and so much smaller diffraction effects were taking place.

The idea is very similar to that of the Rutherford scattering experiment with α particles that revealed the presence of the nucleus inside the atom.

The SLAC results have since been refined using neutron targets and also neutrino beams. The physicists involved have been able to show that the three objects have the charges expected of the u and d quarks, and in the correct ratios. For this pioneering work, which really started the general acceptance of the quark idea, the team leaders Friedman, Taylor, and Kendall were awarded the Nobel Prize in 1990.

Partons

Feynman named the objects inside the proton **partons**. For a while it was not obvious that the partons were the same as Gell-Mann's quarks. Indeed there is a subtle difference as the gluons (see spread 9.16) that exist inside hadrons to carry the strong force between the quarks are also counted as partons.

The November revolution

When Gell-Mann and Zweig originally presented their ideas, only three types of quark were required to explain all the known hadrons. On 11 November 1974 two teams of physicists from America announced the discovery of a new meson. The team from SLAC named the new meson the ψ (psi) but the Brookhaven name was the J meson. To this day it is known as the J/ψ (jay/psi). At the time this was a remarkable discovery as the new meson had a mass of 3.1 GeV/c^2 (greater than any other known hadron) and a lifetime that was 2000 times longer than most other hadrons.

Within a very short time the new meson had been 'explained' as the charm/anti-charm meson (c$\bar{c}$) – a fourth quark had been discovered. Any lingering doubts about this interpretation went away when in 1975 and 1976 the charmed mesons D$^+$ (c$\bar{d}$) and D^0 (c$\bar{u}$) were discovered. Samuel Ting and Burton Richter were awarded the 1976 Nobel Prize for the J/ψ discovery.

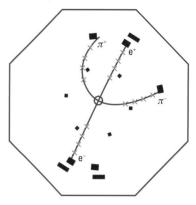

The first ψ event was named after the pattern of particles on this electronic 'photo'.

Jets

Some very nice evidence for the existence of quarks came up in 1978 from the Positron–Electron Tandem Ring Accelerator (PETRA) at the German laboratory DESY (Deutsches Elektronen Synchrotron). This machine was colliding electrons and positrons by accelerating them in opposite directions. When the collision energies rose to about 15 GeV per beam the physicists started to observe hadrons being produced after the collisions. These particles showered out from the collision point in two very narrow cones in opposite directions. Since then these jets of particles have been seen in a variety of experiments and at different energies.

The quark model provides a very neat explanation for these results. When the electron and positron collide they annihilate each other into a burst of electromagnetic energy (a very high-energy photon). This energy is then able to rematerialize back into particles, provided they are of opposite electrical charge (to conserve electric charge). As a result, an electron and a positron might reappear:

$$e^- + e^+ \rightarrow e^- + e^+$$

or a pair of muons might be produced:

$$e^- + e^+ \rightarrow \mu^+ + \mu^-$$

However, if the energy available is great enough, then a quark and its anti-quark might be produced, for example

$$e^- + e^+ \rightarrow u + \bar{u}$$

The quark and the anti-quark are produced moving in opposite directions with the same speed (to conserve momentum). When they get about a proton's diameter away from each other the strong force between them produces a shower of quarks that combine to form the jets of hadrons. This effect (known as **dressing the quarks**) is very complex and not well understood. However, the particle jets do follow the original path of the quarks. The original quarks are never seen directly – they are lost among the others produced when the dressing takes place.

> **A date to remember**
>
> Because it coincided with the anniversary of the Russian revolution, the J/ψ discovery is known as the **'November Revolution in physics'**.

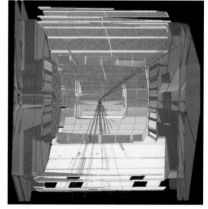

A typical jet event reveals the presence of quarks. This computer image was obtained with the DELPHI detector at CERN.

PRACTICE

1 What is the momentum in kg m s^{-1} of an electron moving with an energy of 30 GeV? What is the de Broglie wavelength of such an electron? Compare this wavelength with the typical size of an atom, the typical size of a nucleus, and the typical size of a proton, and comment on the comparison.

2 The Rutherford scattering experiment used naturally occurring α particles of 10 MeV energy. What is the momentum of such a particle? Calculate the de Broglie wavelength of 10 MeV α particles. Compare this wavelength to the typical size of an atom and the typical size of a nucleus (mass of α particle = 6.64 × 10^{-27} kg).

3 Which of the following four reactions can take place? (Explain your answer.)

a $\nu_e + d \rightarrow e^- + u$

b $\nu_e + u \rightarrow e^- + d$

c $\bar{\nu}_e + d \rightarrow e^+ + u$

d $\bar{\nu}_e + u \rightarrow e^+ + d$

Explain how you could use a neutrino beam and an anti-neutrino beam to count the number of u and d quarks inside a proton.

EXCHANGE FORCES – 1

Standard model

The quark–lepton picture of matter and the quantum theory of the fundamental forces are together known as the standard model of particle physics.

In the macroscopic world in which everyone lives, it is very easy to think of forces as being pushes or pulls. On becoming slightly more knowledgeable about physics you will have started to think in terms of force fields and energy transfers. However, even these ideas start to fail when you consider the forces that act between particles in the subatomic world. Here you will find that the very word 'force' does not adequately describe what is going on: particle physicists generally prefer to talk about 'interactions'.

Feynman's contribution

In the 1950s the physicist Richard Feynman developed a method for dealing with the electromagnetic force between charged particles that has since been adapted for use with the other fundamental forces. Feynman's approach built on the work of Dirac and Pauli, among others, who had attempted to bring relativity and quantum mechanics together. When this is done a new picture of what forces are emerges. The actual physical push or pull becomes less important than the energy and the momentum that is passed from one object to another. Feynman provided a 'simple' way of calculating how likely a particular interaction was by considering the energy and momentum to be carried by quanta of the force field. The approach is very mathematical and requires an understanding of advanced quantum theory. However, the outline of his technique can be seen in the (now famous) **Feynman diagrams**.

Feynman diagrams

In a simple situation, such as two electrons scattering off each other, the event is completely described by stating the energy and momentum of each electron before and after. These are the only measurements that can be made without ruining the reaction. In the diagram below this is symbolized by the shaded 'bubble of ignorance' in the middle. All we can tell is what went in (*before*) and what came out (*after*).

Feynman diagrams 'fill in' the region between the start and end of a reaction. Each diagram is a term in a mathematical series. Only the total series represents the interaction.

Feynman worked out how to calculate the likelihood of a given *before* leading to a given *after*. He did this by imagining all the possible things that might have happened inside the 'bubble' to get from the *before* state to the *after*. He learnt to represent each of the possible events as a 'doodle' or Feynman diagram. The figure above shows the first five simple Feynman diagrams that can be drawn for electron–electron scattering.

In a Feynman diagram the straight lines represent the electrons and the wavy line is the energy and momentum being passed from one to the other in the reaction. As the reaction is happening because of the electromagnetic force between the charged electrons, the energy and momentum is symbolized as a photon. Each Feynman diagram represents a precise mathematical formula. Adding these together in a certain way enables physicists to calculate the likelihood of the reaction.

Decoding doodles

1. The lines in a Feynman diagram do not represent the actual directions in which the particles are moving – it is not a velocity–time graph or anything similar. The arrows on the lines can only be read as 'coming in' and 'going out'.

2. No one diagram tells the whole truth. You will often see only the first Feynman diagram drawn (this is normally all that is required at this level), but do not make the mistake of thinking that the one diagram is all that is 'actually' happening.

3. Each diagram is a 'term' in an approximation. The formula $(1 + x)^n$ can be written as a line of mathematical terms:

$$(1 + x)^n = 1 + nx + \frac{n(n-1)x^2}{2} + \frac{n(n-1)(n-2)x^3}{6} + \ldots$$

which goes on for ever. Each Feynman diagram represents a term in a similar, but more complicated, line. The more you add up, the more correct the answer.

Applying Feynman's ideas to other forces

Following Feynman's basic principle, the likelihood of any reaction for any force can be calculated by using Feynman diagrams. The only difference is that you have to use the different exchange particles for the different forces.

The exchange particles.

Force	Particle	Charge	Mass
electromagnetic	photon (γ)	0	0
strong	gluon (g)	0	0
	W^+	$+e$	$89m_p$
weak	W^-	$-e$	$89m_p$
	Z^0	0	$99m_p$

Just as a photon is a packet, or quantum, of energy moving through the electromagnetic field, a gluon (for example) is a packet of energy moving through the field of the strong force. It is the exchange of gluons between quarks that gives rise to the strong force between them.

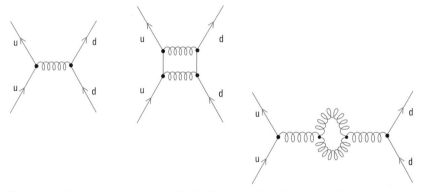

> **QED and QCD**
>
> Feynman's theory of the electromagnetic force is called **quantum electrodynamics (QED)**. The theory of the strong force that is on similar lines is called **quantum chromodynamics (QCD)**.

Gluons passing between quarks inside hadrons produce the strong force.

The weak force has three different exchange particles. Two of them are charged. The exchange of these particles results in the reacting particles changing type. This is characteristic of the weak force.

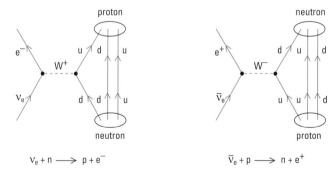

$$\nu_e + n \longrightarrow p + e^-$$

$$\overline{\nu}_e + p \longrightarrow n + e^+$$

*The weak force has two charged exchange particles. In these diagrams the two quarks that pass straight through are called **spectator quarks**. Notice that in the reaction $\nu_e + n \rightarrow p + e^-$, the W^+ could not have been absorbed by a u quark as conservation of charge would have made the particle produced have a charge of $(+2/3 + 1) = +5/3$ and no such quark exists. For the same reason the W^- in the other reaction could only have been absorbed by a u quark.*

The weak force is so short-range and so weak because of the great mass of the exchange particles. In order to react via the weak force, particles must get very close to each other so that there is a great deal of potential energy in the weak field (think about pushing like charges together). This energy is needed in order to stand any chance of creating a W or a Z.

- decays

- unifying the forces

- electroweak unification

Backwards in time

The reversal of particle lines and the switching of particle with anti-particle is often summarized as 'a particle moving backwards in time is the same as an anti-particle moving forwards in time'. Really all this means is the reversing of directions on the Feynman diagram.

EXCHANGE FORCES – 2

Decays

One of the strangest aspects of particle physics is the way in which fundamental forces are responsible for the decay of unstable particles. Nothing could be further from the simple idea of a force that is a push or a pull. The Feynman diagram gives some insight into how the two aspects of a fundamental interaction – forces and decays – can be brought together.

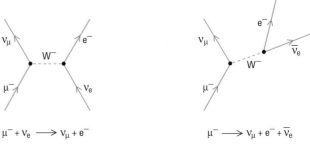

$$\mu^- + \nu_e \longrightarrow \nu_\mu + e^- \qquad \qquad \mu^- \longrightarrow \nu_\mu + e^- + \overline{\nu}_e$$

Fundamental interactions.

In the left-hand diagram above a muon interacts with an electron-neutrino via the exchange of a W^-. In the right-hand diagram a muon has decayed into a muon-neutrino by emitting a W^- which has then gone on to convert into an electron and an anti-electron-neutrino. Notice that in rearranging the left diagram into the right diagram the direction of the arrow on the electron-neutrino has had to be reversed (it is moving away from the W^-, not towards it) and it has also been changed into an anti-neutrino. In general, reversing a particle line and changing into the anti-particle go hand in hand on Feynman diagrams. The same mathematical ideas work for both diagrams – the theory that covers one will also account for the other.

Strong-force decays

The Δ^+ and the proton have the same quark content (uud). However, the Δ^+ has a much greater mass than the proton. This is because the quarks in the Δ^+ have a higher energy than those in the proton.

One of the quarks can get rid of some of this excess energy by emitting a gluon. This is very similar to an electron in a high atomic energy level emitting a photon and dropping into a lower energy level. In this case, however, the gluon contains so much energy that it rapidly materializes into a quark–anti-quark pair (see spread 9.9). The result is the decay of the Δ^+.

In the left-hand diagram below the quarks that materialize from the gluon bind together to form a π^0 meson (which will rapidly decay itself). In the right-hand diagram, one of the gluon's quarks has bonded with an original quark from the Δ^+, while the other has bonded with the two spectator quarks to form a neutron.

Helpful hint

Remember, more energy means more mass.

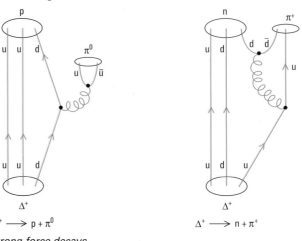

$$\Delta^+ \longrightarrow p + \pi^0 \qquad \qquad \Delta^+ \longrightarrow n + \pi^+$$

Strong-force decays.

Unification

Physicists are prejudiced people. They believe that Nature is simple and elegant. Einstein believed that it was better to have beauty in equations than for them to agree with experiment. Modern-day particle physicists are searching for beauty in the idea that the four fundamental forces can be united into one theory that shows them to be different aspects of the one force (in just the way that electricity and magnetism can be brought together in electromagnetism). In 1979 Glashow, Salam, and Weinberg received the Nobel Prize for bringing this goal a step nearer with their combined theory of the weak and electromagnetic forces.

The electroweak force

In 1961 Sheldon Glashow showed how the weak force could be explained by Feynman diagrams and linked with the electromagnetic force. For 30 years physicists had known about the weak '**charged current**' reactions such as

$$\bar{v}_e + n \rightarrow p + e^- \qquad W^+ \text{ exchange}$$
$$\bar{v}_e + p \rightarrow n + e^+ \qquad W^- \text{ exchange}$$

Glashow predicted the existence of another weak interaction (the 'neutral current') that involved the Z^0 exchange particle:

$$\bar{v}_\mu + e^- \rightarrow \bar{v}_\mu + e^- \qquad Z^0 \text{ exchange}$$

Evidence for such a reaction came from a giant CERN bubble chamber in 1973. Once the neutral current had been established the link with the electromagnetic force was more promising – in many ways the Z^0 and the photon were similar objects. The final problem was that the W and Z exchange particles are very massive and the photon is massless. Any theory seemed to require all the exchange particles to be massless.

The next step was taken in 1967 when Steven Weinberg and Abdus Salam applied an idea first suggested by Peter Higgs to the electroweak theory. Higgs's idea was to introduce another type of force field (now called the **Higgs field**) into the theory. By carefully selecting the right sort of field to add, Weinberg and Salam made the Higgs field interact with the W and Z in such a way as to give them mass, but not the photon.

The Higgs field is very important – without it you cannot explain why the W and Z exchange particles have mass. If the Higgs field exists, then **Higgs particles** should exist as well (see spread 14.7).

Unification spreads

Since the Nobel Prize was awarded for the electroweak theory, physicists have been trying to bring the other forces into line as well. There are several promising theories that unite the strong, weak, and electromagnetic forces (know as **Grand Unified Theories** – or GUTs). Again, the next generation of accelerators will be looking for evidence to show which one is correct. A feature of these theories is that they predict that baryon number is not conserved – so the proton should decay. Experiments have, so far, failed to detect this, indicating that the proton's lifetime (if it does decay) must be greater than 10^{32} years!

Sheldon Glashow, Abdus Salam, and Steven Weinberg at their Nobel Prize ceremony, 1979.

The problem with gravity

Gravity has proved to be very difficult to unify with the other forces. There is a theory which may succeed in doing this, but it is mathematically very complicated and at the moment it is hard to make any predictions using it.

PRACTICE

1 Draw diagrams for the following strong force decays:

  a $\Delta^{++} \rightarrow p + \pi^+$

  b $\Delta^0 \rightarrow p + \pi^-$

  c $\Delta^- \rightarrow n + \pi^-$

2 Draw a diagram for the weak force decay:

  $\Omega^- \rightarrow \Xi^0 + e^+ + v_e$

3 Draw a diagram for the neutral current reaction:

  $\bar{v}_\mu + e^- \rightarrow \bar{v}_\mu + e^- \quad Z_0 \text{ exchange}$

PARTICLE PHYSICS IN THE NUCLEUS

In the current era of the universe's evolution the most common forms of matter are hydrogen (about 78%) and helium (about 22%). The rest of the universe is made up of trace 'impurities' such as the carbon that makes up our bodies and the iron inside our planet. In cosmological terms these materials hardly add up to a single percentage point. As these atoms are predominantly made from protons and neutrons, which are composed in turn of up and down quarks, that makes the up and down quarks by far the most common in the current universe. Strange quarks are also to be found in the particles that compose cosmic rays. The heavier quarks are virtually extinct in the low-energy environment of the current universe.

Protons and neutrons in the nucleus

The neutron is an unstable particle when it is found in isolation away from a nucleus. This instability arises mainly because there is a lower-mass particle into which it can decay – the proton. The decay has to proceed by the weak force (the strong force cannot change particle types), giving it a long lifetime. The lifetime is further extended by the small mass difference between the proton and neutron – there is not much energy released in the reaction. At the quark level the decay proceeds in the following manner:

$$d \rightarrow u + e^- + \bar{v}_e$$

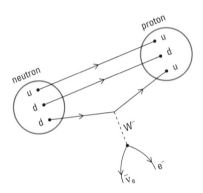

The weak decay of a neutron proceeds by the emission of a W^- which then converts into an electron and an anti-electron–neutrino.

Inside a nucleus the effects of the strong nuclear force stabilize the neutron. The strong nuclear force derives from the strong force between quarks, but its effects are different.

A close analogy is with the van der Waals forces between molecules, which are due to the electrostatic forces between electron distributions in the molecules. The force is complicated and varies with distance in a far more complex way than the electrostatic force. When molecules are distant the force is attractive as one electron distribution polarizes the other, which allows the positive charge of the nuclei to be 'seen'. As the molecules get close to each other the electron distributions overlap, giving rise to a strong repulsive force. The van der Waals force is nothing more than the electromagnetic force in a complex situation.

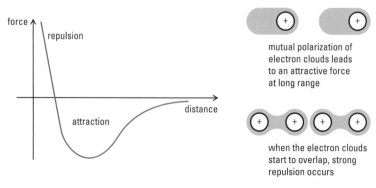

Attractive and repulsive forces between molecules.

Protons and neutrons are equally complex structures of quarks and gluons which affect one another in close proximity. This gives rise to an interaction between the nucleons which we know as the strong nuclear force. The effects of this force can be modelled in a similar manner to the other fundamental forces (although it is not a fundamental force itself) using pions as the exchange particles. It must be emphasized that the interaction between protons and neutrons in the nucleus is a highly complex affair and simple pion exchange barely touches this.

Exchange diagrams, such as those illustrated, show that every proton spends part of its time being a neutron and vice versa. This gives rise to the stabilizing effect that prevents neutrons from decaying in many nuclei. It can also destabilize the proton, leading to nuclear β⁺ decay:

$$p \rightarrow n + e^+ + \nu_e$$

or at the quark level:

$$u \rightarrow d + e^+ + \nu_e$$

Without this stabilizing effect, all nuclei would be β⁻ decaying.

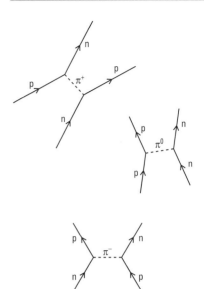

The strong nuclear force can be modelled at a simple level as the exchange of pions between the nucleons.

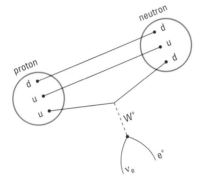

In the nucleus proton decay can proceed via W⁺ emission.

In the nucleus the stability (or otherwise) of individuals is down to the collective stability of the nucleus as a whole. A nucleus is more than just a bag of protons and neutrons – it is a structure, similar in many ways to a drop of liquid.

Proton decay

The most modern theories of how the fundamental forces relate to one another – and a possible unification of them – suggest that protons are not in fact stable objects. They may decay, but only with enormous lifetimes. Various predictions have been made and experiments have established lower bounds to the lifetime. Currently it is thought that a proton must have a lifetime in excess of 10^{32} years. There are various modes by which proton decay can take place – for example into a positron and a pion – and they all, of course, violate baryon number conservation.

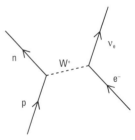

Some nuclei can achieve the same shift in the n/p ratio as β⁺ decay by 'capturing' an orbital electron, as above.

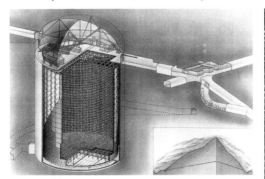

Super-Kamiokande is a joint Japan–US collaboration to construct the world's largest neutrino observatory. It is located in the Kamioka mine in Japan and comprises a vast tank (40 m tall by 40 m diameter) of ultra-pure water viewed by thousands of sensitive phototubes. Neutrino oscillation was detected in 1998. in 2001, 6600 photomultiplier tubes imploded; the observatory was later replaced by Super-Kamiokande III and then SK-IV which is collecting data on natural sources of neutrinos. Left: how the tank was built within the mine. Right: inside the tank.

PRACTICE EXAM QUESTIONS

1 a i State what is meant by the β-decay of a nucleus.

    ii Describe the results of experiments in which the energies of particles emitted in β-decay have been measured.

  b Explain the evidence for the belief that two particles are emitted in a β-decay process rather than one.

  c Why is one of these particles very difficult to detect directly?

2 a i State the difference between a hadron and a lepton in terms of the type of force experienced by each particle.

    ii Give *one* example of a hadron and *one* example of a lepton.

    iii Hadrons are classified as either baryons or mesons. In terms of quark composition, explain the difference between a baryon and a meson.

  b i State the quark composition of a neutron.

    ii Describe, in terms of quarks, the process of β⁻ decay when a neutron changes into a proton. Sketch a Feynman diagram to represent this process.

3 a i State the quark composition of the proton.

    ii State the charge, baryon number and strangeness of the proton.

    iii The π^+ meson has a charge of 1 and strangeness of 0. State its composition in terms of quarks and antiquarks.

  b In β⁺ decay, a proton changes into a neutron because one type of quark in the proton changes into a different type of quark. The Feynman diagram for this process is shown in Figure 9.1.

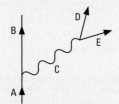

Figure 9.1

    i What type of quark is represented at A?

    ii What type of quark is represented at B?

    iii What particles are represented at C, D and E respectively?

    iv Which of the particles A, B, C, D and E are leptons?

4 a State *one* difference and *one* similarity in the principle of operation of a cloud chamber and a bubble chamber.

  b Figure 9.2 represents a photograph of two events, labelled X and Y, from a bubble chamber in which pair production has occurred as a result of a gamma photon creating an electron and a positron. In event X an electron has also been ejected from an atom. The track created by this electron is labelled 'atomic electron'. The tracks created by pair production are labelled A, B, C and D.

    i Which *two* of the tracks A, B, C and D were produced by positrons?

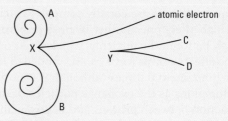

Figure 9.2

    ii Explain why gamma photons do not leave visible tracks.

    iii Explain why the tracks created by pair production in event X are much more curved than those produced in event Y.

5 β⁻ decay of a free neutron may be described by

$$n \rightarrow p + e^- + \nu_e^-$$

  a Draw the Feynman diagram which represents this process.

  b Describe each stage of the decay, including the exchange particle.

  c Table 9.1 below gives properties of quarks.

Table 9.1

type of quark	charge	baryon number	strangeness	approximate rest mass
	Q	B		GeV/c^2
u	+2/3	+1/3	0	0.005
d	−1/3	+1/3	0	0.01
s	−1/3	+1/3	−1	0.2

Use the table to deduce the quark transformation which takes place during β⁻ decay. Explain your reasoning.

6 a Explain the difference between fixed target experiments and colliding beam experiments in particle physics.

  b Use conservation of linear momentum and energy to show that the proportion of the initial kinetic energy available for the production of new particles is greater in the colliding beam case.

  c Discuss any other advantages or disadvantages which one of these techniques has over the other.

  d Indicate briefly how colliding beam experiments can be set up. Include a diagram in your answer.

7 a i Calculate the force of repulsion between two protons separated by a distance of 1.0×10^{-14} m in an atomic nucleus.

    ii The Rutherford scattering experiment was used to investigate the structure of the atom. Deep inelastic scattering provided evidence about the structure of protons and neutrons. Outline the similarities and differences between these two experiments.

    What conclusion was drawn from the deep inelastic scattering experiment?

    iii The quarks in a nucleon are held together by the strong interaction. What exchange particle mediates this interaction?

    iv State the mass and charge of this exchange particle. What evidence is there that the range of this exchange particle is less than 10^{-14} m?

b i State one piece of evidence that supports the Big Bang Theory and explain how it does so.

ii The very early Universe was made up of a collection of quarks, leptons and exchange particles. Through what stages has it passed in order to reach its present state?

iii Explain way atoms were unstable when the Universe was at a temperature greater than about 4000 K.

8 In answering this question you might wish to use the following:

(i) Hadrons with zero baryon number B are known as mesons; those with $B = 1$ are known as baryons.

(ii) In strong and electromagnetic interactions, charge Q, strangeness S and B are conserved.

(iii) Baryons can be constructed from three quarks and mesons from one quark and one antiquark.

(iv) The particles K^-, π^-, and π^+ are mesons. Both K^- and the Σ^- particle have strangeness –1, while the proton p, π^- and π^+ have strangeness zero.
The reaction $K^- + p = \pi^+ + \Sigma^-$
and the decay $K^- = \pi^+ + \pi^- + \pi^-$ both occur

a i Which one of the above two processes is caused by a weak interaction? Justify your answer.

ii Name and use a conservation law to discover whether the Σ^- is a meson, a baryon or an antibaryon.

b Table 9.2 below shows the properties of the three quarks originally proposed to describe the structure of mesons, baryons and antibaryons. Use it to find the quark content of K^-, π^+ and p.

Table 9.2

quark	symbol	spin	Q	B	S
up	u	1/2	2/3	1/3	0
down	d	1/2	–1/3	1/3	0
strange	s	1/2	–1/3	1/3	–1

9 a i State why it is possible to achieve much higher energies using circular accelerators than using linear ones of a similar size. Explain why the particles in an accelerator must be contained within a vacuum tube.

ii What other two factors in the design of a circular accelerator limit the maximum energy available?

iii A charged particle moving in a plane perpendicular to a uniform magnetic field follows a circular path. Show that, if the speed of the particle is increased, the radius of the circular path increases but the period remains constant.

b In 1962 the existence of a particle with strangeness –3 was predicted. The particle Ω^- was identified in 1964 in an experiment involving a strong interaction between a K^- meson of strangeness –1 and a proton in a hydrogen bubble chamber. The interaction involved was

$$K^- + p = \Omega^- + K^+ + K^0$$

i Is the Ω^- particle a baryon or a meson? Give two reasons for your answer.

ii Using the information given in Table 9.3, deduce the quark composition of all the particles involved.

Table 9.3

Type of quark	Charge	Strangeness
u	+2/3	0
d	–1/3	0
s	–1/3	–1

iii The Ω^- (lifetime 8.2×10^{-11} s) subsequently decayed in a three stage process to a proton and a number of pi mesons. Pi mesons have zero strangeness. What fundamental interaction must be involved at some stage in this process? Give two reasons for your answer.

iv What exchange particle(s) could be mediating this process?

10 a Figure 9.3 shows the track of a proton in a magnetic field.

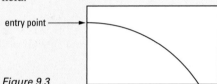

entry point

Figure 9.3

i Deduce the direction of the magnetic field.

ii Describe how the track would change if a proton of *higher* energy in the *same* magnetic field were used. How would it change if a proton of the *same* energy in a *stronger* magnetic field were used?

iii Name *one* type of particle detector that might be used to record these tracks. Explain why this detector cannot record the passage of neutrons.

b The Feynman diagram in Figure 9.4 represents the β^- decay process.

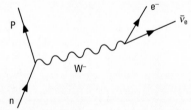

Figure 9.4

With reference to the diagram, describe each stage of the decay. Then redraw the diagram so that it shows the quark transformation that occurs in β^- decay.

c The K^+ is a meson with strangeness +1. One of its common decay modes is $K^+ \rightarrow \pi^+ + \pi^0$. Pions are not strange particles.

i Name the type of interaction responsible for the $K^+ \rightarrow \pi^+ + \pi^0$ decay.

ii Table 9.4 gives the properties of the relevant quarks. Deduce the possible quark content of the K^+, the π^+ and the π^0.

Table 9.4

Type of quark	Charge	Baryon number	Strangeness
u	+2/3	1/3	
d	–1/3	1/3	0
s	–1/3	1/3	–1

iii The rest mass of a proton is $938\,\text{MeV}/c^2$. The K^+ rest mass = 0.53 proton masses, and the π^+ rest mass = π^0 rest mass = 0.15 proton masses.

Assuming that the K^+ is stationary when it decays, show that the total energy of each pion produced in the decay is 249 MeV.

iv Given that $E^2 = m_0{}^2c^4 + p^2c^2$, calculate the momentum of each pion. Express your answer in units of MeV/c.

11 a i Explain why there was a critical temperature during the cooling of the Universe above which hydrogen atoms were unlikely to be stable.

Calculate an approximate value for this temperature. You may assume that the typical energy of a photon in black body radiation at temperature T is kT.

(The Boltzmann constant, $k = 1.38 \times 10^{-23}\,\text{J K}^{-1}$, electronic charge, $e = 1.60 \times 10^{-19}\,\text{C}$, the ionisation energy of a hydrogen atom = 13.6 eV.)

ii The equation for the production of a helium nucleus from the fusion of a tritium nucleus with a deuterium nucleus is

$$^3_1\text{H} + {}^2_1\text{H} \rightarrow {}^4_2\text{He} + {}^1_0\text{n}$$

Explain, in terms of the appropriate fundamental interactions, why this reaction could only take place at temperatures above about $10^8\,\text{K}$.

b Use the laws of conservation of lepton number, baryon number and charge to decide which of the following reactions is/are possible and which is/are not. Show clearly how you applied the laws in each case. (Both π and K particles are mesons.)

i $\text{n} \rightarrow \text{p} + \text{e}^- + \nu_e$

ii $\pi+ \rightarrow \mu^+ + \nu_\mu$

iii $\pi^- + \text{p} \rightarrow K^- + \pi^+$

c Figure 9.5 represents a simple particle interaction. State the nature of the interaction and describe what happens to the particles involved. State three properties of the exchange particle.

Figure 9.5

12 a Table 9.5 shows the charge, baryon number and strangeness of each of the u, d and s quarks.

Table 9.5

Type of quark	Charge	Baryon number	Strangeness
u	+2/3	1/3	0
d	−1/3	1/3	0
s	−1/3	1/3	−1

There are nine possible ways of combining u, d and s quarks and their associated antiquarks to make nine

different mesons. List all the possible combinations. From your list select any strange mesons and state the charge and strangeness of each of these.

Three of the mesons in the list have zero charge and zero strangeness. What will distinguish these mesons from each other?

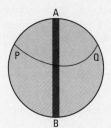

Figure 9.6

b Figure 9.6 shows the track of a charged particle in a magnetic field. The field is at right angles to the plane of the paper, and its direction is out of the plane of the paper. AB is a thin sheet of lead through which the particle passes.

Deduce the direction of movement of the particle and the sign of the charge on the particle. Explain clearly how you made your deductions.

c In a colliding beam experiment a greater proportion of the colliding particles' total kinetic energy is available for producing new particles than in a fixed target experiment. Explain why this is so. When might a fixed target experiment be preferable?

A moving proton of kinetic energy 2.12 GeV collides with a stationary antiproton. These particles annihilate and a new particle of rest mass $2.74\,\text{GeV}/c^2$ and momentum 2.91 GeV/c is produced. Show that no other new particles have been produced as a result of this collision.

(Rest mass of proton = $0.938\,\text{GeV}/c^2$.)

* 13 In the synchrotrons in the Large Hadron Collider (LHC) protons are accelerated to speeds of approximately $0.9999c$. The circumference of the LHC is 27 km. The protons travel around the accelerator in bunches of 1.2×10^{11} particles, each proton having a final energy of about $7 \times 10^{12}\,\text{eV}$.

a Calculate the total energy in MJ of each bunch of protons in the LHC.

b Calculate the time taken for a proton to travel round the LHC:

i in the experimenter's frame of reference

ii in the proton's frame of reference.

c To reduce energy loss, the LHC has to be evacuated so that there are only 400 million atoms of gas, each with a mass of about $4.8 \times 10^{-26}\,\text{kg}$, in each cubic centimetre of the chamber. The temperature inside the collider is only 1.9 K.

i Calculate the root-mean-square speed of the gas atoms that remain in the LHC.

ii Calculate the pressure of the gas that remains in the LHC.

*Pre-U-style question

14 This question is about the operation of particle accelerators.

a The simplest accelerator is an ion gun as shown in Figure 9.7.

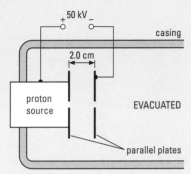

Figure 9.7

Protons drift with negligible speed into the region between the parallel plates. The plates are 2.0 cm apart and a p.d. of 50 kV is maintained between them.

i Calculate the electric field strength between the plates.

ii What will be the energy, in J, of the protons emerging from the hole in the negative plate?

iii Show that the speed of the protons emerging from this hole is approximately $3.1 \times 10^6 \, \text{m s}^{-1}$.

Charge on an electron $e = -1.6 \times 10^{-19} \, \text{C}$.

Mass of a proton $= 1.7 \times 10^{-27} \, \text{kg}$.

iv What would be the speed of deuterons, which are particles consisting of a proton and a neutron, when accelerated using this apparatus?

v Why is the apparatus evacuated?

b A linear accelerator produced particles of higher energy. Further acceleration is produced by using a 50 kV alternating p.d., with a frequency of 10 MHz, applied to a succession of tubular electrodes (WX, YZ and so on) shown in Figure 9.8.

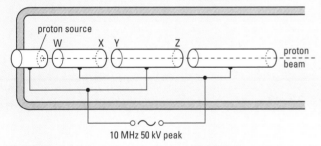

10 MHz 50 kV peak

Figure 9.8

The tube lengths are such that, in the time taken to travel the length of a given tube, the polarity changes and further acceleration takes place.

Therefore, each tube length is such that the time to travel down the tube is half the period of the alternating p.d.

i Explain why the particles travel at constant speed along the length of an electrode even though the electrode is at a high potential.

ii Calculate the length WX of the first tube so that the protons arrive at the right time to be accelerated in the gap XY.

15 Figure 9.9 illustrates a cyclotron.

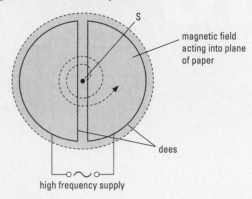

Figure 9.9

An ion source S is placed between two hollow 'dees' in an evacuated chamber. The dees are connected to a high frequency alternating voltage supply. A uniform magnetic field acts normally into the plane containing the dees.

a State the sign of the charge on the particles emitted by the ion source.

b Explain why the dees are situated in an evacuated chamber.

c State the region in which the particles emitted by the ion source gain energy.

d i The magnetic field has a uniform flux density (field strength) B.

By consideration of the force acting on the particles due to the magnetic field, show that the frequency f of the supply voltage connected to the dees is given by

$$f = \frac{Bq}{2\pi m}$$

where q/m is the charge-to-mass ratio of the particles.

ii The magnetic flux density B is 1.3 T. Each particle has a charge-to-mass ratio of $4.8 \times 10^6 \, \text{C kg}^{-1}$.

Calculate the frequency of the supply voltage.

e This cyclotron is capable of accelerating particles to an energy of 20 MeV.

i Calculate a value in joules for this energy.

ii Show that the kinetic energy E_k of a particle moving in the cyclotron in an arc of radius r is given by

$$E_k = \frac{q^2 B^2 r^2}{2m}$$

iii Use your answers to **e i** and **ii** to calculate a value for the minimum diameter of the dees used in this cyclotron. Assume that the particles, of charge-to-mass ratio $4.8 \times 10^6 \, \text{C kg}^{-1}$, are singly-ionised.

f Name one other type of accelerator which is capable of accelerating charged particles to energies much higher than that obtained in this cyclotron.

10

The Physics of **MATERIALS**

At first sight, copper and diamond do not appear to have much in common. However, if you could reduce yourself to the size of an atom and wander through both structures, you would find a fair degree of similarity. In both cases the atoms stretch off as far as you can see. Flaws in these regular structures are very few and far between. Some materials, like glass and many polymers, do not have a regular crystalline structure. We know this because structures can be analysed by passing X-rays or electron beams through them. This is useful because material properties depend on the microscopic arrangements of atoms and molecules. The properties of pure materials can sometimes be improved by heat treatment or the addition of impurities, or by combining them into composites. It is through our understanding of the structure and properties of various materials that we are able to construct enduring and functional monuments such as the Golden Gate Bridge.

In this chapter you will look at the structure of various types of material, such as crystals, metals, polymers, glasses, and ceramics. You will study how their mechanical properties are affected by stress, strain, and wear and tear. Other properties, for example magnetic, electrical, and optical, are then examined, and the chapter ends with sections on lasers and their applications.

Looking down from the east tower of the Golden Gate Bridge, which hangs above the entrance to San Francisco Bay (San Francisco itself can be seen to the upper left).

• stress and strain

• the Young modulus

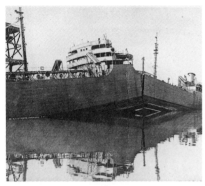

Not understanding the properties of materials can have disastrous consequences. The SS Schenectady broke in two while lying in dock in 1943. The principal causes of this failure were poor design and bad workmanship.

Measuring the Young modulus *E* of metal wire

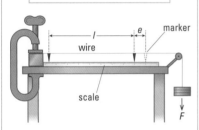

Tips:

• Wear safety glasses.

• Measure length *l* and load *F* as *F* is varied.

• Mark two points on the wire.

• Measure the length between markers for each value of *F*.

• Calculate the extension *e* for each value of *F*.

• Measure the diameter *d* using a micrometer screw gauge.

• Use a long wire (e.g. 2.0 m) to increase extensions.

• Pre-load the wire to straighten it.

• Plot the load *F* versus the extension *e*.

• Measure the gradient in the region obeying Hooke's law.

• The gradient is equal to $\left(\dfrac{\pi d^2 E}{4 l_0}\right)$ where l_0 is the initial length.

• Beware of units! Convert to SI units.

• Estimate uncertainties in the final result.

MECHANICAL PROPERTIES OF MATERIALS

Stress and strain

A fine steel spring may stretch and bend more easily than a rubber brake block, but this does not mean that rubber is stiffer than steel. To compare the properties *of materials* the effects of size, shape, and construction must be removed. It is then possible to weigh up the advantages and disadvantages of using different materials for the same job. For the time being we shall restrict our attention to:

• **Tensile** (stretching) forces or compressive forces, *F*, along the axis of the object.

• Rod-like objects, for example, cylinders (like wires) or rectangular blocks of length *l* and cross-sectional area *A*.

• Loads for which Hooke's law is obeyed.

The extension (or compression) *e* is related to *l*, *A*, and tensile force, *F*, in the following way:

$e \propto l$ since lengths are added in series.

$e \propto \dfrac{1}{A}$ since extra cross-sections are in parallel with one another.

$e \propto F$ since Hooke's law is obeyed in this region.

Therefore:

$$e \propto \frac{Fl}{A} \quad \text{or} \quad F \propto \frac{Ae}{l}$$

This relates the force per unit area to the fractional increase in length. This is usually written:

$$F = \frac{AeE}{l}$$

where *E* is a constant called the **Young modulus**.

Definitions

Tensile stress $(\mathrm{N\,m^{-2}}) = \dfrac{\text{axial force (N)}}{\text{cross-sectional area } (\mathrm{m^2})}$ $\sigma = \dfrac{F}{A}$

Tensile strain (no unit) $= \dfrac{\text{extension (m)}}{\text{original length (m)}}$ $\varepsilon = \dfrac{e}{l}$

Young modulus $(\mathrm{N\,m^{-2}}) = \dfrac{\text{tensile stress } (\mathrm{N\,m^{-2}})}{\text{tensile strain}}$ $E = \dfrac{\sigma}{\varepsilon}$

Hooke's law (for materials):

    stress $(\mathrm{N\,m^{-2}}) \propto$ strain   or $\sigma \propto \varepsilon$

The advantage of using stress and strain instead of force and extension is that the ratio of stress to strain is the same regardless of the size and shape of the object. This ratio, the Young modulus, *E*, is a *property of the material*. It is a measure of the material's stiffness; that is, its resistance to stretching or bending. The value of the Young modulus is usually very large because it is equal to the stress that would cause unit strain (a doubling of length) if the material could stretch that far, but most materials cannot.

Mild steel is about three times as stiff ($E_{\text{steel}} = 2 \times 10^{11}\,\mathrm{N\,m^{-2}}$) as aluminium ($E_{\text{Al}} = 7 \times 10^{10}\,\mathrm{N\,m^{-2}}$), so an aluminium rod would have about three times the strain of a steel rod if both were subjected to the same tensile stress. Notice that this is true even if the rods do not have the same dimensions.

Worked example

What is the strain and extension of a 1.4 m steel wire of diameter 2.0 mm when it supports a load of 500 N?

$$\sigma = Ee \quad \text{so} \quad \varepsilon = \frac{\sigma}{E} = \frac{F}{AE} = \frac{500\,\text{N}}{\pi r^2 E} = \frac{500\,\text{N}}{\pi \times (0.001\,\text{m})^2 \times 2 \times 10^{11}\,\text{N m}^{-2}}$$

$$= 8 \times 10^{-4} \;(= 0.08\%)$$

$$\varepsilon = \frac{e}{l} \quad \text{so} \quad e = \varepsilon l = 8 \times 10^{-4} \times 1.4\,\text{m} = 10^{-3}\,\text{m} \;(1\,\text{mm})$$

Properties of some materials.

Material	Mild steel	Copper	Aluminium	Glass	Concrete	Wood
E/N m^{-2}	20×10^{10}	13×10^{10}	7×10^{10}	5×10^{10}	4×10^{10}	10^{10} varies
σ_{max}/N m^{-2}	11×10^8	5×10^8	2×10^8	10×10^8	2–$4 \times 10^8 (C)$	$2 \times 10^7 (P)$
ε_{max}/%	15	30	30	2	not applicable	
ρ/kg m^{-3}	7700	8920	2700	3500	2300	600

E is the Young modulus; σ_{max} is the **breaking stress**; ε_{max} is the **breaking strain**; ρ is the density; C gives the compressive strength, P is the strength parallel to grain.

Describing the mechanical properties of materials

Terms such as **strength**, **stiffness**, and **toughness** are used in a special way by physicists: some common terms are explained in the margin. When choosing a material it is always important to consider how the material behaves under the conditions in which it will be used. For example, concrete is very strong under compression, but is weak in tension, where it breaks by brittle fracture as cracks open up. It is good for walls but not so good for ceilings. Aluminium has a similar stiffness to glass, but glass is usually **brittle** and shatters for very small strains, whereas aluminium undergoes plastic deformation before breaking – in other words it is tough. Most metals reach their limit of proportionality for quite small strains, typically $\varepsilon = 0.001$. (This is often quoted as a percentage, 0.1%).

There are many other factors to consider including density, ease of working, and cost, and most engineering projects will need to find a compromise solution to a complicated problem. In many cases this will involve the use of **composite materials**, for example, reinforced concrete or fibreglass. These give a combination of the most useful properties of all the materials used.

> **Describing mechanical properties**
>
> **Strong:** a large stress is needed to break it.
>
> **Stiff:** small strains for large stresses (large Young modulus); not stretchy or bendy.
>
> **Elastic:** returns to unstretched form when stresses are removed.
>
> **Plastic:** undergoes permanent deformation under large stress rather than cracking.
>
> **Tough:** undergoes considerable plastic deformation before it breaks. Consequently, tough materials absorb a great deal of energy before they break.
>
> **Brittle:** breaks suddenly as cracks travel through it; little or no plastic deformation.
>
> **Hard:** resists indentation on impact.

PRACTICE

1 What is the stress in a copper wire of diameter 0.4 mm when it supports a 2 kg mass?

2 A mass of 0.20 kg is suspended from an aluminium wire of diameter 0.80 mm and length 2.0 m, and extends it by 110 μm.

   a What is the Young modulus for aluminium?

   b What would be the extension of an aluminium wire of twice the length and half the diameter?

   c What would be the extension in a copper wire of the same dimensions under the same stress as **a**?

   d Did you need to know the length or diameter of the copper wire to answer part **c**? Explain.

3 The yield strength of mild steel is 2.5×10^8 Pa. What is the minimum diameter mild steel wire that could safely support a 75 kg adult?

4 **a** Give examples of materials that are **i** strong, **ii** stiff, **iii** tough, **iv** elastic, and **v** brittle.

   b Use these categories to describe **i** rubber, **ii** lead, **iii** stone, and **iv** glass.

5 On a common set of axes sketch graphs of stress versus strain for the materials in the above table. What features of the graph show stiffness, strength, and plastic/brittle behaviour?

6 Compare the mechanical properties of biscuit, steel, stone, plasticine, glass, and bread.

7 Assume the atoms in a solid are held together in a cubic lattice by bonds of length r and spring constant k. Derive an expression for the Young modulus of the material in terms of k and r.

8 The table below gives values of tension and extension for a steel rod of radius 6.4 mm and length 100 mm. Plot a suitable graph to determine the Young modulus and the limit of proportionality. Account for the shape of the graph.

T/kN	0.0	5.0	10.0	15.0	20.0	25.0	30.0	35.0	37.5
e/μm	0	18	37	55	74	92	114	155	240

9 Compare the legs of an elephant and a mouse. Do you think a mouse the size of an elephant would be able to support itself? Explain your answer.

STRESS–STRAIN GRAPHS

Stress and strain are used when we are more interested in the properties of a material (e.g. steel) than the properties of a particular object (e.g. the suspension spring in a car). Graphs of stress versus strain tell us a great deal about the material.

Units and axes

Tensile stress is measured in $N\,m^{-2}$. This is the same as the unit for pressure, the pascal (Pa), so pascals are often used instead. Stresses in metals are usually very high (e.g. $10^8\,N\,m^{-2}$) so it is common to use MPa ($10^6\,Pa$) and GPa ($10^9\,Pa$) to record stresses. Strain is a dimensionless ratio so it has no units, however it is fairly common to find strains expressed as a percentage. For example, a strain of 0.02 is the same as 2% strain.

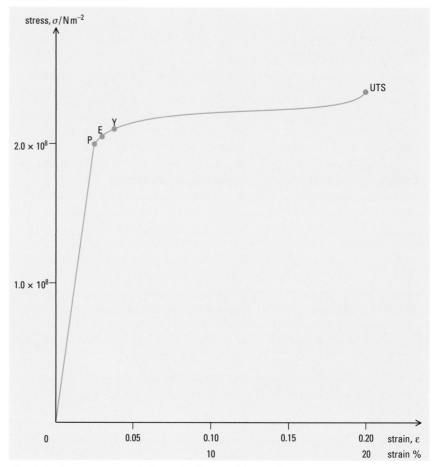

Graph 1: stress versus strain for a ductile metal (copper).

The gradient of the graph (above) in the initial linear section is equal to the Young modulus of the metal. The steeper the graph the stiffer the material. This follows from Hooke's law expressed in terms of stress and strain: $\sigma = E\varepsilon$, so E is the gradient of a graph of σ versus ε.

Beyond P, the **limit of proportionality**, the graph is curved. Stress and strain are no longer directly proportional and Hooke's law does not apply.

The material continues to behave elastically up to point E, the **elastic limit**. If the load is removed below E the material returns to its original shape and size. Beyond E it will be permanently deformed.

At a certain stress the strain begins to increase rapidly with increasing stress. This is the **yield point**, Y.

The **ultimate tensile strength** (**UTS**) of the material is the stress at which it breaks. With copper, plastic deformation occurs prior to fracture, so a wire narrows or **necks** before it breaks. This means the cross-sectional area of the sample at the fracture point is less than elsewhere in the material.

Ductile metals (e.g. copper) undergo a great deal of plastic deformation before they break. This appears as a long shallow region beyond the yield point. Materials that behave like this are described as **tough**.

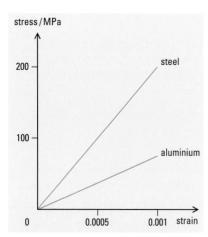

Graph 2: comparing the elasticity of steel and aluminium for small strains (Hooke's law is obeyed).

Brittle materials undergo little or no plastic deformation before fracture, so their stress–strain graphs are virtually straight lines. Ceramics are an example of strong (high UTS), stiff (steep gradient), brittle (no curved section) materials.

The area under a stress–strain graph is equal to the work done per unit volume of material that had to be supplied to stretch it. If the material is behaving elastically this energy is stored as elastic potential energy, and can be retrieved by making the material do work as it contracts when the original stretching forces are removed. Tough materials have a large area under their stress–strain graphs up to the point at which they fail. This means a lot of work must be done before a tough material breaks.

Rubber has a complicated stress–strain graph, being hard to stretch at low strains, becoming easier for a while and then hard again when the strain is very high (perhaps 3 or 4). It is highly elastic but its stress–strain graph for unloading lies beneath its stress–strain graph for loading. The stress–strain graph for a loading–unloading cycle forms a closed loop called a **hysteresis loop**. The area inside the loop is the energy per unit volume transferred to heat inside the rubber.

Brittle fracture of a ceramic.

Necking

The UTS is calculated at the point of fracture so, strictly speaking, this reduced cross-sectional area must be used in calculations rather than the cross-sectional area of the wire before loading, or away from the fracture point.

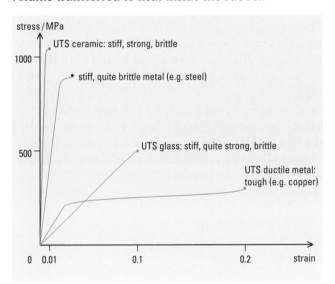

Graph 3: comparing ceramic, glass, and metal.

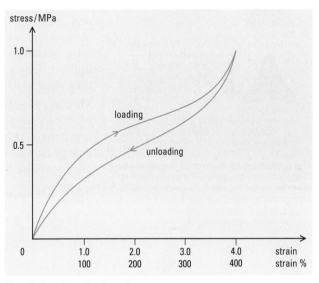

Graph 4: rubber hysteresis.

PRACTICE

1 Sketch stress–strain curves for the following materials, labelling them to indicate significant properties: glass, plasticine, gold.

2 Find a rubber band. Stretch it quickly two or three times and immediately touch it to your lip. What do you feel? Why?

3 Use graphs 1, 2, and 3 above to estimate the Young modulus of steel, copper, and aluminium.

4 Explain why copper is an ideal material to use as a conductor in electrical circuits.

5 Ceramics are used in the rotor blades of some jet engines. What properties of the ceramic material make it suitable? Use graph 3 to estimate the Young modulus of the ceramic.

6 Compare the properties of the four materials shown on graph 3. Suggest a use for each material.

7 What does the area inside the hysteresis loop on graph 4 represent? Use this graph to estimate the percentage of the energy needed to stretch the rubber that is transferred to heat when it contracts.

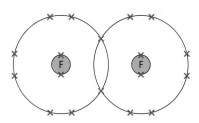

The fluorine molecule. The only true covalent bonds are between atoms of the same type. In all other cases the bonds are to some extent ionic.

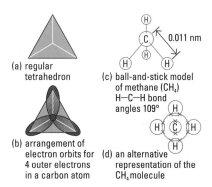

(a) regular tetrahedron

(c) ball-and-stick model of methane (CH_4) H—C—H bond angles 109°

(b) arrangement of electron orbits for 4 outer electrons in a carbon atom

(d) an alternative representation of the CH_4 molecule

The tetrahedral structure of the methane molecule can be explained by assuming that the outer four (n = 2) electrons in carbon form symmetric orbitals as in (b). Each hydrogen atom then shares one of these electrons.

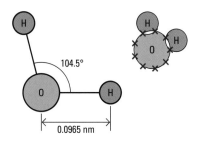

104.5°

0.0965 nm

If the hydrogen atoms in the water molecule did not interact with one another, the expected bond angle would be 90°. The larger angle of 104.5° is caused by the repulsion of the two partially exposed protons.

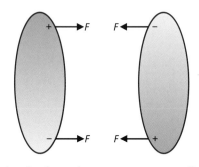

London forces between two temporarily induced dipoles.

BONDING AND PACKING

To understand and manipulate the large-scale (macroscopic) properties of materials you first need a good understanding of their small-scale (microscopic) structure. All materials are made from atoms, but atoms can bond together in various ways and the nature of these bonds has a crucial effect on mechanical, electrical, and optical properties.

Primary (interatomic) bonds

Covalent bonds   These form when one or more electrons are shared between adjacent non-metallic atoms. Unlike all the other bonds described here, covalent bonds are strongly directional and the angles between them are responsible for the geometry of molecules such as methane CH_4, water H_2O, and ammonia NH_3. The tetrahedral structure of diamond is directly related to the angles between the covalent bonds formed between carbon atoms. Single covalent bonds allow rotation about the bond direction, double bonds do not.

Ionic bonds   These form when electrons from one atom are completely transferred to another so that the donor atom becomes a positive ion and the recipient atom becomes a negative ion. The ions then bond by electrostatic attraction. These are the strongest bonds between atoms and ionic materials tend to be very stiff (they have a high **Young modulus**) and brittle (since deformation pushes like ions close to one another and they then repel strongly).

Metallic bonds   Metals form extended structures in which bonding electrons are not fixed in any particular place. They are effectively free to move through the structure as long as they remain within certain allowed energy bands. These '**delocalized**' electrons are responsible for the high electrical and thermal conductivity of metals.

Bond energy

The energies required to break a mole of each type of bond are compared below:

- covalent (single bonds)   100–500 kJ mol^{-1}
- ionic   500–3000 kJ mol^{-1}
- metallic   60–600 kJ mol^{-1}

Secondary (intermolecular) bonds

Van der Waals bonds   These are weak forces between molecules that arise because of electrostatic attraction either between permanent molecular dipoles or temporary dipoles induced on the molecules as a result of the fluctuating electron clouds in the molecule (known as **London forces**). If one part of a molecule becomes slightly positive it will induce a small negative charge on the closest part of an adjacent molecule and attract it. (A full explanation of these forces requires quantum theory.)

Hydrogen bonds   These occur when molecules have hydrogen atoms attached to either nitrogen, oxygen, or fluorine atoms. All of these atoms are highly electronegative; that is, they tend to drag electrons away from other atoms, in this case the hydrogen, leaving positive regions near the hydrogen and negative regions on the other atom. If the hydrogen bond were weaker, water would be a gas at room temperature and life on Earth would be impossible!

Intermolecular bond energies are in the range 1 to 50 kJ mol^{-1}.

Crystalline materials

Crystalline materials have a regular geometric arrangement of particles. The regularity or periodicity of a crystal means its entire structure can be generated by repeating small cells, rather like the way a brick wall is made up of many identical bricks. These **unit cells** fix the geometry and symmetry of the whole crystal. The simplest example of this is a cubic lattice. Its unit cell has a single particle at each corner of a cube and this cube repeats over and over again throughout the structure of the material – the ionic lattice of salt (NaCl) is like this.

The clear-cut geometry of large single crystals comes about because they cleave (split) easily along the planes of atoms. This is why their large-scale symmetry reflects their microscopic atomic arrangement. In metals this is less obvious because they are usually poly-crystalline and the tiny individual crystals within them have grown together in a haphazard way (like a pile of stones rather than a regular pyramid).

Amorphous materials

Crystals have highly ordered atomic arrangements. **Amorphous** materials have little or no long-range order – this means that it is impossible to predict the positions of distant atoms from the positions of those nearby, although there may be some limited form of order over short distances. Glass and soot are two good examples. The lack of a large-scale structure also shows up when their microscopic structures are analysed using X-ray diffraction techniques. In some ways the pattern for an amorphous material is rather like that of a liquid. Glass can be considered a super-cooled liquid which has solidified too quickly to allow crystallization to take place. A diffraction photograph would therefore have much fuzzier diffraction rings or be even more featureless.

Polymers

These contain long-chain molecules; rubber and polythene are two examples. They can be considered as semicrystalline materials since the molecules can align with one another to different extents depending on how the material has been prepared or treated. In rubber the individual molecules are held together by weak intermolecular forces allowing the coiling and uncoiling of molecular chains which account for rubber's great elasticity. Polythene has regions of crystalline structure and regions of amorphous structure. In a **thermosetting polymer** the long-chain molecules bind together with covalent bonds – this creates a material which is stiff, can be quite brittle, and which does not soften on heating. Thermosetting polymers are often used for light switches and other similar electrical fittings. **Thermoplastic polymers** are not cross-linked by covalent bonds and do soften on heating.

Ceramics

These are based on covalently bonded oxides of elements such as silicon, aluminium, and magnesium. They can be formed into various structures and are usually brittle, strong, and resistant to high temperatures. In clay, used for ceramic pottery, the oxides form small plates which slide over one another when wet but which lock together when the water is driven off by heating. The rotor blades in some jet engines are made from ceramic materials.

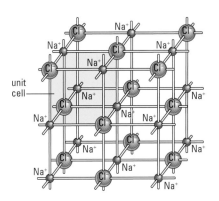

The cubic ionic lattice of sodium chloride (NaCl), showing the unit cell. Each atom is shared by 8 unit cells which meet at these points, so really there is 1/8 atom at each of the 8 corners of the unit cell and therefore 1 atom per unit cell.

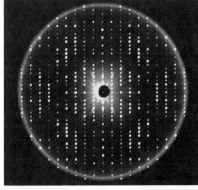

X-ray diffraction photographs of a single crystal (above) and polycrystalline material (below).

Composites

Sometimes two materials are combined to form a composite that exploits the best properties of both. GRP (glass-reinforced plastic) consists of glass fibres in a polymer matrix. The plasticity of the plastic combined with the strength of the glass makes the material light, strong, and flexible. Carbon fibres can also be used in this way to make a composite material that is weaker but stiffer than GRP.

PRACTICE

1 Materials scientists use common words in a special way. Explain what they mean by each of the following terms: strong; tough; brittle; plastic; elastic; hard.

2 Classify each of the following as: crystalline, amorphous, polymeric, ceramic, or composite.

   a steel-reinforced concrete   b copper

   c fibreglass   d bone china

   e DNA.

3 In practice covalent bonds between different atoms always have a partially ionic character. Why is this?

CRYSTALLINE MATERIALS

In 1848 Bravais showed that there are just 14 different kinds of unit cell, so all crystalline materials are classified according to which of the 14 **Bravais lattices** they belong.

Metals usually crystallize in one of three forms: **face-centred cubic** (f.c.c.), **body-centred cubic** (b.c.c.), or **hexagonal close pack** (h.c.p.). The unit cell for each of these is shown in the diagram.

Metallic structures.

Metallic structure	Number of atoms per cell	Number of nearest neighbours for each atom (coordination number)
b.c.c.	2	8
f.c.c.	4	12
h.c.p.	6	12

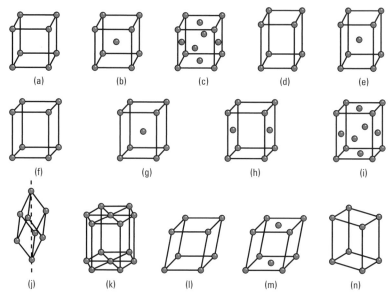

The 14 Bravais lattices: (a) simple cubic, (b) body-centred cubic, (c) face-centred cubic, (d) simple tetragonal, (e) body-centred tetragonal, (f) simple orthorhombic, (g) body-centred orthorhombic, (h) end-centred orthorhombic, (i) face-centred orthorhombic, (j) rhombohedral, (k) hexagonal, (l) simple monoclinic, (m) end-centred monoclinic, (n) triclinic.

A continuous region of the same structure inside a material is called a **phase**. Some materials can crystallize in more than one phase, often dependent on the conditions of crystallization. For example, iron has a b.c.c. structure at room temperature but changes to f.c.c. above 912 °C; titanium, which is h.c.p. at room temperature, adopts the more open b.c.c. structure (with a 3.3% increase in volume and reduction in density) above 880 °C. Even ice has three distinct phases (labelled ice I, II, and III).

Worked example

Calculate the packing density in an f.c.c. crystal. There are 4 atoms per f.c.c. unit cell so the fraction of space that is filled by the atoms is the ratio of 4 atomic volumes to the volume of the cell. To calculate this assume that the atoms are spherical and close-packed, so that their centres are separated by two atomic radii ($2r$). The diagonal line between opposite corners of one face of the unit cell is equal to $4r$. This can be related to the length of side of the cubic cell, a, using Pythagoras's theorem (see diagram):

$$(4r)^2 = 2a^2 \text{ or}$$

$$r = \frac{a}{2\sqrt{2}}$$

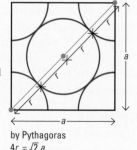

by Pythagoras
$4r = \sqrt{2}\,a$

volume occupied by 4 atoms $= 4 \times \frac{4}{3}\pi r^3 = \frac{16}{3}\pi \times \frac{a^3}{16\sqrt{2}} = 0.74a^3$

Now the volume of a cell $= a^3$, so the packing density is 74%

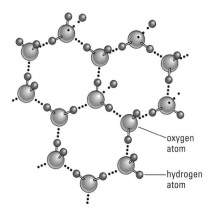

The structure of ice. Each oxygen atom is tetrahedrally surrounded by four hydrogen atoms. Two of the hydrogen atoms are close, forming H_2O molecules, while the other two are further away. The distribution of hydrogen atoms in these two types of position is completely random.

oxygen atom

hydrogen atom

Ice is unusual, having a more open structure than water (for water below 4 °C). This gives it a lower density and explains why icebergs and icecaps float. Carbon also has several phases: diamond and carbon have been known for a very long time; buckminsterfullerine and its related structures were discovered more recently.

Another crystalline structure that has been of particular interest in recent years is the **perovskite** structure. 1-2-3 Yttrium-barium-copper oxide, a ceramic, is an unlikely sounding compound, but it has this structure and is a high-temperature superconductor. The mechanism by which these strange compounds become superconducting is not fully understood but is linked to the distribution of charge in this structure.

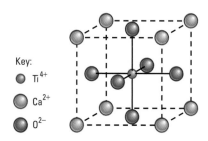

Key:
● Ti^{4+}
○ Ca^{2+}
● O^{2-}

The perovskite structure.

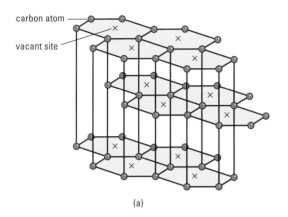
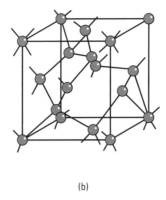
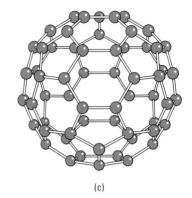

Three different forms of carbon: (a) graphite, (b) diamond, (c) buckminsterfullerene.

Molecular crystals

Linear polymers can also crystallize, although it is most common to find a mixture of crystalline and amorphous (disordered) phases in any particular polymeric material. This crystalline structure shows up in X-ray diffraction patterns and can be used to determine lattice spacings just as for metals and other crystals. Even deoxyribonucleic acid (DNA) molecules (the biological molecules that carry the genetic code) can be crystallized, and X-ray diffraction evidence helped to unravel their regular structure.

Crystal defects

Real crystals usually contain many defects, and these have a crucial influence on mechanical, electrical, and optical properties.

- **Vacancies** These are holes in the structure where an atom is missing. Vacancies are an example of a point defect.

- **Dislocations** These are line defects, and can take many forms. Edge dislocations are formed when an extra plane of atoms extends partway through a crystal lattice. At the end of this incomplete plane there is a disruption in the structure that extends in a line through the crystal. This weakens the crystalline structure and allows planes of atoms to slip over one another much more easily than in a perfect crystal. This explains why the actual strength of metals is an order of magnitude lower than the theoretical strength based on the assumption of an ideal crystal lattice.

The double helix

Maurice Wilkins and Rosalind Franklin used X-ray diffraction techniques with crystallized DNA to show that a previous model of its structure, proposed by Linus Pauling, was incorrect. Wilkins passed his data to James Watson and Francis Crick at Cambridge, and Wilkins, Watson, and Crick shared the 1962 Nobel Prize for physiology. Franklin's role was played down, and controversially she did not share in the prize, although her X-ray diffraction photographs were the first to imply a helical structure and she was also the first to show that the phosphate groups in DNA lie outside the chain (the crucial piece of evidence which brought about the downfall of Pauling's model). Watson and Crick correctly identified the double helix and the arrangement of base pairs that carries the genetic code.

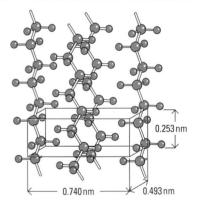

Polythene as a molecular crystal.

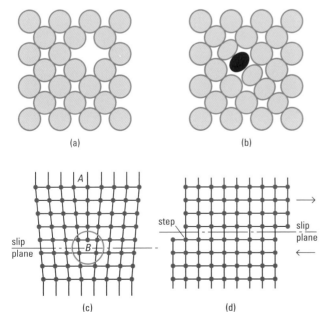

Crystal defects: (a) a vacancy, (b) an interstitial impurity, (c) atomic packing near an edge dislocation (at B); far from the dislocation (e.g. at A) the atoms are in their normal configuration, (d) dislocations allow slip to occur at lower stresses than if the structure were perfect.

PRACTICE

1 Draw a diagram of a simple cubic lattice – the unit cell has one atom at each corner of a cube, and each of these atoms is shared by 8 similar unit cells.

  a What is the coordination number for this lattice?

  b Calculate the packing fraction for the lattice.

2 How would you expect the structure of rubber, which consists of long-chain molecules, to change as the rubber is stretched?

METALLIC STRUCTURES

Metallic bonds are not directional, so metal atoms tend to join together in a close-packed way, rather like apples stacked on a market stall or marbles in a box. This kind of packing can be modelled in a simple experiment. Take a bag of marbles or similar small spheres and tip them out into a small box or tray. Lift one end of the box and shake it slightly – the spheres will pack together at the lower end. You will notice some particularly close-packed regions in which there is an obvious hexagonal arrangement. This is the tightest packing of spheres in a plane (they tend to fall into this pattern because it packs more of them closer to the bottom of the box and so minimizes the gravitational potential energy). If you add more and more spheres so that several layers form you will see that the hexagonal pattern can be repeated over and over to form a three-dimensional close-packed structure.

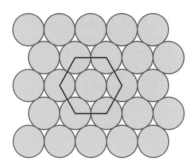

Two-dimensinal close-pack. All atomic planes in both h.c.p. and f.c.c. structures are packed like this. Notice the hexagonal symmetry and the fact that each atom has 6 nearest neighbours in the plane.

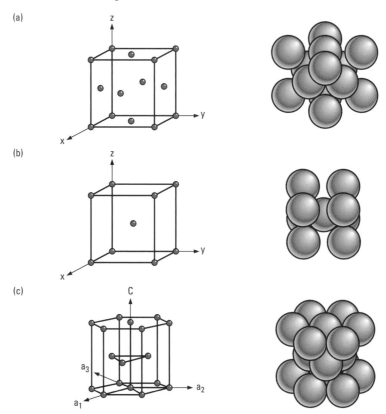

The principle crystal structures of metals: (a) f.c.c., (b) b.c.c., (c) h.c.p.

Close-packing

There are two different ways in which close-packed planes can stack to form a close-packed three-dimensional crystal. One repeats the original layer in every second plane of atoms (*abab* etc.) whereas the other repeats in every third plane (*abcabc* etc.). These close-packed structures are the densest way of filling space with spheres and each of the atoms in both alternative close-packed arrangements has 12 nearest neighbours (6 in its own plane and 3 in each of the planes above and below it). This is called the **coordination number** for the structure and is greatest for these close-packed arrangements. Spherical atoms in close-packed structures fill about 74% of the available space.

Although these arrangements give equally close-packed structures they have different crystal symmetries.

• **Hexagonal close-packed (h.c.p.)** The *abab* structure has a hexagonal symmetry.

• **Cubic close-packed (f.c.c)**. The *abcabc* structure has a cubic unit cell with an atom at the centre of each cubic face – hence the term 'face-centred cubic' (f.c.c.) for this structure.

The unit cells for each structure are shown in the diagram. Copper and aluminium have f.c.c. structures (a) whereas zinc and cadmium are h.c.p. (c).

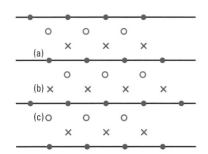

h.c.p. and f.c.c. If the first two atomic layers are in positions (a) and (b) above, then the third layer can go either directly above the first (above (a)) or in a new position (c). The first alternative gives h.c.p, the second f.c.c.

Body-centred cubic

Some metals adopt a slightly more open structure which also has cubic symmetry but is not close-packed. Iron has this this body-centred cubic (b.c.c.) structure at room temperature but changes to f.c.c. above 800 °C. The b.c.c. unit cell is also shown in the diagram above (b).

Crystals and grains

When a metal solidifies crystals grow as atoms settle into place on each layer. The conditions under which the metal solidifies control how large the crystals become, and most metals contain a very large number of tiny crystals called **grains**. These may be no more than 0.01 mm across, although they can be much larger than this. The arrangement of atoms in one grain is not related to their arrangement in adjacent grains, and the grains and grain boundaries show up clearly when suitably prepared specimens are observed under a microscope.

Metal grains.

Bubble rafts

Metal structures can be modelled using a **bubble raft**. If a low-pressure gas or air jet is blown through a syringe into a shallow dish containing soap solution, a steady stream of regular bubbles forms on the surface of the liquid. The bubbles have a long-range attraction and short-range repulsion to one another with no preferred direction for these forces, so they tend to pack together like metal atoms. The bubble raft that forms usually contains **grains**, **grain boundaries**, and various **crystal faults** such as **dislocations** (where the crystalline structure is flawed), **vacancies** (gaps where a bubble is missing), and **impurities** (bubbles of the wrong size that distort the structure).

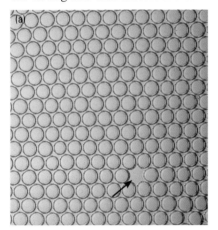
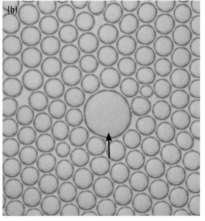
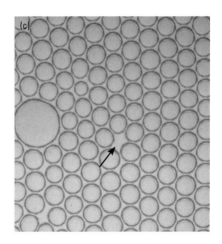

Crystal defects modelled using a bubble raft: (a) vacancy, (b) impurity, (c) dislocation.

The structures of some metals.

Element	Atomic number	Structure(s)	Interatomic distances/nm	Atomic radius /nm
sodium	11	b.c.c.	0.37	0.19
magnesium	12	h.c.p.	0.32	0.16
aluminium	13	f.c.c.	0.29	0.14
calcium	20	f.c.c.	0.39	0.197
		h.c.p.	0.40	0.200
iron	26	b.c.c.(α)	0.248	0.124
		f.c.c.(γ)	0.252	0.126
copper	29	f.c.c.	0.26	0.13
silver	47	f.c.c.	0.29	0.14
tungsten	74	b.c.c.(α)	0.27	0.137
		cubic(β)	0.28, 0.25	0.141
gold	79	f.c.c.	0.29	0.144
lead	82	f.c.c.	0.35	0.18

EXPLAINING METAL PROPERTIES

The mechanical properties of metals can be explained in terms of their microscopic structure. Most metals are **ductile** (they can be drawn out into thin wires), **malleable** (able to be beaten into thin sheets), **tough** (able to undergo plastic deformation before breaking), and **elastic** (able to return to their original shape when an applied force is removed). These properties can be enhanced or diminished by treating the metal in different ways – if the microstructure is changed the large-scale properties can be manipulated to suit particular purposes.

Plastic deformation and dislocations

Ductility, malleability, and toughness all depend on the ability of the metal to undergo **plastic deformation**. This involves planes of atoms sliding over one another. If a single crystal of a metal is deformed plastically, ridges form on its surface where individual planes of atoms have slipped. However, if you calculate how much stress is needed to slide one perfect layer over another (which requires enough force to break all the bonds at once) the result is much higher than the stress that is actually required to deform real metal samples (the stress needed to make a single crystal undergo plastic deformation is much greater than for a polycrystalline metal sample).

The reason metals undergo plastic deformation so easily is because plastic deformation results from the movement of flaws in the crystal structure, called **dislocations**. These allow atomic planes to slip past one another without having to overcome all the bonds at once. This mechanism is similar to moving a carpet across a floor by introducing a hump and kicking that across – this is much easier than sliding the whole carpet. There are various different kinds of dislocation depending on the way in which the structure is flawed. The type shown in the crystal diagram (left) is called an **edge dislocation**, and is formed where an extra half row of atoms is squeezed in between two other complete planes. When the sample is stressed the dislocation moves *by breaking one bond at a time* and migrating sideways as shown. The net effect is that the planes of atoms move past one another and the metal is plastically deformed.

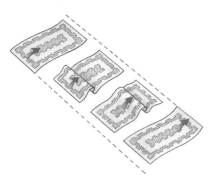

It is much easier to move a large carpet across a floor by moving successive 'humps' than by dragging the whole carpet in one go.

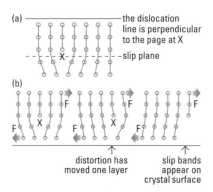

(a) An edge dislocation. (b) Dislocation movement in plastic deformation.

Controlling metal properties

Dislocations cannot cross grain boundaries, so metals with small grains will not deform plastically as readily as those with larger grains. This makes them more **brittle** and **harder**. Dislocations can also be trapped by atoms that don't fit the crystal structure, so adding even small amounts of impurities can drastically change properties – a fact that is utilized in the production of carbon steels. **Cold-working** a metal (e.g. by repeated bending or hammering) introduces more and more dislocations, and eventually these get so 'tangled up' that they are again trapped, making the metal harder and more brittle (this process is also known as **work-hardening**). You can easily feel the effects of work-hardening if you take a steel coat hanger and bend the wire backwards and forwards until it snaps.

Some common treatments and their main effects are summarized below.

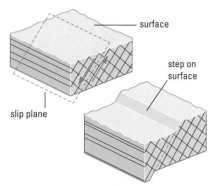

Slip planes leave steps on the surface of a single crystal.

• **Cold-working** The metal is hammered, beaten, or bent repeatedly – more dislocations form and lock each other in place. The metal becomes harder, stiffer (its Young modulus increases), and more brittle.

• **Annealing** The metal is heated to red heat and then slowly cooled – grains grow larger, making the metal softer, more easily bent, hammered out, or scratched.

• **Tempering** The metal is heated to red heat, cooled slowly to a particular temperature, and then **quenched** by plunging it into cold water, oil, or brine to freeze in a particular grain structure – this gives control over brittleness and hardness. The higher the quenching temperature the smaller the grains and the harder and more brittle the resulting metal.

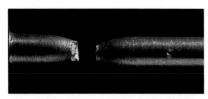

The ductile fracture of a tough material. This shows the ends of a copper wire that has failed under tensile stress. Being tough, a considerable amount of plastic deformation occurred before it finally broke. Notice the characteristic 'neck'.

When a liquid metal cools in a casting it begins to crystallize from many points, but mainly from the walls of the casting, where it cools first. This process is called **nucleation**, and long branching crystalline arms called **dendrites** grow inwards into the liquid from these nuclei. As the liquid continues to cool, the rest of the liquid crystallizes onto the interlocking matrix of dendrites, forming a polycrystalline grain structure. The irregular shapes of the grains come about because of the arbitrary distribution of dendrites and because of their irregular growth. This is why the grains themselves do not reflect the underlying symmetry of the metallic crystals.

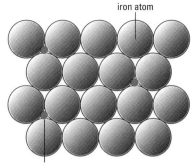

iron atom

carbon atom

*The addition of even a small proportion of carbon atoms to iron radically alters its mechanical properties. The impurity atoms which squeeze into gaps in the lattice are called **interstitial impurities**.*

Impurities

Another way of altering metal properties is to introduce **impurities**. Steel is about 99% iron, but its mechanical properties can be varied over a wide range by altering the amount of impurities it contains, and changing its heat treatment. About 0.1 to 1.5% of steel is carbon: the smaller carbon atoms get in between the layers of iron atoms, making it harder for them to slide over one another. This makes the steel stronger and more elastic because it will deform further before any slip occurs. Pure iron, on the other hand, is flexible but not elastic; it undergoes plastic deformation fairly readily. Cast iron, which has 3–4% carbon, is neither flexible nor elastic: it is quite brittle. The excess of carbon forms flakes inside the structure and these separate iron crystals so that the material breaks apart fairly easily if stressed. Carbon is not the only impurity added to iron – chromium, nickel, manganese, titanium, and molybdenum are other examples. Stainless steel (which resists the formation of iron oxide, or rust) contains chromium and nickel as well as carbon impurities.

Alloys

The ability of molten metals to mix together to form **alloys** gives yet another way of controlling metal properties. Often it is possible to adjust the proportions of different metals to get a useful compromise between their properties. Duralumin, for example, which is an alloy of aluminium with 4% copper (plus small amounts of manganese, magnesium, and silicon), takes advantage of aluminium's low density, but is much stronger. It has been very important in the aerospace industry. Bronze (about 90% copper, 4% tin) is harder and stronger than copper alone. Brass is about 70% copper, 30% zinc, and is often used for cheap decorations and ornaments as it can be polished to a smooth golden lustre. Tungsten carbide 'steel' is used for the edges of cutting tools because it is very hard. However, it is actually an alloy of about 90% tungsten carbide with about 10% cobalt, and does not contain any iron!

Sintering

Sometimes metals or ceramics are made from powders rather than a molten mixture. In **sintering** a compacted powder is heated to a temperature just below its melting point. The individual atoms now have sufficient thermal kinetic energy to diffuse through the structure via vacancies in the crystal, and the material becomes denser, stronger, and polycrystalline. However, it is usually of lower density than a metal formed from a melt and is often absorbent (e.g. brass bearings formed in this way can be soaked in oil to give them a longer service life). Tiles on the Space Shuttle were sintered.

Creep

The descriptions of material properties given so far have ignored **time-dependent effects**, but engineers must take these into account in their designs. For example, if a load heavy enough to cause plastic deformation is suspended from a copper wire the extension of the wire gradually increases. This is **mechanical creep**. The atoms inside the copper are in a state of thermal agitation and are under stress. Every now and then a dislocation is freed and two planes of atoms slip over one another. The slip is always in the direction of the applied stress so the strain increases with time. Turbine blades are constantly in tension as they rotate, so they creep. Too much creep would be disastrous, so alloys or ceramics with great creep resistance are used.

Superalloys

Materials used in parts of the aerospace and power industries must be extremely strong and hard and must withstand creep and corrosion under extreme conditions of temperature and in hostile chemical environments. A new class of high performance alloys (or 'superalloys') based on nickel, cobalt, or nickel–iron has been developed for these applications. Superalloys are used, for example, in turbine blades, for aircraft skins, inside reactors, and for spacecraft and have influenced the development of new orthopaedic and dental prostheses.

PRACTICE

1 Why can't dislocations cross grain boundaries?

2 Suggest and explain how increased temperature might affect the rate at which a loaded metal wire creeps.

3 How would you expect the mechanical properties of a metallic single crystal to differ from those of the bulk, polycrystalline metal?

4 Discuss the suitability of each of the following materials for a car body: aluminium; steel; lead; plastic; fibre glass; glass; rubber.

Consider hardness, stiffness, plasticity, brittleness, strength, cost, and density, and present your answer in the form of a table.

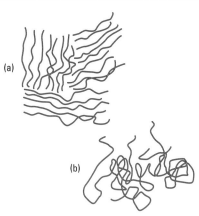

Two-dimensional representation of (a) semicrystalline and (b) amorphous polymers.

POLYMERS

Imagine a tangled mess of slippery cooked spaghetti. This is rather like the molecular structure of an amorphous **polymer**. While the molecules are slippery the whole bundle can be stretched out so that the strands align, adopting a semicrystalline structure. Cool the spaghetti down and the stickiness congeals linking individual strands together in a more permanent configuration. This is what happens to amorphous polymers below a certain critical temperature; all their molecules stick together and the material becomes hard, glassy, and brittle. Of course, this is just a simple model, but it reproduces some of the behaviour of natural and synthetic polymers and shows how manipulating structure allows materials scientists to create materials with particular properties.

The defining characteristic of all polymers is that they are hydrocarbons made of long-chain molecules. Natural polymers include rubber, cellulose, and proteins, and synthetic polymers (plastics) include nylon, polythene, perspex, PVC, and bakelite. Their properties depend on three main things:

- **Degree of crystallinity** – the extent to which their molecules are arranged in an ordered way.
- **Nature and extent of cross-linking** – atoms within the polymers are joined together by covalent bonds, but the molecules themselves may be joined by weak (van der Waals) or strong (covalent) bonds.
- **Temperature** – the higher the temperature the more thermal energy the molecules have, and if this is enough to allow molecules to move past one another the material's properties are more like those of a super-viscous liquid than a true solid.

Amorphous polymers

Synthetic polymers are made by joining lots of short molecules (**monomers**) together to form long chains (**polymerization**). Although the backbone of these chains consists of many carbon atoms bonded together, there are many other elements and chemical groups that can be attached to the chains to alter the properties of the plastic that is formed. These plastics are usually amorphous polymers, and are particularly useful because they have a low density, are cheap, can be made in any colour, and can be moulded into any shape.

Thermosetting polymers   When these are first heated strong covalent bonds form cross-links between the molecules and lock them into fixed positions. The whole structure is like an extended macromolecule and the plastic is strong (it has a large breaking stress), stiff (it has a high Young modulus), brittle (it does not deform plastically before breaking), and heat-resistant (it does not soften on heating). This means it must be formed in its final shape since it cannot be moulded later. In fact heating it is likely to break down the molecules before it melts the plastic. Melamine (used for dishes) and epoxies (used for glues) are both thermosetting polymers. The cross-links in an epoxy resin are formed by chemical reactions when the glue is mixed with a hardener.

Thermoplastic polymers   These soften on heating and can be moulded into any desired shape. They harden again on cooling. They differ from thermosetting polymers in that their molecules are only bound by weak van der Waals forces between adjacent atoms, and moderate heating is sufficient to break these bonds. They have lower values of Young modulus, and are softer and tougher than thermosetting polymers. Polypropylene (used for plastic sheets and gutter piping) and polythene (used for drinks containers and clear sheeting) are both thermoplastic.

The mechanical properties of two themosetting and two thermoplastic polymers are compared in the following table.

Comparing polymers.

Name	Type of polymer	Cost £/kg	Tensile strength /MPa	Maximum strain (%)	Relative hardness (Rockwell scale*)	Young modulus /MPa	Density /g cm⁻³
polyesters	thermosetting	0.7	30	0	100	7	1.1
epoxies	thermosetting	2.1	70	0	90	7	1.1
polythene (low density)	thermoplastic	0.6	14	90–800	10	0.17	0.92
polypropylene	thermoplastic	0.6	35	10–700	90	1.4	0.91

*A standard hardness scale which measures the degree of penetration when a diamond-tipped indenter is pressed on to the material surface.

Glass transition temperature

At low temperatures, amorphous thermoplastic polymers are glassy and brittle. There is a critical temperature, the **glass transition temperature**, T_g, above which thermal agitation is sufficient to loosen the molecules and the material becomes rubbery. Rubber itself becomes glassy below about −200 °C. The graphs show how this transition affects the density and elasticity of the polymer.

Effect of time

When strain occurs molecules must move and this takes time. If stresses are applied very suddenly there may not be enough time for all the molecules to respond and so the material becomes stiffer and more brittle, and responds more like a glassy polymer; it may even shatter. However, a gradual stress applied to the same material may result in rubbery behaviour. Silly Putty is an extreme example of this – it bounces like a ball if dropped, but gradually flows out like a liquid if left on a flat surface for long enough.

Semicrystalline polymers

Thermoplastic polymers like polythene are not entirely amorphous. As they form there is a tendency for the linear molecules to line up parallel to one another and the material ends up with amorphous and crystalline regions, so it is best described as semicrystalline. The degree of crystallinity affects its mechanical properties because the crystalline part behaves like other covalently bonded atomic crystals whereas the amorphous part is **viscoelastic** (molecules flow over one another under stress). High-density polythene (used to hold packs of beer cans together) has a significant degree of crystallinity and exhibits some rather strange properties because of this. If you take a strip of this material and try to stretch it, it will behave elastically at first (up to about 10% strain) and then it 'gives' quite easily and irreversibly, more than doubling its original length, but not narrowing in the way a metal would. This is called **cold-drawing**. The molecules have become partially aligned and now any further stress is opposed by the covalent bonds in the molecules. This dramatically increases the strength and stiffness of the polymer.

The efffects of temperature on a thermoplastic polymer.

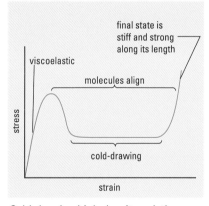

Cold-drawing high-density polythene.

PRACTICE

1 Suggest molecular explanations for the following observations about rubber.

   a At room temperature rubber can undergo elastic strains of several hundred percent.

   b Below about −70 °C rubber becomes a brittle solid.

   c If a rubber band supporting a load is heated it will raise the load slightly.

2 Both metals and polymers can undergo plastic deformations. In what respects are these processes **a** similar and **b** different?

3 Define each of the following terms as simply as you can:

   polymer; amorphous; crystalline; glassy; brittle; plastic; cross-link; heat-resistant; thermoplastic; thermosetting; semi-crystalline; monomer.

RUBBER

The *Challenger* Space Shuttle disaster in 1986 was apparently caused by the failure of one component, a rubber O-ring used as a seal. At the senate hearings the physicist Richard Feynman demonstrated that the O-ring would have lost most of its elasticity in the freezing temperatures that prevailed at the time of launch, resulting in the failure of the seal and ultimately the tragedy of the Space Shuttle explosion.

The property of rubber that makes it suitable for seals, car tyres, balls, and so on is its ability to withstand enormous elastic strains without breaking. It belongs to a class of materials called **elastomers**. Elastomers are polymers that return to their original length after being stretched repeatedly to a strain of 100% or more. Note that it is not just the large elastic strain but also the **resilience** to repeated strains that make elastomers so useful.

Rubber is a natural polymer whose long-chain molecules are randomly tangled at room temperature, but which align when under an applied stress. This process of molecular movement is called **viscoelastic flow**. The material changes from amorphous to crystalline when stretched, as can be seen from X-ray diffraction patterns of stretched and unstretched rubber samples.

 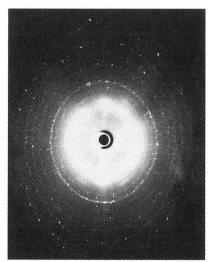

(a) Electron diffraction pattern for unstretched natural rubber. The lack of detail implies a random amorphous structure. (b) Electron diffraction pattern for highly stretched natural rubber.

Vulcanized rubber

The amazing ability of rubbers to undergo **elastic** (i.e. reversible) strains of several hundred percent should not be confused with the ability of thermoplastic polymers to undergo similar **plastic** (i.e. irreversible) strains. The key is in the degree of cross-linking. Before Charles Goodyear (1800–1860) natural rubber was too soft and weak to be of much use for anything other than pencil erasers. Goodyear discovered that if rubber is heated with sulphur it becomes harder and stronger and yet retains its elasticity. This process is called **vulcanization**. The sulphur atoms bond to carbon atoms on adjacent chains, binding them together. The more vulcanization occurs, the harder and stiffer the rubber becomes. Nowadays some thermoplastic elastomers have been produced whose molecular chains have both elastic and stiff regions (**domains**). These avoid the need for the labour-intensive vulcanization treatment.

Elastic hysteresis

If you plot force against extension for a strip of rubber as you add and remove loads you will see that it does not obey Hooke's law. The loading and unloading lines form a closed **hysteresis loop** on the graph.

Elastic bands

You may have noticed that old elastic bands sometimes get stiffer and more brittle. This is because further cross-linking occurs by exposure to oxygen in the atmosphere.

advanced **PHYSICS**

The shape of the loading curve below can be understood by thinking about the molecular processes involved in stretching. Initially the molecules are held in their random tangled positions by van der Waals forces (OA). Once these are overcome the molecules begin to align with one another and the rubber becomes less stiff (AB). In this region the effect of an applied force is to unravel the molecules by making them rotate about single carbon–carbon bonds. It is during this process that the large strain of an elastomer occurs. Eventually the molecules are more or less aligned and unravelled and the applied force is resisted by covalent bonds: the rubber now becomes stiff (BC).

The forces exerted by the rubber when it contracts (as unloading takes place) are less than the forces exerted when it is stretched, so the work done by the rubber in contraction is less than the work done on the rubber to stretch it. Total energy is conserved so some of the work done stretching the rubber must have been transferred to thermal energy inside it. This can be calculated from the graph if you remember that work done is equal to the area under a force–extension graph. The work done to stretch the rubber is the area under the loading curve, and the work done by the contracting band is the area under the unloading curve, so the energy transferred to thermal energy is the area inside the hysteresis loop.

Tyre war! Before 1997 Goodyear supplied tyres to all Formula 1 teams. A rival tyre company, Bridgestone, entered Formula 1 for the first time at the start of the 1997 season. As a result of the competition to produce more effective race compounds, lap times were lowered by several seconds.

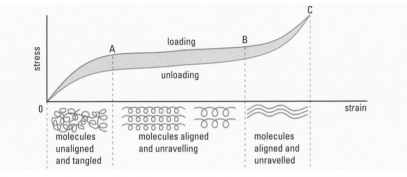

Stress versus strain for rubber. The area contained within the hysteresis loop is the energy per unit volume transferred to thermal energy in one loading and unloading cycle.

Properties of rubbers.

Name	Tensile strength /MPa	Maximum strain (%)	Resistance to oil, gas, etc.	Useful temperature range/°C
natural rubber	21	800	poor	−51 to 82
silicone	4.8	300	poor	−117 to 315
urethane	35	600	excellent	−54 to 115

Rubber is highly resilient because it can stand up to repeated distortion around the hysteresis loop. Think about the treatment a car tyre must withstand and you will see how important this resilience is.

- The heating effect of hysteresis can be felt directly in a simple experiment. Take a rubber band, stretch it suddenly to two or three times its original length and hold it against your lip (which is sensitive to quite small changes in temperature). It will be noticeably warmer.

- Another rather surprising experiment uses the reverse effect. If you warm a stretched rubber band it will contract. The thermal excitation increases the amount of random rotation about the single bonds and the molecules ravel up and become shorter. This can be used to extract work from thermal energy to drive a small 'heat engine' (see above right).

A rubber band heat engine. As bands on one side are heated, they all contract and the centre of mass of the wheel is displaced. This creates a turning effect and the wheel begins to rotate, moving other cooler bands in front of the heat source. The process continues.

PRACTICE

1 Sketch stress–strain curves for copper, steel, glass, and rubber onto a common set of axes. Explain how the shape of each graph relates to the mechanical properties of the material concerned.

2 Give a concise explanation of each of the following terms: elastic hysteresis; viscoelasticity; vulcanization; elastomer.

10.9

O B J E C T I V E S

- natural materials
- manufactured materials
- composites

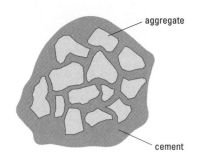

The structure of concrete.

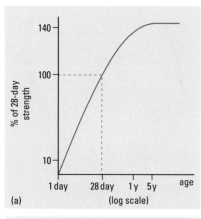

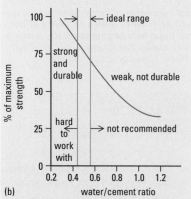

(a) The strength of setting concrete increases with time, taking several years to reach its maximum value.
(b) Increasing the water/cement ratio makes concrete weaker and less durable.

BUILDING MATERIALS

Natural stone is very strong in compression, but weak in tension. This makes it excellent for supporting things but not much use if it has to span a large gap. However, large natural stones free of cracks and imperfections (that make them weak) are not always readily available and a number of artificial stone-like building materials have been developed over the centuries. The two most important are **bricks** and **concrete**, and these are held together by **cement**.

- **Bricks** Clay is formed when rocks such as granite weather into very fine grains. The grains stick together when wet to form a viscous mud similar to plasticine. In hot dry places wet clay is moulded into bricks which harden when they are baked in the sun. In wetter climates these bricks would soften when it rained, so house bricks are **fired** (placed in a kiln at a high temperature) to make them hard, stony, and resistant to water.

- **Concrete** This is a **composite material** – a mixture of gravel or broken stones and pebbles is mixed with sand, cement, and mortar and dries to form a hard stony material in any desired shape. Solid stone would be as strong as concrete, but concrete is far more adaptable. Its tensile strength is limited because it has a biscuit-like structure, made up of many small particles cemented together, and it tends to crack and break where these join. However, concrete is very strong in compression and much cheaper than other alternatives.

- **Cement** This is the essential ingredient for gluing bricks or aggregate together. Portland cement is made by burning clay and limestone together and crushing the resultant **clinker** into a fine powder. When this is mixed with sand and water, hydration reactions initiate crystallization. As the crystals grow they make contact with one another and form bonds. This is a slow process, taking up to an hour to really get going and many weeks to complete. The reactions are exothermic and as the concrete sets they release about 300 kJ kg^{-1}. This can cause a significant increase in temperature in a large concrete structure and can be a problem, since the centre of a large block gets quite hot and contracts as it cools, sometimes cracking the block. In some cases (e.g. large dams) pipes are cast into the concrete and chilled water is pumped through them to keep the concrete cool while it is setting.

Properties of building materials.

Material	Density/kg m^{-3}	Tensile strength/MPa	Compressive strength/MPa
concrete	2400	0–5	10–70
steel	7860	360	360
cast iron	7200	250	850
aluminium	2800	150	150
wood	500–700	90–100	50–60

Other properties of concrete

- Low thermal conductivity – so it provides good thermal insulation.
- High specific heat capacity – so it is suitable for fire resistance.
- Retains its properties at low temperatures – so it can be used to store industrial quantities of very cold substances (e.g liquid methane at $-160\,°C$), as metals become very brittle at low temperatures.
- Concrete expands by 1.2 parts per million for every °C increase in temperature.

The strength of concrete changes as it sets and building designs are usually based on the **28-day strength** of the concrete. However, the concrete continues to strengthen after this time and may reach 150% of the 28-day strength after a few years, so older buildings are stronger than new ones. The final strength of building concrete can be adjusted by changing the water/cement ratio, but very dry concrete is hard to work with, so a compromise between strength and ease of working is chosen. This is usually in the range 0.45–0.55 (see graph (b)).

Reinforced concrete

Concrete can be made more resilient to tensile stress by reinforcing it with steel rods (steel is strong in tension and compression). In road building a steel wire mesh is embedded into the concrete. **Pre-stressed concrete** is made by inserting steel rods into the setting concrete. The rods are placed in tension while the concrete sets and then the tension is released. Hence the rods put the concrete in compression so it can survive tensions applied to it. One problem with reinforced concrete is the effect of corrosion on the reinforcing material which can result in the concrete cracking and weakening – various coatings and treatments are used to try to prevent this.

Wood

Wood remains one of the most important building materials. In the USA, for example, the mass of wood used per year (3×10^8 tonnes) is greater than the combined masses of steel and concrete used. As well as a wide range of natural woods, many composites are also used, such as plywood, chipboard, and paper. Although wood is made of polymers, its structure is a complex honeycomb of different cells whose cell walls are strengthened by arrays of cellulose fibres (which have high tensile strength and low stiffness) 'glued' together by lignin (which has low tensile strength and high stiffness). In this respect it is already a composite biological material deriving its overall properties from a combination of the properties of its components. For all the other materials under consideration here the **microstructure** is the key factor that fixes mechanical properties. This is not the case for wood, in which the **macrostructure** is critical. In wood, many properties vary by a factor of up to 20 depending on the position in a particular log and the direction of testing. Natural wood is divided into two classes.

- **Softwoods**   These come from evergreen trees such as pine, and are usually softer and weaker than hardwoods. However, there are important exceptions as some 'softwoods' are harder and stronger than some hardwoods.

- **Hardwoods**   These come from deciduous trees (which lose their leaves every autumn) and are usually harder and stronger than softwoods.

Both types of wood have a grain structure parallel to the long axis of the branch from which the wood is taken. This means the properties of wood are different in different directions – it is an **anisotropic** material. For example, the Young modulus for a piece of Douglas fir may be 13 000 MPa parallel to the grain but only 670 MPa perpendicular to it. You can feel the difference if you try to cut or saw a piece of wood along, and then across the grain.

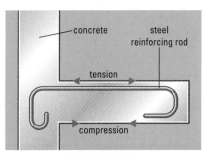

A cantilevered concrete beam. The steel reinforcing rod is placed on the side that will be in tension. Concrete alone would be too weak in tension.

Concrete radiation shields

Concrete containing lumps of lead is not especially strong, but is used as a radiation shield around nuclear power reactors, waste dumps, and at installations such as the European Laboratory for Particle Physics (CERN – for Conseil Européen pour la Recherche Nucléaire).

Wood composites

One way to overcome the problems of warping (as wood dries out) and weakness across the grain is to use wood composites.

- **Plywood** This is a composite material made by gluing several thin sheets of wood together so that their grains are at right angles to one another on adjacent layers. This makes its properties more **isotropic** (same in all directions). 'Three-ply' and 'five-ply' refer to the number of layers.

- **Chipboard (particle board)** This cheap alternative to natural wood is made by mixing wood shavings with glue and compressing them. Although the surface of chipboard is itself rather unattractive, a thin layer of natural wood can be stuck onto it – this is called a **veneer**.

Look at the photographs of buildings and list the most significant materials used for each one. Explain how the properties of the materials you have listed suit each to its purpose in the buildings.

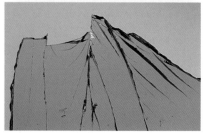

Above: ductile fracture – note the necking that results from plastic deformation before fracture. Below: brittle fracture – there is no plastic deformation; the material shatters as cracks move rapidly through it.

Griffiths cracks

The weakness of glass was explained by A. A. Griffith in 1920. He pointed out that stress concentrates at the tips of tiny cracks in the surface and body of glass when it is subjected to a distorting force. The stress at these points soon reaches the theoretical value at which the glass fractures, and then the tip of the crack travels through the rest of the glass at high speed causing brittle fracture. The more cracks present in the sample, the lower its breaking stress.

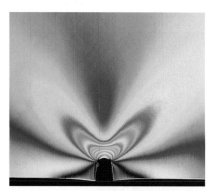

Polarized light reveals the strains resulting from a stress concentration at the tip of a crack in a transparent material.

GLASS AND CERAMICS

Pliny, the Roman historian, writing just after the birth of Christ, claimed that glass was discovered by accident when merchants cooking a meal on the sandy banks of the River Belus in Phoenicia (now in Israel) propped their pots on lumps of soda. These melted and mixed with the sand and transparent streams of liquid were formed. Whatever the true story, glass has been an important material throughout most of recorded history. Ordinary window glass is made by mixing sand (silica), soda, and limestone and heating them to about 1400 °C. The molten glass is then rolled into a flat sheet and allowed to cool slowly and evenly (**annealed**). The molecular structure is amorphous, but the molecules are fixed in position so that glass is stiff, hard, and brittle. However, if it is reheated, it flows and can be blown (for bottles, etc.) or moulded into any shape. The giant amorphous structure of glass has no internal boundaries to scatter light, so it is transparent.

Fracture mechanisms

The main problem with glass is that it is very brittle and will shatter easily. **Brittle fracture** occurs because stresses are concentrated at the tips of cracks and scratches on the glass surface. Metals also have these imperfections, but they undergo plastic deformation near the point of the crack, blunting it and reducing the maximum stress. When a metal breaks it is often by **ductile fracture**, and a lot of plastic deformation occurs before the material fails. This does not happen in a brittle material and the crack moves rapidly through it, causing it to shatter. Brittle and ductile fractures leave quite different surfaces at the point of failure, as you can see in the photographs. Glass and tile cutters take advantage of brittle fracture. If the surface of a sheet of glass or ceramic tile is scratched using a very hard sharp edge the glass or tile will break easily along this line when it is tapped.

Strengthened glass and glass composites

Windscreen glass   Have you ever noticed the coloured pattern on car windscreens when you look at them through polarized sunglasses? This is a stress pattern linked to the pattern of cooling jets used in making the windscreen. The sheet of glass has been heated until it softens and then both the surfaces are cooled by blowing jets of cold air at them. This makes the outer surfaces contract and so the central layer is held in compression (this inhibits crack formation in much the same way as in pre-stressed concrete). As a result the windscreen is more impact-resistant, and when it shatters it breaks into smaller blunt pieces that are less hazardous than large glass splinters.

Laminated glass   This is made of thin glass layers stuck together by a resin. It is used to make bullet-proof glass (car windscreens are also laminated). When the glass breaks the pieces are held together by the resin.

Reinforced glass   This is a cheaper alternative way to strengthen glass and provide greater security. A fine wire mesh is embedded in the glass.

Glass fibre   Glass is actually quite strong – it has a large maximum tensile stress. Unfortunately it usually breaks far below this value because of cracks on the surface and in the bulk of the glass. If thin glass fibres are encased in a resin (plastic material), a tough crack-resistant material is formed. Fibre glass has low density, high strength, is tough, fire-proof, and rot-proof. It is also an extremely good thermal insulator. A similar approach – embedding carbon fibres in a polymer resin – results in a material which is less dense but stronger than steel, known as **carbon fibre**.

Ceramics

Brick, earthenware, stoneware, and porcelain are all ceramics. These are traditionally made by mixing two or more forms of fine powdered clay with sand or powdered rock and water, and then firing the paste that is formed in a kiln at high temperature. Chemically, ceramics are based on oxides of silicon, aluminium, and magnesium, and the strong ionic bonds inside them result in a material that is hard, strong, brittle, and resistant to heating and chemical degradation.

Ceramics have a microcrystalline grain structure like metals, but cannot deform plastically because the bonds are strong and directional, and slip would require ions of like charges to slip past one another (unlike the non-directional bonds in metals that can relatively easily switch from one atom to another when stress is applied). The smaller the grain size in the ceramic the greater its strength, but this is at the expense of brittleness, and most ceramics shatter easily on impact.

The main drawback of traditional ceramics is their brittleness, and a great deal of research has gone into increasing the fracture toughness of ceramics. To increase resistance to brittle fracture something must happen inside the material to absorb energy at the root of a crack (in the way plastic deformation does in a metal). The detailed way in which this is achieved is beyond the scope of this book, but the basic approaches are described on the right.

The aim of this research is to develop new materials that can replace metals in many applications. For example, ceramics have greater strength, hardness, creep resistance, and heat resistance than metals. If they are tough enough they can be used inside engines and turbines to allow them to operate at higher temperatures (the higher the operating temperature of a heat engine the greater its maximum efficiency). They are already used for turbine blades in some jet engines. The development of tough, fracture-resistant ceramics also provides a better raw material for future composite materials.

A ceramic turbine.

Increasing fracture toughness

Transformation toughening The ceramic is formed in a two-phase structure in which one phase transforms, under stress, to a new lower-density phase. This transformation will occur where the stress is concentrated at the tips of cracks and absorbs strain energy, stopping the crack from opening any further and making the material tougher.

Microcrack formation Whereas large cracks travel through a material and cause it to fail, if the tip of a large crack creates more and more microcracks these absorb the strain energy and stop the large crack from moving on, again making the material tougher.

Crack deflection This is what happens in a composite material. Cracks are unable to cross the boundaries between grains, so they do not cause failure. Two-phase microstructures achieve the same effect on a much smaller scale.

Properties of glass and ceramics.

Material	Density/ kg m^{-3}	Young modulus/ GPa	Refractive index at 589 nm	Composition
sheet (window) glass	2.46	70	1.51	73% SiO_2 13% Na_2O
high lead glass	4.28	53	1.69	35% SiO_2 58% PbO
light crown glass	2.90	73	1.54	57% SiO_2 27% BaO
dense crown glass	3.56	79	1.61	36% SiO_2 45% BaO
bone china	2.80	90	opaque	–

PRACTICE

1 Sketch a graph of stress versus strain for a ceramic tile and label the point at which it fractures. What are the main differences between this graph and one for a copper wire?

2 Thin glass fibres that have just formed are remarkably elastic. However, with a little handling they soon become brittle. Why does this happen?

3 a Describe the differences between ductile and brittle fracture.

  b Explain why glass does not undergo any significant plastic deformation before it fractures.

4 The theoretical strengths of metals and glasses are much higher than their actual experimental strengths. What is it that weakens each type of structure?

5 Which of the following describe the mechanical properties of glass?

  stiff; strong; elastic; plastic; ductile; brittle; hard; soft.

6 Before the regulations were tightened up, it was possible to stop a crack in a car windscreen from growing by drilling a small hole at the crack tip. Explain how this works. (Damaged windscreens must now be replaced immediately.)

- friction

- lubrication

- metal fatigue

- hardness

WEAR AND TEAR

Sliding contact between metal surfaces produces very high stresses at the points of contact. This can result in temporary welding of the two surfaces and then shearing as they move apart. It can also result in intense heating and one metal may melt locally. An extreme example of this is used in friction welding, where one half of a shaft is rotated at very high speed and the other half is brought into contact with it. Enough thermal energy is generated to weld the two halves together. However, in many situations this heating effect is undesirable. The effect on the surfaces and the frictional force that results depend on the materials. If the weld is stronger than either metal the weaker metal will tear. On the other hand, whereas local heating of lead is likely to melt it, heating steel may actually harden it. All these changes could occur when a machine is first used, so many machines (for example car engines) must be 'run in' for a while before being used at full power.

Lubricants are used to reduce frictional forces. There are two approaches to lubrication.

- **Hydrodynamic lubrication** This approach tries to keep a fluid (e.g. oil) film between the surfaces so they do not touch. Several factors reduce the effectiveness of hydrodynamic lubrication: intermittent operation, breakdown of the lubricant molecules, and large forces applied at low speeds.

- **Boundary lubrication** This approach involves applying a coating of lubricant molecules to the metal surface. The metal reacts with the lubricant to produce a **metal soap** that allows easier shear for lower applied forces.

It is not possible to predict resistance to abrasion simply from the hardness and strength of a material because of the complex nature of the surface interactions and the effect of bumps and impurities on the surface. The most reliable approach is to test representative samples of the materials under realistic loads.

Fatigue

If you have ever taken a metal coat hanger and bent it backwards and forwards repeatedly to break it you have experienced **work-hardening** followed by **metal fatigue** and failure. The repeated strains generate large numbers of new dislocations in the metal and these eventually get tangled up. In this condition dislocations are unable to move to relieve stress so the material becomes brittle and fails. This is an important consideration in the design of machines whose parts will be subjected to vibration or cyclic stresses.

The design of most structures allows a significant safety margin between the largest expected stress and the maximum tensile strength of the material, so it is very unusual for anything to fail in this way. The most common cause of failure is as a result of repeated stress leading to fatigue. For example, if a steel bar is repeatedly loaded to 80% of its **yield strength** it will eventually fail. This kind of failure is evident from the nature of the fracture that occurs. Unlike tensile failure, where the steel would elongate by about 30% before breaking, there is no noticeable elongation in a fatigue fracture.

Engineering materials that are going to suffer repeated cyclic stresses must undergo fatigue testing before use. The simplest test applies a stress to a sample and then releases the stress, measuring how many cycles are required to break it. This generates an **S–N curve** (S is stress required for failure and N is number of cycles). The lower the applied stress the more cycles are required. The following graph shows an S–N curve for

| Static fatigue |

Some ceramics exhibit **static fatigue**. If they are subjected to a high static load for a long time they may suddenly shatter. This does not happen in a vacuum or in dry air; it is initiated by a chemical reaction between water in the air and the stressed ceramic surface.

a ferrous material. Notice that the curve levels off after 10^6 cycles. If the sample survives 10^6 cycles it will not fail thereafter – this is called its **fatigue strength** or **endurance limit**. For non-ferrous materials there is no endurance limit; the curve continues to fall.

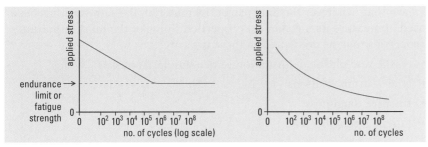

In a simple fatigue test a sample is subjected to cyclic stresses and the number of cycles it takes for fracture to occur is plotted against stress. (a) Testing iron. For stresses below the endurance limit it will never fail. (b) Testing aluminium. Aluminium does not have an endurance limit and will eventually fail under repeated stress, however low the stress.

The Comet disasters

The de Havilland Comet was one of the first aircraft to have a pressurized fuselage. It was also one of the first aircraft made by de Havilland to use aluminium alloys rather than fatigue-resistant wood. In 1953 and 1954 three Comets crashed as a result of metal fatigue. In all three cases cracks started from the same point in the fuselage and spread as a result of repeated strains during pressurization and depressurization for each flight. Eventually the cracks reached a critical '**Griffith length**' and the structure failed by brittle fracture. Nowadays aircraft fuselages are designed in such a way that they will not fail even if quite long cracks (>0.5 m) develop. They are also subject to regular detailed inspections.

Testing an aircraft for metal fatigue. The fuselage of an aircraft is repeatedly pressurized and depressurized to replicate normal operating conditions. Sensors detect any effects.

The remains of an early Comet airliner.

Hardness

Hardness measures resistance to penetration or scratching, and most hardness tests (e.g. the Vickers and Rockwell tests) involve pressing a diamond pyramid or small steel ball into the surface of the material and measuring the indentation produced. These tests depend on resistance to plastic deformation, so hardness increases if the material is cold-worked because this locks up the dislocations and restricts plastic deformation.

TESTING, SQUEEZING, AND BENDING

There are agreed standards for materials testing. These specify the way a sample must be prepared and how the test must be carried out. This means that, whatever the final use of the material, the same test has been used to measure its mechanical properties. Some of the most important properties that are measured in these tests are:

- elastic and plastic deformation under standard stresses;
- stress–strain relations;
- fracture toughness;
- fatigue strength;
- wear and abrasion.

However, the single most important property for most applications is the ability to withstand applied stress without failing or undergoing permanent deformation. An overview for several different materials is given in the table below.

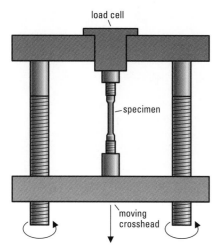

A tensile test machine.

Properties of materials.

Material type	Example	Stress/MPa for strain of 0.01%	Maximum stress/MPa	Maximum plastic strain (%)	Fracture mechanism
metal	annealed copper	11	210	45	ductile
metal	hardened steel (1% carbon)	21	1700	3	ductile (but not very!)
ceramic	glass (Pyrex)	7	70	0	brittle
ceramic	alumina	40	400	0	brittle
polymer	thermoplastic	0.07	20	100	ductile
polymer	thermoset	0.7	70	1	brittle
polymer	elastomer	0.007	20	1	brittle

A standard tensile test for a metal is shown in the diagram. The specimen is manufactured to an exact shape and size and then screwed into the test jig. The lower part of the jig is then pulled downwards and the force is measured by the load cell (this allows stress to be calculated), and the extension (and hence strain) is recorded continuously. A stress–strain graph is then plotted.

A rather different approach is taken when testing the tensile strength of ceramics. A standard tensile strength usually results in values well below those expected theoretically because small misalignments can lead to stress concentrations and early brittle fracture. Tests based on three-point or four-point bending are preferred since these limit the region under maximum stress and so reduce the effects of misalignment or of flaws in the structure.

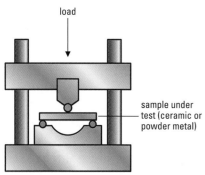

A three-point bend test for a ceramic or sintered (compacted powder) metal sample.

Other kinds of elasticity

Tensile stress is not the only kind of stress. An object might be subjected to an increased pressure or sheared by opposite forces acting along different lines. Pressure is called a **bulk stress** and its effect is to reduce the volume of an object. **Bulk strain** is defined as the proportional change in volume, and is proportional to pressure over a significant range for most materials.

$$p = K\frac{\Delta V}{V}$$

where K is the **bulk modulus** ($N\,m^{-2}$) and $\Delta V/V$ is the fractional change in volume. The **bulk modulus** is sometimes called the **modulus of rigidity**.

The **shear modulus** (μ) is a measure of how hard it is to slide one layer of a solid over another. Think of a cube of material. A **shear stress** would be applied if forces in opposite directions, but parallel to the faces, were exerted on opposite faces of the cube. This is shown in the diagram. The cube would tend to distort. If the shear modulus is high there is little distortion for a given shear stress. If the modulus is low then there is a lot of distortion. For shear the best measure of distortion is the angle in radians through which the cube tips.

$$\text{shear stress (N\,m}^{-2}\text{)}, \sigma = \frac{\text{shearing force (N)}}{\text{area being sheared (m}^2\text{)}}$$

shear strain (dimensionless), ε = angle of shear (rads)

$$\text{shear modulus (N\,m}^{-2}\text{)}, \mu = \frac{\sigma}{\varepsilon}$$

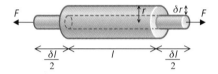

(a) Bulk stress. (b) Shear stress.

The elastic moduli affect many of the mechanical properties of a material, for example, how fast sound travels through it and how much elastic energy is stored in the material when it is distorted.

Poisson's ratio

If you take a cylinder of plasticine and pull it along its axis it will get both longer and thinner. The ratio of the fractional decrease in radius to the fractional increase in length is called **Poisson's ratio**, v.

$$v = \frac{-\delta r/r}{\delta l/l}$$

For many metals it is about 1:3 or 0.33.

This ratio depends on the relation between different elastic moduli for a particular material. When a sample is subjected to tensile stress it elongates as a result of stretched bonds parallel to the applied force. If this was all that happened then the sample volume would increase. This increase in volume is opposed by the bulk modulus so that atoms are pulled inwards towards the axis introducing local shear strains. A small value of Poisson's ratio indicates a large resistance to shear; a large Poisson's ratio corresponds to something that shears relatively easily. The maximum value of Poisson's ratio is for a fluid, which has no resistance to shear.

longitudinal strain = $\frac{\delta l}{l}$

lateral strain = $\frac{\delta r}{r}$

Poisson's ratio = $\frac{-\delta r/r}{\delta l/l}$

Poisson's ratio is the ratio of change in radius to change in length when a sample is subjected to tensile stress.

Some typical values for Poisson's ratio.

Material	Poisson's ratio
concrete	0.20
glass	0.23
metals	0.33
vulcanized rubber	0.40
soft rubber	0.50

PRACTICE

1 Water has a bulk modulus of 2.0×10^9 Pa, and a density of 1000 kg m^{-3} at atmospheric pressure (10^5 Pa).

a What volume does 1.0 kg of water occupy at atmospheric pressure?

b What is its change in volume when the external pressure is increased to 5 atmospheres?

2 a A rectangular swimming pool has uniform depth of 1.2 m. How much does its depth change as a result of an increase in atmospheric pressure of 1%?

b Discuss whether an average decrease in atmospheric pressure of 1% would have a significant effect on global sea level.

3 a Prove that Poisson's ratio is 0.5 if the volume of a material remains constant when it is stretched (assume the sample under test is cylindrical).

b What happens to the volume of a metal sample when it is subjected to tensile stress? (Take Poisson's ratio to be 0.33.)

MAGNETIC MATERIALS

The magnetic fields generated by electric currents are not very strong. To generate a 0.1 T field in a typical solenoid of, say, 500 turns and 10 cm long would require a current of nearly 16 A! A modern superconducting coil could maintain such a current with no significant heat production. However, superconductors have to be cooled to very low temperatures, which is expensive and impractical in many situations. Practical electromagnets rely on the properties of magnetic materials, that is, materials that become magnetized when exposed to a magnetic field. There are three classes of magnetic material. Two of them – **paramagnetic** and **diamagnetic** materials – can only produce very weak magnetic fields and are of little use technologically. **Ferromagnetic** materials, on the other hand, can produce fields that are up to 10^5 times stronger than the applied field. This enables strong electromagnets to be produced by wrapping solenoids round lumps of ferromagnetic material.

Paramagnetism

Electrons in orbit round the nuclei of atoms are moving charges and so generate magnetic fields. These fields are due partly to the orbital motion of the electrons, and partly to the electron's own spin. In most atoms the magnetic fields produced by the motions of the individual electrons cancel out, leaving the atom with no net magnetic field. Paramagnets (for example, chromium) have electron motions that do not totally cancel out, leaving a small net magnetic field. On its own this effect does not lead to a material that is magnetic. The individual fields are very weak and there is no tendency for all the magnetic fields to be pointing in the same direction – the net field for the material would be zero.

However, if a paramagnet is placed in an external magnetic field, the individual magnetic atoms tend to line up with the field, producing a weakly magnetized material.

In 1895 Pierre Curie discovered that the magnetization of a paramagnetic material depended on both the absolute temperature (T) and the strength of the external field (B). This is **Curie's law**:

$$\text{magnetization} \propto \frac{B}{T}$$

The alignment of the atomic magnets tends to be disrupted by their thermal motion. Even with external fields as strong as 3.0 T, room temperature thermal motion will disrupt the alignment sufficiently to prevent a field of more than 3 mT from being produced. However, at temperatures of a few kelvin some paramagnetic materials can be fully aligned or **saturated**.

Diamagnetism

A paramagnetic material placed near to a magnetic field would be weakly attracted because of the lining up of the atomic magnets. A diamagnetic material would be weakly repelled in the same situation. The field produced by a diamagnet is always in the opposite direction to that of the applied field. All materials show diamagnetic effects. However, if the material is also paramagnetic or ferromagnetic these effects will always dominate.

A rough understanding of diamagnetism can be gained from the diagram. Two electrons are orbiting a nucleus in opposite directions, so that their atomic currents produce magnetic fields in opposite directions, and so they cancel. If an external magnetic field is turned on, both electrons will experience a magnetic force, but in opposite directions. The effect of this force is to break the symmetry between the two electrons, so their magnetic fields no longer cancel. The atom gains a magnetic field opposite to that of the external field.

Again the effect is extremely weak. Generally the material gains a field that is less than the external field by a factor of 10^5.

The magnetic force on an electron orbiting a nucleus distorts the electron's motion. If two electrons are orbiting in opposite directions, then the forces will act differently and the magnetic fields due to the electrons will no longer cancel.

Ferromagnetism

Ferromagnets are the most useful magnetic materials. The name derives from the Latin word for iron which, along with cobalt and nickel, is one of the most commonly used ferromagnets.

The atoms within ferromagnetic materials have their own weak magnetic fields (like paramagnets). A quantum mechanical effect known as **exchange coupling** tends to align the magnetic fields of nearby atoms, producing regions known as **domains**. An individual domain may contain many thousands of atoms, but will not be as large as an individual crystal.

Above the Curie temperature (about 1040 K for iron) the thermal motion of the atoms breaks up the alignment and the ferromagnet becomes paramagnetic.

Within a ferromagnet (below the Curie temperature) each individual crystal will be subdivided into domains. Without an external field, the domains tend to be randomly distributed so the crystal has no net magnetic field, but in an external field the domains become aligned, giving the material a magnetic field.

Hysteresis

The total field inside a solenoid wound round a ferromagnetic core is

$$B_{tot} = B_{sol} + B_{core}$$
$$= \mu_0 n I + B_{core}$$

which can be written as

$$B_{tot} = \mu_0 \mu_r n I$$

where μ_r is the **relative magnetic permeability** of the core.

However, μ_r is not constant but depends on the **magnetic history** of the material. (Compare this with the electrical case in which ε_r is a constant for the material.)

The diagram below right shows how the magnetic field in a sample of steel varies with the externally applied field. Such curves are known as **hysteresis loops**.

Starting from an unmagnetized sample, increasing the magnetization field tends to cause domain boundaries to shift (AB) – those domains that are nearly parallel to B_{sol} grow at the expense of the others. This effect is reversible. Reducing B_{sol} to zero will restore the domain walls to their initial positions.

As the domain walls continue to grow in stronger fields the effect starts to become irreversible (BC). Impurities and dislocations within crystals block the growth of domains so that stronger fields are required (the curve gets steeper along BC). When the domain walls force themselves past the blockage, a shock wave passes through the crystal and some magnetic energy is lost as heat. Reducing B_{sol} does not restore the domain walls; there would be some residual magnetism even if B_{sol} is zero (CD). The magnetic field of the core can only be reduced to zero by applying a **reverse** magnetic field (the **coercive field**).

Eventually the domain walls have moved to the edges of crystals. To increase the magnetization further the domain directions have to be rotated to come into line with the external field (CE). This effect is partially reversible. Finally all the domains are parallel to B_{sol} and the material is saturated (E).

Ferromagnetic materials are specified in terms of their **remanence** (B_r) and **coercivity** (B_c) along the saturation curve.

- Hard magnetic materials have high coercivity, are difficult to demagnetize, and are used in permanent magnets (for example, steel).
- Soft magnetic materials have low coercivity, are easy to demagnetize, and are used as the core for coils, transformers, etc. (for example, iron).

<table>
<tr><td>Curie temperature</td></tr>
</table>

The Curie temperature of a ferromagnetic material is the temperature above which the material becomes paramagnetic. Estimates of the temperature of the Earth's core suggest that it is above the Curie temperature of iron. This is how we know that the Earth's magnetic field is not due to magnetic rocks.

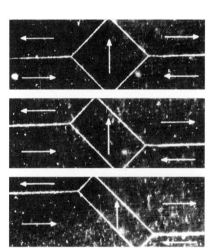

Photomicrographs of domain walls in a crystal of iron. As a magnetic field is applied (top: zero field; bottom: greatest field) the magnetic domains change in shape. Those domains that are favourably oriented with the magnetic field grow at the expense of those that are unfavourably oriented. The arrows indicate the magnetization direction of each domain.

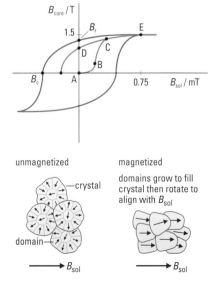

What happens when an external magnetic field is applied to a sample of steel.

OBJECTIVES

- valence and conduction bands
- energy gap
- conductors and insulators

ENERGY BAND THEORY

Electrons in atoms can only occupy certain discrete energy states. However, when atoms come together in molecules or in solids the electrons in different atoms interact. The atoms also have thermal vibrations. These effects shift the allowed energies up or down a little bit so that the electrons can exist in a band of energies around the original level. The **energy band** structure of solids can explain many of their electrical and optical properties.

When atoms bond together to become molecules their energy levels merge and split – this results in the splitting of spectral lines in molecular spectra. In a solid this process takes place between large numbers of atoms, and the energy levels divide into bands of closely spaced levels with large energy gaps between the bands. In the upper energy band (the **conduction band**) the electrons are free to move between atoms and become charge carriers. In the lower energy band (**valence band**) electrons are tightly bound to their atoms and are not free to move about. In some circumstances it is possible for an electron in the lower band to gain enough energy to jump into the higher band and become a charge carrier.

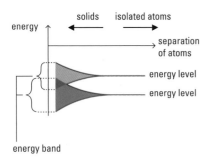

Spreading of energy levels to form energy bands.

Insulators

In an insulating material the outer electrons are all shared between atoms to form bonds, so they are not available as charge carriers (the inner electrons are, of course, tightly bound to individual atoms). These bonding or valence electrons have a range of allowed energies which form a 'valence band'. This valence band is completely filled, so there are no vacant allowed energy levels which would allow the electrons to gain energy from an applied electric field and move from an occupied state in one atom to an empty state in another. There *is* another allowed energy band above the valence band and if electrons could somehow get into this empty band they *could* skip from atom to atom through the structure. Thermal vibration might give individual electrons energy boosts, but the energy gap is much greater than the typical size of these thermal excitations so the material is an insulator.

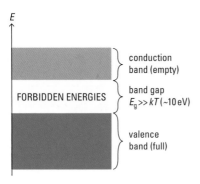

Energy band structure for an insulator.

$$\text{typical thermal excitation} \approx kT = 0.025 \, \text{eV}$$

$$\text{typical insulator band gap} = 10 \, \text{eV}$$

Thermal excitation is a random process, so electrons will sometimes get energy boosts a lot greater than kT, but the band gap in an insulator is so great that the probability of an electron jumping into the conduction band and becoming a useful charge carrier is virtually zero. In the valence band the electrons are unable to gain energy from an applied electrical field as there are no vacant energy levels for them to jump into. This explains the extremely high resistivity and low conductivity of many insulators.

Conductors

In a conducting material the energy levels of adjacent atoms have spread to form bands that overlap. The top of the valence band is above the bottom of the conduction band. This means that electrons in the valence band can easily move to vacant energy levels in the partially filled conduction band. In practice this means that an applied electric field (e.g. created by a cell connected across the conductor) supplies the tiny amount of energy needed to move electrons from one atom to a vacant state in the conduction band of an adjacent atom and then accelerates the electrons through the material.

Semiconductors

A semiconductor has a filled valence band, and an empty conduction band, just like an insulator. In fact, all semiconductors are insulators at

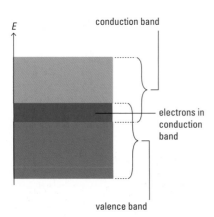

Energy band structure for a conductor (metal).

very low temperatures. The band gap, however, is much smaller than in materials that are insulators at room temperature. As a result, some electrons will be thermally excited into the conduction band where they can move freely through the material. When this happens the electrons leave '**holes**' in the valence band.

These holes behave just like positive charge carriers and can also move through the material. When a potential difference is connected across a semiconductor, holes and electrons drift in opposite directions and both contribute to the current that flows. The concentration of charge carriers in a typical semiconductor at room temperature is about $10^{21}\,\mathrm{m^{-3}}$. The concentration of charge carriers in a metal is about ten million times greater; this is where semiconductors get their name.

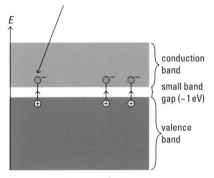

Energy band structure for a semiconductor. Thermal excitation creates pairs of electrons and holes, both of which act as charge carriers in the material.

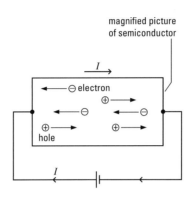

Current flow in a pure (intrinsic) semiconductor. There are equal numbers of holes and electrons but they do not make an equal contribution because they have different mobilities.

Group in periodic table	Crystal	Band gap /eV
IV	C	5.3
	Si	1.1
	Ge	0.7
Semiconductor compounds		
III–V compounds	GaAs	1.4
	InAs	0.4
IV–VI compounds	PbS	0.4
	PbSe	0.3

Energy gaps for a number of semiconductor materials.

PRACTICE

1 Why is light unable to promote electrons to the conduction band in an insulator?

2 LEDs emit light when electrons in the conduction band drop into holes in the valence band. What can you say about the band gaps of semiconductors used to make green and red LEDs?

3 Use the energy band theory to explain why electrons in metals are effectively 'free' even though they are trapped inside the metal.

4 Look at the table of semiconductor band gaps. Which of these will have the smallest number of free charge carriers at room temperature? Explain your answer.

5 Negative temperature coefficient thermistors (i.e. ones in which the resistance reduces with increasing temperature) are made from semiconductors. Why? What type of material could be used to make a positive temperature coefficient thermistor?

- intrinsic semiconductors

- p- and n-type extrinsic semiconductors

- band theory of semiconductors

This innocent-looking material that gets into sandwiches is the key to the modern electronics industry ...

... it helped make Bill Gates a multi-millionaire.

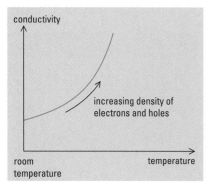

A graph of conductivity versus temperature.

A large heat sink.

SEMICONDUCTOR MATERIALS

One of the most important materials in the modern world is sand. Sand contains silicon, and silicon is the key to the microelectronics industry which has made complex electronic circuits small enough to fit into a watch. Silicon is an example of a semiconducting material. Such materials are vital to the industry as their conducting properties can be very precisely controlled.

Intrinsic semiconductors

Pure semiconducting materials like silicon and germanium are called **intrinsic semiconductors**. They have very few charge carriers compared with metals ($4.3 \times 10^{21}\,\text{m}^{-3}$ for germanium, $8 \times 10^{28}\,\text{m}^{-3}$ for copper), so the charge carriers must move faster when the same current flows in a semiconductor as in a metal of similar cross-sectional area. This is clear from the equation for drift velocity (v is inversely proportional to n if everything else is the same):

$$v = \frac{I}{nAe}$$

where v is the drift velocity, I the current, n the density of charge carriers, A the cross-sectional area, and e the charge.

Thermal effects

The density of charge carriers (mobile electrons and holes) in a semiconductor increases rapidly with temperature. This can cause problems in electronic devices. If a semiconductor begins to overheat its conductivity also increases, so the current passing through it becomes bigger, increasing the temperature still further. This causes a further increase in conductivity and further heating. This positive feedback loop is called **thermal runaway** and it can destroy the device. Large integrated circuits, such as microprocessors, have built-in heat sinks and fans to prevent this happening.

The Hall effect and semiconductors

The **Hall voltage** (see spread 10.16) is given by $V_H = Bvd$, where v is the drift velocity of charge carriers in the magnetic field. v is much higher for semiconductors than metals, so semiconductors are used in Hall probes to detect and measure magnetic fields. In the semiconductor industry the Hall voltage is used to analyse the electrical behaviour of semiconductors – in particular to work out charge carrier density, from

$$I = nAve.$$

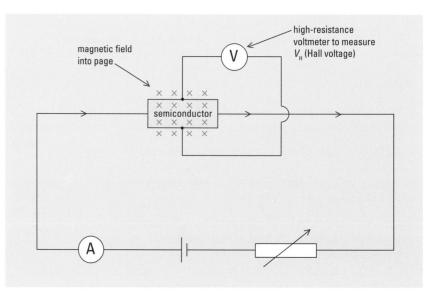

Extrinsic semiconductors

These are materials that have had their semiconducting properties artificially modified by the addition of impurity atoms to the crystal. This is called **doping**.

In a **p-type** semiconductor tiny amounts of boron (a three-valent group III element) are added to increase the number of charge carriers (one boron atom in 10^6 silicon atoms will decrease the resistivity by a factor of 10^5). Boron has a very strong grip on its outer electrons. When it tries to bond with other atoms in the crystal it grabs an electron from a nearby atom, creating a hole on that atom. The boron holds this electron so tightly that none of the surrounding silicon atoms can pull it back. For every boron atom added a free hole is created. In such a crystal there are more free holes than free electrons and the **majority carriers** are holes – hence the name p-type (for positive). The boron is called an **acceptor atom** since it takes one electron from the surrounding material.

In an **n-type** semiconductor the impurity is an atom such as arsenic (a five-valent group V element) which has a very weak grip on its outer electron. This electron easily breaks free, by picking up thermal energy, and becoming a charge carrier. For every arsenic atom added a free electron is created. In such a crystal there are more free electrons than free holes and the majority carriers are electrons – hence the name n-type (for negative). The arsenic is called a **donor atom** since it donates one extra electron to the conduction band.

In an n-type semiconductor the impurity atoms add extra levels near to the conduction band, allowing easy promotion and leaving the holes in the donor energy levels. In a p-type semiconductor the impurity atoms add acceptor levels near to the valence band. Electrons are easily promoted into these levels, leaving holes in the valence band. However, these electrons are not free to move as the acceptor levels are localized at the impurity atoms.

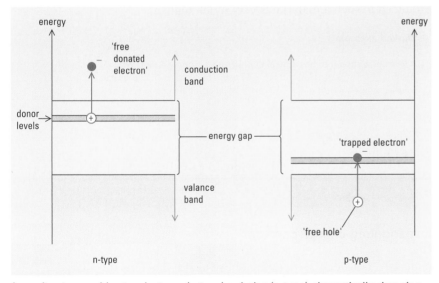

Impurity atoms add extra electrons (n-type) or holes (p-type), dramatically changing the conductivity of semiconductor materials.

Extrinsic semiconductors

Both p- and n-type semiconductors are electrically neutral – each has equal numbers of electrons and holes. p-types have more free holes than free electrons (the difference being the trapped electrons). n-types have more free electrons than free holes (the difference being the trapped holes).

Holes

Holes may seem to be an invention used to explain certain types of conduction, but their reality as charge carriers can be demonstrated by the Hall effect (see spread 10.16). A p-type semiconductor used as a Hall probe will have a Hall voltage polarity that is the reverse of that for an n-type. This implies that the charge carriers are of opposite sign.

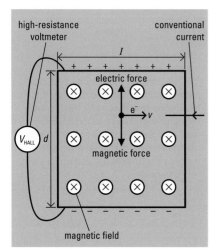

The Hall effect.

THE HALL EFFECT

The **Hall effect** is another example of the magnetic force on a moving charge carrier.

Charge carriers drifting through a block of material in a magnetic field will experience a force that deflects them from their path. In the diagram (left) electrons flow in a block of conductor with a magnetic field passing through one of the larger faces at 90°. The electrons drift downwards because of the magnetic force acting on them. After a short period negative charge will accumulate on the bottom face of the block, leaving a positive charge on the upper face. As the amount of charge on the faces increases, so does the electrical force due to the charges which is tending to repel the electrons against the magnetic force. The charge build-up will stop when the electric and magnetic forces are equal:

electric force on electrons = magnetic force on electrons

$$Ee = Bev$$

Now

$$E = \frac{V_{\text{Hall}}}{d} \quad \text{and} \quad I = nAev = n(dw)ev$$

$$\therefore \quad e\frac{V_{\text{Hall}}}{d} = \frac{BeI}{ndwe}$$

$$\therefore \quad V_{\text{Hall}} = \frac{BI}{nwe}$$

To achieve a measurable Hall voltage the conductor must be very thin (w must be small), and have as few charge carriers as possible (n must be small). With a metal, n is comparatively large ($\sim 10^{28}$) so the Hall voltage is typically a few millivolts for magnetic fields that are easily produced in a lab. On the other hand, in a **semiconductor**, there are fewer charge carriers than in a metal (by a factor of about 10^9), and the Hall voltage produced in such a material is very much higher.

Worked example

A piece of germanium has dimensions 10 mm × 5 mm × 1 mm. When it is carrying a current of 150 mA the Hall voltage is 57 mV. The number of charge carriers per unit volume in germanium is 4.3×10^{21} m^{-3}. What is the magnetic field strength?

$$V_{\text{Hall}} = \frac{BI}{nwe}$$

$$\therefore \quad B = \frac{nweV_{\text{Hall}}}{I}$$

$$= \frac{(4.3 \times 10^{21}\,\text{m}^{-3}) \times (1 \times 10^{-3}\,\text{m}) \times (1.6 \times 10^{-19}\,\text{C}) \times (57 \times 10^{-3}\,\text{V})}{150 \times 10^{-3}\,\text{A}}$$

$$= 0.26\,\text{T}$$

Investigating charge carriers

One use for the Hall effect is to investigate conduction processes in materials. Consider the diagram above again; this time the charge carriers are positively charged and moving in the same direction as the conventional current. The magnetic force on the charge carriers would also be downwards (both charge and direction of motion have been reversed) but the polarity of the Hall voltage would be reversed (the lower surface of the slab is positive). The sign of the Hall voltage is then an indication of the polarity of the charge carriers. For example, the sign of the Hall voltage shows that the charge carriers in copper are negative (electrons), and that the charge carriers in semiconducting materials can be either negative or positive (see spread 10.15).

The Hall probe

Traditional and quite fiddly methods of measuring magnetic fields have been replaced in recent times by the **Hall probe**. This device is cheap, it is simple to use, and it is small in size, which enables accurate maps to be drawn of non-uniform magnetic fields. This is especially important if the field is used to bend particle paths in order to determine momentum.

The Hall probe consists of a small rectangular block of semiconducting material. It is placed into the magnetic field so that the field lines are passing though one of the larger faces at 90°. A current is then passed through the block in a direction parallel to the long edges. As a result, a p.d. is produced between the smaller faces. This p.d. is measured by a sensitive voltmeter and is proportional to the strength of the field passing through the probe.

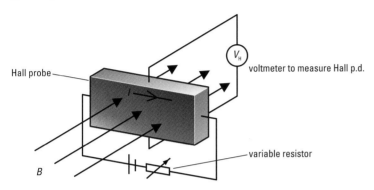

A Hall probe.

The probe needs to be calibrated by measuring a magnetic field of known strength. One way of producing such a field is to use a pair of **Helmholtz coils**.

This pair of flat, tightly wound coils is set so that the distance between the two is the same as their radii. They will produce a field that is very nearly uniform midway between the coils, if they have the same number of turns and the same current passing through them.

The strength of the field between the coils is accurately given by:

$$B = \frac{8\mu_0 NI}{5\sqrt{5}r}$$

where N is the number of turns on each coil, I the current in each coil (A), and r the separation between the centres of the coils (m).

The Hall probe is placed at the centre of the Helmholtz coils and the meter reading compared with the calculated value. As the measured p.d. is linearly proportional to the field strength it is sufficient to measure one field accurately in order to calibrate the whole scale.

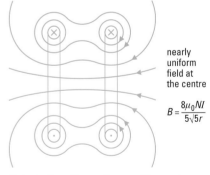

Helmholtz coils can be used to produce uniform magnetic fields for calibrating Hall probes.

PRACTICE

1 A strip of metal is 10.0 mm wide and 2.0 mm thick. It carries a current of 6.0 A, and is placed so that a magnetic field of 0.09 T is passing at right angles through its surface. The metal has 8.0×10^{28} charge carriers per cubic metre.

Calculate the velocity of the charge carriers, and the Hall voltage that would be produced. Comment on the size of this voltage.

2 A block of semiconductor material of the same size as the metal in question 1 has a Hall voltage of 60 mV in the same magnetic field. The number of charge

carriers in the material is 8.0×10^{20}. What is the current in the semiconductor?

3 A thin strip of semiconductor material carries a current parallel to its long edge. A magnetic field passes through the strip at 90° to its surface. Describe how you can use the Hall effect to tell if the current is produced by positive or negative charge carriers within the semiconductor. (Note: it is possible to design semiconductor materials which have either positive or negative charge carriers.)

- absorption/opacity

- transmission/transparency

- scattering

A variety of metals.

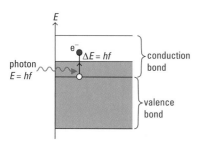

Metals can absorb photons of all wavelengths.

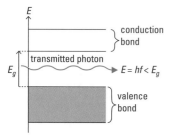

Insulators cannot absorb visible photons.

Good sunglasses transmit visible light and absorb the harmful ultraviolet.

OPTICAL PROPERTIES OF MATERIALS

Light is an electromagnetic disturbance. When it strikes electrons in atoms they interact with the oscillating electromagnetic fields, and one of three things may happen.

- Photons may be absorbed.

- Photons may be scattered.

- Photons may be unaffected.

Which of these actually happens depends on the energy (and hence wavelength and frequency) of the photons, and on the energy band structure of the material and its purity. If the light contains a wide range of wavelengths they may be affected in different ways, so that the reflected or transmitted light is coloured. A material that is transparent to visible light may be opaque to ultraviolet light (e.g. glass), or opaque to visible and transparent to infrared (e.g. many semiconductors).

Metals

Metals have overlapping valence and conduction bands. This means that electrons have a continuous range of vacant available energy levels above them. They are able to absorb photons of any energy at all and jump to higher energy levels. This means light cannot pass through metals (because all the photons are absorbed) so all metals are opaque. If the surface of the metal is smooth it is a good reflector. This is because the excited electrons re-radiate photons as they lose energy. This process is not 100% efficient, so some of the radiated energy is transferred to thermal vibrations, heating the metal. The amount that heats the metal is dependent on the wavelength, so the reflected light is usually coloured (e.g. the reddish colour of copper, the yellow of gold).

Insulators

Insulators have a fixed energy gap between the valence and conduction bands and electrons are bound to atoms in the structure (which is ionic or covalent). Low-energy photons do not have enough energy to promote valence electrons to the conduction band, so they are not absorbed and pass right through the material. This makes pure insulators transparent. However high-frequency photons (e.g. ultraviolet or X-ray) do have enough energy to promote these electrons, so these photons are absorbed and the insulator is opaque to these short wavelengths.

In practice most insulators contain impurities which may absorb certain wavelengths (an **absorption band**) so that the transparent insulator is coloured. For example, ruby contains chromium ions which absorb in the blue and green regions of the visible spectrum. This is why rubies have a red colour. The presence of impurity ions creates **colour centres** in the material structure where electrons are more easily excited, so photons of a lower energy can be absorbed. Also, most crystals have many grain boundaries and structural faults, all of which can scatter light, resulting in a white opaque appearance (e.g. salt, sugar, etc.).

Semiconductors

Semiconductors have a similar energy band structure to insulators. There is, however, one very important difference: the energy gap between the two bands is quite small (around 1 eV). This means that photons of visible light may have enough energy to promote electrons to the conduction band, whereas infrared photons with lower energy may not. The table gives energy gaps for two group IV materials (diamond is, of course, a form of carbon) and one group VI material (sulphur) and the wavelength of photons with the same energy. Each material will be transparent to photons with longer wavelengths (lower energies) and opaque to shorter wavelength. This explains their appearance.

Band gaps and photon wavelengths.

Material	Energy gap/eV	Corresponding photon wavelength /nm	Comments	Appearance
diamond	5.6	220 (UV)	transparent to visible light, absorbs hard UV	colourless
sulphur	2.2	560 (orange/green)	transparent to red/yellow, absorbs UV, blue/green	yellow
silicon	1.1	1100 (IR)	transparent to far IR, absorbs UV, visible light, near IR	opaque

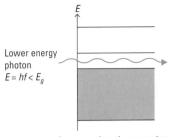

long wavelengths transmitted

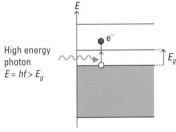

short wavelengths transmitted

Semiconductors transmit long-wavelength photons and absorb short-wavelength photons.

Photoconductivity

When electrons are promoted to the conduction band the conductivity of the material increases (its resistivity falls). This is used to make electronic components like LDRs (light-dependent resistors). It is also used to detect heat radiated by missiles and aircraft even when they are hundreds of kilometres away. This is possible because semiconductors can be made with an energy gap that corresponds to the energy of photons of a particular wavelength, in this case the infrared radiation emitted by the source to be tracked. Photons with a longer wavelength will not be absorbed, while photons with a shorter wavelength will be absorbed so strongly that they only affect the surface of the semiconductor and so have a small effect on overall conductivity. Photons with energy very close to the energy of the gap (i.e. with the critical wavelength) penetrate the crystal and cause a large increase in conductivity.

The same effect is used in photographic light meters, where cadmium sulphide is used as the light detector because of its high sensitivity in the visible region.

Scattering

Shafts of sunlight in a forest can be seen because suspended dust motes scatter light from them into our eyes. Scattering by small particles is related to both reflection and diffraction.

If the particles are larger than the wavelength of light then each one causes diffuse reflection of the light and all wavelengths are affected equally, so if the light source is white then so is the scattered light. This is why milk is white: light is scattered by droplets suspended in the milk and the droplets are larger than the wavelengths of visible light.

If the particles are small compared to the wavelength of light then they act like point sources of diffracted light, and circular waves travel out from each particle. In 1871 Lord Rayleigh analysed this kind of scattering and showed that the intensity of the scattered light varies as $1/\lambda^4$, so it is much more intense for short wavelengths (blue end of the spectrum) than long wavelengths (red end). **Rayleigh scattering** explains the blue colour of the sky. It is mainly due to molecular scattering. As sunlight passes through the atmosphere the blue part of the spectrum is scattered more intensely (about 10 times more) than red light. So if you look up your eye will receive mainly blue light. On the other hand, at sunset you are looking toward the sun and receiving light that has travelled a long way through the atmosphere. It has lost the blue end of the spectrum and so has a bias toward the yellows and reds.

Blue light is scattered more strongly than red; this explains blue skies (seeing the scattered light) and sunsets (what remains when the blue end is scattered).

PRACTICE

1 Check the wavelengths quoted in the table above (band gaps and photon wavelengths).

2 Why are mirrors 'silvered'?

3 Cadmium sulphide has an energy gap of 2.42 eV. What is the corresponding photon wavelength? Where is this in the electromagnetic spectrum?

4 Show that Rayleigh scattering of white light in the atmosphere results in the scattered violet light (400 nm) having about 10 times the intensity of the scattered red light (720 nm).

10.18

The warning sign on the Albert Bridge.

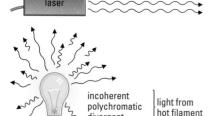

Coherent and incoherent light.

Laser beams can be powerful enough to produce a light spot on the Moon. Here a laser is being used to create an 'artificial star' in the atmosphere for testing astronomical equipment.

Using lasers

The science fiction image of powerful lasers being used as weapons of mass destruction is not entirely ridiculous, and it is true that they have been used with devastating effect to guide bombs and missiles, but the majority of their uses are far more diverse and positive. They are used to read CDs and bar codes, for accurate range-finding and surveying work, as scalpels in keyhole surgery, to adjust the curvature of the cornea to correct defects of vision, and to form images in laser printers. They are also widely used as transmitters in communication systems based on optical fibres.

LASERS

There is a curious old sign on the Albert Bridge which crosses the river Thames near Battersea Park in London. It says that soldiers must break step when crossing. This seems strange when you consider that the bridge is quite secure carrying cars and buses. However, the effect of marching soldiers is to apply many small impulses in phase with one another and these impulses might set up a resonant vibration in the bridge, resulting in damage to its structure. There is no problem when the soldiers break step because the impulses are then **incoherent** and no particular frequency dominates, so the vibrations induced in the bridge are small.

The reinforcing effect of coherent vibrations is used in **lasers**. Lasers emit light in which all the photons are 'in step', making it coherent and **monochromatic** (a single frequency). This gives a far more intense beam than the incoherent light that is emitted from a filament lamp of similar power.

Photon emission

LASER stands for Light Amplification by Stimulated Emission of Radiation. This is a concise description of what a laser does. The key to the 'light amplification' is in a process that is very rare in nature, the stimulated emission of a photon by an atom in an excited state. You will recall that electrons in the outer shells of atoms can be excited to higher vacant energy levels if they absorb a photon of just the right energy. Normally they only stay in the excited state for a very short time (typically less than a microsecond) before dropping back to a lower level (perhaps the one from which they came) and emitting one or more photons in the process. This process occurs at random and is called **spontaneous emission of radiation**.

There is no reason why photons emitted spontaneously from different atoms should have any particular phase relation, so light emitted by spontaneous emission (as in a filament lamp) is usually incoherent. It is also obvious that spontaneous emission cannot result in amplification – if the atoms are 'pumped' into excited states by irradiating them by photons the best that could be achieved is for every absorbed photon to result in a photon of equal energy being emitted a little later.

There is another way in which excited atoms emit photons. If a photon of exactly the right frequency passes an atom in an excited state it can induce the atom to de-excite and emit a second photon in phase with the first. This process is called **stimulated emission**. It is uncommon in nature because it is unlikely that the right photons will pass excited atoms in the brief time before they undergo spontaneous emission. They are much more likely to pass atoms in their ground state and get absorbed. In lasers, however, everything is arranged so that stimulated emission dominates. One photon can then create a copy of itself and these two can go on to stimulate further atoms to emit more photons and so on. There is a kind of optical chain reaction resulting in a burst of photons which are all in phase and monochromatic.

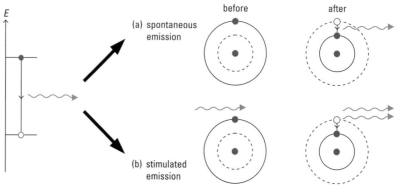

Spontaneous emission and stimulated emission.

The population inversion

If energy is supplied to a material it distributes itself among the atoms. The chance of a particular atom getting a large amount of energy is small, so the expected distribution is **bottom heavy**: lots of atoms in low-energy states and few in high-energy states. If a photon passed through this material it would be much more likely to be absorbed than to stimulate emission of radiation. The amplifying materials used in lasers are chosen because they have a particular characteristic, an energy level that is almost stable. These **metastable states** can temporarily (e.g. for a millisecond) trap electrons above the ground state. This means that as energy is supplied to the medium more and more electrons accumulate in the metastable state. This can result in a situation where there are more atoms in this excited state than in the ground state – a **population inversion**.

Of course, some of these metastable states will decay quite rapidly (it is a random process), and when they do the photons they emit have just the right frequency to stimulate other similarly excited atoms to emit similar photons in phase with them. This is how laser light is generated. Various materials have been used as amplifying media in lasers, but all of them possess these metastable states that allow a population inversion to be set up. They include synthetic ruby (the first lasers), a mixture of helium and neon (He–Ne lasers), and various semiconductors.

The laser cavity

The process described above achieves a significant amount of light amplification, but a laser takes this even further. The amplifying medium is placed between two parallel mirrors (which may be the polished ends of the medium itself). One mirror is a near-perfect reflector, the other is a partial reflector. The light that leaks through the partial mirror is the laser output. The purpose of the reflectors is to bounce photons back and forth along the tube so that a great deal of stimulated emission occurs and the amplification is increased (this is an example of positive feedback). The space between the two mirrors is called the **laser cavity**, and as well as increasing amplification it has two other advantageous effects.

- Only photons travelling almost parallel to the axis of the cavity will remain within it as they bounce back and forth. This means that the beam intensity is concentrated in one direction and has very little divergence.

- The mirrors themselves act like the fixed ends of a guitar string in that they select certain optical standing waves from the cavity. The amplifying medium usually emits photons over a small range of frequencies; the cavity selects an even narrower range from within this, improving the spectral purity of the beam.

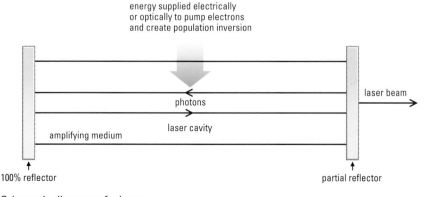

Schematic diagram of a laser.

A, B, C are energy levels:

A → B   'pumped'

B → C   rapid spontaneous

C → A   delayed stimulated

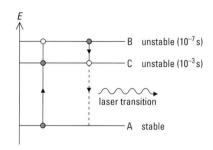

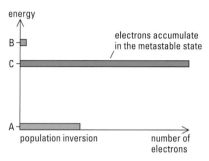

Creating a population inversion.

Pumping two-, three-, and four-level systems

Optical or electrical 'pumping' is used to create a population inversion. This requires an input energy equal to the energy difference between the ground state and the level to which the electrons are being 'pumped'. However, the probability of electrons jumping back to the ground state increases as more enter the excited state, and with a two-state system this approaches an equilibrium state with equal numbers of electrons in both states however hard the system is pumped; therefore a population inversion cannot be achieved with such a system. For this reason lasers are based on three- or four-level systems (the diagram above shows a three-level system). Population inversions can be achieved in these systems because there is a greater probability that the pumped level will decay to an intermediate level than to the ground level. Four-level systems are more efficient than three-level systems because the third level (i.e. the base level for the laser transition) is initially empty, which is not the case for a three-level system.

USING LASERS

The world's first laser was demonstrated by Theodore H. Maiman at Hughes Research Laboratories in Malibu, California, on 16 May 1960. Many new companies were set up to build and market the new device, but most of the obvious applications, drilling holes or transmitting messages through the atmosphere, could be done more cheaply and effectively by other means. Many of the companies collapsed and one observer branded the laser 'a solution looking for a problem'. However, from the mid-1970s onwards lasers have found a multitude of uses in domestic, medical, and industrial situations, and you are probably less than a couple of metres from one as you read this page.

Bar codes

One of the first mass markets for lasers was as bar-code readers. In 1973 the Super Market Institute in the United States adopted the **Universal Product Code** (**UPC**) that assigns a unique symbol (marked by a series of parallel black bars) to every product sold in its members' grocery stores. The UPC is now used on almost all manufactured goods. It is interesting to note that bar codes, which were originally developed to keep track of which cars went with which engines on the American railroad system, were invented long before bar code readers. The code itself is quite simple: the thickness of each line codes for a digit from 0 to 9. At a checkout the laser scans across the bar code and variations in reflected light (less on the black lines) are picked up by a suitable photodetector. This generates an electrical signal that varies with the same pattern as the bar code and so the product can be identified. Bar codes are not only used in supermarkets, and not only to aid sales. For example, they are also used to record book issues from libraries, and to keep track of stock in warehouses.

Laser communication

Laser beams can be used to communicate directly through the atmosphere. A signal can be carried if the beam is modulated or pulsed, but such systems rely on direct line-of-sight transmission and good atmospheric conditions. In practice it is usually too unreliable. However, in 1966, Charles Kao and George Hockman of Standard Telecommunications Laboratories suggested that optical fibres could be used as light guides. Initially this seemed unlikely because all 'transparent' materials scatter and absorb light to some extent, and the glass available at the time would not have been clear enough to allow signals to travel a significant distance. New kinds of extremely clear glass have since been manufactured and it is now possible to transmit signals hundreds of kilometres with no need to amplify them along the way. These signals are generated by semiconductor lasers which emit light in the near infrared, at wavelengths which are only very weakly absorbed by the glass.

Medical lasers

When laser light hits living tissue some of its energy is absorbed. This can have two main effects.

- **Chemical** – the absorbed photons excite electrons and promote certain chemical reactions.

- **Thermal** – the tissue is heated rapidly and then vaporized (**photoablation**).

Both are used medically, for example to remove benign tumours or perform **keyhole surgery**. Lasers are particularly good for microsurgery because the high power density at the point of absorption means they vaporize tissue rapidly before heat conducts to surrounding healthy tissues and damages them. **Pulsed lasers** (**excimers**) are usually used for microsurgery so that there is no cumulative heating effect. One important example is laser ablation of the cornea. Ultraviolet light is used because

Laser ablation of the cornea enables short and long sight to be corrected by 'scoring' the cornea's surface and so altering the position of its focus.

this selectively breaks covalent bonds, leading to explosive evaporation from a tiny region of the surface of the eye. The cut depth on each pulse is less than half a micrometre, and the laser can be scanned across the cornea to adjust its curvature, and hence focal length, to correct defects of vision. The laser is being used to form an extra lens on the surface of the cornea that will take over the role of a contact lens or glasses!

Laser printers

Laser printers are usually quieter, faster, and sharper (but also more expensive) than inkjet printers. Laser printing uses a similar technique to photocopying, in that light is used to form an electrostatic image of the picture to be copied on a light-sensitive plate or drum. The drum or plate is charged and only conducts where light falls on it (a bit like a light-dependent resistor (LDR)). The narrow beam and fine scanning control enable images with very high resolution to be formed. Once the electrostatic image is in place a fine toner powder will stick to it and can be transferred to paper by contact. It is fixed in place by heating (this melts small plastic particles mixed in with the black powder, making it stick to the page, and explains why the copies are warm when they emerge from the printer).

Lasers for information storage and retrieval

Perhaps the most obvious modern use of lasers is to read compact discs (CDs) and digital versatile discs (DVDs). In both cases information is stored on a track that spirals out from the centre of the disc (and back to the centre or back out from the centre on the second layer in the case of DVDs) and contains a series of pits. The laser is focused onto the disc surface and made to follow this spiral track. As it passes over the step at the edge of a pit, light reflected from the top and bottom of the step interferes. The steps are all one-quarter of the wavelength of the laser light, so the extra path involved in travelling to the bottom and back is half a wavelength. This means that light from the lower surface is 180° out of phase with light reflected from the upper surface, and destructive interference occurs, reducing the intensity of the reflected signal. The reflected light falls on a photodiode where the pattern of high- and low-intensity light is converted to a binary digital electronic signal. This is then converted back to analogue sound using a digital-to-analogue converter. Lasers are ideal because they produce monochromatic coherent light. Most players use an aluminium gallium arsenide semiconductor laser which emits light with a wavelength of 780 nm. Since the wavelength of the light determines the minimum size for the pits, this also limits the amount of information that can be stored on a disc of any particular diameter. The development of blue-light lasers may allow more information to be stored. Alternatively the information can be compressed so that the same amount of storage space results in longer playing time (this is the technique used with DVD technology).

The laser lens inside a CD player.

PRACTICE

1 What properties of laser light make it suitable for use in communications?

2 Why is ultraviolet laser light able to break covalent bonds in matter whereas infrared laser light is not?

3 What is the main advantage of using an excimer (pulsed laser) rather than a continuous laser when performing laser surgery on the cornea?

4 a Estimate the maximum number of bits of information that can be stored on an audio CD (assume each bit occupies a minimum area equal to the square of the laser wavelength).

   b CD audio is sampled at 42 kHz. This means 42 000 audio levels are recorded each second. Each of these levels is coded as an 8-bit word. Estimate the maximum playing time for an audio CD. (In practice, extra information must also be stored on the disc (error correction codes and synchronization information, etc.) and the audio information is spread out in such a way that a realistic maximum is about one-third of the value you have calculated.)

PRACTICE EXAM QUESTIONS

1 Table 10.1 gives corresponding values of load and extension when masses are hung on a wire of length 1.5 m and diameter 0.30 mm.

Table 10.1

load/N	0.0	2.0	4.0	6.0	8.0	10.0	11.0	11.2
extension/mm	0.0	1.0	2.1	3.1	4.2	5.4	7.3	9.0

  a **i** Plot a graph of load (vertical axis) against extension (horizontal axis).

      ii Indicate on your graph the region over which Hooke's Law is obeyed.

  b Use your graph to calculate a value for the Young Modulus of the material from which the wire is made.

2 **a** The Young modulus of an elastic material is defined as *tensile stress ÷ tensile strain*. Explain the meaning of the words in italics.

A playground swing consists of a plastic seat of negligible mass supported by two nylon ropes, each of unloaded length 2.5 m. The area of cross-section of the rope is 7.5×10^{-5} m^2. The rope obeys Hooke's law and has a Young Modulus of 7.0×10^7 Pa. A child of mass 32 kg sits on the swing seat.

  b Calculate the distance by which the seat drops when the child sits on the seat.

  c The child's mother pulls the seat aside until the rope makes an angle of 30° with the vertical.

      i Calculate the distance by which the child and seat rise vertically. Assume that the additional change in length of the ropes caused by moving the swing from the vertical is negligible.

      ii The mother then releases the seat. Find the speed of the child and seat as they swing through the position where the ropes are vertical.

3 This question is about the stretching of a rubber band.

Figure 10.1 shows a graph of the extension of a rubber band and the force applied to it, as the force is increased (line OAB) and then reduced (line BCO).

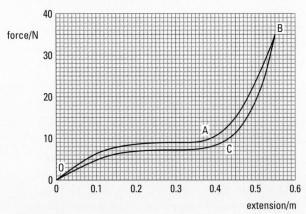

Figure 10.1

  a Explain, in terms of the molecular structure of rubber, why there are two distinct regions along the stretching graph, that is, OA and AB.

  b Estimate how much work was done on the band to stretch it to its maximum extension. Explain how you made the estimate.

  c What does the enclosed area OABCO of the graph represent?

4 Figure 10.2 shows a section through a possible crystalline structure for copper. Each circle represents a copper atom.

Figure 10.2

  a Show that the spacing, d, between the centres of adjacent atoms is approximately 2×10^{-10} m.

    molar mass of copper = 6.4×10^{-2} kg

    the Avogadro constant = 6.0×10^{23} mol^{-1}

    density of copper = 8.9×10^3 kg m^{-3}

  b Figure 10.3 shows how the force, F, between a pair of atoms in a solid varies with the distance, x, between the atoms. Copy the figure and mark on it the distance, d, you have calculated in **a** and explain why you have chosen the distance you indicate.

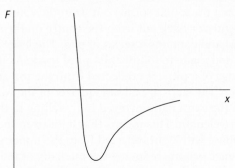

Figure 10.3

5 The mechanical behaviour of many polymers shows a strong dependence on temperature.

  a What is meant by the term polymer?

  b **i** Explain, in terms of microstructure, why acrylic (perspex) is a brittle material at room temperature.

      ii If the temperature of acrylic is slowly increased from room temperature the material will lose its brittle property and become rubbery. State the name given to the temperature at which this happens and explain the change in behaviour in terms of microstructure.

6 **a** Solids can be classified as crystalline, amorphous and polymeric. Explain the meanings of these terms by referring to the structures of different types of solids.

  b **i** Sketch a typical force–extension diagram for a thin rod of a ductile material when it is gradually loaded to breaking. Clearly indicate any important points and/or regions of the diagram.

      ii Explain what is meant by the terms ductile and brittle.

  c **i** When a specimen of rubber is gradually loaded and then unloaded it may show elastic hysteresis and permanent set. What do these terms mean?

    Illustrate your answers with a sketch of the load–extension diagram which would be obtained.

ii By referring to the molecular structure of rubber explain why:

(I) rubber has a low value of the Young Modulus compared with metals;

(II) the value of the Young Modulus increases with rise in temperature.

7 a Explain what is meant by the term unit cell.

b Figures 10.4 and 10.5 represent two ways in which spheres may be packed together in crystal models so that they touch one another.

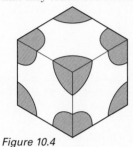
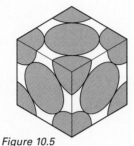

Figure 10.4   *Figure 10.5*

i State the type of packing in each case.

ii A student stated that both represented close-packed unit cells. Criticise this statement and give reasons for your answer.

c The unit cell for copper is represented by Figure 10.5.

The length of the side of the unit cell was found by X-ray diffraction to be 0.3615 nm.

i By considering one face of the unit cell show that the radius of a copper atom is 0.1278 nm.

ii Show that the density of a unit cell for copper is $8.934 \, \text{kg m}^{-3}$.

the relative atomic mass of copper = 63.54
the Avogadro constant = $6.022 \times 10^{23} \, \text{mol}^{-1}$

iii The actual density of copper may differ from that calculated for the unit cell. State two reasons why this may be so. Describe the way in which the density of the material will be affected in each case.

8 a Explain what is meant by

i crystal lattice spacing,

ii grain morphology.

In each case name one practical investigative technique.

b A metal X has a density of $1.13 \times 10^4 \, \text{kg m}^{-3}$ and the atoms have a mass number of 183. The structure of X is simple cubic with an atom of X placed at each intersection of a regular cubic lattice, as shown in Figure 10.6 below. The distance between the centres of nearest-neighbour atoms is $3.0 \times 10^{-10} \, \text{m}$.

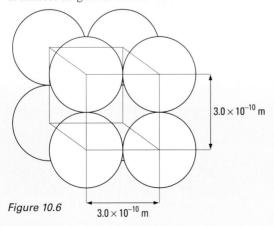

Figure 10.6 $3.0 \times 10^{-10} \, \text{m}$

Use the above information to obtain a value for the Avogadro constant.

9 a An increasing load is applied to each of the following specimens until it breaks:

i a glass fibre (a brittle material)

ii a rubber cord (an elastomeric material)

Sketch force–extension graphs on the same axes for each specimen, labelling clearly which graph is which. Describe the main features of and differences between the two graphs in terms of the microscopic structural differences of the two materials.

b The interaction energy U between nearest neighbours in a long-chain molecule composed of equally spaced identical atoms is given by:

$$U = -A/x^6 + B/x^{12}$$

where A and B are positive constants and x is the distance between the centres of neighbouring atoms. The graph of this relationship is shown in Figure 10.7. The interaction force is given by:

$$F = 6A/x^7 - 12B/x^{13}$$

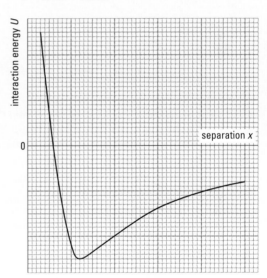

Figure 10.7

Calculate the equilibrium separation r_0 of atoms in the chain and the energy, E, needed to break a bond in the chain. Assume that only nearest neighbour interactions are important and that kinetic energy effects may be neglected.

Take $A = 1.4 \times 10^{-77} \, \text{J m}^6$

and $B = 1.0 \times 10^{-135} \, \text{J m}^{12}$.

Copy the graph of U against x marking the values of r_0 and of E on the axes.

10 This question is about an experiment in which two materials are stretched.

Figures 10.8 and 10.9 are stress/strain graphs for copper and glass, drawn to different scales.

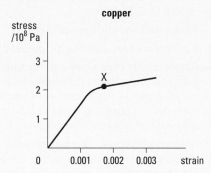

Figure 10.8

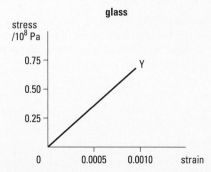

Figure 10.9

a i Name a quantity which can be found from the part of the graph near the origin in Figure 10.9. State how you would find its value from the graph.

ii Figure 10.9 shows both *elastic* and *plastic* behaviour of copper. Explain the italicised words with reference to the graph.

iii Copy Figure 10.9 and sketch on it the shape you would expect the graph to have when, having reached point X, the stress is gradually reduced to zero.

b i In Figure 10.9, Y represents the point at which the glass specimen fractures. What property of glass causes this failure?

ii Is the deformation of glass shown by Y elastic or plastic? Justify your answer.

iii Name one difference between the stretching of glass before it fractures, Figure 10.9, and that of copper over the initial straight portion of Figure 10.8.

iv Suggest a reason why fracture occurs in glass at such a low strain.

11 a Glass is a brittle material. The graph in Figure 10.10 shows that the tensile strength of glass in the form of rods and fibres varies with the diameter of the fibre. For very thin fibres the graph levels out at about 8 GPa.

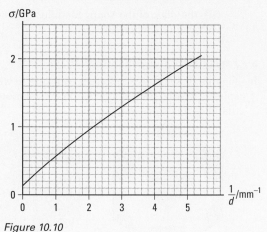

Figure 10.10

i What is meant by a brittle material?

Define the term tensile strength.

Use the graph to estimate the greatest mass which can be hung from a glass rod of diameter 2 mm.

ii For fibres of fixed length, how does the surface area of a fibre vary with the fibre diameter?

Explain why very thin fibres have a greater tensile strength than thicker ones and suggest why there is a maximum value.

iii 'Fibre-glass' is a widely used composite material. State what is used in its manufacture and name one object which is made from it.

b Sketch a graph of potential energy against separation for two atoms. Indicate an approximate scale on the separation axis and mark, with the letter E, the equilibrium separation.

Explain, with reference to your graph, the expansion of solid materials with increasing temperature.

12 a Figures 10.11 and 10.12 show typical fatigue curves for (i) steel, and (ii) aluminium alloy. (Both diagrams are drawn to the same scale.)

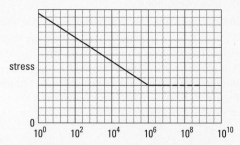

Figure 10.11

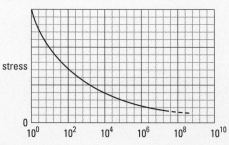

Figure 10.12

Describe the difference in behaviour of the two materials with increasing number of stress cycles. Why is steel a more suitable material for a component that is expected to suffer many stress reversals?

The Comet (an early jet airliner) was one of the first to have its passenger compartment pressurised, and had an aluminium alloy body. Suggest why this combination led to a number of accidents in flight after several years' service.

b The cabin floor panels of an aircraft may be constructed from a composite material such as that shown in Figure 10.13.

Figure 10.13

What is the difference between an alloy and a composite material?

What type of composite is shown in the diagram? In what *two* ways does it satisfy the requirements of a suitable flooring material for an aircraft?

Initial attempts to design a carbon fibre composite material for aircraft engine fan blades were unsuccessful. The blades were unable to withstand the impact of small birds, and had to be replaced by titanium ones. State one property that titanium possesses, but that a carbon fibre composite does not, that makes it able to withstand such impacts.

c A strain gauge, attached to a large steel cylinder, consists of an aluminium wire 2.0 m long and 2.0×10^{-6} m^2 in cross-section. If its resistivity $\rho = 2.8 \times 10^{-8} \, \Omega$ m, calculate its resistance.

When a longitudinal force is applied to the cylinder the resistance of the strain gauge becomes 3.0×10^{-2} W. Calculate the strain in the wire, and the stress experienced by the cylinder. (The Young modulus for steel = 2.0×10^{11} N m^{-2}.) Assume that the change in cross-sectional area of the wire is negligible.

13 A new polymer is being considered for use in car seats. It is subjected to a tensile test in which an applied stress is gradually and uniformly increased over a period of one minute. The stress is then maintained at 0.6 MPa for one minute before being reduced uniformly to zero over a further period of one minute. The results of this test are shown in the graph in Figure 10.14.

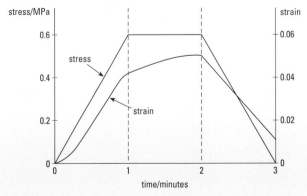

Figure 10.14

a i *Sketch* a stress/strain graph for the material, showing the values of strain along the horizontal axis and values of stress on the vertical axis.

ii Indicate the portions of your sketch graph corresponding to loading and unloading of the test sample.

b The strain of the material is seen to increase under conditions of constant stress.

i What is the name given to this process?

ii Attempt an explanation of the process in terms of the behaviour of the molecules of the material.

iii What additional information might be expected to accompany data for a test of this sort? Explain why this information would be considered essential.

c The maximum tensile stress in the material covering the seat of a car is estimated to be 0.6 MPa when an adult is sitting in it.

Discuss the consequences of using the polymer described above for the covering of a car seat by considering a typical loading/unloading cycle, stating any assumptions you make.

14 This question is about the Hall effect.

a A slice of n-type semiconductor carries a current I in the positive x-direction as shown in Figure 10.15. A uniform magnetic field B is applied perpendicular to the plane of the slice as shown.

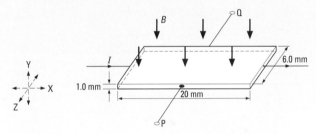

Figure 10.15

i The moving electrons, creating the current through the semiconductor, experience a force caused by the presence of the magnetic field. Along the line of which of the three axes x, y, or z will this force act?

ii Why is an electric field established parallel to the z-axis?

iii A voltage, called the Hall voltage, caused by this electric field can be measured across PQ. For a semiconductor slice with the dimensions shown in Figure 10.15 show that the electric field which causes a Hall voltage of 17 mV is about 3 N C^{-1}.

b i The strength of the magnetic field required to maintain the Hall voltage of 17 mV is 4.5×10^{-2} T. Show that the mean drift velocity of conducting electrons through the slice is more than 60 m s^{-1}. Electronic charge = 1.6×10^{-19} C.

ii The current I in the slice is 2.6×10^{-2} A. Calculate the number of conducting electrons per m^3 of semiconductor.

c The density of conducting electrons in a metal is many million times greater that that in n-type semiconductor. Explain how the Hall voltage expected across a slice of metal would compare with that across a semiconductor slice of the same dimensions carrying the same current and subjected to the same magnetic field.

11

The Physics of the BODY

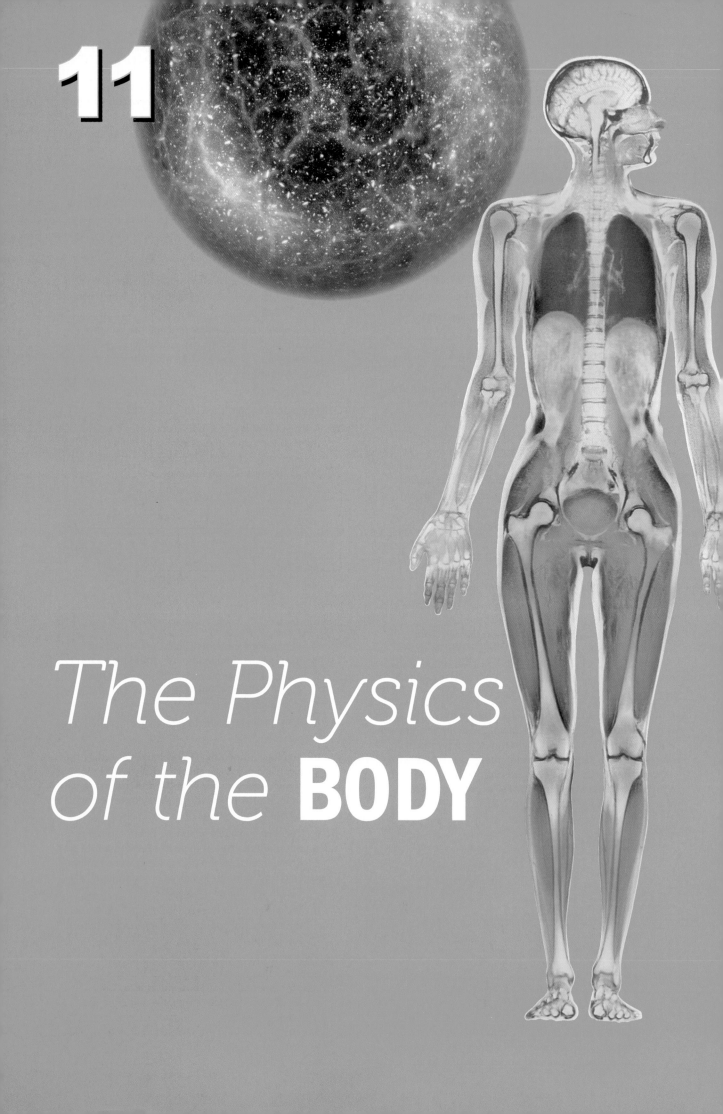

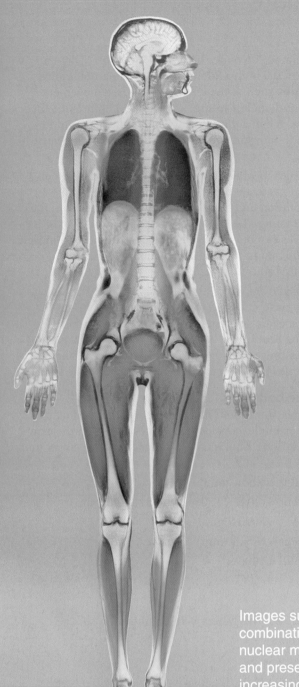

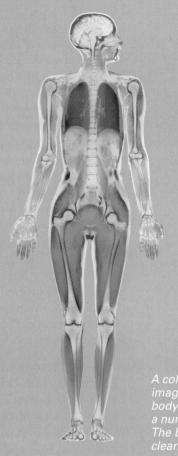

A coloured magnetic resonance imaging (MRI) scan of the whole body of a woman made up from a number of separate MRI scans. The bones of the skeleton can be clearly seen, as can the internal organs. The brain is pink in the skull; the lungs are dark in the chest; the lobes of the liver are blue/green in the abdomen; and the bladder is the rounded orange feature in the pelvis.

Images such as this of the entire human body are prepared with a combination of techniques that rely on physical phenomena (here nuclear magnetic resonance) and powerful computers to interpret and present the data obtained. These processes are becoming increasingly common in medicine, with doctors relying more and more on information from such images. Every hospital has a medical physics section responsible for producing radioactive elements to be used to trace chemicals around the body. Some even have their own particle accelerators to produce radiation for treatment and also to make artificial isotopes. Highly complex scanning and imaging equipment allows sections of the body to be imaged without cutting open a patient. In today's hi-tech world, physics is playing an increasingly important role in helping doctors to diagnose and treat illnesses.

In this chapter you will first study the structure and movement of the body as a whole, before focusing on two sensory organs, the eye and the ear, and the nervous system. The second half of the chapter deals with modern diagnostic and therapeutic techniques in which physics plays a key role.

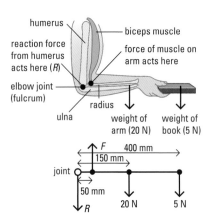

The system of bones and muscles in the arm, with a free-body diagram showing the forces acting on the arm while it carries an object.

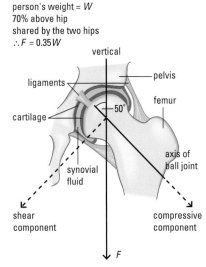

The force on the head of the hip bone is not parallel to the bone, so it puts the bone under both compressive and shear stresses.

THE BODY – STRUCTURE

The human skeleton contains 206 bones linked by 186 joints. There are two main types of joints in the body which allow a considerable range of movement. **Hinge joints**, such as the knee, provide free movement in one plane, while **ball-and-socket joints**, such as the hip, provide the maximum range possible. When a joint fails to function properly due to disease the restriction in movement and the resulting pain can cause considerable discomfort.

Ligaments hold the bones together at the joints, and muscles attach to the bones via tendons to activate the movement. A muscle is a fibrous tissue that can contract and become thicker at a signal from the nervous system. By contracting, the muscle applies a force at each of its connection points. Muscles can only contract. It requires a second muscle acting in opposition to the first to pull the bone back to its original position and restore the first muscle to its initial state.

Bone is an ideal material from which to build a structure. It is living, so self-repairing, light in mass, stiff, and strong. Some of the bones in the body have evolved a specific structure to make them more suited to their job. For example, the **femur** (leg bone) is thick-walled, hollow, and contains an internal cross-brace. This makes it ideal for supporting the strong forces (up to 30 times body weight) that are applied to the leg.

Analysing forces in the body

The body's system of bones, joints, and muscles can be analysed using free-body diagrams. For example, consider the forces acting on the arm while it is carrying a book (left). This is easier to analyse if the elbow joint is bent at 90°.

Taking moments about the elbow joint, we get

$$\text{clockwise moment} = (20\,\text{N} \times 150\,\text{mm}) + (5\,\text{N} \times 400\,\text{mm})$$
$$= 5000\,\text{N}\,\text{mm}$$
$$\text{anticlockwise moment} = (F \times 50\,\text{mm})$$
$$\therefore \quad 50F = 5000\,\text{N}$$

hence

$$F = 100\,\text{N}$$

Vertical forces must balance if the system is in equilibrium, so

$$R + 20\,\text{N} + 5\,\text{N} = F = 100\,\text{N}$$
$$\therefore \quad R = 75\,\text{N}$$

Note that as the force applied by the muscle acts at a point close to the pivot, the moment of the force is not very large. This means that the force applied by the muscle is very much greater than the weight of the object being lifted. This is the price that humans pay for having a range of movement – having the muscle linked to the bone further from the pivot would do some very strange things to our anatomy!

The **mechanical advantage** of a joint system is defined as

$$\text{mechanical advantage} = \frac{\text{load}}{\text{effort}}$$

which in this case is

$$\text{mechanical advantage} = \frac{(20\,\text{N} + 5\,\text{N})}{100\,\text{N}} = 0.25$$

Forces on the hip joint

The hip is a ball-and-socket joint which allows the freedom of motion required in walking, sitting, bending, etc. When a person is standing with his weight evenly balanced on both legs, each hip must support half of the weight of the body structure above the hips. Typically this is 70% of the total body weight. The diagram on the left is a free-body diagram of the force at the hip joint in these circumstances. The head of the femur normally makes an angle of 50° to the vertical at the socket. The load acts vertically on the head of the femur, but because of the angle causes both compression and shear on the bone (if the bone entered the joint vertically then there would only be compression).

$$\text{shear force} = F \sin 50° = 0.35W \times \sin 50° = 0.27W$$
$$\text{compressive force} = F \cos 50° = 0.35W \times \cos 50° = 0.22W$$

The backbone

The 33 vertebrae that together make up the backbone, or **spinal column**, provide the main support for the body. The top 24 vertebrae are separated from each other by fibrous pads called **discs** between them. This gives the upper back some flexibility. The lower 9 vertebrae are fused together. The **spinal cord**, which is a vital part of the central nervous system, runs down through the spinal column. The discs are under considerable compressive and shear stresses. In time they lose their toughness. Under certain circumstances a disc can slip out of position so some of the load falls directly on a vertebrae. This is an extremely painful condition.

The most likely time for damage to occur to the back is when bending and lifting loads. The forces involved can be estimated by considering the back to be a rod jointed at the pelvis. A set of muscles links the vertebrae of the spine to the pelvis and controls the spine which can pivot about its joint with the pelvis.

Forces in the body when bending and lifting

Bending

Lifting

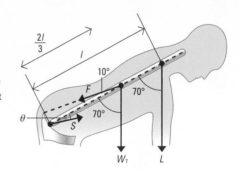

W_T = weight of trunk ≈ 60% of body weight. The centre of gravity of the upper body is about 2/3 of the way along the spine.

The angle between the spine and the line of action of the muscles is about 10°. As the spine is in equilibrium under the action of three forces, their lines of action must meet at a single point (see spread 3.4), so the reaction force S from the pelvis acts along the spine.

Resolving forces at 90° to the line of the spine gives

$$\text{component due to } S = 0$$

as S acts along the spine

$$\therefore \quad W_T \sin 70° = F \sin 10°$$

or

$$F = \frac{W_T \sin 70°}{\sin 10°}$$

So, for an average body weight of 700 N, W_T = 420 N and F is 2.27 kN (considerably more than the body weight!)

Now there is a fourth force acting (the load L), so S no longer acts along the spine but at some small angle θ to the spine.

Taking moments about the joint, we get

$$\text{clockwise moment} = l \times L \sin 70° + \frac{2l}{3} \times W_T \sin 70°$$

$$\text{anticlockwise moment} = F \times \frac{2l}{3} \sin 10°$$

$$\therefore \quad F = \frac{(L + \frac{2}{3}W_T)\sin 70°}{\frac{2}{3}\sin 10°}$$

So, when a person of average body weight (700 N) lifts half her own weight (350 N), W_T = 420 N and F = 5.11 kN.

S has two components, a compressive one (along the spine) of 1.64 kN and a shear (90° to the spine) of 196 N.

PRACTICE

1 What load can be held by the arm bent at 90° if the force exerted by the muscle is 120 N? What is the reaction force acting on the joint? What would be the mechanical advantage of the system?

2 By resolving forces parallel to the spine and perpendicular to the spine, confirm the compressive and shear components of S. What is the magnitude of S? At what angle to the spine does S act?

THE BODY – MOVEMENT

The intricate mechanical engineering of the body and its system of joints, muscles, and nerves does not simply have to cope with supporting the mass of the rest of the body while at rest. Human beings tend to move about. Some of them move about in very strange ways and the skeletal system has to support the forces involved in these movements. Not only that, but the muscles have to provide the forces required to initiate the movement in the first place.

Standing

The sensation of weight that we have while standing still on the ground is not our body directly detecting the force of gravity pulling us towards the centre of the earth. That force of course exists, but we do not have any 'force-detecting' cells situated at our centre of gravity to spot that it exists. The sensations that we associate with our own weight are actually the compressions of the various joints in the body.

Two types of force are acting upon a stationary person – there is the pull of gravity as described above, but there is also the reaction force exerted by the ground upwards. Without this force the body would not be in equilibrium and would be accelerating downwards (drilling a hole in the Earth in the process!). In fact this is exactly what does happen for a very brief period of time until the material under our feet has deformed enough to exert a force up on us equal to our weight. If the material breaks before it can exert this force, then we will fall through.

The combination of these two sets of forces puts the body into a state of compression, squeezing the joints and making us aware of our own weight. Astronauts in free fall do not have this compressive effect acting on them and so their joints expand slightly.

Walking

Even a simple motion such as walking is revealed to be quite a complex process when it is fully analysed.

As a simple start consider that each leg will require a horizontal force to accelerate it forwards. The leg then rotates about the hip joint, so a turning moment is also required to produce the angular acceleration. Finally the leg must be decelerated again when it reaches the end of the pace. This time, however, the body does not have to provide all of the force as gravity helps.

The diagram on the next page shows the forces acting during the three stages of walking. The vertical reaction force of the ground on the foot changes. Assuming that there is only one foot in contact with the ground at a time, then:

a During the support moment $R_2 = W$ (the weight).

b At **heel-strike** $R_1 > W$, as the ground is reacting to the extra deformation caused by the decelerating foot over that due to the weight.

c At **toe-off** $R_3 > W$, as the ground is reacting to the thrust of the calf muscles as well as the weight.

Frictional forces act during stages **a** and **c** which must be less than the limiting friction for that surface, i.e. $F_1 < \mu R_1$ and $F_2 < \mu R_2$.

During heel-strike and toe-off the resultant force of the ground on the foot is given by $G = \sqrt{R^2 + F^2}$, which acts at an angle θ to the ground where

$$\tan \theta = \frac{F}{R}$$

Now, if the walker is not going to slip:

$$F < \mu R$$

hence

$$\tan \theta < \mu$$

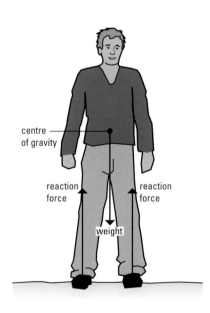

centre of gravity

reaction force   reaction force

weight

The sensation of weight.

This places a maximum angle at which the foot can contact the ground for a given surface. Consequently, there is a maximum stride length for every surface.

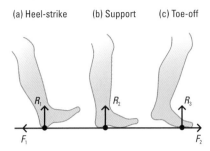

(a) Heel-strike　　(b) Support　　(c) Toe-off

The three stages of walking and the forces acting on the foot.

Another feature of the walking motion to consider is the change in height of a person's centre of gravity. This is at its lowest point when the legs are in mid-stride, with one in front and one behind the vertical centre line of the body. The highest point of the centre of gravity comes when both legs are in line.

A tallish adult male has a leg length of about 1 m and a comfortable stride of 80 cm. Consequently the change in height of the centre of gravity will be:

$$\Delta h = 100\,\text{cm} - \sqrt{100^2 - 40^2}\,\text{cm} = 100\,\text{cm} - 92\,\text{cm} = 8\,\text{cm}$$

With a weight of around 750 N this corresponds to 60 J of work having to be done against gravity. However, the estimation of Δh carried out above is inaccurate as the legs bend as we walk, so the variation in height of the centre of gravity is somewhat less.

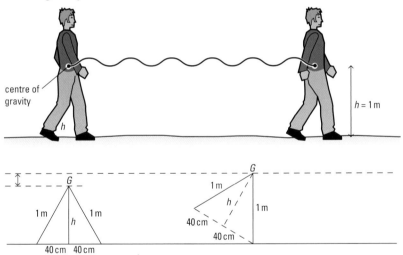

PRACTICE

1 a Draw a simple diagram of a person standing in a lift supported by a single cable (the lift is not moving). Now draw free-body (force) diagrams of **i** the lift and **ii** the person.

b The lift now accelerates upwards. Which forces on your free-body diagrams have now changed and in what way have they changed? By referring to these diagrams explain why the person experiences a sensation of increased weight.

c A simple weighing machine consists of a metal plate supported by a vertical spring underneath. When someone stands on the plate, the spring is compressed and the extent of the compression is a measure of the person's weight. The person in the lift is now standing on such a weighing machine. When the lift is stationary, the machine correctly reads the person's weight. Draw a free-body diagram showing the forces acting on the plate of the weighing machine. What happens to the reading once the lift starts to accelerate upwards? Justify your answer from the free-body diagram.

2 Measure the length of your leg. Imagine that your leg is a simple pendulum pivoted at the hip. What would be the period of this pendulum? Measure a comfortable stride length for one foot. Using this information estimate a comfortable walking pace for you. Is your answer reasonable?

THE EYE

The mammalian eye could be regarded as a miracle of evolution. In order to produce an image the eye has to act as an optical refracting system (to focus a sharp image) and a photodetector (to analyse the image and send information to the brain).

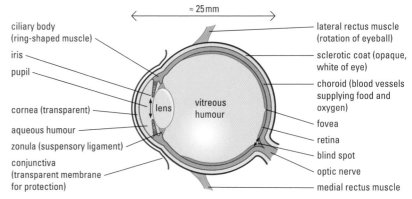

A cross-section through the human eye. The aqueous and vitreous humours are liquids that provide nutrients without interfering with vision.

The human eyeball is filled with fluid, which exerts an outward pressure that keeps the eyeball spherical. At the front of the eye is the optical system, made up of the transparent curved **cornea** and the adjustable biconvex **lens**. The photodetector is the layer of light-sensitive cells at the back of the eye, called the **retina**.

The optical system

The amount of light entering the eye is controlled by the **iris**. If the light is too bright the circular muscles of the iris contract, making the **pupil** smaller. In dim light the radial muscles contract, opening out the pupil.

Light rays are strongly refracted as they pass through the cornea, which focuses the rays on the retina. Sharp focusing is achieved by altering the shape of the lens. The shape of the lens is controlled by the ring of **ciliary muscle** which runs round the outside of the lens. The lens itself is linked to this muscle by radial ligaments. When the ciliary muscle relaxes the ligaments are stretched and the lens is pulled into a thin shape. This reduces its refractive power, allowing the eye to bring distant objects into focus. When the ciliary muscle relaxes, the tension in the ligaments is reduced and the lens bulges outwards. This shape has a greater refractive power and the eye can focus on nearer objects. The process of changing the shape of the lens is called **accommodation**.

The image distance in the eye (see spread 6.20) is fixed by the size of the eyeball. In principle, for any shape of the eye lens, there is only one distance at which an object can be brought into focus on the retina. In practice, objects at a range of different distances can be brought to an acceptable focus without having to adjust the lens. This range of object distances is called the **depth of field**. The eye will also accept images that are focused either just in front of or just behind the retina. This is known as the **depth of focus**. Both depth of field and depth of focus are influenced by the size of the pupil. If the pupil is wide open, light can enter the eye at quite large angles, which makes it more difficult to bring all the light from one object to a focus – this reduces the depth of field and the depth of focus. The converse is true when the pupil contracts. People tend to 'screw up' their eyes to read small print.

The photodetection system

The retina contains light-sensitive cells with diameters of a few micrometres. When light falls on them it causes photochemicals to decompose, sending electrical signals to the brain. The rod-shaped cells are not sensitive to colour but are capable of working in low-light conditions (**scotopic vision**). The cone-shaped cells do register colour (**photopic vision**) but require quite bright light in order to operate. This is why objects seem to lose their colour in dim light. There are about 6 million **cones** and 125 million **rods** in the retina. As there are only about a million nerve fibres running from the eye to the brain, a degree of sharing is required. The extent to which cells have to share nerve fibres varies; rods share more than cones. At the **fovea**, directly behind the lens, there is one nerve per cone (this is the region most sensitive to fine detail), while at the edge of the retina 600 rods share a single fibre.

Resolution

Diffraction at the edge of the pupil spreads light out as it enters the eye. This means that a point light source will always produce a slightly diffused image on the retina. The width of the patch of light produced is such that, applying the Rayleigh criterion (see spread 6.10), two patches cannot be resolved unless they are at least 2 μm apart. Two point sources placed 10 m from the eye and 1 mm from each other would just be resolved. At the fovea the diameter of a cone is 1.5 μm and each cone has its own nerve connection to the brain. Two points of light cannot be distinguished by the retina unless there is at least one unstimulated cell between them. So, at the fovea the minimum separation between patches of light must be about 3 μm. This is of the same order as the optical resolution, so the fovea is capable of exploiting the maximum performance of the optical system.

Scanning

The image produced in the brain is only detailed at the centre, but we do not notice this because our eyes are continually moving back and forth over the whole viewing area. This process, known as **scanning**, is rather like 'refreshing' a picture on a TV screen. If the image on the retina were fixed, then the cells would stop firing as their illumination would be constant. The image would slowly fade. If the source of light is removed, the image takes about one-fifth of a second to fade. This **persistence of vision** is used in cathode-ray displays. A fast-moving spot will seem to form a line. If a bright light flashes on and off, it will seem to be continually on provided it is flashing at somewhere in the region of 60 times a second. It is no coincidence that 50 frames per second is the refresh rate for TV.

Colour vision

There are three types of cone in the retina. One responds to a range of wavelengths in the red part of the spectrum, another detects green wavelengths, and the third is sensitive to blue, so red, green, and blue are regarded as primary colours. If the eye is illuminated by orange light, the cells fire to different extents. The exact pattern of firing can be reproduced using a mixture of red, green, and blue lights. This mixture would be seen as orange. Consequently there is **true orange** (part of the visible spectrum), and **psychological orange** which is an 'optical illusion'. This optical illusion is exploited in TV.

Persistence of vision

It is commonly accepted that the retina retains an image for a fraction of a second (about 0.04 s) after removal of the stimulus that produced it. This persistence of vision theory is often used to explain how a rapidly shifting sequence of still frames is perceived as motion (e.g. in film and TV). However, in reality the persistence of vision merely explains why we do not see the black frames between images. The illusion of movement results from the brain/eye system interpolating information between frames.

The edge of the eye

At the edge of the retina the resolution drops off due to the multiple sharing of nerve cells by the rods. However, in compensation the gathering of information from many rods into one nerve means that the edge of the eye is quite sensitive in dim light. Astronomers using telescopes try to look 'out of the corner' of their eyes when trying to see dim objects.

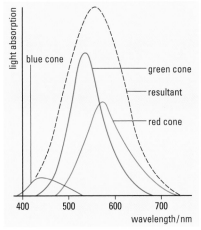

Relative sensitivities of the three types of cone.

- ray diagrams of image formation

- near point, far point, range of accommodation

- defects and their correction

The power of a lens

The power of a lens is defined as 1/focal length; if the focal length is in metres this gives a power in **dioptres** (D).

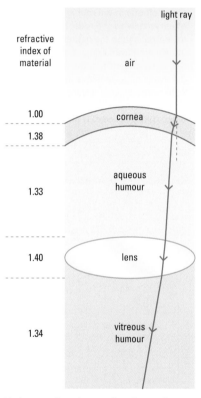

Light entering the eye is refracted most at the cornea.

Refracting power

The curvature of the cornea and the change in refractive index between the air and the material of the cornea produce a refracting power of about 46 D at the boundary. There is a slight defocusing as the light moves from the cornea into the aqueous humour as the refractive index drops slightly (about –5 D).

THE EYE'S OPTICAL SYSTEM IN DETAIL

Most of the refraction produced by the eye's optical system is achieved by the cornea. It has an optical power of about 46 D. In comparison the lens has very little ability to bend light, but its role is absolutely vital as it allows the eye to accommodate for objects at different distances from us. With an object at infinity (in practice, anything on the horizon) the rays of light arriving at the eye can be assumed to be parallel and so the lens can be at it thinnest – in which case it has a power of 18 D. With the lens at its fattest its power goes up to about 29 D. Of course it is not the thickness of the lens that is important, it is the curvature that determines the amount by which it will refract light (along with its refractive index). Curved lenses have to be fat to accommodate the curvature!

The maximum refraction achievable by the cornea and the lens in combination determines the minimum distance from the eye at which an object can be viewed. The nearer the object, the greater the angle between the rays of light arriving at the eye, and so the more powerful the optical system has to be in order to bring these rays into focus on the retina. With most young people the **near point** (minimum focusable distance) is about 25 cm and the **far point** (maximum distance from which an object can be focused) is at infinity. However, as people get older the lens and its muscles become less flexible and so less able to accommodate. Consequently the near point recedes until only distant objects can be focused correctly. This degrading of the eye's ability to accommodate is called **presbyopia** and is a normal part of the ageing process.

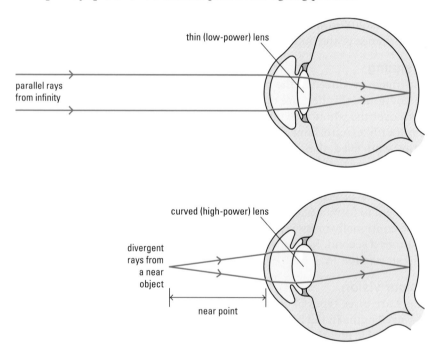

The principle of accommodation.

Defects in the eye's optical system

1 **Long sight** (**hypermetropia**) People with this condition can see distant objects more clearly that ones that are nearby. This is usually because the cornea is not curved enough, which reduces the angle of refraction, and the lens cannot accommodate sufficiently to bring the diverging rays from near objects into sharp focus on the retina. The situation can be corrected by placing a converging lens in front of the eye (by means of spectacles or contact lenses). In extreme circumstances laser surgery can be used to correct the shape of the cornea.

Worked example 1

A man has a near point which is 1.5 m from his eyes. What type of lens must be used to restore the near point to 25 cm?

The distance between the eye lens and the retina is generally about 0.02 m, which is the image distance of the eye's optical system. Using the lens formula

$$\frac{1}{f} = \frac{1}{u} + \frac{1}{v}$$

gives the power of his optical system ($1/f$):

$$P = \frac{1}{f} = \frac{1}{1.5} + \frac{1}{0.02} = 50.7\,\text{D}$$

whereas for a normal eye, with a near point of 25 cm, the power would be

$$P_\text{n} = \frac{1}{0.25} + \frac{1}{0.02} = 54\,\text{D}$$

Hence a converging lens of 54 D – 50.7 D = 3.3 D must be used to correct this defect.

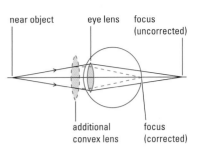

Long sight is the inability to focus the light from near objects. This can be corrected by adding a converging lens in front of the eye.

2 **Short sight (myopia)** Near objects are clearly seen, but distant objects cannot be brought into focus. If the cornea is too curved, the light from distant objects is refracted too much and comes to a focus in front of the retina. The situation can be corrected by placing a diverging lens in front of the eye.

Worked example 2

A woman has a far point of 1 m. What type of lens must be used to enable her to see objects at infinity?

Using the lens formula again shows that the power of her optical system when fully relaxed is

$$P = \frac{1}{1} + \frac{1}{0.02} = 51\,\text{D}$$

For the far point to be at infinity a power of 50 D is required, so the woman needs to use a pair of diverging lenses of –1 D.

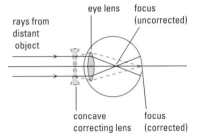

Short sight is the tendency to focus the light from a distant object to a point before the retina. It can be corrected with a diverging lens.

Spectacle lens prescriptions

A spectacle or contact lens prescription lists numbers under three headings:

OD (*oculus dextrus*): right eye;

OS (*oculus sinister*): left eye;

OU (*oculus uterque*): both eyes.

The power of a lens used to correct long sight will be shown by a positive number (in dioptres); a lens to correct sort sight will have a negative number.

For astigmatism the prescription will be in the format $S \times C \times$ axis, where S is the spherical near/far power as described above, C is the cylinder (the degree of astigmatism in dioptres), and the axis is an angle 0–180° showing the orientation of the astigmatism.

---

3 **Astigmatism** If the cornea is curved in an uneven manner, it will refract light that enters the eye in different planes to different extents. The eye will focus light in some planes better than others (i.e. verticals might be in focus, but horizontals not). This can be corrected by using a cylindrical lens (i.e. one that only curves in one plane).

4 **Presbyopia** This is the natural reduction in the eye lens's accommodating power with age. It can be corrected by using bifocal lenses, which have different curvatures in different regions. While looking at close objects (such as text) the eye tends to look down, so the lower region of the lens is manufactured to correct for defects in short sight. When we look at distant objects we tend to look up, so the upper part of the lens caters for long-sight problems.

PRACTICE

1 In Worked example 2 above, the lenses in the woman's eyes have an accomodation of 6 D. What will be the distance of her near point:

a while she is not using her correcting lenses;

b while she is using the correcting lenses?

2 Prescribe correct lenses for the following conditions:

a far point at infinity, near point at 1 m, accommodating range 10 D;

b far point at 5 m, near point at 0.5 m, accommodating range 1.8 D.

THE EAR

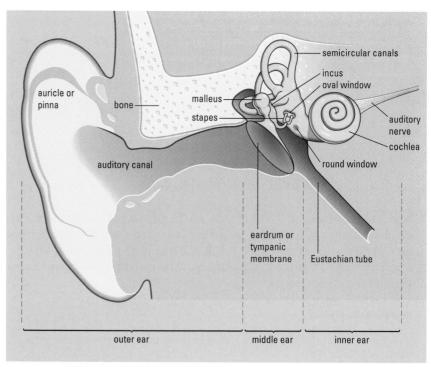

A cross-section through a human ear.

The structure of the ear is divided into three sections – the outer, middle, and inner ear.

The outer ear collects air vibrations and channels them to the eardrum (**tympanic membrane**) in the middle ear. The outer ear is a confined passage which will resonate at a certain frequency (usually about 3000 Hz). Resonances also occur in the middle ear. Together they help shape the frequency response of the ear (see below).

The eardrum vibrates when sound waves strike and the vibrations are passed on to the three bones of the middle ear. These in turn set a membrane across the **oval window** vibrating and the energy is passed on to the fluid in the inner ear. To reduce the acoustic impedance mis-match between air and the fluid (see spread 11.13) the ear has evolved an impedance-matching system. The **round window** linking the inner ear to the middle ear is covered by a membrane which moves in response to pressure from the inner ear. This prevents the inner ear becoming pressurized and allows some 'give' in the system – effectively reducing the impedance of the fluid. The three bones in the middle ear act as a lever. The lever is geared to multiply the force at the oval window. As the oval window is about 20 times smaller than the eardrum the pressure on it is greatly increased. This higher pressure helps put the dense liquid in the inner ear into motion which reduces the effect of the acoustic mis-match.

In the inner ear the very complex **cochlea** turns the pressure waves in the fluid into nerve signals for the brain. The exact process is not understood, but above 20 Hz the pressure waves set a membrane in the cochlea moving. This membrane in turn causes a line of hairs linking it to a more rigid membrane to bend back and forth, which sends electrical impulses to the brain. The **semicircular canals** of the inner ear are responsible for detecting movement of the body and balance. They do not take part in the hearing process.

The ear's ability to detect sound

The **intensity** of a sound is defined as the amount of energy per metre squared per second carried by the wave. The lowest sound intensity that can be heard at a frequency of 1 kHz is known as the **threshold of hearing**. It corresponds to an intensity of about 10^{-12} W m^{-2} (which would be produced by a pressure variation between the compressions and rarefactions in the wave of 2×10^{-5} Pa). At the other end of the scale the most intense sound that the ear can be subjected to without damage is 100 W m^{-2}.

The **intensity level** of a sound is the ratio between the intensity of the sound (I) and the threshold intensity (I_0), measured in decibels (see spread 6.2). Consequently

$$\text{intensity level} = 10\ln\left(\frac{I}{I_0}\right)\text{dB}$$

The intensity level of the maximum intensity referred to above is

$$10\ln\left(\frac{I_{max}}{I_0}\right) = 10\ln\left(\frac{100}{10^{-12}}\right) = 322\,\text{dB}$$

An adaptation of the intensity level scale, known as the dBA scale, compensates for the ear's frequency response by electronically weighting the contributions to a sound at different frequencies to match the ear's response.

Loudness is a sensation produced by the combination of the ear and the brain and so there is no reason why loudness (which is subjective) and intensity (which is objective) should correspond to one another. Experiments show that people believe two sounds to be equally loud even though they are of different intensities provided that they are also of different frequencies. In order to measure loudness a standard source at a frequency of 1 kHz is used. This can be placed next to a test sound source and its intensity adjusted until it is perceived as being the same loudness as the test source. The intensity of the standard source is then measured. If this intensity is n decibels over the threshold intensity, then the loudness of the sound is said to be n **phons**.

Not only is loudness not directly related to intensity, but the perception of a change in loudness is not simply related to a change in intensity either. Experiments have shown that a change from 1×10^{-6} W m^{-2} to 2×10^{-6} W m^{-2} is not the same increase in loudness as an increase of 2×10^{-6} W m^{-2} to 3×10^{-6} W m^{-2}. However, it is perceived to be the same as an increase of 2×10^{-6} W m^{-2} to 4×10^{-6} W m^{-2} – which is a doubling of intensity in each case. This leads to the idea that equal loudness increases are produced by the same fractional increases in intensity:

$$\text{loudness change} \propto \frac{\text{change in intensity}}{\text{initial intensity}} \propto \frac{dI}{I}$$

Hearing defects

There are two main categories of hearing problem. In **conductive loss** the sound vibrations do not reach the inner ear. **Nerve loss** is the failure of the cochlea to pass nerve impulses on to the brain.

Disease and the ageing process reduce the ability of the inner ear bones and the oval window to respond to pressure waves. The ability to perceive high frequencies reduces with age. An infection can fill the inner ear with fluid, leading to conductive loss. A more serious problem is damage to the eardrum, which can be ruptured by a high-intensity sound (190 dBA). A blow to the head can also damage the inner ear. Conductive losses can sometimes be corrected by surgery, but there is no cure for nerve loss. Using a hearing aid can only compensate for the loss of sensitivity.

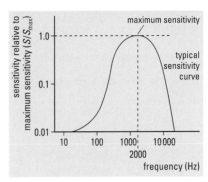

The frequency response of the ear.

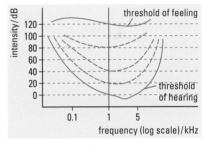

The ear's perception of loudness depends on intensity and frequency. Along each of these lines a sound would have the same perceived loudness. A person's equal-loudness curve is known as an audiogram, produced by testing their hearing through earphones in a soundproof room.

Turn the music down!

Your hearing may deteriorate if you subject your ears to continuous loud sound. People exposed to loud music or noise e.g. in factories can experience hearing loss, especially around 4 kHz.

- electrical conduction in nerve cells

- detecting the electrical activity of the heart

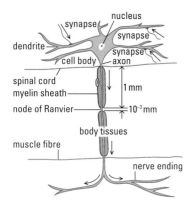

The structure of a nerve cell.

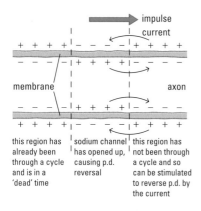

The action potential of a nerve axon.

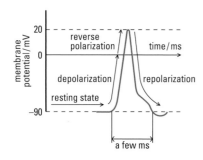

Variations in p.d. across a nerve axon due to the action potential.

THE BODY ELECTRIC

Nerve cells

The diagram (left) shows a typical nerve cell running from the spinal cord to a muscle fibre. Such a nerve cell would be used to control the contraction of the muscle fibre. Nerve cells communicate with each other via connections between them made at the **synapses**. When a cell is stimulated, an electrical signal runs the length of the **axon** to the muscle. The process by which the signal passes down the axon is not fully understood, but it is known that it relies on the concentration of ions inside and outside the cell.

A chemical process known as the **sodium–potassium pump** forces Na^+ ions out of the axon and K^+ ions in from outside. As a consequence of this pumping action Cl^- ions also drift from inside the cell to outside. The high concentration of Na^+ outside the axon contributes to a p.d. across the membrane of about 70 mV (the outside being positive). Most of this p.d., however, is due to negatively ionized organic molecules inside the axon (they are too big to pass through the membrane).

Running across the membrane are long-chain molecules which curl up into helical shapes. These form tunnels along which sodium and potassium ions can drift. A triggering voltage from the main body of the cell causes the first few sodium tunnels to open up (the inner radius of the molecules increases) and Na^+ ions start to flow in from the higher concentration outside (pushed by the p.d.). This flow of charge reverses the p.d. in the region of the tunnels. The reversed p.d. (known as the **action potential**) triggers sodium tunnels in the next part of the membrane to open as well, starting the process further down the line.

When the original reversed p.d. reaches a peak the sodium tunnels close and the potassium tunnel molecules slowly open up (stimulated by the reversed p.d.). K^+ ions flow out of the axon. This flow helps trigger the sodium–potassium pump to start restoring the original p.d. and the axon slowly returns to its original state. Meanwhile, the open sodium tunnels further down the membrane are triggering the next section – and so on. As a result, a region of reversed p.d. moves down the axon – this is the nerve impulse.

Once a region has gone through the full cycle, a period of time must elapse before it can happen again. This ensures that the impulse can only travel one way along the axon.

The heart

The heart is about one-third of a kilogram of muscle which powers two separate pumps. The pump on the right receives oxygenated blood from the lungs and forces it out into the arterial system at very high pressure. Blood which has been deoxygenated and loaded with waste products arrives at the left pump, via the veins, where it is forced into the lungs. Waste gases are extracted and exhaled, replaced by oxygen, and the process starts again.

Typically a heart beats at about 60–100 beats per minute. The pumping action is controlled by an action potential produced by a cluster of cells called the **sinoatrial node** in the upper right region of the heart. As the electrical signal spreads through the muscular tissues via nerves, it causes depolarization and muscular contraction. The signal arrives first at the atria, causing them to contract and force blood through one-way valves into the ventricles. The signal then reaches the ventricles, making them contract, which forces the blood out of the heart and into the circulatory system.

Measuring the electrical signals of the heart

The p.d.s caused by the spread of the action potential can be detected by placing electrodes on the skin. Many of the materials in the body have a high resistivity (blood plasma about $0.6\,\Omega\,m$, fatty tissue about $25\,\Omega\,m$) so the currents involved are very small. The site for placing the detecting electrodes must be very carefully chosen, and rubbed with an abrasive pad to remove the outer layer of skin cells. A conductive paste is sometimes used on the skin before the electrode is taped in place. Even with this preparation, the signal needs amplification before it can be recorded. The resulting display is known as an **electrocardiogram** (**ECG**).

There are two ways of recording ECGs:

- A pair of electrodes (**bipolar leads**) are placed on the body, and a differential amplifier is used to measure the potential difference between the sites. There are three favoured connections:

  a right arm and left arm;

  b right arm and left leg;

  c left arm and left leg.

  It is pointless to connect to the right leg as it is too far from the heart. However, a neutral lead is often run to the right leg to minimize interference by earthing the person.

- One electrode is kept at zero potential, while the other measures changes in potential (**unipolar leads**). Usually the two arms and left leg are connected together to the zero potential lead, while the other is placed on the chest over the heart.

A normal ECG

An ECG from a heart under normal conditions is shown below right. The main features are listed below.

- The **P-wave** signals the contraction of the atria due to their depolarization. Their subsequent repolarization would be visible on an ECG if not for the next feature which swamps the signal.

- The **QRS-wave** indicates depolarization of the ventricles leading to their contraction.

- The **T-wave** indicates that repolarization of the ventricles has occurred.

Various problems with the heart can be diagnosed from their effect on the normal ECG trace.

- The normal heart rate is 60–100 beats per minute, depending on the level of exertion. The heart rate can easily be measured on an ECG. A slow rate can sometimes occur in a very fit athlete who has a large heart capacity. The cyclist Bradley Wiggins has a resting heart rate of about 35 beats per minute, while that of the racing driver Jenson Button is in the 40s.

- An irregular trace indicates an irregular pumping action (**arrhythmia**).

- A blockage of the heart due to disease may be diagnosed if part of the trace fails to appear. Muscular damage (due to a heart attack) can be seen in a reduced waveform height.

- If the heart goes into fibrillation (beating fast and irregularly, and pumping little blood) because of an electric shock for example, the result is a very irregular trace.

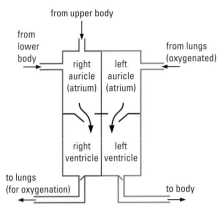

The basic structure of the heart.

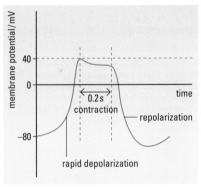

The action potential of the heart.

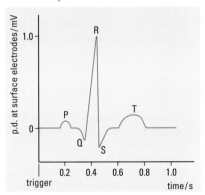

A typical ECG.

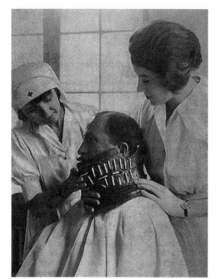

In the early days of research into radioactivity people were allowed to carry out some highly dangerous activities! The tubes contain small samples of radium and are being applied as a cancer cure.

Methods of exposure

It is important to distinguish between **contamination** and **irradiation**. If an object is exposed to ionizing radiation it has been irradiated. Only in very extreme circumstances (e.g. neutron exposure) will this make the object radioactive. Irradiation by gamma rays is used as part of the sterilization procedure for medical instruments. A move to irradiate vacuum-packaged food produced a public outcry as people thought, wrongly, that it would make the food radioactive. Contamination is exposure to the radioactive material itself. This is far more dangerous, especially if the material manages to get inside the body (e.g. by breathing in a gas, or if a small particle is absorbed into the blood stream from a cut). Workers in the nuclear industry sometimes wear protective clothing – this is to prevent contamination, not irradiation.

RADIATION AND THE BODY

When radium was first discovered its medical possibilities were widely advertised and on sale. Unfortunately, many of the suggested uses of radium, such as adding radium to one's bath water, were 'quack' medicine and carried a significant amount of risk to anyone using them.

How ionizing radiation affects biological systems is not yet fully understood. Clearly the ionization of molecules can lead to significant damage. In some cases molecules that are essential to the operation of cells (such as enzymes and DNA) are damaged directly, but damage may also occur indirectly. For example, ionization of water molecules in cells produces H and OH free radicals which are chemically highly reactive, and having such powerful components free in a cell produces chemical havoc. Extensive cellular damage can lead to the formation of cancers or leukaemia. High levels of exposure can destroy the body's ability to repair itself or to defend itself from infection. In extreme cases burns appear on the skin, and blindness can result, as well as nausea, loss of hair, and eventual death.

Some of the effects described (e.g. leukaemia) can be produced by any exposure to radiation although the risks increase with the amount of exposure. These are referred to as **stochastic effects**. Other damage, such as burns, will not occur unless the exposure is greater than some minimum amount (**non-stochastic effects**). Another useful distinction is that between **somatic effects**, which only happen to the person exposed to the radiation, and **hereditary effects**, which are passed on to children of the exposed person.

Units used in nuclear medicine

The rate at which ionizing radiations are produced by a sample is called the **activity** of the sample. This can be very high if the sample is large and the half-life is short, or low for a small sample containing long-lived isotopes. However, the activity alone does not determine how dangerous a sample can be to a person's health; the degree of danger depends on the nature of the radiation. When radiation passes through matter it loses energy by ionizing atoms. The **dose** of radiation received is the amount of energy that is absorbed from the radiation per kilogram of body mass. The greater the dose, the greater the amount of energy deposited in the body, and consequently the more damage that is done.

The body's ability to repair damaged tissue depends on the extent of the damage and its localization. When alpha particles pass through body tissue they deposit energy in a short straight path. Beta particles, on the other hand, spread their energy over a longer and more crooked path. The same dose of each would not necessarily produce the same damage. For this reason the **dose equivalent** is a commonly used unit which takes into account the nature of the radiation:

dose equivalent (Sv) = dose (Gy) × quality factor

The ironically named **quality factor**, Q, is a numerical constant that accounts for the nature of the radiation. For example, Q for neutrons is 10, Q for alpha particles is 20, and Q for X-rays is 1. In principle, exposure to the same dose equivalent should have the same medical implications whatever the type of radiation.

Radiation quantities.

Quantity	Definition	Unit
activity	number of decays per second in a sample	becquerel, Bq 1 Bq = 1 decay per second
dose	energy absorbed from ionizing radiation per kg of target mass	gray, Gy 1 Gy = 1 J kg^{-1}
dose equivalent	equivalent energy absorbed per kg	sievert, Sv

Exposure

It is difficult to measure the amount of energy deposited in a material by ionizing radiation. When X-rays or gamma rays pass through a material the amount of ionization produced is directly related to the energy absorbed. The amount of energy absorbed per kilogram in the material depends on the energy of the photons and the atomic number, Z, of the atoms in the material. Air has an effective Z of 7.6, which closely resembles that of body tissues. For this reason, ionization in air is an effective way of monitoring X-ray and gamma-ray exposure.

If Q is the total amount of either positive or negative charge produced in a mass, m, of air by the passage of X-rays or gamma rays, then the exposure, X, is defined as:

$$X \ (\text{C kg}^{-1}) = \frac{Q}{m}$$

(This applies only to X-rays and gamma rays in air.)

Exposure and dose

In practice, an ionization chamber can be used to monitor the likely X-ray or gamma-ray dose received by people working in the region of the chamber. To make sense of the information the exposure recorded by the chamber must be related to the effect that the radiation has on the body.

An exposure of $1 \ \text{C kg}^{-1}$ implies that $1/e$ electrons have been produced per kilogram of air. The average ionization energy of the molecules in air is $34 \ \text{eV}$, so the energy absorbed by the air must be:

energy = number of electrons produced × ionization energy

$$= \left(\frac{1}{e}\right) \times (34 \ \text{eV}) = \left(\frac{1}{e}\right) \times (34 \times e \ \text{J}) = 34 \ \text{J}$$

An exposure of $1 \ \text{C kg}^{-1}$ is a dose of $34 \ \text{J kg}^{-1}$ or $34 \ \text{Gy}$.

Of course, this is not necessarily the same dose as a person would receive, as the energy absorbed depends on the material. However, the relative energy absorption properties of different materials are well known and shown in the graph.

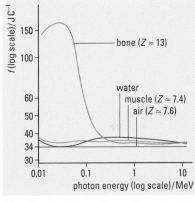

The absorbed dose, D, can be calculated from the exposure, X, by D = fX where f is a function of photon energy as shown in the graph.

Measuring dose

Film badges are the most common way of measuring a person's dose. A film badge is a piece of photographic film pinned to the clothing and worn for a few weeks. After this it is removed and developed. The blackening of the film measures the dose received. Film badges have the advantage of being cheap and easy to use as well as providing a permanent record of the strength and nature of the dose. They are being replaced by **thermoluminescent dosimeters** (**TLDs**) which contain a material, such as lithium fluoride, that can store energy absorbed by radiation. The energy is then released as light when the material is heated (~200 °C).

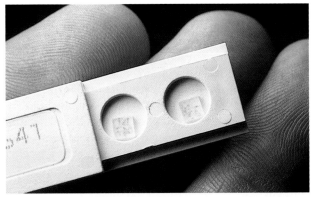

A radiation detection tag. The small squares inside the tag are affected by radiation over time. They are periodically analysed to check the exposure levels over a set time for the wearer of the tag.

MEDICAL USES OF RADIATION

Provided that a proper degree of control is exercised, ionizing radiations can be medically beneficial. **Diagnosis** (the attempt to find out what is wrong) can be aided by the patient ingesting a radioactive source and the radiation being traced round the body, or by measuring the amount of radiation in a sample of body fluids extracted from the patient, or by producing an image from gamma rays that escape the body (the gamma camera). Radiation treatments are used in **therapy** (the attempt to cure) either by subjecting the patient to a dose of radiation from a source outside the body, or by the patient ingesting the source.

Sources for ingestion

Several factors must be taken into account when selecting an isotope to be ingested by a patient.

- **The nature of the radiation produced** For diagnostic purposes the radiation must be detectable outside the body. This requires penetrating radiations such as X-rays or gamma rays. In therapy a beta emitter is generally employed as the particles are absorbed within a small distance, localizing the dose (alpha particles would not penetrate enough). Some beta emitters decay into excited states that produce gamma rays which penetrate the body, allowing dosage to be measured.

- **Energy of radiation** For diagnosis it is preferable to use radiation with uniform energy of at least 100 keV and less than about 500 keV. Below 100 keV the photons get scattered about too much to produce a sharp image, and above 500 keV the detectors lose efficiency. In therapy the amount of energy is not important provided it is sufficient to allow a reasonable penetration.

- **Half-life** The **radioactive half-life**, T_r, is the average time it takes for half the number of atoms of a given isotope to decay. The **biological half-life**, T_b, is the average time taken for biological processes (respiration and excretion) to remove half the number of atoms from the body. The **effective half-life** of an isotope within the body, T_e, is given by:

$$\frac{1}{T_e} = \frac{1}{T_r} + \frac{1}{T_b}$$

  In diagnosis it is important that the dose is minimized. If the biological half-life is long, then an isotope with a radioactive half-life similar to the time needed for the investigation must be used. If the body disposes of the isotope relatively quickly, then a long half-life source can be used. In therapy effective half-lives of several days are generally used.

- **Biological behaviour** The majority of the material ingested must arrive at the correct place. Chemical compounds are chosen which will concentrate in the part of the body of interest quite naturally. Radioactive iodine has been useful in studying the thyroid as the gland takes up iodine from the bloodstream to produce hormones. An overactive thyroid absorbs more of the iodine and therefore will produce more radiation. The most commonly used isotope is one of technetium ($^{99}Tc^m$). Compounds of technetium are taken up by bone and red blood cells when used in heart and blood-flow studies. Genetic engineering techniques allow antibodies tagged with radioactive isotopes to be targeted to attack particular types of cell.

External radiotherapy

Radiotherapy is used to kill cancer cells without damaging healthy cells. Healthy cells are generally more resistant to radiation than cancerous cells. However, it is still important to localize the dose at the cancer. If this cannot be done chemically, then beams of radiation must be applied from outside the patient's body.

Doses can be localized either by using multiple beams from several directions (the combined dose at the cancer is considerable, but little energy is deposited along each beam), or by using a single beam and rotating the patient (so that the beam is not continually passing through the same part of the body). The radiation can be artificially produced (X-rays or gamma rays from particle accelerators), or come from an unstable isotope. ^{60}Co is frequently used as it has a long half-life (5.3 years), so it does not often need replacing, and it produces gamma rays of sufficiently high energy.

Generating radioisotopes

Almost all the isotopes used in medicine are artificially produced. Short half-life isotopes have to be manufactured on site at the hospital.

Reactors are used to produce artificial isotopes by bombarding a target material with neutrons that have been produced by the fission process:

$$^{130}_{52}\text{Te} + \text{n} \rightarrow {}^{131}_{52}\text{Te} + \gamma$$

The particular Te reaction above is frequently used as ^{131}Te decays by beta decay into iodine-131, which is a very useful tracer isotope.

$$^{131}_{52}\text{Te} \rightarrow {}^{131}_{52}\text{I} + \text{e}^- + \overline{\nu}_e$$

Iodine-131 is an important tracer isotope used in thyroid diagnosis and radiotherapy. It has a long half-life (8 days) and produces additional beta radiation (increasing the patient's dose) as well as gamma rays so iodine-123 (which only produces gamma rays) is more often used when imaging alone rather than imaging with radiotherapy is required.

Milking the cow

Some very important isotopes happen to be the daughters of more long-lived parents. ^{99}Mo (molybdenum) decays by beta emission with a half-life of 6.7 hours producing ^{99}Tc^m, which decays by emitting a gamma ray with a half-life of 6 hours, making it a very useful radioisotope.

A steady supply of ^{99}Tc^m can be produced using a **radioisotope generator**. ^{99}Mo is obtained from a nuclear reactor by an (n, γ) reaction, and is used to make the compound ammonium molybdenate. This is then adsorbed onto a column of alumina stored in a glass or plastic tube.

As the ^{99}Mo decays, its ions are replaced by those of the technetium. When a sample needs to be extracted a saline solution is passed through the column. Some of the Cl$^-$ ions in the salt water exchange places with the technetium ions. The solution extracted from the bottom of the column will contain a high proportion of the ^{99}Tc^m isotopes that have been produced. This process is known as '**milking the cow**'. The column has been **eluted** by the salt water (the **eluent**) producing the sample, which is the **eluate**. The solution is sterile and can safely be injected into the blood stream.

After each elution the generator must be left for a period of time so that the concentration of ^{99}Tc^m can build up again. The optimum time between milkings depends on the half-lives of the daughter and parent. The concentration of ^{99}Tc^m will build up to a maximum and then reduce as the isotopes decay.

Which iodine?

Iodine-131 is a beta and gamma emitter. The beta emissions can be used to destroy thyroid-cancer tissue and the gamma rays can be detected by a gamma camera (see spread 11.11) so iodine-131 can be used as a diagnostic imager delivering simultaneous beta treatment. In pure imaging situations iodine-123 which emits only gamma rays is more commonly used.

---

^{99}Tc^m

'm' stands for **metastable**. Normally, a gamma emission would happen very soon after the beta decay, but in this case it has a half-life of 6 hours, which is a very long time on the atomic scale. This makes the state seem virtually stable or 'metastable'.

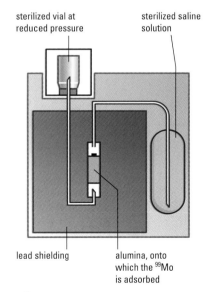

sterilized vial at reduced pressure

sterilized saline solution

lead shielding

alumina, onto which the ^{99}Mo is adsorbed

A ^{99}Tc^m generator.

MEDICAL X-RAYS

X-rays are an ionizing radiation. Their most well-known medical use is the production of X-ray photographs (**radiograms**). Dense material, like bone, absorbs more X-rays than softer tissues, so bones cast a 'shadow' onto a photographic film. X-rays are also used for therapeutic purposes in the treatment of tumours. X-ray production is discussed in spread 8.33.

X-rays passing through matter

Any point source will emit radiation equally in all directions. If the radiation is not absorbed by the material through which it is passing then its intensity will fall off as:

$$\text{intensity } (I) = \frac{1}{d^2}$$

(where d is the distance from the source). If the material does absorb some of the radiation, then an exponential reduction in intensity is produced.

In passing through a thickness of material Δx the intensity, I, of radiation falls by a fraction proportional to the thickness:

$$\frac{\Delta I}{I} = -\mu \Delta x$$

where μ is the **linear attenuation coefficient** – i.e. the probability of a single photon of radiation being absorbed in 1 m of material. The value of μ depends on the nature of the radiation (in the case of X-rays this is the photon energy) and on the material through which it is passing.

This equation is very similar to the equations for radioactive decay and the discharge of a capacitor. It can be solved to produce a relationship showing how the intensity of radiation falls as it penetrates the material:

$$I = I_0 e^{-\mu x}$$

where I is the intensity of radiation a distance x into the material, and I_0 the intensity entering the material.

In radiology it is useful to work with the **mass attenuation coefficient, μ_m**, which is the attenuation of the radiation per unit mass of material:

$$\mu_m = \frac{\mu}{\rho}$$

where ρ is the density of the material.

The half-value thickness of the material, $x_{\frac{1}{2}}$, is the thickness required to reduce the intensity of the radiation to half the initial value. This can be found by analogy with half-life in the case of radioactivity:

$$t_{\frac{1}{2}} = \frac{\ln 2}{\lambda} \quad \text{and} \quad x_{\frac{1}{2}} = \frac{\ln 2}{\mu}$$

^{60}Co is a commonly used gamma-ray source which produces 1 MeV photons. Its radiation has a half-value thickness of 10 mm in common lead.

Attenuation processes

Various processes take place when a beam of X-rays passes through a material. They all contribute to the mass attenuation coefficient of the material, although they do not contribute equally at all energies. Their

Attenuation processes.

Process	Variation of μ_m with photon energy, E	Variation of μ_m with atomic number of material, Z	Energy range in which process dominates in soft tissue
scattering	$1/E$	Z^2	1–20 keV
photoelectric effect	$1/E^3$	Z^3	1–100 keV
Compton scattering	falls slowly with E	independent	0.5–5 MeV
pair production	rises slowly with E	Z^2	>5 MeV

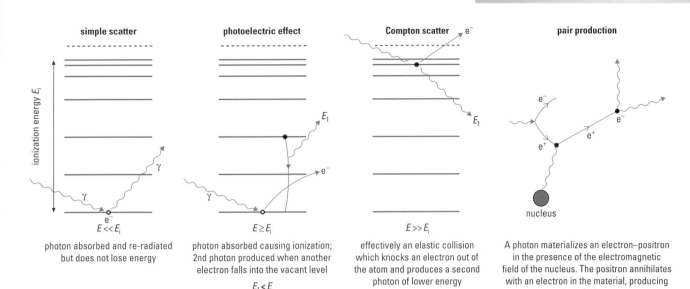

simple scatter	photoelectric effect	Compton scatter	pair production
photon absorbed and re-radiated but does not lose energy	photon absorbed causing ionization; 2nd photon produced when another electron falls into the vacant level $E_1 < E$	effectively an elastic collision which knocks an electron out of the atom and produces a second photon of lower energy	A photon materializes an electron–positron in the presence of the electromagnetic field of the nucleus. The positron annihilates with an electron in the material, producing two photons.
$E \ll E_i$	$E \geq E_i$	$E \gg E_i$	

The four mechanisms of X-ray attenuation in matter. The first three are shown on energy-level diagrams for an atom within the material.

contributions also strongly depend on the nature of the material. It is possible to characterize this by a rough relationship with Z, the atomic number of the atoms in the material.

When using X-rays for *diagnostic* radiography, the best beam energy to use is about 30 keV. At this energy the photoelectric effect is the most important mechanism. Consequently, μ_m varies with Z^3 and as the average values of Z for fat, muscle, and bone are respectively 5.9, 7.4, and 13.9, the bone will absorb radiation 11 times more effectively than the surrounding tissue. This gives good contrast on the radiogram. However, X-ray therapy uses beam energies in the MeV range which Compton scattering dominates. As this mechanism is independent of Z, bones do not absorb too much energy from the beam on its way to the target area.

Producing images

X-ray machines used in medical applications use an angled target to increase the area struck by the electron beam (and so reduce the temperature rise in the tissue). The X-rays pass through an adjustable diaphragm and conical beam definer. A narrow beam is preferable as this reduces the scattering which always tends to blur the image. A further technique to reduce blurring is to remove the scattered photons by placing a lead grid under the patient and above the film. Any photons arriving at the grid off-line will be absorbed by the lead. To prevent the grid casting shadows it can be slowly moved across the film while the exposure is taking place.

The dose of X-rays can be reduced by using an **intensifying screen**. Zinc sulphide in the intensifying screen absorbs the X-ray photons and re-radiates the energy as visible light. This exposes the film with the same image as the X-rays. X-rays passing through the film can produce light in the rear screen. The intensity can be increased by a factor of 40, but with some loss of detail in the image.

Alternatively an **image-intensifier tube** can be used, in which X-rays strike a fluorescent screen and the light produced excites electrons from a photo cathode. The electrons are then accelerated, as in a TV tube, and strike another fluorescent screen to produce an image.

In some cases the contrast of the image must be improved. For example, in investigations of the gut all the tissues are of similar material and so contrast is low. The available contrast can be increased by introducing a high-Z radio-opaque barium compound (a **barium meal**) into the digestive tract.

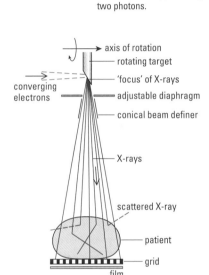

A medical X-ray machine. Rotating the target avoids localized heating, so an electron beam of a much higher intensity can be used. The accelerating voltage used on the electron beam determines the X-ray wavelength.

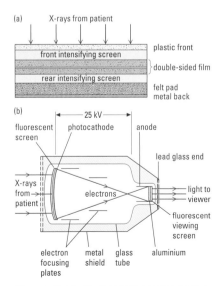

Two methods for intensifying the image and reducing the required dose. (a) An image-intensifying screen. (b) An image-intensifier tube.

Tumours – medical issues

A tumour is a mass of abnormal tissue produced when normally healthy cells start to reproduce at too great a rate. Benign tumours are generally slow-growing and do not spread to other parts of the body. They only tend to cause problems if their growth interferes with the correct operation of vital organs in the body. A malignant tumour, on the other hand, tends to be fast-growing and spreads to other parts of the body. The treatment that can be employed depends on the location of the growth. Generally the options are:

surgery – cutting the growth out of the body;

chemotherapy – introducing chemicals designed to destroy cells into the body;

hormone therapy – altering the hormone balance in the body can have an effect on some hormone-dependent growths;

radiotherapy – using ionizing radiation to destroy cells within the growth.

RADIATION THERAPY

The purpose of radiation therapy

It is ironic that one of the characteristics that makes cancerous cells dangerous, their ability to reproduce rapidly, also makes them susceptible to being selectively destroyed by ionizing radiation. **Radiation therapy** (often abbreviated to **radiotherapy**) aims to deliver a destructive dose of ionizing radiation to the cancerous cells without damaging surrounding healthy cells. Of course it is impossible to completely achieve this aim, but several strategies can be used to help.

- Multiple beams of radiation can be directed into the body from several different directions. Each beam carries less than the required dose, but the collective effect at their crossing point is to deliver a large enough dose.

- A single beam can be used to deliver the required dose, provided that it is rotated around the patient while always passing through the tumour site. This ensures that the radiation is not always passing through the same region of the body on its way to the target area.

- Absorbers manufactured from wax or lead foil of varying thickness can be placed on the body to compensate for the fact that its surface is not flat. This allows a much more uniform dose of radiation to enter the body, by compensating for areas where there is less tissue in the direction of the beam.

- Thin lead wedges can be used to protect regions of the body that are especially sensitive to radiation damage, or are better at absorbing radiation than others.

The dose used is very critical – a variation of between 5 and 10% in absorbed dose can lead to very different results. Too much and the healthy tissue suffers damage, too little and the cancer survives to re-grow later. It is therefore very important that the treatment is carefully planned, monitored, and controlled.

Megavoltage therapy

X-rays that are used for diagnosis are generally restricted to the kilovolt energy range, where their absorption is strongly dependent on the proton number of nuclei in the tissue, and so produce clear contrasts on the exposed film. When X-rays are used for therapy, it is important to ensure that the energy is much higher – in the megavolt range. At energies similar to that used in diagnosis the dependence of absorption on Z would be a problem as bone would tend to absorb radiation and be damaged at the doses required. Not only that, but the distance to which X-rays can penetrate into the body depends on their energy, so to reach tumours within the body higher energies are required. The maximum dose is delivered at a distance into the body (in cm) roughly $\frac{1}{4}$ of the beam energy (in MeV). Consequently, to deliver a dose 5 cm into the body a beam energy of 20 MeV would be required.

Lower-energy beams (in the kilovolt range) can be used to treat skin cancers while not penetrating deep enough to damage other tissue. However, this form of treatment is slowly being replaced by electron beam therapy.

Treatment devices

The earliest therapy devices used naturally produced γ rays from a cobalt-60 source. This isotope has a comparatively long half-life for a medical source (5.3 years) and so does not need replacing very frequently. The energy of the γ rays it produces (1.25 MeV) is in the correct range for therapy and the source is compact enough to permit short exposure times. However these devices have disadvantages (principally the fact that the radiation cannot be switched off so permanent shielding is required,

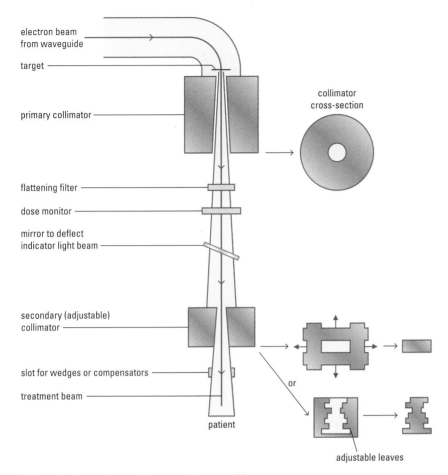

The key features of a radiotherapy X-ray machine.

but the beam definition is also inferior) which has led to their being
phased out in favour of linear accelerator X-ray machines.

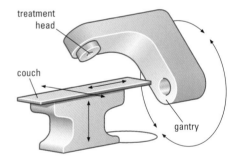

Linear accelerator X-ray sources are now the primary means of
delivering radiotherapy. The diagram shows the key features of the
treatment head of such a device.

A linear accelerator produces a beam of electrons with an energy
between 4 and 25 MeV. This beam is directed onto a **target**, with a high
atomic number, which is thick enough to absorb the majority of the
electrons. Collimators are used to adjust the beam profile and to reduce
the amount of X-ray scatter (by absorbing rays that are not in the beam
direction). The **primary collimator** is a large lead block that produces a
conical beam. A **flattening filter** has a specially designed cross-section
that helps to define a uniform intensity across the beam. The secondary
collimator can either be in the form of thick lead blocks that slide back
and forth (to define the exact beam profile) or a multi-leaf collimator with
up to 40 individually moveable thin lead strips.

A mirror deflects a light beam into the X-ray beam path so that a
visible indication of the beam's position can be displayed on the patient.
In addition a dose monitor is used to measure the cumulative dose and to
cut off the device at the appropriate time.

The treatment head is mounted on a gantry that allows it to rotate
about the couch on which the patient is positioned. Various accessories
can be attached to the treatment head to allow the alignment of the beam
on the patient to be verified. A simple light beam (as mentioned above)
can be used, or a video system, or a laser beam. Lower-energy X-rays
can be used to provide a picture of the beam profile on the patient by
exposing a suitable film.

- images from gamma rays produced inside the body

- making nuclei resonate to produce an image

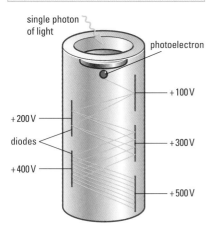

A photomultiplier tube works by using the incoming light to knock electrons out of a photocathode. The series of secondary electrodes (dynodes) are at higher positive potentials and the electrons are accelerated from one to another. Each time they strike a dynode more electrons are emitted. By the time the current pulse has reached the anode at the end of the tube an amplification of at least 10^5 has taken place. With careful design a current pulse proportional to the intensity of light arriving can be created.

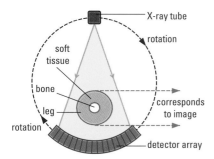

The principle of the CAT scanner.

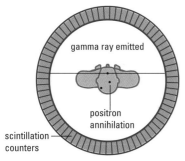

The principle of the PET scanner.

ELECTROMAGNETIC IMAGING

The ability to examine what is happening inside a patient without having to cut the body open is invaluable in diagnosis. To do this we have to either pass waves (ultrasound or electromagnetic waves) through the body, or detect them originating in the body.

The gamma camera

A radioisotope such as $^{99}Tc^m$ is used to tag a chemical which is absorbed by part of the body. The isotope emits gamma rays. The gamma camera detects the gamma rays and relays infomation about the relative numbers of photons produced to a computer that can create an image of that part of the body from the information.

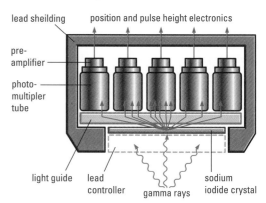

The gamma camera.

Gamma photons from the body pass through a collimator that consists of a circle of lead into which has been drilled a regular sequence of holes. A photon arriving at the collimator at an angle is absorbed by the lead. Any photons that get through the collimator pass into a large sodium iodide crystal covering the back of the lead block. The photons striking the sodium iodide promote electrons into higher energy levels. When the electrons fall back into lower levels, visible-light photons are emitted. In this way a single gamma photon can be converted into many visible-light photons. These photons are detected by **photomultiplier tubes** (see left) arranged in an array behind the crystal. The photomultipliers convert the photons into a pulse of charge that can be electronically monitored. The size of the pulse is proportional to the intensity of light entering the tube. Light emitted from a point in the crystal will travel to each of the photomultipliers, but the intensity arriving will depend on the distance from the point in the crystal to the tube. By monitoring the intensity at each tube, a computer can reproduce the position and intensity of the emission from the crystal, and locate the position of the ray source in the body.

The computerized axial tomography (CAT) scanner

The use of sensitive radiation detectors to collect X-rays and computers to analyse the resulting information produces greatly improved X-ray images. Although simple radiograms still have their uses, the resulting image suffers from poor contrast between different types of soft tissue and shows no depth information.

The most common type of computerized axial tomography (CAT) scanner has the patient on a moveable table surrounded by a gantry housing the X-ray tube and detectors. A wedge-shaped beam of monochromatic X-rays passes through a slice of the body to the detector array, which contains about 1000 devices similar to those used in the gamma camera. The X-ray tube and detectors then rotate about the patient, emitting and absorbing pulses at each angle. Several hundred pulses are used to assemble a section during a scan time which is typically 4–5 seconds. The detectors relay information about the level of X-radiation reaching them to a computer that assembles the data into the resulting image.

The positron emission tomography (PET) scanner

Positron emission tomography (PET) is a technique that has revolutionized research into the activity of the brain. A patient inhales carbon monoxide containing some carbon-11 isotopes. Carbon monoxide is very good at attaching itself to haemoglobin molecules in red blood cells.

When areas of the brain are active the blood flow to them increases, so the concentration of carbon-11 in that part of the brain increases. The ^{11}C isotope of carbon is artificial and decays by β⁺ (positron) emission. Within about 1 mm of its emission point a positron will annihilate with an electron to produce two gamma-ray photons. As the positrons are not moving very quickly when they annihilate with an electron, the two photons emerge virtually back-to-back, which conserves momentum. The patient is surrounded by a ring of **scintillation counters** which detect the emerging gamma-ray photons (scintillation counters are photomultiplier tubes, each with its own sodium iodide crystals). A computer processes this information to reconstruct, very accurately, the point inside the patient from which the photons originated. The result is a map of the blood flow in the brain. If the patient is asked to carry out some activity such as reading, the PET scanner can detect the change in blood flow as parts of the brain become active. One disadvantage of the technique is that it cannot record the activity of parts of the brain that are constantly active – only *changes* in blood flow can be detected.

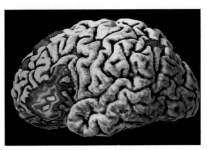

A PET scan showing activity in the speech centre of the brain.

Carbon monoxide and haemoglobin

Haemoglobin molecules transport oxygen round the body. If you inhale too much carbon monoxide the haemoglobin is unable to do this and you suffocate. This is why carbon monoxide is such a dangerous gas.

Producing the CAT scan image

A CAT scan produces an image of a section through the body. The section is divided into small-volume elements called **voxels** and the X-ray attenuation coefficient for each voxel calculated.

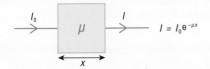

$$I = I_0 e^{-\mu x}$$

A cubical voxel of side x attenuates a beam of X-rays.

There are various techniques for extracting the attenuation information; they all rely on combining data from the X-rays passing through the body from a variety of different directions. The intensity at the detector gives information about the sum of the attenuation coefficients μ for the tissues that the X-rays passed through.

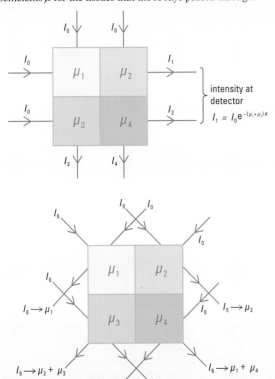

intensity at detector
$$I_1 = I_0 e^{-(\mu_1 + \mu_2)x}$$

X-rays passing through a sequence of voxels are attenuated by each one. Comparing $I_1 \rightarrow I_0$ tells us $\mu_1 + \mu_2$, $I_2 \rightarrow I_1$ gives $\mu_3 + \mu_4$, etc.

The process can be illustrated with a 4-voxel cube as illustrated below.

Detector intensity	Gives	Detector intensity	Gives
I_1	$\mu_1 + \mu_2$	I_6	$\mu_1 + \mu_4$
I_2	$\mu_3 + \mu_4$	I_7	μ_3
I_3	$\mu_1 + \mu_3$	I_8	μ_1
I_4	$\mu_2 + \mu_4$	I_9	$\mu_2 + \mu_3$
I_5	μ_2	I_{10}	μ_4

Step 1 Start with a set of 'empty' voxels and add in the vertical and horizontal intensity information.

$\mu_1 + \mu_2$	$\mu_1 + \mu_2$
$\mu_3 + \mu_4$	$\mu_3 + \mu_4$

$\rightarrow \mu_1 + \mu_2$
$\rightarrow \mu_3 + \mu_4$

$2\mu_1 + \mu_2 + \mu_3$	$\mu_1 + 2\mu_2 + \mu_4$
$\mu_1 + 2\mu_3 + \mu_4$	$\mu_2 + \mu_3 + 2\mu_4$

Step 2 Add in the diagonal information for each voxel.

μ_1

$3\mu_1 + \mu_2 + \mu_3$	$\mu_1 + 3\mu_2 + \mu_3 + \mu_4$
$\mu_1 + \mu_2 + 3\mu_3 + \mu_4$	$\mu_2 + \mu_3 + 3\mu_4$

$\mu_2 + \mu_3$ μ_4

$4\mu_1 + \mu_2 + \mu_3 + \mu_4$	$\mu_1 + 4\mu_2 + \mu_3 + \mu_4$
$\mu_1 + \mu_2 + 4\mu_3 + \mu_4$	$\mu_1 + \mu_2 + \mu_3 + 4\mu_4$

μ_2

μ_3 $\mu_1 + \mu_4$

Step 3 Subtract the total attenuation coefficient $(\mu_1 + \mu_2 + \mu_3 + \mu_4)$ from each voxel.

$3\mu_1$	$3\mu_2$
$3\mu_3$	$3\mu_4$

Then divide by (number of directions – 1) to give the individual attenuations.

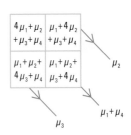

μ_1	μ_2
μ_3	μ_4

The data are used to calculate $\mu_1 - \mu_4$.

Finally each pixel in the image is given a greyscale value dependent on the attenuation coefficient within the voxel.

MAGNETIC RESONANCE IMAGING

Magnetic resonance imaging (MRI) is a technologically complex technique that can provide detailed images of slices through the body. It does not require the patient to ingest a radioisotope and it can distinguish between tissues with similar X-ray absorptions. It is ideal for detecting tumours and neurological diseases.

Nuclear spin

Many nuclei spin about an axis. With the protons in the nuclei being charged, this nuclear spin acts like a tiny circulating electric current generating a small magnetic field, similar to that of a bar magnet, aligned along the spin axis. If a collection of these nuclei is placed in an external magnetic field (which will inevitably be much stronger), then they will tend to align their spin (and magnetic) axes along the direction of the external magnetic field. Most of them will have their magnetic field in the same direction as the external field, which is a low-energy state, but some of them will be aligned anti-parallel to the external field, which is a high-energy state. There is a tendency for the nuclei in a high-energy state to 'flip' back into the low-energy state by emitting energy to the surrounding nuclei. However, there is generally enough thermal energy in a material to ensure that some of the nuclei are flipped back.

Although there are many nuclei that possess spin, the most useful in the context of imaging the human body is the hydrogen nucleus (a proton). Approximately 10% of the body mass is hydrogen, about 70% of which is contained in water. The rest is mostly contained in fats, but there is a small amount of hydrogen in proteins. Coupled to that, protons can be induced to give a comparatively large signal in a MRI scanner.

Precession

With any collection of nuclei in the presence of an external magnetic field the nuclear magnets do not arrange themselves parallel or anti-parallel to the field with perfect alignment. Rather they tend to **precess** about the magnetic field lines rather like a spinning top precessing about a vertical line (precession is the movement of the axis of a spinning body around another axis). This precession takes the nuclei in a complete loop about the magnetic field line in a definite period corresponding to a fixed frequency that is proportional to the strength of the external magnetic field, B_0:

$$f = \frac{\gamma B_0}{2\pi}$$

where γ is a constant called the **gyromagnetic ratio** and f is the **Larmor frequency**. For a proton, γ is $42.57\,\text{MHz}\,\text{T}^{-1}$.

With a large magnetic field, of the order of 1 or 2 tesla, as used in MRI, the Larmor frequency is about 50 MHz – in the radio-frequency part of the spectrum.

The precession of the nuclear magnets has an effect on the total magnetic field produced in the material. This consists of the external applied field plus the sum of all the small magnetic fields due to the proton's spin. The contributions that the proton's fields make to the total is best analysed by dividing the field into a longitudinal component (parallel to the external field axis) and a transverse component (at right angles to the external field). More of the protons will be oriented parallel to the external field than anti-parallel, so there will be a net longitudinal field to add to the external one. All the protons will be precessing about the external field axis at the same frequency, but they will all be at different phases of their motion. Consequently their transverse fields will all tend to cancel each other out. As a result there is a net small increase in the total field along the external field's axis, but no additional field at right angles to this.

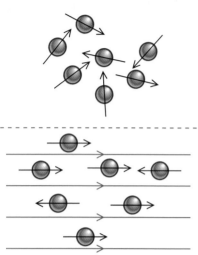

magnetic field applied externally

In the presence of an external field the nuclei tend to line up with their spin axes along the magnetic field lines. Some will be in a high-energy state (anti-parallel to the external field) but most will be in the low-energy state (parallel to the external field).

external field

the proton's magnetic field axis precesses about the external field at the Larmor frequency

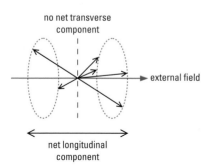

no net transverse component

external field

net longitudinal component

Resonating the protons

If the protons can be made to precess about the external field lines in phase with each other, then their transverse field components will add up to give a small net transverse field that rotates about the external field axis at the Larmor frequency. It is this magnetic field that can be detected and which produces the information from which the image is generated in a MRI scanner.

The trick is to make the protons precess in phase with each other. This can be achieved by making them absorb radio-frequency radiation of the same frequency as the Larmor frequency of the precession. When protons in the low-energy state absorb this energy they flip into the high-energy state (with their axes anti-parallel to the external field lines) and, crucially, they also start precessing in phase with the applied signal – and hence with each other. The number of protons that flip depends on the duration of the radio-frequency pulse that is applied.

Producing a magnetic resonance image

The patient is placed in a uniform magnetic field of 1–2 T. To locate a part of the body three smaller magnetic fields are added to the constant field. These fields are not uniform. One increases steadily along the length of the patient's body (the z axis). This is used to define slices through the body. The other two fields increase along x and y axes within the plane of a slice. The magnetic field strength at any point is the sum of the four fields which sets a unique Larmor frequency and phase at every point in the body.

The next step is to send a pulse of electromagnetic radiation through the body at a set radio frequency. Those protons with a Larmor frequency equal to this will absorb energy from the pulse and flip into the high-energy state. This produces a rotating transverse field at the Larmor frequency for that part of the body. Sensitive detectors pick up this signal – a rotating magnetic field (consequently a changing field) will induce an electric current in the detector. The image is built up by sending a sequence of radio-frequency pulses through the body at different frequencies to pick out the Larmor frequencies at different locations. The signals produced and the **relaxation time** involved are processed by computer to assemble the image.

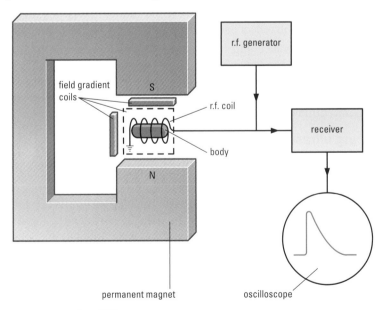

Component parts of an MRI scanner.

| field gradient coils | S | r.f. coil | body | N | permanent magnet | r.f. generator | receiver | oscilloscope |

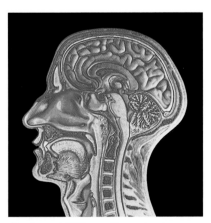

A coloured MRI scan through a human head.

Acoustic impedance

The acoustic impedance Z of a material is frequency related and determines the sound pressure produced by molecular vibrations in a material. It is defined as:

$$Z = \frac{p}{vS}$$

where p is the sound pressure, v the particle velocity, and S the surface area through which the sound wave propagates.

The specific acoustic impedance z is found by:

$$z = \frac{p}{v} = ZS = \rho c$$

where ρ is the density of the material and c the velocity of sound in the material.

Piezoelectric effects

When certain non-conducting crystals are subjected to mechanical stress between opposite faces, a p.d. is generated across these faces. This is known as the **direct piezoelectric effect**.

The **reverse piezoelectric effect** is the opposite: if a voltage is applied across opposite faces in a piezo crystal, the crystal changes its shape.

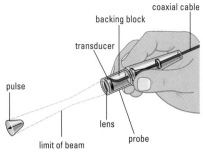

An ultrasound transducer.

ULTRASOUND IMAGING

The advantages of being able to produce images of the internal structure of the body without invasive surgery are obvious. The electromagnetic techniques discussed in the previous sections are either highly expensive (magnetic resonance imaging), or involve the patient ingesting a radioactive source (gamma camera and positron emission tomography scan). Ultrasound offers the ability to produce images in real time, relatively cheaply, and without the need for exposure to radiation. As far as can be determined there are no harmful side-effects due to exposure to ultrasonic waves.

Comparing the major techniques for medical imaging.

Device	Image resolution	Cost	Safety
Ultrasound scan	Useful but no detailed contrast and low resolution.	Comparatively cheap	No ionizing radiation or contrast material, so very safe.
CAT scan	Good resolution and good contrast in soft tissue. Ingested contrast agent required.	Expensive	Uses ionizing radiation. Adverse reactions to contrast agents may occur.
MRI scan	Excellent contrast in soft tissues. Can distinguish between tissues that have similar X-ray absorptions. Very good resolution.	Very expensive	No ionizing radiation, and contrast material used tends not to cause adverse reactions. Cannot be used for patients who have implanted metal e.g. pacemaker.

Principle of ultrasonic imagery

Ultrasound is very high-frequency sound. A typical ultrasound device produces waves at 1–3 MHz (the range of human hearing being roughly 20 Hz–20 kHz).

Whenever a wave passes from one medium into another in which the wave speed is different, reflection and refraction take place. The reflected waves bounce back, while the refracted waves pass into the second medium. The extent to which a wave reflects back rather than passing through is determined by the change in wave speed. If the change is very big, then a lot of the wave energy will reflect back (see margin note on acoustic impedance).

An ultrasound device creates an image by firing a pulse of ultrasound into the body. When the waves in the pulse meet a bone or organ in the body they reflect back. The imaging device measures the **intensity** of the reflection and the **time delay** from the instant when the pulse was emitted, and builds up a picture from this information.

Producing ultrasound

Ultrasound transducers are used to turn electrical energy into sound, and vice versa, by using **piezoelectric effects**. Some crystals will change their physical dimensions when an electric field is applied between opposite faces. This is due to an interaction between the charges in the crystal and the external field. Regularly switching the polarity between the faces of the crystal will set it vibrating. When the frequency at which the polarity is switched corresponds to the natural vibration frequency of the crystal, resonance occurs and the crystal produces an intense sound wave. This effect can be created using quartz crystals, but for ultrasound the most commonly used crystal is a ceramic: lead zirconate titanate. The frequency at which the crystal will resonate can be tuned by selecting its thickness.

Creating an image

The velocity of sound in air is very much less than that in skin, so the two media are said to be **acoustically mis-matched**. This produces a high proportion of reflection. So when an ultrasound transducer is placed against the skin a proportion of the energy produced is immediately reflected back if there is an air gap. To ensure that there is no air gap between the transducer and the skin, a gel is smeared on the skin and the transducer placed on the gel. The sound velocity in the gel is midway between that of air and skin, ensuring a smooth change in sound velocity and less reflection. This technique is known as **impedance matching**.

When sound waves pass from one medium to another with a different speed of sound, refraction will take place. With light the extent of the refraction is determined by the refractive index – with sound the acoustic impedance of the medium determines the refraction. There is also some reflection of energy at the surface. The ratio of reflected energy to refracted energy is the intensity reflection coefficient, α.

$$\alpha = \left(\frac{Z_2 - Z_1}{Z_2 + Z_1}\right)^2$$

where Z_1 and Z_2 are the acoustic impedances of the media.

If $Z_1 = Z_2$ there is no reflection and the media are impedance matched.

If one impedance is much bigger than the other, then α is very large and most of the energy is reflected. Typically this is the case for a wave travelling from a gas to a liquid (as the speeds of sound are very different). Often a gel is used on a patient's skin to couple the media more effectively and prevent much reflection.

Once the transducer has sent out one pulse of sound waves, it can be switched into a different mode in which reflected ultrasound sets the crystal vibrating, producing an electrical signal that can be detected. The transducer sends information about the time delay between the pulses, and the strength of the reflected signal, to a computer. In an **A scan** the information is used to display a graph of distance into the body (calculated by the time delay) against strength of reflection. This sort of scan requires some skill in interpretation and cannot readily display movement. In a **B scan** a sequence of dots is placed on the screen along a line, the distance between them being proportional to the distance into the body. The brightness of the dots is proportional to the strength of the reflection. By moving the transducer about, many such lines can be assembled into an image which represents a section through the body.

A more sophisticated ultrasound device uses an array of transducers that send out pulses in many different directions at once. This provides many image lines together without the instrument having to be moved, and a picture is rapidly built up. Such instruments are widely used to monitor babies in the womb, as they can display movement in real time.

Doppler measurements

The **Doppler effect** is the shift in frequency of a wave when the source and/or the receiver is moving (see spread 6.18). A wave reflecting off a moving object is Doppler-shifted as if the object was a moving source of waves. A ultrasound device can measure this shift in frequency to detect movement within the body. One important use for this is in the monitoring of blood flow. A continuous set of waves has to be used, so two transducers are required – one to produce the ultrasound and one to receive it. Any changes in the blood flow along a vein or artery due to a clot (**thrombosis**) or a constriction on the walls (**atheroma**) show up as a change in the frequency shift.

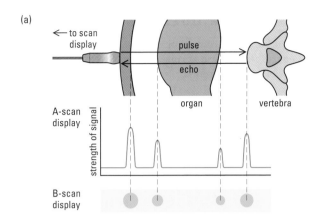

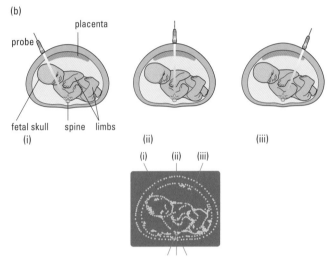

(a) A- and B-scan ultrasound imaging.
(b) Building a picture from a sequence of B-scan lines.

Image resolution

Axial resolution is the ability to distinguish closely spaced structures along the direction of the ultrasound pulse. The higher the frequency the better the axial resolution. However, high-frequency ultrasound is strongly attenuated in the body, so can only be used to resolve structures close to the surface. Axial resolution is (spatial length of pulse)/2 so the shorter the duration of the pulse, the better the axial resolution.

Lateral resolution is the ability to distinguish closely spaced structures at 90° to the pulse direction. It is influenced by the width of the beam, so that high frequency and close focusing improve the resolution. Pulse duration does not affect lateral resolution. Lateral resolution is generally not as good as axial resolution.

Doppler shift

The equation:

$$\frac{\Delta\lambda}{\lambda} = \frac{v}{c}$$

is used to relate the Doppler shift of ultrasound to the velocity of blood flow (spread 6.18).

PRACTICE EXAM QUESTIONS

1 a By reference to refraction at the cornea and at the lens, explain how a normal eye focuses an image of an object on the retina.

   b A patient is able to see clearly text which is nearby but finds that the same text appears blurred when moved further away.

      i State the eye defect from which the patient is likely to be suffering.

      ii Copy Figure 11.1 and on your copy draw rays to show how the light from a distant object passes through the eye to the retina for this patient.

      iii State the name of the corrective lens for the defect you have stated in **i** above.

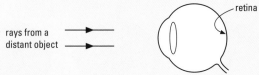

Figure 11.1

   c A second patient suffers from astigmatism.

      i Describe a test which could be carried out to diagnose this defect.

      ii Explain how astigmatism can be corrected using an appropriate lens.

   d A third patient cannot see text clearly unless the text is held at least 100 cm from the eye. Calculate the minimum power of the lens required in order that the patient may see the text clearly when placed 25 cm from the eye. State any assumption you make.

2 a When assessing radiation damage to human beings, the quality factor of alpha particles is given as 20. Explain what quality factor means, and how it affects the radiation dose equivalent which a person receives.

   b Explain why alpha radiation from small particles of a radioisotope which has been inhaled by a person is considered to be particularly dangerous while alpha radiation from an external source is considered less hazardous.

3 a Describe the two main processes by which a beam of X-rays from a typical X-ray tube used for diagnostic work is attenuated by a thin slab of material.

   b The average value of Z, the atomic number, for human tissue lies in the range 5 to 7.5 and is close to 14 for bone. Justify the claim that the optimum energy of the X-ray photons is around 30 keV for radiographic purposes.

   c Explain how, instead of allowing the X-radiation to form an image directly on a photographic emulsion, it is possible to reduce the dose absorbed by the patient and still obtain a photographic image of the same density.

4 In an investigation into plasma volume and plasma iron turnover, a patient was injected with iron citrate containing a radioisotope, ^{59}Fe. Four blood samples were taken from the patient at different times after the injection and their count-rate measured. Table 11.1 gives the corrected count rates in counts per minute per ml.

Table 11.1

Time after injection/minutes	Corrected count rate/ counts minute^{-1} ml^{-1}	ln(corrected incount rate)
15	697	
30	621	
45	554	
60	496	

   a Copy Table 11.1 and complete the ln (corrected count rate) column in your table.

   b Use the values from the table to plot a graph of ln (corrected count rate) against time.

   c Use your graph to find **i** the initial corrected count rate and **ii** the effective half-life of the isotope.

   d Give *two* factors, apart from half-life, which would have to be considered when selecting the isotope for this investigation.

5 a Endoscopes are commonly used in medicine for the purpose of viewing some internal organs. In conjunction with a laser, the endoscope may also be used for surgery.

      i Explain how an endoscope allows a surgeon to view the stomach.

      ii What properties of laser light make it suitable for use as a scalpel?

      iii Suggest a physical process which occurs in soft tissue exposed to laser light to enable the light to act as a scalpel.

   b Figure 11.2 illustrates laser light being used to destroy unhealthy tissue.

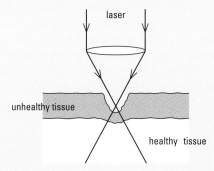

Figure 11.2

      i Explain why only some tissue in the laser beam is destroyed.

      ii Suggest one advantage of using a lens of short focal length for the removal of a thin layer of tissue.

   c In a process known as spot welding, a pulsed laser is used to repair small tears in the retina. The power of the laser light prior to entry into a bundle of optical fibres is 125 W. 40% of the incident power is not transmitted through the fibres.

      i Calculate the light intensity in W mm^{-2} of the laser beam when it is focused on to an area of 1.5×10^{-3} mm^2 of the retina.

      ii The duration of each pulse of the laser light is 0.5 ms. Calculate the energy supplied to the retina in one pulse.

6 Tables of sound intensities, such as the one which is given in Table 11.2, are often published in textbooks dealing with human hearing.

Table 11.2

Sound	Intensity /W m^{-2}	Intensity level /dB
(Rupture of ear-drum)	10^4	160
Jet engine	10	130
(Threshold of pain)	1	120
Loud thunder	10^{-1}	110
Heavy street traffic	10^{-4}	80
Normal conversation	10^{-6}	60
Whisper	10^{-10}	20
(Threshold of hearing)	10^{-12}	0

a What is meant by the term intensity?

b By inspection of the table deduce

  i the intensity of the sound from a motor cycle when the intensity level is 100 dB,

  ii the intensity level of bird song with an intensity of 10^{-8} W m^{-2}.

c The ear of a particular person collects sound from an effective area of 15 cm^2. What power of sound is the ear detecting when at the threshold of hearing?

d Explain how the power of the sound entering a person's ear is used to maintain resonant frequency oscillations in the inner ear.

e State three factors which make obtaining data for a table such as this, very difficult.

f Suggest additional information which needs to be given in order to make use of the values given in the table.

7 a People who work in radiation areas often wear a film badge as a personal monitor. Typically, two pieces of film are enclosed in a plastic holder in which there are three openings. A sensitive emulsion (fast film) is immediately below the openings and a less sensitive emulsion (slower film) is underneath it in the holder. Two of the openings have filters over them. Figure 11.3 shows a top view and Figure 11.4 shows a cross-section of the film badge.

  i Explain why the badge is constructed in this way, indicating the type of information which can be deduced from the blackening of film under the three openings in the holder.

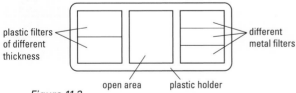

Figure 11.3

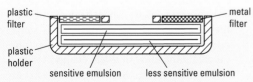

Figure 11.4

  ii What is the purpose of having two films of different sensitivity in the holder?

  iii State *two* advantages and *two* disadvantages of this type of radiation detector as a safety monitor.

b The intensity of a certain monochromatic beam of X-rays is reduced by 25% of its initial value after passing through 1.5 mm of lead. What will the transmitted intensity be if the thickness of lead is increased to 7.5 mm?

8 A person jumps down from a wall and lands feet first exerting a force which causes compression along the axis of each lower leg. The average reaction force on each leg on landing is 8.0×10^4 N.

Estimate the stress in one lower leg bone on landing and show that the bone is likely to fracture. Discuss two factors which would normally make it possible for people to jump off this wall without breaking their legs.

cross-sectional area of lower leg bone = 3.5×10^{-4} m^2
compressive breaking stress for bone = 1.5×10^8 N m^2

9 a When a person is standing upright and still, the load on the head of one femur (thigh bone) is vertical. The neck of the femur makes an angle of 50° to the vertical as shown in Figure 11.5.

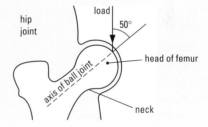

Figure 11.5

The weight, W, of the person is 700 N and the load on the head of one femur is $0.35W$. This load will produce a compressive force along the axis of the ball joint and a shear force normal to this axis.

  i Indicate on a copy of Figure 11.5 how the compressive and shear components of the load act on the head of the femur.

  ii Calculate the values of these two components.

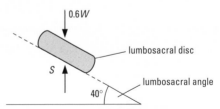

Figure 11.6

b Figure 11.6 shows the forces acting on the lumbosacral disc of the same person still standing upright. The vertical downwards load, $0.60W$, and the upwards support force, S, are equal in magnitude but produce compressive and shear forces on the disc.

  i If the effective area of the disc which supports the compressive load is 3.0×10^{-4} m^2, calculate the compressive pressure on the disc.

  ii A person with a poor posture may curve the spine and change the angles of the vertebrae relative to each other. Describe the consequences of such poor posture over many years for the lumbosacral disc and for the person.

10 A cardiac pacemaker is a device which is used to ensure that a faulty heart beats at a suitable rate. In one pacemaker the required electrical energy is provided by converting the energy of radioactive plutonium-238 ($^{238}_{94}$Pu). The atoms of plutonium decay by emitting 5.5 MeV alpha particles. The daughter nucleus is an isotope of uranium (U). The plutonium has a decay constant of 2.4×10^{-10} s^{-1}.

a **i** How many neutrons are in the nucleus of the radioactive plutonium?

  ii Write down the equation representing the decay indicating clearly the atomic (proton) number and the mass (nucleon) number of each nucleus.

b **i** Define the term half-life.

  ii Calculate the half-life of plutonium-238.

  iii State why alpha particles are more suitable than either beta particles or gamma radiation for use in the power source.

c A new pacemaker contains 180 mg of plutonium.

  i Determine the number of radioactive atoms in a new pacemaker.

  ii Show that approximately 1×10^{11} disintegrations occur each second when the power supply is new.

  iii Determine the initial power, in W, of the source when it is new.

  Avogadro constant, $N_A = 6.0 \times 10^{23}$ mol^{-1}.

  Charge on an electron, $e = 1.6 \times 10^{-19}$ C.

11 Blood contains ions in solution. Figure 11.7 shows a model used to demonstrate the principle of an electromagnetic flowmeter which is used to measure the rate of flow of blood through an artery.

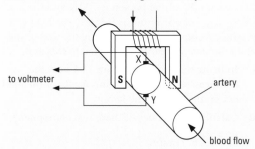

Figure 11.7

When a magnetic field of 2.0 T is produced by the electromagnet, a potential difference (p.d.) of 600 μV is developed between the two electrodes X and Y. The cross-sectional area of the artery is 1.5×10^{-6} m^2. The separation of the electrodes is 1.4×10^{-3} m.

a Write down an expression for the force on an ion in the blood which is moving at right angles to the *magnetic* field, defining the symbols you use.

b An ion has a charge of 1.6×10^{-19} C.

  Show that the force on the ion due to the electric field between X and Y is 6.9×10^{-20} N.

c Given that a p.d. of 600 μV is developed when the electric and magnetic forces on an ion are equal and opposite, calculate:

  i the speed of the blood through the artery;

  ii the volume of blood flowing each second through the artery.

12 a There are two types of light-sensitive receptor found in the human eye.

  i Write down the name of each receptor and state the light conditions under which each operates.

  ii What is the approximate wavelength range over which the eye can see?

  iii Draw a graph of relative light absorption (y axis) as a function of wavelength (x axis) for the more sensitive receptors indicating clearly the wavelength, in nanometers, at which these receptors are most sensitive.

  iv The other receptors take part in the process of colour vision. How do these receptors differentiate colour?

b In an experiment to determine the sensitivity of the eye to a flash of light it was found that an energy of 2.1×10^{-17} J, measured in front of the cornea, could just be detected.

  The wave-length of light used was 510 nm.

  i How many photons reached the cornea?

  ii Only about 10% of these photons arrived at the retina. What may have happened to the others?

c **i** What is meant by the resolution of the eye?

  ii Two objects which are close together will produce two images on the retina.

  Explain, in terms of the receptors, how these two images should be positioned for them to be perceived as separate.

  the Planck constant, h, = 6.63×10^{-34} J s

  speed of light in vacuum, c, = 3.00×10^8 m s^{-1}

13 Table 11.3 contains data relating to the rate of loss of heat energy from a person.

Table 11.3

air temperature /°C	activity	person's rate heat loss/W	% rate of heat loss A	B	C	respiration
32	sunbathing	400	10	10	78	2
0	walking	400	8	50	2	40

Columns A, B and C relate to the mechanisms 'conduction and convection', 'evaporation' and 'radiation' in no particular order.

a Identify each column, giving a reason for each answer.

b The average rate of energy expenditure of a person for a game of tennis lasting 90 minutes is 500 W. The loss of 1.0 kg of body fat releases 38 MJ of energy.

  i Calculate the mass of body fat lost during the game, assuming that all of the energy expended was at the expense only of body fat.

  ii The average efficiency of the muscles during the game is 25%. All of the energy not converted into mechanical work becomes heat energy: 40% of the heat energy produced during this game is removed from the body through perspiration. Calculate the mass of sweat evaporated as a result of the game, assuming that 2.4 MJ are required to evaporate 1.0 kg of sweat.

14 This question is about the generation and use of ultrasound in medical diagnosis.

The ultrasound required for medical diagnosis is normally generated and detected using a piezoelectric transducer. Short pulses of ultrasound are transmitted into the body from the transducer via a coupling medium, and pulses which are reflected back are detected and analysed. Echoes or reflections occur from interfaces between media in the body.

The intensity reflection coefficient for reflection at an interface between two media is given by

$$\alpha = \left(\frac{Z_2 - Z_1}{Z_2 + Z_1}\right)^2$$

where Z_1 and Z_2 are the acoustic impedances of the two media.

Table 11.4 gives data about some typical biological materials.

Table 11.4

Medium	Density ρ/kg m^{-3}	Ultrasound velocity c/m s^{-1}	Acoustic impedance Z/
air	1.3	330	
oil	950	1500	
water	1000	1570	
soft tissue	1065	1530	
bone	1500	4080	

a Give a simple description of the piezoelectric effect in an ultrasonic transducer.

b Describe the main features in the design of a piezoelectric transducer and explain the reasons for their inclusion.

c Why is it important to know values of acoustic impedances?

d i Use the data contained in the table to calculate the values of Z for air, bone and soft tissue. Copy the table and enter in it your values and the unit for Z. Hence estimate the percentage of sound intensity reflected at air/tissue and tissue/bone interfaces.

ii Use the results you obtained in part **d i** to justify the need for a coupling medium.

iii From your results, explain why echo-sounding can be used to locate bone in soft tissue.

e The time delay between transmitted and reflected pulses for an echo from an interface in soft tissue is 0.15 ms. Calculate the depth of the interface below the skin.

15 a A typical metabolic rate for a 20-year-old man of mass 60 kg is 66 W when he is asleep. This is called the basal metabolic rate. The metabolic rate might increase to 1200 W when the man is running although the mechanical power output he develops in running is only 350 W.

i State three uses for the 66 W of power when the man is asleep.

ii Calculate the efficiency of the man in converting energy input to the muscles into work done when running.

iii When the man is running, there is an 850 W difference between the mechanical power output and the metabolic rate. What is the effect on the body of this difference?

iv How does the body cope with the effect mentioned in **iii**?

b It has been suggested, for all animals, that r, the basal metabolic rate per unit mass, varies inversely with the fourth root of m, the mass of the animal, i.e.

$$r \propto \frac{1}{\sqrt[4]{m}}$$

i Calculate the basal metabolic rate per unit mass for the man in **a**.

ii Assuming the relation given above to be valid, show that the basal metabolic rate of a horse of mass 960 kg is about 530 W.

12

The Physics of **SPACE**

Lie out under the night sky on a warm, clear night and you will be able to see about 3000 stars, each with its own distinct colour, the bright, broad band of the Milky Way – our own galaxy seen edge on – and the occasional bright line flaring and dying as a fragment of rock burns up in the atmosphere. You are looking out into space and back into time. It is awesome and magnificent. Understanding what the stars are and how they shine only adds to this sense of beauty and wonder. The development of ever-more sophisticated tools has enabled us to investigate and observe the planets, our own galaxy, the galactic neighbourhood, strange objects such as pulsars, quasars, and black holes, and the large-scale structure of the universe itself.

In this chapter you will learn how we can examine objects beyond the Earth using everything from X-rays to radio waves, how to estimate truly astronomical distances, how we are able to determine the temperature and composition of the stars, and how long our Sun may yet shine. You will also find out when and how we believe the universe came into being and how Einstein's theory of gravity has changed forever our understanding of space, time, and motion.

Trapped in a mutual galactic embrace NGC 2207 (on the left) and IC 2163 (on the right) will continue to distort and disrupt one another for billions of years until they merge into a single, more massive galaxy. It is believed that many present-day galaxies, including the Milky Way, were assembled in this way.

- Aristotle's model

- Ptolemy's model

- Copernicus's model

- Newton's model

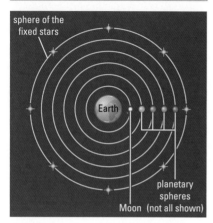

Aristotle's universe was linked to his philosophical and religious ideas. Other Greek philosophers had proposed that the Earth rotates on its axis and even that it orbits the Sun, but Aristotle's authority meant that these ideas were forgotten for two millennia.

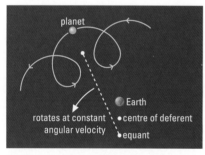

Ptolemy's universe. The stars are fixed once again on a large outer sphere. Though the Ptolemaic system explained the observed motions of the planets very accurately, from a philosphical point of view it seemed rather contrived, lacking the beauty and simplicity of Aristotle's model.

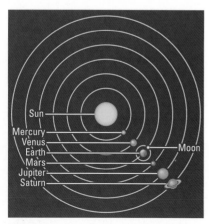

The Copernican Sun-centred model.

THEORIES OF THE UNIVERSE

Ancient astronomers and philosophers looked up at the sky and saw patterns and regularities. The Babylonians divided the sky into the constellations of the Zodiac and recorded how the planets moved against this background. They invented myths to explain what they saw and used skilful mathematical techniques to predict how the patterns change with time. The ancient Greeks invented philosophical systems to describe the same observations, and one of these survived for 2000 years.

Aristotle's universe

One of Plato's pupils, Eudoxus (about 408–355 BC), proposed that the Earth is at the centre of the universe and is surrounded by a set of concentric spheres which rotate about it. The outermost sphere carries the stars, while the inner spheres carry planets and move about various inclined axes. This model was developed further by Aristotle. The Aristotelian universe was:

- finite in size;

- **geocentric** (having Earth at the centre);

- based on perfect mathematical forms (circles).

The Ptolemaic system

One particular problem bothered the Greek astronomers. Planets do not always move in the same direction (as they would be expected to do if they were fixed on rotating spheres). Sometimes a planet changes direction for a short time before continuing its motion. This is called **retrograde motion**. Claudius Ptolemy, an astronomer of the 2nd century AD, adapted the Aristotelian universe to take into account the retrograde motions. He did this by using several clever geometric constructions.

- He displaced the centres of the spheres slightly from the centre of the Earth; this meant the circular orbits (**deferents**) were eccentric (off-centre).

- He allowed the planets to move around circles whose centres ride on the rotating sphere. These small circles are called **epicycles**.

- He introduced another off-centre point, called an **equant**, about which the centre of the epicycle moved at a constant angular rate. This had the effect of varying the speed of motion around the deferent.

By adjusting the positions, sizes, and rate of motion of deferents, epicycles, and equants, Ptolemy was able to fit the observed motions very accurately.

The Copernican revolution

Nicolaus Copernicus (1473–1543) was a canon of the Catholic Church. He was aware that a Greek philosopher, Aristarchus in the 3rd century BC, had proposed a Sun-centred (**heliocentric**) model of the universe that had been ignored by Ptolemy. Copernicus was unhappy with Ptolemy's model because of its awkwardness and complexity. He thought that a heliocentric system would provide a simpler explanation of planetary motions and could return cosmology to the perfection of circular motions, getting rid of deferents, epicycles, and equants. In Copernicus' universe the Sun is at the centre, the Moon moves in a circular orbit around the Earth, and the planets, including the Earth, move in large circular orbits around the Sun. The stars remain on a sphere at great distance. Copernicus's universe was:

- finite in size;

- heliocentric (Sun at centre);

- based on perfect mathematical forms (circles).

In his attempt to get the motion right, Copernicus ended up with a system almost as complex as that of Ptolemy.

Brahe and Kepler

Tycho Brahe (1546–1601) was a Danish astronomer who made incredibly careful and accurate observations of planetary positions using the best apparatus available (prior to the use of telescopes). He also speculated on the structure of the universe, but it was his successor as Imperial Mathematician of the Holy Roman Empire, Johannes Kepler (1571–1630), who inherited and used his data and constructed one of the most important models of all. Kepler accepted Copernicus's basic heliocentric sytem but realized that it could only be used if the planets moved in elliptical rather than circular orbits. He proposed three laws of planetary motion that would later be explained by Newton's theory of gravitation.

1 Planets move in elliptical orbits with the Sun as one focus.

2 A straight line from the Sun to the planet sweeps out equal areas in equal times.

3 The cube of the planet's mean distance from the Sun divided by the square of the orbital period is the same for all planets (r^3/T^2 = constant).

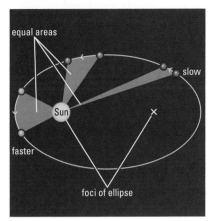

Kepler's laws. Planets move in elliptical orbits sweeping out equal areas in equal times.

Galileo

Galileo did not invent the telescope, but he was probably the first to use it for systematic astronomical observations. His early discoveries included mountains and craters on the Moon, and moons orbiting Jupiter. The first confirmed that astronomical bodies are not all perfect spheres (as had been suggested), and the second that the Earth is not the only centre for rotation (as in the geocentric theories). He promoted the Copernican heliocentric theory but was attacked by the Catholic Church and placed under threat of torture. In the end he was forced to recant, although it is quite clear he in no way changed his opinion on the matter.

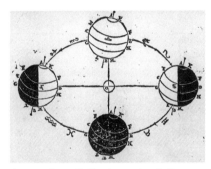

A page from Galileo's notebook. It shows the orbit of the Earth according to Copernicus.

Newton

Isaac Newton built on the work of Galileo and developed his laws of mechanics. When he tried to explain the motions of the planets he did so by asking what forces would be needed to give them their observed motions. This led to the idea of **gravity**, an attractive force exerted by all masses on all other masses. The Moon, for example, would move in a straight line at constant velocity if no resultant force acted on it. In fact it moves in a near-circular path. Gravitational attraction toward the Earth acts as a centripetal force for this circular motion. More detailed analysis showed that the attraction depends on the product of masses and the inverse square of their separation.

$$F = \frac{Gm_1m_2}{r^2}$$

He used this equation to derive Kepler's laws and predict planetary motions in great detail. Newton's law allowed detailed calculations to be made, and led to many discoveries. For example, slight irregularities in the orbit of Uranus suggested the existence of an undiscovered planet. Calculations told astronomers where to point their telescopes and Neptune was discovered (1845). Pluto was discovered in a similar way. Nowadays periodic variations in the motions of distant stars are used to search for other planetary systems.

Newtonian gravity is a purely attractive force. He realized that the Sun must be the centre of the Solar System but saw no reason to place it at the centre of the universe; in fact, in an infinite universe there is no centre. So Newton's universe was:

• centreless

• infinite

• based on mathematical laws of motion and gravitation.

OBJECTIVES

- atmospheric effects
- windows
- using the electromagnetic spectrum
- space telescopes

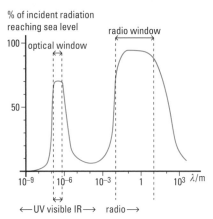

Atmospheric 'windows'.

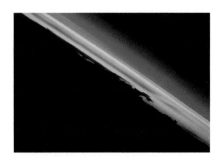

Light interacts with our atmosphere (shown here towards the horizon, edge on) in a variety of ways. This is a real problem for ground-based optical astronomy.

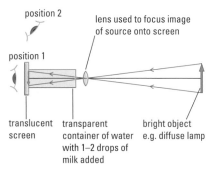

A simple experiment to show that small particles scatter blue light more strongly than red. The transmitted beam (position 1) has had most of the blue light scattered out of it and so appears red or orange. Light scattered to position 2 appears more blue.

OBSERVATIONAL ASTRONOMY

The night sky is awesome, but few of us are privileged enough to see it these days. Light pollution from street lamps in cities and towns reflects off the sky and obscures the stars. Pollution from factories, cars, and power stations obscures ever more of the beauty – but it is still there. On a clear night the view from a remote place, preferably on top of a hill or mountain, can be breathtaking, but even in ideal conditions the atmosphere itself limits what a person can see. The graph on the left shows what percentage of radiation hitting the top of the atmosphere gets down to sea level.

There are two '**windows**' in this spectrum; the **optical window** and the **radio window**. This is why you can see the stars, and why optical and radio astronomy can work with ground-based observatories. Of the two, optical astronomy is the most affected by atmospheric effects and pollution so optical observatories are usually situated at high altitude (to reduce the amount of atmosphere above them) in remote places (away from sources of pollution) where the climate is favourable (plenty of clear nights). As radio waves are longer than visible light waves, they are hardly affected by the small water droplets and dust particles that scatter visible light, so the location of radio telescopes is not quite so critical.

The way different wavelengths of light are scattered by particles depends on the ratio of the wavelength λ to particle size d. (Optical wavelengths are around 500 nm.)

- $d > \lambda$ particles act like tiny mirrors causing diffuse reflection – this is why light scattered by clouds or mist (drop size about 10–100 μm) appears white.

- $d \approx \lambda$ (**Mie scattering**) amount of scattering is proportional to $1/\lambda$ – i.e. blue light is scattered about twice as strongly as red light.

- $d < \lambda$ (Rayleigh scattering) amount of scattering is proportional to $1/\lambda^4$. This explains why the sky is blue – blue light has a shorter wavelength than red so the scattered light is predominantly blue. On the other hand, sunsets are red because the light coming directly to us from the setting Sun has to travel a longer distance through the atmosphere than when the Sun is overhead, and has its shorter-wavelength components scattered out of the beam.

Another problem for optical astronomy is **scintillation** – the 'twinkling' of stars as light passes through moving regions of the atmosphere with different refractive index. Scintillation can smear the apparent point-like image of a star into a disc about a thousandth of a degree across. Even from a well-situated mountain observatory the image size or '**seeing disc**' is unlikely to be less than 0.0001°. This places a more severe limitation on the resolving power of a telescope than the Rayleigh criterion derived from diffraction effects in the aperture (see spread 6.10). Radio telescopes, which do not suffer from these effects, can come closer to achieving their theoretical resolving powers, so their resolving power is said to be **diffraction limited**.

The Hubble Space Telescope

The ultimate solution to the scattering and absorption of light by the atmosphere is to mount a telescope on an orbiting spacecraft. The **Hubble Space Telescope** (**HST**) project was started in 1979 and launched in 1990. The telescope orbits about 600 km above the Earth's surface and its detectors are sensitive to waves from the near infrared through to the ultraviolet (about 100–1000 nm). The telescope has a 2.4 m mirror and it carries a wide-field planetary camera and a faint-object camera for studies of distant galaxies. Its optical resolving power is many times greater than any terrestrial telescope at present. To achieve sharp images it has to be pointed at particular objects to within about 2×10^{-6} degrees over a 24 hour period. This degree of accuracy is equivalent to the width of a hair at a distance of about a kilometre! The HST has been used to check theories of galaxy and star formation, to measure the **Hubble constant** (see spread 12.19) and to look for other planetary systems around distant stars.

Other parts of the electromagnetic spectrum also bring us useful information about their sources but cannot penetrate the Earth's atmosphere. Infrared, ultraviolet, and X-ray astronomy must be carried out using either high-altitude balloons or orbiting telescopes.

IRAS and Rosat

The advantage of space-based astronomy was emphasized by the success of **IRAS**, the **Infrared Astronomy Satellite** launched in 1983. It was designed to make observations in the far infrared at wavelengths in the range 10–100 nm and found more new infrared sources *each day* than were previously listed in all astronomical catalogues! The instruments themselves had to be cooled by an on-board **cryostat** using liquid helium, otherwise the thermal radiation from the equipment would dwarf the signals from astronomical sources.

Although visible, infrared, and radio waves can all be focused fairly easily, anything that can refract X-rays significantly also absorbs them. It is also difficult to build an X-ray reflecting telescope because the X-rays only reflect from metal surfaces if they strike them at a glancing angle. However, it can be done: the X-rays reflect from the inside of a series of nested metal 'cylinders' to focus beyond them. At this point they collide with a detector and create showers of electrons. The **Röntgen satellite (ROSAT)** was launched in 1990 and was used to survey the sky at X-ray (0.6–10 nm) and hard ultraviolet (5–25 nm) wavelengths. The advantage of using X-rays is that they record the energetic photons emitted in violent events – e.g. matter falling into black holes – so they give very different information from infrared or visible surveys.

COBE and WMAP

The **Cosmic Background Explorer (COBE)** was used to study the cosmic microwave background radiation (CMB) across the whole sky. The COBE results included a measurement of tiny fluctuations in the brightness (and temperature) of the background radiation coming from different points in the sky. These fluctuations had been predicted and correspond to tiny differences in density in the early universe from which the galaxies formed. These results caused a sensation when they were first published in 1992, producing an image of the universe as it was less than 400 000 years after the Big Bang and showing that the temperature of the universe is 2.735 K above absolute zero. (See also spread 14.6.)

The **Wilkinson Microwave Anisotropy Probe (WMAP)** was launched in 2001 and carried out even more detailed measurements of the distribution of CMB. It confirmed the existence of fluctuations in the CMB, measured the age of the universe to be 13.7 billion years to within 1%, and provided evidence to support the Standard Model of Cosmology.

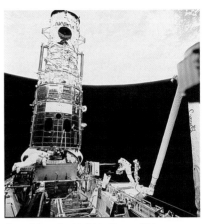

Repairing the Hubble Space Telescope. At first, problems with an incorrectly-ground mirror blurred and distorted all the images and there were fears that the HST might be an expensive flop. However, an amazing repair job was carried out by astronauts on a maintenance visit in 1993 and since then the beautiful images it has sent back have impressed everyone.

An incredible HST image of Betelgeuse – the first time the disc of a star other than our Sun has been resolved.

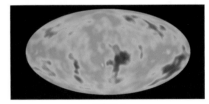

The distribution of cosmic microwave background radiation across the whole sky. The colours indicate temperatures, which fluctuate by only a few parts per million from the average. The upper image shows COBE data; the lower one shows WMAP data.

PRACTICE

1 Why do stars appear to twinkle but planets do not?

2 a Calculate the orbital period of the Hubble Space Telescope if its altitude is 600 km and the Earth has a radius of 6400 km and a mass of 6.0×10^{24} kg.

   b Suggest a reason for placing the HST in such a low Earth orbit rather than a geostationary orbit.

3 Explain why the resolving power of a radio telescope is diffraction-limited whereas the resolving power of an optical telescope is atmosphere-limited.

4 a What are typical wavelengths for radio, infrared, visible, ultraviolet, and X-ray astronomy?

   b Compare photon energies at these wavelengths.

   c How might the sources of X-rays differ from those emitting visible or radio waves?

12.3

O B J E C T I V E S

- the laws of planetary motion and Newton's theory
- derivation of the third law
- dark matter

KEPLER'S LAWS

Johannes Kepler (left), astronomer and mathematician (1571–1630).

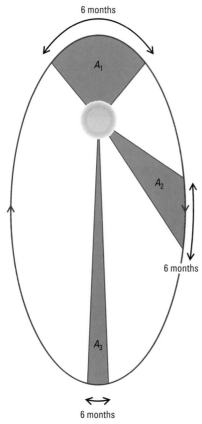

Kepler's 2nd law in more detail.

Laws of planetary motion

Johannes Kepler devoted a considerable portion of his life to the study of planetary motions. He succeeded where others had failed for two reasons.

- He was working from observations of extraordinary accuracy provided by Tycho Brahe.
- He rejected the philosophical assumptions that had led so many earlier workers astray.

Kepler regarded his work as a problem set by God: to decode the workings of the Solar System. In the process he discovered a set of laws that describe the motion of any orbital body, whether it be a communications satellite in orbit around the Earth, a star in orbit around a giant black hole at the centre of the galaxy, or a planet in orbit around the Sun. These laws were discovered directly from the observations of positions of planets in the sky without a computer or any other calculating aid.

Kepler's three laws of planetary motion

1 All the planets move in elliptical orbits with the Sun at one of the focal points of the ellipse.

2 A line drawn from the Sun to a planet will sweep out equal areas in equal intervals of time as the planet moves in its orbit.

3 The period of a planet's orbit squared is proportional to its mean distance from the Sun cubed: T^2 is proportional to r^3.

Kepler and Newton

Kepler's laws are empirical, that is, they are mathematical summaries of the observations. They make no attempt to *explain* these observations. In this respect there is no common link between the three laws. Newton used his inverse-square law of gravitation to derive Kepler's laws, so they all follow from this one fundamental principle.

Maths box: derivation of Kepler's third law

It is quite straightforward to prove Kepler's third law for a circular orbit from the inverse-square law. The gravitational attraction between a planet of mass m and the Sun of mass M provides centripetal force to move the planet in a circular path.

$$F = \frac{GMm}{r^2} = -\frac{mv^2}{r}$$

$$v^2 = \frac{GM}{r} \qquad (1)$$

(The fact that m cancels means T is independent of m.) The time for one orbit is orbital circumference divided by speed:

$$T = \frac{2\pi r}{v}$$

$$\therefore T^2 = \frac{4\pi^2 r^2}{v^2} = \frac{4\pi^2 r^3}{GM}$$

(substituting from equation (1))

so

$$T^2 \propto r^3 \quad \text{(Kepler's third law)}$$

This can also be proved for a general elliptical orbit.

Dark matter in the galaxy

Kepler's laws apply whenever smaller masses orbit an object whose mass is much greater than the total mass of the objects in orbit around it. In a galaxy, the further out towards the edge you go the more stars there are between you and the galactic centre. If the density of stars in the galaxy is more or less constant this number should increase in proportion to the volume contained between you and the galactic centre. The stars at greater radius can be ignored (there is no field inside a hollow spherical shell because of the mass of the shell itself). This implies that the mass causing stars to orbit at radius r should increase as r^3. If this is substituted into equation (1) in the maths box it gives

$$v^2 = \frac{Gkr^3}{r} \quad (k \text{ is a constant})$$

$$\therefore v \propto r$$

Therefore a plot of v against r for stars at the outer edges of a galaxy might be expected to be linear. The graph shows that this is not the case. The velocity distribution levels off towards the edge of the galaxy, suggesting there is something wrong with our simple model of mass distribution in galaxies.

Turning the argument around, if v is constant then equation (1) implies that the mass increases in proportion to radius; a result that contrasts strongly with the observed distribution of luminous sources in galaxies, as their density is seen to fall away quite rapidly with distance. The conclusion is that there must be some form of **dark matter** in the outer regions of galaxies (**galactic haloes**) that contributes mass but not light. Prime candidates for this dark matter are **brown dwarf** stars which have relatively low mass and so radiate very little. This makes them difficult to observe, but recent surveys have found some in our own galaxy. However it is too early to say whether this will solve the problem of the 'missing mass'.

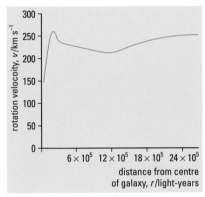

Graph of v vs. r.

Practice

1 Use the data in the table to plot a suitable graph to test Kepler's third law for planets in our Solar System and use the graph to estimate the mass of the Sun.

Planet	Mean orbital radius r/m	Period T/s
Venus	$1.0^8 \times 10^{11}$	1.94×10^7
Earth	1.50×10^{11}	3.15×10^7
Mars	2.28×10^{11}	5.93×10^7
Jupiter	7.78×10^{11}	3.74×10^8
Saturn	1.43×10^{12}	9.29×10^8
Uranus	2.87×10^{12}	2.65×10^9

2 Use Newton's law of gravitation and the conservation of angular momentum to verify Kepler's second law for the points of closest approach (perihelion) and furthest distance (aphelion) for a comet. (The magnitude of angular momentum is mvr where m is the comet mass, v is its orbital speed, and r is the distance at aphelion or perihelion.)

3 If the Sun is at one focus of a planet's elliptical orbit, what is at the other?

4 Newton's law of gravitation explains Kepler's laws of planetary motion, but does it explain gravity?

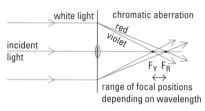

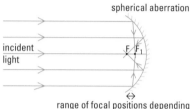

Chromatic and spherical aberration.

Degrees, seconds, radians

The objects of interest to astronomers usually subtend only small angles, so it is convenient to measure them in seconds of arc, an old unit based on the degree. 1 second of arc (1″) is 1/60 of 1 minute of arc (1') which is 1/60 of 1 degree (1°), so 1 second is equal to about 2.8×10^{-4} degrees. The small-angle approximations, however, are only valid if angles are measured in radians, so care must be taken when calculating limits of resolution. 1 rad = 180/π degrees.

The two Keck telescopes on Mauna Kea are the largest optical telescopes in the world. The primary mirror of each has a diameter of 10 m and is constructed from 36 hexagonal mirror segments each about 1.8 m across and 7.5 cm thick. Signals from both telescopes can be combined to make an optical interferometer (see spread 12.5) with an effective diameter of about 100 m!

OPTICAL ASTRONOMY

Light from any point on a distant object arrives as virtually parallel rays at the eye or telescope. This is because rays passing through opposite sides of the receiving aperture have only diverged by a small amount (mm, cm, or m) on their entire journey from the source (perhaps many light-years away). So telescopes are designed to focus parallel rays. This can be done either by passing the light through lenses (**refractors**) or bouncing light off curved mirrors (**reflectors**). A large objective lens or curved mirror is used to collect as much incident light as possible and form an intermediate image, which is then viewed through an eyepiece that acts as a magnifying glass or is directed at light-sensitive equipment that records an image electronically. The effect of the telescope is to produce a brighter image with a larger angular size than if it were viewed with the naked eye.

The angular magnification of an astronomical telescope is equal to the ratio of focal lengths of the objective lens/mirror to the eyepiece lens:

$$M = \frac{\text{angle subtended at eye by image in telescope}}{\text{angle subtended by object at the unaided eye}} = \frac{f_o}{f_e}$$

The secondary mirror in a reflecting telescope blocks some of the incident light. This is a major problem for small reflecting telescopes, so small telescopes tend to be refractors. However, the main drawback with refractors is **chromatic aberration** – different wavelengths (colours) have different speeds in glass and so refract different amounts. This means white light from a star will not be focused in one plane. The blue light is focused closer to the lens than the red light. If this image is captured on a film it will never be sharp. Chromatic aberration does not occur with reflectors so all large optical telescopes are reflectors.

Spherical aberration limits the performance of both reflectors and refractors. Rays far from the optic axis are focused to a different position from those close to it. It is possible to correct for this by using a parabolic reflector or shaping the lens in a special way. However, off-axis objects are then distorted (they are said to suffer from **coma**).

Resolving power

Diffraction at the aperture of the telescope means that point objects cannot be focused to point images. The theoretical limit to resolution is then given by the **Rayleigh criterion**:

$$\theta_{min} = \frac{1.22\lambda}{D}$$

where θ_{min} is the smallest angular separation of point-like objects that results in distinguishable images. λ is the wavelength of light and D the diameter of the objective lens or mirror.

The resolving power of terrestrial optical telescopes is limited still further because they receive light that has passed through the atmosphere. Convection currents and temperature gradients in the air change the refractive index slightly from moment to moment and from place to place. This causes 'twinkling' or **scintillation** and limits ground-based resolution to about 0.0003°. The next generation of optical telescopes will use 'adaptive optics' to correct for atmospheric disturbances. The mirrors (like those in the Keck telescopes) will be made from a large number of separate segments and the surface of each one will be continually flexed to keep the image as sharp as possible.

Seeing the stars

Forming images is one thing, recording and analysing them is the next step. It is important to understand the response characteristics of your detector and its limitations. The sensitivity of a light detector is called its **quantum efficiency**: this is the ratio of **information bits** recorded to the total number of photons absorbed. An information bit is a picture element or **pixel** making up one part of an image (like an individual dot in a printed photograph or on a TV screen).

$$\text{quantum efficiency } Q = \frac{\text{number of bits of information recorded}}{\text{number of photons absorbed}} \times 100\%$$

The quantum efficiency of a particular detector is not constant; it usually depends on the wavelength and intensity of the light absorbed.

- **Using film** ($Q < 4\%$) Photographic emulsions record the light that falls on them by undergoing a chemical reaction when they absorb photons. However, the images they produce need careful interpretation. For example, film is relatively more sensitive to bright light than faint light so that the magnitudes of faint stars are underestimated unless the film is carefully calibrated. Also, the spectral response varies with the manufacturer and has to be allowed for when doing spectral analysis.

- **CCDs (charge-coupled devices)** ($Q > 70\%$) These are slices of silicon that store the electrons freed by incoming photons and build up an image as a pattern of pixels. The number of electrons liberated is proportional to the intensity of the light. They are faster and more sensitive to light than film and have the advantage that the image information can be read automatically in digital form for processing and measurement using computers.

A CCD camera in an optical telescope.

- **Photomultipliers** ($Q \approx 20\%$) Incident photons eject electrons from a cathode in a vacuum tube (photoelectric effect) and this is accelerated toward another electrode where it ejects several electrons. After about 10 similar stages of amplification an avalanche of electrons produces an electric pulse in an external circuit which is counted. Photomultipliers detect individual photons.

- **Photodiodes** ($Q < 1\%$) Although not especially sensitive, these can respond at MHz frequencies, so they are useful for detecting very rapid changes. Incident light frees electrons in a reverse-biased photodiode and a current of a few microamps flows for a very short time.

- **Optical fibres** These are not detectors in their own right, but bundles of optical fibres can be used to direct light from a collection of stellar images to separate spectrometers so that they can all be analysed at the same time, thus making more efficient use of telescope time.

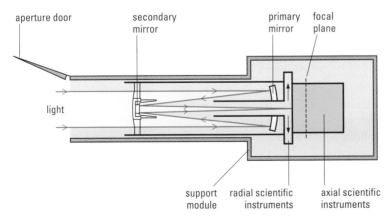

aperture door   secondary mirror   primary mirror   focal plane

light

support module   radial scientific instruments   axial scientific instruments

The Hubble Space Telescope is an example of a Cassegrain-type reflector. The 2.4 m primary mirror collects light from a region of sky 28 arcminutes across (comparable to the angle subtended by a full Moon from Earth) and directs it onto a 0.3 m secondary mirror. This focuses the beam back through a 0.6 m hole in the primary to project it onto an area about the size of a frisbee in the focal plane 1.5 m behind the primary mirror. From there it is divided up and sent to each of five instruments.

*The European Southern Observatory's **Very Large Telescope (VLT)** is at Paranal in Chile. It consists of four 8.3 m diameter telescopes connected together to form an interferometer. Its resolving power of milli-arcseconds is equivalent to being able to separate car headlamps at the distance of the Moon!*

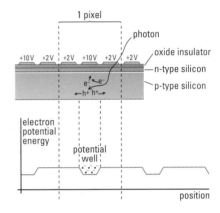

CCDs. Each pixel consists of three electrodes across which an undulating pattern of voltages is applied, as shown above. Photons release holes and electrons in the semiconductor lattice. Electrons are stored beneath the central electrode of each pixel (since they have the lowest potential energy here – in a potential well close to the most positive potential). Holes move away from these positions. An electron pattern is built up which is identical to the image formed on the CCD. The pattern of charge is then read out sequentially from electrodes on the surface of each pixel.

PRACTICE

1 The densest cluster of optical and near infrared telescopes on Earth is on top of Mauna Kea (altitude 4200 m) in Hawaii.

  a Why is this such a good site?

  b Compare the advantages and disadvantages of the Mauna Kea site with space-based astronomy using satellites such as the Hubble Space Telescope.

The 100 m fully-steerable radio telescope at Effelsberg in Germany.

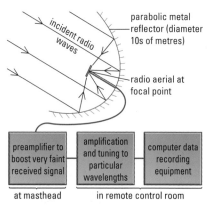

A radio telescope.

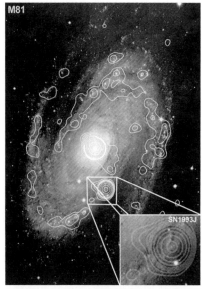

A radio contour map at 5.0 GHz observed with the Very Large Array superimposed on an optical image of the galaxy M81. The brightest source is the nucleus of M81 and the second brightest source is SN1993J, which has been enlarged in the inset to show detail.

RADIO ASTRONOMY

Radio waves were first demonstrated by Heinrich Hertz in 1887, and just three years later the great electrical pioneer Thomas Edison suggested that astronomical objects like the Sun might also be sources of radio waves. He even proposed carrying out an experiment with a huge radio receiver to try to convert incoming waves to sound. No record of the experiment exists. Others did try it and failed until 1932, when Karl Jansky noticed that the radio noise in an antenna he was testing varied throughout the day and was always most intense from the direction of the galactic centre (the direction in which there is the greatest concentration of stars).

For some years after Jansky's discovery a radio amateur, Grote Reber, was the only radio astronomer in the world. One of the most striking things about a radio map of the sky is that most of the strongest radio sources do not match up with obvious optical sources (in Reber's original maps only the Sun could be identified). This underlines the importance of radio astronomy: it gives us *different* information to optical astronomy. Reber's papers and the results of research into radar during the Second World War prepared the way for an explosion of interest in radio astronomy after the war. One scientist, J. S. Hey, made three of the most important discoveries in radio astronomy.

• The Sun emits radio waves in the metre wavelength range. These waves are from active sunspots and kick-started the investigation of the Sun using radio waves.

• Meteors leave trails of ionization in the upper atmosphere that reflect radio waves. This led to the use of radio waves to investigate both meteors and the atmosphere.

• Hey also discovered the first discrete source of radio waves, from the radio galaxy Cygnus A.

The single-dish radio telescope

The **radio window** in the atmosphere lies between wavelengths of about 10 mm and 10 m, a range over which the atmosphere is almost transparent. Longer waves are reflected back into space by the ionosphere; shorter waves are absorbed by water molecules and oxygen in the atmosphere.

The parabolic shape of the reflecting surface of the single-dish radio telescope focuses incoming parallel rays onto a small radio receiver and prevents spherical aberration.

There are two advantages to having a large dish diameter – the signal strength and the resolving power are both increased, so weaker sources can be detected and sources with smaller angular separations can be distinguished. The power received by a radio telescope is proportional to the area of its collecting dish, so doubling the diameter quadruples the power received.

Unlike optical telescopes, whose practical resolving power is limited by the atmosphere, radio telescopes are **diffraction limited** (because the atmosphere is so much more transparent to radio waves), so they come much closer to the theoretical resolving power of $1.22\lambda/D$. However, radio wavelengths are much longer than optical wavelengths, so the resolving power of even a large single-dish radio telescope does not compare with that of an optical telescope, or the unaided eye!

Making a dish larger also makes it heavier, and eventually a point is reached when distortions in the structure caused by its own weight begin to distort the image and also make the dish harder to steer. To reduce this effect the solid sheet-metal reflector can be replaced by a fine-wire mesh (but the mesh spacing must be less than 1/20 the wavelength of the radio waves).

Interferometers

If the signals from two or more radio telescopes are combined they interfere, and the pattern of interference can be used to construct radio images that show much finer detail than any of the individual telescopes used. Such an arrangement is called an **interferometer**. The simplest type of interferometer consists of two telescopes placed some distance apart with their output signals added at a common receiver. The path difference between parallel rays from a distant point source arriving at each telescope will vary as the Earth rotates and carries the telescopes with it. This variation produces a characteristic interference pattern which shows up as a series of maxima and minima in the combined signal. The diagram on the right shows a simple radio interferometer consisting of two radio telescopes separated by a distance d (this is called the **baseline** of the interferometer) and tuned into radio waves of wavelength λ coming from a distant point-like source.

The signal moves from one maximum to the next when θ changes by λ/d (this is just like the angular separation of maxima from a double-slit experiment, except this time it is a pattern for reception). If an individual source has an angular size larger than λ/d it will not produce interference fringes (because maxima and minima caused by different points on the source will all overlap) and a radio interferometer can reject it. This allows the interferometer to pick out detail missed by a single dish by emphasizing signals from sources of small angular size. The limit of resolution for such an interferometer is about λ/d, which is equivalent to that of a single dish antenna of diameter d, but is much cheaper and simpler to build. Of course, this resolution is only possible along the line parallel to the baseline of the interferometer.

Early interferometers used baselines of a few kilometres. There is no theoretical limit to the size of baseline that can be used, however, and later interferometers, such as MERLIN, used hundreds of kilometres. These are called **very long-baseline interferometers (VLBIs)**. In principle, the resolving power of a dish with diameter equal to the diameter of the Moon's orbit could be simulated by having one telescope on Earth and another on the Moon.

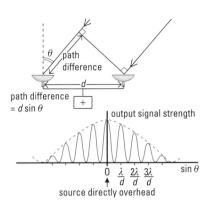

A baseline interferometer and the signal it receives as the Earth rotates and a radio source passes overhead.

Interferometer techniques

Measuring the angular size of a source If the separation of a pair of telescopes is changed, the interference pattern from a particular source disappears when the angular size of the source corresponds to the resolving power of the telescope. This can be used to measure the angular size of the source. The first radio interferometer was used in this way to measure the size of active sunspots.

Aperture synthesis The performance of a very large-diameter radio dish can be simulated by an array of individual telescopes inside the perimeter of the imaginary large telescope. This technique is called **aperture synthesis**. An alternative to a two-dimensional array is to have many telescopes along a particular baseline and then to rely on the rotation of the Earth to rotate them. To synthesize the signal from the filled aperture, signals collected at different times (when the array or line is in different positions) must be added together.

Telescopes in the Very Large Array near Socorro, New Mexico. There are 27 dishes arranged in a Y-shaped array, with each 'arm' of the Y about 20 km long.

PRACTICE

1 Explain why:

  a Radio telescopes are much larger than optical telescopes.

  b Radio astronomy can be carried out from terrestrial telescopes, even in locations where optical astronomy would be impossible.

  c The resolving power of a large single dish radio telescope is less than that of the unaided human eye.

2 Compare the theoretical limits of resolution and total power collected for the 100 m radio telescope at Effelsberg in Germany with the 76 m diameter Jodrell Bank radio telescope in Cheshire (assume they are tuned to the same wavelength).

3 If a very large single-dish radio telescope of diameter d could be built, in what ways would it be superior to a long baseline interferometer with baseline d?

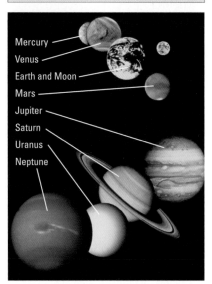

Mercury
Venus
Earth and Moon
Mars
Jupiter
Saturn
Uranus
Neptune

The Solar System, showing all eight planets. Pluto was reclassified as a dwarf planet in 2006.

command module

service module

lunar module (within hull)

launch vehicle

Apollo 11 *at launch.*

A mosaic of Viking Orbiter *photos showing ancient river channels running from the cratered highlands to the east of* Mangalla Valles *on Mars.*

EXPLORING THE SOLAR SYSTEM

A great deal of data can be collected using Earth-based instruments, but the study of our own Solar System has been changed beyond recognition by observations and measurements made by various space probes. **SOHO**, the Solar and Heliospheric Observatory, is measuring the solar magnetic field, looking at the solar wind, and 'listening in' to sound waves bouncing about inside the Sun. On Christmas day 2004 the *Cassini Huygens* mission arrived at Saturn and the Huygens Probe detached from the main spacecraft in order to descend through the atmosphere of Saturn's moon Titan and land safely on the surface. This was the first successful landing in the outer Solar System and Titan sent back incredible images of an alien coastline with rivers and lakes. However, this is no Earth-like world – the temperature at the surface of Titan is –180° C and the oceans are probably liquid methane! Robotic spacecraft have also visited minor Solar System bodies such as comets and asteroids. In 1986 Giotto made a close fly-by of Halley's comet and in 2005 the Deep Impact mission fired a projectile into the nucleus of comet 9P/Tempel in order to analyse the ejected materials and improve our understanding of the structure and composition of comets – information that might well be important if we ever need to deflect one that is on a collision course with the Earth. One of the most exciting current missions is *Kepler*. This was launched in 2009 and is surveying 100 000 main sequence stars in the Milky Way, trying to detect the tiny variation in brightness caused by a transiting exoplanet. At the time of writing Kepler has confirmed over 100 exoplanets with a further 2700 requiring further analysis. These discoveries have confirmed that solar systems are commonplace and the search continues for Earth-like worlds that might be able to (or do) support life. (See also spread 14.5.)

Some important early missions are described briefly below.

Apollo

The ***Apollo*** programme consisted of 17 missions, and between July 1969 and December 1972 there were six Moon landings. Each *Apollo* craft consisted of three sections, a command module (CM), a service module (SM), and the lunar module (LM). The CM and SM remained in orbit around the Moon while two of the three astronauts went down to the Moon's surface in the LM. On return to Earth the SM was jettisoned just before re-entry. The 12 astronauts who landed on the Moon brought back a total of 400 kg of Moon rock and carried out a variety of experiments while they were there.

The surface of the Moon and other planets suggests that there have been periods of intense meteoritic bombardment during the 5 billion year history of the Solar System. Volcanic activity on Earth has concealed much of the evidence here.

Viking

Science fiction has often suggested that Mars is or was once inhabited by intelligent aliens. The main quest of the ***Viking*** mission was to establish whether the Martian soils show any signs of micro-organisms. *Vikings 1* and 2 were launched in 1975 and arrived at Mars about nine months later. They made a complete photographic survey of the Martian surface, recording details to a resolution of about 150–300 m, sampled the Martian atmosphere and carried out meteorological and geological experiments. Whilst Viking did not find conclusive evidence that Mars has ever supported life more recent missions have confirmed the existence of water on Mars and of flowing water in the past. In 2013 NASA claimed that this water would have been drinkable. Robotic rovers such as Spirit and

Discovery landed on Mars in 2004 and the Curiosity Rover, a larger, more sophisticated, robot geologist landed in 2012. These mobile laboratories collect and analyse surface and subsurface samples as they search for the organic molecules that would be the signature of life past or present.

Voyager

Voyagers 1 and *2* were sent on a grand tour of the Solar System. The mission relied on the **gravity assist** (**slingshot**) principle to propel the craft from one planet to the next and was only possible because of a fortunate alignment of Jupiter, Saturn, Uranus, and Neptune that occurs less than once a century. The spacecraft are now headed out of the Solar System and carry information about humans for any interested aliens!

Magellan

Magellan was a Venus orbiter launched from the Space Shuttle *Atlantis* in 1989. It reached Venus in 1990 and used radar reflections to map the surface to a resolution of a few 100 m. This is possible because the thick atmosphere is transparent to radio frequencies. The images showed details of volcanic activity along with enormous impact craters.

Galileo

Galileo mission used the gravity assist principle to reach Jupiter. Its mission was to investigate the Jovian system and to launch a probe into the atmosphere to test theories of planetary formation. Remarkable images were received from the probe, which entered the Jovian atmosphere in December 1995.

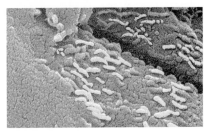

Evidence for life on Mars? This coloured scanning electron micrograph shows tube-like structures (coloured yellow) on a meteorite which originated from Mars. Despite a great deal of excitement further analysis in 1997 suggested that the formations might be geological rather than biological.

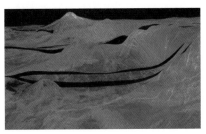

An image from the Magellan *mission. The* Magellan *spacecraft surveyed the surface of Venus by bouncing radio waves from it. It constructed the detailed images using the time delay for the return pulse and the Doppler shift due to the velocity of the spacecraft over the surface.*

Planetary data (including the dwarf planet Pluto).

	Mercury	Venus	Earth	Mars	Jupiter	Saturn	Uranus	Neptune	Pluto
Mass (relative to the Earth)	0.055	0.815	1	0.108	318	95.2	14.6	17.2	0.1
Diameter (km)	4880	12 104	12 756	6787	142 800	120 000	51 800	49 500	2300
Density ($kg\,m^{-3}$)	5400	5200	5500	3900	1300	700	1200	1700	2000
Surface gravity (relative to the earth)	0.37	0.88	1	0.38	2.64	1.15	1.17	1.18	0.07
Mean surface temperature (°C)	350 (day) −170 (night)	−33 (cloud) 480 (surface)	22	−23	−150 (cloud)	−180 (cloud)	−210 (cloud)	−220 (cloud)	−230?
Atmospheric pressure ($10^5\,Pa$)	10^{-12}	90	1	0.006	?	?	?	?	?
Atmospheric gases	none	carbon dioxide	nitrogen oxygen	carbon dioxide argon	hydrogen helium	hydrogen helium	hydrogen helium methane	hydrogen helium methane	none?
Maximum distance from Sun ($10^6\,km$)	69.7	109	152.1	249.1	815.7	1507	3004	4537	7375
Minimum distance from Sun ($10^6\,km$)	45.9	107.4	147.1	206.7	740.9	1347	2735	4456	4425
Mean distance from Sun (relative to the Earth)	0.387	0.723	1	1.524	5.203	9.539	19.18	30.06	39.44
Orbital period (days)	88	224.7	365.26	687	11.86	29.46	84.01	164.8	247.7
Rotation period	59 days	−243 days retro	23 h 56 min 4 s	24 h 37 min 23 s	9 h 50 min 30 s	10 h 14 min	−11 h retro	16 h	6 days 9 h

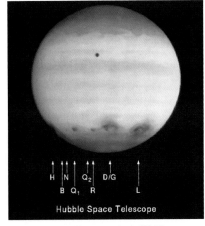

Hubble Space Telescope

It could never happen here(?). The collisions of the Shoemaker–Levy comet with Jupiter.

Dwarf planets

Pluto was discovered in 1930 by Clyde Tombaugh and was initially classified as the ninth planet in the Solar System. In 2006 the International Astronomical Union voted to reclassify Pluto as a dwarf planet along with several other newly discovered members of the Kuiper belt. Pluto orbits at the inner edge of the Kuiper belt which contains other comet-like bodies made of rock and ice, some of which (e.g. Eris) are more massive than Pluto. Ceres, the largest body in the asteroid belt, is also classified as a dwarf planet.

The Earth–Moon system photographed by Galileo on its way to Jupiter.

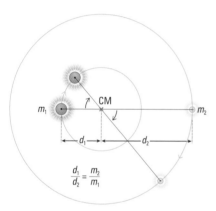

A binary system. Both partners orbit their common centre of mass (CM).

The dense band of stars rising up through the night sky is an edge-on view of our own galaxy, the Milky Way.

OUR PLACE IN SPACE

When the Earth solidified from its molten state, isotopes of elements were fixed in the rocks of the crust. Some of these isotopes are radioactive, and as time goes on their abundance falls and that of the elements they decay into rises. For example, uranium-238 decays to thorium-234 with a half-life of 4.5 billion years. Thorium-234 is also radioactive and decays, via several other isotopes, to lead-206 which is stable and gets trapped in rocks. The isotopic ratio of lead-206 to uranium-238 increases with time and if measured can be used to estimate the age of the rocks. This age can be checked using other isotopic ratios.

When this is done for rocks on Earth the most ancient are found to be about 3.8 billion years old. Moon rocks are up to 4.3 billion years old (the Moon, being smaller than the Earth, would have solidified earlier) and samples of meteorites which came from the asteroid belt are a little older, about 4.5 billion years. This suggests that the Solar System formed about 4.5 billion years ago when the Sun itself condensed out of an enormous cloud of galactic gas and dust.

Binary stars

Many stars are formed in binary systems orbiting around their combined centre of mass, rather like the Earth and Moon. Since the forces on each star are equal (Newton's third law) they orbit at distances d_1 and d_2 from the centre of mass (and rotation) which are inversely proportional to their individual masses m_1 and m_2. Centripetal forces are equal, so

$$m_1 d_1 \omega^2 = m_2 d_2 \omega^2$$

$$\therefore \frac{m_1}{m_2} = \frac{d_2}{d_1}$$

The total mass of the binary system can be calculated from its period by using a generalization of Kepler's third law (see spread 12.3):

$$T_2 = \frac{4\pi^2 r^3}{G(m_1 + m_2)}$$

This works for circular or elliptical orbits, but for ellipses r must be replaced by the length of the semimajor (longer) axis.

The galaxy

Look up at the stars on a clear night far away from city lights, preferably somewhere in the southern hemisphere, and you will see a dense band of stars across the night sky. This is the **Milky Way**. Our Sun is situated in one arm of a spiral galaxy which is about 100 000 light-years across and contains about 200 billion stars. The Milky Way is our best view of our own galaxy from Earth. When we look towards it we are looking along the plane of the galaxy towards the galactic centre. The darker sky on either side is when we look away from that plane through the narrower region of stars in the spiral arms and out into the almost empty space that surrounds us. The galaxy rotates and we orbit the galactic nucleus with a period of about 2×10^8 years at a radius of about 30 000 light-years. If these figures are substituted into Kepler's third law an estimate of the mass of the galaxy can be made – about 2×10^{11} solar masses.

The galaxy actually consists of two parts, the **disc** and the **halo**. The Milky Way is our view of the disc. The halo is spherical, with a diameter of about 200 000 light-years, and is centred on the galactic nucleus. In its central region it contains densely packed stars that form the **nuclear bulge** of the galaxy. Further out it consists of low-density gas, widely spaced stars and about 120 globular **clusters**. The globular clusters each contain hundreds of thousands of stars and move around the galactic centre in elliptical orbits. There is a distinction between the types of stars found in the disc and halo. Disc stars are usually relatively young (**population I**) stars (like our Sun) whereas those in the halo tend to be

older (**population II**) stars and were formed when the galaxy was young. The nuclear bulge contains a mixture of both populations.

Galaxies formed when regions of above average-density intergalactic gas and dust collapsed under the influence of internal gravitational forces a few billion years after the Big Bang. The Hubble Space Telescope provides images of young galaxies in various stages of evolution and these are supplemented by images from the next generation of terrestrial telescopes (e.g. Keck I and II and the VLT in Chile (see spread 12.4)).

Radio waves can penetrate the dust clouds between us and the galactic centre far more effectively than visible light. The intense radio emissions that originate there suggest that extremely energetic events are occurring, and that there is something very massive at the centre. Perhaps matter is falling into a huge black hole and radiating energy as it falls.

Other galaxies

Early models of the universe assumed that stars are scattered more or less randomly through space. The band of the Milky Way suggests that this is not actually true; local stars at least are grouped in a systematic way. However, in 1924 Edwin Hubble used the powerful 100 inch telescope on Mount Wilson to observe faint **nebulae** (thought to be gas clouds within our own galaxy) and realized that they are complete systems in their own right – galaxies at great distances from our own Milky Way. His subsequent work on the spectra, redshifts, and distances of these galaxies made an extraordinary impact on our view of the universe (leading to the ideas of expansion and the Big Bang) and he invented the classification system which we still use today. This may be depicted in what has become known as Hubble's 'tuning fork' diagram, because of its shape. Hubble believed that this showed how galaxies evolve in time. We now know that this is not the case, but the classification system is still useful.

Spiral galaxies similar to our own Milky Way, seen edge on: NGC 4565 (top) and obliquely: M81 (bottom).

Hubble's tuning fork diagram.

Clusters of galaxies

Most galaxies are members of **clusters**, some of which are roughly spherical (**regular clusters**) and some of which have rather arbitrary shapes (**irregular clusters**). Our own galaxy is a member of the Local Group, an irregular cluster of about 20 galaxies. The nearest great irregular cluster is the Virgo cluster containing about 1000 galaxies, and the nearest regular cluster is the Coma cluster. The amount of matter we observe from Earth in these clusters seems insufficient to hold them together in gravitationally bound states. This is more evidence for the presence of dark (i.e. unseen) matter (see spread 12.3). In some cases the observed mass of galactic clusters is only about 10% of the mass required to bind them. On the largest scale the universe has a kind of frothy structure with galaxies, clusters, and super clusters lying on the surface of enormous 'bubbles' or voids of almost empty space.

step 1 $_1^1H + {}_1^1H \rightarrow {}_1^2H + {}_1^0e^+ + {}_0^0\nu$

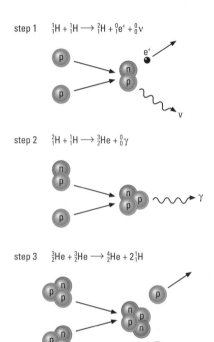

step 2 $_1^2H + {}_1^1H \rightarrow {}_2^3He + {}_0^0\gamma$

step 3 $_2^3He + {}_2^3He \rightarrow {}_2^4He + 2{}_1^1H$

The proton–proton cycle (p–p cycle).

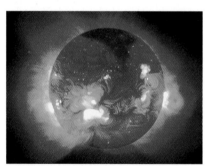

The Sun in X-rays. The Sun's radiant power (about 4×10^{26} W) comes from nuclear fusion reactions at its core.

THE ORIGIN OF THE ELEMENTS

How old are you? Seventeen? Eighteen? No, billions of years old. We are all made from atoms and, apart from hydrogen (there is no helium in our bodies) the nuclei of these atoms were made in nuclear fusion reactions inside stars that exploded billions of years ago. We are, literally, 'stardust'.

Power from the Sun

The Sun has been radiating at a power of about 10^{26} W for about 4.5 billion years and is thought to be about halfway through this part of its life. There is no way that chemical or gravitational energy can be responsible for generating such an enormous amount of energy. All stars are gigantic self-sustaining fusion reactors which balance the immense outward pressure generated by reactions in their cores against the continual inward pressure of gravitation due to their large mass. The Sun consists almost entirely of hydrogen and helium, and the Sun's power comes from the fusion of hydrogen into helium in its core.

The details of the nuclear reactions were worked out by several physicists including Sir Arthur Eddington (the British astronomer who first measured the deflection of starlight by gravity) and the American Nobel Prize winner Hans Bethe. Nearly 80% of the Sun's power is generated by the so-called proton–proton reaction. The net effect of this process is to convert four hydrogen to one helium-4 nucleus and release a large amount of energy. The overall reaction is

$$4{}_1^1H^+ \rightarrow {}_2^4He^{2+} + 2{}_0^0\nu + 2{}_1^0e^+$$

Of course, the probability of four protons meeting and undergoing this reaction is virtually zero. In practice it proceeds by a series of simpler steps called the **proton–proton cycle**:

step 1: $_1^1H + {}_1^1H \rightarrow {}_1^2H + {}_1^0e^+ + {}_0^0\nu$

step 2: $_1^2H + {}_1^1H \rightarrow {}_2^3He + {}_0^0\gamma$

step 3: $_2^3He + {}_2^3He \rightarrow {}_2^4He + 2{}_1^1H$

The positron created in step 1 soon collides with an electron and annihilates, releasing energy as high-frequency gamma-ray photons. These photons and others created from the energy of the reactions are absorbed and re-emitted countless numbers of times before they eventually result in the emission of visible, infrared, and ultraviolet photons from the surface of the Sun. This gradual diffusion towards the surface takes many millions of years.

The neutrinos escape far more easily. This is because they only interact with matter through the weak nuclear force and this has only a tiny probability of an interaction in each collision. The flux of solar neutrinos reaching the Earth can be calculated using theoretical models of the fusion reaction in the Sun. This leads to an expected neutrino flux about three times higher than the flux that is actually observed. This **solar neutrino problem** has yet to be satisfactorily resolved.

Mass and energy

The energy released in nuclear fusion reactions comes from the mass defect (Δm) of the reactions themselves. The products of the fusion reactions have slightly less rest mass than the reactants, so the products have an energy $\Delta E = c^2 \Delta m$. For the proton–proton cycle:

$$\text{proton mass} = 1.007\,276\,\text{u}$$
$$\text{helium nucleus mass} = 4.001\,505\,\text{u}$$
$$\text{positron mass} = 0.000\,549\,\text{u}$$
$$\text{mass defect} = (4 \times 1.007\,276\,\text{u}) - (4.001\,505\,\text{u} + 2 \times 0.000\,549\,\text{u})$$
$$= 0.027\,501\,\text{u}$$
$$1\,\text{u} = 1.663 \times 10^{-27}\,\text{kg}$$

$$E = 0.027\,501 \times 1.66 \times 10^{-27}\,\text{kg} \times (3.0 \times 10^8\,\text{m s}^{-1})^2$$
$$= 4.2 \times 10^{-12}\,\text{J} = 26\,\text{MeV/helium nucleus.}$$

The Sun radiates with a power of about 4×10^{26} W; this means it is producing helium nuclei at a rate of about 9.5×10^{37} nuclei per second and losing mass at a rate of about 4×10^9 kg s^{-1}. It has a total mass of about 2×10^{30} kg, most of which is hydrogen. If it continued to shine at the present rate until all its hydrogen had been turned to helium it would last more than 10^{13} years. However, fusion reactions at the core are expected to die away when just over 0.0003 of its remaining mass has been lost. This gives an expected lifetime (from now) of about:

$$\text{lifetime} = \frac{\text{mass to be lost}}{\text{rate of mass lost}}$$

$$= \frac{0.0003 \times 2.0 \times 10^{30}\,\text{kg}}{4 \times 10^9\,\text{kg s}^{-1}} = 1.5 \times 10^{17}\,\text{s} \approx 5 \times 10^9\,\text{years}$$

The Solar System was formed just under 5 billion years ago, so the Sun is almost exactly halfway through its 'life'.

The origin of the elements

The Big Bang was such a violent event that the only atoms that formed and survived were simple ones, mainly hydrogen with about 25% helium by mass (i.e. one helium atom for 12 hydrogen atoms). The process by which nuclei are created by a sequence of fusion reactions is called **nucleosynthesis**. The Sun, as we have seen, is busy converting hydrogen to helium, but where did the other elements come from? The first clue was found in stellar spectra which contained spectral lines characteristic of a wide range of elements; the idea grew that the elements are created by fusion reactions inside heavy stars. The reason they must be heavy stars is because the conditions needed at the core to bring about fusion of heavier nuclei are more extreme than those required in helium synthesis inside the Sun. The details of these fusion reactions are quite complex, but they are capable of building all nuclei up to and including iron-56. Then the process stops. Heavier elements cannot be formed because iron-56 has the largest binding energy of all the elements, so fusion beyond this point needs a net energy input rather than releasing energy.

Where could this energy come from? In 1957 it was realized that heavy nuclei could be formed explosively in a **supernova** (exploding star). The idea is quite simple – when a heavy star reaches the end of its life, fusion reactions at the core, which is now mainly nickel and iron, die away. The star can no longer support itself against gravitational collapse. The rapid release of gravitational potential energy as the outer parts of the star fall inwards provides the energy for endothermic reactions that fuse nucleons in the core to form all the heavy elements. These reactions generate a huge surge of neutrinos whose outward pressure blows the star apart in an explosion that makes it, briefly, brighter than an entire galaxy. It also spreads the elements through space, where they may be drawn together to form new stars and planets (like our solar system, which is a second- or third-generation system).

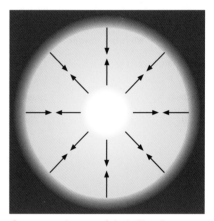

Stars spend most of their 'lives' balancing gravitational forces against the outward pressure generated by fusion reactions in their cores.

PRACTICE

1 Use data in the section on 'mass and energy' to confirm that the Sun is losing mass at about 4×10^9 kg s^{-1}.

2 Why will iron-56 not be formed by fusion reactions in the Sun?

3 Elements whose nuclei are multiples of the helium nucleus (e.g. carbon-12, oxygen-16) are particularly abundant. What does this suggest about the process of nucleosynthesis in stars?

4 There are no stable nuclei with mass numbers 5 or 8. Why does this make it difficult to explain nucleosynthesis as a systematic construction process involving the addition of protons to existing nuclei?

5 Why are such high temperatures needed to make nuclei fuse?

6 Why will the Sun stop fusing hydrogen long before all its hydrogen has been converted to helium?

These two images are of 61 Cygni and were taken 32 years apart. Its motion relative to the star field is obvious.

Memory jogger

Remember Kepler's third law:

T^2/r^3 = constant

Knowing T and r for one planet allows us to calculate r for all the others if we know their orbital periods.

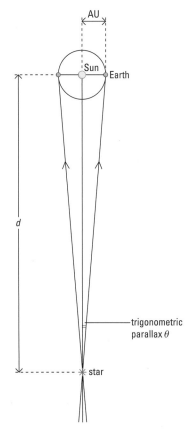

Trigonometric parallax.

HOW FAR TO THE STARS?

When you look at the night sky there are no obvious labels to tell you how far away things are, but there *are* subtle clues, and astronomers have developed a range of techniques based on these clues to estimate the distances to stars and galaxies. Some methods, particularly for objects at great distances, are rather unreliable and depend on controversial assumptions.

The astronomical unit

By timing the return journey of pulses of radio waves bounced off the surface of the Moon and correcting for delays due to atmospheric refraction, its distance has been measured to an accuracy of a few centimetres. The same technique has been used to measure distances to the inner planets. However, the most important distance (to us) is the distance to the Sun. This is called the **astronomical unit** or **AU** and it is measured indirectly by making radar measurements to determine the distance to Venus as it orbits the Sun. The orbits of planets can be derived from Newton's laws, so measuring our distance to any one of them fixes a distance scale for the solar system.

Recent measurements give

$1\,\text{AU} = 1.495\,978\,92 \pm 0.000\,000\,03 \times 10^{11}\,\text{m}$

This local distance scale can be linked to distances measured by other means and forms part of a 'ladder' of distance scales starting on Earth and reaching out to the limits of the observable universe.

Proper motions

Nothing is fixed in space because there is nothing to fix it to. The Earth orbits the Sun, the Sun swings in an arc about the galactic centre, the galaxy is full of stars and teeming with motion and is itself moving under the influence of gravitational forces dragging it toward other clusters of moving galaxies. And yet, look up on a starry night and the stars seem fixed in place: indeed, you can admire constellations named by the Ancient Greeks and Babylonians. Why can't you see the motion? It is because the stars and galaxies are very far away and very far from one another, and because human history is a brief instant in astronomical time.

When someone walks in front of you the line from you to them sweeps through most of 180° as they pass. Now imagine they are far off in the distance and again move the same distance perpendicular to your line of sight. Now their motion subtends a much smaller angle at your eye and, if they are far enough away, it may be difficult to decide if they moved or not. Many ancient astronomers predicted that the stars should have their own '**proper motions**' but they were first measured in 1718 by Edmund Halley (of Halley's comet). A few of the closest stars shift their position by about one thousandth of a degree per second but the majority of proper motions are far smaller than this.

Astronomical parallax

Look past something close to you and focus on a distant object, then move your head from side to side. Both objects will change their position relative to you and the change will be greater for the closer object. In the same way, the Earth's own orbital motion should cause an annual shift in the apparent positions of the stars. The amplitude of this annual angular motion is called the **trigonometric** or **astronomical parallax** (p) and is usually measured in seconds of arc (see the Mathematics toolbox). Its possible existence had even been suggested by Aristotle! However, the shift is tiny, even for relatively close stars, and the first measurement was made by Bessel in 1838, observing the double star 61 Cygni. He chose this because of its unusually large proper motion, a shift of about 0.0015° or 5 seconds of arc per year. He reasoned that a large proper motion probably meant a star was close, and so the trigonometric parallax should be relatively large and easier to detect. In the end it amounted to just ± 0.33 seconds of arc.

The width of the Earth's orbit is fixed, so the trigonometric parallax is a measure of distance and is used to measure distances to nearby stars. The **parsec** (pc) is a unit of distance based on this method of measurement. If a star has a trigonometric parallax of 1 second of arc its distance is 1 parsec. This leads to a very simple formula:

$$\text{distance (pc)} = \frac{1}{\text{trigonometric parallax (seconds of arc)}}$$

The parsec is obviously related to the astronomical unit (since this fixes how far the Earth moves). From trigonometry

$$\tan \theta = \frac{1\,\text{AU}}{d} \quad \text{(with } d \text{ in AU)}$$

$$d = \frac{1\,\text{AU}}{\tan \theta}$$

$$1\,\text{pc} = \frac{1\,\text{AU}}{\tan 1''} = 2.06 \times 10^5\,\text{AU} = 3.09 \times 10^{16}\,\text{m} = 3.26\,\text{light-years}$$

The more distant a star, the smaller its trigonometric parallax. Distances of about 25 pc are the practical limit for ground-based measurements of single stars (θ = 0.04 seconds of arc). At greater distances the uncertainties in individual measurements become comparable to the parallax itself. However, if the stars are all in a cluster, individual parallaxes can be averaged to give a reasonable accuracy for the parallax of the group as a whole whose distance can then be calculated.

Moving clusters

If the distance d to a globular cluster (close-knit group of stars) or galaxy is known then its angular size (radians) when seen from Earth can be used to measure its size x. These angles are usually small, so the small-angle approximation can be used and

$$x = \alpha d$$

If a cluster is moving its velocity can be found from the Doppler shift of stellar spectra. If it is approaching or receding from us its apparent size also changes. If measurements of velocity and rate of change of apparent size are combined they can be used to estimate distance. This is called the **moving-cluster method**.

Although orbiting telescopes will increase distances over which trigonometric parallax methods can be used, the enormous distances to most astronomical objects mean that this method is severely limited (our own galaxy is about 30 kpc across). Geometrical methods can be used up to about 200 pc.

The 10 nearest stars to the Earth.

Star	Distance in AU	Distance in l.y.	Distance in pc
Sun	1	1.6×10^{-5}	4.9×10^{-6}
Alpha Centauri	2.7×10^5	4.3	1.3
Barnard's Star	3.8×10^5	6.0	1.8
Wolf 359	4.8×10^5	7.6	2.3
Lalande 21185	5.1×10^5	8.1	2.5
Sirius	5.4×10^5	8.6	2.6
Luyten	5.6×10^5	8.9	2.7
Ross 154	5.9×10^5	9.4	2.9
Ross 248	6.4×10^5	10.3	3.2
Epsilon Eridani	6.7×10^5	10.7	3.3

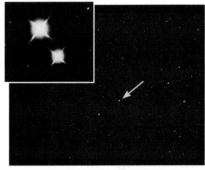

Proxima Centauri, the closest star to our Sun. The inset shows Alpha Centauri, a binary star too far away from Proxima to be seen in the main picture.

PRACTICE

1 The distances to the Sun and Moon are about 1.5×10^{11} m and 3.8×10^8 m respectively. Both subtend an angle of about 0.5° from Earth. Use this information to estimate their radii.

2 Convert 200 pc to:

 a metres

 b light-years

 c astronomical units.

3 Express the mean radii of orbits for planets in our solar system in AU (refer to the table in spread 12.6).

4 Barnard's star has a trigonometric parallax of about 10 seconds of arc. How far is it from Earth?

5 If the apparent size of a globular cluster reduces by 1 part in 10^4 in 1 year and it is known to be moving away from us at 0.01 times the speed of light, estimate its distance.

12.10

OBJECTIVES

- surface temperature and colour
- luminosity and size of stars
- radiation flux

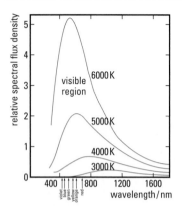

Radiation from black bodies of various temperatures.

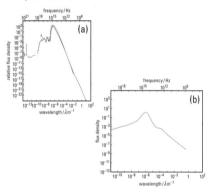

Broad-band spectra of the Sun (a) and a spiral galaxy (b), showing their general similarity to a black-body spectrum.

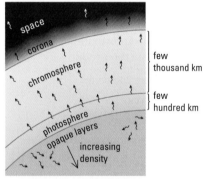

The outer layers of a star like our Sun.

Photosphere and chromosphere

The **photosphere** is a layer just a few hundred kilometres thick near the surface of a star. Outside this is a low-density hot gas called the **chromosphere**. This is transparent and because of its low density it produces relatively little radiation, so it does not significantly distort the black-body spectrum. Beyond this is a layer of even lower density, containing ionized gas. This is the **corona**. The Sun's corona is visible outside the disc of the Moon during a solar eclipse.

STARS AS BLACK BODIES

If you watch the filament of a lamp as the current through it is gradually increased you will see it begin to glow a deep dull red, become brighter and more orange-yellow, and eventually appear white. It has gone from warm to 'red hot' to 'white hot'. As its temperature increases two things happen.

- The intensity of the radiation it emits increases at all wavelengths.
- The spectrum of radiation shifts towards higher frequencies. This is why the colour changes through red to white as more and more of the visible spectrum is included.

An ideal thermal radiator is called a **black body** (see spread 7.16) and emits a characteristic broad-band spectrum of radiation that depends only on its temperature. Surprisingly, stars and filament lamps both behave like ideal black-body radiators. This is very important in astronomy because it means a great deal can be found out about stars simply by measuring their spectra. In particular:

- The colour of a star is related to its surface temperature (through **Wien's law**), red stars being cooler than yellow stars which are cooler than white or blue stars.
- The intensity of radiation leaving the surface of a star is related to its temperature (through the Stefan–Boltzmann law) and so to its colour.

How can stars be 'black'?

A black body absorbs all the radiation that falls upon it, but if it is hot it will also emit all wavelengths. If this was not the case black objects could never be in thermal equilibrium with their surroundings because they would absorb more energy than they emitted at certain wavelengths and so continue to increase in temperature. To understand how a star can be 'black' you have to look at its structure. The main body of the star consists of a hot gas dense enough to be opaque to radiation. This means that any photons entering from outside the star will be absorbed and scattered many times by particles in the gas which makes it an ideal absorber.

Any photons coming out from the core of the star (where the nuclear reactions take place) scatter and are absorbed and re-emitted countless numbers of times before escaping from the **photosphere**, a thin layer near the surface of the star where the gas density drops fairly rapidly and it becomes transparent. A million years might elapse between the creation of a photon in the core of a star and the emission of its energy into space! These multiple scatterings ensure that the photons come to thermal equilibrium with the gas and make their spectrum close to that of a true black body. Without the scattering there would be an abundance of very high-energy, high-frequency photons characteristic of nuclear fusion energies.

Colour and temperature

Wien's law states that for a black-body spectrum the product of peak wavelength λ_p and temperature T (in kelvin) is constant:

$\lambda_p T$ = constant = 2.898×10^{-3} m K

This means we can work out the surface temperature of stars simply by looking at the spectrum of light they emit. Even with the naked eye it is clear that some stars are red-orange in colour and others are yellow or white. Red stars are relatively cool, with surface temperatures around 3000 K, yellow stars are closer to 6000 K, and white stars are around 10 000 K. Some blue stars have surface temperatures in excess of 20 000 K.

Worked example 1

Estimate the surface temperature of a red-orange star whose spectrum peaks at $\lambda \approx 650$ nm (red light).

$$T = \frac{2.898 \times 10^{-3} \, \text{m K}}{6.5 \times 10^{-7} \, \text{m}} \approx 4500 \, \text{K}$$

---

advanced **PHYSICS**

In practice astronomers have to measure the apparent magnitude of a star at more than one wavelength in order to determine the shape of its radiation curve and hence its surface temperature. The human eye responds most strongly to light in the yellow part of the spectrum, as do photoelectric photometers or charge-coupled devices (CCDs), which mimic the human eye, so this 'visual magnitude' measurement is supplemented by measurements in the blue (and possibly ultraviolet) region.

This photograph of star trails shows the variety of stellar colours.

Temperature, luminosity, and size

The power radiated per unit area (**intensity**) of a black body is given by the Stefan–Boltzmann law

$$I = \sigma T^4$$

I is the energy radiated per second per unit area (intensity in $W\,m^{-2}$). T is the temperature of the source (in K). σ is the **Stefan constant** which is equal to $5.7 \times 10^{-8}\,W\,m^{-2}\,K^{-4}$.

The total power radiated by a black body of surface area A at temperature T is then

$$P = A\sigma T^4$$

The total power radiated by a star is called its **luminosity**, L.

> **Terminology**
>
> **Luminosity** (or **flux**) – total power radiated by a star (W).
>
> **Flux density** (or **luminous flux**) – power per unit area of surface perpendicular to the radiation at a distance r from the star ($W\,m^{-2}$).

Worked example 2

The peak wavelength in the spectrum of a white dwarf star is 500 nm. Its luminosity is $5.0 \times 10^{23}\,W$. Calculate: **a** its surface temperature T and **b** its radius r.

a Use Wien's law. $\quad \lambda T = 2.898 \times 10^{-13}\,K \quad$ giving: $\quad T = 5800\,K$

b Use Stefan's law. $\quad P = 4\pi r^2 \sigma T^4 \quad$ so $\quad r = \sqrt{\dfrac{P}{4\pi\sigma T^4}} = 2.5 \times 10^7\,m$

Radiation flux and the inverse-square law

If radiation is emitted uniformly in all directions and is not significantly absorbed or scattered in space, its intensity will obey an inverse-square law. That is, if you double your distance from the source the *intensity* of radiation that reaches you (also called the **flux density**) falls to one quarter of its previous value. It is easy to see why this must be the case. The total power L radiated by a star (its luminosity or **flux**) passes through spherical surfaces of ever greater radii (see diagram). The surface area of one of these spheres is $4\pi r^2$ so the intensity I or flux density at a distance r is

$$I = \frac{L}{4\pi r^2} \quad \text{(clearly proportional to } 1/r^2.)$$

If the luminosity L of a star is known, its distance r can be worked out using this and the intensity I of radiation arriving at the Earth.

> **The solar constant**
>
> The solar constant is the flux density from solar radiation at the edge of the Earth's atmosphere. It is almost constant at $1400\,W\,m^{-2}$ with a variation of only 0.1% over the solar cycle.

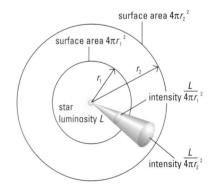

Calculating power radiated.

PRACTICE

1 Alkaid (in the Plough) has a surface temperature of $17\,000\,K$ and a luminosity of $6.1 \times 10^{29}\,W$.

  a Calculate the peak wavelength in its broad-band emission spectrum.

  b What colour would the star appear to be?

  c Calculate its radius.

2 The background radiation is a remnant of the Big Bang and corresponds to thermal radiation from a black body at temperature $2.7\,K$.

  a What is the peak wavelength in this spectrum?

  b What kind of detector would be needed to pick up the background radiation?

3 Estimate the surface temperature of the Sun from its colour.

4 Betelgeuse is 10^5 times more luminous than the Sun. How far would Betelgeuse have to be from Earth (in AU) if the flux density of radiation arriving from it was to equal that from the Sun?

5 On the same diagram sketch graphs to show the broad-band emission spectra from the Sun and Betelgeuse (surface temperatures about $6000\,K$ and $3000\,K$ respectively).

CLASSIFYING STARS BY MAGNITUDE

A powerful light a long way off may seem brighter than a weak lamp nearby. The same is true for stars, so it would be foolish to think that the brightest stars in the night sky are also the closest (although most of them are). It would also be foolish to assume that the brightest stars have the greatest luminosities (as very often they do not). To make accurate comparisons you would need to compare them all at the same standard distance.

Magnitudes

The first person to classify stars according to their apparent brightness was the Greek astronomer Hipparchus, over 2000 years ago. He divided visible stars into six classes, called **magnitudes**, the first (magnitude 1) representing the brightest stars and the last (magnitude 6) the faintest that could be seen. This turns out to be a logarithmic scale because the eye's response to light intensity is logarithmic. This means that going from one magnitude to the next **multiplies** the intensity received by some constant number a (rather than going up in a series of equal steps). In practice, the ratio of received intensity between a magnitude 6 star and a magnitude 1 star is one hundred times. This is a difference of 5 magnitudes, so

$$a^5 = 100 \quad \text{and} \quad a = 100^{1/5} = 2.512$$

So a magnitude 3 star is 2.512 times more intense than a magnitude 4 star and about 6.3 times more intense than one of magnitude 5 ($6.3 \approx 2.512^2$). In general the ratio of intensities of two stars is given by

$$\frac{I_1}{I_2} = a^{m_2 - m_1}$$

Apparent and absolute magnitude

In Hipparchus's time the stars were thought to be in a fixed celestial sphere at a constant distance from the Earth, so it would follow that a star of greater apparent magnitude would also have greater luminosity. We now know that the stars are all at different distances, so this is not the case. To compare the magnitudes of different stars their **apparent** magnitudes need to be adjusted to the values they would have if the stars really were all at the same distance. The distance used is 10 parsecs and the standardized values are called **absolute magnitudes**. They can be calculated using the inverse-square law.

Apparent magnitudes m are converted to absolute magnitudes M by using the equation:

$$M = m - 5\log\left(\frac{d}{10}\right)$$

where d is in parsecs and the log is to base 10.

A large absolute magnitude means the star has a low luminosity; a small value implies a large luminosity. If the absolute and apparent magnitudes for a particular star are both known this equation can be used to calculate its distance:

$$m - M = 5\log\left(\frac{d}{10}\right)$$

Because of its use to calculate distances, the difference between apparent and absolute magnitudes ($m - M$) is called the **distance modulus**.

Maths box: deriving the absolute magnitude

Distances are measured in parsecs (pc). Intensity from a star at distance d is denoted I_d. Using the inverse-square law we get

$$\frac{I_{10}}{I_d} = \left(\frac{d}{10}\right)^2$$

Now, if the apparent magnitude at distance d is m and the absolute magnitude (at 10 pc) is M then the ratio of intensities received at these two distances will be given by

$$\frac{I_{10}}{I_d} = a^{m-M}$$

where $a = 100^{\frac{1}{5}}$

and

$$\log_{10} a = \frac{2}{5} \text{ (useful later)}$$

so

$$a^{m-M} = \left(\frac{d^2}{10}\right)$$

Now take logarithms (base 10) of both sides:

$$\log_{10} a^{m-M} = \log_{10}\left(\frac{d}{10}\right)^2$$

$$(m - M)\log_{10} a = 2\log_{10}\left(\frac{d}{10}\right)$$

$$m - M = 5\log_{10}\left(\frac{d}{10}\right)$$

$$M = m - 5\log_{10}\left(\frac{d}{10}\right)$$

Worked example 1

Deneb, a star in the constellation of Cygnus, has $m = 1.25$ and is 500 pc from Earth. What is its absolute magnitude?

$$M = m - 5\log\left(\frac{d}{10}\right) = 1.25 - 5\log\left(\frac{500}{10}\right) = -7.2$$

This is a very luminous star.

Worked example 2

Calculate the absolute magnitude of the Sun, $m = -26.74$, at a distance of 4.8×10^{-6} pc.

$$M = m - 5\log\left(\frac{d}{10}\right) = -26.74 - 5\log\left(\frac{4.8 \times 10^{-6}}{10}\right) = 4.9$$

At a distance of 10 pc the Sun would be so faint that it could not be seen under full moonlight.

Magnitude and luminosity

In worked examples 1 and 2 the absolute magnitudes of Deneb and the Sun are −7.2 and +4.9 respectively. This means that Deneb has a much greater luminosity than the Sun. The ratio of their luminosities can be calculated from the difference in their absolute magnitudes. Each unit on the magnitude scale changes the received intensity by a factor of 2.512, so if both stars were 10 pc from Earth the intensity of radiation from Deneb would be $2.512^{(4.9 - (-7.2))} = 63\,000$ times greater than that from the Sun. This means that the luminosity of Deneb is also 69 000 times greater than the luminosity of the Sun.

In general the ratio of luminosities for two stars of absolute magnitudes M_1 and M_2 is

$$\frac{L_1}{L_2} = 2.512^{M_2 - M_1}$$

Notice that it is $M_2 - M_1$ rather than the other way round because *larger* absolute magnitudes correspond to *lower* stellar luminosities.

The 10 brightest stars as seen from Earth (in order of apparent magnitude).

Star	Apparent magnitude	Distance (pc)	Absolute magnitude
Sirius	−1.46	2.65	1.42
Canopus	−0.72	70	−5
Alpha Centauri	−0.01	1.33	4.37
Arcturus	−0.04	10.3	−0.10
Vega	0.03	7.5	0.65
Capella	0.08	12.5	−0.40
Rigel	0.12	265	−7
Procyon	0.38	3.4	2.71
Achernar	0.46	27	−1.7
Betelgeuse	0.50 (variable)	320	−7

Sirius, the brightest star in the sky, seen here with its white dwarf companion (the small white dot immediately to the right), some 10 magnitudes fainter.

PRACTICE

1 Two stars have apparent magnitudes 3 and 6.

   a Which appears to be brighter?

   b Explain whether it necessarily true that this is also the closest?

   c What is the ratio of luminosities of the two stars if they are at the same distance from Earth?

2 Use the table above and the data given in worked example 2 to arrange the list of stars below in order of:

   a increasing brightness in the night sky;

   b increasing luminosity;

   c increasing distance modulus.

Sun, Betelgeuse, Rigel, Vega, Procyon, Canopus.

3 Use the table above to calculate the ratio of luminosities of Vega and Procyon.

4 The luminosity of the Sun is about 4×10^{26} W and its apparent magnitude is −26.74 at a distance of 4.8×10^{-6} pc. Use this data and the table above to work out a value for the luminosity of Rigel.

5 **a** What is the distance modulus of Canopus?

   b Bellatrix has a distance modulus of 2.72. How far from the Earth is it?

   c Pollux has apparent magnitude 1.14 and absolute magnitude 1.00. How far is it from the Earth?

The origin of absorption lines

How can a star approximate to a black body and still have sharp absorption lines? The essential characteristic for a black-body spectrum is thermal equilibrium, achieved by continual absorption and re-radiation of photons, so you might think that these photons would 'fill in' the gaps due to absorption. The reason this does not happen is that photons with the right energy to excite a quantum jump are much more likely to be absorbed than those with other nearby frequencies, so they can only be radiated from the photosphere if they originate in the cooler regions *very near its surface* (they have a short **'mean free path'**). Photons at other frequencies will have less chance of absorption and travel, on average, farther between absorptions. These photons can therefore escape from deeper inside the photosphere, where the temperature is higher and the radiation more intense (Stefan's law). The net effect is a sharp reduction in emitted intensity very close to the frequency of the absorption line.

An objective prism spectrogram of stars in the Hyades. A low-angle prism is placed over the objective lens of a telescope, spreading stellar images to form stellar spectra. Notice the differences in intensities and in the distribution of light radiated at different wavelengths. Photographs like this were used to make quick visual classifications of thousands of different stars.

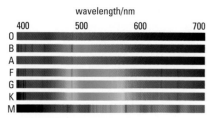

The sequence of spectra emitted from O- to M-type stars.

STELLAR SPECTRA

The broad-band spectrum of visible light emitted by the Sun is crossed by thousands of dark absorption lines, each of which can be related to quantum jumps in particular kinds of atoms.

Magnitudes and colours are very useful for determining surface temperatures and luminosities of stars but the detailed spectra of the light they emit reveal a great deal more about their structure and history. The familiar continuous spectrum of the Sun is crossed by a large number of dark **absorption lines** and by 1815 Joseph von Fraunhofer had measured and classified more than 300 of them – thousands have now been catalogued. Absorption lines are produced when light from the hot surface passes through the cooler gases in the outer part of the Sun's photosphere and excites **quantum jumps** in atoms, ions, and molecules. The photons absorbed by atoms and ions correspond to quantum jumps between electron energy levels, while those absorbed by molecules make the molecules vibrate or rotate.

These quantum jumps have distinct energies characteristic of the particular elements and transitions involved, so particular frequencies of photons are absorbed (since $E = hf$ for a photon) removing them from the radiated light and leaving dark bands in the spectrum when viewed from Earth. The positions of these absorption bands in the broad-band spectra of astronomical objects make it possible to identify elements and compounds present in space. The element helium (from 'helios', meaning 'Sun') was actually identified by the pattern of its absorption lines in the solar spectrum before the element was discovered on Earth.

Spectra and temperature

For an absorption line to occur certain conditions must be satisfied.

• There must be a plentiful supply of atoms, ions, or molecules in reasonably low-energy states with available excited states.

• There must be a large number of photons with enough energy to excite the atoms, molecules, or ions to their vacant excited states.

Exactly what is available depends very much on temperature. At low temperatures the first criterion is satisfied for most atoms, ions, and molecules present, but there will only be low-energy photons available. These will only be able to excite vibrational states in molecules as this requires less energy than ionization or atomic excitation. At higher temperatures collisions between the molecules will excite them and break them up so the molecular lines disappear. However, there are more energetic photons around too, and these will excite neutral atoms. At very high temperatures collisions between atoms result in many more of them being in excited states, or ionized, and so ionic absorption lines and atomic hydrogen lines dominate. At extremely high temperatures even the hydrogen is ionized and the dominant lines are due to absorption in ionized helium.

Classification by spectra

When stellar absorption spectra are compared with the spectrum of the Sun, it is clear that there is a very wide range of elements and compounds present in stars. Spectra from hot white stars like Vega and Sirius are dominated by hydrogen lines, blue stars also show prominent helium lines, whereas cooler, orange-red stars have less hydrogen but strong absorption bands from titanium dioxide and carbon compounds. To simplify this complexity stars are put into alphabetical groups according to the strength of their hydrogen absorption lines. The seven main spectral classes are:

O, B, A, F, G, K, M

The characteristic of each spectral class is given in the following table.

The seven main spectral classes.

Spectral class	Intrinsic colour	Temperature (K)	Prominent absorption lines
O	blue	25 000–50 000	He$^+$, He, H
B	blue	11 000–25 000	He, H
A	blue-white	7500–11 000	H (strongest), ionized metals
F	white	6000–7500	ionized metals
G	yellow-white	5000–6000	ionized and neutral metals
K	orange	3500–5000	neutral metals
M	red	< 3500	neutral atoms, TiO

Some important spectral lines

Molecular absorption bands Cooler stars (below 4900 K) have molecular absorption bands and are orange or red in colour.

Metal lines F and G class stars (4900–7400 K) like our Sun have strong absorption lines corresponding to the excitation of electrons in metals and are yellow or yellow-white in colour.

Balmer lines Any jump from the ground state ($n = 1$) requires so much energy that the photons absorbed are all in the ultraviolet region of the spectrum. However, A (and B) stars (7400–28 000 K) have many hydrogen atoms already in the $n = 2$ state and the smaller energy jumps made when these atoms are excited absorb visible photons and result in visible absorption lines. These are called the **Balmer lines**, after the Swiss schoolteacher who first noticed that their wavelengths form a mathematical series. The hydrogen lines are labelled H_α, H_β, H_γ, etc. in order of increasing frequency and decreasing wavelength.

Neutral helium lines (He I) At higher temperatures collisions between hydrogen atoms become so violent that most of the atoms are ionized. This removes the electrons and so the hydrogen lines weaken and disappear. There is then enough energy to excite helium atoms, producing the He I lines in B stars.

Ionized helium lines (He II) O class stars are very hot (28 000–50 000 K) and collisions between helium atoms result in a great deal of ionization. Helium ions are like hydrogen atoms with twice the charge on the nucleus. Their characteristic spectrum produces the He II absorption lines.

Thermal broadening

The atoms that emit or absorb radiation in the atmosphere of a star are in rapid thermal motion. Their thermal kinetic energy depends on the temperature according to the equation

$$\frac{3}{2}kT = \frac{1}{2}m\langle v^2 \rangle$$

where $\langle v^2 \rangle$ represents the average squared velocity.

This means that atoms in the line of sight may be moving toward or away with a range of speeds dependent upon the temperature, and the spectral line will be broadened either side of the natural wavelength (i.e. instead of a sharp line a range of absorption or emission frequencies will be detected). The width of the line depends on the range of velocities and the most representative value for this range is $v_{\text{r.m.s.}}$. So the line width is given by

$$\frac{\delta\lambda}{\lambda} \approx \pm \frac{v_{\text{r.m.s.}}}{c} = \pm \frac{\sqrt{\langle v^2 \rangle}}{c} = \pm \frac{\sqrt{3kT/m}}{c}$$

so

$$\delta\lambda \propto \sqrt{T}$$

This means that stellar temperatures can be estimated from the width of the spectral lines.

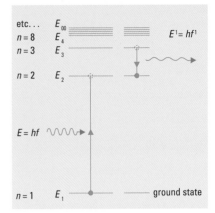

Absorption and emission by an isolated atom. The left-hand side of the sketch shows how a photon of energy $E = E_2 - E_1$ can excite an electron to make a quantum jump from the ground state to the first excited state. On the right it shows how an excited atom in the $n = 3$ state can decay to the $n = 2$ state by emitting a photon of energy $E' = E_3 - E_2$.

Some important spectral lines.

Wavelength λ/nm	Absorber (symbol)
656.3	hydrogen (H)
589.3	neutral sodium (Na I)
587.6	neutral helium (He I)
527.0	neutral iron (Fe I)
516.7 517.3 518.4	neutral magnesium (Mg I)
495.5	titanium oxide (TiO)
486.1	hydrogen (H)
468.6	ionized helium (He II)
438.4	neutral iron (Fe I)
430.0	CH molecule
434.0	hydrogen (H)
422.7	neutral calcium (Ca I)
410.1	hydrogen (H)
396.8	ionized calcium (Ca II)
393.4	ionized calcium (Ca II)

Star stuff

Analysis of stellar spectra shows that the roughly 75% of the visible universe by mass is hydrogen and 24% helium, with small quantities of other elements.

THE HERTZSPRUNG–RUSSELL DIAGRAM

The hotter something is the more intensely it radiates. According to the Stefan–Boltzmann law its luminosity (total power radiated) is proportional to its surface area and the fourth power of its absolute temperature. This means that stars might be very luminous because they are very hot, or because they are very large, or a combination of both. So it is surprising to discover that, for a representative sample of stars, there is a clear relationship between spectral class (which depends on temperature) and absolute magnitude (proportional to luminosity). This was discovered independently by the Danish astronomer Ejnar Hertzsprung and the American astronomer Henry Norris Russell, both of whom plotted absolute magnitude against spectral class. The main features of this plot are:

• Most stars fall inside a narrow band called the **main sequence** within which luminosity rises with surface temperature.

• There is a smaller but significant number of stars that lie in another band that leaves the main sequence and shows an increasing luminosity with *reducing* temperature. The only way for a cooler star to be more luminous than a hot one is if it is bigger, so this branch is called the '**giant branch**'.

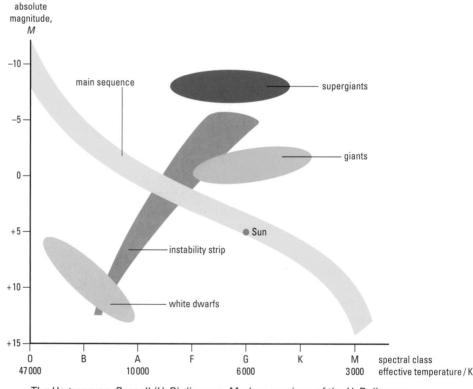

The Hertzsprung–Russell (H–R) diagram. Modern versions of the H–R diagram contain many branches corresponding to distinct types of stars, many of which will be discussed in the following pages. The fact that the majority of stars lie close to the main sequence implies that this represents some kind of stability and it turns out that most stars spend most of their lives here, but exactly where on the sequence depends on their mass. At the end of their lives, when they stop fusing hydrogen, they evolve away from the main sequence: most of the other stars on the H–R diagram are old stars in various stages of 'decline'.

Mass and luminosity

The mass of each star in a binary system can be worked out from the rotation period and radius. If this information is combined with a measurement of the spectral class and luminosity an approximate relationship between mass, M, and luminosity, L, for main-sequence stars can be derived:

$$L \propto M^{3.5}$$

Why should more massive stars be more luminous?

The gravitational pressure inside a star increases with increasing mass, so, when it is in equilibrium, fusion reactions in the core of a star must generate a greater pressure than in a less massive star. This means its reaction rate must be greater and it must burn at a higher temperature and generate more power – giving it greater luminosity.

The amount of fuel available to a star depends on its mass, but the rate at which it consumes this fuel depends on its luminosity, and the luminosity increases as $M^{3.5}$. This means that more massive stars have shorter lifetimes. The Sun will remain on the main sequence for about 10 billion years whereas Vega, a dwarf star of about 2.5 times the solar mass, will remain on the main sequence for about 1/10 as long. Giant stars have much shorter lives.

Stars range in mass from about 8% of the Sun's mass (this is the minimum required to 'ignite' the star), to about 120 times its mass. Extremely massive stars are very rare.

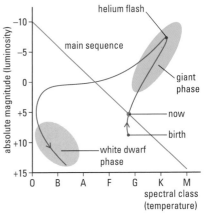

H–R diagram showing the approximate trajectory of our Sun during its lifetime.

- **Red giants** Above and to the right of the main sequence are some very luminous stars that are relatively cool. This is surprising. A cooler star radiates far less power per unit area (Stefan–Boltzmann law) so the fact that these stars are *more* luminous than main-sequence stars of similar temperature suggests that they are *much* larger. They are giant stars – the **red giants**.

- **White dwarfs** Below and to the left of the main sequence are hot stars that are not very luminous. As with red giants, by the same reasoning, these stars must be much smaller than main-sequence stars of similar temperature. They are called **white dwarfs**.

The instability strip

The luminosity of some stars varies periodically. Mira, in the constellation Cetus, varies in brightness by 7 magnitudes (a change in visual luminosity of about 600 times) with a period of about 330 days. This variation is clearly visible to the naked eye. Mira is a **Cepheid variable**, one of the most important types of star in astronomy (see spread 12.17). Variable stars congregate in a great band of instability that sweeps up through the middle of the H–R diagram. These stars pulsate, expanding and contracting repeatedly as they struggle to strike a balance between the pressure of gravitational collapse and the pressure of radiation from internal fusion reactions. Their luminosity changes with the change in surface area of the photosphere.

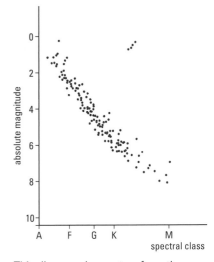

This diagram shows stars from the Hyades plotted on the H–R diagram. The main sequence is obvious.

PRACTICE

1 Explain why:

   a A cool star must be big to have a large luminosity.

   b A very massive star is likely to have a large luminosity.

   c Stars with large absolute magnitudes have relatively short lives.

   d There is a lower limit to stellar mass.

   e Variable stars become more luminous as they expand.

2 Sirius B is a white dwarf star of absolute magnitude 11.4, smaller than the Earth, but its mass is about equal to that of the Sun.

   a Does this star obey the mass–luminosity relation for main sequence stars? (Compare it with the Sun, absolute magnitude 4.8.)

   b How is it possible for such a massive star to be so dim?

3 Sirius A is 2.3 times the mass of the Sun and about 20 times as bright.

   a Is this consistent with the mass–luminosity law?

   b What, roughly, is the absolute magnitude of Sirius A?

   c Suggest a value for its effective temperature and say what absorption lines are likely to be prominent in its spectrum.

   d Will Sirius spend more or less time on the main sequence than our Sun? Give your reasons.

- stages in stellar evolution
- relation to the H–R diagram
- supernovae

Stars form in giant clouds of gas and dust where regions of above-average density collapse under their internal gravitational forces.

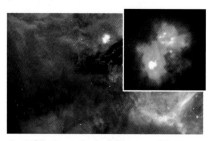

An HST view of a turbulent cauldron of star birth, called N159, some 150 light-years across, taking place 170 000 light-years away in our satellite galaxy, the Large Magellanic Cloud (LMC).

A rare type of butterfly-shaped ionized nebula, known as a 'papillon' nebula, less than 2 light-years across at upper centre is shown in more detail in the inset.

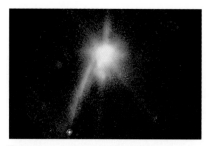

There is growing evidence that many other stars have planetary systems. The tilted disc of Beta Pictoris suggests the existence of a large planet.

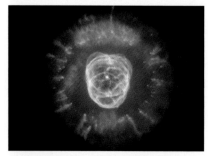

A planetary nebula, NGC2392, the Eskimo Nebula in Gemini. The central star is still heating up.

THE LIFE AND DEATH OF A STAR

There are several major phases in the life and death of a star.

The nursery Stars are born inside enormous interstellar gas clouds as they collapse under the influence of their own gravitational attraction. These clouds are almost entirely hydrogen and helium left over from the Big Bang, plus 1–2% by mass of heavier elements created in previous generations of stars. The trigger for starbirth seems to be shock waves set up inside the gas by galactic motion or even by intergalactic collisions. This means that it is more likely that large numbers of stars are born together in a great '**stellar nursery**', rather than one or two here and there in isolated regions.

Gravitational collapse As the gas collapses it spins at an ever increasing rate. This can be the end of many a promising young star as its rapid rotation tears it apart before it has a chance to ignite the nuclear fusion reactions in its core that signal its existence to the universe. However, there are ways the spinning disc of gas can lose some of this momentum to the surrounding gas and slow down enough to become a star. If this does happen, then loss of gravitational potential energy as the **protostar** shrinks generates more and more thermal energy that radiates out from the core.

Ignition The core temperature continues to rise and at about 106 kelvin deuterium begins to fuse to helium. The energy generated helps to 'blow away' some of the outer layers of gas. But the deuterium soon runs out and collapse continues until the core is hot enough to ignite hydrogen fusion reactions. It now has access to an enormous energy reserve and undergoes a brief convulsive struggle to reach a stable state in which the forces of gravity are balanced by the pressure of radiation from its core. It is now a **T Tauri-type star**, an irregular varying star surrounded by gas and dust. As it stabilizes, a few tens of millions of years after the initial collapse, the star joins the H–R diagram on the main sequence, and it will remain here fusing hydrogen to helium for billions of years.

Planetary discs As the embryonic star collapses, rotation forces it to flatten into a disc, and in this disc some of the elements heavier than hydrogen and helium will join together to form tiny dust grains. Over time these collide and stick together to form meteoroidal rock and ice. Driven by gravitational forces, their growth may continue forming them into larger bodies like planets. For several hundred million years after reaching this state, a struggle goes on between the star and the debris from its formation. It strips the inner planets of their atmospheres and gathers in the smaller **planetisimals** and **meteoroids**, causing intense bombardment of the planets. Scars from this era are clearly visible on the Moon (especially the dark side) and planets such as Mercury and Mars. Some of the planetisimals are actually captured by planets and become their moons.

Many ways to die After a star has consumed its supply of hydrogen its core is mainly helium. Gravitational forces are no longer balanced so it begins to collapse once again. This ignites hydrogen fusion in a shell surrounding the core, and energy released by gravitational collapse and fusion in the shell causes the star to swell up. It now leaves the main sequence and becomes a red giant. But its days are numbered – it has little hydrogen left and can only continue to radiate if it fuses heavier and heavier elements all the way up to iron (after that fusion requires a net energy input). The process begins to speed up, but exactly what happens and how quickly depends on the mass of the star. The fates of stars of mass M less or greater than 1.4 times the solar mass, M_s, are described below. The critical mass $1.4M_s$ is called the **Chandrasekhar limit**.

- **$M < 1.4M_s$** The core continues to collapse until its density and temperature are sufficient to smash helium nuclei together so hard that they fuse. This condition is reached suddenly at a temperature of about 10^8 K and the explosive reaction, called the **helium flash**, blasts away the outer envelope, leaving just the central bright core. This continues to shrink to become a white dwarf star surrounded by some ejected gases called a **planetary nebula**. This is thought to be the fate that awaits the Sun in about 5 billion years' time. White dwarfs then continue to shine with gradually diminishing luminosity as they cool over several billion years. They are prevented from further collapse by **electron degeneracy pressure**, i.e. the fact that the **Pauli exclusion principle** forbids two electrons from being in exactly the same quantum state.

- **$M > 1.4M_s$ – supernovae** More massive stars remain on the main sequence for a shorter time, burning their hydrogen at a much more rapid rate. These too become red giants, but the core continues to contract, becoming hotter and denser until it can fuse helium to carbon and oxygen and become even more luminous. At a core temperature above about 3×10^9 K and a density in excess of a tonne per cubic centimetre, carbon and oxygen undergo fusion reactions to create progressively heavier elements like neon, magnesium, silicon, and so on up to nickel and iron. The structure of the star is now layered like an onion with more extreme conditions and reactions taking place closer to the centre. However, these reactions generate relatively little extra energy, and beyond iron, fusion requires a net input of energy, so the star is reaching a critical condition. At the same time enormous numbers of neutrinos are being created in the nuclear reactions in the collapsing core, and the neutrino flux eventually carries away more energy than electromagnetic radiation. There is no way back: as the core runs out of energy, gravitational collapse accelerates and even *supplies* the extra energy to create elements heavier than iron. The intense neutrino flux blasts the mantle of the star off into space and the collapsing core crushes the remaining nuclei to protons and neutrons. This incredible explosion is called a **supernova** and for a short time the exploding star is more luminous than all the stars in a galaxy. The signature of such an event is a sudden increase in neutrino flux followed by the visible flaring of a distant star. The material blasted off into space seeds the galaxy with heavier elements which can be incorporated into future stars and planets as they too 'collapse out' of the interstellar gas and debris of old supernovae.

Supernovae

There are two main types of supernova.

- **Type I** These occur in binary systems when one star goes into its red giant phase and the other is a white dwarf. As the giant swells up gas is drawn from its surface onto the white dwarf, increasing its mass above the Chandrasekhar limit. The carbon core of the white dwarf is now unable to support itself against the increased gravitational forces and it collapses. This re-ignites the core and the star explodes. The net effect is to burn about one solar mass of carbon and oxygen to produce nickel-56. This is a radioactive isotope which decays to cobalt-56 and eventually iron-56. The brightness of Type I supernovae decays exponentially as a result of this, and the characteristic 60-day half-life makes them easy to identify.

- **Type II** These are the type described earlier. Their luminosity increases by a factor of about a billion, increasing their brightness by about 23 magnitudes. This increases the distance at which the star can be seen by about 30 000 times. If Type II supernovae are used as **'standard candles'** they can be used to estimate distances out to about 1000 Mpc.

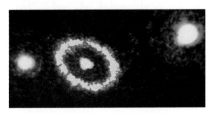

Supernova 1987A was an amazing stroke of luck for astronomers. It appeared suddenly on 23 February 1987 at all wavelengths from gamma to radio. It was heralded, 20 hours earlier, by a burst of neutrinos – several million million of which passed through your body.

The Crab Nebula, M1, is the most famous of all supernova remains. It is the brightest supernova on record and was recorded by the Chinese as a 'guest star' in 1054. The filaments are moving outwards at about 0.2 seconds of arc per year.

Starbirth

A comparison of numbers of young stars in local galaxies with those seen forming in the young galaxies photographed by the Hubble Space Telescope show that the rate of star formation is falling. As the universe continues to expand starbirth will become increasingly rare.

PRACTICE

1 Suggest a reason why the rate of star formation now is much lower than it was a few billions of years after the Big Bang. Does this mean the universe is dying?

2 A song by Joni Mitchell has the line, 'We are stardust, billion-year-old carbon …'. How good was Joni Mitchell's astrophysics?

3 It was stated that the luminosity of a Type II supernova changes by about a billion times. Show that this corresponds roughly to an increase of 23 magnitudes and would increase the distance over which the star was visible by a factor of about 30 000.

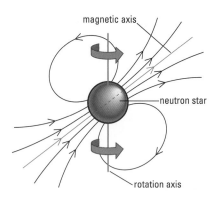

A neutron star has an intense magnetic field.

COLLAPSING STARS

Neutron stars

What happens to the collapsing core of a supernova? If it is more than about 1.4 times the solar mass the gravitational forces in its core are so strong that electrons are forced into protons to make neutrons and the star becomes a **neutron star**. This is like a massive neutral nucleus composed entirely of neutrons, a few tens of kilometres across, and held together by gravity. The density of nuclear matter is around $10^{17}\,\mathrm{kg\,m^{-3}}$ – a neutron star with the same mass as the Earth would have a radius of just a few hundred metres. This is possible because ordinary matter made of neutral atoms is virtually all empty space, the nucleus itself occupies less than one part in 10^{15} of the atomic volume. Therefore forcing nucleons (i.e. neutrons) close together creates amazingly dense matter. One teaspoonful of matter from a neutron star would have a mass of about 100 million tonnes!

Neutron stars are prevented from collapsing further by the **Pauli exclusion principle** which forbids the neutrons from crowding together in their lowest energy state. This opposition to collapse is called **neutron degeneracy pressure**.

Pulsars

At Cambridge, in 1967, Jocelyn Bell was monitoring rapid changes in the signal strengths of certain point-like radio sources when she noticed something amazing. One of the sources was pulsing at a very precise rate, once every $1.337\,011\,\mathrm{s}$. At first she thought this might be a message from an alien civilization, some kind of beacon radiating in space to attract attention. After all, it was unlikely to be a star because it was difficult to imagine how something as large as a star could vary so quickly. The only kind of system that could create such a strong stable and regular signal would be a massive rapidly rotating body – something like a neutron star.

- As the core of a supernova collapses it increases its rate of rotation. This is because of conservation of angular momentum and is just like a pirouetting ice skater who draws in her arms and spins even faster.

- The magnetic field of a star is linked to its rotation, so this too is intensified by the collapse and increased angular velocity. The polar field strength of a neutron star is about 10^{12} times stronger than that of the Earth and, like the Earth's poles, the north and south poles of the neutron star often do not line up with the rotation axis.

- The intense rotating field induces an intense electric field that accelerates charged particles toward the poles of the neutron star and makes them emit a concentrated beam of radiation. This is like an intense directional beacon that rotates with the star. As it crosses our 'line of sight' we receive a pulse of radiation – the signal Jocelyn Bell detected.

- In some cases the beam can be seen from both poles, but this depends on the inclination of the magnetic axis to the rotation axis.

These rapidly rotating, pulsing neutron stars are called **pulsars** and have time periods whose regularity is comparable to that of an atomic clock. Extremely rapid, millisecond pulsars have been discovered. These are usually members of a binary system and are whipped into even more rapid rotation by matter that falls onto them from their companion.

Black holes

No known physical law prevents supernova remnants above about 2.5 solar masses collapsing towards a point of infinite density (called a **singularity**) and the star ends its life as a **black hole**. The name itself defines what we mean by a black hole. It is 'black' because even light cannot escape its

---

The magnetic field of a star

You may be surprised to learn that a star consisting entirely of neutrons has an intense magnetic field. However, a neutron star has a much more complex structure than has been suggested so far and includes layers of superfluid protons and electrons which can maintain a very strong field. The magnetic field of a collapsing star is frozen in place by a process called **magnetic flux freezing** and increases as the inverse-square of the radius of the collapsed star.

Explaining singularities

The general theory of relativity predicts that singularities must exist, but it cannot explain what they are, as the theory breaks down at a singularity. It is hoped that quantum gravity theory may avoid this problem.

intense gravitational field. According to relativity theory nothing can travel faster than light, so if light is unable to escape nothing else can either, and anything that falls into a black hole is lost forever. (Stephen Hawking has predicted a quantum mechanical process that allows black holes to radiate, but so far no one has detected this '**Hawking Radiation**'). A proper mathematical treatment of the black hole requires the use of Einstein's **general theory of relativity**, which is complex; fortunately, many of the same results can be derived by using Newton's theory.

For something to become a black hole it must trap light. In Newton's theory this means the escape velocity from the surface of the object (see spread 5.5) is at least as great as the speed of light.

The escape velocity from a sphere of mass M and radius R is

$$v_{esc} = \sqrt{\frac{2GM}{R}}$$

For a black hole to form, $v_{esc} \geq c$, or

$$\sqrt{\frac{2GM}{R}} \geq c$$

The minimum radius inside which the mass of the star must be compressed to form a black hole is called the **Schwarzschild radius**, R_s. If the equation above is rearranged, we find that

$$R \leq \frac{2GM}{c^2}$$

$$R_s = \frac{2GM}{c^2}$$

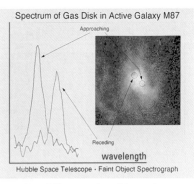

Spectrum of Gas Disk in Active Galaxy M87

Approaching

Receding

wavelength

Hubble Space Telescope · Faint Object Spectrograph

Doppler shifts reveal rapid rotation at the centre of galaxy M87. This can be explained if there is an extremely massive but tiny object at the centre of the galaxy – a black hole.

Worked example

What is the Schwarzschild radius for the Earth?

$$R_s = \frac{2 \times 6.7 \times 10^{-11}\,\mathrm{N\,m^2\,kg^{-2}} \times 6.0 \times 10^{24}\,\mathrm{kg}}{(3.0 \times 10^8\,\mathrm{m\,s^{-1}})^2} = 8.9 \times 10^{-3}\,\mathrm{m}$$

All the mass of the Earth would have to be compressed into a sphere smaller than a golf ball!

Once a star has collapsed inside its Schwarzschild radius there is no hope for it. The core is now invisible to the outside universe and continues collapsing to become a singularity. From the outside the spherical shell of radius R_s is the final place from which messages from the collapsing star can escape to the outside. For this reason the surface is called an **event horizon**. No information about events inside the event horizon gets out.

Time and black holes

According to the general theory of relativity time runs slow in a strong gravitational field. If you watch a clock falling into a black hole it will appear to 'tick' slower and slower as it approaches the event horizon. It would also get fainter and fainter as the photons struggle against gravity to escape and become more and more redshifted (this reduces photon energy and shifts more of the radiation out of the visible range). Seconds on the falling clock would last minutes and then days and years of your time. In fact it would take an *infinite* amount of your time for the clock, or anything else, to just reach the event horizon. You could never see anything (including the surface of the collapsing star) actually fall through the event horizon, so the inside of a black hole is in one sense beyond the end of time for the outside universe!

PRACTICE

1 Use the law of electromagnetic induction to explain how the rotating magnetic field of a neutron star can result in the acceleration of charged particles in an intense electric field.

2 The Pauli exclusion principle prevents the collapse of atoms and neutron stars.

  a Explain how it applies in each case and say what would happen if it did not apply.

  b How tall would you be if you were scaled down to have the density of matter in a neutron star? (Take this density as $10^{-17}\,\mathrm{kg\,m^{-3}}$.)

3 a Calculate the Schwarzschild radius for the Sun ($M_S = 6.0 \times 10^{30}\,\mathrm{kg}$).

  b If the Sun were suddenly replaced by a black hole of the same mass, would this affect the Earth's orbit?

4 a Write down an expression for the mass of a spherical object in terms of its radius and density.

  b Substitute this into the equation for escape velocity, and rearrange it to find an expression for the density at which a spherical object becomes a black hole (you will need to set $v = c$).

  c Comment on how this density depends on the size of the object concerned. Is it true that things have to be extremely dense to become black holes?

- distances to stars and star clusters

SPECTROSCOPIC PARALLAX

Trigonometric parallax is limited by our ability to detect small changes in the position of distant stars as the Earth moves in its orbit. The H–R diagram provides a powerful alternative method for measuring distance, known as **spectroscopic parallax**, that can be extended to more distant stars. The crucial point is that it allows us to predict absolute magnitude from spectral class. The difference between absolute and apparent magnitude ($m - M$) then gives the **distance modulus**, from which distance can be calculated. The technique is summarized below.

- Identify the spectral class of the star and measure its apparent magnitude m.
- Use the H–R diagram to predict its absolute magnitude M.
- Substitute the distance modulus, $m - M$, in the distance equation to give:

$$m - M = 5 \log \frac{d}{10}$$

$$10^{\frac{m-M}{5}} = \frac{d}{10}$$

$$d = 10^{\frac{m-M+5}{5}}$$

Misnomer

'Spectroscopic parallax' is a rather inappropriate name for this technique since no parallax measurements are actually involved.

Worked example

Sirius A is an A-type star with apparent magnitude –1.5. Estimate its distance.

From the H–R diagram we can read off an absolute magnitude of about 1.5 for a main-sequence A star.

$$d = 10^{\frac{-1.5 - 1.5 + 5}{5}} \approx 10^{0.4} = 2.5 \, \text{pc}$$

More careful measurements give a value of 2.65 pc.

Care has to be taken when using this method because dust in the line of sight will absorb some of the star's radiation and make it seem dimmer and more distant than it actually is. Another problem is that dust scatters blue light more than red light, making the star appear redder (and so cooler) than it actually is. This means you are likely to classify it incorrectly on the H–R diagram and underestimate its distance.

For accurate work the spectral classes have to be subdivided. This is usually done on the basis of the widths of certain spectral lines. This broadening of spectral lines is another Doppler effect caused by the thermal motion of the radiating atoms (see spread 12.12). The higher the temperature of the star the greater the mean velocities and the broader the spectral line. The line spreads above and below its mean frequency since atoms are moving in all directions randomly relative to us.

Using clusters

Star clusters come in two distinct types, and both types are important in defining astronomical distances.

- **Open clusters** These are found in the disc of the galaxy and contain hundreds of young stars; they are often surrounded by the gas clouds from which they were born.
- **Globular clusters** These are spherical in shape and contain hundreds of thousands of old stars.

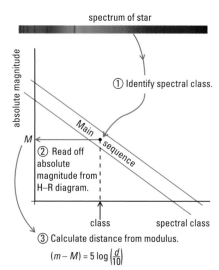

① Identify spectral class.

② Read off absolute magnitude from H–R diagram.

③ Calculate distance from modulus.

$$(m - M) = 5 \log \left(\frac{d}{10} \right)$$

How to use a star's spectral class and apparent magnitude to determine its distance.

Some clusters are close enough to Earth for their distance to be measured by geometric methods. If it is assumed that a particular cluster is 'typical' it can be used as a '**standard candle**' to estimate distance. By comparing the brightness of one cluster against another similar one you can use the inverse-square law to determine their relative distances. For example, if cluster A appears 4 times dimmer than cluster B it must be twice as far away. The nearest open cluster is the Hyades and it contains about 200 stars at a distance of about 140 light-years.

The distance scale can be extended by a method called **main-sequence fitting**. All stars in a particular cluster are about the same age and were born from the same gas cloud, so they have the same composition. If their *apparent brightnesses* are plotted against their surface temperatures onto an H–R diagram they will run parallel to the main sequence some distance below it. The vertical distance between the two plots is the distance modulus $m - M$ and can be used in the magnitude equation to determine the distance to the cluster.

For more distant clusters the apparent magnitude is very low and often only giant stars are visible. To determine the distance these can be compared directly with the giant branch on the H–R diagram. This method is calibrated by comparing the distant clusters against the H–R diagram for stars in the Hyades. The distance to the Hyades is measured using geometric methods (moving-cluster and parallax) and so the accuracy of spectral parallax methods depends on the accuracy of this geometric method.

Spectroscopic methods are useful for distances up to about 50 kpc.

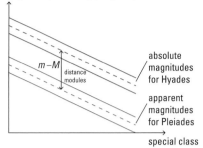

Main-sequence fitting. The top plot shows an H–R diagram for the Hyades and the lower one shows one for the Pleiades. The absolute magnitudes of the Hyades stars are compared with the apparent magnitudes for those in the Pleiades. If we assume the two clusters are similar the distance between them on the plot below gives the distance modulus from which we can calculate distance.

Summary: using clusters to do astronomical distances

- Measure m and spectral classes of stars in cluster.
- Plot m versus T (or spectral class) on H–R diagram.
- Find average distance modulus ($m - M$) from main sequence.
- Use the distance modulus equation to find distance.

PRACTICE

1 Use the distance modulus for the stars listed below to calculate their distances. Give your answers in both parsecs and light-years.

Sirius $M = 1.42$ $m = -1.46$
Deneb $M = -7.2$ $m = 1.25$
Altair $M = 2.30$ $m = 0.77$
Arcturus $M = -0.10$ $m = -0.04$

2 What will be the effect on a distance measurement of:

   a not allowing for the presence of dust clouds in the line of sight to a distant star;

   b not allowing for reddening due to scattering of shorter wavelengths by dust particles?

3 What is the apparent magnitude of a star of absolute magnitude 1 at a distance of 100 pc?

4 Explain what is meant by a 'standard candle'. Suggest different structures that can be used as standard candles for distance measurements.

5 A particular galaxy is at distance d. Another galaxy similar to the first appears 70 times dimmer. What is the distance to the second galaxy? State any assumptions.

- calibrating distance scales
- Cepheid variables

USING VARIABLE STARS

Cepheid variables

The luminosity of many stars varies with time – you will have already seen that a great band of instability crosses the main sequence of the H–R diagram. Among these variable stars are the **Cepheid variables**, which are particularly useful for making distance measurements. Polaris, the pole star, is a Cepheid whose apparent magnitude varies between 1.95 and 2.05 with a period of 4 days. Other variables change in brightness more dramatically than this. What makes Cepheids so important is that their period of variation is related to their absolute magnitude (and luminosity) in a well-understood way. This is called the **period–luminosity law**, and makes it possible to work out their absolute magnitude from their measured period. They are also very luminous, so they are visible at great distances.

To determine the distance to a star you need both its apparent and absolute magnitude. Then the distance modulus can be calculated and used in the distance equation. Typical periods of Cepheids are a few days or more and so are readily measurable. Once this has been done you can read off the absolute magnitude from the period–luminosity law and then calculate distance.

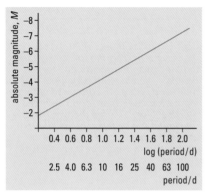

The period–luminosity law. The period of a variable star can be measured from Earth by observing how its intensity varies with time. Its absolute magnitude can then be read from the graph.

How to use Cepheids to determine distance

- Measure apparent magnitude m.
- Measure period T.
- Use period–luminosity law to find M.
- Use distance modulus $m - M$ to find d.

Worked example

The graph shows a light curve for δ Cephei in the constellation Cepheus. What is its distance?

From the light curve δ Cephei has a period of about 5.4 days and an apparent magnitude of about 5.3. From the period–luminosity law the absolute magnitude is about –3.3.

$$m - M = 5.3 - (-3.3) = 8.6$$

$$m - M = 5 \log \frac{d}{10}$$

$$\log \frac{d}{10} = \frac{8.6}{5} = 1.72$$

$$\frac{d}{10} = 52$$

$$d = 520 \, \text{pc}$$

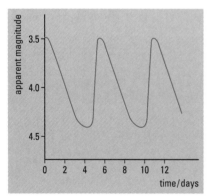

The light curve for δ Cephei.

The period–luminosity law

It may have occurred to you that you need to know the distance to at least some Cepheids in order to work out the period–luminosity law in the first place. To do this you need Cepheids which are relatively close to Earth so that their distance can be measured by other means.

Although Cepheids are relatively rare stars, some are found in open clusters whose distances can be measured by main-sequence fitting with Hyades. In this way, absolute magnitude can be calibrated against period for local Cepheids and the period–luminosity law can be used to read off the absolute magnitude of more distant Cepheids once their period is known.

advanced **PHYSICS**

Cepheids are not the only variable stars and another type, RR-Lyrae stars, are also used in a similar way. Variable stars allow the measurement of distances up to 4 Mpc using terrestrial telescopes and up to about 40 Mpc using the Hubble Space Telescope.

Far out

At very great distances Cepheids are too faint to be used for distance measurements – you can no longer see them. Supernovae, on the other hand, increase their brightness by more than 20 magnitudes as they explode and could be used to measure distances up to hundreds of Mpc. The essential assumption is that supernovae of the same type have similar maximum luminosities and so can be used as 'standard candles'. The distance can then be found by measuring the apparent magnitude of a supernova and using the inverse-square law. The method is calibrated against the Cepheid variables using supernovae in closer galaxies.

You can also estimate the distances to very distant galaxies using Hubble's law (see spread 12.18).

Why do Cepheids vary?

The luminosity of a star varies when the star is unstable and pulsates. Cepheids are very luminous stars and their atmospheric gases, in particular singly ionized helium, absorb radiation emitted from the core and heat up. Collisions between helium ions become more violent and doubly ionized helium ions become more abundant. These now absorb even more energy and the atmosphere expands rapidly. During this phase the surface temperature and area both increase, resulting in an increase in luminosity. However, its increased luminosity means that it is now losing energy at a phenomenal rate so the atmosphere cools, the helium becomes singly ionized, and gravitational forces cause the star to shrink. The cycle then begins again.

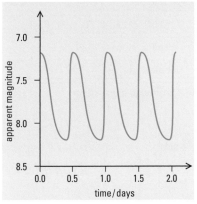

The light curve for an RR-Lyrae star. Whilst these are significantly dimmer than cepheids they also have a well defined period luminosity relationship and are usually found in globular clusters.

PRACTICE

1 The apparent magnitude of Polaris varies from 1.95 to 2.05 with a period of about 4 days.

  a Use the period–luminosity law to estimate its absolute magnitude.

  b What is its distance modulus?

  c Use the distance modulus to estimate its distance.

2 The graph above right shows how apparent magnitude varies with time for a particular star in the Andromeda galaxy.

  a Use this data to estimate the distance to Andromeda.

  b If you were attempting to find the distance to Andromeda by using Cepheids, what else would you do?

3 Explain why a Cepheid variable can be considered a 'standard candle' for making distance measurements.

4 Cepheids appear at the high-luminosity end of the instability strip on the H–R diagram.

  a Why does this make them especially useful for distance measurements?

  b Why are Cepheids of little use beyond about 50 Mpc?

  c Explain why supernovae may take over from Cepheids for measuring even greater distances.

  d What are the assumptions that must be made if supernovae are to be used as 'standard candles'?

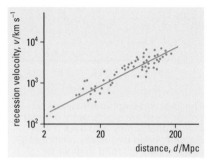

Hubble's law.

Hubble's law

The recession velocity v of a distant galaxy is directly proportional to its distance d.

$$v \propto d \quad \text{or} \quad v = H_0 d$$

where H_0 is the Hubble constant.

The best current measurements give $H_0 = 70 \pm 7 \, \text{km s}^{-1} \, \text{Mpc}^{-1}$

or (in SI units)

$2.2 \times 10^{-18} \pm 0.23 \, \text{s}^{-1}$

Helpful hint : spacetime expansion

As these ideas are so radical, it will help to understand them if you pause to think about a simplified model.

Imagine a large sheet of rubber, so large that you can't see the edges – this represents all of space as it is right now. Small coins are scattered all over the sheet – these represent galaxies in their present positions. If, somewhere out of sight, something is pulling outwards on the rubber sheet in all directions and it is stretching, what happens to the coins? Obviously they 'move' further apart, but this is due to the expansion of the rubber, not their motion across it. And where is the centre of this expansion? A little thought shows that there isn't one. Whatever coin you focus on, all the other coins move away from it, and the rate at which they move away increases in proportion to their distance – Hubble's law applies. If you are still not convinced make the argument quantitative. How does the distance to nearest neighbours change when the scale of the sheet doubles? How does the distance to second nearest neighbours change? Which recedes fastest?

THE EXPANDING UNIVERSE

Member of cluster in	Approximate distance / Mpc	Redshifts and velocities/ km s^{-1}
Virgo	16	1200
Ursa Major	200	15 000
Corona Borealis	300	22 000
Bootes	500	39 000
Hydra	800	61 000

 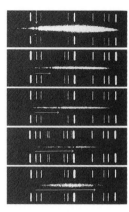

To the right are photographs and spectra of the galaxies listed in the table above. The spectra show that there is a relation between distance and redshift and hence recession velocity. A rate of expansion of 75 km s^{-1} Mpc^{-1} has been used to calculate the distance.

Vesto Slipher was a dedicated spectroscopist who spent his entire career (from 1901 to 1952) at the Lowell Observatory in Arizona. His most significant discovery was made when measuring the Doppler shifts in spectral lines from the arms of spiral galaxies in order to work out their rotation rates. In addition to the effect of rotation he discovered a Doppler shift due to motion. This in itself was not surprising, but by 1923 he had discovered that 36 of the 41 galaxies he had studied had **redshifts** – the observed wavelengths were slightly longer than those from similar sources on Earth. If this was caused by a Doppler shift due to relative motion it implied that the majority of the observed galaxies must be moving away from the Earth, and some of the velocities he calculated were enormous – hundreds of kilometres per second. It was *as if* the Earth is situated at the centre of some cosmic explosion.

At the same time Edwin Hubble had been developing and improving techniques to measure the distances of galaxies (he was the first to show that Andromeda is outside our own galaxy). When Hubble's distances and Slipher's redshifts are put together there is a very interesting link – the distance increases in direct proportion to the redshift.

According to the theory of the Doppler shift (see spread 6.18) the fractional change in wavelength z due to relative motion is given by:

$$z = \frac{\Delta \lambda}{\lambda} \approx \frac{\Delta f}{f} \approx \frac{v}{c}$$

This shows that redshifts are directly proportional to velocities, and leads on to the derivation of Hubble's law (see box).

Spacetime expansion

There are two common misunderstandings about Hubble's law:

- It does not necessarily mean that the Earth is in a very special position at the centre of the universe – in a sense everything could be. Every position in the universe is equivalent. If you look out from any galaxy all the others recede from you with velocities proportional to their distance, and the velocities and distances obey Hubble's law.

- It does not necessarily mean that the galaxies are moving through space as they recede from us. If *space itself* is expanding, the separation between galaxies increases and the light transmitted across this expanding gap gets stretched. In this model of an expanding universe redshifts *do not arise because of a Doppler shift*, but because the scale of the universe has increased since the light left its source. In fact the redshift z is directly linked to this change of scale. The ratio of received wavelength to emitted wavelength is equal to the ratio of the scale of the universe at the time of reception to the scale it had when the light was emitted.

The discovery of Hubble's law combined with Einstein's model of space and time indicates that we are living in an *expanding universe*. And if the scale of the universe is increasing now, it must have been smaller in the past. Run time far enough backwards and there must have been a moment in the far-distant past when the scale of the universe was so small that everything – all the space, time, matter, and energy – was packed into a vanishingly small volume. Everything started from the same place; everything was at the 'centre' at time zero in a point of infinite density and temperature – a singularity. Now imagine time running forward from zero: the universe explodes from that moment and begins to expand out to its present scale. That explosion is called the **Big Bang** and physicists are confident that our understanding of physics is now good enough to explain the broad outline of events that occurred from a tiny fraction of a second after the Big Bang to the present day, and to make predictions about the future evolution and possible fate of the universe. The study of the universe as a whole is called **cosmology**.

- **Before the Big Bang?** It is tempting to think that there was an enormous infinite empty space *before* the Big Bang and that the singularity that exploded to create the universe was one point in this space. This is *not* the model used by cosmologists. The Big Bang is the origin of space and time – there was no 'before' and the explosion did not take place in space – space itself burst out of the Big Bang.

- **Open or closed universe?** If gravitational forces are strong enough to halt the expansion and make the universe collapse it is a closed universe which is finite but unbounded (this means that there are no edges or boundaries, rather like the two-dimensional surface of the Earth). If the expansion cannot be stopped the universe is open and infinite.

- **The cosmological principle** The idea that the universe should look the same from all points (apart from local small-scale irregularities) is called the **cosmological principle**. It implies that every place is equivalent and the universe looks the same in all directions and from any point – it is isotropic and homogeneous. Although such an assumption is little more than an article of faith among cosmologists, it seems to work quite well as a starting point for theories of the universe as a whole.

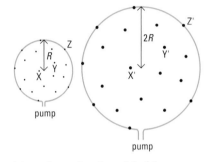

A two-dimensional model of the expanding universe is like the surface of a balloon as it is inflated. Dots on the surface represent galaxies. All dots recede from all others and the recession velocity is proportional to distance. In this case the scale (radius) has doubled so $X'Y' = 2XY$ and $X'Y' = 2XZ$, etc.

PRACTICE

1 Why was Slipher surprised that the majority of galaxies had redshifts?

2 Quasars are the most distant objects whose redshifts have been measured. This is done by comparing their spectra with a reference spectrum from a terrestrial source and calculating the fractional shift in prominent spectral lines. Quasar 3C 273 has a prominent emission line Hδ at 475.0 nm corresponding to a particular quantum jump in hydrogen atoms. The normal Hδ line is at 410.2 nm.

   a Is this a redshift or a blue shift? Explain your answer.

   b Why do hydrogen lines play an important role in determining redshifts?

   c What is the fractional shift z in wavelength?

   d Use the Doppler formula to calculate the velocity of 3C 273 and say in which direction it is moving relative to the Earth.

   e If the present scale of the universe is taken to be 1 and the present age of the universe is 16 billion years, what was the scale and age of the universe when this light was emitted from the quasar? (Assume a constant expansion rate.)

3 The latest value of the Hubble constant, H_0, is 70 km s^{-1} Mpc^{-1} (± 10).

   a Explain why there is a connection between the time since the Big Bang and the Hubble constant.

   b If the present value of H_0 is an underestimate will it lead to an underestimate or overestimate of the age of the universe?

   c Explain what 1 km s^{-1} Mpc^{-1} means.

4 What will determine whether the universe continues to expand forever or eventually slows down and re-collapses?

12.19

O B J E C T I V E S

- calibrating Hubble's law
- measuring H_0
- distant objects
- quasars

THE HUBBLE CONSTANT

In principle the Hubble constant can be measured from observations of a single galaxy. Its distance can be measured using Cepheid variables and its redshift measured by spectroscopy. These values could then be substituted into Hubble's law to calculate a value for the Hubble constant. Once this has been done the distances to all other galaxies could then be read off the graph of redshift versus distance or calculated by substituting into Hubble's law. Of course, using a single galaxy would be bad experimental practice, so data from many galaxies is used. This technique can be used to calibrate a distance scale out to about 500 Mpc.

However, although redshifts can be measured quite accurately, there is still some uncertainty surrounding distance measurements, so estimates of H_0 have a large uncertainty. Use of the Hubble Space Telescope in the late 1990s extended the range over which variable stars can be used to calibrate Hubble's law, and pinned down the Hubble constant to $70 \pm 7 \, \text{km s}^{-1} \, \text{Mpc}^{-1}$.

Worked example 1

The Ursa Major cluster has a redshift $z = 0.05$. Estimate its distance assuming the Hubble constant lies between 60 and 80 km s^{-1} Mpc^{-1}.

$$z \approx \frac{v}{c} \quad \text{so} \quad v \approx zc \quad \text{(Doppler shift)}$$

$$v = H_0 d \quad \text{(Hubble's law)}$$

$$d_{\text{max}} = \frac{v}{H_0(\text{min})} = \frac{zc}{H_0(\text{min})} = \frac{0.05 \times 3.0 \times 10^5 \, \text{km s}^{-1}}{60 \, \text{km s}^{-1} \, \text{Mpc}^{-1}} = 250 \, \text{Mpc}$$

$$d_{\text{min}} = \frac{v}{H_0(\text{max})} = \frac{zc}{H_0(\text{max})} = \frac{0.05 \times 3.0 \times 10^5 \, \text{km s}^{-1}}{80 \, \text{km s}^{-1} \, \text{Mpc}^{-1}} = 190 \, \text{Mpc}$$

NB The speed of light has been used in km s^{-1} to be consistent with the units used for H_0.

Variable stars are not the only way to determine the Hubble constant. For example, the distances to galaxies of similar types can be estimated by using their apparent brightnesses (or the apparent brightnesses of globular clusters they contain) and applying the inverse-square law. At double the distance a similar object appears only one quarter as bright. This method can be calibrated using the Virgo cluster of galaxies whose distance of 16 Mpc has been calculated using a different method.

The age of the universe

The rate of expansion can be used to estimate the time since the Big Bang and hence the age of the universe. The faster the present expansion rate, the more recently all the galaxies were very close together, and the younger the universe. All you need to do to get a rough estimate of the age is to assume that the expansion rate has not changed since the Big Bang and to work out how long ago any two galaxies would have been together. For example, if a sprinter is running away from you at 10 ms^{-1} and is presently 5 m away he or she must have passed you 5 m/10 ms^{-1} = 0.5 s ago. Similarly, if a galaxy is now a distance d from you and is moving at velocity v then it must have been right next to you a time $t = d/v$ in the past. Since distance and velocity are linked in Hubble's law, the age of the universe is linked to the Hubble constant:

$$v = H_0 d$$

so

$$t \approx \frac{d}{v} = \frac{1}{H_0}$$

This time period is called the **Hubble period**.

Slowing expansion

It is unrealistic to assume that the universe has always expanded at its present rate. Gravitational attraction between the galaxies is thought to have been slowing the expansion ever since the Big Bang, so the expansion rate was faster in the past and the Hubble period is an overestimate of the age of the universe. In 1998 new observations of the redshifts of distant supernovae surprised everyone. They implied that, rather than slowing down, the expansion of the universe may actually be accelerating! This remarkable conclusion is neither definitely confirmed nor fully explained at present, but it may lead to exciting new physics.

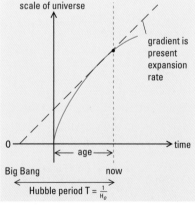

Variation of the scale of the universe with time. If we assume the universe has always expanded at its present rate we will overestimate its age.

Worked example 2

Calculate the range of possible ages for the universe assuming the Hubble constant lies in the range 60–80 km s^{-1} Mpc^{-1}.

To get an age in seconds, H_0 must be expressed in s^{-1}:

$$1\,\text{Mpc} = 3.08 \times 10^{22}\,\text{m}$$

so

$$1\,\text{km s}^{-1}\,\text{Mpc}^{-1} = \frac{10^3\,\text{m s}^{-1}}{3.08 \times 10^{22}\,\text{m}} = 3.25 \times 10^{-20}\,\text{s}^{-1}$$

This conversion factor is used below.

$$t_{max} \approx \frac{1}{H_0(\text{min})} = \frac{1}{60 \times 3.25 \times 10^{-20}\,\text{s}^{-1}} = 5.1 \times 10^{17}\,\text{s} = 1.6 \times 10^{10}\,\text{y}$$

$$t_{min} \approx \frac{1}{H_0(\text{max})} = \frac{1}{80 \times 3.25 \times 10^{-20}\,\text{s}^{-1}} = 3.9 \times 10^{17}\,\text{s} = 1.2 \times 10^{10}\,\text{y}$$

This gives an age between about 12 and 16 billion years.

This calculation shows how important the measurement of the Hubble constant is to our understanding of the evolution of the universe.

Quasars

The largest redshifts ever observed are all associated with mysterious objects called **quasars**. These are exceptionally luminous for their size and typical redshifts are in the range 0.5–4.0. If these are caused by expansion of the universe the most distant quasars are more than 10 billion light-years away and the light that reaches us was emitted when the universe was less than half its present age. The name quasar comes from quasistellar radio source, because the first quasars were very luminous in the radio-frequency part of the electromagnetic spectrum. Most of the thousands of quasars now known are stronger in optical frequencies so they are now more often referred to as **quasistellar objects (QSOs)**.

Great luminosity requires a huge power source. The present model of quasars is as **active galactic nuclei** – probably massive black holes in the centres of large galaxies. As matter falls into the black hole it loses gravitational potential energy and radiates. Black holes are the only known objects that are small enough and massive enough to account for the small size and great luminosity of a quasar.

Eternal expansion?

Studies of high redshift supernovae since the late 1990s suggest that the universe is now expanding at an accelerating rate. To explain this, physicists assume that there is a form of 'dark energy' that acts like a repulsive gravitational force to drive the expansion. However, the future fate of the universe remains a topic of controversy and alternative models are hotly debated. Precise measurements of cosmological parameters, such as the Hubble constant, help to distinguish between these models.

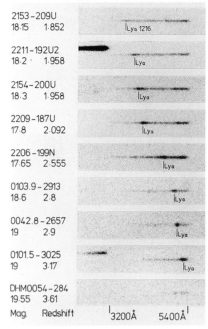

Mag.	Redshift	
2153–209U		
18·15	1·852	Lyα 1216
2211–192U2		
18·2	1·958	Lyα
2154–200U		
18·3	1·958	Lyα
2209–187U		
17·8	2·092	Lyα
2206–199N		
17·65	2·555	Lyα
0103.9–2913		
18·6	2·8	Lyα
0042.8–2657		
19	2·9	Lyα
0101.5–3025		
19	3·17	Lyα
DHM0054–284		
19.55	3·61	
		3200Å 5400Å

Many quasars have redshifts greater than 1, showing that they are at enormous distances and moving away with extremely high velocities. The label 'Lyα' refers to a particular spectral line (in the Lyman series). Notice how this line is (red-) shifted further and further to the right (longer wavelengths) in the quasar spectra shown above. The greater the redshift, the further away the quasar is and the faster it is moving.

---

z > 1?!

If z is greater than 1, this seems to imply that $v > c$. However, the simple formula $z \approx v/c$ breaks down for velocities approaching c. A more useful equation relates the redshift to the change in scale (or size), R, of the universe since the light was emitted:

$$z = \frac{R_2}{R_1} - 1$$

So a redshift of 4 means that $R_2/R_1 = 3$, and the universe is now 3 times larger than when that light was emitted.

PRACTICE

1 If the distances to galaxies are to be measured by comparing their apparent brightness to that of the Virgo cluster we need to know the distance to the Virgo cluster.

  a How do you think the distance to the Virgo cluster is measured in practice?

  b Two objects similar to the Virgo cluster have apparent brightnesses 1% and 0.1% as great. How far away are they? (Distance to Virgo cluster is 16 Mpc.)

2 **a** Use Hubble's law to estimate the distance to Ursa Major, whose redshift is 0.05. Take the value of the Hubble constant to be 42 km s^{-1} Mpc^{-1}.

  b How is this estimate affected if the value of the Hubble constant turns out to be an underestimate?

3 Why do people think quasars are:

  a far away; **b** extremely energetic;

  c more common in the early universe than now?

In a dense forest all lines of sight end on trees.

The darkness of the night sky suggests that all lines of sight do not end on stars.

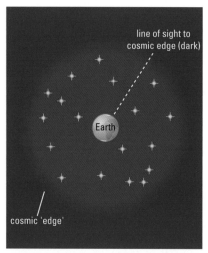

If the universe is small and finite the darkness at night is where we look between the stars toward the 'cosmic edge'.

OLBERS'S PARADOX

The night sky is dark; at least, the spaces between the stars are dark. This isn't surprising at first thought. At night we are on the side of the Earth that points away from the Sun, so it is bound to be dark. Isn't it? Actually it is not so obvious as that. In fact it is very difficult to explain why the sky at night is dark and this question has taxed the brains of many astronomers.

Olbers's paradox

To understand why the night sky might not be dark, imagine you are standing in a forest. What do you see around you? Trees. If the forest is large enough and you are not near an edge then whichever (horizontal) direction you look in, your line of sight will stop at a tree trunk. Now, think about the night sky. If the universe is infinite or even just very large and there are stars scattered through it like trees in a forest, then every direction you look in should point at a star. This means light should arrive from all directions with equal intensity and the night (and daytime) skies should be as bright as the surface of a star. If this was the case we would not be here to see it; the intensity of the stellar radiation would be too great for life to survive. Kepler was the first to discuss this problem (although it is now known, incorrectly, as **Olbers's paradox**).

Kepler's solution

The universe is not infinite but finite, and the dark spaces are where we look between the stars to a dark outer boundary or cosmic edge. This solution fails if the universe is very large, as it seems to be, or infinite. Worse than that, the idea of a cosmic edge is in conflict with both physics (as we know it) and observation.

Cheseaux and Olbers's solution

Starlight is absorbed as it travels through space. Unfortunately this conflicts with the law of conservation of energy. In 1848 Herschel pointed out that any medium that absorbed the radiation would inevitably warm up and re-radiate it. Once equilibrium was reached the night sky would, once again, be intensely bright. So this cannot resolve the paradox either.

A finite spherical universe?

General relativity allows space to curve, and it is possible that the space of the universe is closed in the same sense that the surface of the Earth is closed – i.e. it has no edges. If this was the case then we could have a finite but unbounded universe. Would this solve Olbers's paradox? Unfortunately not. Light could pass round the 'sphere' so we might see some stars more than once or from both sides. Either way our line of sight would always end on a star.

Finite lifetimes

One thing has been completely neglected so far: the average lifetime of a star. Stars do not shine forever. A star like our Sun lives for about 10 billion years. If this is taken as a representative luminous lifetime for stars in the universe then it can resolve Olbers's paradox. Imagine all the stars started shining at the same time. For the first 10 billion years the intensity of starlight reaching Earth would gradually increase as rays from more distant stars arrived. This radiation would reach a maximum value at the end of this time. Then the nearest stars would begin to go out. However, this would not reduce the brightness of the night sky because light from stars a little further away (beyond 10 billion light-years) would now be reaching us (although their sources would actually have stopped shining). They supply exactly as much light as those in a similar shell closer to us because the area of the shell increases as r^2 (surface area of a sphere is $4\pi r^2$) and the intensity of light from each star falls off as an inverse-square law with distance. This means that a shell of stars 10 times farther away contains 100 times more stars and the light from each one is 100 times weaker.

Of course, the stars did not all come into being at the same time, but this does not affect the argument. At any moment we are only receiving light from a finite number of stars. Light from others is still on its way to us, and others stopped shining long ago. The overall effect is that we are continually seeing light from the equivalent of a spherical shell of stars that is not wide enough to cover the entire sky. This explains why the gaps between the stars are dark.

A neat way to resolve Olbers's paradox was suggested by E. R. Harrison. He imagined a star surrounded by a cosmic box with reflecting walls. The box has a volume equal to the average volume occupied by a star in the universe. As the star shines photons flow into the box and its temperature rises. In equilibrium it would reach the same temperature as the surface of the star. Calculations show that it would take about 10^{23} years to reach equilibrium. The star is luminous for about 10^{10} years, so the sky is much cooler, on average, than the surface of the stars.

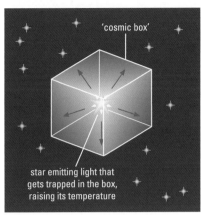

Harrison's resolution of Olbers's paradox.

The cosmological principle

We have been assuming that the universe is the same in all directions (**isotropic**) and the same at all places (**homogeneous**). This is a very convenient assumption as it allows us to use local information to construct large-scale models. For example, when resolving Olbers's paradox we assumed that the mean density of stars in the universe is the same everywhere, even if the universe is infinitely large. This assumption is one of the most important in cosmology, and is called the **cosmological principle**:

The universe has the same large-scale structure when viewed from any point within it.

This principle implies that there is no centre or special location within the universe; all places are equivalent. There is also a **perfect cosmological principle** which goes one step farther and states that the universe looks the same at all times and all places. This has been refuted by the discovery of **expansion**. The past is different from the future.

Expansion

Modern observations show that the universe is expanding. This means that distant galaxies and the stars they contain are all moving away from us. Light emitted by a source that moves away has its wavelength 'stretched' or redshifted. This reduces its frequency and the energy of its photons. The intensity is also reduced. Some people have suggested that this resolves Olbers's paradox. It does not. If the universe was infinite and the stars eternal an expanding universe would still have a bright sky. However, expansion does make the sky slightly darker than it would be if the universe were static.

Doesn't an expanding universe contradict the cosmological principle? Hubble's law states that distant galaxies recede with a speed proportional to their distance. This seems to imply that we are at the centre of some great explosion and therefore at a very special place in the universe. A little thought shows that this is not the case. Imagine stretching a metre ruler so it doubles in length in 1 second. The cm markings are now 2 cm apart. Imagine a tiny observer sitting on one of the marks. He will see the mark nearest to him move from 1 cm to 2 cm in 1 s, receding at $1\,\text{cms}^{-1}$. He will see marks initially 2 cm from him move from 2 cm to 4 cm in 1 s, receding at $2\,\text{cm s}^{-1}$. Those 10 cm from him move to 20 cm in the same time, receding at $10\,\text{cm s}^{-1}$. It is clear that the recession speed is directly proportional to distance, as in Hubble's law. It is also clear that it doesn't matter from which mark the observer is looking out; he will still see the same uniform expansion. The same is true in the universe at large. Wherever you are you see other galaxies receding, and their speeds obey Hubble's law, so the cosmological principle is upheld.

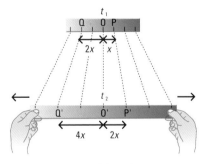

Simulating Hubble's law. Between t_1 and t_2 the ruler above has expanded to double its length. In this time P, initially a distance x from O, has 'moved' to a position 2x away. In the same time Q, twice as far from O as P was, has increased its distance from 2x to 4x. The 'recession velocity' of Q from O is twice that of P from O as in Hubble's law.

O B J E C T I V E S

- the equivalence principle
- gravitational time dilation
- the deflection of light by gravity

GENERAL RELATIVITY

The laws of physics in a uniform gravitational field are the same as in a uniformly accelerating laboratory.

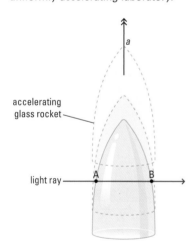

accelerating glass rocket

light ray

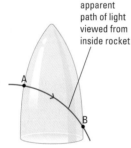

apparent path of light viewed from inside rocket

The equivalence principle predicts that light will be deflected by gravity.

Einstein's masterwork was, without doubt, the **general theory of relativity**. In it he described gravity not as a force of attraction between masses but as a distortion of space and time that affects everything. Although general relativity uses a lot of advanced mathematics it is possible to get some important results using simple thought experiments.

The equivalence principle

Special relativity is based on the **principle of relativity**:

The laws of physics are the same for all uniformly moving observers.

This led to ideas like time dilation, length contraction, and the speed of light as a limit. The problem with special relativity is that it does not say how the laws of physics appear to observers who are not moving uniformly (e.g. accelerating) or who are inside a gravitational field.

One day, while Einstein was sitting at his desk looking out of the window, it occurred to him that a freely falling observer in a gravitational field does not experience gravity (at least not until she hits the ground). Imagine standing inside a freely falling lift. You and everything in the lift accelerate downwards *at the same rate* relative to the outside world. However, from inside the lift, objects appear to hang suspended in space before you or, if pushed, they move away at constant speed in a straight line. This is exactly how they would behave in deep space or in a uniformly moving spacecraft (and is the reason why NASA trains astronauts for weightlessness in jets undergoing free-fall dives). Einstein described this as the most beautiful thought of his entire life – it was the link he needed between uniform motion, acceleration, and gravity.

Imagine now that you are inside a closed room holding a tennis ball. When you let go of the ball it falls to the ground; so does anything else you drop. Can you assume your closed room rests on the surface of a planet and that you are in a gravitational field? No! You may have been transported into deep space, far from gravitational effects, and aliens may have attached a tow rope to your room and pulled it along with an acceleration of $9.8\,\mathrm{m\,s^{-2}}$. Inside the room all the same things would happen. You cannot distinguish acceleration from gravitation.

It is clear from these thought experiments that accelerations can be used to switch gravitational effects on or off. Einstein turned this into **the principle of equivalence**:

The laws of physics in a uniform gravitational field are the same as in a uniformly accelerated reference frame.

This was one of the most important steps toward general relativity. It means that anything that happens in an accelerated reference frame must also happen in a gravitational field.

Deflection of light by gravity

Here is a beautiful example of the equivalence principle in action. Imagine a light ray passing through a glass spacecraft. It enters on one side of the craft and leaves on the other. If the spacecraft is accelerating the light ray will appear to follow a curved (parabolic) path inside it (although an outside observer could confirm that the light travelled in a straight line). Now imagine the spacecraft is at rest on the surface of a planet and again the light ray passes through it. By the equivalence principle it must behave in exactly the same way as it did when the spacecraft was accelerating; it moves in a parabolic curve and 'falls' toward the surface of the planet. The conclusion is obvious – light is deflected by gravity! And there is more: a freely falling observer would still see the light travel in a straight line.

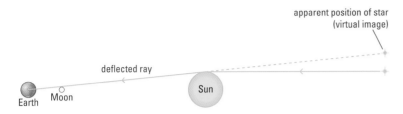

apparent position of star
(virtual image)

deflected ray

Earth   Moon   Sun

Deflection of starlight (a schematic representation not to scale).

But is it true? Does light really get deflected in a gravitational field? Einstein suggested a way to test the idea. The deflection in ordinary gravitational fields is tiny, but light from distant stars passing close to the Sun should be deflected enough for us to measure a small change in their apparent positions. The problem with this is that the brightness of the Sun itself usually prevents us seeing them. In the 1919 total eclipse of the Sun the astronomer, Arthur Eddington, led an expedition to photograph the positions of stars close to the Sun's disc during totality. When these photographs were compared with others taken at night in the same region of sky they confirmed a small systematic shift in apparent positions (about 0.0005 degrees), in agreement with Einstein's prediction.

More recently the effect has been confirmed using radio waves emitted by astronomical radio sources. These are not obscured by the Sun's radiation, so the experiment can be carried out whenever the Sun passes in front of the source. We don't have to wait for an eclipse.

Gravitational time dilation

Elsewhere we showed that moving clocks run slow. So do clocks in strong gravitational fields. This can be shown by another simple thought experiment. Imagine two clocks, one near the surface of a star, the other at a distance from the star in a weaker gravitational field. Both clocks can be calibrated using light from a standard spectral source. This is done by ensuring the same number of oscillations of the light correspond to one 'tick' on both clocks.

Now imagine someone standing beside the surface clock beaming their calibration signal up to an observer beside the other clock. Light is affected by gravity so the photons increase their gravitational potential energy as they move upwards. This must reduce their electromagnetic energy and also their frequency (since $E = hf$). The upper observer will see that they have a lower frequency and longer period than the photons from the source he used to calibrate his clock. He will conclude that the period between ticks on the lower clock is longer than on his own: it loses time. But it is not just the clock that loses time; time passes more slowly on the surface (strong field) than at height (weak field). This is called **gravitational time dilation** and is a separate effect from the time dilation caused by high-speed motion.

Time slows down in strong gravitational fields.

Advance of perihelion

The planets move in relatively weak gravitational fields as they orbit the Sun. In weak fields general relativity leads to almost the same results as Newtonian gravitation. However, in Newton's theory the planets follow closed ellipses. General relativity predicts that the axis of these ellipses should rotate around the Sun very slowly. The effect is largest for Mercury, because it is closest to the Sun and therefore in the strongest field. In fact, Mercury's **perihelion** (closest approach to the Sun) had already been observed to rotate in this way, leading some astronomers to speculate that there must be another undiscovered planet close to the Sun causing the disturbance. The rate of advance of perihelion (just 1.76 seconds of arc per century) was accurately explained by general relativity.

Time travel

Gravitational time dilation opens an intriguing possibility for time travel. A rocket in orbit around a neutron star or black hole (not too close!) could remain in a strong gravitational field where time passes slowly and then re-emerge to find that a great deal of time had passed in the outer universe. This would be a way of travelling into the future.

GPS corrections

The clocks used in global positioning system (GPS) satellites are set to run fast on Earth. This is to offset the combined effects of gravitational and special relativistic time dilation (making them run fast and slow respectively). Once in orbit the 'offset' clocks keep perfect time with atomic clocks on the Earth's surface.

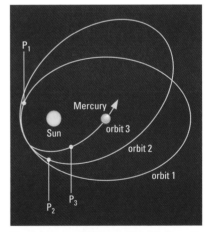

The advance of Mercury's perihelion. P_1, P_2, and P_3 mark the successive points of perihelion. The amount of the advance has been greatly exaggerated here.

12.22

OBJECTIVES

• gravity and geometry

• the fate of the universe

• critical density

• dark matter

Albert Einstein predicted that gravity would deflect light. In 1919 Arthur Eddington (above) measured the apparent shift in the positions of stars close to the edge of the Sun's disc during a total eclipse. The tiny shift was consistent with Einstein's prediction. This has now been tested much more accurately using radio waves.

Geodesics

Great circles drawn onto a globe are geodesics because they are the shortest routes between any two points.

A gravitational lens: a massive cluster of galaxies, Abell 2218, imaged by the HST. The cluster is so massive that its enormous gravitational field has produced a magnifying effect, providing a zoom lens for viewing galaxies that are 5 to 10 times further away and that appear here as distorted arcs.

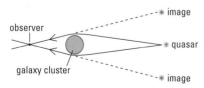

The principle of a gravitational lens.

CURVED SPACE

Think back to the rubber-sheet model that was used to explain the expanding universe (see spread 12.18). The rubber represented space and coins scattered onto it were galaxies. Imagine pressing down on the rubber at some point so that a depression forms round your finger. If someone rolls a marble across the sheet it will be deflected by the depression; it may even fall into it and become trapped. If it was projected along a tangent at the right speed it might go into orbit around the centre of the depression. If the marble represented a planet or comet and your finger represented a star the distortion would be like the star's gravitational field. In Einstein's theory of gravitation (the general theory of relativity) gravity is simply the distortion of geometry, or the curvature of spacetime geometry. The closer you get to a massive object the more severe the distortion. The paths followed by free bodies (e.g. stars, galaxies, planets, etc.) are all the straightest possible paths through curved space – these paths are called **geodesics**.

Far away from large masses these paths look like the normal straight lines you might draw with a ruler. Close to massive bodies they look curved. The Earth's orbit around the Sun is actually its shortest path through spacetime near the Sun – the Earth follows a geodesic.

This remarkable idea can be summarized as follows.

• Matter tells space how to curve.

• Space tells matter how to move.

One of Einstein's first predictions from this theory was that light, which follows the shortest path between two points, should be deflected by gravity just like anything else. This deflection was first observed in 1919 and news of the discovery helped make Einstein into an international celebrity. Nowadays there is dramatic confirmation of the effect in **gravitational lenses**. These are usually massive galaxies that deflect the light of very distant sources like quasars so that the rays bend around the galaxy and reach us from more than one direction, forming multiple images of the source.

The geometry of the universe

The flat rubber-sheet model of the universe raises one very important question – what happens at the edge of the sheet? In other words, is space infinite so that there is no end to it – or is it finite, stopping somewhere? And if there is an edge of space, what is beyond it? All these questions are problematic, but spacetime curvature offers one more possibility. The rubber sheet might look flat *here*, just like the surface of the Earth looks flat locally, but it might actually be just a small part of some enormous sphere so that it joins up with itself and has no edge. This sort of space would be *finite but unbounded* and the universe would be *closed*. Could the universe as a whole have this sort of geometry? It turns out that it can, although in four-dimensional spacetime rather than on the two-dimensional surface of a sphere – in fact Einstein's equations allow for many different geometries and this is just one of them.

So what decides whether the universe is closed or open? Remember that *matter tells space how to curve*, so it must depend on how much matter is packed into the universe. There is a critical density ρ_0 which must be exceeded to close the universe – all that needs to be done is to calculate the value and then measure the density of the universe and there will be the answer!

The geometry of the universe is also linked to its fate. The expansion of the universe is opposed by the gravitational attraction of all the things in it to each other. Whether or not these gravitational forces are sufficient to halt the expansion eventually or even reverse it depends on the density of matter ρ:

- **Less than critical density** $(\rho < \rho_0)$ Expansion continues forever, the universe is eternal and open.
- **Critical density** $(\rho = \rho_0)$ There is just enough matter to halt the expansion, but only after an infinite time when it reaches infinite size – the universe is eternal and open (but cosmologists call this a 'closed' universe!).
- **Greater than critical density** $(\rho > \rho_0)$ Expansion will stop in a finite time and the universe will re-collapse to a Big Crunch. The universe is not eternal and is closed, finite, and unbounded (i.e. it has no edge).

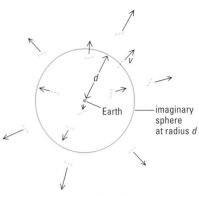

Considering critical density.

The critical density

Although Einstein's full general theory of relativity is needed for a thorough derivation of the critical density, the same result can be derived from Newton's theory. For expansion to continue forever the galaxies must be receding from Earth at escape velocity. But what are they escaping from? It is certainly not just the Earth's gravity. If you draw a line from the Earth to a particular galaxy then it is escaping from all the matter contained inside a sphere that has this radius. Take a galaxy at distance d from Earth.

The mass contained within a sphere of radius d and density ρ is

$$\frac{4}{3}\pi d^3 \rho$$

The escape velocity from the surface of such a sphere is

$$v_{esc} = \sqrt{\frac{2GM}{d}} = \sqrt{\frac{2G\frac{4}{3}\pi d^3 \rho}{d}} = \sqrt{\frac{8G\frac{4}{3}\pi d^2 \rho}{3}}$$

The velocity of the galaxy is given by Hubble's law, $v = H_0 d$. For a closed Universe the galaxies do not escape,

so $v \leq v_{esc}$. Hence

$$H_0 d \leq \sqrt{\frac{8G\pi d^2 \rho}{3}}$$

Rearranging this gives $\qquad \rho \geq \dfrac{3H_0^2}{8\pi G}$

So the critical density is $\qquad \rho_c = \dfrac{3H_0^2}{8\pi G}$

Effect of other galaxies

You might be worried about ignoring the effect of all the galaxies beyond the one we have chosen, but remember that the net gravitational field strength inside a hollow sphere is zero, even in an infinite hollow sphere. The other galaxies play no part in deciding whether a particular galaxy can escape or not.

Measuring the critical density is rather harder than deriving an expression for it. All estimates of mass based on things so far detected have lead to a density far below the critical density. However, detailed observations of the motions of galaxies suggest that they are being influenced by the gravitational fields of objects not yet seen, and one of the most important problems of modern cosmology is to detect and identify this **dark matter**. Some of it is predicted to be cool dark matter made up of hadrons, maybe some comprises small failed stars (**brown dwarfs**), or maybe neutrinos have a tiny mass. But there is also a strong suspicion that some of the 'missing mass' is in new forms of exotic matter that are as yet undiscovered. The question of whether the universe is open or closed must itself remain open, but there are some clues that suggest it may be balanced exactly on the critical mass.

Accelerating expansion

The calculations on this spread include the gravitational effects of all the mass in the universe, whether this is in ordinary or dark matter. However, the discovery that the universe is undergoing an accelerated expansion has changed things, and cosmologists speculate that some kind of 'dark energy' is driving the expansion. Current evidence suggests that the universe will continue to expand forever.

PRACTICE

1 Explain how the two-dimensional surface of the Earth can be 'finite but unbounded'.

2 Lines of longitude are geodesics on the Earth's surface, but only one line of latitude is a geodesic.

 a Identify the one line of latitude that is a geodesic.

 b Explain why the others are not geodesics.

3 a What are the limits on critical density for values of the Hubble constant that lie between 30 and 100 km s^{-1} Mpc^{-1}?

 b How many hydrogen atoms per cubic metre would have the same density as the values you have calculated? (Mass of a hydrogen atom is about 1.7×10^{-27} kg.)

 c Explain why the critical density varies so much with the value of the Hubble constant.

4 What is meant by:

 a an 'open' universe;

 b a 'closed' universe;

 c cold dark matter?

A BRIEF HISTORY OF THE UNIVERSE

One of the most awe-inspiring claims of modern physics is that it can explain the evolution of the universe from less than 1 microsecond after the Big Bang up to the present day. However, there would be no reason to believe this claim unless it could be checked against detailed evidence – but what evidence remains of something that happened more than 10 billion years ago? Surprisingly, there is rather a lot – and it is very convincing too – but before weighing up the evidence look at the model below.

Quark soup

The further back you look, the denser and hotter you find the early universe was, and the more violent the collisions between particles. Our knowledge of high-energy interactions on Earth comes from particle accelerators that reach up to a few teraelectronvolts (1 TeV = 10^{12} eV). At these energies protons and anti-protons shatter when they collide and quarks fly out. (Free quarks have never been observed, but 'jets' of new particles emerge.) About a billionth of a second after the Big Bang this was the *average energy of all particles* in the Universe and they were packed together with a density greater than 10^{25} kg m^{-3}. At this density the mass of the entire Earth would fit inside a suitcase. There would have been no stars or galaxies, no atoms or molecules, no protons or neutrons, just a violently energetic soup of quarks in the exploding universe.

The hadron era

As the universe expanded it cooled and the collision energy dropped. About one millionth of a second after the Big Bang the quarks briefly formed into clusters before being broken apart again by new collisions. At about 10 μs these clusters began to survive in large numbers forming **hadrons** (particles made of quarks) like protons, neutrons, and pions and their anti-matter counterparts. There were also large numbers of electrons, positrons, neutrinos, and photons, but hadrons dominated. Particles and anti-particles continually collided and annihilated, and in that dense seething world the gamma rays emitted in annihilation created yet more particles and anti-particle pairs.

The origin of matter

Then something strange happened. As the universe cooled, nearly 1 second after the Big Bang, the expansion of spacetime redshifted the gamma-ray photons so far that they no longer had enough energy to materialize new particles. At this point particle annihilation outstripped creation and the hadrons began to disappear. If the hadrons had been present in equal numbers pretty well all matter would have annihilated with anti-matter at this stage and no atoms, stars, or galaxies would ever have formed. However, for reasons still not totally understood, there seems to have been a slight excess of matter over anti-matter, perhaps 1 part in 10^{18}, so that after the great annihilation a little matter remained – we are made of it.

For a while, lighter particles, the leptons (electrons, positrons, and neutrinos) continued to be created and annihilated, but as the universe continued to expand and cool they too virtually wiped one another out, leaving just a trace of matter and a huge amount of radiation.

Helium synthesis

The protons created during the hadron era were nuclei of hydrogen. The other dominant hadrons were neutrons. It is possible to predict quite accurately how the proportions of protons and neutrons changed after the Big Bang and then how they combined to form deuterons (nuclei of the hydrogen isotope deuterium containing one proton and one neutron). For a while the universe was hot enough and dense enough to fuse deuterons to form helium-4 and traces of some other isotopes like

Matter from energy. A very high-energy (about 15 000 GeV) cosmic ray iron nucleus collides with a nucleus in photographic film emulsion, producing a tremendous 'jet' of about 850 mesons.

helium-3 and lithium, but between 100 s and 1000 s after the Big Bang helium synthesis finished and, theory predicts, about 25% of matter had been converted to helium. (This is in close agreement with estimates of the ratio of helium to hydrogen based on astronomical observations, and is another strong piece of evidence for the Big Bang.)

The radiation era

There had been so much annihilation that the universe had become dominated by gamma-ray photons; about a billion of them for every proton or neutron. At first they continued to interact strongly with charged particles and the universe was opaque, but with continued expansion and cooling, electrons and protons combined to form neutral hydrogen atoms and photons **decoupled** from matter (i.e. they stopped interacting with it).

At about 100 000 years the universe became transparent. (This is the earliest time we could ever hope to see back to with our telescopes.) The universe was then filled with photons which shared in the expansion of space, getting stretched to longer and longer wavelengths, but always remaining as a fading echo of the Big Bang itself. More than 10 billion years later this radiation blast is still present as the **microwave background radiation** that is detected from all directions in space – the single most important piece of evidence supporting the Big Bang model. The radiation follows a black-body spectrum corresponding to an absolute temperature of about 2.7 K. However, recent measurements show slight fluctuations in radiation intensity from different directions (see spread 12.2). These were predicted because the present universe is not uniform – there are galaxy clusters in some places, huge gaps elsewhere – so the radiation distribution could not have been uniform when it decoupled from matter.

The last few billion years

At about a million years old the universe was full of a brilliant yellow light and its density was a few hundred hydrogen atoms per cubic centimetre. Tiny density fluctuations were frozen into the universe and these caused matter to begin to clump together in some places under the influence of gravitational attraction. From one hundred million to one billion years after the Big Bang galaxies formed and stars burst into life as gas clouds collapsed and ignited nuclear fusion reactions in their cores. The density of the universe was just a few hundred hydrogen atoms per cubic metre and falling. We are now able to see back to this time with the Hubble Space Telescope. Star formation continues now wherever the intergalactic gas clouds are large and dense enough. Expansion also continues, and it is still not known whether the mutual gravitational attraction of matter in the universe will eventually stop it, or if the universe will continue to expand forever.

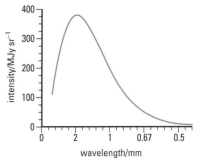

The cosmic microwave background spectrum. The curve is the theoretical spectrum of a black body at 2.175 K, as predicted by the Big Bang theory. Data from the Cosmic Background Explorer (COBE) satellite match the curve so exactly, with error uncertainties less than the width of the black-body curve, that it is impossible to distinguish the data from the theoretical curve. The results show that the radiation matches the predictions of the Big Bang theory to an extraordinary degree.

Inflation

The picture of the Big Bang and the expansion of spacetime is good at accounting for the state of the universe from about 10^{-35} s onwards, but what happened before this? According to an influential idea suggested by the American physicist, Alan Guth, the present expansion is as nothing compared to what came before. He thinks that the universe underwent a period of unbelievably rapid inflation from about 10^{-44} s to 10^{-35} s during which it increased its scale by a factor of about 10^{50}! After this it settled into the present expansion. Amazingly, this theory helps to explain some puzzling properties of the universe and is supported by quantum theories that allow the vacuum of space to exist in different states – the transition from inflation to expansion corresponding to a vacuum transition from an excited state to a ground state.

Summary: Big Bang theory

Evidence for the Big Bang theory

- Redshifts and Hubble's law imply cosmic expansion.

- Background radiation – redshifted radiation from matter/anti-matter annihilation.

- Helium abundance agrees with calculated value from Big Bang model.

Problems with the Big Bang theory

- Pinning down the Hubble constant and the cosmic distance scale.

- What happened before 10^{-40} s – quantum gravity?

- How did the matter/anti-matter asymmetry come about?

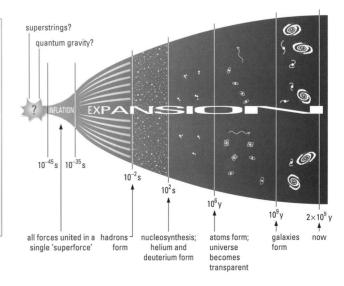

PARTICLE PHYSICS AND COSMOLOGY

The size of the universe

The size of the universe is extremely difficult to define – it may even be infinite! Cosmologists prefer to work in terms of the **scale parameter**, R. The idea is to take a reference distance (it does not matter what) and compare its size as the universe expands. If the reference length is x, then at any other time it will be Rx. For convenience R is taken to be 1 at the current time in the universe's history. In the future R will be greater than 1 (as the universe expands); in the past it was less than 1 (as the universe was smaller).

The temperature of the universe

Whenever an accelerator causes two particles to react at high energy, a situation first seen in the early moments of the Big Bang is re-created. Although energy is the term often used in particle physics, when applying these ideas to the universe as a whole it is more convenient to work in terms of temperature:

$$\text{temperature of universe} = \frac{\text{average energy of particles}}{k}$$

or

$$\langle E \rangle = kT$$

The temperature of the universe depends on R and so also on time, t. As the universe expands so the wavelengths of any photons within it get longer – this is the **Hubble redshift** (see spread 12.18). A similar effect applies to the de Broglie wavelength of other particles. Detailed calculations show that

$$T \propto t^{-\frac{1}{2}}$$

$R \propto t^{\frac{1}{2}}$ while the universe is dominated by relativistic particles

$R \propto t^{\frac{2}{3}}$ when the universe has cooled below relativistic energies

$$T \propto \frac{1}{R}$$

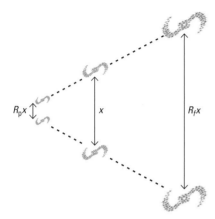

If the distance between two galaxies is taken as reference distance, x, and the scale parameter fixed at $R = 1$ for today, then the distance between the galaxies in the future will be $R_f x$ where $R_f > 1$. At some time in the past the distance between the galaxies was $R_p x$ where $R_p < 1$. As you wind time backwards R_p gets smaller and the universe approaches the conditions of the Big Bang.

Missing factor

You may have expected the factor $\frac{3}{2}$ to appear in the equations in the examples, in comparison with the kinetic theory of gases. The exact factor that should appear will depend on the type of particle, but in any case it is always ~1, so it can be replaced by 1 when estimating the temperature of the whole universe and averaging over all particles.

Expanding infinity

Many people get confused by the idea of the universe getting bigger, or smaller, and yet being infinite. Imagine a chess board with white and black squares stretching off into infinity. If R is getting bigger then the sizes of the squares will get bigger – but there are still an infinite number of them. At the moment no one knows if the universe is infinite or not, but physicists and cosmologists are quite sure that it is expanding in this way.

Worked example 1

If the universe is now 10^{15} years old, how much smaller was it when it was 1 second old?

$$10^{15} \text{ years} = 10^{15} \times 365 \times 24 \times 60 \times 60 \text{ seconds} = 3.15 \times 10^{22} \text{ seconds}$$

If $R = 1$ now, then the scale parameter after 1 second, r, is given by

$$\frac{r}{R} = \frac{(1 \text{ second})^{\frac{1}{2}}}{(3.15 \times 10^{22} \text{ seconds})^{\frac{1}{2}}} = 5.63 \times 10^{-12}$$

This will be an estimate as it has been assumed that the universe has been relativistic up to the present day – this is, of course, not true.

Worked example 2

What was the temperature of the universe when it was 1 second old?
If you take the temperature of the universe now to be $3\,\text{K}$ (the temperature of the background radiation), then

$$\text{temperature at 1 second} = \frac{R}{r} \times (3\,\text{K}) = 5.33 \times 10^{11}\,\text{K}$$

Stages in the history of the universe

The quick estimates in worked examples 1 and 2 show that the early universe was a place with a density very much greater than it has now, and was ablaze with a temperature of billions of kelvin. Yet, incredibly, the history of this place can be traced down to a few fractions of a second after it was created (see spread 12.18). As the universe has evolved it has passed through several distinct stages characterized by the temperature and the type of physics that was taking place.

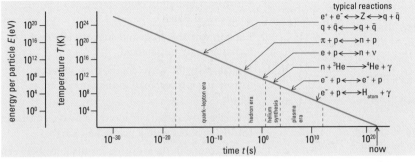

The variation of temperature with time in the history of the universe. Some of the exact details depend on theories which are not completely worked out, but physicists are confident of the overall features in this history.

$t = 10^{-43}$ s, $T > 10^{32}$ K This is the starting point – before this time we cannot be sure of the physics involved. At this period in the universe's history, the particles had so much energy that the GUT forces were in operation (see spreads 9.16 and 14.7). The universe was full of quarks, gluons, leptons, photons, Ws, Zs, and the more exotic particles of the GUT forces. By the end of the period a slight imbalance between the amount of matter and anti-matter in the universe had appeared (in every 10^{18} particles there was one more matter particle than anti-matter particle).

$t = 10^{-35}$ s, $T = 10^{27}$ K Next came the era of the quark plasma. The weak and electromagnetic forces were still unified, but the strong force had separated out. The quarks were too close to each other for the strong force to bind them into hadrons. The universe was a sea of quarks and leptons. Towards the end of this period the universal expansion had moved the quarks far enough apart for the strong force to bite – hadrons started to form. The small excess of quarks over anti-quarks from the GUT time now produced a small excess of baryons over anti-baryons.

$t = 10^{-12}$ s, $T = 10^{15}$ K At this point the weak force separated out from the electromagnetic force.

$t = 7 \times 10^{-7}$ s, $T = 1.1 \times 10^{10}$ K Protons and anti-protons started to annihilate each other (as did neutrons and anti-neutrons). This had been happening before this time, but the energy of the photons in the universe had been high enough to materialize new ones. At this stage the photons had cooled so that pair creation of baryons could not take place. After the general annihilation a small excess of matter was left over – this is the matter that evolved into what we see in the universe now.

$t = 3$ s, $T = 5.9 \times 10^9$ K Electrons and positrons started to annihilate as photons were, by then, too cool to recreate them. Eventually just a small excess of electrons remained.

$t = 3.2$ mins, $T = 9 \times 10^8$ K At this stage **nucleosynthesis** started. The energy in the photons was no longer big enough to blast apart any protons and neutrons that had bound together to form deuterium. Minute amounts of heavier nuclei were also formed (e.g. helium and lithium). The exact proportions of the various nuclei formed would have depended on the details of the expansion rate and the lifetime of the neutron. The calculations are in impressive agreement with the amounts observed in the oldest stars.

$t \approx 300\,000$ years, $T = {\sim}4000$ K Re-combination followed. By now the universe was cool enough to allow free electrons to bind into atoms with nuclei. Before this time the photon energy was high enough to ionize atoms. After this time the universe became 'transparent' – i.e. the photons no longer interacted with matter to any extent as there were hardly any free electric charges. From this point the photons in the universe washed around being slowly redshifted by the expansion – today they are detected as the microwave background.

The future of the expanding universe

Until the late 1990s cosmologists argued over whether or not the mass of all the matter in the universe was great enough to eventually stop expansion and then cause re-collapse to some 'Big Crunch' in the far future. Even if there was not enough matter to stop expansion they expected the universally attractive nature of gravitation to cause expansion to slow. This led to three possible scenarios – a closed universe with a Big Crunch, a universe of critical density in which expansion continued but slowed ever closer to zero, and an open ever-expanding universe.

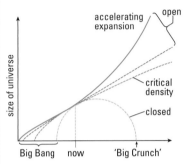

Alternative histories of the universe. The dotted lines represent the alternative scenarios considered before the discovery of dark energy. The solid line shows the onset and subsequent increase of accelerated expansion caused by dark energy.

However, measurements of the redshifts of extremely distant (and therefore ancient) supernovae using the HST in 1998 provided startling evidence that the rate of expansion in the distant past was lower than it is now. This implied that, rather than slowing, the rate of expansion is accelerating. This was completely unexpected but has been supported by further observations. Saul Perlmutter, Brian Schmidt, and Adam Reiss shared the 2011 Nobel Prize for Physics for this discovery.

What could be causing the accelerating expansion? We don't know – but it does have a name – dark energy, a universal repulsion acting over great distances and pushing the galaxies apart. We also know how much of it there is – it must account for 70% of the mass–energy content of the universe! Since dark matter is thought to account for a further 25% that leaves just 5% for all the matter we see in stars and galaxies! The nature of dark matter and dark energy is one of the most important unsolved problems in physics.

PRACTICE EXAM QUESTIONS

1 a Write an account of the evolution of stars. In your account, include specific reference to

   i the formation of a star,

   ii the processes that occur during a star's lifetime,

   iii how the later stages of a star's evolution are affected by its mass.

   b Suggest ways in which the conservation of energy and the conservation of momentum are involved in the processes you have described.

2 a Starting with the largest first, and working in descending order of size, write a sentence about each of the following astronomical terms, showing that you know their relation to one another in space.

   STAR, PLANET, GALAXY, MOON, UNIVERSE

   b Explain how observations made by Hubble led to the idea of the expanding Universe.

   c When a graph is plotted of the recessional speed of galaxies against their distance from the Earth, it is found to be a straight line through the origin. The Hubble constant, H_0, is the gradient of this graph and has an approximate value of $2.4 \times 10^{-18}\,\text{s}^{-1}$. Use this information to deduce

   i the speed of recession of a galaxy which is a million light years away from the Earth,

   ii the approximate age of the Universe.

   d Suggest why there is an observational edge to the Universe.

   e Suggest what the blue shift is and what might cause it.

3 a A typical Milky Way star, such as Vega, has a speed within our Galaxy of $20\,\text{km s}^{-1}$.

   Estimate the maximum shift of a line of wavelength 656 nm in the hydrogen spectrum from this star which results from such a speed.

   b The spectrum of the Andromeda galaxy (the nearest spiral galaxy beyond the Milky Way) shows blue shift.

   i Why is this observation unusual?

   ii With the aid of a diagram, suggest an explanation for this observation.

4 a Trigonometrical parallax is a method of measuring stellar distances.

   i Explain the principle of the method.

   ii State the measurements made and show how the distance to a star is obtained from these measurements.

   iii State a correction which has to be made to the measurements.

   b Describe the mode of operation of a charge coupled device (CCD) used as a detector in a telescope. Give two advantages of using a CCD rather than a photographic plate as a detector.

5 a Emission from stars and galaxies covers most of the electromagnetic spectrum.

   i Radio waves of certain frequencies emitted by galaxies cannot be observed on Earth. Explain why this is so.

   ii Ultraviolet radiation from stars cannot be detected on Earth.

   Explain why this is so and suggest how such radiation may be detected.

   b i Explain why the wavelengths used in radar astronomy are confined to the range 1 cm to 20 cm.

   ii Describe how radar astronomy is used to measure the distance between a planet and Earth.

   c i In the course of their orbits, the distance between the Earth and Venus changes. The radii of the orbits of the Earth and Venus are $1.5 \times 10^8\,\text{km}$ and $1.1 \times 10^8\,\text{km}$ respectively. If a radio pulse is transmitted from Earth towards Venus, calculate the greatest and least times that would be measured before the return pulse was detected.

   ii Explain why radar cannot be used to take measurements of a star such as α *Centauri* which is at a distance of 1.3 parsec from Earth.

6 a Regions of the electromagnetic spectrum are listed in Table 12.1.

Table 12.1

Region of radio electromagnetic spectrum	γ-ray	X-ray	ultraviolet	visible	infrared
Degree of transparency of Earth's atmosphere					

Copy Table 12.1 and indicate the degree of transparency of the Earth's atmosphere to e.m. radiation in each region of the spectrum. Use 'N' for not transparent, 'P' for partially transparent and 'H' for highly transparent.

   b i Observatories using large modern optical telescopes are never sited in cities.
   Suggest three reasons, apart from cost, for this.

   ii X-ray observations were used to help to confirm the existence of black holes.

   Suggest a suitable type of location for an X-ray telescope and justify the cost involved.

7 In the absorption spectrum of a particular faint galaxy the wavelength of the calcium K line is found to be greater than the wavelength of the same line when produced by a laboratory source.

   a Explain why there is this increase in wavelength.

   b Explain how this 'red-shift' in the wavelength can be used to measure the distance of the galaxy from Earth, provided the Hubble constant is known.

   c If the two measured values of the wavelength are $3.934 \times 10^{-7}\,\text{m}$ and $4.733 \times 10^{-7}\,\text{m}$, calculate the speed of the galaxy with respect to the Earth.

   d Hence calculate the distance in light-years of the galaxy from Earth.

   Take the Hubble constant to be $5.0 \times 10^4\,\text{m s}^{-1}\,\text{Mpc}^{-1}$.

8 a Measurements made in 1994 using the Hubble Space Telescope indicate that a cluster of galaxies in the constellation Virgo is 51 million light-years away.

i What other information about this cluster is needed in order to calculate a value of the Hubble constant?

These observations give a value of about $80\,\text{km}\,\text{s}^{-1}\,\text{Mpc}^{-1}$ for the Hubble constant.

ii What age of the Universe does this value of the Hubble constant indicate?

Comment on your answer.

b The critical density ρ_0 of matter in the Universe and the Hubble constant H_0 are related by the equation

$$\rho_0 = \frac{3H_0{}^2}{8\pi G}$$

i From your knowledge of the derivation of this equation, suggest why π appears in the denominator.

ii Use base units to check the homogeneity of this equation.

9 Hubble's law relates the speed of recession of a galaxy to its distance from Earth. The Hubble constant is taken to have a value within the limits of $50\,\text{km}\,\text{s}^{-1}\,\text{Mpc}^{-1}$ and $100\,\text{km}\,\text{s}^{-1}\,\text{Mpc}^{-1}$.

a Give *one* reason why there is such a large uncertainty in the value of the Hubble constant.

b Show how the maximum distance to the edge of the Universe may be obtained from Hubble's law.

c When estimating the age of the Universe, what is the significance of the lower limit of the Hubble constant?

10 a i The wavelength of the calcium H line in the laboratory is $3.968 \times 10^{-7}\,\text{m}$. The wavelength of the same line in the spectrum of a distant galaxy is found to be greater by $0.198 \times 10^{-7}\,\text{m}$. Explain the reason for this increase.

ii The wavelength of the calcium K line from the same source is observed to be greater than the wavelength of the K line in the laboratory by a different amount. Explain why the respective increases in wavelength for the H line and K line are different.

If the wavelength of the K line is $3.934 \times 10^{-7}\,\text{m}$, calculate the increase in wavelength for the K line.

iii Calculate the speed of the source with respect to Earth.

b A single narrow line appears in the spectrum of a non-spinning star. In the spectrum of a similar spinning star, the same line is broader. Account for this difference in appearance of the spectral line.

11 a Figure 12.1 shows part of the visible region of the 'continuous' emission spectrum of a star, in which dark lines are observed against a bright background.

Figure 12.1

i Explain the occurrence of the dark lines in this spectrum.

ii Intensity of this spectrum is greatest in the yellow region.

What does this indicate about the star?

b Figure 12.2 shows qualitatively the spectrum of the microwave background radiation.

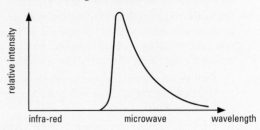

Figure 12.2

What does this indicate about the Universe?

12 a State what is meant by parallax.

b i Define the parsec.

ii With the aid of a diagram explain the principles underlying the trigonometric parallax method of stellar-distance measurement.

iii State a correction which has to be made to the measurements.

13 a Sketch a Hertzsprung–Russell diagram, showing the positions of Main Sequence stars, Red Giant stars and White Dwarf stars. Indicate values of absolute magnitude and temperature on the axes.

b Use the diagram to explain why Red Giant stars must be much larger than White Dwarf stars.

14 a Outline the evidence which has led to the general acceptance of the 'Big-Bang' theory of the origin of the universe. Explain how notions of density and temperature are involved.

b From what evidence can an estimate be made of the age of the Universe?

c Give a brief account of the changes which are supposed to have taken place in the structure of matter in the Universe in the first 10 000 years of its existence.

15 a Give a qualitative account of the evolution of galaxies and galactic clusters from about ten milliseconds after the Big Bang.

b Describe the structure of our own Galaxy. Include in your description an indication of the Sun's position and the galactic motion of stars.

c Other galaxies have structures different from that of our own Galaxy. Outline these structures.

16 a Explain why black holes are so named.

b i Explain what is meant by an event horizon.

ii The radius of the event horizon for a star of mass M is given by

$$R = \frac{2GM}{c^2}$$

where G and c have their usual meanings.

Calculate the radius to which the Sun would have to collapse for it to become a black hole.

Mass of the Sun $= 2.0 \times 10^{30}\,\text{kg}$

17 a Explain what is meant by

  i apparent magnitude of a star,

  ii absolute magnitude of a star.

b In the Hertzsprung–Russell diagram, stars are grouped according to their spectral class. State what you understand by spectral class and give the prime reason for the range of spectral classes which occurs.

c i Outline the method of spectroscopic parallax for determining the distance of a Main Sequence star from the Earth.

  ii Explain, without calculation, why this method can lead to an error as large as 50% in the calculated distance to the star.

  iii Distances to some stars are measured by trigonometric parallax; distances to other stars are measured by spectroscopic parallax.

  What determines which of these methods is more appropriate for a particular star?

d The star Polaris is a Cepheid Variable. Outline a method of determining the distance to Polaris using a property of Cepheid Variables.

18 a An astronomical telescope, consisting of a thin converging objective lens of focal length f_o and a thin converging eyepiece lens of focal length f_e, is set up in normal adjustment by an observer to view a distant object.

  i Show, by means of a labelled ray diagram, how a magnified image is formed by the telescope. The diagram should show the paths through the telescope of at least two rays from a non-axial point on the object.

  ii Defining the angular magnification M as

  $$M = \frac{\text{angle subtended at the eye by the final image}}{\text{angle subtended at the unaided eye by the object}}$$

  show that $M = \dfrac{f_o}{f_e}$

  iii If the separation of the lenses in the above telescope is 26 cm and $M = 12$, calculate the values of f_o and f_e.

b i State, without the aid of a ray diagram, how the eye-ring in an astronomical telescope is formed. Calculate its position for the telescope described in **a iii**.

  ii Show that f_o and f_e are related to d_1 and d_2, the respective diameters of the objective lens and eye-ring, by the expression

  $$\frac{f_o}{f_e} = \frac{d_1}{d_2}$$

  iii Assuming the observer's eye is placed at the eye-ring, estimate the minimum diameter of the objective lens of the telescope necessary for maximum light to enter the eye.

c Give *two* similarities between a dish radio telescope and an optical reflecting telescope.

19 a Distinguish between a planet and a comet.

b The critical density ρ_0 of matter in the Universe is given by the expression

$$\rho_0 = \frac{3H_0^2}{8\pi G}$$

  i A Explain the significance of the constant H_0.

  B Discuss the possible fate of the Universe if the density ρ of matter in the Universe is

  I greater than ρ_0,

  II less than ρ_0.

  ii Figure 12.3 shows possible curves for the variation with time of the size of the Universe.

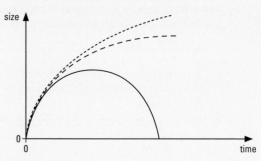

Figure 12.3

  Copy Figure 12.3 and on your copy

  A label the curve corresponding to $\rho > \rho_0$,

  B label the curve corresponding to $\rho < \rho_0$,

  C mark on the diagram a point to represent the Universe at the present time.

  iii Discuss your answer to **ii** with reference to the possible fate of the Universe.

c Telescopes, such as the Hubble telescope, which orbit the Earth have enabled scientists to photograph astronomical objects. These photographs can provide far more information than photographs obtained using Earth-based telescopes. State reasons for this improvement.

20 a i The mean distance of the Earth from the Sun is 1.496×10^{11} m. The distance is called the astronomical unit. Explain briefly the particular significance of this distance for astronomers.

  ii The star Sirius A has a parallax of 0.377 seconds of arc. Explain, with the aid of a labelled diagram, the meaning of the term parallax.

  iii Calculate the distance of Sirius A from the Earth.

  iv Explain how the velocity of Sirius A towards or away from the Earth can be found.

b i Sirius A, a massive main sequence white star, is the brightest star in the sky. Describe the fusion process known as the pp chain which accounts for some of the immense amount of energy pouring out from Sirius A.

  ii Sirius A has a faint companion, Sirius B, which is a white dwarf. Explain the term white dwarf. How would you account for the differences between Sirius A and Sirius B assuming that the two stars are roughly the same age?

21 a Astronomers have discovered from their observations of the star Capella that:

- its surface temperature T is 5200 K
- its distance from the Earth is 4.3×10^{17} m
- at the Earth's surface the intensity of the radiation received from Capella is 1.2×10^{-8} W m^{-2}

   i Explain briefly how the surface temperature is determined.

   ii Describe, in outline only, the parallax method for finding the distances of the star from the Earth.

   iii Calculate the radius r of Capella, given that its luminosity L can be found by using

$$L = 4\pi r^2 \sigma T_4$$

where σ is the Stefan-Boltzmann constant which is 5.7×10^{-8} W m^{-2} K^{-4}.

b i Sketch and label an HR diagram showing the main sequence and the regions occupied by red giants and white dwarfs.

   ii Explain why main sequence stars of large mass have higher luminosities and shorter lives than main sequence stars of low mass.

   iii Describe the processes which occur within a star similar in mass to the Sun when it leaves the main sequence.

22 a i Stars emit radiation at all wavelengths, but most regions of the electromagnetic spectrum cannot be observed from the Earth's surface. Explain why this is so for γ-rays, X-rays, ultraviolet radiation and infrared radiation.

State how observations at these wavelengths can be made.

   ii Radio waves of certain frequencies emitted from galaxies cannot be observed on Earth. Explain why this is so.

   iii Radiation in the visible region of the electromagnetic spectrum is scattered in the atmosphere. State what causes this scattering and under what conditions the scattering will be most significant.

b i Radar waves of wavelength 0.20 m are sent from Earth towards Venus. The signal reflected from Venus is detected after a lapse of 276 s. Determine the distance of Venus from Earth.

   ii The reflected signal has a total spread of wavelength = 2.40×10^{-9} m about the original wavelength. Explain what causes this spread.

If the radius of Venus is 6050 km, calculate the period of rotation of the planet. Neglect any spread in wavelength due to motion of the planet either away from or towards the Earth.

23 a i There are two types of photoreceptors in the eye. Discuss the difference between them with respect to their response to the intensity of light falling on them. Hence explain why stars of low apparent brightness all appear to be the same colour when observed by the unaided eye, but can be seen to be of different colours when seen through a telescope.

   ii Treating the pupil of the eye as a circular aperture of diameter 2.0 mm, calculate the smallest angular separation between two objects which the eye can resolve when light of wavelength 550 nm falls on it from the two objects.

Light from two point objects which the eye can just resolve activates an area on the retina. Calculate how many photoreceptors there are in this area of the retina, given that each photoreceptor has a diameter of 1.1 μm and that the retina is 18 mm from the pupil. Assume that the diameter of the pupil does not change.

   iii For a point source, emitting light of wavelength 550 nm, to be detectable by an eye, the minimum energy falling per second on the eye must be 1.0×10^{-16} J. Calculate the energy of each photon and hence the number of photons arriving at the eye per second. If only 5 of these photons excite the retina per second and thus produce a nerve impulse, calculate the quantum efficiency of the eye.

Explain why a photographic film can record much weaker sources of light, although its quantum efficiency is not much greater than that of the eye.

b i Explain what is meant by the apparent magnitude and the absolute magnitude of a star.

   ii Cepheid Variables are stars whose brightness increases and decreases periodically. The graph in Figure 12.4 shows the relationship between the mean absolute magnitude and the pulsation period for Cepheid Variables.

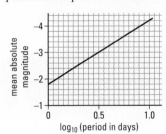

Figure 12.4

Delta Cephei has a period of pulsation of 5.2 days.

During this time its apparent magnitude varies between 3.6 and 4.2. Use these data to *estimate* the distance of Delta Cephei from the Earth.

13

The Physics of
COMMUNICATIONS

This communications satellite was hanging in orbit just minutes before this shot was taken of it being recaptured by astronauts on the Space Shuttle. Although they look rather like the inside of a washing machine, such satellites are in fact extremely hi-tech devices. They receive, amplify, and re-transmit radio, television, and digital communications signals. The information revolution that is taking place at the moment may well be more significant than the microchip revolution.

In this chapter some of the most important ideas of information technology are developed. The basic terminology of communications is defined and the principles of radio discussed. Finally, you will look at some applications of digital technology, including CD/DVD players and the Internet.

INTELSAT VI being recaptured by astronauts on the Space Shuttle Endeavour *in 1992, some two years after the satellite was first launched. Owing to a malfunction, INTELSAT VI was not raised into a geosynchronous orbit, and the main goal of this mission was to capture the satellite and install a new motor to propel it into the correct orbit.*

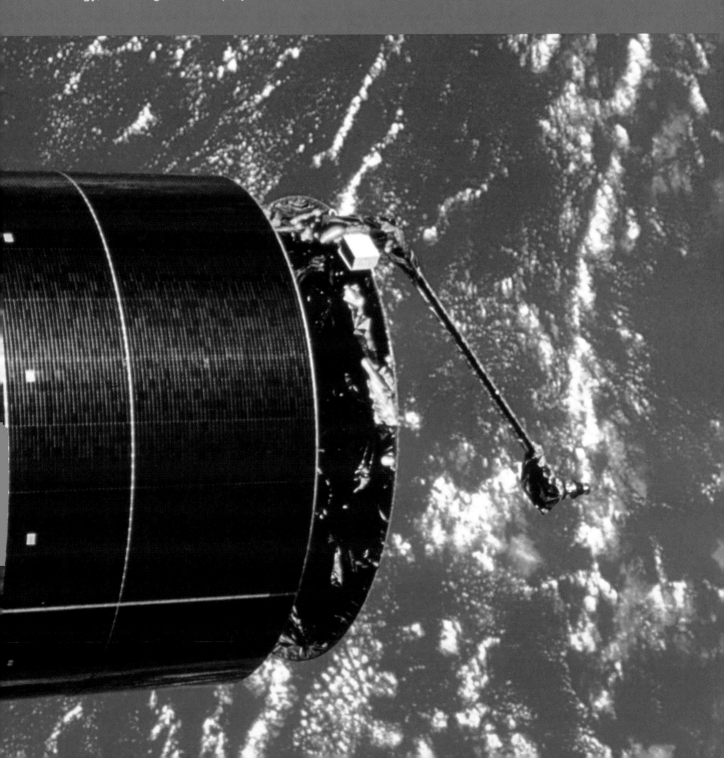

COMMUNICATIONS SYSTEMS

Information is power. Knowing what is happening or about to happen can be vital. Political, military, and financial decisions frequently rest on having up-to-date information about events. People who study future trends suggest that the real gap between the developed world and the developing countries is their access to information.

Telecommunications science centres on developing technology to transmit information over long distances, and is one of the fastest-growing areas of industry.

A telecommunications dictionary

Communications engineers employ a wide variety of systems and have to be familiar with a certain amount of terminology.

• **Signal** A transfer of information in a certain medium i.e. radio, telephone cable, or fibre optic.

• **Encoding** The process of transforming the information into a form suitable for transmission.

• **Decoding** The process of extracting the information from a received signal.

Each different medium requires a different technology, but the basic process is the same in each case – the information is encoded, in a way suitable for transmission via the medium, and then decoded again on reception.

Passing information invariably involves encoding at one end and decoding at the other.

The process of encoding and decoding is invariably electronic.

• **Modulation** In order for a transmission to carry information – i.e. to be a signal – it must vary with time. Encoding devices are used to produce a pattern of variation that represents the information that is to be transmitted – for example, a telephone encodes speech as a variation in electrical signals. Frequently the encoded variations are superimposed on a steady signal or **carrier**. This process is called **modulation**. The reverse process at the reception end of the chain is called **demodulation**. **Amplitude modulation (AM)** and **frequency modulation (FM)** are both used in radio transmissions (see spread 13.4).

• **Transducer** Encoding may involve the transformation of energy from one form into another – for example, sound energy must be transformed into electrical energy for radio transmission. Devices that do this are called **transducers** – for example, a microphone in this case. Telecommunications systems almost always involve transducers.

• **Channel** A transmission path is referred to as a **channel**. One-way communication requires only one channel, two-way communication requires two channels.

• **Analogue** and **digital signals** These are the two main types of signal used in telecommunications.
An **analogue** quantity is one that can assume any value – for example, the height of a person.
A **digital** quantity can only assume one of a set of values – for example, the sex of a person.
Electronic digital values are restricted to being either an electronic 1 (normally +5 V) or an electronic 0 (normally 0 V). Digital information is carried as a transmission of a sequence of **binary numbers**.

Building blocks

Any communications system can be broken down into a series of components, each of which is designed to carry out a specific job.

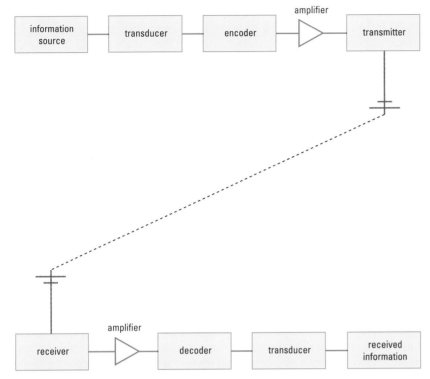

Block diagram of a typical communications system. This example uses radio waves, but the link could be by wire in a telephone system.

1 **Encoder** Converts the information into a form suitable for transmission.

2 **Transmitter** Produces the signal. This may involve characterizing the signal in some way so that the receiver can select it – for example, a radio station transmits at a known **frequency**.

3 **Receiver** Accepts the signal. This may involve selecting the appropriate signal from a range available – for example, tuning a radio.

4 **Decoding** Extracts the information from the received signal.

5 **Storage** The information is placed in a medium that preserves it for an extended period of time (this may involve encoding it in a different way).

6 **Retrieval** Extracting the information from storage.

7 **Transducers** Devices for converting energy from one form to another, which are frequently needed at each end of the communications channel.

Example 1 *Ship-to-ship signalling*

Encoder	person sending message
Transmitter	signal (Aldis) lamp i.e. a lamp with an opening shutter
Receiver	eye
Decoder	brain
Storage	paper (written message)

Example 2 *Radio*

Receiver	antenna (wire containing electrons which move in response to the electromagnetic wave)
Decoder	diode (crystal set)
Amplifier	increases the size of the electrical current to the size that can drive the loudspeaker
Output transducer	loudspeaker

A	•—	N	—•
B	—•••	O	———
C	—•—•	P	•——•
D	—••	Q	——•—
E	•	R	•—•
F	••—•	S	•••
G	——•	T	—
H	•••••	U	••—
I	••	V	•••—
J	•———	W	•——
K	—•—	X	—••—
L	•—••	Y	—•——
M	——	Z	—•••

Morse code. In this code the letters of the alphabet are represented by combinations of long and short light or sound signals. This code used to be used extensively in shipping and military communications.

COMMUNICATION CHANNELS

The decibel

Telecommunications engineers often have to compare power levels in different parts of the system. Sometimes these power levels differ by very large factors, making the standard units of measurement inconvenient. To allow for this the *decibel* is almost always used (see spread 6.2). It is defined in terms of the ratio between two power levels P_2 and P_1:

ratio in decibels (dB) = $10 \log \left(\dfrac{P_2}{P_1} \right)$

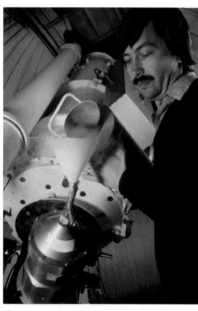

One way of minimizing the thermal noise in electronics is to cool the equipment to very low temperatures (for instance, with liquid nitrogen). This is especially important if the signal power is very weak – for example, when receiving signals from distant space probes, or imaging very faint objects with the CCD detector of a telescope, as shown here.

Noise and distortion

Noise is an unwelcome addition to the signal. It will still be present even if no signal is being sent. Distortion is a corruption of the original signal.

There is an old story about a general trying to communicate from the front line. In order to get an important message through to his Headquarters he employed a series of runners to relay the message from person to person. According to the story the message started out as 'send reinforcements, we are going to advance'. By the time it had been passed from person to person several times, and each person had slightly misheard the message, what actually arrived was 'send three [shillings] and four pence, we are going to a dance'.

The story does not go on to explain what Headquarters made of this message, but it does illustrate the difficulties experienced in communications. Every communication channel suffers from **noise**, **distortion**, and **attenuation** to some degree. Various things can be done to minimize the problems, but it is important to use the right communications channel for the information.

Noise

Noise is an unwelcome, random, extra signal added to the signal that is being transmitted. For example, there is often a background hiss on telephone conversations or when listening to the radio – the hiss is noise. It is not really distortion of the original signal as noise tends to occur in the system even if there is no signal being sent. Thunderstorms and atmospheric conditions can add to the amount of noise in a radio signal. Noise from some artificial sources, such as electric motors and switches for large currents, can also be picked up by radio and TV receivers.

Noise is always produced in electronic circuits. The random thermal motion of electrons in a conducting material is an electronic noise source. As this tends to increase with temperature many high-quality amplifiers, transmitters, and receivers are cooled to very low temperatures to reduce the noise. This is especially important if the signal is very weak as then it can be about the same power level as the noise produced in the electronics at room temperature.

The **signal-to-noise ratio** (**S/N**) measures the power of the noise in a system compared with the signal power. Every communications channel has a minimum acceptable signal-to-noise ratio. If there is too much noise compared to the signal, then the channel decoder will have difficulty separating the signal from the noise.

Worked example 1

A communications channel has a minimum signal-to-noise ratio of 35 dB. If the signal power to be transmitted is 3 W, what is the maximum amount of noise allowable in the channel?

$$S/N = 35 \text{ dB} = 10 \log \left(\frac{\text{signal power}}{\text{noise power}} \right) = 10 \log \left(\frac{3.0 \text{ W}}{P_N} \right)$$

$$\therefore \quad \frac{3.0 \text{ W}}{P_N} = 10^{3.5} = 3162$$

$$\therefore \quad P_N = 9.5 \times 10^{-4} \text{ W}$$

Attenuation

All signals get weaker the further they have to travel. This is called **attenuation**. In the case of a radio or TV signal the signal power is attenuated according to an approximate inverse-square law. In telephone systems the resistance of the wires will lead to attenuation. Optical fibres are very efficient at transmitting signals over long distances, but the small amount of scattering that takes place each time the light is internally reflected will also give rise to attenuation.

The amount of attenuation is normally quoted as a decibel loss per kilometre. One way of compensating for attenuation is to put amplifiers at various points along the channel. With a digital signal the wave shape of the pulse can also be 'cleaned up' by using a **regenerator**.

Worked example 2

The diagram shows a communications system. What is the output power of the system?

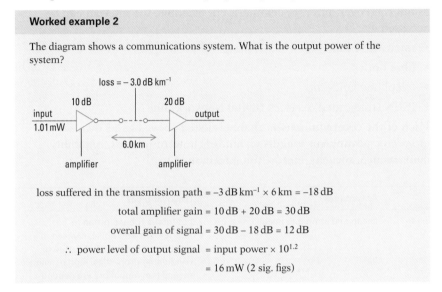

loss suffered in the transmission path = $-3\,\mathrm{dB\,km^{-1}} \times 6\,\mathrm{km} = -18\,\mathrm{dB}$

total amplifier gain = $10\,\mathrm{dB} + 20\,\mathrm{dB} = 30\,\mathrm{dB}$

overall gain of signal = $30\,\mathrm{dB} - 18\,\mathrm{dB} = 12\,\mathrm{dB}$

∴ power level of output signal = input power × $10^{1.2}$

= $16\,\mathrm{mW}$ (2 sig. figs)

Bandwidth

The **signal bandwidth** is the range of frequencies that must be transmitted in order to make a signal intelligible. For example, the range of frequencies transmitted over a telephone system is 300–3400 Hz – a bandwidth of 3.1 kHz. The **channel bandwidth** is the range of frequencies that can be sent over a particular channel. Of course this is normally far greater than the signal bandwidth.

Multiplexing

The most efficient way of using a communications system is to send more than one signal at the same time. This is known as **multiplexing**. Clearly the various signals have to be kept separate in order to prevent confusion at the receiving end. There are two basic ways in which signals can be multiplexed.

1 **Frequency-division multiplexing (FDM)** This is currently the most common method of multiplexing. It is used in telephone systems. Each telephone signal requires a signal bandwidth of 3.1 kHz, but this need not be 300 Hz to 3400 Hz. One of the processes that can be done at encoding and decoding is to shift the frequencies used – any block 3.1 kHz wide will do. Consequently, if the cables used can support a channel bandwidth of 100 kHz, around 30 conversations can be sent at the same time, each occupying its own 3.1 kHz slice of the channel. In practice a 4 kHz signal bandwidth is used to ensure that the signals do not overlap.

2 **Time-division multiplexing (TDM)** As digital systems have become more common, this has increasingly taken over from FDM. It has the advantage over FDM of keeping the system bandwidth low. The idea is to break up the digital signals into 'bursts' of information. If there are three conversations going on at once, a burst of 1 can be sent, followed by a burst of 2, then 3, and then 1 again, etc. These bursts are typically 8 μs long and consist of digitally encoded information.

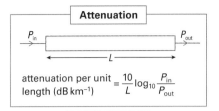

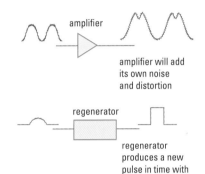

Attenuation can be compensated for by using amplifiers. A digital signal can be cleaned up into a sharp-sided pulse by using a regenerator which generates a new pulse in time with the old one.

Helpful hint

An amplifier's **gain** is the ratio of its output to its input. Gains and losses measured in dB can be added like ordinary numbers. Gains would normally be multiplied together, but the definition of the decibel requires the log of the gain, and the log of two numbers multiplied together can be found by adding the logs:

total gain = $G_1 \times G_2$

total gain in dB = $10 \log (G_1 \times G_2)$

= $10 \log (G_1) + 10 \log (G_2)$

= gain G_1 in dB + gain G_2 in dB

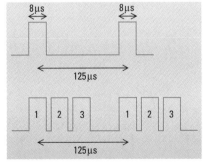

Time-division multiplexing.

Internet and intranet

An **intranet** is a set of information (normally as web pages) that can only be accessed by computers within the company or within a particular building that set it up. The **internet** is a set of information that people have made available to anyone who wishes to access it from all over the world.

COMMUNICATIONS NETWORKS

Even within a comparatively 'simple' situation such as an office building, there is a bewildering variety of different communications channels in use, such as voice mail (internal messenger service), telephone, email, intranet, internet, fax, cellphone, etc. Many of these possibilities can be established over different channels. For example, the internet can be accessed via:

- microwave link
- fibre optics
- telephone land-line
- ISDN (Integrated Services Digital Network) line.

Each of the communications channels listed above has its own set of specific advantages and disadvantages in terms of cost, reliability, convenience, security, bandwidth, and ease of use.

Comparing channels.

Channel	Description	Advantages	Disadvantages
wire pair	a thin pair of copper wires used to relay telephone messages	cheap and simple to install	not totally robust, prone to interference and noise, low security, small bandwidth (500 kHz) due to conducting properties, high resistance means that amplifiers are needed regularly along long lines
coaxial cable	single solid copper core surrounded by insulator then a copper sheath (used as second conductor)	greater bandwidth (50 MHz) than wire pair, less prone to interference as outer sheath provides electrical screening	more bulky than wire pair, harder to install, less resistance (thicker) but amplifiers still needed, more expensive
fibre optic	thin glass fibre – signals relayed as light pulses	much higher bandwidth, more secure, much less prone to interference and noise (mostly digital transmission), comparatively cheap	regenerators needed, but much less frequently than with electrical cables (km lengths)
radio	electromagnetic waves	comparatively cheap, portable, extensively used	restricted bandwidth, AM prone to interference and noise, not secure – can be received by anyone tuned to the right frequency (digital systems allow encoding, but FM requires complex circuitry for encoding and decoding)
microwave	shorter-wavelength electromagnetic waves	higher frequency so greater bandwidth possible	line of sight (but that can mean greater security), used for digital transmissions

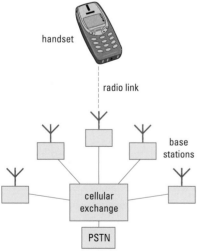

Mobile phones are linked to the PSTN network via a base station and cellular exchange.

The cellular phone network

In traditional telephone networks the handsets are essentially fixed in place with an individual wire running to each one. Calls are then routed electronically from one device to another via an exchange. This system, known as the public switched telephone network (PSTN), had to be supplemented during the 1970s and 80s as mobile telephone devices were developed and became popular, increasing demand on the PSTN.

A mobile phone is effectively a handset that contains a radio transmitter/receiver. When a call is being made or received, the handset establishes a radio (microwave) link to a nearby base station detected by its mobile phone aerial. The base station is linked to a cellular exchange via a cable and from there to the standard PSTN system, if required.

The cellular phone network divides the country up into a set of regions or **cells**. These cells can be quite wide where the density of phone traffic is low (e.g. 30 km in the countryside) and quite small where there are a large number of calls going on at once (e.g. 2 km in the city). Each cell contains a set of transmitter/receiver stations using a range of frequencies in the region of 900 MHz (which puts them in the microwave range of the electromagnetic spectrum). Cells are arranged so that no two cells next to each other use the same range of frequencies, to prevent interference.

When a mobile phone user attempts to make a call, the cellphone broadcasts a signal requesting a service on a special control channel. The nearest transmitter/receiver station will receive the signal and act on it. Once the call is connected the cell transmitter sends a signal back to the cellular phone instructing it to tune to one of the available frequencies for the two-way conversation. If the phone is being used in a moving car then the network can transfer the call to a new region as the car moves from one cell to another. This will involve the phone tuning to a new frequency as it moves from cell to cell.

Handset construction

At the core of a mobile phone handset is a microwave radio transmitter/receiver system. When the user speaks into the microphone the pattern of pressure waves formed by the voice is digitized by the analogue-to-digital converter (ADC). The digital number representing the voltage level at each sample is then converted into a bit stream by the parallel-to-serial converter. The resulting sequence is used to modulate the carrier which is then amplified and transmitted via the aerial.

The received signal is routed to a tuning circuit that isolates the specific carrier frequency allocated to the handset by the computer

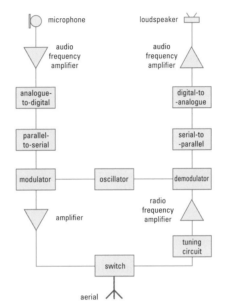

Block diagram of a mobile phone handset.

in the cellular base station. A demodulator extracts a bit stream from the tuned carrier, and the bit stream is passed to the serial-to-parallel converter to be restructured as a series of samples. The digital-to-analogue converter reconstitutes a pattern of voltages to drive the loudspeaker via an audio-frequency amplifier.

The worldwide web

The worldwide web (internet) is a distributed network of computers. This means that no one computer routes information across the network – the job is distributed over a variety of computers so that if any one of them malfunctions, information (**traffic**) can still flow via other routes. When a user wants access to the internet they must sign up with a provider. The company involved will provide the user with access to the internet via a telephone link or ISDN line. The user then 'logs in' to the provider's system, through which they can have access to web pages (including social networking, blogs, photo sharing, etc.), email, Twitter, and other facilities.

Access to the internet is fast becoming ubiquitous with the development of 3G and 4G networks for mobile browsing, wireless hotspots in shops and cafes, as well as faster telephone networks into businesses and homes.

WAP, 4G, and bluetooth

The wireless application protocol (WAP) is a global standard developed to make the internet available to mobile phone users. Although based on standard internet technology, WAP and the main internet sit side by side, with users having to produce mobile versions of their websites that are more easily used on devices with small screens.

4G is the fourth generation of mobile phone communication standards, replacing the previous third generation (3G), and designed to bring mobile ultra-broadband internet access to smartphones, laptops with wireless modems, and other mobile devices. Faster access and data compression could bring better gaming, high definition TV, and even 3D TV to mobile devices.

Bluetooth is a wireless technology and standard designed for the exchange of data over short distances (up to 30 m). It uses the 2400–2480 MHz band to create a local network with a high level of security. A frequency-hopping technology is employed to send chunks of data over 79 different 1 MHz sub-bands. Bluetooth can be used to link devices such as keyboards to computers, or mobile phones to in-car entertainment systems.

The internet is a system that allows the easy exchange of information between computers across the world. The system was developed at the European Laboratory for Particle Physics (CERN – for Conseil Européen pour la Recherche Nucléaire) in Switzerland, for exchanging information between physicists in different countries.

- AM and FM
- frequency ranges
- ground waves
- sky waves
- space waves

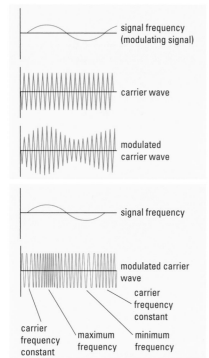

Amplitude modulation (top) and frequency modulation (bottom).

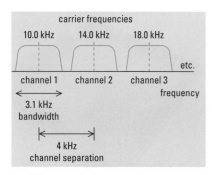

An FM transmission occupies a range of frequencies either side of the carrier frequency.

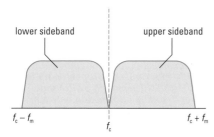

The frequency spectrum of an AM transmission has two sidebands about the carrier frequency, f_c. The width of the sidebands is defined by the maximum modulation frequency, f_m.

RADIO TRANSMISSION

The majority of radio signals are encoded by the modulation of a **carrier wave**. The carrier wave is a fixed signal at some frequency that defines the channel that is in operation – i.e. you have to tune to the frequency of a particular radio station. The information is then encoded by variations in that carrier signal. The pattern of variation that is imposed on the carrier represents the information to be transmitted. There are several ways in which radio signals are modulated.

Amplitude modulation (AM)

When a carrier is **amplitude modulated**, the amplitude of the carrier wave varies in response to the information signal. A positive signal increases the carrier's amplitude. When the signal goes negative, the carrier's amplitude reduces. Any audio signal (e.g. voice or music) can be used to amplitude-modulate an r.f. (radio-frequency) carrier wave for radio transmission – this is AM radio.

The frequency of the carrier needs to be a lot greater than the frequency of the modulating signal, otherwise the amplitude modulation is not very well defined.

The standard range of audio frequencies transmitted in AM radio is 50 Hz to 4.5 kHz, rather less than the complete range that the human ear can hear, which is 20 Hz to 20 kHz. This is why AM produces poor quality in comparison with FM radio, in which the frequency range extends up to 15 kHz. AM radio could transmit the range of audio frequencies that FM does. However, since anything that affects the amplitude of the signal (loss of strength, noise, etc.) is interpreted as part of the signal by the decoder, the system is inherently low-quality, so nobody bothers to transmit the full range.

Frequency modulation (FM)

In this system, the audio signal modulates the carrier by varying the **frequency** either side of the basic carrier frequency. The more positive the signal, the greater the increase in carrier frequency. Negative signals reduce the frequency. By convention, the maximum allowed modulation is set at ±75 kHz. Over-modulated signals will be distorted when decoded unless the receiver has a wide bandwidth.

In order to accommodate the wide bandwidths of broadcast FM signals, **VHF (very high-frequency)** radio frequencies must be used. VHF transmission must be by line of sight as the wavelengths are too short to diffract round obstacles. This is both an advantage and a disadvantage. It means that national radio stations have to have many repeater stations to re-transmit the signal, but it also enables local stations some distance apart to use the same frequencies as there is little chance of interference between them.

Frequency ranges

The radio part of the electromagnetic spectrum is split, by international convention, into the following ranges.

Frequencies and wavelengths for the classifications of radio waves.

Classification	Frequency range	Wavelength range
VLF (very low)	3–30 kHz	100–10 km
LF (low)	30–300 kHz	10–1 km
MF (medium)	300–3000 kHz	1000–100 m
HF (high)	3–30 MHz	100–10 m
VHF (very high)	30–300 MHz	10–1 m
UHF (ultrahigh)	300–3000 MHz	100–10 cm
SHF (superhigh)	3–30 GHz	10–1 cm
EHF (extremely high)	30–300 GHz	10–1 mm

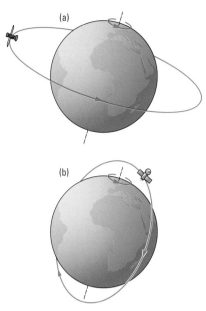

The full range of frequencies shown in the classification table is not used for public service broadcasting. Those frequencies that are used are listed below.

Public service wavebands.

Waveband	Frequency range	Wavelength range
LW (long-wave)	150–285 kHz	2.00–1.05 km
MW (medium-wave)	525–1605 kHz	571–187 m
SW (short-wave)	5.95–26.1 MHz	50.4–11.5 m
FM/VHF	87.5–108 MHz	3.43–2.78 m

Global radio

Setting up a global communications network depends critically on understanding the way in which radio waves travel from the transmitter to the receiver.

1 Ground (surface) waves These electromagnetic waves travel in close contact with the ground. The better the conductivity of the surface the further the wave will travel – transmission over water is especially good. The waves are sustained by inducing small electrical currents in the surface. These currents produce heat in the ground, sapping some of the energy of the wave. Long-wavelength ground waves can diffract round hills and buildings and follow the curvature of the Earth, giving them a long range.

2 Sky waves These waves travel in straight lines with no diffraction and do not hug the ground. Sky waves in the correct frequency range can be totally internally reflected off the ionosphere, which effectively increases their range.

3 Space waves Space waves are sky waves of a high enough frequency to pass through the ionosphere, i.e. greater than 30 MHz. These frequencies can be beamed through the ionosphere to satellites in orbit as space-wave transmissions which can then be beamed back to ground from the satellite. The small wavelengths are an advantage in this form of transmission as there are few diffraction effects to spread out the beam, and so it can be aimed accurately at the satellite and the receiving station.

Satellites in geostationary orbits (a) appear stationary above the equator. Hence transmission and reception aerials do not have to track a path across the sky. Satellites in polar orbits (b) will eventually pass over each point on the Earth's surface as the Earth turns on its axis.

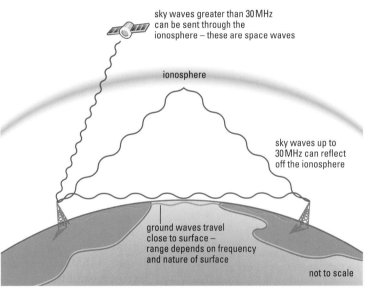

sky waves greater than 30 MHz can be sent through the ionosphere – these are space waves

ionosphere

sky waves up to 30 MHz can reflect off the ionosphere

ground waves travel close to surface – range depends on frequency and nature of surface

not to scale

How radio waves travel.

Ground waves and sky waves.

Frequency	Waveband	Ground waves	Sky waves
up to 3 MHz	LW and MW	long enough wavelength to diffract round hills and buildings little attenuation by ground range up to 1000 km	little importance as the ground wave has a greater amplitude
3–30 MHz	SW	range ~100 km some signal loss due to interference with the reflected space wave of the same frequency	transmissions can be by line of sight, or by reflecting off ionosphere giving longer range
30–300 MHz	VHF and UHF	range <100 km used for telephones, police, etc.	pass through ionosphere so must be line-of-sight, or be reflected off satellites (space waves)
>1 GHz	microwaves	range <100 km	space wave reflection off satellites, or line-of-sight microwave links for telephones, etc.

OBJECTIVES

- tuned circuits
- a simple MW radio
- the ferrite aerial
- the dipole aerial

TUNED CIRCUITS AND AERIALS

Imagine walking into a crowded room. Lots of conversations are going on at once. The sound waves reaching your ears are a complicated mixture of all the individual sound waves beings produced by the people in the room. In order to carry on a conversation with a particular person you have to concentrate on the part of the sound being produced by them – you need to 'tune in' to the sound waves they are producing.

The same is true when you select a particular TV or radio station. The aerial is receiving waves from *all the stations at the same time*. The receiver has to sort out the chaos and concentrate on the selected signal. In order to do this the receiver contains a **tuned circuit** that selects the station required.

A simple tuned circuit

The simplest type of tuned circuit contains a **variable capacitor** and an **inductor**. Variable capacitors work by having fixed plates between which are moving plates mounted on a common shaft. As the shaft is turned the amount of overlap between the fixed and moving plates changes. The capacitance of a parallel-plate capacitor depends on the area of overlap between the plates, so turning the shaft alters the capacitance.

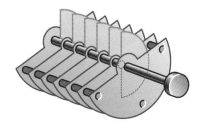

A variable capacitor works by adjusting the amount of overlap between fixed and moving plates.

Aerials (antennae) are made from conducting materials. Electromagnetic waves produce forces on the free electrons within the aerial, causing them to move back and forth. Each transmission arriving at the aerial produces its own pattern of current. If the transmission is AM then the tuned circuit needs to pick out the carrier frequency used by the station. All the other carrier frequencies will be producing alternating current (a.c.) in the aerial at the same time, but at different frequencies. The total current is passed into the tuned circuit.

Reactance

Reactance is a.c. resistance. An inductor has a resistance due to the resistivity of the wire that it is made from. When a.c. is passing through an inductor it will also have a reactance – the back e.m.f. induced will act as an extra 'resistance'. A.c. makes a capacitor charge and discharge rapidly – the capacitor appears to conduct charge. This gives it an a.c. reactance. (See spread 5.26 for more details.)

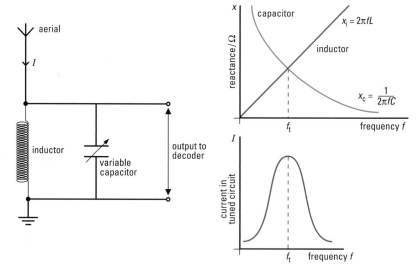

This simple tuned circuit relies on the a.c. resistance (reactance) of an inductor and capacitor.

The **reactance** of an inductor *increases* as the frequency *increases*. The reactance of a capacitor *decreases* as the frequency increases.

When the inductor and capacitor are connected in parallel the combined reactance of the circuit is a maximum at a particular frequency which can be selected by adjusting the capacitance of the variable capacitor. Currents produced by low-frequency carriers are conducted to earth through the inductor. Currents from high-frequency carriers flow to earth through the capacitor. At the tuned (resonant) frequency the current meets a large reactance, so a large p.d. is produced across the circuit.

Selectivity and sensitivity

A sensitive tuned circuit can receive very weak signals. This can be improved by including a radio-frequency (r.f.) amplifier with the aerial. Selectivity is the ability to separate stations with very similar carrier frequencies. Circuits sometimes have to compromise one to optimize the other.

A simple AM receiver

Once the AM carrier has been selected by the tuned circuit the signal needs to be demodulated. In a very simple AM receiver this can be done with a single diode.

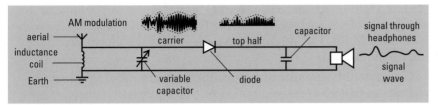

The diode selects half of the signal and then the headphones can only react to the lower-frequency modulation signal.

The diode trims away half of the signal, leaving the top half of the carrier which is modulated by the sound signal. The second capacitor is chosen to have a very low reactance at the frequency of a typical carrier. It will conduct the carrier part of the signal to earth and charge and discharge in time with the modulating signal. This ensures that a varying direct current (d.c.) is passed through the headphones which will reproduce the sound signal. Alternatively, an amplifier based round an op-amp can be used to amplify the signal and pass it to a loudspeaker. (For a suitable op-amp circuit see spread 13.8.)

Aerials

1 **The ferrite aerial** An AM radio uses a coil aerial wound round a ferrite core. This forms the inductor in the tuned circuit. Such an aerial detects the magnetic part of the electromagnetic wave (by electromagnetic induction). Alternatively, a long wire could be used as an aerial and the current from it passed into a coil, but to be efficient this aerial would have to have a length half the wavelength of the signal. BBC Radio 4 transmits at 198 kHz which is a wavelength of 1.5 km, requiring an aerial of 750 m. This is why medium-wave radios use ferrite aerials.

2 **The dipole aerial** The efficiency of the tuning process can be improved if the aerial itself is only sensitive to a narrow band of frequencies. One way of achieving this is to make the aerial a specific length. A dipole aerial consists of two pieces of wire which together are half the wavelength of the station. This requires impractically large wires for medium-wave reception, but VHF FM stations only require dipoles of about 3 m in length. The aerial will pick up other stations in the same band, so a tuned circuit is still required to eliminate their signals. Dipole aerials detect the electric field of the electromagnetic wave, so they must be aligned parallel to the plane of polarization or very little signal will be received: if you rotate the aerial you will notice the signal decrease (see spread 6.19).

3 **The multi-element aerial** Multi-element aerials are used to receive TV signals but the design is also used in high-quality VHF FM reception. Basically it is a dipole aerial with other elements added to increase the amount of energy extracted from the electromagnetic wave. The short director wires receive parts of the signal and the currents produced in them effectively re-radiate the signal. The directors are designed to be of such a size and length that their signals add to the wave received at the dipole. The reflector works in a similar manner. It is placed a distance $\lambda/4$ behind the dipole. The signal reflected from it has an extra path length of $\lambda/2$ to get to the dipole which, combined with the 90° phase shift on reflection, ensures that the reflected signal arrives in phase and adds to the energy at the dipole.

4 **Car aerial** A car aerial is half a dipole. As it is mounted on a large conducting surface (the car body) an 'image' of the aerial is produced. This makes up the other half of the dipole.

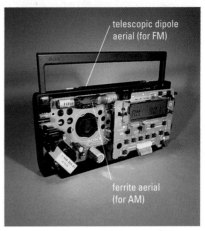

A typical multiband radio in which can be seen an internal ferrite aerial, for AM reception, and an external, telescopic dipole aerial, for FM reception.

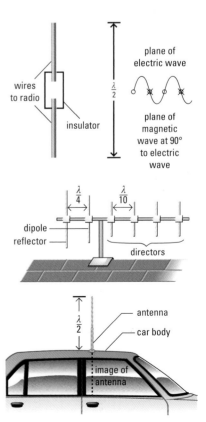

The dipole, multi-element, and car aerial.

DIGITAL INFORMATION

Although electronics has played an increasingly important role in our lives over the last 20 years, the real electronic revolution has only just started. Digital electronic technology promises a new information age in which data of any sort can be accurately and instantly retrieved from any part of the world. The internet is only the start of this global sharing of information.

Central to this revolution has been the development of simple techniques that allow information to be transformed into digital numbers that computers can understand.

Analogue and digital

With an **analogue** quantity the difference between two values can be very small or large, and can take on any value in between. For example, two lamps can be adjusted so that one is very much brighter than the other, or so that they are exactly the same brightness, or any difference in between these two. The brightness of the lamps is an analogue quantity and its size varies continuously, that is, smoothly with no gaps or jumps in value.

A **digital** quantity is all about gaps or jumps. With a digital quantity there is always a small difference between two values that are not the same. Digital quantities are often represented by binary numbers in which the smallest piece of information (the bit) is represented by either ON (1) or OFF (0) states.

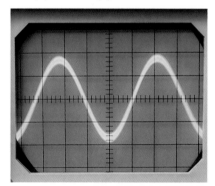 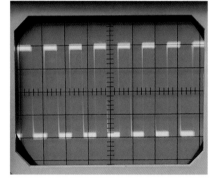

Information can be transmitted and stored in digital and analogue forms.

Advantages of digital information

The invention of reliable electronic processes for converting analogue quantities to digital quantities has caused a steady revolution in electronics, and many tasks that used to be carried out using analogue electronics are now done digitally. Digital information has several advantages over analogue information.

• Digital information is more easily stored – the storage medium (for example tape, computer memory, or laser disks) remembers a series of 1s and 0s rather than a range of values.

• Digital information is more easily transmitted – the transmitter and receiver need only tell the difference between 1 (ON) and 0 (OFF), rather than being sensitive to a range of values. This also makes it easier to amplify or 'boost' a signal en route without producing interference.

• Digital information is less prone to interference – random electrical signals (noise) are often confused with part of an analogue signal that is being transmitted. This is far less likely to occur with a digital signal because the noise is analogue and is very unlikely to be confused with a string of 1s and 0s.

The key to the digital revolution has been the conversion of signals from analogue to digital, their processing and transmission in digital form, and their eventual conversion back into an analogue form.

Analogue-to-digital conversion

One of the most common forms of **analogue-to-digital conversion** (A–D) is the digital recording of music. Music recording uses a technique known as **pulse-code modulation** (**PCM**).

PCM works by measuring the amplitude of the analogue music signal at regular intervals (a process known as **sampling**). This amplitude becomes a voltage in the electronics of the recording device and this is compared with a fixed set of voltages preset within the machine. For a compact disc player, the fixed set of voltages is numbered using binary numbers which are 16 bits long. There are 2^{16} (65 536) different voltage values that are numbered in the system (0000 0000 0000 0000 to 1111 1111 1111 1111). The number of the voltage nearest to the sampled value is then stored electronically. In this way the analogue signal is converted into a set of stored 16-bit digital numbers.

The number of times a second that this is done (the **sampling rate**) is fixed by the maximum frequency of signal that is recorded (the **Nyquist theorem**). To record all the information in a signal, the sampling rate must be at least twice the maximum frequency in the signal. The human ear can only hear up to 20 kHz, so the minimum sampling rate would have to be 40 kHz. In fact, 44.1 kHz is used because at the time the standard was set this slightly higher frequency made it easier to design the electronics that processed the signal before and after the digital conversion.

For every second of music a digital recorder has to store just over 44 thousand digital numbers that are 16 bits long.

A–D conversion

Assigning the actual voltage to a given level in the process of converting an analogue signal to a digital signal does introduce **quantization distortion**. This effect, due to the 'rounding off' of the signal, is especially serious at low volumes, as the gap between levels is a large fraction of the actual signal level. To overcome this it is normal to use close levels at low voltages and more spread-out levels at higher voltages (i.e. not uniform as in the diagram below).

Altering the sample rate also involves potential problems. Too low a sample rate will mean that high frequencies in the analogue signal will be missed out. A problem will also be caused at the D–A stage. The squared-off signal produced by the converter will have signals in it that are higher than the sample frequency – but if that sample frequency is too low then these false signals will be in the audible part of the spectrum. This problem is known as **aliasing** and is avoided by ensuring that the sample frequency is beyond the range of hearing.

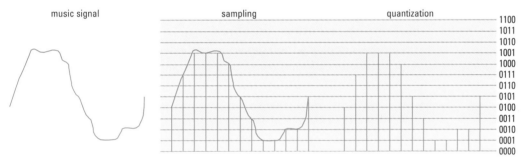

(0100) (0111) (1001) (1001)
(1001) (1000) (0101) (0011)
(0001) (0001) (0010) (0010)
(0101)

stored digital strings of numbers representing the music signal

Pulse-code modulation.

Digital-to-analogue conversion

A CD player needs to read the digital data off the disc and recreate the music using **digital-to-analogue conversion**. The digital numbers are read at the rate of 44.1 kHz. The voltage represented by each number is then generated, giving an odd-looking 'squared-off' representation of the original music waveform. If this is then treated as an analogue signal it is found to contain all the frequencies of the music, plus many more that are higher than the 20 kHz maximum in the original. The 'squared-off' signal is passed through a filter that removes all frequencies higher than 20 kHz, and this results in an accurate reproduction of the original signal.

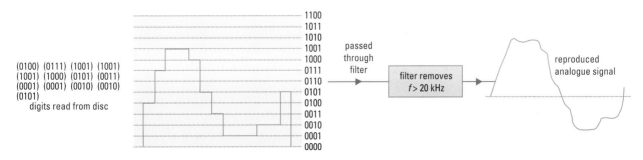

(0100) (0111) (1001) (1001)
(1001) (1000) (0101) (0011)
(0001) (0001) (0010) (0010)
(0101)

digits read from disc

passed through filter

filter removes *f* > 20 kHz

reproduced analogue signal

Digital-to-analogue conversion.

- compact disc format
- the laser head

Compact discs, shown here being manufactured, were the first large-scale digital storage system to be introduced to the domestic user. Although they were originally designed to store musical information in digital form they rapidly found applications in computer and video systems.

THE CD PLAYER

Compact disc (**CD**) players were first introduced to the European market in 1983. Their design and development were the result of a joint venture between the Dutch Phillips company and the Japanese Sony corporation. Phillips had done the preliminary work on the system and had announced in 1978 the intention of bringing it to the mass market. Sony's contribution was to help in the marketing and development, especially the techniques for correcting any mistakes in reading the information from the disc. Although the two industrial giants agreed on the basic format of the disc and how the information was to be encoded onto it, they developed different systems for converting the digital information back into sound. In the early 1990s a new technology was developed that effectively merged the two systems. In addition, further applications of the disc in computer storage and video recording started to be developed.

Optical disc technology has been extended further since the introduction of the CD. **Digital versatile discs** (**DVDs**) were introduced in 1995 using a smaller-wavelength laser and improved error-correction technology, which allowed a higher density of data to be stored (4.7 GB rather than 700 MB). In 2006 Blu-ray discs advanced the technology to the double-layer 50 GB capacity needed for high-definition video. The primary development behind this technology was the production of 405 nm (blue) semiconductor lasers.

The format of the disc

A compact disc is a plastic disc of 120 mm diameter. The sound is encoded onto the disc in digital form using the pulse-code modulation (PCM) system outlined in spread 13.6.

- Sound is sampled 44 100 times per second: each sample is compared with a reference sound level set by the system.
- There are 65 536 reference levels.
- Each reference level is given a digital code number between 0000 0000 0000 0000 and 1111 1111 1111 1111 – this is called **16-bit encoding**.
- So, there are 44 100 16-bit digital numbers that have to be recorded onto the disc for every second of music.

The disc The digital numbers are recorded onto the disc as a series of pits. A metal stamper impresses these pits onto the plastic disc which is then coated with a reflective layer of aluminium or gold. Finally, the disc is overcoated with a transparent protective film.

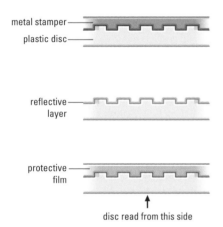

The three stages in making a compact disc. Note that the disc is read from the underside in this diagram (the label you see on a disc is on the protective film).

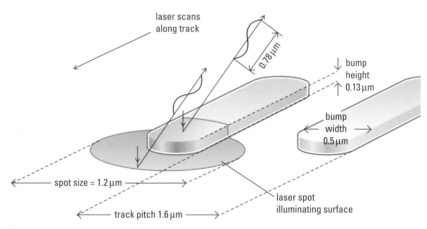

Waves reflected off the top and bottom of a bump are out of phase and interfere destructively – this is read as a digital zero.

The compact disc player

There are two major parts to a compact disc player. The **transport** reads the digital information from the disc and sends it to the **converter** which is responsible for the digital-to-analogue conversion. A CD-ROM drive is a transport section of a CD player. In some of the most expensive players the transport and converter are housed in separate boxes so that their electronic systems do not interfere with each other, but most compact disc players have transport and converter in the same box, isolated from each other by a Faraday cage (see spread 5.10).

The transport The information is read from the disc by reflecting laser light from the back surface. This surface contains a pattern of bumps which mirror the pits produced by the pressing.

Compact discs are read from the inside outwards as the pattern of bumps follows a spiral path from the centre of the disc outwards. The information must be taken from the disc at a constant rate, so the laser tracks over the disc at a constant linear speed. To achieve this the angular velocity of the disc increases from about 200 r.p.m. at the start to 500 r.p.m. when the transport is reading information from the outside edge.

The laser head A semiconductor laser produces a beam of red light (780 nm wavelength) that passes through a **beam splitter**. On the way up the light passes through the splitter to the lens where it is focused onto the reflective layer. The beam is unfocused when it strikes the transparent protective layer on the disc. This helps the laser to pass through any minor scratches on the surface. Light reflected back from the disc is passed to the photodetector by the beam splitter.

The height of one of the bumps on the disc is set to be approximately 1/4 of the laser wavelength. Hence there is a path length difference between light reflected from a pit and from a bump of $\lambda/2$. Consequently, when the laser spot illuminates the edge of a bump, the light from the top surface destructively interferes with the light from the pit and no reflected intensity is detected (binary 0). If the laser spot illuminates a patch of flat surface (top surface of a bump or between bumps) that is wider than it, then all reflections will be in phase and constructive interference takes place (binary 1).

Correcting for errors in reading the disc

With so much information packed into such a small space it is inevitable that mistakes will be made in reading some of it from the disc. There are several ways in which the manufacturers have compensated for such errors.

- The music does not play in sequence along the track. The digital numbers corresponding to a snatch of music are scattered round the disc. This means that if a portion of the disc is damaged, only part of any section of music is lost (a fraction of a second). This is known as **interleaving**. Consequently the laser head has to be accurately steered round the disc and is shuttling to and fro all the time as the disc is playing. This means that instead of increasing smoothly from inside to outside the disc speed is continually changing. All this calls for very high-quality motors and good control electronics. Other digital information is included on the disc to allow the transport to keep track of its location.

  The transport is very sensitive to being knocked as a jolt can move the laser head the wrong way – it loses its place! This may cause the player to go back to the start of the track or jam completely, so portable CDs and in-car CDs have very sophisticated designs to prevent vibration disturbing the laser head.

- If a small fraction of a second is lost due to damage to the disc the player can 'guess' what the missing part was by looking at special **'error correction codes'** on the disc, and also at the pattern of the data before and after the missing part.

'Cutting' a master CD.

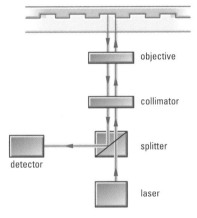

A cross-section through a laser head which reads the digital data from the disc. Light reflected from the disc is sent to the detector by the beam splitter.

- amplifiers in general
- the ideal amplifier
- the operational amplifier
- the inverting amplifier

AMPLIFIERS

Producing music, either by instruments or by playing back recordings, has become one of the most popular of leisure activities. In recent times MP3 players and home hi-fi system have become vital, in sales terms, to the consumer electronics industry, rivalled only by computer games consoles and tablet computer systems. All of these devices use a combination of digital and analogue electronics. The preferred means of storing music, however, is digital. Digital data has, at some point, to be converted into an analogue waveform and sent to loudspeakers so that the electronic signal can be converted back into sound.

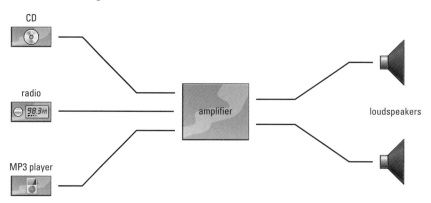

A typical modern hi-fi system.

A vital link in the stage between replaying the signal that is stored digitally and producing the sound is the **amplifier**. This device takes the analogue signal and increases its amplitude until it is strong enough to drive the loudspeakers.

Amplifiers of different types are used in a wide variety of applications other than in hi-fi. Telephones, radios, and televisions, for example, all use amplifiers.

The ideal amplifier

An ideal amplifier would have several characteristics that engineers try to design into real amplifiers.

- **Gain** – this is the ratio of input to output:

$$G = \frac{V_{\text{out}}}{V_{\text{in}}}$$

or

$$\frac{I_{\text{out}}}{I_{\text{in}}}$$

  if the amplifier was designed to amplify current, not voltage.

  Some amplifiers need to have a large gain (for example, $G = 10^5$), while others have a small gain ($G = 2$). In either case the gain should not depend on the frequency of the signal.

- **Low distortion** – the output signal should look exactly like the input signal, except bigger! Any difference between the two is distortion. A good hi-fi amplifier will have 0.01% distortion or less.

- **Input resistance** – this should be as large as possible so that the amplifier does not draw too much current from the source of the signal.

- **Output resistance** – this should be as low as possible so that the amplifier can deliver high currents to the device connected to the output, for example, the loudspeakers.

R_{in} and R_{out} are not real single resistors, but the effective resistance of the whole circuit input and output

The ideal amplifier.

The operational amplifier (op-amp)

Operational amplifiers were designed to be nearly ideal amplifiers for use in the first types of computer. Since then they have been modified and miniaturized, and can now be bought as an individual integrated circuit. They are a general-purpose amplifier with a high specification and performance.

Typical op-amp specifications.

amplifier	R_{in}	R_{out}	open-loop gain (G)
741	2 MΩ	75 Ω	10^5

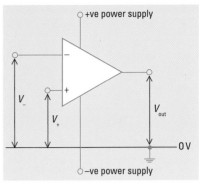

Some operational amplifiers use dual-rail power supplies to allow them to amplify positive and negative voltages.

The **open-loop gain** is to the gain of the integrated circuit itself without any external components.

- The power supply is **dual-rail**, that is, there is both a positive and a negative supply line, and a centre tap which is connected to earth (0 V). This allows the amplifier to deal with a.c. signals, as well as d.c. signals with a varying amplitude.

- The op-amp has two inputs: '–' (inverting) and '+' (non-inverting). The inverting input swaps negative for positive on the input signal as it is amplified.

$$V_{out} = G\,(V_+ - V_-)$$

As G is very large $(V_+ - V_-)$ must be very small or the amplifier will **saturate** (see below).

The inverting amplifier

The diagram on the right shows a circuit for a typical low-gain amplifier made from an op-amp. In this circuit the non-inverting input is not used and so is connected to 0 V.

Note that generally the power supply connections are missed off such diagrams to save cluttering them up.

The **closed-loop gain** of the circuit is

$$g = \frac{R_2}{R_1}$$

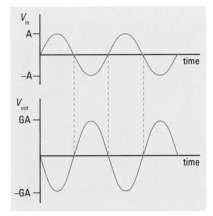

Input and output signals for an inverting amplifier.

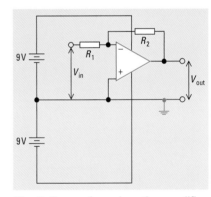

Circuit diagram for an inverting amplifier.

Worked example

In the circuit on the right, for an inverting amplifier, $R_1 = 2.7\,\text{k}\Omega$ and $R_2 = 10\,\text{k}\Omega$. The input signal is also shown. Sketch the output signal.

The gain of the circuit is

$$g = \frac{R_2}{R_1} = \frac{10\,\text{k}\Omega}{2.7\,\text{k}\Omega} = -3.7$$

The amplitude of the signal is 4 V, so the maximum output voltage is

$$V_{out} = g \times V_{in} = 3.7 \times 4\,\text{V} = 14.8\,\text{V}$$

The power supply for the circuit is only ±9 V, so this is the maximum ouput that the amplifier can provide. Under these conditions the amplifier is saturated and will remain so as long as the size of the input voltage is greater than 9 V/3.7 = 2.4 V.

The resulting output is also shown in the diagram below right.

Feedback

Part of the output of the inverting amplifier is **fed back** to the input via R_2. This is a common technique used for reducing distortion and for improving the general behaviour of amplifiers. Distortion causes the output to be different to the input. This difference is also passed back via R_2, which alters the input and compensates for the distortion. Negative feedback reduces the overall gain of the amplifier, but reduces distortion and makes the gain more stable if the power supply or temperature of the circuit varies.

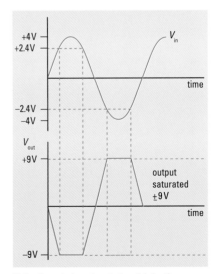

If the input signal gets too high, the output will saturate to the supply voltage.

13.9

OTHER OP-AMP CIRCUITS

The operational amplifier has become the general workhorse of analogue electronics. Op-amps are cheap, easy to use, and have a wide range of applications. The examples in this section are just a few of the different circuits in which these amplifiers are used.

The comparator

This circuit uses both inputs to the op-amp. The output voltage, which is driving the lamp, is given by $V_{out} = G(V_+ - V_-)$ and because G is very large ($\sim 10^5$) the circuit is very sensitive to differences between V_+ and V_-. If $(V_+ - V_-) > 90\,\mu V$ the output of the op-amp will saturate and the lamp will come on at full brightness. If $V_- > V_+$, the output of the op-amp will be forced down to $0\,V$. (In this circuit there is no negative supply line.)

When the light-dependent resistor (LDR) is dark, V_+ is high, and when it is light, V_- is low. The variable resistor is used to set the light level at which the op-amp will switch on the lamp. The LDR could be replaced with a thermistor or another sensing device.

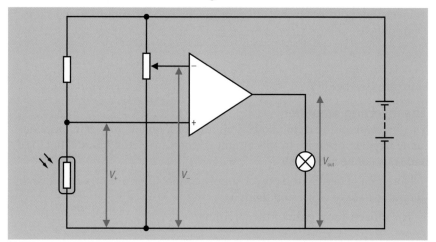

The comparator circuit.

The non-inverting amplifier

This circuit uses negative feedback, but does not invert the output signal. The gain of the circuit is:

$$g = 1 + \frac{R_2}{R_1}$$

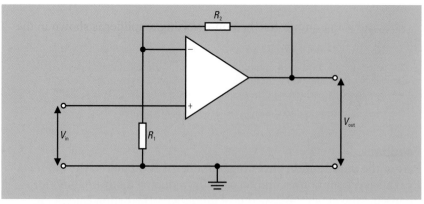

The non-inverting amplifier. NB This would use a dual-rail power supply that is not shown on the diagram.

As a special case of this circuit, a gain of 1 can be achieved by setting R_2 to zero and R_1 to infinity (i.e. a wire and a gap respectively). This is known as the **voltage follower** because the output is locked to the same voltage as the input. The advantage of such a circuit is that the op-amp has a very high input resistance and so it can 'measure' V_{in} without drawing any current.

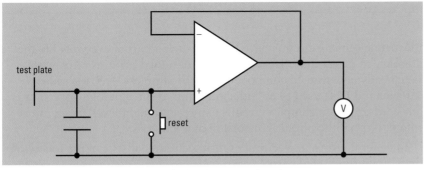

The voltage follower as used as a device for measuring charge.

The diagram above shows a voltage follower with a capacitor to make a charge-measuring device. If a charged object touches the test plate, it will transfer charge to the capacitor, and the p.d. between its plates will rise. If the capacitor were connected directly to a voltmeter this charge would drain away through the meter and a proper reading would not be obtained. The op-amp has such a high resistance that practically no charge is removed from the capacitor, yet the voltage can be measured.

Analysing op-amp circuits

Circuits that involve op-amps can be easily analysed provided that two approximations are used:

1 The input resistance is so high that the op-amp does not draw any current.

2 The gain is so high that V_+ and V_- have to be at virtually the same p.d.

The second approximation is sometimes known as **virtual earth**. In some circuits either V_+ or V_- is 0 V (as the input is connected to earth), in which case the other input must also be virtually at earth potential.

Using these ideas the circuit of the inverting amplifier can be reduced to that shown in the diagram on the right.

As there is no current into the input of the op-amp, the current in R_1 is the same as that in R_2. If V_{in} is positive, then V_{out} is negative (it is an inverting amplifier) so:

$$I = \frac{V_{in}}{R_1} = -\frac{V_{out}}{R_2}$$

$$\therefore\ g = \frac{V_{out}}{V_{in}} = -\frac{R_2}{R_1}$$

Obtaining the result is simple once the correct circuit has been obtained.

The equivalent circuit for the non-inverting amplifier is shown in the diagram below.

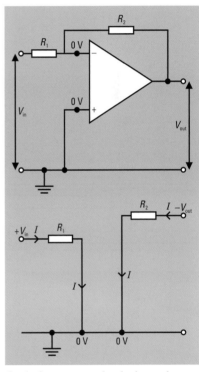

Analysing op-amp circuits is much simpler if the op-amp is replaced by an equivalent circuit.

PRACTICE

1 An op-amp with an open-loop gain of 2×10^6 is to be used in a comparator circuit. The power supply is ±18 V. What voltage difference will cause the comparator to switch from one state to another?

2 Use the diagram on the right to prove the equation for the gain of a non-inverting amplifier.

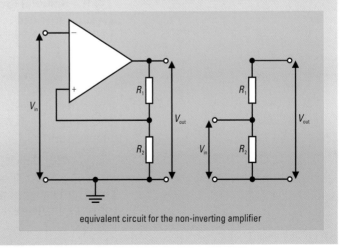

equivalent circuit for the non-inverting amplifier

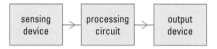

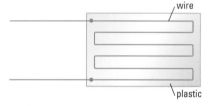

Block diagram of a remote sensing device.

SENSING CIRCUITS

Electronic sensing circuits have become such a part of everyday life that it is easy to forget that they are specially designed features often with specifically constructed components. At the simplest level, the sensor can be a red LED used as an indicator lamp to show that an appliance is powered up. More complex circuits can react to light level, temperature, sound, or even the physical distortion of a material.

In general terms an electronic sensing circuit consists of the following three main functional parts.

- A **sensing device** takes the form either of a transducer, that converts some physical input (such as sound) into an electrical signal, or of a component that changes its electrical characteristics when one of its physical properties changes (such as temperature, illumination level, or size).

- A **processing circuit** reacts to changes in the sensing device and produces an appropriate output depending on the change. An operational amplifier set up as a comparator circuit would be an appropriate simple processing circuit.

- An **output device** displays some indication of the change that has taken place.

Sensing light or temperature

Light-dependent resistors (LDRs – see spread 4.9) drop their resistance as the illumination level rises and thermistors (spread 4.9) become less resistive as their temperature increases. This makes them ideal sensing components, coupled with a potential divider circuit (spread 4.10) to convert their change in resistance to a p.d. This p.d. can be detected by an operational amplifier configured as a comparator (spread 13.9). The output of the comparator can then be used to control an indicator or to trigger an appropriate response (lighting a lamp in low light levels, or switching on a cooling circuit when the temperature rises above a set point, for example).

Sensing strain

It is frequently necessary to measure the physical distortion of an object subjected to forces. This may be to check that the material is operating within its design tolerances, as in aerospace, machine tooling, automotive, architecture, and other engineering applications. Biometric, medical, and dental applications also benefit from the use of a simple **strain gauge**. Such devices consist of a length of fine wire that is sealed into a thin plastic sheet, which can be easily bonded to the surface of the test material.

If the material is distorted, the plastic of the strain gauge experiences a strain (provided it is well bonded to the material), which in turn alters the length of the wire within it. The relationship between resistivity and resistance:

$$R = \frac{\rho l}{A}$$

shows that the resistance of the wire will vary in proportion to its length (assuming that the cross-sectional area does not change significantly under strain) so:

$$\Delta R = \frac{\rho \Delta l}{A}$$

The strain gauge can be placed into a potential-divider circuit, which converts the distortion of the material into a detectable p.d.

wire

plastic

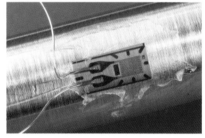

The resistance of a simple strain gauge varies as the device deforms under stress.

Sensing sound

In a piezoelectric crystal such as quartz, SiO_2, the structure contains polar bonds with a greater electron density around the atoms with the most attraction for electrons. Overall the crystal does not display a charge distribution because the arrangement of dipoles cancels out. However, if the crystal is subjected to stress, and as a result physically deforms, then the dipole moments separate out and a p.d. develops across the crystal. The polarity of the p.d. depends on the compression or expansion of the crystal. Painting a pair of crystal faces with metal allows an electrical contact to be made, and so a current is produced as a result of the stress.

Piezoelectric transducers are used in microphones where the pressure variations of a sound wave deform the crystal and generate an electrical signal. Piezoelectric transducers are also used to create and detect ultrasound signals in scanners, and they can be used to detect audible sound waves and hence be used in sensing applications.

Output devices

Operational amplifiers are excellent general-purpose amplifiers used in a variety of applications; however they are unable to produce output currents greater than 25 mA or so. This is insufficient to drive output devices such as electrical heaters, spotlights, and other high-current devices. The relay (spread 5.18) provides an easy solution to this, with the output of the amplifier feeding the relay coil that magnetically closes a contact, routing a much higher current to the output device.

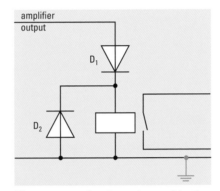

Electromagnetic relays triggered by the output of an operational amplifier can be used to route large currents to output devices such as heaters and spotlights.

The amplifier is capable of producing positive and negative p.d.s. Diode D_1 ensures that the relay operates only when the output is positive and switches off when the output turns negative. When the relay opens a back e.m.f. produced by electromagnetic induction could damage the amplifier by driving a large current back through it. Diode D_2 protects the amplifier by drawing the current away.

In a simpler application an indication of the positive or negative nature of the amplifier's output can be shown by a pair of LEDs arranged as shown on the right. D_1 lights up if the p.d. is positive and D_2 if the p.d. is negative.

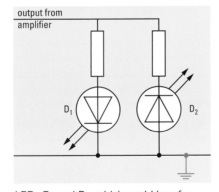

LEDs D_1 and D_2, which could be of a different colour, indicate the polarity of the amplifier's output.

A voltmeter connected across the potential divider or to the output from an amplifier provides a recordable output, for example in response to a temperature or illumination sensor. Interpreting the reading requires calibration as the sensor's response is generally not linear. The sensor could be used to measure a range of known temperatures, for example, and the p.d. on the voltmeter recorded. From this a calibration curve of temperature against p.d. could be plotted allowing the reading to be interpreted in other contexts.

13.11

OBJECTIVES

- multipath dispersion
- types of fibre
- material dispersion
- choice of material for fibres

FIBRE-OPTIC COMMUNICATIONS

Digital communication via optical fibres has several distinct advantages over other communications systems.

1 The fibres themselves are capable of an enormous information-carrying capacity. In a telephone system a pair of fibres can carry 8000 conversations at once. A typical bundle of fibres making up a cable can support 370 000 simultaneous conversations. Old-style coaxial cables can only support 2000 conversations at best.

2 The fibres are much lighter and easier to handle than copper cable, although connecting them can be trickier.

3 Optical fibre systems are much more noise-resistant than cables that carry information by patterns of electric current.

4 Cross-talk (the process by which information from one channel can leak onto another) is considerably reduced by using optical fibres.

5 Information can be transmitted over greater distances without degrading to the point at which it has to be boosted or regenerated.

How an optical fibre works

Information is transmitted down an optical fibre, digitally encoded as pulses of electromagnetic radiation in the infrared part of the spectrum (see spread 6.6). In order for a digital pulse to be correctly received, and hence decoded, it is important that it have very sharply defined edges. Without this, the exact moment at which the transition takes place in the pulse becomes obscured and the distinction between 1 and 0 harder to read in a rapidly changing sequence of pulses.

Optical fibres are designed to allow the pulse to travel with the smallest amount of '**rounding**' possible. The chief cause of this rounding is **multipath dispersion**. The infrared waves travel along the fibre by totally internally reflecting off the boundary between the core and the cladding of the fibre (see spread 6.6). This will happen as long as the angle at which an individual ray strikes the boundary is greater than the critical angle. Clearly there are many different paths that a ray can take along the fibre by totally internally reflecting in this way.

A pulse of infrared light signifying a digital 1 shone into a fibre will produce a vast number of rays following different paths along the fibre. Each of these rays will have a different distance to travel, depending on the path taken, before it can reach the other end. Consequently, although all the rays start from one end at the same moment, and they all travel at the same speed (the speed of light in the material of the fibre), they do not all arrive at the other end together. This has the effect of turning a sharp increase in intensity at the transmission end (pulse on) into a more gentle increase at the receiving end – the pulse is broadened.

Types of optical fibre

The first type of fibre to be developed is now known as a **step index** fibre. The central core is typically 50 μm in diameter and surrounded by cladding which takes the whole fibre to 125 μm. Such fibres suffer from multipath dispersion (30 ns km^{-1} is a typical dispersion for such a fibre) and consequently cannot be used for rapidly changing signals over long distances. However, they are cheap, and can be used for some low-capacity local systems.

One way of solving the mutipath dispersion problem is to use a **graded index** fibre. These have cores and cladding of about the same diameters as the step index fibre, but rather than using just two slightly different materials, the core is made so that there is a gradual change in its refractive index across the diameter. This is arranged so the speed of light increases with diameter across the core. Consequently, rays that follow paths which give rise to a greater transmission distance also spend more

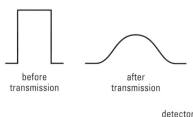

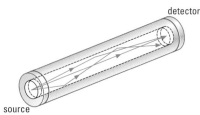

Multipath dispersion.

before transmission

after transmission

detector

source

of their time towards the outside of the core, and so they pass mostly through the material in which their speed is greater. As a result, they can travel the greater distance in about the same time as the rays that stay nearer to the centre of the core. This reduces dispersion (typically to $1\,\text{ns km}^{-1}$). The graded index fibre was widely used in early high-capacity systems, but has since replaced by **monomode fibres**.

Monomode fibres cut down multipath dispersion by using a significantly reduced core diameter ($5\,\mu\text{m}$). Although this significantly reduces the amount of energy that can be transmitted down the fibre, it also reduces the difference in distance between the rays that travel straight down the fibre, and those that internally reflect off the boundary with the cladding. In essence, as the diameter of the core is similar in magnitude to the wavelength of the waves, all the rays travels more or less directly down the centre of the core.

Material dispersion

As the core of a monomode fibre is so thin, a laser generally has to be used as a light source. Semiconductor lasers are now very cheap and small, and are consequently ideally suited to this sort of work. They also have the advantage of having a smaller range of frequencies in their radiation than a light-emitting diode (LED) (for example). This also reduces dispersion in the signal – refractive index changes with frequency and a signal of many frequencies will be dispersed due to the different speeds. This is known as **material dispersion**.

Choice of material

There are three main mechanisms by which light can be attenuated when passing through a material.

1 Impurities in the material scatter light passing through.

2 The material can absorb some of the light.

3 Crystalline defects in the structure of the material scatter the light.

The last of these is called **Rayleigh scattering** and is proportional to $1/\lambda^4$ where λ is the wavelength of the light. The graph on the right shows the absorption curve for silica (used in manufacturing fibre optics). At low wavelengths the absorption is dominated by Rayleigh scattering, at high wavelengths absorption is the most significant factor. The large maximum in the middle is due to absorption by OH^- radicals, which are almost impossible to eliminate. The two minima at $1.3\,\mu\text{m}$ and $1.55\,\mu\text{m}$ correspond to infrared wavelengths and are generally chosen for signalling work.

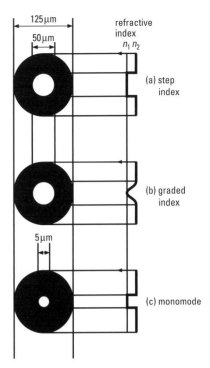

Types of optical fibre.

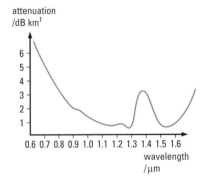

Endoscopes

An **endoscope** is a thin flexible optical instrument used for imaging inside the body or inside machinery. The optical system primarily consists of two light guides using fibre-optic bundles. One bundle carries light from the device to the imaging area. This bundle carries only diffuse illumination and is **non-coherent**, which means that the individual fibres are not aligned along the length of the bundle. The other bundle carries image information back to the operator. The image consists of a collection of pixels and each fibre carries the light data for just one pixel. To transmit the image the fibres must be aligned along the length of the bundle so that there is a one-to-one correspondence between the position of the fibre at the imaging end and its position at the operator's end (this is a **coherent** bundle). A small lens routes the light into the bundle at the imaging end, and an eyepiece is used to view the image at the operator's end.

Modern endoscopes carry a camera at the imaging end which relays the image back wirelessly. These devices need only one non-coherent fibre bundle to provide the illumination, so they can be thinner and cheaper.

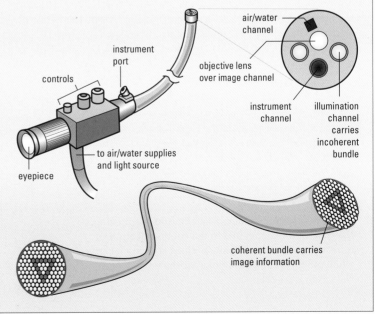

PRACTICE EXAM QUESTIONS

1 a With the aid of a diagram, describe the nature of electromagnetic waves in a vacuum in terms of electric and magnetic fields.

b Hertz discovered how to produce and detect radio waves, as illustrated in Figure 13.1.

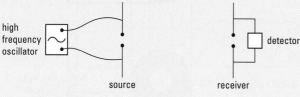

Figure 13.1

i He found that no radio waves could be detected if a large metal plate was placed between the radio wave source and the receiver. Use your knowledge of electromagnetic waves to explain this observation.

ii He also found that no radio waves could be detected if the receiver was turned through 90° about the line between the source and the receiver. Use your knowledge of electromagnetic waves to explain this observation.

2 Figure 13.2 shows the simplest possible way of transmitting the human voice by electromagnetic radiation. A microphone picks up the voice signal which is passed into an audio amplifier. The amplifier output is connected directly to a dipole aerial matched to an average voice frequency. Some of the electromagnetic radiation from the transmitting dipole is picked up by a receiving aerial of equal size. The signal received is amplified by an audio amplifier and passed to a loudspeaker.

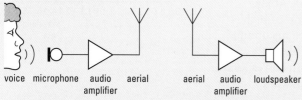

Figure 13.2

a Suggest reasons why radio stations do not broadcast audio signals in this simple manner, and explain how, in practice, the information is broadcast.

b With the aid of a block diagram, explain how a radio receiver operates. You should include sketch graphs of the signal at various stages in the receiver.

c Modern portable telephones are essentially small radio transmitters and receivers. In what respects do the transmission systems of portable telephones differ from those of commercial radio stations? Comment on the social implications of the availability of portable telephones.

3 Considerable use is made of satellites in monitoring the state of the planet and in the communication of information. Some of these satellites are in polar orbits while others are in geostationary orbits.

a Explain, with the aid of suitable diagrams, how the position of polar-orbiting and of geostationary satellites changes with time relative to the Earth below.

b Discuss the advantages and disadvantages of the use of such satellites over earlier methods of data gathering and communication.

c Discuss the impact on society of the use of satellites.

4 Telephone messages are transmitted using light as a carrier, along glass fibres which may be many kilometres in length.

a State *two* factors associated with the microstructure of glass that will reduce the distance over which light may be transmitted along a single fibre.

b Semiconducting lasers are used to generate the light in optical-fibre systems. State four properties that make this type of light source suited to this application.

5 A sinusoidal signal of frequency 1.0 kHz and amplitude 0.50 V is to be transmitted by means of a carrier wave of frequency 100 MHz and amplitude 50 V. The carrier wave is to be either amplitude modulated (AM) or frequency modulated (FM).

a What do you understand by a carrier wave?

b Describe, giving numerical values where possible, the form of the carrier wave when it is

i amplitude modulated,

ii frequency modulated, the variation of the carrier wave frequency being at a rate of 8.0 kHz per volt of the modulating wave.

c i By reference to an amplitude modulated wave, explain what is meant by bandwidth.

ii Explain the implications of bandwidth for radio reception.

6 Figure 13.3 shows the frequency spectrum of the signal from a radio transmitter.

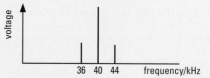

Figure 13.3

a What name is given to this type of radio transmission?

b Describe quantitatively how this signal has been formed from its components.

c Copy Figure 13.4 and sketch a graph of this radio signal as a function of time for a time.

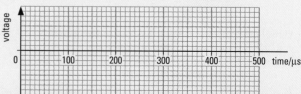

Figure 13.4

d State the bandwidth of this signal.

e In which radio waveband does this signal belong?

7 **a** State *two* advantages which can be gained by transmitting a signal digitally rather than in analogue form.

b Explain the following disadvantages of digital signals and state how they are overcome.

    i An analogue signal, reconstructed after digital transmission, has a stepped shape rather than that of the original smooth analogue signal.

    ii A digital signal requires a high bandwidth for transmission.

c An analogue signal, as shown in Figure 13.5, is to be converted into digital form and transmitted along an optic fibre.

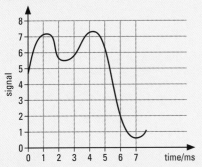

Figure 13.5

Assuming that it is digitised into three-bit words at times 0, 1, 2, …, 7 ms, what digital signal needs to be transmitted? Explain how you have obtained the three-bit digital word at $t = 0$.

d A digital signal of initial power 70 mW is transmitted along a fibre of length 3000 m. If the absorption coefficient a is $4.2 \times 10^{-4}\,\mathrm{m}^{-1}$, calculate the power received at the end of the fibre.

(The power falls exponentially as given by the equation $P = P_0 e^{-ax}$.)

e Draw sketches to show how the shape of a typical digital signal changes between the input and the output of the optic fibre considered in **d**.

8 **a** Explain the need for multiplexing in communication systems, and describe two different methods of multiplexing.

b Explain the need for modulation in communication systems. Illustrate your answer by describing the processes of amplitude modulation (AM) and pulse-code modulation (PCM).

c Discuss the relative advantages and disadvantages of pulse-code modulation over amplitude modulation for the communication of information.

9 **a** State three advantages of optical fibres over electrical cables for long-distance communication.

b Explain how several telephone conversations can be transmitted at the same time along a single optical fibre.

c An optical fibre is to be used for the transmission, in digital form, of a number of television channels. One such television channel carries 25 pictures per second and each picture is made up of 625 lines. Each line is composed of 600 colour points (pixels), and an eight-bit binary code is used for each point in order to control the brightness and colour at that point.

Calculate, for each television channel,

    i the rate at which colour points are sampled,

    ii the rate at which bits are transmitted.

d The light entering the optical fibre in **c** is produced by a laser which can be switched on and off at a frequency of 600 MHz.

    i What is the maximum number of television channels which could be transmitted along a single fibre?

    ii Suggest why, in practice, the actual maximum number of television channels is less than the number you have calculated in **i**.

10 **a** A radio enthusiast communicates with a friend on the opposite side of the Earth by means of a short-wave radio link operating at a frequency of approximately 15 MHz.

    i Explain, with the aid of a diagram, how the ionosphere may be used to propagate radio waves from one side of the Earth to the other.

    ii Calculate the length of the dipole aerial which could be used as a receiving antenna for the radio waves.

b A television broadcasting system operates at frequencies in the range 3–20 GHz by means of a satellite in geostationary orbit.

    i Explain why television broadcasting is achieved by means of satellites rather than by reflection from the ionosphere.

    ii Describe what is meant by a geostationary orbit and explain the advantage for communication of satellites placed in such orbits.

    iii Describe the structure of a reflecting dish aerial suitable for use as a receiving antenna.

    iv Suggest a reason why a dish aerial is used with satellite television but a dipole aerial is used for terrestrial television broadcasts.

11 **a** What is meant by the term modulation?

b Why is it necessary to use a modulated carrier wave in the transmission of a radio signal?

c Describe, using diagrams, the two common types of modulation used for analogue radio transmission. State the name of each type of modulation and label your diagrams to show relevant frequencies.

d What is meant by the term bandwidth?

e Identify and explain three problems which arise because of the need for bandwidth.

12 **a** With the aid of a labelled diagram, distinguish between the methods of propagation of

    i ground (surface) radio waves,

    ii sky waves,

    iii space waves.

State the typical range of frequencies usually transmitted by each of the three methods.

What do you understand by an active satellite?

b What do you understand by noise in a communications link? In what way does this noise create problems? Name one natural and one artificial source of noise.

Explain what is meant by signal to noise ratio. For accurate decoding the signal to noise ratio must exceed 20 dB. A 10 W signal suffers a loss of 4 dB km^{-1} on its route to a receiver, and accumulates noise of power 1.0×10^{-21} W. What is the maximum distance the signal can be propagated if it is to be satisfactorily decoded at the receiver?

13 A High Definition Television (HDTV) signal is to be transmitted digitally. The transmission is composed of 30 pictures (frames) per second, 1250 lines per picture and 1440 phosphor points per line, with a 4-bit code controlling the brightness and colour of each phosphor point.

  a Explain why such a signal could not be transmitted through any appreciable distance by a pair of wires.

  b Explain why such a signal could not be transmitted on the VHF (or FM) waveband.

  c The HDTV signal is to be transmitted by a laser feeding an optic fibre. The laser is switched on and off to generate a digital, pulsed signal.

    i Calculate the minimum frequency at which the laser must be capable of being switched in order to transmit this signal.

    ii The laser actually produces pulses at a frequency of 2.2 GHz. Calculate the length of time required to transmit the signal for a single picture (frame).

    iii Estimate how many separate HDTV transmissions could share the same laser and optic fibre. Explain your reasoning.

14 As telephone systems developed, it became necessary to transmit many separate calls along a single coaxial cable. This was achieved by a process known as frequency division multiplexing. In this system, the analogue signal from each call modulates a separate high frequency carrier which is transmitted along the cable. At the other end of the cable, there are circuits, each tuned to a different carrier frequency, to extract, or demodulate, each call from its carrier.

  a If a single call is allocated a bandwidth of 4 kHz and if coaxial cable can transmit frequencies up to 60 MHz, calculate the maximum number of telephone calls which could be multiplexed in a single cable.

  b In practice, the maximum numbers of calls actually multiplexed is less than the theoretical maximum. Suggest one reason for this.

  c There is a limit to the length of uninterrupted cable along which a signal can be transmitted.

    i Explain why there is this limit.

    ii What must be done before this limit is reached?

  d A cable has a loss of 2.5 dB km^{-1}. A 4.5 W signal is fed into this cable. Calculate the maximum uninterrupted length along which the signal can be transmitted if the noise power in the cable is 5.0×10^{-6} W and the signal-to-noise ratio must not fall below 20 dB.

  You may find the following equation useful:

$$\text{number of dB} = 10 \lg \frac{P_1}{P_2}$$

  where $\dfrac{P_1}{P_2}$ is the ratio of two powers.

  e State two advantages of transmitting telephone signals on a microwave link rather than on coaxial cable.

15 Figure 13.6 illustrates a satellite in geostationary orbit receiving TV signals from a ground station and re-transmitting them back to Earth. The ground station transmits on a frequency of 14 GHz and the satellite transmits on a frequency of 12 GHz. About 90% of the signal power output of the satellite is contained in a narrow beam which has a diameter of coverage on the Earth's surface of 800 km.

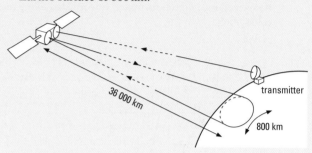

Figure 13.6

  a Explain what is meant by a geostationary orbit.

  b Calculate the time delay between the transmission of information from the Earth to the reception of the signal by a dish aerial on Earth.

  c Explain why the frequency of the signal transmitted to the satellite is different from the frequency of the signal broadcast from it.

  d Calculate the length of a dipole aerial tuned to receive signals from this satellite.

  e Explain, with the aid of a diagram, why a parabolic dish must be used with the receiving aerial.

  f If the signal power transmitted by the satellite is 2.2 kW, calculate the maximum power that can be received by a 1.5 m diameter parabolic dish on Earth.

  g There is a 190 dB loss in the signal received by the satellite compared with the 17 kW signal transmitted to it. Use the equation below to determine the actual power received by the satellite:

$$\text{number of dB} = 10 \lg \frac{P_1}{P_2}$$

  where $\dfrac{P_1}{P_2}$ is the ratio of two powers.

16 a A radio station transmits an amplitude-modulated wave with a carrier frequency of 200 kHz. The transmission consists of speech and music within the frequency range 50 Hz–9 kHz.

    i Explain the terms amplitude-modulated and carrier frequency.

    ii Calculate the value of the maximum and of the minimum sideband frequency.

    iii Hence, or otherwise, explain what is meant by bandwidth.

  b Draw and label a block diagram of the elements in an amplitude-modulated radio receiver, suitable for receiving the signal in **a**.

c Tuning of the radio is achieved by means of resonance in a circuit containing a coil of inductance L (and low resistance) and a capacitor of capacitance C. The resonant frequency f_0 is given by the expression

$$f_0 = \frac{1}{2\pi\sqrt{LC}}$$

The coil has a fixed inductance of $4.0\,\text{mH}$.

i What is the value of the capacitance required in order to tune into the broadcast described in **a**?

ii Why is it necessary that the coil should have low resistance?

17 a What is meant by a digital signal?

b The variation in light output from an optical fibre is sampled every millisecond and this output is converted into a 4-bit number, the most significant bit coming first. A series of consecutive 4-bit numbers is given below:

0001 0010 0100 0110 1010 1100 1010 0110 0010

0001 0001 0010 0110

i Plot a graph showing the variation with time of the output.

ii Estimate the frequency at which the pulses of light are being transmitted along the fibre.

c An optical fibre transmission system consists of a transmitter, an optical fibre of length 30 km and a receiver. The minimum detectable power leaving the fibre and entering the receiver is $1.0 \times 10^{-8}\,\text{W}$.

i Calculate the minimum power entering the fibre from the transmitter, given that the power P is related to the distance x along the fibre by the expression

$$P = P_0 e^{-ax},$$

where P_0 and a are constants and the value of a is $0.15\,\text{km}^{-1}$.

ii List *two* possible sources of power loss associated with an optical fibre.

18 This question is about a simple a-m radio receiver. Figure 13.7 shows the circuit of a medium wave radio receiver with sections of the circuit in dotted boxes.

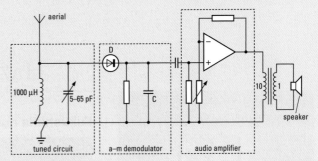

Figure 13.7

a The first section contains the aerial and tuned circuit of capacitor and inductor in parallel.

i Explain how the tuned circuit selects a small band of frequencies to be passed to the second section, the a-m demodulator.

ii The circuit shown has a fairly high selectivity. Explain, using an appropriate diagram if you wish, the meaning of the words in italics.

iii Find the range of resonance frequencies of this tuned circuit.

b The second section is the a-m demodulator, where the audio signal which modulates the high-frequency carrier is separated from the carrier. How do the diode D and capacitor C demodulate the signal? Draw suitable sketches to illustrate your answers.

c Why is the third section present? What is the purpose of the transformer between the audio amplifier and the loudspeaker?

New frontiers

HAZARDS OF NUCLEAR ENERGY

Fukushima nuclear disaster

Explosion at Fukushima unit 3. This is not a nuclear explosion; it is a hydrogen explosion which has blasted through the containment of the reactor, releasing radioactive material into the environment.

At 2.46 p.m. local time on Friday 11 March 2011 a magnitude 9 earthquake struck Japan. The earthquake killed over 16 000 people and destroyed over a million buildings. The earthquake was followed by a huge tsunami which caused severe damage at the Fukushima Daiichi nuclear power complex, resulting in a major nuclear disaster. The social and environmental consequences have raised serious questions about the safety of nuclear fission programmes, leading many nations to reassess their energy policies, in some cases (such as Germany) deciding to phase out all nuclear power plants.

Failure of the cooling circuits

When the earthquake struck the three units that were generating electricity automatically shut down. However, even in shut-down, the core continues to generate about 15% of its power as a result of the radioactive decay of short half-life fission products (mainly beta-emitters), and this has to be removed by circulating a coolant (water) in the core.

Cooling ponds at a nuclear reactor. Spent fuel rods are placed here for years after they are removed from the reactor core.

The coolant passes through a heat exchanger which transfers excess heat to sea water. However, the earthquake destroyed the external generators that powered the coolant pumps and then the tsunami swamped the back-up generators in each reactor. The temperature in the core rose, and eventually parts of the core melted. The high temperatures released hydrogen from the water and resulted in explosions that breached the containment of at least two of the three reactors. Radioactive isotopes

---

Reactor safety

1 To control reactivity: control rods can be lowered into the core. These are made of materials such as boron or cadmium that absorb neutrons and stop the chain reaction.

2 To cool the fuel in the reactor core: back-up emergency cooling systems switch on if the primary coolant fails. These prevent the fuel rods from overheating and melting.

3 To contain the radioactive materials: the reactor core is enclosed in a steel containment vessel. This is surrounded by a thick outer concrete building designed to withstand the effects of earthquakes, terrorist assault, or even a crashing airliner.

Despite the risks of nuclear power it has been responsible for fewer direct deaths than any other source of energy over the past few decades. The figures below show the number of fatalities in severe accidents per TWy (terawatt-year) from different energy sources between 1969 and 2000.

Energy source	Associated fatalities
coal	754
natural gas	196
hydroelectric	10 288
nuclear	48

(mainly of iodine and caesium) were released into the atmosphere and leaked through the containment (2.6 m of reinforced concrete) into the sea. The total release of about 200 PBq of radioactive material was about 15% of the amount released in the Chernobyl disaster.

Damage to the cooling ponds

Spent fuel rods are highly radioactive. When they are removed from the reactor core they are placed in water-filled cooling ponds for several years until their activity has fallen to a more manageable level. After this they are removed and stored with convective cooling systems. New fuel rods are also placed in these ponds prior to insertion in the reactor core. Each reactor at Fukushima had its own cooling pond. When power was lost the water that is usually pumped through the ponds began to heat up and boil off, exposing fuel rods. The pond at unit 4 (which had not been operating) contained both spent fuel rods and 548 new fuel assemblies ready for insertion. The total amount of radioactive material, if released into the environment, would dwarf the release at Chernobyl and some scientists have suggested that if this did happen it would require the complete evacuation of Japan. This event is still a possibility because of the precarious state of the reactor pool and its location in an earthquake zone. In April 2012 the Japanese Government drew up plans for the evacuation of Tokyo (39 million people) in the eventuality that reactor pool 4 collapsed. Fallout from the release of this material in the case of a fire could contaminate much of the western coast of the USA.

Environmental and social consequences

The number of immediate deaths attributable to the Fukushima nuclear disaster was tiny and in most cases was an indirect consequence (such as from stress linked to evacuation). However, estimates suggest that over 1000 deaths might result from exposure to radiation already released to the environment. In order to minimize these consequences the Japanese Government ordered a complete evacuation of the area within 20 km of the plant and a voluntary evacuation up to 30 km. By December 2012, a year and a half after the disaster, large numbers of people were still displaced from their homes. The greatest long-term threat may well be to mental health as a result of disrupted lives.

The coastal location of the facility and the weather systems in place at the time of the disaster meant that about 80% of the radioactive material that was released fell over the Pacific Ocean and almost all of the rest over Japan. Whilst levels of radioactivity in sea water close to the plant rose dramatically during the disaster, dispersion in the ocean means that the average levels now are not significantly above those elsewhere. However, the levels of radiation from fish in the waters close to Fukushima have been monitored and remain high, particularly among bottom feeders, suggesting that there remain sources of caesium on the sea bed or that there are continued releases from the stricken nuclear plant. It seems unlikely that commercial fishing will be able to resume in these waters in the near future.

Although most radioactive material went into the ocean there are many thousands of square kilometres of land that have been contaminated. The policy of the Japanese Government is to attempt to decontaminate much of this land. This involves spraying and scraping buildings and removing layers of topsoil. The consequence will be a large quantity of contaminated material that will have to be stored securely. This approach contrasts with the response to the disaster at Chernobyl where the government enforced long-term evacuation of large areas. The difference in approach is related to the amount of habitable land available in these two countries. One interesting consequence of the Ukrainian approach has been the way in which animals and birds have recolonized the evacuated areas, raising questions about the ability of life to tolerate increased levels of radiation.

The ferris wheel at Pypiat. 350 000 people were evacuated following the Chernobyl disaster. Towns such as Pypiat remain empty 25 years after the disaster.

Safe handling and disposal of nuclear waste

Low-level waste, such as radiation suits and laboratory equipment used in the handling of radioactive sources, is stored for about 10 years and then disposed of as normal waste. Low level contaminated water can be released into the sea.

Intermediate-level waste includes metal cladding for fuel rods and chemical sludges. These materials are encased in a resin or concrete and sealed into steel drums. The drums can then be buried in concrete chambers and covered with a concrete roof.

High-level waste such as spent fuel rods is stored underwater for years and then vitrified (made into a solid glassy material). This will probably be stored deep underground in stable geological formations but its ultimate fate is still subject to discussion and negotiation. It has to be isolated from the environment for thousands of years.

THE FUTURE OF NUCLEAR ENERGY

Cascade of gas centrifuges used to produce enriched uranium. This photograph is of the US gas centrifuge plant in Piketon, Ohio from 1984.

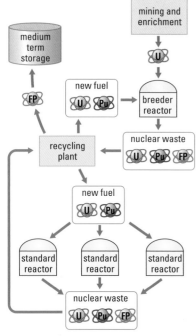

The use of breeder reactors can extend the supply of nuclear fuel for tens of thousands of years. The diagram shows how fuel bred in a fast breeder reactor can be used to fuel both this reactor and several standard reactors whilst generating less waste than standard reactors alone. FP = fission products.

Fertile and fissile material

A **fissile** nuclide is one that undergoes a nuclear fission reaction when it absorbs a neutron.

A **fertile** nuclide can be converted to a fissile nuclide by absorption of a neutron and subsequent decays.

Proliferation of nuclear technology – the ambiguity of enrichment

The development of civil nuclear power has often gone hand in hand with the development of nuclear weapons. The same technologies (e.g. the enrichment of uranium) are involved in both undertakings so it is difficult to judge whether any particular project is purely civil or has military ambitions. The Iranian nuclear industry is a particularly relevant example. The Iranians claim that they are enriching uranium for use in their own nuclear reactors but some external observers (e.g. from the USA and Israel) claim that they are developing an independent nuclear weapons programme that could further destabilize the region. To understand why it is difficult to distinguish civil and military nuclear projects we must look at the need for, and method of, enrichment.

Nuclear fission reactors harness energy from a chain reaction in which neutrons emitted during fission go on to induce further fissions in other uranium-235 nuclei. However, many neutrons are either lost from the material or absorbed by the non-fissile isotope uranium-238. In order to achieve a critical assembly natural uranium (which is 99.3% U-238 and only 0.7% U-235) must be enriched until it consists of about 2–4% U-235. This can be done by passing the gas uranium hexafluoride through a cascade of thousands of centrifuges, each of which causes a slight separation of the two isotopes (U-238 atoms have slightly greater mass than U-235 atoms so gas containing U-238 atoms is more concentrated at the outside of the centrifuge). However, this process can be continued until weapons grade uranium (>80% U-235) is produced. (It is also possible to make a weapon with just 20% enrichment.)

Iran has thousands of centrifuges in place, hundreds of which are operational. Many are in the Fordar enrichment facility which is buried under about 80 m of rock beneath a mountain, providing security from air strikes. Iran's claims that it is not developing a military nuclear weapon cannot be verified because inspectors from the International Atomic Energy Authority (IAEA) have not been granted full access to the country's facilities and there has been considerable discussion of the possibility of a pre-emptive Israeli air strike if 'the programme' is not halted. However, it is very difficult to make clear judgements on these issues. Israel is thought to have a large number of nuclear weapons itself but has never actually admitted to being a nuclear power.

Breeding fuel

Uranium is a common metal in the Earth's crust and is also present in sea water (at about 0.003 ppm). The most common uranium ores are uraninite (UO_2) and pitchblende (U_3O_8) and the largest reserves are in Canada, Australia, and Kazakhstan. Estimates based on current rates of use suggest that present reserves will last about 200 years. However, this can be extended enormously by:

• breeding more fissionable fuel from present resources (30 000 years), or

• extracting uranium from sea water (60 000 years at current levels of use).

Neither of these options is economically viable at the moment, particularly as more reserves have been identified over the past few decades. However, an increase in the cost of uranium could make them more attractive and there has been a considerable amount of research into the construction and operation of breeder reactors.

A **breeder reactor** makes more fissile material (material that can sustain a chain reaction) than it consumes. It does this by converting 'fertile' material, such as uranium-238 or thorium-232, into fissile material inside the reactor core while it is operating. The conversion occurs when the fertile material captures neutrons and undergoes radioactive decay to become a fissionable isotope.

For example, the conversion of U-238 to Pu-239 involves two beta-minus decays:

$$^{238}_{92}\text{U} + ^{1}_{0}\text{n} \rightarrow ^{239}_{92}\text{U} \rightarrow ^{239}_{93}\text{Np} + ^{0}_{-1}\beta + ^{0}_{0}\overline{\upsilon}$$

$$^{239}_{93}\text{Np} \rightarrow ^{239}_{94}\text{Pu} + ^{0}_{-1}\beta + ^{0}_{0}\overline{\upsilon}$$

The fissile material that accumulates in the core (Pu-239 or Th-233) is then extracted by reprocessing the fuel once the spent fuel rods have been removed from the reactor. It can then be used as fuel in a nuclear reactor. Thorium is more abundant than uranium so this massively increases the potential supply of nuclear fuels. The use of breeder reactors also reduces the amount of radioactive waste (since it is reprocessed) making the nuclear fuel cycle more like a renewable resource.

Whilst several nations have built and run prototype breeder reactors in the past, many of these have now been closed down because of cost and fears over safety. However, India is now developing a programme of fast breeders which use a mixture of uranium and plutonium as fuel. Fast-breeder reactors have a denser, hotter core than conventional thermal nuclear reactors and harness fast neutrons rather than moderated thermal neutrons in the chain reaction. This means that there is a much higher power density in the core, so a more efficient coolant must be used. Current designs pump liquid sodium through the core and this is one of the aspects of fast-breeder technology that raises most concerns about safety.

Harnessing nuclear fusion

The challenge of generating electricity from nuclear fusion has been tantalizing nuclear scientists for more than half a century. If this can be done economically and safely it will solve our energy problems for the foreseeable future. However, the conditions needed for nuclear fusion to take place are extreme (as in the core of a star) and difficult to achieve in a controlled way. It is also extremely difficult to control matter under these conditions. Two different approaches are currently front runners but both are decades away from success.

The world's largest and most advanced fusion reactor, the International Thermonuclear Experimental Reactor (ITER), is under construction at Cadarache, France. This is a multi-billion-dollar project and is a collaboration between China, the European Union, India, Japan, Korea, Russia, and the United States. Its purpose is to prove that nuclear fusion is a viable energy source for the future and to test possible designs for the first electricity-producing fusion reactors. The reactor is based on Andrei Sakharov's Tokamak design in which a plasma is confined in a toroidal magnetic field and heated to millions of kelvin by electric currents circulating through the plasma. Under these extreme conditions nuclei begin to fuse and energy is released. The reactor is expected to be switched on by 2020.

The National Ignition Facility (NIF) at the Lawrence Livermore Laboratories in the USA uses a different approach, called inertial confinement. The idea is simple. A pellet of hydrogen fuel (a mixture of deuterium and tritium) is heated and compressed by an intense burst of laser light to such an extent that the conditions necessary for nuclear fusion are achieved and more energy is emitted by fusion reactions than was supplied by the lasers. The fuel pellets are tiny, just millimetres across, but the lasers trained on them are the most powerful in the world: 192 beams providing 500 TW for 20 ns. During this brief, intense irradiation the inertia of the target particles prevents them from escaping (inertial confinement) and some of the nuclei in the fuel fuse to form helium. To get an idea of the extreme conditions it is worth noting that the power delivered to this tiny target is over 1000 times the average power consumed by all of the USA! The aim of this facility is to achieve ignition – the point at which nuclear fusion in the fuel emits more energy than is supplied to it by the lasers. It also provides a laboratory where scientists can investigate the behaviour of matter under conditions that exist in the core of a star or at the heart of a nuclear explosion.

Key equations

Induced fission reaction, for example:

$$^{235}_{92}\text{U} + ^{1}_{0}\text{n} \rightarrow ^{236}_{92}\text{U} \rightarrow ^{139}_{56}\text{Ba} + ^{94}_{36}\text{Kr} + 3\,^{1}_{0}\text{n}$$

Creation of U-233 from Th-232 in a breeder reactor:

$$^{232}_{90}\text{Th} + ^{1}_{0}\text{n} \rightarrow ^{233}_{90}\text{Th} \rightarrow ^{233}_{91}\text{Pa} + ^{0}_{-1}\beta + ^{0}_{0}\overline{\text{v}}$$

$$^{233}_{91}\text{Pa} \rightarrow ^{233}_{92}\text{U} + ^{0}_{-1}\beta + ^{0}_{0}\overline{\text{v}}$$

Nuclear fusion of deuterium and tritium:

$$^{2}_{1}\text{H} + ^{3}_{1}\text{H} \rightarrow ^{4}_{2}\text{He} + ^{1}_{0}\text{n}$$

NIF target chamber at the Lawrence Livermore Laboratories. A fuel pellet just a few millimetres across will sit at the centre of this chamber and the light from 192 powerful lasers will be focused on it.

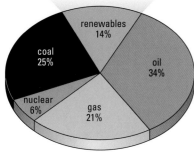

Fuel shares of the world total primary energy supply. The 'other' category includes tidal, solar, and wind generation. Source: International Energy Agency (2007).

Finite resources – too much or too little?

Current reserves of oil and natural gas will last at least 50 years. Coal could last for 250 years. However, as reserves dwindle and demand grows the price of these commodities (particularly oil) rises, making it economical to use more expensive extraction methods such as:

- fracking – releasing oil reserves from underground rocks by cracking them with high-pressure water jets and letting them accumulate in reservoir rocks

- shale oil extraction – mining the shale and then converting the kerogen it contains (a mixture of organic and non-organic chemicals found in sedimentary rocks) into shale oil in a chemical processing plant.

Both technologies could vastly extend the life of oil as a primary energy source but both have an environmental impact, and continuing to burn the oil increases the concentration of carbon dioxide in the atmosphere.

It has been argued that we are approaching 'Peak Oil' – the point at which the rate of oil extraction reaches its maximum value and after which it enters terminal decline. From an environmental point of view this might be irrelevant. If we were to burn even a fraction of the remaining oil reserves we would put so much carbon dioxide into the atmosphere that we would never be able to halt a runaway greenhouse effect. In this context one could argue that we have too much oil!

ENERGY CHALLENGES AND ALTERNATIVES – 1

The demand for primary energy sources has increased approximately 2.5% per year for the past 20 years. The rapid growth and industrialization of countries such as China and India make continued strong growth in energy demand almost certain. In 2011 the International Energy Agency predicted an increase in primary energy consumption of up to 40% by 2030, with the biggest increase in energy used to generate electricity.

Current energy sources

At the moment 80% of the world's energy comes from burning fossil fuels in the form of coal, oil, or gas. All of these are finite resources. Furthermore, combustion of fossil fuels results in the emission of greenhouse gases (primarily carbon dioxide) which cause global warming and climate change. They also generate local atmospheric pollution. Fifty-six per cent of the world's oil reserves are concentrated in the Middle East, an area of political instability, making oil prices fluctuate significantly and affecting energy security, the world economy, and international conflicts. For all of these reasons we need to reduce our dependence on fossil fuels and particularly on oil.

At present just 6% of the world's energy is generated from nuclear sources. Nuclear reactors do not produce greenhouse gases but uranium (particularly as it is used in conventional thermal reactors) is a non-renewable fuel. However, prior to the Fukushima disaster of 2011 many countries were considering an increase in nuclear capacity. Since then there has been a global reassessment of nuclear technology and of its political impact which has resulted in some proposed developments being scrapped or scaled back and others being put on hold.

The greenhouse effect and climate change

The temperature of the Earth's atmosphere is determined by the balance of radiant energy coming in from the Sun and re-radiation by the Earth back into space. Short-wavelength radiation from the Sun penetrates the Earth's atmosphere and warms the surface. The surface radiates longer-wavelength radiation back toward space. However, these longer wavelengths are absorbed strongly by greenhouse gases such as water vapour, carbon dioxide, and methane and are re-radiated in all directions, including about 30% back towards the Earth. This reduces the radiation that returns to space so that the temperature of the atmosphere rises. The higher the concentration of greenhouse gases, the higher the equilibrium temperature of the atmosphere. If it were not for this natural greenhouse effect the surface of the Earth would have an average temperature of about –18 °C, too cold to sustain life.

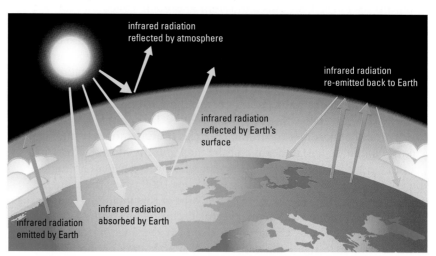

The natural greenhouse effect is increased with rising greenhouse gas concentrations.

The combustion of fossil fuels, coupled with deforestation (reducing the amount of carbon stored in trees) and the release of methane from landfill sites, has increased the proportion of greenhouse gases in the atmosphere. Carbon dioxide released from the combustion of fossil fuels is the main problem – in 2012 the concentration of carbon dioxide in the atmosphere had reached 391 parts per million (ppm) compared with 280 ppm in pre-industrial times. Carbon dioxide stays in the atmosphere much longer (hundreds of years) than other greenhouse gases, and so has long-term irreversible effects on the atmosphere and climate. Even if emissions stopped tomorrow it would take many centuries to reverse the trend and restore pre-industrial lower concentrations.

During the past century average global temperatures have increased by about 1 °C. Whilst there is still discussion about the detailed link between human activity, carbon dioxide concentrations, and climate change, the strong consensus amongst climate scientists is that human activity is causing global warming and climate change. Future projections vary enormously because of the difficulty of modeling the Earth's climate, but there is great concern that if we do not act to reduce carbon emissions temperatures could rise disastrously. Temperature increases of between 1 and 5 °C over the next century have been predicted by the US Environmental Protection Agency. Increasing global temperatures would result in:

- melting ice sheets over Antarctica, Greenland, and the Arctic seas
- reduced snow cover and permafrost
- changing patterns of precipitation
- more frequent floods and droughts
- rising sea levels (between 20 and 60 cm by 2100)
- increased acidity of the oceans
- raised temperatures, particularly over land, making some parts of the Earth uninhabitable
- changing ecosystems and some extinctions.

What can be done?

There are compelling reasons to reduce our dependence on fossil fuels, particularly on oil. However, in the short term at least we will continue to burn them, so there is a need to develop techniques that can reduce their impact on the environment. These could include:

- develop more efficient power stations – it is theoretically possible to increase from the present 35% to about 50% efficiency by adopting different methods of combustion (such as coal gasification)
- develop carbon-capture systems
- develop more efficient and cleaner engines for motor vehicles
- encourage lifestyle changes to reduce carbon emissions, such as fewer long-haul flights, car sharing, eating only locally grown food
- encourage the use of more efficient consumer devices (such as low-energy lighting) thus reducing demand for electricity
- build more nuclear power stations
- (longer term) switch to renewable sources such as wind, solar, or hydroelectric power
- (farther into the future) develop commercial nuclear fusion reactors.

There are currently 1199 proposed new coal-fired power stations worldwide with 76% of these in India and China.

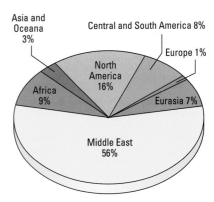

World oil reserves by region. Source: US Energy Information Administration (2007).

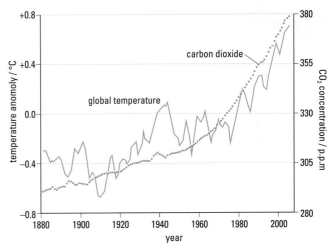

Data such as those shown above have convinced most scientists that global warming is real and is linked to the emission of carbon dioxide by humans into the atmosphere.

Photochemical smog, Beijing.

- renewable resources

- fuel cell technology

Power output of different types of power station.

Energy source	Example	Power output /GW
hydroelectric power	Tarbela dam, Pakistan	3.5
wind turbines	Thanet offshore wind farm, UK	0.30
solar	Gujarat solar power station, India	0.60
coal	Ratcliffe-on-Soar, UK	2.0
nuclear	Palo Verde nuclear power station, USA	3.9

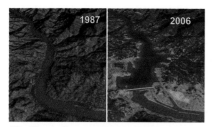

The Three Gorges Dam, built across the Yangtse river in China, has the world's largest installed hydroelectric capacity of 22.5 GW. The water level on the upstream side of the dam is 110 m higher than downstream. The construction of the dam resulted in the displacement of 1.3 million people and has had a major ecological impact on the area and on the river itself.

When the Siemens SWT-6.0-154 wind turbine was constructed it had the largest rotor blades in the world – 75 m long. This wind turbine can generate 6 MW, enough to power over 5500 households.

ENERGY CHALLENGES AND ALTERNATIVES – 2

An ideal primary energy source would be endlessly renewable, would generate no greenhouse gases, have little environmental impact, and would involve simple safe technologies. Furthermore it would be free at source. Many would argue that such sources are all around us, particularly in the form of wind, solar, and hydroelectric power. However, fossil and nuclear fuels represent compact energy-dense sources and we have grown to depend on large centralized power stations connected to an electrical grid to distribute the energy. Whilst hydroelectric power stations can be used in this way, the energy harvested from the sun and wind is more dispersed and often better suited to localized uses.

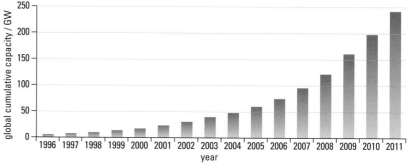

The global capacity for wind-generated electricity is growing at an increasing rate as governments look to renewables to reduce their dependence on fossil fuels.

In the past decade, against a background of growing concern over greenhouse gas emissions from fossil fuels, governments in many countries have provided economic incentives to encourage investment in renewables. The percentage of global electricity generated by the three most important renewables in 2011 was approximately:

- 18% from hydroelectric plants
- 2.5% from wind turbines
- 0.5% from solar panels (photovoltaics).

Hydroelectric power

The conventional method of generating hydroelectricity is to dam a river and then let the trapped water fall through turbines to a lower level beyond the dam. In the process gravitational potential energy is transferred to electricity in turbines whose output connects to transmission lines for long-distance distribution. In many ways a large hydroelectric power plant is comparable in output and utility to a large thermal reactor. However, the energy produced is much cheaper and there is a much lower emission of greenhouse gases. One might expect no greenhouse gases to be emitted from a hydroelectric power station, but in fact rotting organic matter in the reservoirs created by hydro projects releases some carbon dioxide, methane, and nitrous oxide.

Wind turbines

When wind moves over the blades of a wind turbine it creates a torque on the turbine causing it to turn and doing work on it. The turbine is connected via a gearbox to an electrical generator. The overall effect is to convert kinetic energy into electrical energy which can be used locally or fed into the grid. Small wind turbines with blade lengths of about 1.0 m can be used to generate a few kilowatts for home use; large off-shore turbines have blades over 50 m in length and their output can reach several megawatts. The UK Government is investing heavily in offshore wind farms. Its target is to supply over 25% of the UK's electricity from wind turbines by 2020. Siting wind farms offshore has a number of advantages – winds are usually stronger and more predictable and there is less visual impact on the environment. However, it is significantly more expensive to generate electricity using offshore rather than onshore wind farms.

Solar power

Solar panels consist of a large number of individual photovoltaic (PV) cells connected together. The photovoltaic effect, originally discovered in 1839 by Henri Becquerel (at the age of 19!) is related to the photoelectric effect discovered by Heinrich Hertz in 1887. Modern PV cells are based on semiconductor materials. When light falls on them photons are absorbed and their energy causes a separation of charge and hence a voltage. If the PV cell is connected to an external circuit then a current will flow. Whilst the efficiency of early solar panels was relatively low, recent technological innovations have shown that efficiencies of 20% or above are now possible. The actual output of any particular panel will vary with the amount of sunlight it receives up to a maximum of about $200\,W\,m^{-2}$ in direct sunlight in clear conditions. This means that large areas of solar panels are required to produce a significant output.

An alternative to using photovoltaics is to use the concentrated solar power (CSP) to heat water, producing high-pressure steam to drive turbines. This method is being used in the world's largest single-unit solar power station, Shams 1, which is nearing completion in Abu Dhabi. This will have an output of 100 MW and is designed to power 20 000 homes locally. The power station itself covers an area of about $2.5\,km^2$. Whilst historically the cost per kilowatt of solar power has been significantly higher than that generated by burning fossil fuels or from nuclear power, it has now fallen below that of nuclear and is expected to continue to fall.

Fuel cells

Whilst fossil-fueled power stations are the main source of greenhouse gas emissions, transportation also produces very significant emissions. In 2010 transportation accounted for 27% of the total US greenhouse gas emissions. There are several strategies for reducing this contribution:

- more efficient vehicle technology
- hybrid vehicles (which combine an internal combustion engine with one or more electric motors)
- electric vehicles (powered by onboard rechargeable batteries)
- fuel cells.

Fuel cells generate electricity by combining a fuel and an oxidant (for example, hydrogen and oxygen) in an electrochemical reaction. Batteries also do this, but whereas a battery contains all of the chemicals it will use inside itself, the fuel cell does not. Fuel cells continue to produce electricity as long as fuel is supplied to them. In 2003 President Bush announced the Hydrogen Fuel Initiative (HFI). The intention is to develop hydrogen-fuel-cell technology and infrastructure so that vehicles using fuel cells are economically viable by 2020. The products of a hydrogen fuel cell are water and electrical energy. No greenhouse gases or atmospheric pollutants are produced. Fuel cells are also efficient, converting about 80% of the fuel's energy to electrical energy. However, the overall efficiency of a fuel-cell-powered vehicle will be lower than this because of the intermediate stages required to actually propel the vehicle (for example, inefficiencies in the electric motor and mechanical connections). Overall the efficiency is around 60%. Battery-powered vehicles are actually more efficient than this (the battery itself is about 90% efficient, and the electric motor 80%, so the overall efficiency of a battery-powered vehicle is about 72%). However, the electricity used to charge the batteries has to be generated somewhere so their overall efficiency in terms of primary energy usage is significantly lower (around 30%). Conventional vehicles powered by internal combustion engines have efficiencies of around 20%. Whilst there are great hopes for the use of fuel cells in future, at the moment they are relatively expensive, there are safety concerns over the fuels (e.g. hydrogen) used to power them, and there is no infrastructure in the form of refuelling stations to support them. However, oil prices are expected to rise significantly in the future so economic and environmental arguments suggest that fuel cells may well play an important part in vehicle propulsion in future.

Advantages and disadvantages of the three main renewables	
Advantages	**Disadvantages**
Hydroelectric power	
free renewable energy source	high cost of construction
power can be varied rapidly	ecological and environmental impact
river flow can be controlled for irrigation	displacement of people (flooded villages)
large centralized power source	huge damage if dam fails
reduced greenhouse gas emissions	
Wind power	
free renewable energy source	output depends on wind speed which varies
no greenhouse gas emissions	aesthetic impact on countryside
land beneath turbines can still be farmed	large wind farms needed to supply towns
wide range of sizes/powers	noise
can supply electricity in remote areas	
Solar power	
free renewable energy source	high initial cost
no greenhouse gas emissions	collection area must be large
non-polluting	only generates electricity during the day
can supply electricity in remote areas	output varies with location/ conditions
	needs linked energy storage

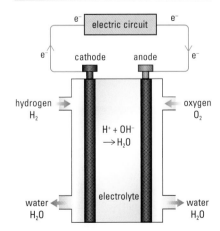

The principle of a hydrogen fuel cell. Hydrogen and hydroxide ions combine in an electrolyte to create water. In the process electrons are pushed around an external circuit and electrical work is delivered to a load (such as a d.c. motor).

14.5

OBJECTIVES

- America in space

- the Chinese and Indian space programs

- extrasolar planets

The conquest of space is worth the risk of life. Our God-given curiosity will force us to go there ourselves because in the final analysis, only man can fully evaluate the Moon in terms understandable to other men.

Virgil 'Gus' Grissom, *Gemini* astronaut. Grissom died in the *Apollo 1* fire, 27 January 1967.

The *Constellation* program

Constellation called for the development of a new manned spacecraft, *Orion*, designed to support a crew of up to six members. Alongside this, *Altair* was a projected lunar lander with four times the volume of the *Apollo* lunar module and a crew of four.

Ares 1 was to be a launch booster for manned missions, and *Ares V* for cargo.

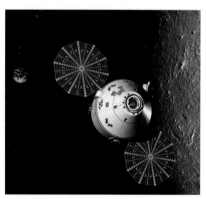

The full Orion *spacecraft, part of the cancelled* Constellation *program. The manned capsule concept is being developed as an ISS lifeboat.*

An artist's rendition of the planned Altair *lunar lander on the Moon.*

THE FINAL FRONTIER

An era of space flight ended on 21 July 2011 when *Atlantis* rolled to a halt after touching down to complete the last ever Space Shuttle mission. With that landing, the country that put people on the Moon for the first time lost the ability to launch a human into space. America currently does not have any manned space flight

Atlantis *landing at the Kennedy Space Center at the end of the final Shuttle mission.*

capability – flights to and from the International Space Station are being serviced by Russian lifting capacity.

Manned space flight is expensive (people, along with the equipment and stores to keep them alive, are heavy) and dangerous. While astronauts like Gus Grissom have always accepted that loss of life is an inevitable consequence of pushing the limits of exploration and technology, politically it is harder to countenance; especially in times when the money would appear to be more usefully spent on domestic issues. There has always been a tension between the manned and unmanned exploration of space. During the *Apollo* era, people were questioning the need to send humans rather than robot probes. In modern times, economic pressures are helping to shape the argument.

Constellation and beyond

In 2004 President George Bush challenged NASA to design new, reuseable manned space vehicles that could replace the Space Shuttle and ultimately develop into the capability to return to the Moon. The *Constellation* program was estimated to cost $97 billion up to 2020, but NASA was unable to provide a complete estimate due to the unsolved technical challenges that would be faced.

In the 2011 budget proposals, President Obama included no funding for the *Constellation* program due to it being "over budget, behind schedule, and lacking in innovation" (costs had now risen to an estimated £150 billion). Instead NASA was encouraged to invest in promoting collaboration with private companies to provide crew and cargo lift capabilities that could be used commercially and by government. The lifetime of the Space Station was extended to 2020 and NASA was encouraged to develop new propulsion technologies that would support manned missions to an asteroid by 2025 and to Mars in the 2030s. The focus is on unmanned space probes, orbiting telescopes, and Earth observation as well as exploration beyond the Moon. This left transportation to low-Earth orbit as an open market for commercial development. A mission to the Moon was dropped as NASA Administrator Charles Bolden emphasized: "as we focused so much of our effort and funding on just getting back to the Moon, we were neglecting investments in key technologies that would be required to go beyond."

China in space

Carrying out space programs is not aimed at sending humans into space per se, but instead at enabling humans to work in space normally, also preparing for the future exploration of Mars, Saturn, and beyond.

Qi Faren, Chinese Academy of Sciences

China became the third manned spacefaring nation on 15 October 2003 when the China National Space Administration launched Yang Liwei into a 21-hour orbital mission. The roots of this achievement can be traced back to the Chinese ballistic missile program, which started in 1950 as a result of perceived American and Soviet threats. In modern times, Chinese politicians see a manned space program as a sign of the country's developing technological and economic power. While they are carrying out planned robot explorations, for example their lunar orbital mission in 2007, the main focus of the five-year plan published in 2011 was manned. China's small orbiting laboratory, *Tiangong-1*, was in orbit since September 2011 and de-orbited in 2013 to be replaced by a sequence of larger *Tiangong-2* and *Tiangong-3* labs leading to a full space station deployment by 2020. Even more ambitiously, development is starting for a manned lunar mission between 2026 and 2030.

Extrasolar planets

While there have been many robot and orbital observatory missions leading to amazing scientific advances, perhaps the most exciting in modern times has been the discovery of planets in orbit around other stars. By the start of 2013, some 863 planets have been discovered in 678 planetary systems, which includes 129 systems of multiple planets. The scientific significance of this is hard to exaggerate. Current estimates suggest that the Milky Way galaxy may contain as many as 400 billion extrasolar planets. Although many of these will be gas giants like Jupiter, and in some cases much larger, a proportion of them will be rocky in nature, and a further fraction exist in a 'temperate zone' the right distance from the central star to allow liquid water to exist on the surface. Such planets may well harbour life.

Early in January 2013, scientists working on the *Kepler* space observatory announced the discovery of KOI-172.02, a rocky Earth-like planet orbiting a star similar to the Sun and in the habitable temperate zone. Although a total of nine planets are known to exist in habitable zones, this is the first confirmed discovery of a planet of Earth size.

The *Kepler* space observatory was launched in March 2009 having been specifically designed to detect Earth-like planets. *Kepler's* orbit around the Sun is designed to allow it an unobstructed view in the direction of the constellations Cygnus, Lyra, and Draco. The observational vector also follows the direction of the Sun's orbit around the galactic core, which means that the stars being observed are all roughly the same distance from the core as our Sun.

Kepler observes the light from stars, looking for the slight drop in illumination (0.01%) when an Earth-like planet crosses in front of the star from our perspective. Repeating the observation confirms that the drop is a repeating event, rather than a random observational glitch. The data have to be carefully analysed to eliminate genuinely variable stars. *Kepler* observes 150 000 stars simultaneously and measures variations in their brightness every 30 minutes. In its 2012 catalogue, the *Kepler* team listed 2740 candidate planets orbiting 2036 stars.

Since December 2010 the 'Planet Hunters' website has provided access to star light curves from the *Kepler* mission so that amateurs can cast an eye over the data looking for planetary candidates. The human mind is still a better pattern-recognizing system than computer programs, especially with planets having long periods, so the dip repeats infrequently. At least 69 planetary candidates missed by the computers have been found in this way.

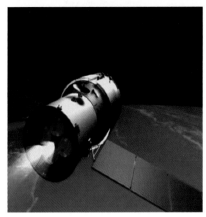

The Chinese Tiangong-1 *space laboratory: 10 m long and 3 m in diameter, with a mass of 8500 kg.*

India in space

The Indian Space Research Organization is working on a two-person orbital vehicle to go alongside their extensive satellite launch capability. There is no fixed date for their first manned mission; planned sometime after 2017.

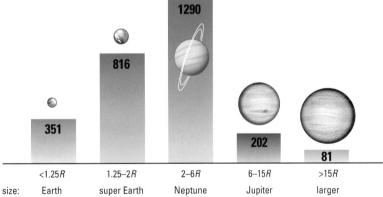

351	816	1290	202	81
<1.25R	1.25–2R	2–6R	6–15R	>15R
Earth	super Earth	Neptune	Jupiter	larger

size:

A size comparison for Kepler *candidates. The chart shows the numbers of candidate planets and their sizes relative to the Earth's radius.*

O B J E C T I V E S

- the cosmic microwave background
- observing temperature fluctuations
- density variations in the early universe
- precision cosmology

ADVANCES IN COSMOLOGY

It is a remarkable testimony to the human mind that theorists can make calculations relating to the first fractions of a second into history, and then experimental scientists can construct satellites and experiments that produce data to confirm these calculations.

The background to it all

According to the Nobel Prize awarding committee, the 'starting point for cosmology as a precision science' was in 1989, when the COsmic Background Explorer satellite (COBE) was placed in orbit. Ground- and balloon-based measurements had already started mapping the microwave background left over from the Big Bang, but COBE provided higher-resolution temperature measurements without interference from the Earth's atmosphere. The data from the COBE mission led to two significant discoveries:

1   The spectrum of the microwave background corresponds very precisely to a black-body curve at 2.7 K.

2   The spectrum shows tiny, but significant, variations in temperature from one point in the sky to another. These patches reflect the density of matter in the universe when the thermal equilibrium between matter and radiation broke down.

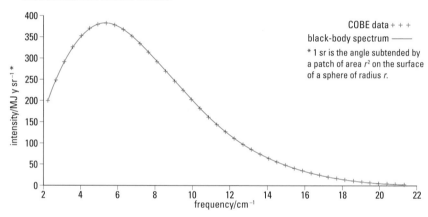

Cosmic microwave background spectrum from COBE. The COBE satellite established that the microwave background followed a black-body spectrum to very high precision. The red points are the measured value, and the green line the black-body spectrum for 2.7 K – not a fit to the data!

In 2001 the Wilkinson Microwave Anisotropy Probe (WMAP) was launched to make a detailed study of these temperature variations. WMAP mapped the background at five distinct frequencies with $0.2°$ resolution across the sky and showed temperature fluctuations $\sim 5 \times 10^{-6}$ K. Nine years of precision measurements served to establish the current Lambda-CDM model of cosmology, which breaks the universe down into the following three constituents.

- **Ordinary (baryonic) matter** – the physical universe we can observe with the naked eye and electromagnetic telescopes. Although it is called baryonic matter, this does include electrons, as they are constituents of atoms, but not other leptons.

- **Cold dark matter (CDM)** – the mysterious matter that does not interact with electromagnetic waves, but which shows up by its gravitational influence on ordinary matter. The best fit to the data comes from the dark matter being 'cold', that is, travelling through the universe at speeds much less than that of light.

- **Dark energy** – the even more mysterious constituent of the universe that exerts a repulsive gravitational effect causing the expansion of the universe to accelerate with time (as confirmed by supernova observations). This constituent is represented in the model as 'lambda', which is a historical reference to the cosmological constant introduced by Einstein when he developed general relativity.

Cosmological parameters (based on WMAP results only).

Parameter	Value	Uncertainty
age of universe (10^9 years)	13.74	± 0.11
Hubble's constant (km s^{-1} Mpc^{-1})	70.0	± 2.2
baryon density	0.0463	± 0.0024
cold dark matter density	0.233	± 0.023
dark energy density	0.721	± 0.025

The amount of each constituent is specified as a fraction of the critical density – the density of matter required to close the universe. The WMAP data shows that the recipe for the universe is 4.6% baryonic matter, 23% cold dark matter, and 72% dark energy.

Density fluctuations

Each patch in the sky, as measured by COBE and WMAP, can be allocated a black-body temperature dependent on the energy spectrum of photons originating from that patch. Overall these temperatures are remarkably uniform, considering that the patches can be sufficiently separated such that light from one patch has not had chance to reach another yet, even in the 13.7-billion-year history of the universe. Cosmologists refer to this as the **horizon problem** – how can different parts of the universe 'over the horizon' from one another have the same temperature to such high precision?

However, there is some variation in temperature, which physicists believe has come about because of small differences in the density of matter at the time when the universe became transparent. Two factors contribute to this: the gravitational red shift imposed on photons escaping from a high-density region; and the Doppler shift of energy when photons scatter off moving electrons. These density variations are very important as the gravity pull of the higher-density regions draws in matter from their surroundings, triggering gravitational collapse that leads to the formation of galaxies and stars.

Theorists calculate that between 10^{-35} and 10^{-32} seconds into history, the universe went through a period of hyper-expansion, doubling in size every 10^{-34} seconds! This **inflationary period** was driven by the gravitational effect of a Higgs field (similar to, but not the same as, the Higgs field in the standard model of particle physics). As inflation ended, the energy density of the Higgs field would cause it to 'break down' into a flood of Higgs particles, and they in turn would rapidly decay into matter and dark matter particles. The original matter in the universe before inflation would now be completely swamped by that originating from the Higgs field, producing large regions of uniform density and explaining the horizon problem.

However, quantum effects would ensure that the strength of the Higgs field varied at different microscopic points in the universe. Inflation would swell these regions into much greater macroscopic sizes producing subtle variations of density inside the very smooth patches. These are the density variations observed by COBE and WMAP. In this amazing picture, galaxy formation can be directly linked to quantum fluctuations in the Higgs field of the early universe!

The era of precision cosmology

The temperature variations in WMAP's sky map are rich in information regarding the cosmological parameters. Inflation predicts that the biggest variations in temperature should be found by averaging over pairs of points separated by 1° across the sky. These temperature differences were set into the microwave background as the universe became transparent. Since then, the expansion of the universe should have changed the size of the patches. you can think of it like this – as the expansion alters the curvature of space, we observe the microwave background as if it were reaching us through a distorting lens. This alters the angular size of the patches we observe. From this information, cosmologists can use computer models to produce 'best fits' to the values of the cosmological parameters. That is how the Lambda-CDM model was confirmed.

After the spectacular success of WMAP, the Planck satellite was launched in 2009 to provide even more detailed information on the microwave background.

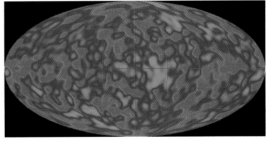

The COBE sky map confirmed the existence of tiny variations in temperature across a generally very uniform background of microwaves. The 'coldest' (blue) and 'hottest' (pink) parts of this whole sky map are 10^{-5} K different from the average 2.73 K. The size of the patches is quite large, indicating the resolution of the COBE satellite.

The much higher resolution whole-sky map from WMAP shows a full range of temperature variations ~200 µK.

O B J E C T I V E S

- the role of mathematical symmetry in theories
- the Higgs mechanism and the discovery of the Higgs particle
- supersymmetry

Professor Peter Higgs was among the physicists who devised a method of incorporating particle mass into the standard model. Particles interact with the Higgs field, which gives them mass.

The Large Hadron Collider at CERN.

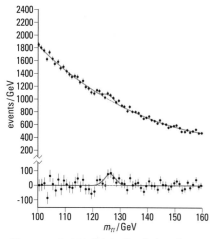

The experimental data that led to the discovery of the Higgs particle, also called the Higgs boson. $m_{\gamma\gamma}$ is the mass of the two photons in the event. The top curve shows the number of two-photon events seen in the experiment as a function of energy. The small bump at 126 GeV is the Higgs. The lower line shows the same data, but with the expected experimental background removed. The significance of the 'Higgs bump' is now much more apparent.

ADVANCES IN PARTICLE PHYSICS

Chapters 8 and 9 detailed the techniques used to explore fundamental particles, the basic constituents of matter. Recent years have seen exciting developments in this field, and an exploration of symmetry has proven to be a very fruitful guide for the production of new theories in particle physics.

In this context, an equation has symmetry if some of its terms can be switched without altering the overall physics described. For example, exploring various symmetries among the phases of wave functions (spread 8.7) led to the development of the mathematical description of the fundamental forces (the strong, weak, gravitational, and electromagnetic forces, spreads 9.1 and 9.15). From this followed the quark–lepton picture of matter and the quantum theory of the fundamental forces – together known as the standard model of particle physics.

Symmetry and the Higgs particle

The underlying symmetries in the equations of the standard model show up only if all particles involved are massless, which is clearly not the case. In 1964 Peter Higgs (along with other physicists working independently) proposed the existence of a new field, now called the **Higgs field**, as a solution to this problem.

Particles passing through this field, which permeates the whole universe, interact with it, causing an energy exchange that gives them mass (from $E = mc^2$). A cunning technique was used to incorporate the Higgs field into the standard model while protecting the underlying symmetry of the equations, and hence preserving the theoretical understanding of the fundamental forces. In this version of the standard model all particles are fundamentally massless, but gain mass due to their interaction with the Higgs field: a process called **spontaneous symmetry breaking**. Photons and gluons do not interact with the Higgs field, so they remain massless.

Because the Higgs field is clearly crucial to the standard model in its current form, confirming its existence experimentally became vitally important. Given that the quantum description of any field implies the existence of a corresponding particle (in the quantum electrodynamic field this is the photon; in the strong field it is the gluon, etc.), the experimental search centred on finding the **Higgs particle**.

From the mid-1990s every technological improvement that could step up the energy of particle accelerators came with the hope that the Higgs particle would be found. Successive failures to find the Higgs raised the lower limit on its mass, and fuelled the need to develop a better accelerator. The Large Hadron Collider (LHC) built at the European Centre for Nuclear Research (CERN) started full operations in 2010, colliding protons accelerated to 3.5 (later 4.0) TeV. In July 2012 two of the experimental teams using the LHC finally announced convincing evidence for the existence of the Higgs particle based on results accumulated from more than a year's worth of data. The Higgs was spotted from its tendency to decay into two high-energy photons creating an excess of photons at around 125.5 GeV. The experimental confirmation of the Higgs particle, and hence the Higgs field, ranks as one of the most significant discoveries of the last 100 years.

Supersymmetry

In the 1970s theorists began to explore the consequences of building equations that were symmetrical if boson particles (photons, Ws, Zs, and gluons) were switched with the quarks and leptons (together known as **fermions**) that make up matter, an idea known as **supersymmetry**.

Immediately, each particle gained a SUperSYmmetric (SUSY) partner: for each quark there is a squark (honestly!), each lepton has a partner slepton, the photon has a photino, the W a Wino and the gluon a gluino. Doubling the number of elementary particles would not normally be a sensible theoretical step, but in this case the benefits of the supersymmetric theories turned out to be very attractive as they brought potential solutions to a number of problems.

- Some of the lower-mass SUSY particles are good candidates for the mysterious dark matter that shows up gravitationally in the universe.
- Including the existence of SUSY particles in the Higgs mass calculations brought its mass down to a sensible value that turned out to be comparable with the 126 GeV estimation from the 2012 LHC results.
- The existence of SUSY particles made the possibility of finding one theory that unifies the fundamental forces as different components of a single underlying quantum field appear more likely.

The last point is of deep significance and worth exploring further.

Physicists have long dreamed of a theory that would bring the fundamental forces together as different aspects of one force (in the way that electric and magnetic fields were discovered to be components of an electromagnetic field). At the sort of energies typical in the universe currently, this seems very unlikely because the fundamental forces have strikingly different strengths. However, the force strengths vary with the interaction energy of the particles involved. Theoretical projections suggest that the strengths become more similar at the stupendous energy scales thought to have been present in the early universe. True unification, however, would require them to be exactly equal at some energy value. Remarkably, the projections show that the strengths of the strong, weak, and electromagnetic forces do become equal if the existence of SUSY particles is taken into account. For this reason, theorists have explored a variety of different Grand Unified Theories (GUTs) that aim to tie together these three fundamental forces (spread 9.16).

In summary, the reasons for believing that some form of supersymmetry is a vital component in the correct description of nature are largely based on the mathematical properties of supersymmetry. Presently there is no decisive experimental confirmation; indeed there are some indications that the theory in its simplest form might not be correct. As yet no SUSY particles have been observed.

SUSY particles that are valid dark-matter candidates would need to have masses ~0.1 TeV. In this case, the energy of the 4 TeV LHC collisions should have been enough to reveal them, if they existed in nature. Of course it may be that some very rare events have not yet shown up clearly in the data, but as time goes on that possibility becomes more remote.

Also the decay of the B_s meson (an $s\bar{b}$ combination) into two muons is very sensitive to the existence of supersymmetric particles (as they affect the decay modes available). Initial results for this decay rate, observed for the first time at the LHC, put it as being too rare for supersymmetry, at least in some of its simplest forms, to be correct. This is bound to be an area of intense theoretical and experimental work over the next decade.

Superstring theory

Aside from the appealing neatness of gathering up all the forces in one theory, some quantum theory of gravity is required to deal with issues such as the fate of matter crushed to a point in the singularity of a black hole. A quantum theory of gravity faces many theoretical challenges, including the tendency to produce nonsensical infinite results when point-like particles are included in the calculations.

One attempt to get round this issue replaces the idea of point particles with that of extended one-dimensional objects called **strings**, which can be closed into tiny loops (~10^{-33} m). With supersymmetric strings, all the fermions and bosons (including the graviton) turn out to be different vibrational modes of strings. The natural appearance of the graviton in this theory, as well as its neat mathematical properties, have made supersymmetric string theory currently the best candidate for a Theory of Everything (TOE).

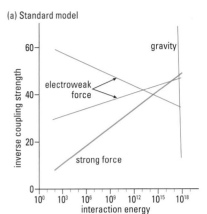

(a) Standard model

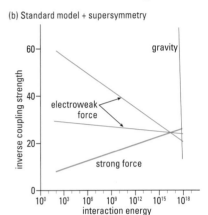

(b) Standard model + supersymmetry

Graph (a) shows how the 'strength' of the fundamental forces varies with interaction energy. Graph (b) shows the same variation, but with the effect of SUSY particles included. The fact that three of the forces meet at the same energy is encouraging to physicists seeking one unified theory of the fundamental forces.

Dimensions in string theory

Early in the history of string theory the need for more dimensions than the three of space plus time was discovered. The exact number of 'extra' dimensions depends on the details of the theory (typical numbers are 10 or 11), but it is necessary to understand why they do not show up in everyday life. The best explanation is that these extra dimensions are 'curled up' on themselves (compactified) so they only extend for ~10^{-33} m – so you can't walk along them!

O B J E C T I V E S

- the Copenhagen interpretation
- the scale of quantum effects
- Heisenberg's uncertainty principle
- Entanglement and teleportation

MODERN QUANTUM THEORY

When physicists put together the **Copenhagen interpretation** (spread 8.7), they were aware of a division between the quantum world and the everyday world. Attempting a quantum-mechanical description of both the subatomic particles being measured and the measuring devices being used led to paradoxes and confusions. To make the theory work, they had to assume that the effects of the quantum world became less significant at some scale and good old ordinary classical physics could be used to describe the measuring devices.

The scale of the classical world

The Planck constant h gives some indication of the scale at which we can expect quantum behaviour to operate. The de Broglie wavelength of an electron moving at $10 \, \text{m s}^{-1}$ can be measured ($6.9 \times 10^{-5} \, \text{m}$), but that of a car at the same speed would be far too tiny to cause any wave-like effects ($\sim 10^{-38} \, \text{m}$).

Anton Zeilenger and his group at the University of Vienna are trying to establish the boundary line between quantum and classical worlds. Their experiments are developments of the Young's slit scheme, which first demonstrated the wave nature of light. Technological improvements have allowed matter waves of sizeable molecules to be detected from their interference patterns: $C_{60}F_{48}$ has been used to successfully demonstrate matter wave interference, and this molecule is $\sim 10^{-10} \, \text{m}$ in size. Perhaps even more significantly, interference has been demonstrated for tetraphenylporphyrin ($C_{44}H_{30}N_4$), an organic molecule related to structures found in chlorophyll and haemoglobin. While this is a very long way from seeing interference in living systems, there is something psychologically significant in detecting the matter waves of an organic molecule.

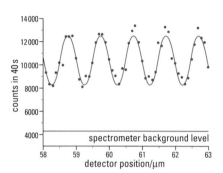

Interference fringes generated by the matter waves associated with the tetraphenylporphyrin molecule.

Uncertainty

The de Broglie relationship $\lambda = \dfrac{h}{p}$ (spread 8.6) relates the wave nature of a quantum object, expressed as the wavelength λ, to the particle nature, expressed as the momentum p. This is a very mysterious equation. It is hard to see how one object can be both a wave (spread out and constantly varying) and a particle (localized and rigid). The Copenhagen interpretation deals with this problem by insisting on the impossibility of designing an experiment that can display *both* wave and particle natures at the same time. This approach is propped up by **Heisenberg's uncertainty principle**:

$$\Delta x \Delta p \geq \frac{h}{2\pi}$$

where Δx is the uncertainty in an object's position and Δp the uncertainty in its momentum (spread 1.5). When Heisenberg first discovered this relationship he was thinking specifically in terms of measurement uncertainties, and so the equation was expressing how any experiment designed to pinpoint the position of an object (small Δx) would necessarily mean that its momentum could not be measured with precision at the same time (large Δp). However, with further thought and discussion (especially with Bohr), Heisenberg came to think of his relationship as expressing something more fundamental about the reality of nature, not simply our ability to experiment on nature. In this later view, Δx expresses the extent to which the particle *has* a position, and Δp the extent to which is *has* a momentum.

Quantum entanglement

Einstein believed in a deterministic universe: that every cause had an effect, and that if we could measure the detailed properties and behaviour of every particle in the universe, the laws of physics would enable us to predict the future with absolute certainty. This was his philosophical position and he was not prepared to compromise on it. As a result he

---

The Copenhagen interpretation

Neils Bohr and Werner Heisenberg, working at Bohr's institute in Copenhagen, devised a way of thinking about quantum theory that allowed physicists to interpret the mathematics and describe experiments. Max Born had already suggested that the 'wave function' used in quantum theory was related to the probability of finding an object, such as an electron, in a small region of space: the probability of finding an electron between x and $x + \Delta x = \psi(x)^2 \Delta x$ or, more simply, 'probability $\propto$ amplitude^2'. This left a problem: to explain how the probability was turned into a reality. The wave function gives an electron a probability of being in many places, but when you make a measurement, it turns up at only one specific location. According to the Copenhagen interpretation, when the electron interacts with the measuring device the wave function 'collapses' so that one of the probabilities becomes an actuality. Nothing is said about *how* this happens. Indeed, if you try to create a wave function for the device itself, then that function becomes entangled with the wave function of the electron, and some other device is needed to collapse both! So, the interpretation has to rely on some devices not being described by a wave function – in other words behaving classically.

was never happy with quantum theory. He felt that the probabilities embedded in the theory had to be a result of a lack of knowledge about the fundamental nature of particles.

In a classic 1935 paper Einstein, working with Podolsky and Rosen, set out to demonstrate this incompleteness in the quantum description of nature. The core of the paper relates to certain special, but not greatly contrived, circumstances in which the wave function of one particle would become 'entangled' with that of another. As a result, any measurement carried out on one of the entangled pair would have a direct influence on the other, even if separated by great distances. The measurement would collapse the wave function of the measured particle, instantaneously influencing the wave function of the other and limiting the outcome of any subsequent measurement on it.

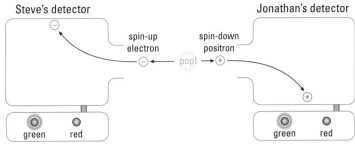

When an electron–positron pair is created their wave functions are entangled. Each wave function describes an equal possibility for the particle to be 'spin up' or 'spin down'. Spin is a quantum mechanical property of particles, which is somewhat like the rotation of a top. When the positron is measured in Jonathan's detector its wave function collapses to be either spin up or spin down. Instantly, the wave function of the electron collapses to be the opposite (to conserve angular momentum). When the electron arrives at Steve's detector its wave function is already collapsed, before the detector acts. This is what Einstein felt was 'spooky'.

Einstein felt that this 'spooky action at a distance' was better explained by assuming that some as yet undiscovered 'hidden variables' were fixed when the particles became entangled, and then influenced the outcome of subsequent measurements. If we did not know the values of these hidden variables, then we would not be able to predict the outcome of an experiment with certainty, and hence quantum theory would have probabilities in it.

In 1964 John Bell constructed a powerful argument showing that unless the hidden variables could communicate between particles at faster than the speed of light, any theory based on their existence could not exactly reproduce the results of quantum mechanics. Experimental measurements on entangled pairs made by Alain Aspect and his team settled the matter in favour of quantum theory in 1981–2. Since then physicists have tried to see how far apart the particle measurements can be and still see the effects of entanglement. The current record stands for pairs of entangled photons, one of which was measured in La Palma (Canary Islands, Spain) and the other 144 km away at an observatory in Tenerife. This experiment is a preliminary test for a possible future plan to send an entangled photon to a satellite or to the International Space Station.

As far as we can see, Einstein's hidden variables do not exist and quantum probabilities are a fundamental part of nature.

Quantum teleportation

Since 1998 physicists have been using entangled photon states to achieve a form of teleportation. This may appear to be the stuff of science fiction, but no actual matter is teleported. In quantum teleportation, the state of one quantum system is duplicated in another at a remote location. To do this, the transmitting station starts with a specific photon in a quantum state. Next a pair of entangled photons are generated, one of which is sent to the receiving station. At the transmitting end the entangled photon is allowed to interact with the other photon and a measurement is made on the combined system. The result of that measurement instantaneously affects the quantum state of the entangled photon at the receiving end. If the transmitting station then sends some simple classical information about the results of their measurements, the receiving station can perform a simple operation on their quantum system to transform it into a duplicate of the original photon state. Teleportation of this nature has been achieved between macroscopic objects, clusters of rubidium atoms, across a distance of 150 m. For photons the record stands at 144 km (again between La Palma and Tenerife.).The dream for the future is to use quantum teleportation as a means of routing quantum information across a quantum internet with significant advantages in encryption and data-processing speed.

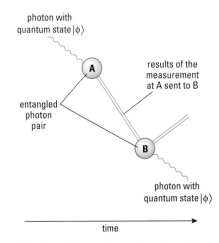

Quantum teleportation allows the state of a photon at a remote station (B) to be converted into a duplicate of that at a transmitting station (A).

MT.1

OBJECTIVES

- the difference between a variable and a constant
- symbols used in mathematics
- manipulating indices

Warning!

When using calculators people can misuse the equals (=) sign. On a calculator, hitting the = button means 'complete the previous calculation and start the next one'. For example, working out

$$\frac{3 + 4}{7}$$

on a calculator would probably involve typing something like

3 + 4 = ÷ 7 =

Using an equals sign in this way when writing mathematics can lead to nonsense. For example, imagine a question that asked you to calculate the total weight of a 70 kg man on a 10 kg bike:

total mass = 70 kg + 10 kg = 80 kg

∴ weight = 80 kg × 9.8 N kg⁻¹ = 784 N

However, one often sees answers written as follows:

total mass = 70 kg + 10 kg

   = 80 kg × 9.8 N kg⁻¹ = 784 N

which is nonsense, since read mathematically, this says that

70 kg + 10 kg = 784 N!

SIGNS, SYMBOLS, AND INDICES

Variables and constants

The power of mathematics, as far as physics is concerned, lies in its ability to represent many different situations in a compact form. For example, Newton's second law of motion can be expressed as $F = ma$. This is a very compact way of writing 'the force applied to a body is equal to the mass of the body multiplied by the acceleration produced by the force'. Furthermore, it allows us to calculate what that force is under any circumstance by replacing the letters with appropriate numbers.

In this expression of Newton's law the letters F, m, and a stand for numbers that can in principle take any value. They are known as **variables** (because they can vary).

Newton's law of gravity can be expressed mathematically as

$$F = \frac{GM_1M_2}{r^2}$$

In this expression all the letters stand for variables, except G, which is a constant of nature (the **universal gravitational constant**), i.e. it is a number with a value that we have to measure in nature and it never takes any other value (as far as we know).

Symbols

There are a variety of different symbols used in mathematics to write statements in a short and simple way. These symbols can be used when answering questions, provided that they are used correctly and that there is sufficient explanation to outline the physical principles involved.

Some mathematical symbols.

Symbol	Meaning	Example of use
=	is equal to	$x = 10$, i.e. the variable x has the value 10; $x = y$, i.e. the variable x has the same value as the variable y
≡	is identical to	$G \equiv 6.67 \times 10^{-11}\,\text{N m}^2\,\text{kg}^{-2}$, i.e. the constant G is $6.67 \times 10^{-11}\,\text{N m}^2\,\text{kg}^{-2}$ and is never anything else; $x \equiv y$, i.e. the variables x and y always have the same value
≈, ≃	is approximately equal to	$x \approx 10$, i.e. x has a value close to 10; $x \approx y$, i.e. x and y have about the same value
≠	is not equal to	$G \neq 5$, i.e. the value of G is not 5; $x \neq y$, i.e. the variables x and y do not have the same value
∝	is proportional to	$x \propto y$, i.e. x is equal to y multiplied by some constant
>	is greater than	$6 > 4$; $x > y$, i.e. x has a value greater than that of y
<	is less than	$4 < 6$; $x < y$, i.e. x has a value less than that of y
≤	is less than or equal to	$x \leq y$, i.e. the value of x is either less than that of y or it is equal to that of y
≥	is greater than or equal to	$x \geq y$, i.e. the value of x is either greater than that of y or it is equal to that of y

Some mathematical symbols (cont.)

Symbol	Meaning	Example of use
$\gg$	is much greater than	$10^6 \gg 1$; $x \gg y$, i.e. the value of x is much greater than that of y
$\ll$	is much less than	$1 \ll 10^6$; $x \ll y$, i.e. the value of x is much less than that of y
$\langle x \rangle$, $\bar{x}$	average or mean	$\langle x \rangle = 10$, i.e. the average value of the variable x is 10
$\sim$	is about the same as	$x \sim 10$, i.e. x is of the same order of magnitude as 10
$\pm$	plus or minus	$x = 5 \pm 3$, i.e. x is probably 5, but it might be as big as 8 or it might be as small as 2

Indices

Mathematicians are always looking for shorthand ways of writing things. For example, $2 \times 2 \times 2 \times 2 \times 2$ can be written as 2^5. In general, x^n means the number x multiplied by itself n times. n is called the index or power, and we would say 'x is raised to the power n', or 'x to the n'. There are rules for handling indices that follow from their meaning.

Rules for indices.

Rule	Example
$x^n \times x^m = x^{m+n}$	$2^3 \times 2^2 = (2 \times 2 \times 2) \times (2 \times 2) = 2^5$
$x^n \div x^m = x^{m-n}$	$2^3 \div 2^2 = (2 \times 2 \times 2) \div (2 \times 2) = 2$
$x^n \times y^m = ?$	this cannot be further simplified
$x^n + x^m = ?$	this cannot be further simplified
$(x^n)^m = x^{m \times n}$	$(2^2)^3 = (2 \times 2) \times (2 \times 2) \times (2 \times 2) = 2^6$

Special values of indices

The idea of indices being whole numbers can be extended to include negative numbers and fractions. These are given special meanings according to the following rules.

More indices.

Value	Meaning	Examples
x^{-1}	$\dfrac{1}{x}$	$2^{-1} = \dfrac{1}{2} = 0.5$
x^{-n}	$\dfrac{1}{x^n}$	$3^{-2} = \dfrac{1}{3^2} = \dfrac{1}{9} = 0.111$
$x^{\frac{1}{2}}$	$\sqrt{x}$	$2^{\frac{1}{2}} = \sqrt{2} = 1.414$
$x^{\frac{1}{n}}$	$\sqrt[n]{x}$	$27^{\frac{1}{3}} = \sqrt[3]{27} = 3$
$x^{\frac{m}{n}}$	$\left(\sqrt[n]{x}\right)^m$	$27^{\frac{2}{3}} = \left(\sqrt[3]{27}\right)^2 = 3^2 = 9$
$x^{-\frac{1}{n}}$	$\dfrac{1}{\sqrt[n]{x}}$	$9^{-\frac{1}{2}} = \dfrac{1}{\sqrt{9}} = \dfrac{1}{3} = 0.333$
x^0	1	$4^0 = 1$

Worked example

What is the kinetic energy of an object of mass 10 kg moving with a speed of 20 m s^{-1}?

$KE = \frac{1}{2}mv^2$

$= \frac{1}{2} \times (10\,\text{kg}) \times (20\,\text{m s}^{-1})^2$

$= \frac{1}{2} \times (10\,\text{kg}) \times (20\,\text{m s}^{-1}) \times (20\,\text{m s}^{-1})$

$= 2\,\text{kJ}$

NB It is only the v that is squared, not the whole expression. If that had been intended, the equation would be

$\left(\frac{1}{2}mv\right)^2$

PRACTICE

1 Write the following expressions in words:

  a $x = y < 100$ **b** $20 \leq x \leq 100$ **c** $10 \leq z < 30$.

2 Write the following in an abbreviated form using mathematical symbols:

  a the variable x is smaller than 5 times the variable y, which is less than or equal to 5;

  b the mean value of z is identical to the mean value of m.

3 Calculate the values of the following:

  a 2^7 **b** $8^2 \times 8^3$ **c** $\dfrac{4^2}{2 \times 8}$ **d** $9^{\frac{1}{2}} \times 3 \times 4^2$

  e $18^{\frac{1}{4}} \times 9^4 \times 3^{-\frac{4}{3}}$

4 Simplify the following:

  a $\dfrac{x^3 \times x^5}{x^7}$ **b** $3p^2 \times 4p^3$ **c** $(3x + 4y + 2x)^2$

  d $\left(x^3\right)^{\frac{1}{3}} + x$ **e** $\left(64x^4\right)^{\frac{4}{16}}$ **f** $\left(x^2 + 3x^2\right)^{\frac{1}{2}}$

EQUATIONS AND APPROXIMATIONS

Equations are extremely important in physics as they enable us to calculate values of variables. For example, the equation $v = u + at$ enables us to calculate the speed achieved by an object that has a starting speed of u and accelerates at a rate a for a time t. In this equation v is the **subject**. We need to have the subject on the left-hand side of the equals sign if its value is to be calculated from the values of the other variables. Of course, it is possible that we might need to calculate the value of one of the variables on the right-hand side. In this case the equation needs to be rearranged so that the variable we need becomes the subject.

Helpful hint

An equation is a balancing act. Look at it as if it were a ruler balanced on a knife edge. The equals sign represents the balance point. In order not to disturb the balance, whatever is done to the equation on one side of the balance point must be done on the other side as well. If this rule is not followed, then the result is bound to be mathematical nonsense.

Worked examples 1–5

1 Rearrange the equation $v = u + at$ to make u the subject of the equation

$$v = u + at$$

To get u on its own we need to remove the at part on the left-hand side. As at is added to u at the moment, the way to do this is to subtract at from both sides:

$$v - at = u + at - at$$
$$\therefore v - at = u \quad \text{or} \quad u = v - at$$

Notice that in the last line we have simply swapped left- and right-hand sides round. This is not strictly necessary, but it is conventional to have the subject on the left-hand side of the equals sign.

2 Make I the subject of the equation $P = IV$.

This time the variable that we want is being multiplied by something that we want to separate it from. The way to do this is to divide both sides of the equation by the unwanted variable:

$$P = IV$$
$$\frac{P}{V} = \frac{IV}{V} = I$$

3 Make r the subject of the equation $B = \mu_0 I / 2\pi r$

This will have to be a two-step process. The variable we are after is on the bottom line of a fraction on the right-hand side. The first step is to get it up to the top line of the equation. The best way to do this is to multiply both sides of the equation by r.

$$B \times r = \frac{\mu_0 I}{2\pi r} \times r$$
$$= \frac{\mu_0 I}{2\pi}$$

Now we have r on the left-hand side (where we want it), but it is multiplied by B. To get rid of the B we divide both sides by B:

$$\frac{Br}{B} = \frac{\mu_0 I}{2\pi B}$$
$$r = \frac{\mu_0 I}{2\pi B}$$

4 Make I the subject of $P = I^2 R$

Separating I^2 so that it is the subject is easy:

$$P = I^2 R$$
$$\frac{P}{R} = I^2 \quad \text{or} \quad I^2 = \frac{P}{R}$$

However, it is I that we want, not I^2. To convert I^2 to I we need to take the square root, but as this is an equation we must take the square root of both sides:

$$I = \sqrt{I^2} = \sqrt{\frac{P}{R}}$$

5 Make g the subject of $T = 2\pi\sqrt{l/g}$

Work through the following steps:

$$T^2 = 4\pi^2 \frac{l}{g}$$
$$gT^2 = 4\pi^2 l$$
$$g = \frac{4\pi^2 l}{T^2}$$

Substitutions

Sometimes it is necessary to combine two or more equations to make the equation that you need. For example, if you wish to calculate the rate at which a wire of length l and cross-sectional area A dissipates energy when a current I is passing through it, the equations $P = I^2R$ and $R = \rho l/A$ will have to be combined. The way to do this is to substitute (replace) the variable R on the right-hand side of the top equation with the right-hand side of the bottom equation. In other words,

$$P = I^2R$$
$$= I^2 \times \frac{\rho l}{A}$$
$$= \frac{I^2 \rho l}{A}$$

Worked example 6

How does the escape velocity of a planet depend on the radius of the planet?

$$v_{escape} = \frac{2GM}{R}$$

which would seem to provide the answer; all one has to do is make R the subject of the equation. However, this overlooks the fact that M is a **dependent variable** – the mass of a planet also depends on its radius as

$$m = \tfrac{4}{3}\pi r^3 \rho$$

if we take ρ to be the mean density of the planet.

So in order to get the full relationship we need to substitute for M in the escape velocity equation:

$$v_{escape} = \sqrt{\frac{2G \times (\tfrac{4}{3}\pi R^3 \rho)}{R}}$$
$$= \sqrt{\frac{8G\pi\rho R^3}{R}}$$
$$= \sqrt{\frac{8G\pi\rho}{3}}\,R$$

Approximations

Approximations are ways of making quick calculations that may not be exactly right but are close enough to the correct answer to make the full calculation a waste of effort.

One of the most useful approximations is

$$(1 + x)^n \approx 1 + nx$$

if x is a small number.

Worked example 7

What is 1.01^2?

$$1.01^2 = (1 + 0.01)^2 \approx 1 + 2 \times 0.01 = 1.02$$

The exact answer is 1.021.

This approximation can be used in a range of different circumstances, as the following examples show.

Expression	Approximation
$\sqrt{1.02}$	$\sqrt{1.02} = (1 + 0.02)^{\frac{1}{2}} \approx 1 + \frac{1}{2} \times 0.02 = 1.01$
$(1 + x^2)^{\frac{1}{2}}$	$(1 + x^2)^{\frac{1}{2}} = 1 + \frac{1}{2} \times x^2$
$\dfrac{1}{R + h}$ ($h \ll R$)	$\dfrac{1}{R + h} = \dfrac{1}{R} \times \dfrac{1}{1 + \frac{h}{R}} = \dfrac{1}{R}\left(1 + \dfrac{h}{R}\right)^{-1} \approx \dfrac{1}{R}\left(1 - \dfrac{h}{R}\right)$

PRACTICE

1 Make the underlined variable the subject of the following equations:

a $v = u + \underline{a}t$

b $R = \dfrac{\rho l}{A}$

c $F = \dfrac{mv^2}{r}$

d $v = \sqrt{\dfrac{T}{\mu}}$

e $F = \dfrac{GM_1M_2}{r^2}$

f $v^2 = u^2 + 2\underline{a}s$

g $E = \dfrac{T_1 - T_2}{T_1}$

h $E = \dfrac{T_1 - T_2}{T_1}$

i $\dfrac{1}{f} = \dfrac{1}{u} + \dfrac{1}{v}$

j $T = 2\pi\sqrt{\dfrac{m}{k}}$

2 Substitute from equation **i** into equation **ii** and simplify the result in the following questions:

a i $v = u + at$

   ii $x = \dfrac{(u + v)}{2}t$

b i $F = \dfrac{mv^2}{r}$

   ii $F = \dfrac{GM_1M_2}{r^2}$

c i $\omega^2 = \dfrac{k}{m}$

   ii $T = \dfrac{2\pi}{\omega}$

O B J E C T I V E S

- converting to decimal degrees
- radian measure
- extending the definition of sin, cos, and tan
- small-angle approximations

Measuring angles

The traditional way of measuring the size of an angle is to take the circumference of a circle and divide it up into 360 equal sections. The angle formed when the ends of one section are joined to the centre of the circle is one degree, 1°. This is the angle measure drawn onto a protractor, which is normally a semicircle with its edge divided into 180 sections. Each degree is then further divided into 60 smaller sections called minutes (60′), and each minute is divided again into 60 seconds (60″). These days, with calculators being used almost universally, it is more normal to work in decimal degrees rather than degrees, minutes, and seconds. Minutes and seconds are still used in astronomy to pinpoint the position of a star in the sky and to measure parallax angles.

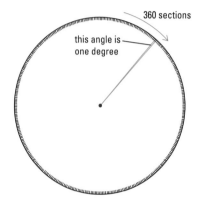

One degree is the angle subtended by one 360th of the circumference of a circle.

Worked example

What is the angle 30° 15′20″ in decimal degrees?

First we need convert 20″ into minutes:

$$20'' = \frac{20}{60} = \frac{1}{3} = 0.33$$

So the total number of minutes is 15.33, which we convert to degrees by dividing by 60:

$$15.33' = \frac{15.33}{60} = 0.26° \text{ (2 d.p.s)}$$

So the angle in decimal degrees is 30.26°

Dividing the circle into 360 sections is an arbitrary choice which seems natural to us only because we have never known anything different. However, there is a mathematically more sensible choice, which is to divide the circle into 2π pieces. After all, the circumference of a circle is $2\pi \times$ radius, so a circle of radius 1 has circumference of 2π. This is the basis of the **radian** measure of angles.

Radians

Consider an angle of $\theta°$ at the centre of a circle, as in the figure below. The same angle can be measured in radians (rad) by calculating the ratio of the arc length to radius:

$$\text{angle in radians} \quad s = \frac{\text{arc length}}{\text{radius}} = \frac{l}{r} \text{ rad}$$

For a complete rotation $l = 2\pi r$

so: $\qquad s = 2\pi$ rad

$\qquad \therefore 2\pi$ rad $= 360°$

$$s = \frac{\theta}{360} \times 2\pi \qquad \text{which is the way to convert from degrees to radians, or}$$

$$\theta = \frac{s}{2\pi} \times 360 \qquad \text{which converts from radians to degrees.}$$

Degrees to radians

Angle in degrees	Angle in rad
0	0
45	$\frac{\pi}{4}$
60	$\frac{\pi}{3}$
90	$\frac{\pi}{2}$
180	π
360	2π

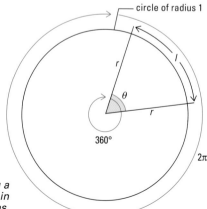

The radian angle measure is defined by dividing a circle up into 2π pieces, not 360°. θ is the angle in degrees, s is the same angle in radians.

Sine, cosine, and tangent

In basic trigonometry the three ratios sin, cos, and tan are defined for a right-angled triangle as in the figure on the right. In physics we often need to calculate the values of these ratios for angles greater than 90°, which cannot fit into a right-angled triangle. To make this possible we extend the way in which the ratios are calculated by using a circle of unit radius.

We imagine a radius line turning like a clock hand from the x axis round the circle in an anticlockwise direction. The angle it has turned through is marked θ on the diagram. This radius forms two right-angled triangles with the x and y axes as shown. As the radius has length 1, the coordinates on the x and y axes mark out the values of $\sin \theta$ and $\cos \theta$. The third ratio, $\tan \theta$, is found by dividing $\sin \theta$ by $\cos \theta$.

We can extend this to negative angles by rotating the radius in a clockwise direction from the x axis.

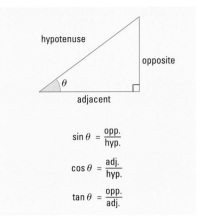

$$\sin \theta = \frac{\text{opp.}}{\text{hyp.}}$$

$$\cos \theta = \frac{\text{adj.}}{\text{hyp.}}$$

$$\tan \theta = \frac{\text{opp.}}{\text{adj.}}$$

Sine, cosine, and tangent in trigonometry. The three definitions work for $0 < \theta < 90°$

The four quadrants of a circle in which the ratios are either positive or negative.

Quadrant	$0 \le \theta < 90°$	$90° \le \theta < 180°$	$180° \le \theta < 270°$	$270° \le \theta < 360°$
$\sin \theta$	+ve	+ve	−ve	−ve
$\cos \theta$	+ve	−ve	−ve	+ve
$\tan \theta$	+ve	−ve	+ve	−ve

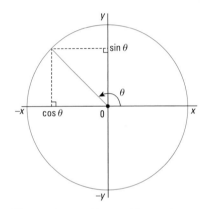

Sine, cosine, and tangent for angles > 90°. In the quadrant shown ($90° < \theta < 180°$) $\cos \theta$ is negative and $\sin \theta$ is positive.

Small angles

Working in radians rather than degrees has a great advantage when it comes to small angles (<0.1 rad). When the angle is very small, the arc length s and distance along the y axis are approximately the same length (see right). The arc length is the size of the angle in radians, and the distance along the y axis is the sin of the angle, so

$$\sin \theta \approx \theta \text{ when } \theta < 0.1 \text{ rad}$$

Cos θ is the distance along the x axis, which is

$$\sqrt{1 - \theta^2} \approx 1 - \frac{\theta^2}{2}$$

So

$$\cos \theta \approx 1 - \frac{\theta^2}{2} \text{ when } \theta \text{ is } < 0.1 \text{ rad}$$

Furthermore

$$\tan \theta \approx \theta \text{ when } \theta < 0.1 \text{ rad}$$

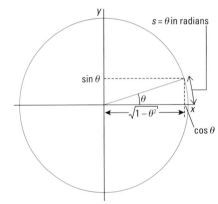

The small-angle approximation.

- vectors and scalars
- resolving into components
- vector addition

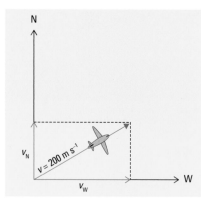

Resolving the velocity of a jet plane flying at 200 m s⁻¹ 25° north of west.

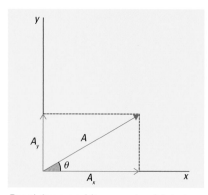

Resolving an arbitrary vector A into two mutually perpendicular components.

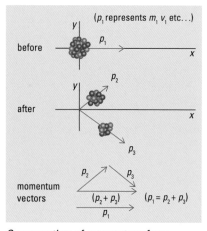

Conservation of momentum for a nucleus emitting an alpha particle.

VECTORS

Scalar quantities have magnitude only. **Vectors** have magnitude and direction. The following are examples.

Scalars: energy; temperature; mass; distance; speed; potential.

Vectors: force; momentum; field strength; displacement; velocity.

Scalars can be added or subtracted like ordinary numbers, but vectors obey different rules. For example, two 1 kg masses together make a mass of 2 kg, but two 1 N forces acting together can result in a force of anywhere between 0 N (opposite directions) and 2 N (parallel directions). Intermediate values occur when they act along different lines.

Resolving vectors into components

Any vector can be split into components parallel to two other directions (in the same plane); for example, a jet flying at 200 m s⁻¹ 25° north of west has components of velocity in the north and west directions. The size of these components can be found using trigonometry:

component west: $v_w = 200 \cos 25° = 180 \, \text{m s}^{-1}$

component north: $v_n = 200 \sin 25° = 85 \, \text{m s}^{-1}$

General case   If a vector of length A is at an angle θ to a particular direction (e.g. an x axis), its component in that direction is A_x, given by

$$A_x = A \cos \theta$$

The component perpendicular to this direction is A_y, given by

$$A_y = A \sin \theta$$

Components can be taken in any directions, but it is usually most convenient to resolve along perpendicular axes, since the components are then independent of one another (like the horizontal and vertical components of velocity). Here we have resolved in a two-dimensional plane. In three dimensions three components would be used, and again it is usually convenient to choose mutually perpendicular axes.

Conservation of momentum

Linear momentum is a vector quantity, so the law of conservation of momentum involves the conservation of a vector. This means that the total linear momentum vector after a collision or interaction (in a closed system) is the same as it was before it. For this to be the case, *every component* of the initial linear momentum must be separately conserved. So in two dimensions conservation of momentum generates two separate conservation equations, one for each of two components. In three dimensions there will be three equations.

An example of this is a moving nucleus that emits an alpha particle. If the initial nucleus is moving in the x direction when it decays, then the sum of the x components of linear momentum for the new nucleus and the alpha particle must equal the original linear momentum. The sum of momentum components perpendicular to the initial direction must be zero. This can be represented graphically by recombining the two final momentum vectors as shown; their sum is the original momentum vector.

Adding vectors

The last example showed how to combine momentum vectors graphically. The sum of two or more vectors is called the **resultant**.

• **Graphical method** Draw each vector to scale pointing in the right direction and place them end to end (so that the 'point' of one touches the 'start' of the other). The sum of all the vectors is equal to a vector drawn from the start of the first vector to the end of the last one in the sequence. (The order in which the vectors are combined does not matter.) The magnitude and direction of the resultant can then be measured straight off the diagram, or it can be calculated using trigonometry.

In the example below, three forces act on a mass. The resultant of the three is found by combining them graphically.

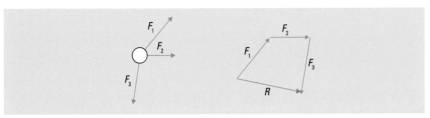

Combining three forces graphically.

• **Parallelogram method** Some people find this a useful way to combine two vectors. Draw both vectors from a common point. Complete the parallelogram defined by these two vectors. The resultant is now the diagonal of the parallelogram, and again this can be measured off a scale drawing, or trigonometry can be used to calculate its magnitude and direction.

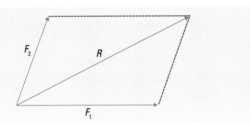

The parallelogram method of combining two vectors.

• **Components method** In more complex examples the most reliable way to add vectors is to resolve them along perpendicular axes and then add their components. The resultant can then be reconstructed using Pythagoras's theorem and trigonometry, as shown below.

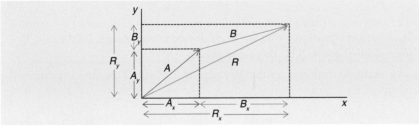

Resolving vectors into components.

Recombining components

Sometimes it is necessary to recombine a resultant vector from its components along mutually perpendicular axes. This is simply an example of vector addition (the components being the vectors to be added) as discussed earlier. In practice it is rather a simple case because the components are perpendicular to one another. In the example above, the components are along the x and y axes and have magnitudes A_x and A_y.

$$A^2 = A_x^2 + A_y^2 \quad \text{with } A = \sqrt{A_x^2 + A_y^2} \quad \text{by Pythagoras}$$

$$\tan\theta = \frac{A_y}{A_x} \quad \text{by trigonometry}$$

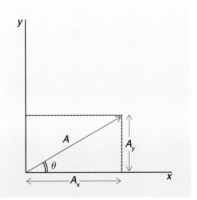

Recombining components.

O B J E C T I V E S

- logarithms to base 10 and base e
- using logarithms to multiply and divide
- logarithms of powers
- logarithms and equations

LOGARITHMS

Logarithms were invented as a means of making the multiplication and division of very large numbers (such as those in astronomy) easier. Now that we all have electronic calculators, the original need for logarithms has disappeared. However, they turn out to be very interesting in their own right and many phenomena in nature are related to each other logarithmically.

What are logarithms?

Multiplying two numbers like 673 and 5490 together is a rather tedious task when it has to be done by hand. However, two numbers like 4096 and 16 384 can be multiplied together rather more easily. All one has to do is to notice that:

$$4096 = 2^{12} \text{ and } 16\,384 = 2^{14}$$

so that

$$4096 \times 16\,384 = 2^{12} \times 2^{14} = 2^{26}$$

So, multiplying the numbers has been reduced to simply adding the indices. This is all very well, provided that the numbers are powers of two and provided that you have some sort of table that converts powers of two into the numbers, so that you can look up $2^{26} = 6710\,8864$. Constructing such a table requires a great deal of work, rather more than multiplying the numbers together in the first place, but the advantage is that it only has to be done once. Once the table exists it can be used many times. Also, it is possible to write any number as a power of 2, provided that you allow the power to be a decimal, not just a whole number:

$$673 = 2^{9.3945} \text{ (4 decimal places)}$$

$$5490 = 2^{12.4226} \text{ (4 decimal places)}$$

In general any number x can always be written as a power of 2, i.e.

$$x = 2^y$$

or

$$y = \text{logarithm (base 2) of } x = \log_2 x$$

> **Definition: logarithms**
>
> The logarithm of a given number to a certain base is the power to which you must raise the base number to in order to obtain the given number, e.g. the logarithm of 5490 to the base 2 is 12.4226 (4 d.p.) as $2^{12.4226} = 5490$

> **Abbreviations**
>
> Rather than having to write $\log_{10}$ to mean 'log to the base 10' all the time, most people just write log. In order to write log to the base e, the abbreviation ln is very popular. These two abbreviations are always found on calculators.

Different base numbers

The number 2 is not the most sensible choice to use as the base of the logarithm; 10 is much more useful since then the table need only be calculated for numbers between 0 and 9.9. Consider

$$1.22 = 10^{0.0864}$$

or

$$\log_{10} 1.22 = 0.0864$$

But, if we want to take the log of a number such as 122, then

$$122 = 100 \times 1.22 = 10^2 \times 10^{0.086\,44} = 10^{2.086\,44}$$

i.e. $\log_{10} 122 = 2.086\,44$, and similarly $\log_{10} 12\,200 = 4.086\,44$.

This would imply that 10 is the most sensible choice of base since any number can easily be obtained once the logarithms of numbers between 0 and 9.9 have been calculated. Indeed, if all we were interested in was the multiplication of numbers, this would be true. However, in nature the very inconvenient-looking number e = 2.718 28... (the so-called **natural number**) is far more often represented. Physicists in particular use log to the base e far more than log to the base 10. The inconvenience of such an odd number is minimal, since logarithms can be found on a calculator.

Worked example 1

What are the logarithms of the following numbers:

(a) 7.89 to the base 10? (b) 7.89 to the base e?

(a) On the calculator, type 7.89 then **LOG** (or, on some calculators, **LOG** then 7.89) and the answer is 0.89708

(b) On the calculator type 7.89 and then **LN** (or **LN** then 7.89) and the answer is 2.0656

Some useful rules involving logarithms

The simple rules for adding and subtracting powers when numbers are multiplied together produce equally simple rules when taking logarithms.

If $a = 10^x$ and $b = 10^y$, then

$$a \times b = 10^x \times 10^y = 10^{x+y} \text{ and } a/b = 10^x/10^y = 10^{x-y}$$

hence

$$\log (a \times b) = \log a + \log b \text{ and } \log (a/b) = \log a - \log b$$

NB These rules apply to any base, not just log to the base 10.

Worked example 2

What is the logarithm of $\rho l/A$?

$$\log \frac{\rho l}{A} = \ln \rho l - \ln A = \ln \rho + \ln l - \ln A$$

Worked example 3

What is the logarithm of 34?

$$\ln 3^4 = \ln (3 \times 3 \times 3 \times 3) = \ln 3 + \ln 3 + \ln 3 + \ln 3 = 4 \times \ln 3$$

In general the logarithm of any power is the power times the logarithm – even if it is a fractional power, e.g. $\log (\sqrt{2}) = \log 2^{1/2} = 1/2 \log 2$.

Logarithms, equations, and graphs

It is possible to take logarithms in an equation, provided both sides of the equation have the logarithm to the same base. For example, the period of a simple pendulum is given by $T = 2\pi\sqrt{l/g}$ so

$$\ln T = \ln \left(2\pi\sqrt{\frac{l}{g}}\right)$$

$$= \ln (2\pi) + \ln \left(\sqrt{\frac{l}{g}}\right)$$

$$= \ln (2\pi) + \frac{1}{2}\ln \left(\frac{l}{g}\right)$$

$$\therefore \ln T = \ln (2\pi) + \frac{1}{2}\ln l - \frac{1}{2}\ln g$$

Worked example 4

The period, T, of a object suspended from a spring is known to depend on the mass of the object, m, and the stiffness of the spring, k, in the following manner:

$$\log T^n = C\frac{m}{k}$$

where n and C are unknown constants. How might one determine the values of n and C from experimental data involving the period measured for several masses?

Taking logs (to either base) gives

$$\log T^n = \log\left(C\frac{m}{k}\right)$$

$$n\log T = \log C + \log m - \log k$$

$$\log T = \frac{1}{n}(\log C - \log k) + \frac{1}{n}\log m$$

So, if one were to plot a graph of $\log T$ against $\log m$, the gradient of the graph would be $1/n$ and the intercept on the y axis would be

$$\frac{1}{n}(\log C - \log k)$$

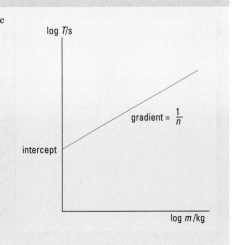

PRACTICE

1 What are the following logarithms to base 10? Try to work them out in your head before you check them on a calculator.
   log 10 log 1000 log 0.01 log 10^{48}

2 Given that log 2 = 0.301 (3 d.p.), work out the following logarithms.
   log 20 log 200 log 0.02 log 4 log 64

3 Write down the algebraic expressions for the following logarithms. $\ln e^2$ $\log 300y^2$ $\log (30\,000/60y^3)$ $\ln [4\pi x^2/(a + b)]$

4 Kepler's third law of planetary motion suggests that the period of a planet's motion is proportional to the average radius of its orbit to some power: $T \propto r^n$

   Using the following data, plot a suitable graph to find n and the constant of proportionality.

Planet	Mercury	Venus	Earth
Average radius of orbit / m	5.79×10^{10}	1.08×10^{11}	1.50×10^{11}
Period	87.97 d	224.7 d	365.3 d

Planet	Mars	Jupiter	Saturn
Average radius of orbit / m	2.28×10^{11}	7.78×10^{11}	1.43×10^{11}
Period	687 d	11.86 y	29.46 y

EXPONENTIALS

Frequently in nature the change in some quantity depends on the amount of it there is to start with. For example, in radioactivity the decrease in the number of radioactive atoms per second (the activity) is proportional to the number of radioactive atoms present in the sample, i.e.

rate of change in number of atoms $\propto$ number of atoms present

$$\therefore \frac{\delta N}{\delta t} \propto N \text{ or } \frac{\delta N}{N} = \text{constant} \times \delta t$$

Worked example 1

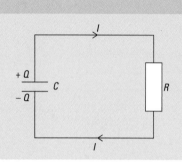

How does the change in charge on a capacitor depend on the amount of charge on its plates?

When a capacitor discharges through a resistor, as in the figure on the left, the voltage drop across the capacitor is $V = Q/C$. Therefore the current is

$$I = \frac{V}{R} = \frac{Q}{CR}$$

The change in charge on the capacitor in time δt is $\delta Q = I\delta t$. Hence

$$\delta Q = \frac{Q\delta t}{CR} \quad \text{or} \quad \frac{\delta Q}{Q} = \text{constant} \times \delta t$$

Worked example 2

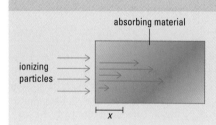
absorbing material

ionizing particles

x

How does the number of ionizing particle decrease with distance as they pass through a material?

If μ is the probability of an ionizing particle being absorbed in a 1 m length of material, and N_0 is the number entering the material, then the number absorbed in a length δx is $\mu N_0 \delta x = \delta N$, so

$$\frac{\delta N}{N_0} = \mu \delta x = \text{constant} \times \delta x$$

The natural number

Consider again the example of radioactive atoms decaying. If the probability of an atom decaying in unit time is λ, then the average number of decays in a time δt is

$$\delta N = \lambda N_0 \delta t$$

So after this period of time, the number of atoms remaining will be

$$N_0 - \delta N = N_0 - \lambda N_0 \delta t = N_0(1 - \lambda \delta t)$$

The number of atoms decaying in the next time interval δt is

$$\delta N = N_0(1 - \lambda \delta t) \times \lambda \delta t$$

so the number remaining is

$$N = N_0(1 - \lambda \delta t) - N_0 (1 - \lambda \delta t)\lambda \delta t = N_0(1 - \lambda \delta t) (1 - \lambda \delta t)$$
$$= N_0(1 - \lambda \delta t)^2$$

and so on. In general, the number of atoms remaining after n time intervals δt is given by

$$N = N_0(1 - \lambda \delta t)^n$$

This is an approximate answer that can be improved by using smaller and smaller time intervals, i.e. by dividing up a time interval, t, into n periods of length δt, so that

$$N = N_0\left(1 - \frac{\lambda t}{n}\right)^n$$

The approximation gets better if smaller and smaller intervals are taken, i.e. for larger and larger values of n. If the calculation is carried out, then as n gets larger, the result gets closer and closer to a fixed quantity. In other words, as n gets larger,

$$N_0\left(1 - \frac{\lambda t}{n}\right)^n \rightarrow N_0 e^{-\lambda t}$$

where e is the infinite decimal number known as the natural number.

Worked example 3

What happens to the value of $\left(1 + \dfrac{1}{n}\right)^n$ as n gets larger?

with $n = 10$ $\quad \left(1 + \dfrac{1}{n}\right)^n = (1 + 0.1)^{10} \quad = 2.593\,742$

with $n = 100$ $\quad \left(1 + \dfrac{1}{n}\right)^n = (1 + 0.01)^{100} \quad = 2.704\,814$

with $n = 1000$ $\quad \left(1 + \dfrac{1}{n}\right)^n = (1 + 0.001)^{1000} \quad = 2.716\,924$

with $n = 10^8$ $\quad \left(1 + \dfrac{1}{n}\right)^n = (1 + 0.000\,01)^{10^8} \quad = 2.718\,282$

Notice as the value of n gets larger, the value of $(1+1/n)^n$ gets closer and closer to a fixed value $2.718\,281\,8\ldots$ which is the value of e.

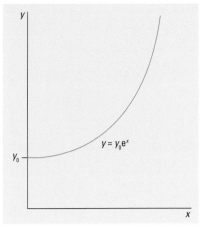

An exponential growth curve.

Exponential graphs

All the examples mentioned so far, radioactive decay, a capacitor discharging, and ionizing particles moving through a material, follow the same pattern. The argument followed to produce the exact equation for the radioactive decay case could work equally well for the other cases, producing

radioactive decay: $\qquad\qquad N = N_0 e^{-\lambda t}$

capacitor discharge: $\qquad\qquad Q = Q_0 e^{-\frac{t}{CR}}$

absorption of ionizing particles: $\quad N = N_0 e^{-\mu x}$

Any relationship that follows this pattern is known as an **exponential** and will produce the same sort of graph if plotted out. Any situation in which the change in some quantity is proportional to the starting amount will always produce an exponential relationship. Such situations are very common in nature. Many physical effects happen in this way, at least to a good approximation. This is why e is known as the natural number and why log to the base e, ln, is so useful in physics.

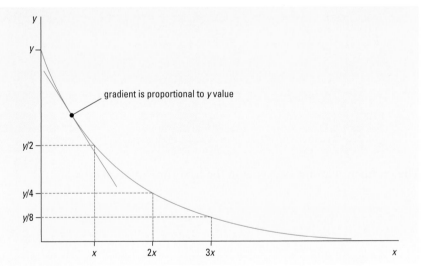

An exponential decay curve.

There are some important features of exponential relationships:

- Constant proportion – if you move in equal steps along the horizontal, x, axis of an exponential graph the corresponding points on the vertical, y, axis move down in equal *ratios*. For example, if moving from 0 to x takes y to $y/2$, then moving from x to $2x$ will take $y/2$ to $y/4$, etc.

- The gradient of the line at any point on an exponential graph is proportional to the y value of that point.

- Plotting $\ln y$ against x will always produce a straight line if the relationship between y and x is exponential.

So far we have only mentioned cases in which the exponential relationship is a decreasing one. It is quite possible for an exponential increase to take place following the relationship $y = y_0 e^{+kx}$.

O B J E C T I V E S

• plotting graphs from data

• sketching graphs

Plotting graphs from data

Plotting a graph from a table of data is a basic skill required of a physicist. There are some fundamental rules that must followed, and examiners expect to see this being done. Sloppy and careless graph-plotting not only leads to the wrong conclusions being drawn from the data, it can also lose many marks in an examination.

Each of the points indicated on the graph in the worked example below would be looked for specifically by an examiner marking a paper.

Worked example 1

Plot a graph of current against voltage for the following data.

Voltage / V	1	2	3	4	5	6	7	8	9	10
Current / mA	0.9	2.1	2.8	4.1	5.0	5.9	7.1	8.0	8.8	9.9

Each current measurement is ± 0.2 mA.

Is the current proportional to the voltage?

What is the gradient of the graph?

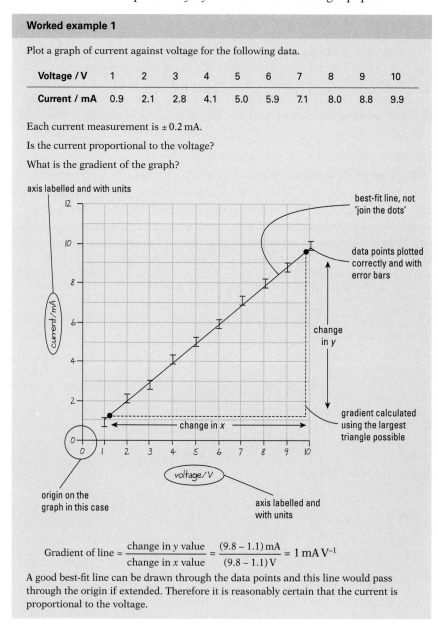

$$\text{Gradient of line} = \frac{\text{change in } y \text{ value}}{\text{change in } x \text{ value}} = \frac{(9.8 - 1.1)\,\text{mA}}{(9.8 - 1.1)\,\text{V}} = 1\,\text{mA V}^{-1}$$

A good best-fit line can be drawn through the data points and this line would pass through the origin if extended. Therefore it is reasonably certain that the current is proportional to the voltage.

Notes on drawing graphs

The following points should be remembered.

• Examiners will check to see that the data points are plotted correctly.

• The origin need not always be on the graph, but in the example above, the question asks if the current is proportional to the voltage – so one must check that the line passes through the origin.

- The data must be plotted with error bars if the uncertainly has been quoted in the question.
- The best-fit line (i.e. the line that passes as close as possible to each data point) has been drawn on the graph above.
- The gradient, if required, must be calculated from the largest possible triangle on the graph paper – this makes reading the scale more accurate.
- The y values and the x values must be read off the scale on the axes, not by simply 'counting the squares'.
- The units must be quoted in the gradient calculation.

Sketching graphs

Another valuable skill for a physicist to acquire is the ability to sketch what a graph looks like given the minimum of data. When faced with this sort of question, students often waste time plotting graphs when a simple sketched curve conveys all the information required.

Worked example 2

Sketch a graph showing how the discharge current changes with time when a $100\,\mu\text{F}$ capacitor charged to $10\,\text{V}$ is discharged through a $100\,\Omega$ resistance.

Notice that the axes have been marked with units, labels, and some indication of the scale on the graph. Although the question did not specifically ask for the initial discharge current and the half-time to be calculated, the information necessary had been provided. Without such scales on the graph it would be impossible to draw any meaningful information from the graph. Examiners will expect this information to be there.

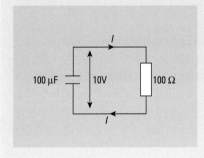 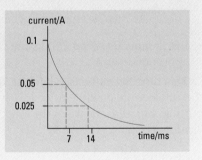

There are a variety of different curve shapes that a student should be able to recognize and sketch. The diagram below shows the most common ones that come up in physical processes besides the exponential curves shown in spread MT.6.

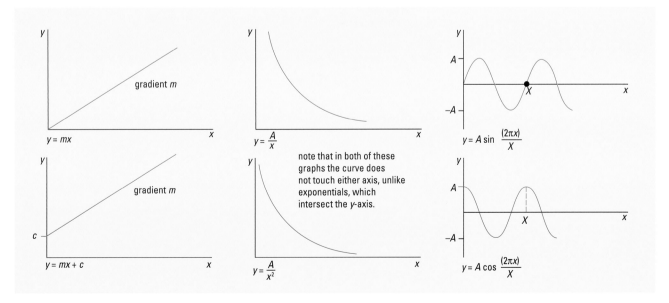

DIFFERENTIAL CALCULUS

Rates of change

If a stone is dropped from rest, the distance s it has fallen after a time t is given by the familiar equation

$$s = \frac{1}{2}gt^2$$

The speed of the falling stone is equal to the rate of change of distance, so it can be derived from this equation. If you wanted to find the speed of the stone after 1 second, there are several ways in which you might do it.

Method 1 Measure the slope of the curve at $t = 1$ s.

Method 2 Find the average speed during a short time δt lasting from just before to just after $t = 1$ s (during which time the stone falls a short distance δs).

Method 3 Differentiate the expression for s and substitute $t = 1$ s.

The accuracy of method 1 will depend on how carefully you draw your tangent to the curve. Method 2 can be made as accurate as you like by taking shorter and shorter time intervals close to $t = 1$ s. As δt is made smaller and smaller, the average speed (given by the ratio $v = \delta s/\delta t$) comes closer and closer to the exact instantaneous value. What differentiation does is provide a way to calculate this ratio in the limiting case when δt approaches zero (often written as $\delta t \to 0$). This gives the *exact* speed.

- Average speed during an interval δt is

$$v = \frac{\delta s}{\delta t} \quad (\delta \text{ means 'small change in'})$$

- Instantaneous speed equals the value approached by $v = \delta s/\delta t$ as δt approaches zero.

This limiting value is called the 'derivative of s with respect to t' or the 'differential coefficient of s with respect to t'. The sounds rather off-putting, but all it means is the *rate of change* of s with t. Sometimes it is written as

$$v = \lim_{\delta \to 0} \frac{\delta s}{\delta t}$$

but usually (thank goodness) we just write

$$v = \frac{\mathrm{d}s}{\mathrm{d}t}$$

- the d.../dt bit is read as 'the rate of change of ... with time'.

- ds/dt simply means 'the rate of change of distance with time'.

You will find plenty of maths books in which the rules for differentiation are derived from first principles. There is not room for that here, so instead there are two tables that bring together most of the functions and equations you are likely to come across. In the first some general forms are shown (these are all differentiated with respect to x rather than t). The second summarizes some common physical expressions that you may well need to differentiate.

Some useful derivatives.

Function	Derivative
A Constant	
$y = A$	$\dfrac{\mathrm{d}y}{\mathrm{d}x} = 0$

NB This is zero because a constant does not change, whatever the value of x.

Powers	
$y = x^n$	$\dfrac{\mathrm{d}y}{\mathrm{d}x} = nx^{n-1}$

Rule: reduce the power by 1, multiply by the old power.

$y = x^2$	$\dfrac{\mathrm{d}y}{\mathrm{d}x} = 2x$
$y = x^5$	$\dfrac{\mathrm{d}y}{\mathrm{d}x} = 5x^4$
$y = x^{\frac{3}{2}}$	$\dfrac{\mathrm{d}y}{\mathrm{d}x} = \dfrac{3}{2}x^{\frac{1}{2}}$
$y = x^{-2}$	$\dfrac{\mathrm{d}y}{\mathrm{d}x} = -2x^{-3}$

NB $y = x^{-1}$ is an exception to this rule for differentiating powers:

$y = x^{-1}$	$\dfrac{\mathrm{d}y}{\mathrm{d}x} = \ln x$

Trigonometric functions	
$y = \sin x$	$\dfrac{\mathrm{d}y}{\mathrm{d}x} = \cos x$
$y = \cos x$	$\dfrac{\mathrm{d}y}{\mathrm{d}x} = -\sin x$
$y = \tan x$	$\dfrac{\mathrm{d}y}{\mathrm{d}x} = \sec^2 x$

Exponentials	
$y = e^2$	$\dfrac{\mathrm{d}y}{\mathrm{d}x} = e^x$

Radioactivity	
$N = N_0 e^{-\lambda t}$	$\dfrac{\mathrm{d}N}{\mathrm{d}t} = -\lambda N_0 e^{-\lambda t} = \lambda N$

SHM

$s = A\cos \omega t \quad v = \dfrac{\mathrm{d}s}{\mathrm{d}t} = -\omega A \sin \omega t$

$$a = \frac{\mathrm{d}v}{\mathrm{d}t} = -\omega^2 A \cos \omega t$$

$$= -\omega^2 x$$

Derivatives in physics.

velocity (speed) is the rate of change of displacement (distance)	$v = \dfrac{\mathrm{d}s}{\mathrm{d}t}$	induced e.m.f. equals rate of change of flux-linkage	$E = -\dfrac{\mathrm{d}(N\phi)}{\mathrm{d}t}$
the acceleration is rate of change of velocity	$a = \dfrac{\mathrm{d}v}{\mathrm{d}t}$	definition of self-induction	$V = L\dfrac{\mathrm{d}I}{\mathrm{d}t}$
current is the rate of flow of charge	$I = \dfrac{\mathrm{d}Q}{\mathrm{d}t}$	resultant force equals rate of change of momentum	$F = \dfrac{\mathrm{d}(mv)}{\mathrm{d}t}$
electric field strength is negative potential gradient	$E = -\dfrac{\mathrm{d}V}{\mathrm{d}x}$	radioactive activity equals negative rate of decay	$A = -\dfrac{\mathrm{d}N}{\mathrm{d}t}$

Worked example 1

What is the velocity of a falling stone 1 s after it is released?

$s = \frac{1}{2}at^2$ where a is constant and equal to $9.8\,\mathrm{m\,s^{-2}}$

$v = \dfrac{\mathrm{d}s}{\mathrm{d}t} = at$ (using the power law for t^2)

$v = 9.8\,\mathrm{m\,s^{-1}} \times 1.0\,\mathrm{s} = 9.8\,\mathrm{m\,s^{-1}}$

Function of a function

Many of the expressions you will meet in physics are not simple functions of x or t. For example, in simple harmonic motion (SHM) the displacement s of a particle that is released from one amplitude at $t = 0$ is given by $s = A\cos\omega t$. This is the cosine of a function of time (t). To differentiate a 'function of a function' like this:

- first differentiate the ωt bit to get ω.
- now differentiate the $A\cos\omega t$ to get $-A\sin\omega t$ (treating the ωt as a fixed lump);
- multiply the first result by the second to get

$$\frac{\mathrm{d}s}{\mathrm{d}t} = \omega \times (-\sin\omega t) = -\omega\sin\omega t$$

This is the velocity of the oscillator.

The acceleration can be found in a similar way:

$$a = \frac{\mathrm{d}v}{\mathrm{d}t} = \frac{\mathrm{d}(-\omega A\sin\omega t)}{\mathrm{d}t}$$

$$= \omega \times (-\omega A\cos\omega t) = -\omega^2 A\cos\omega t$$

The chain rule

The example above dealt with a cosine function of a function of time. The method of solution can be written much more generally as the '**chain rule**' for differentiation.

If g is a function of time (for example ωt) and f is a function of g (for example cosine), then the complete expression y can be written as

$$y = f(g(t))$$

$$\frac{\mathrm{d}y}{\mathrm{d}t} = \frac{\mathrm{d}f}{\mathrm{d}g} \times \frac{\mathrm{d}g}{\mathrm{d}t}$$

Notice that if this was ordinary algebra dg would be cancelled out.

For the SHM example,

$$f = A\cos\omega t, \quad g = \omega t$$

$$f = A\cos g$$

$$\frac{\mathrm{d}f}{\mathrm{d}g} = -A\sin g, \frac{\mathrm{d}g}{\mathrm{d}t} = \omega$$

$$\frac{\mathrm{d}y}{\mathrm{d}t} = \frac{\mathrm{d}f}{\mathrm{d}g} \times \frac{\mathrm{d}g}{\mathrm{d}t} = -\omega A\sin\omega t$$

as before.

Worked example 2

Derive an equation for the activity at time t of a radioactive source of decay constant λ. At $t = 0$ there are N_0 nuclei of the source present.

$N = N_0 e^{-\lambda t}$ (function of a function)

$A = -\dfrac{\mathrm{d}N}{\mathrm{d}t} = (-\lambda) \times (-N_0 e^{-\lambda t}) = \lambda N_0 e^{-\lambda t} = \lambda N$

Notice that this implies that the activity will decay with the same half-life as the source.

Newton and Leibniz

Calculus was invented twice, independently, by Isaac Newton in England and Gottfried von Leibniz in Germany. Leibniz and Newton started off with a friendly rivalry, but when Newton heard that in Europe Leibniz was getting the credit for inventing calculus, he reacted petulantly and accused Leibniz of stealing his ideas. Leibniz responded with counter-accusations, and so the two most gifted mathematicians of their generation were alienated from one another. The reason calculus was invented at all was to help solve problems which involved continuous change, for example the motion of the Moon orbiting the Earth or the motion of a falling stone, and soon Newton, Leibniz, and the next generation of mathematical physicists were using it throughout physics.

Second derivatives

Acceleration is the rate of change of velocity. Velocity is the rate of change of displacement, so acceleration is the rate of change of the rate of change of displacement, in other words a derivative of a derivative – acceleration is the second derivative of displacement with respect to time:

$$a = \frac{\mathrm{d}v}{\mathrm{d}t} = \frac{\mathrm{d}}{\mathrm{d}t}\left(\frac{\mathrm{d}s}{\mathrm{d}t}\right) = \frac{\mathrm{d}^2 s}{\mathrm{d}t^2}$$

Note that, although this looks like a squared term, it really indicates that you have carried out the operation of differentiating twice. The $\mathrm{d}^2/\mathrm{d}t^2$ bit cannot be broken up or cancelled out in any way.

Products and quotients

Two useful rules can be used to differentiate expressions that are products or quotients of simpler expressions:

The product rule

If $y = u(t) \times v(t)$ then

$$\frac{\mathrm{d}y}{\mathrm{d}t} = v\frac{\mathrm{d}u}{\mathrm{d}t} + u\frac{\mathrm{d}v}{\mathrm{d}t}$$

The quotient rule

If $y = \dfrac{u(t)}{v(t)}$ then

$$\frac{\mathrm{d}y}{\mathrm{d}t} = \frac{v\dfrac{\mathrm{d}u}{\mathrm{d}t} - u\dfrac{\mathrm{d}v}{\mathrm{d}t}}{v^2}$$

INTEGRAL CALCULUS

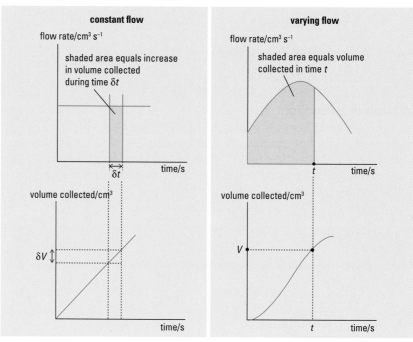

constant flow

flow rate/cm³ s⁻¹

shaded area equals increase in volume collected during time δt

δt time/s

volume collected/cm³

δV

time/s

varying flow

flow rate/cm³ s⁻¹

shaded area equals volume collected in time t

t time/s

volume collected/cm³

V

t time/s

If water drips into a cup, it will eventually fill it. The volume of water in the cup is equal to the sum of volumes of all drops that have fallen into it. If the same cup is filled from a freely flowing tap, the volume of water supplied is again equal to the sum of drops supplied, but now the drops have merged together into a single stream, so they cannot be counted separately. However, if you know the flow rate is, say, $10\,\mathrm{cm^3\,s^{-1}}$ and the cup takes 20 seconds to fill, then the volume would be given by

$$V = \text{flow rate} \times \text{time of flow} = 10\,\mathrm{cm^3\,s^{-1}} \times 20\,\mathrm{s} = 200\,\mathrm{cm^3}$$

This is equal to the area under a graph of flow rate versus time. In this case the area would be a rectangle of height $10\,\mathrm{cm^3\,s^{-1}}$ and length $20\,\mathrm{s}$.

If the flow is continuous but variable, the simple product cannot be used. However, the volume collected is still equal to the area under the graph, so the graph could be used to calculate this if you counted squares or used some other more sophisticated approximation. One way might be to:

- divide the time of flow into very short intervals of duration δt (during which the rate is more or less constant);

- work out how much water flowed during each interval (from flow rate times δt);

- sum the volumes of flow from each short interval.

The accuracy of this method depends on how short the time intervals are. In the limit that δt approaches zero the method is exact. Integration is a mathematical trick for carrying out sums with an infinite number of terms, each of which is vanishingly small.

Writing down integrals

Charging a capacitor is a bit like filling a container with water. As charge flows into the capacitor, the voltage rises in direct proportion and the energy stored also increases (but not in direct proportion). A graph of voltage versus charge is a straight line through the origin, and the area under this graph represents the energy stored on the capacitor. This is easy to understand if we break up the charging process into a sequence of very short intervals, during each of which an amount of charge δq is deposited onto the capacitor while its voltage has a more or less constant value of V. The work done to increase the charge by this amount is simply $\delta W = V\delta q$. But this is just the area of a narrow strip, as shown on

voltage/V

$V = \dfrac{q}{c}$

area of strip is approximately equal to $V\delta q$. This is the work done to increase charge by an amount δq
$\delta W = V\delta q$

δt

q charge/C

voltage/V

to charge from $q = 0$ to $q = Q$ requires an amount of work equal to the sum of many narrow 'strips'

$W = \sum\limits_{q=0}^{q=Q} V\delta q$

0 Q charge/C

the graph. The sum of all these narrow strips gives the total area under the graph and so the total energy stored.

The work δW done to increase by δq is

$$\delta W = V \delta q$$

This is equal to the area of one of the small vertical strips on the graph. So the total work W done to charge the capacitor is

$$W = \sum_{q=0}^{q=Q} V \delta q$$

(This means the sum of all the contributions δW as the capacitor charges.)

In the limit of many very small charging steps, that is, when $\delta q \rightarrow 0$, the value of W will be exact:

$$W = \lim_{\delta q \rightarrow 0} \sum_{q=0}^{q=Q} V \delta q$$

This is what is meant by the integral of V with respect to q and is written as

$$W = \int_{q=0}^{q=Q} V \delta q$$

The values above and below the **summation** and **integration** signs are the **limits** for the integration, that is, the range of values of q over which the summation takes place. In this case the capacitor starts off empty and ends up with charge Q, so appropriate limits are $q = 0$ and $q = Q$. When an integration is carried out between fixed limits it is called a **definite integral**. If no limits are defined, it is an **indefinite integral**.

Integrating

The process of integration is the inverse of differentiation. This means that if the derivative of f is g, then the integral of g is f. For example,

$$x^2 \rightarrow 2x \quad \text{(differentiation)}$$

$$2x \rightarrow x^2 \quad \text{(integration)}$$

This assumes that we are dealing with definite integrals.

You can use the table of derivatives (see spread MT.8) to work out most of the integrals you will need to use (just read the table backwards!). However, a table of useful integrals is included here for convenience.

Evaluating integrals

Once you have successfully integrated an expression, you need to work out its value. This is done by subtracting the value of the integral at the lower limit from its value at the upper limit. Now we can finish off the capacitor problem and derive an equation for the energy it stores when charged.

$$E = \int_{q=0}^{q=Q} V dq$$

First the capacitor equation must be used to express V in terms of q. This is very important because we are integrating with respect to q, and if we do not do this then V's dependence on q is not taken into account and the final result is incorrect.

$$V = \frac{q}{C} \quad \therefore \quad E = \int_{q=0}^{q=Q} \frac{q dq}{C}$$

C is a constant so it will carry straight through to the integral. This means we are effectively integrating q with respect to q so the result is $\frac{1}{2}q^2$.

This is written as follows:

$$E = \int_{q=0}^{q=Q} \frac{q dq}{C} = \left[\frac{q^2}{2C} \right]_{q=0}^{q=Q}$$

The integral appears in square brackets. Next we must put in the limits:

$$\left[\frac{q^2}{2C} \right]_{q=0}^{q=Q} = \frac{Q^2}{2C} - \frac{0^2}{2C} = \frac{Q^2}{2C}$$

This is the energy stored on the charged capacitor.

Some useful integrals.

Expression	Definite integral
powers	
x^n	$\dfrac{1}{n+1} x^{n+1}$

This is the inverse of the rule for differentiation.

x	$\frac{1}{2} x^2$
x^2	$\frac{1}{3} x^3$
x^{-4}	$-\frac{1}{3} x^{-3}$
$x^{\frac{1}{2}}$	$\frac{2}{3} x^{\frac{3}{2}}$

NB x^{-1} is an exception to this rule.

x^{-1}	$\ln x$
Trigonometric functions	
$\sin x$	$-\cos x$
$\cos x$	$\sin x$
$\sin \omega t$	$-\dfrac{1}{\omega} \cos \omega t$
$\cos \omega t$	$\dfrac{1}{\omega} \sin \omega t$
Exponentials	
e^x	e^x
$e^{-\lambda t}$	$-\dfrac{1}{\lambda} e^{-\lambda t}$

Worked example

How much work must be done to stretch a spring from an extension x_1 to an extension x_2? Assume the spring obeys Hooke's law and has spring constant k.

Work done to increase extension by a small amount δx is δW:

$\delta W = F \delta x$ and $F = kx$ (Hooke's Law)
so $\delta W = kx \delta x$

Taking limits and integrating:

$$W = \int_{x=x_1}^{x=x_2} kx \delta x = \left[\frac{kx^2}{2} \right]_{x=x_1}^{x=x_2}$$

$$= \frac{1}{2}k(x_2^2 - x_1^2)$$

If none of this work is transferred to thermal energy this expression is equal to the increase in elastic potential energy stored in the spring.

MT.10

OBJECTIVES

- using a spreadsheet to analyse data

Spreadsheets

Spreadsheets were first devised as a means of doing financial calculations more easily. Since then they have developed in sophistication and now have some very useful scientific features.

A spreadsheet is divided into **rows**, labelled with a number, which run horizontally, and **columns**, labelled with a letter, which run vertically. The place where a row and a column meet is called a **cell**. For example, column B meets row 3 at cell B3.

	A	B	C	D	E
1					
2					
3					
4					
5					
6					

It is possible to type information, in the form of text or numbers, into a cell and then to work on this information using **functions** and **formulae**. For example, in the diagram below, some text and some numbers have been typed into cells.

	A	B	C	D	E
1	length of pendulum	time for 50 swings			
2	10	33			
3	20	45			
4	30	57			
5	40	62			
6	50	71			
7	60	77			

Using different software

The details of how data, formulae, and functions are entered into the cell of a spreadsheet vary slightly from one version to another. The examples used here are based on Microsoft® Excel spreadsheet software.

Cell B4 contains the number 57; cell A1 contains the text 'length of pendulum'.

The most useful aspect of spreadsheets is the ability to work with formulae and functions. In this example a student is processing data from an experiment to measure the value of g using a pendulum. Fifty swings have been timed for a range of different pendulum lengths, and the next step is to calculate the period for each length. This repetitive calculation can be simplified by using a spreadsheet. The student types the formula = B2/50 into cell C2:

Using spreadsheets

Spreadsheets can be used for a variety of different purposes as well as analysing data. It is possible to model various systems by getting the spreadsheet to go through a sequence of repetitive calculations.

	A	B	C	D	E
1	length of pendulum	time for 50 swings			
2	10	33	=B2/50		

The moment the student hits the return key the formula disappears, to be replaced by a number which is the value in the cell B2 divided by 50:

	A	B	C	D	E
1	length of pendulum	time for 50 swings			
2	10	33	0.66		

The number appears in the cell, but the software remembers the formula. If the contents of cell B2 are changed, the spreadsheet will automatically alter cell C2 to give the right answer. Spreadsheet allow formulae to be copied into other cells without you having to type them in every time. If the formula from cell C2 is copied into cell C3, then it will automatically be changed into =B3/50. To the software the formula reads 'take the contents of the cell one place to the left and divide by 50'.

Formulae can be more complicated than this. In this example, the value of g can be obtained from the period by using

$$T = 2\pi\sqrt{\frac{l}{g}} \qquad g = \frac{4\pi^2 l}{T^2}$$

	A	B	C	D
1	length of pendulum	time for 50 swings	period T	g
2	10	33	0.66	=4×π²(A2/100)/(C2×C2)
3	20	45	0.90	
4	30	57	1.14	

Notice that in the formula A2 has been divided by 100, to make sure that it is metres, not centimetres. This formula can then be copied into the cells of column D, producing the measured value of g for each length:

	A	B	C	D
1	length of pendulum	time for 50 swings	period T	g
2	10	33	0.66	9.06
3	20	45	0.90	9.75
4	30	57	1.14	9.11

Finally, the average value of g can be calculated:

	A	B	C	D
1	length of pendulum	time for 50 swings	period T	g
2	10	33	0.66	9.06
3	20	45	0.90	9.75
4	30	57	1.14	9.11
5	30	57	average	=(D2+D3+D4)/3

Formulae can be made more powerful by using some of the functions built into the spreadsheet. For example, the average can be worked out directly by typing into cell D5 the function =AVERAGE (D2:D4). The term in brackets specifies the range of cells that are to be averaged. D2:D4 means 'all the cells from D2 to D4'. There are a wide range of functions available, including all the common mathematical functions such as SIN, COS, TAN, LOG, LN, etc.

Spreadsheets will also allow you to produce graphs or charts of the data, which can then have best-fit lines calculated and drawn through the points. This is an area where the student has to take some care as the tendency is for the software to draw a graph by 'joining the dots', which is not appropriate in physics. The best way of plotting a graph on a spreadsheet is to produce an 'XY scatter' and then put on the best-fit (trend) line. For details of how to do this it is best to refer to the software instructions.

Appendix

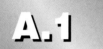

USEFUL DATA

Constants of nature	Symbol and value
speed of light in a vacuum	$c = 2.998 \times 10^8 \, \text{m s}^{-1}$
gravitational constant	$G = 6.672 \times 10^{-11} \, \text{N m}^2 \, \text{kg}^{-2}$
Planck constant	$h = 6.626 \times 10^{-34} \, \text{J s}$
permittivity of free space	$\varepsilon_0 = 8.854 \times 10^{-12} \, \text{F m}^{-1}$
permeability of free space	$\mu_0 = 4\pi \times 10^{-7} \, \text{H m}^{-1}$
electron (elementary) charge	$e = 1.602 \times 10^{-19} \, \text{C}$
electron mass	$m_e = 9.110 \times 10^{-31} \, \text{kg}$
proton mass	$m_p = 1.673 \times 10^{-27} \, \text{kg}$
neutron mass	$m_n = 1.675 \times 10^{-27} \, \text{kg}$
Bohr radius	$a_0 = 5.292 \times 10^{-11} \, \text{m}$
Avogadro constant	$N_A = 6.022 \times 10^{23} \, \text{mol}^{-1}$
Boltzmann constant	$k = 1.381 \times 10^{-23} \, \text{J K}^{-1}$
molar gas constant	$R = 8.314 \, \text{J K}^{-1} \, \text{mol}^{-1}$
Stefan constant	$\sigma = 5.670 \times 10^{-8} \, \text{W m}^{-2} \, \text{K}^{-4}$
Wien constant	$W = 2.90 \times 10^{-3} \, \text{m K}$

Useful numbers	Value
typical radius of an atom	$\sim 10^{-10} \, \text{m}$
typical radius of a nucleus	$\sim 10^{-15} \, \text{m}$
mean radius of the Earth	$6.378 \times 10^6 \, \text{m}$
mass of the Earth	$5.974 \times 10^{24} \, \text{kg}$
gravitational field strength close to the Earth's surface	$9.81 \, \text{N kg}^{-1}$
acceleration of free fall close to the Earth's surface	$9.81 \, \text{m s}^{-2}$
atmospheric pressure at sea level	$1.013 \times 10^5 \, \text{Pa}$
pressure due to 10 m of water	$\sim 1 \, \text{atmosphere}$
density of the atmosphere at s.t.p.	$1.29 \, \text{kg m}^{-3}$
mass of the Moon	$7.35 \times 10^{22} \, \text{kg}$
gravitational field strength close to the Moon's surface	$1.62 \, \text{N kg}^{-1}$
mean radius of the Sun	$6.96 \times 10^8 \, \text{m}$
mass of the Sun	$1.99 \times 10^{30} \, \text{kg}$
power output of the Sun	$3.9 \times 10^{26} \, \text{W}$
solar constant	$1360 \, \text{W m}^{-2}$
atomic mass unit	$1 \, \text{u} = 1.661 \times 10^{-27} \, \text{kg}$
electron-volt	$1 \, \text{eV} = 1.602 \times 10^{-19} \, \text{J}$
kilowatt hour	$1 \, \text{kWh} = 3.600 \times 10^6 \, \text{J}$
astronomical unit	$1 \, \text{AU} = 1.496 \times 10^{11} \, \text{m}$
light-year	$1 \, \text{ly} = 9.461 \times 10^{15} \, \text{m}$

Useful numbers	Value
absolute zero	$-273.15 \, \text{K}$
density of mercury	$1.360 \times 10^4 \, \text{kg m}^{-3}$
density of water	$1.000 \times 10^3 \, \text{kg m}^{-3}$
specific heat capacity of water	$4.190 \times 10^3 \, \text{J kg}^{-1} \, \text{K}^{-1}$
latent heat of fusion of water	$3.344 \times 10^5 \, \text{J kg}^{-1}$
latent heat of vaporization of water	$2.261 \times 10^6 \, \text{J kg}^{-1}$
triple point of water	$273.16 \, \text{K}$
ice point	$273.15 \, \text{K}$
speed of sound in air at s.t.p.	$3.314 \times 10^2 \, \text{m s}^{-1}$
half-life of carbon-14	$5.730 \times 10^3 \, \text{y}$
half-life of a free neutron	$11.7 \, \text{min}$
half-life of uranium-238	$4.51 \times 10^{19} \, \text{y}$
refractive index of glass	1.5
typical drift speed of electrons	$\sim 10^{-2} \, \text{m s}^{-1}$
number of charge carriers per unit volume in copper	$8 \times 10^{28} \, \text{m}^{-3}$
$1/4\pi\varepsilon_0$	$\sim 9 \times 10^9 \, \text{m F}^{-1}$
π^2	~ 10
number of seconds in a year (a quirk of nature)	$\sim \pi \times 10^7$
range of visible wavelengths	400–$700 \, \text{nm}$
range of human hearing	$20 \, \text{Hz}$–$20 \, \text{kHz}$

The quarks

Name	Symbol	Charge/+e	Mass/GeV/c^2	Stable
up	u	$+\frac{2}{3}$	0.33	yes
down	d	$-\frac{1}{3}$	~0.33	no
charm	c	$+\frac{2}{3}$	1.58	no
strange	s	$-\frac{1}{3}$	0.47	no
top	t	$+\frac{2}{3}$	180	no
bottom	b	$-\frac{1}{3}$	4.58	no

The leptons

Name	Symbol	Charge/+e	Mass/GeV/c^2	Stable /lifetime/s
electron	e^-	−1	5.31×10^{-4}	stable
electron neutrino	ν_e	0	$<10^{-10}$	oscillates
muon	μ^-	−1	0.106	2×10^{-6}
muon-neutrino	ν_μ	0	$<10^{-10}$	oscillates
tau	τ^-	−1	1.78	3×10^{-13}
tau-neutrino	ν_τ	0	$<10^{-10}$	oscillates

Some baryons

Name	Symbol	Charge/+e	Mass/GeV/c^2	Lifetime/s	Quarks	Strangeness	Year of discovery
proton	p	+1	0.938	$>10^{39}$	uud	0	1911
neutron	n	0	0.940	900	udd	0	1932
lambda	Λ	0	1.115	2.6×10^{-10}	uds	−1	1947
sigma plus	Σ^+	+1	1.189	0.8×10^{-10}	uus	−1	1953
sigma minus	Σ^-	−1	1.197	1.5×10^{-10}	dds	−1	1953
sigma zero	Σ^0	0	1.192	6×10^{-20}	uds	−1	1956
xi minus	Ξ^-	−1	1.321	1.6×10^{-10}	dss	−2	1952
xi zero	Ξ^0	0	1.315	3×10^{-10}	uss	−2	1952
omega minus	Ω^-	−1	1.672	0.8×10^{-10}	sss	−3	1964
lamda chi	$\Lambda\chi$	+1	2.28	2.3×10^{-13}	udc	+1 charm	1975

Some mesons

Name	Symbol	Charge/+e	Mass/GeV/c^2	Lifetime/s	Quarks	Strangeness	Year of discovery
pi zero	π^0	0	0.135	0.8×10^{-16}	$u\bar{u}$, $d\bar{d}$	0	1949
pi plus	π^+	+1	0.140	2.6×10^{-8}	$u\bar{d}$	0	1947
pi minus	π^-	−1	1.115	2.6×10^{-10}	$d\bar{u}$	0	1947
K zero	K^0	0	0.498	*	$d\bar{s}$	+1	1947
K plus	K^+	+1	0.494	1.2×10^{-8}	$u\bar{s}$	+1	1947
K minus	K^-	−1	0.494	1.2×10^{-8}	$s\bar{u}$	−1	1947

The K^0 and the $\overline{K^0}$ are capable of 'mixing', since the weak force can change one into the other – this makes it impossible to measure their lifetimes separately.

Some more mesons

Name	Symbol	Charge/+e	Mass/GeV/c^2	Lifetime/s	Quarks	Strangeness	Year of discovery
psi	Ψ	0	3.1	10^{-20}	$c\bar{c}$	0	1974
D zero	D^0	0	1.86	4.3×10^{-13}	$c\bar{u}$	+1	1976
D plus	D^+	+1	1.87	9.2×10^{-13}	$c\bar{d}$	+1	1976

USEFUL FORMULAE

Mechanics

Motion with uniform acceleration

$$v^2 = u^2 + 2s$$

$$s = ut + \frac{1}{2}at^2$$

$$s = \frac{u+v}{2}t$$

Resultant force and acceleration $\quad \Sigma F = ma$

Resultant force and momentum $\quad \Sigma F = \frac{\Delta p}{\Delta t} \qquad \Sigma F = \frac{dp}{dt}$

Weight $\quad W = mg$

Work and Power $\quad \Delta W = F\Delta x \qquad W = \int F dx$

$$P = \frac{\Delta W}{\Delta t} = \frac{\Delta E}{\Delta t} = Fv \qquad P = \frac{dW}{dt}$$

Kinetic energy $\quad KE = \frac{1}{2}mv^2$

Gravitational PE change (constant g) $\quad \Delta GPE = mg\Delta h$

Work done $\quad F\Delta x = \Delta\left(\frac{1}{2}mv^2\right)$

Moment of force F about origin O $\quad = $ (size of F) $\times$ (perpendicular distance from O to the line of action of F)

Angular speed $\quad \omega = \frac{\Delta\theta}{\Delta t} = \frac{v}{r} \qquad \omega = \frac{d\vartheta}{dt}$

Radial acceleration $\quad a = r\omega^2 = \frac{v^2}{r}$

Newton's law of gravity $\quad F = -\frac{GM_1 M_2}{r^2}$

Simple harmonic motion

Displacement $\quad x = x_0 \sin(\omega t)$

Maximum speed $\quad v_{max} = \omega x_0$

Acceleration $\quad a = -\omega^2 x$

Period of a simple pendulum $\quad T = 2\pi\sqrt{\dfrac{l}{g}}$

Period of a mass on a spring $\quad T = 2\pi\sqrt{\dfrac{m}{k}}$

Electricity

Charge and current $\quad I = \frac{\Delta Q}{\Delta t} \qquad I = \frac{dQ}{dt}$

$$I = nAQv$$

Electrical energy $\quad \Delta W = IV\Delta t$

Electrical power $\quad P = I^2 R$

Terminal potential difference $\quad V = \varepsilon - Ir$

Resistances in series $\quad R_T = R_1 + R_2 + R_3 + \ldots$

Resistances in parallel $\quad \dfrac{1}{R_T} = \dfrac{1}{R_1} + \dfrac{1}{R_2} + \dfrac{1}{R_3} + \ldots$

Electromagnetism

Electric field strength $\quad E = -\frac{\Delta V}{\Delta x} \qquad E = -\frac{dV}{dx}$

Field of a point charge $\quad E = \frac{Q}{4\pi\varepsilon_0 r^2}$

Potential of a point charge $\quad V = \frac{Q}{4\pi\varepsilon_0 r}$

Coulomb's law $\quad F = \frac{Q_1 Q_2}{4\pi\varepsilon_0 r^2}$

Magnetic force $\quad F = Bqv$

Force on a current carrying wire $\quad F = BIl$

Field inside a long solenoid $\quad B = \mu_0 nI$

Field surrounding a long wire $\quad B = \frac{\mu_0 I}{2\pi r}$

Flux $\quad \phi = BA$

Faraday's law $\quad \varepsilon = -\frac{\Delta\phi}{\Delta t} \qquad \varepsilon = -\frac{d\phi}{dt}$

Capacitors

Charge stored $\qquad Q = CV$

Energy stored $\qquad E = \frac{1}{2} QV = \frac{1}{2} CV^2 = \frac{1}{2}\frac{Q^2}{C}$

Parallel plate capacitor $\qquad C = \frac{\varepsilon_0 \varepsilon_r A}{d}$

Discharge time constant $\qquad \tau = CR$

Discharge $\qquad Q = Q_0 e^{-\frac{t}{RC}}$

Capacitances in series $\qquad \frac{1}{C_T} = \frac{1}{C_1} + \frac{1}{C_2} + \frac{1}{C_3} + \dots$

Capacitances in parallel $\qquad C_T = C_1 + C_2 + C_3 + \dots$

Waves

Intensity $\qquad I \propto A^2 \qquad\qquad I = \frac{P}{4\pi r^2}$

Young's slits $\qquad \lambda = \frac{xs}{D}$

Diffraction grating $\qquad d \sin\theta = n\lambda$

Refraction $\qquad n_1 \sin\theta_1 = n_2 \sin\theta_2$

Heat

Celsius temperature $\qquad \theta\,/\,^\circ C = T\,/\,K - 273.15$

Practical temperature scale $\qquad \theta = \frac{X_\theta - X_0}{X_{100} - X_0} \times 100\,^\circ C$

First law of thermodynamics $\qquad \Delta U = \Delta W + \Delta H$

Rate of energy conduction $\qquad H = kA\frac{\Delta\theta}{\Delta x}$

Energy for a temperature change $\qquad E = mC_s\Delta T$

Energy for a phase change $\qquad E = mL_s$

Ideal gas equation $\qquad PV = nRT$

Kinetic theory $\qquad PV = \frac{1}{3} nm\langle c^2 \rangle = \frac{1}{3}\rho\langle c^2 \rangle$

Heat engine theoretical maximum efficiency $\qquad \varepsilon = \frac{T_1 - T_2}{T_1}$

Modern physics

Mass–energy $\qquad E = mc^2$

Activity $\qquad A = \frac{\Delta N}{\Delta t} = \lambda N \qquad\qquad A = \frac{dN}{dt}$

Radioactive decay $\qquad N = N_0 e^{-\lambda t}$

Half-life $\qquad t_{\frac{1}{2}} = \frac{\ln 2}{\lambda}$

Photons $\qquad E = hf$

Maximum energy of photoelectrons $\qquad KE_{max} = hf - W$

de Broglie wavelength $\qquad \lambda = \frac{h}{p}$

Relativity

gamma factor $\qquad \gamma = \frac{1}{\sqrt{1 - \frac{v^2}{c^2}}}$

Length contraction $\qquad l = \frac{l_0}{\gamma}$

Time dilation $\qquad t = \gamma t_0$

Relativistic momentum $\qquad p = \gamma m_0 v$

Relativistic energy $\qquad E^2 - p^2 c^2 = m_0^2 c^4$

Materials

Hooke's law $\qquad F = k\Delta x$

Stress and strain $\qquad \sigma = \frac{F}{A} \qquad\qquad e = \frac{\Delta x}{x}$

Young's modulus $\qquad Y = \frac{\text{stress}}{\text{strain}}$

Energy stored $\qquad E = \frac{1}{2} F\Delta x = \frac{1}{2} k\Delta x^2$

NUMERICAL ANSWERS TO SELECTED PRACTICE AND EXAM QUESTIONS

Please note: the following suggested answers are intended as a guide for the reader. The respective exam boards have not supplied the answers to past exam questions nor are they in any way responsible for their accuracy or correctness.

CHAPTER 2

Practice questions

Spread 2.2

1 $1\,N = 1\,kg\,m\,s^{-2}$

2 **a** $ML^{-1}T^{-2}$ **b** ML^2T^{-3}
   c MLT^{-1} **d** I
   e $ML^2I^{-1}T^{-3}$ **f** $ML^2I^{-2}T^{-3}$

3 $kg\,m^{-1}\,s^{-3}$

Spread 2.6

1 2.00 ± 0.02

2 4.7 cm; The 0.1 cm (half smallest scale division is 0.5 mm on each end)

3 ~0.5 kg, yes as mean = 4.6 ± 0.5
   ∴ skewed away from 4.0 kg

Spread 2.7

1 $1.01 \pm 0.02\,s$

2 $(2.71 \pm 0.3) \times 10^3\,kg\,m^{-3}$

3 $\sin(30 \pm 0.10) = 0.500 \pm 0.002$

4 $1.476 \pm 0.008\,s$

CHAPTER 3

Practice questions

Spread 3.1

1 $8.4\,m\,s^{-1}$

2 $10.18\,m\,s^{-1}$

5 $465.4\,m\,s^{-1}, 385.7\,m\,s^{-1}$

6 $1.5 \times 10^7\,s$

Spread 3.2

1 $2.13\,m\,s^{-1}$

2 0.85 m

3 $0.5\,m\,s^{-2}$ 225 m

4 $9\,m\,s^{-2}$ upwards

6 $5.5 \times 10^7\,m\,s^{-1}$

Spread 3.3

1 4.1 km north, 2.87 km east

2 $5.37\,m\,s^{-1}$, 21.4° below horizontal

3 $14.1\,m\,s^{-1}$ horizontal,
   $5.1\,m\,s^{-1}$ vertical

4 60°

5 $8.06\,m\,s^{-1}$, 29.7° from north

6 2.75

7 386.4 N

Spread 3.4

1 30 N

2 **a** 5400 N **b** 5400 N
   c 5400 N

3 15.3 N

4 57.7 N

5 1562.8 N up the hill

7 72.8 N

8 3.50 N, 1.52 N

Spread 3.5

1 66.67 N

4 **a** 12 kN **c** 19 kN
   d 45 kN

5 45°

Spread 3.6

3 **a** mass: 48 kg, weight: 470.4 N,
   b mass: 48 kg, weight: 470.4 N,
      density: $108.7\,kg\,m^{-3}$
   c weight changes to 81.6 N

4 $1.4 \times 10^{11}\,kg\,m^{-3}, 7 \times 10^{-12}\,m^3$

5 Si: $1.84 \times 10^{17}\,kg\,m^{-3}$
   Ca: $1.69 \times 10^{17}\,kg\,m^{-3}$
   Co: $1.94 \times 10^{17}\,kg\,m^{-3}$
   Sr: $2.29 \times 10^{17}\,kg\,m^{-3}$
   Au: $2.41 \times 10^{17}\,kg\,m^{-3}$

6 $8653\,kg\,m^{-3}$

7 **a** 2.3 m above the ground

Spread 3.8

taking $g = 9.8\,N\,kg^{-1}$

1 **a** $500\,N\,m^{-1}$ **b** 2.94 mm
   c 16 mm

2 **a** $10^6\,N\,m^{-1}$ **b** 0.735 mm

3 **a** 0.125 m **b** 85.75 mm
   c 50 mm

4 1/3 m, $12\,N\,m^{-1}$, 5 N

Spread 3.9

taking $g = 9.8\,N\,kg^{-1}$

1 **a** 78.4 m **b** $39.2\,m\,s^{-1}$

2 **a** 0.14 s

3 **a** 1.02 s **b** 1.28 m

6 **a** 0.64 s **b** 6.4 m

7 **a** 7.2 kW **b** 7.35 m

Spread 3.10

2 $26.3\,m\,s^{-1}$, 40.6° from horizontal; 35 m

3 **a** 0.11 m **b** 0.31 s
   c 0.8 m

Spread 3.12

1 **a** 58.8 J **b** 0.011 25 J
   c 1915.1 J

3 640 N

4 16.3 N

Spread 3.13

5 5.88 MJ

6 1.28 m

8 **a** $63.3\,m^3$ **b** $3.3 \times 10^{13}\,J$
   c 92.3%

Spread 3.14

1 **a** 20 W **b** 80 W
   c 6 kJ

2 **a** 26.25 kW **b** 52.5 kW

4 1 MW

5 23.2 kJ   19%

6 **a** $7.35 \times 10^{12}\,J$ **b** 4.08 GW

8 746 W

Spread 3.15

1 9.8 kPa, 39.2 kPa, 1.96 kPa

3 **b** 2.25 kN

Spread 3.16

3 **b** $200\,m\,s^{-1}$

4 **a** $1.98\,m\,s^{-1}$ **b** $1\,cm^3\,s^{-1}$

Spread 3.21

1 4.5 kN

4 $2.68\,m\,s^{-2}$

5 $1.09\,m\,s^{-2}$ (50 kg down, 40 kg up)

7 $30\,\text{km s}^{-1}$

Spread 3.22

1 a woman: $585\,\text{kg m s}^{-1}$,
daughter: $75\,\text{kg m s}^{-1}$
  b woman: $117\,\text{kg m s}^{-2}$,
  daughter: $25\,\text{kg m s}^{-2}$

3 $0.36\,\text{m s}^{-1}$

Spread 3.23

6 a $55\,\text{N}$ **b** $50\,\text{N}$
  c $50\,\text{N}$

Spread 3.24

1 a $0.5\,\text{kg}$ mass moves back at
$0.66\,\text{m s}^{-1}$, $1\,\text{kg}$ mass moves
forward at $1.33\,\text{m s}^{-1}$
  b $0.67\,\text{m s}^{-1}$ forward

2 a $18.5\,\text{m s}^{-1}$ **b** $5.2\,\text{kJ}$

3 a $240\,\text{N s}$, $400\,\text{N}$
  g $480\,\text{J}$ per pass, $213.3\,\text{J s}^{-1}$

Spread 3.25

1 a $171.5\,\text{N}$ **c** $171.5\,\text{N}$

4 a $4 \times 10^{-5}\,\text{m s}^{-1}$ **e** $0.62\,\text{mm}$

Spread 3.26

2 a $3 \times 10^{-23}\,\text{kg m s}^{-1}$
  b $6 \times 10^{-23}\,\text{kg m s}^{-1}$
  c 1.66×10^{27} collisions $\text{s}^{-1}\,\text{m}^{-2}$

Spread 3.27

1 $1{:}2{:}2$

2 a $1.1 \times 10^{-9}\,\text{N}$

3 a $21.4 \times 10^{3}\,\text{m}^{3}$ **b** $63.1\,\text{km s}^{-1}$

Spread 3.28

1 $3.33\,\text{Hz}$ and $20.9\,\text{rad s}^{-1}$

2 $13.3\,\text{rad s}^{-1}$, $2.12\,\text{Hz}$

3 b $0.04\,\text{Nm}$
  c $0.027\,\text{rad s}^{-2}$
  d $0.16\,\text{rad s}^{-1}$

Spread 3.29

2 $160\,\text{MN}$

5 a $1.99 \times 10^{-7}\,\text{rad s}^{-1}$
  b $5.95 \times 10^{-3}\,\text{m s}^{-2}$
  c $3.57 \times 10^{22}\,\text{N}$
  d $6.7 \times 10^{6}\,\text{m}$

Spread 3.31

1 a $0.83\,\text{Hz}$, $5.2\,\text{rad s}^{-1}$
  b $0.61\,\text{J}$
  c $0.23\,\text{J s}$
  d $0.035\,\text{rad s}^{-2}$
  e $1.57 \times 10^{-3}\,\text{N m}$

4 a $0.44\,\text{rad s}^{-1}$
  b initial KE $= 1225\,\text{J}$,
  final KE $= 153\,\text{J}$

Spread 3.32

1 a $0.02\,\text{s}$ **b** $0.36\,\text{Hz}$
  c $250\,\text{Hz}$

Spread 3.33

2 $0.32\,\text{s}$

3 $9\,192\,631\,770\,\text{Hz}$

4 b $0.377\,\text{m s}^{-1}$ **c** $3.55\,\text{m s}^{-2}$

Spread 3.34

3 a $10\,\text{N m}^{-1}$ **b** $0.36\,\text{Hz}$

5 $7.12 \times 10^{12}\,\text{Hz}$

6 $2.23\,\text{Hz}$

Spread 3.35

1 $0.011\,\text{J}$, $0.21\,\text{m s}^{-1}$

8 a 9.75%
  b 77% of the original

Exam questions

2 c i $22.6\,\text{N}$ **ii** $11.3\,\text{N}$
  iii $21°$

---

3 a $6600\,\text{N}$ **b** $86\,400\,\text{W}$

4 a i force at A $= 675\,\text{N}$;
force at B $= 1375\,\text{N}$
  b $3.5\,\text{m}$

5 c i $87\,\text{m}$ **d** $6.7 \times 10^{3}\,\text{N}$

6 a $7.5\,\text{kN}$ **b** $15\,\text{kJ}$

7 a $25\,\text{J}$ **b** $24.5\,\text{m s}^{-1}$

8 a $45.9\,\text{m}$ **b** $360\,\text{J}$

9 a i AB: 0.063; BC: 0; CE: -0.056;
EF: 0.083
  b $0.18\,\text{m}$

10 a ii $3.0 \times 10^{-4}\,\text{N}$
  iii $0.12\,\text{mW}$
  b $0.032\,\text{m}$

11 a $2.5\,\text{m s}^{-2}$
  b i $4000\,\text{N}$ **ii** $8000\,\text{N}$

12 a $0.15\,\text{m}$

13 b i $2.3\,\text{Nm}$ **ii** $14.1\,\text{J}$
  c $2.0 \times 10^{5}\,\text{Pa}$

14 a $3.1\,\text{s}$
  b i $15\,\text{m s}^{-1}$ **ii** $1.5\,\text{s}$

15 b $0.56\,\text{m}$
  c ii $15–20\,\text{m s}^{-1}$

16 c i $44\,\text{kJ}$ **ii** $1.1\,\text{kN}$

17 a $6.3\,\text{m s}^{-1}$ **c** $16.4\,\text{N}$

19 a i $2\pi L\rho r\delta r$ **ii** $2\pi L\rho r^{3}\delta r$
  b i $0.0159\,\text{m}$ **ii** $0.13\,\text{s}$
  iii $63.7\,\text{rad s}^{-1}$

20 a ii $92\,400\,\text{kg m s}^{-1}$
  iii $37\,400\,\text{N}$

21 b i $0.13\,\text{Hz}$ **ii** $2.45\,\text{m s}^{-1}$
  iii $7500\,\text{J}$

22 b i $30\,\text{s}$ **ii** $18\,000\,\text{N}$
  iii $450\,\text{kW}$

23 b i $0.04\,\text{J}$ **iii** $0.14\,\text{m}$

24 a $14\,\text{m s}^{-1}$
  b i $710\,\text{J}$ **ii** $160\,\text{J}$
  iii $15.4\,\text{m s}^{-1}$

25 b i $5000\,\text{N}$ **ii** $5000\,\text{N}$
  iii $32\,000\,\text{N}$ **iv** 6

26 b i $1100\,\text{N}$ **ii** $2.7\,\text{rad s}^{-1}$

27 b $31.4\,\text{rad s}^{-1}$
  c iii $9.9 \times 10^{-3}\,\text{m}$

28 b i $38.4\,\text{rad s}^{-1}$ **ii** $184\,\text{J}$

30 b ii $10.5\,\text{m s}^{-1}$ **iii** $33 \times 10^{-6}\,\text{m}^{3}\,\text{s}^{-1}$

31 a ii $0.6\,\text{J}$
  b i $0.12\,\text{rad s}^{-1}$ **ii** $0.72\,\text{J}$

33 b i $9ML^{2}/12$ **ii** $4.8 \times 10^{7}\,\text{kg m}^{2}$
  iii $1.9 \times 10^{7}\,\text{J}$

Chapter 4

Practice questions

Spread 4.1

1 The charges are the same sign, but
nothing can be said about the sizes.
Newton's third law tells us that the
forces must be the same no matter
what the size of the charge.

2 10 h 40 min, 2 ms

3 6.2×10^{18}

6 $8.93 \times 10^{-4}\,\text{kg}$, 8.36×10^{21},
5.35×10^{23}

Spread 4.2

1 $2 \times 10^{-5}\,\text{m s}^{-1}$

2 10 h 40 min. $3.8 \times 10^{4}\,\text{s}$

3 drift speed is four times faster in the
thinner wire

4 $5 \times 10^{5}\,\text{m s}^{-1}$

Spread 4.3

1 $2.3\,\text{mA}$

3 $170\,\Omega$

Spread 4.5

1 2 A, 1 A
  a 1 V, **b** 2 V, **c** 1 V, **d** 0 V, **e** 2 V,
  f 2 V, **g** 2 V, **h** 3 V, **i** 3 V

2 $10\,\text{V}$, $1\,\text{k}\Omega$

3 $4.5\,\text{V}$, $10\,\Omega$ each

4 series combination: $3\,\text{V}$, $2\,\Omega$
parallel combination: $3\,\text{V}$, $1\,\Omega$
the second combination is better – the
terminal p.d. will be nearer to $3\,\text{V}$

Spread 4.6

1 $0.5\,\Omega$

2 a 3 resistors – two $3\,\Omega$ in series
together in parallel with a single $3\,\Omega$
  b 5 resistors – three $4\,\Omega$ in series in
  parallel with a single $4\,\Omega$ and this
  combination in series with another
  $4\,\Omega$

3 low: all three in series, $1.44\,\text{k}\Omega$, $0.16\,\text{A}$;
medium: two in parallel in series with
the third, $720\,\Omega$, $0.32\,\text{A}$; high: all three
in parallel, $160\,\Omega$, $1.4\,\text{A}$

Spread 4.7

1 $1 \times 10^{-3}\,\Omega$

3 $0.94\,\Omega\,\text{m}^{-1}$, $1.6\,\text{m}$

4 $1.4\,\Omega$, $1.9\,\Omega$, $3.0\,\Omega$, about $7000\,\text{K}$

Spread 4.8

1 $1032\,\text{s}$ – just over 17 minutes,
$25\,800\,\text{s}$ – just over 7 hours

2 $6.9\,\text{kW}$, $3.5\,\text{kW}$

3 $26\,\Omega$

4 $3564\,\text{MJ}$, $790\,\text{W}$

5 33 hours

Spread 4.10

1 a $5\,\text{V}$ **b** $6.7\,\text{V}$
  c $8.3\,\text{V}$

2 a $5.3\,\text{V}$ **b** $9.6\,\text{V}$

4 $1.5\,\text{V}$, because of the internal
resistance of the $2\,\text{V}$ cell

Spread 4.11

1 a 1 A, 1 V **b** 0.073 A, 1.46 V
  c 0.030 A, 1.49 V

3 $5\,\text{V}$, $3\,\Omega$

Spread 4.12

1 6.5

2 $1.2\,\mu\text{m}$ (assuming a d.c. current of
$42\,\text{A}$ in a half cycle), $0.86\,\mu\text{m}$
(assuming a $30\,\text{A}$ average current
during a half cycle).

3 $4.2\,\text{A}$

4 a $18.4\,\text{A}$ **b** $18.4\,\text{A}$
  c $-14.9\,\text{A}$, $6.7\,\text{ms}$

Spread 4.15

1 $96\,\text{mC}$, $48\,\text{A}$

2 $100\,\text{pF}$

3 $600\,\mu\text{C}$, $3.6\,\text{mJ}$

4 b $15\,\mu\text{C}$ **c** $10\,\mu\text{F}$

Spread 4.16

1 $150\,\mu\text{C}$, resistor: $3\,\text{V}$,
capacitor: $3\,\text{V}$, $6\,\mu\text{C}$

2 $4.2\,\mu\text{A}$, $1.2\,\text{ks}$ or just under 20 minutes

3 about $32\,\text{s}$

Spread 4.17

1 series: $1\,\mu\text{F}$, parallel: $11\,\mu\text{F}$

2 $0.75\,\mu\text{F}$

3 0.36 J, 100 μF, 120 V, 0.72 J; 400 μF, 60 V, 0.72 mJ

Spread 4.18

1 400 V, 179 V

2 100 μC on each capacitor, 0.5 mJ in total

3 80 μC on 20 μF capacitor, 120 μC on 30 μF capacitor, 0.4 mJ in total

4 1 nC, 0.01 μJ, 40 V, 0.02 μJ, 3.6×10^{-7} N (0.36 μN)

5 5 mJ, 10 mJ, 1 GJ

Exam questions

1 a 0.47 A **b** 0.16 A
  c 1.53 V

3 a i 0.8 A **ii** 3.2 W
  b i 1.6 A **ii** 0.8 A

5 b 0.011 W
  c i 1090 A
  d i 180 kC **ii** 1710 s

7 b 10.2 mΩ **c** 30.6 mV

9 a 0.065 W

10 a ii 15 Ω

11 a 0.6 V
  b i 1.69 A **ii** 0

13 b i 60×10^{-9} C
  ii 12 V **iii** 3.6×10^{-8} J

14 c i 4.5 J **ii** 4.5×10^{-3} J
  iii 2.3×10^{-3} J

15 a i 0.79 C **ii** 0.95 J
  b ii 0.46 C **iii** 128 h
  c 1.7×10^{-6} W

16 a i 600 V **ii** 200 μF
  iii 360 J
  b i 3.8 kV **ii** 20 s
  iii 1.4 mΩ

CHAPTER 5

Practice questions

Spread 5.1

3 a Earth 740 N, Mars 230 N, Jupiter 1900 N (to 2 sig. figs.)
  b Earth 7.4 kJ, Mars 2.3 kJ, Jupiter 19 kJ (to 2 sig. figs.)

Spread 5.2

2 47 mN

3 a i $\sqrt{2}R_E$ **ii** $10R_E$
  b false

6 a 10^9 **b** 10^{27}
  c 10^{27} **d** 10^{36}

8 a 740 N (to 2 sig. figs.)
  b 130 N (to 2 sig. figs.)
  c 1900 N (to 2 sig. figs.)

Spread 5.3

1 25 N kg^{-1}

4 a 3.8×10^7 m
  b Moon's gravity at Earth = 34 μN kg^{-1}
    Sun's gravity at Earth = 5.9 mN kg^{-1}

6 a 1.49×10^{13} N kg^{-1}
  b 1.1×10^{15} N
  c 0.1% less
  d yes – differential force $\approx 1.5 \times 10^{14}$ N

Spread 5.4

1 6.3×10^7 J kg^{-1}

2 $\Delta Vg = 9.8$ J kg^{-1} m^{-1} = 9.8 N kg^{-1}

3 a 235 MJ **b** 2.25 MJ

Spread 5.5

1 a 0.15 N kg^{-1} **b** 85 kJ kg^{-1}
  c 410 m s^{-1}

Spread 5.6

1 geostationary: 42 Mm, 100 mins orbit: 7 Mm

Spread 5.7

1 5.14 N C^{-1}, 8.2 μN

2 2.45 mC

Spread 5.8

1 1.9×10^7 m s^{-1}

2 60 V, 0.12 μJ, 0.24 μJ

3 4.4×10^{-17} J, 2.2×10^{-17} J (=13.6 eV)

4 2.3×10^{-14} m

Spread 5.9

3 potential at 1.0 m: 18.0 V, potential at 1.1 m: 16.3 V, average potential gradient: 1.6 V m^{-1}, field strength at 1.05 m: 16.3 V m^{-1}

Spread 5.11

1 1.8 pF

2 0.47 m^2

3 3 V, 27 pC

4 5 mJ, 300 μF, 3.3 V, 1.7 mJ

Spread 5.14

2 a 0 **b** 8×10^{-12} N
  c 6×10^{-12} N **d** 4×10^{-12} N

Spread 5.15

2 C: 2.8×10^{-5} T, directed at 45° to the top right of the diagram,
  D: 2.8×10^{-5} T, directed at 45° to the bottom right of the diagram

3 2 μN

4 between the wires, 0.24 m from the wire carrying the 2 A current

5 a 7.2×10^{-5} N
  b 2.2×10^{-4} N

Spread 5.16

2 58 μT at 72° to the horizontal, 0.28 A

3 85 μN, forces attract

Spread 5.17

1 1.1 mN m

3 102 μN, 51 μN (opposite directions)

Spread 5.19

1 1.8×10^6 m s^{-1}, 2.0×10^7 m s^{-1}, 3.2 cm

3 3.2×10^{-17} J, 8.4×10^6 m s^{-1}
  2.7×10^{-13} N, 2.4×10^{-4} m
  4.5×10^{-11} s

Spread 5.21

1 1.7 mV

2 3.1×10^{-9} N, 6.3×10^{-8} W, 1 Ω

3 4.8×10^{-5} T

4 using values for the Boeing 747-400: wing span = 64.4 m, typical cruising speed = 910 km h^{-1}, then the e.m.f. induced ~ 0.8 V

Spread 5.22

1 8.3 μV

2 600 μWb, 5 mV

Spread 5.24

1 π rad s^{-1}, $\pi/30$ rad s^{-1}

2 10 mV

4 0.2 H, 1.5 A s^{-1}

5 1 H = 1 kg m^2 A^{-2} s^{-2}

Spread 5.25

1 a ~2.8×10^4 As **b** 5.6 V
  c 6.3 V

2 3 V, 2.7 W

Spread 5.26

2 a 4.1 Ω **b** 41 Ω
  c 410 Ω

Spread 5.27

1 V_C = 126 V, V_L = 124 V, X = 5.07 Ω, ϕ = −9.4°, f_0 = 252 Hz

Spread 5.28

1 5:1

2 0.25 mA

Exam questions

1 a i 2.6 A **ii** 88 Ω
  iii 3.7 A
  b i 32 kΩ **ii** 7.2 mA

2 a 4.2×10^{-9} F **b** 3.0×10^{-7} J

4 b 6.6×10^{-15} m

5 b i 2.6×10^{-9} C
  ii 1.6×10^{10}
  c 1.2×10^{-9} C

7 a ii 6.4×10^{-15} N
  b ii 0.17 N

8 a i 537 kV **ii** 1.5×10^5 V m^{-1}

9 a i 5.5×10^{-10} F
  ii 2.8×10^{-6} J
  iii 5.5×10^{-9} C

11 b iii 8 T **iv** 0.080 m s^{-2}

12 a ii 5.6 pF
  b i 24 000 V m^{-1}

15 b i 9 V **ii** 45 A s^{-1}
  iii 0.2 A

16 a 72.6 V **b** 90 V

17 a 5200 V **b** 1.95×10^4 V m^{-1}
  c i 0.53 T **ii** 10.4 mm
  e 58.77

18 a i 90 **ii** 0.45 A
  iii 200 W
  b i 17 V

19 a i 4.2×10^7 km
  iii 0.28 s
  b i 3.5×10^9 J **ii** 4.3×10^{10} J
  iii 8.1×10^7 J

20 c i 60 Ω **ii** 0.19 H
  d i 18 W **ii** 12 W
  iii 67% **iv** 0.75 Ω

23 b ii 0.19 N
  c i 3.6×10^2 m s^{-2}
  ii 0.57 A

25 c i 9.83 N **ii** 0.034 N
  iii 9.80 N
  d i 9.83 m s^{-2}
  ii 9.80 m s^{-2}

26 c i 1.62 m s^{-2}
  ii 60.6 m

27 c 7.3×10^{22} kg

29 a ii 2.5×10^{-12} kg^{-1}
  b 5×10^{18} kg

CHAPTER 6

Practice questions

Spread 6.1

1 7.5×10^{14} Hz (400 nm) to 4.3×10^{14} Hz (700 nm)

2 a 3.2 m s^{-1}
  b i 0
  ii 180° or π or 1/2 cycle
  iii 90° or $\pi/2$ or 1/4 cycle
  iv 180° or π or 1/2 cycle

3 a few μW

Spread 6.2

3 a 1800 km

Spread 6.4

1 1.3×10^{11} m

Spread 6.5

3 a 30° to normal
b 40.8°

Spread 6.6

1 a 24.4°
b increases to 32.5°
2 1.51
3 b 98 m   **c** 500 ns
d i yes   **ii** no

Spread 6.7

2 a 1 m
b i 2 A, 4 I
ii 0, 0
iii 2 A, 4 I

Spread 6.8

2 a 570 mm   **c** 1.03 mm

Spread 6.10

3 a UV, IR respectively
b 3.5×10^{-8} rad
c for 110 nm: 5.6×10^{-8} rad, for 1100 nm: 5.6×10^{-7} rad

Spread 6.11

1 b N
2 a 589 nm at 17.13°, 36.09°; 589.5 nm at 17.15°, 36.13°
b 3
c 3rd
3 3rd-order, 450 nm missing

Spread 6.12

3 b 2.86 eV
c 2.86 eV, 435 nm (visible)
d 91 nm
4 a 124 µm, microwave
b 622 nm, visible
c 1.24 nm, X-ray
d 2.49×10^{-11} m, X-ray
e 1.24×10^{-12} m, X-ray/gamma ray

Spread 6.13

1 a 3.0×10^8 m s⁻¹
2 c 3.1×10^8 m s⁻¹
3 2930 Hz

Spread 6.14

1 $V_{al} = 5090$ m s⁻¹, $V_{steel} = 5190$ m s⁻¹

Spread 6.15

2 b 75 m
3 a 267 m s⁻¹
b 0.60 m
c 445 Hz, 891 Hz
4 a 512 Hz, 768 Hz, 1024 Hz

Spread 6.16

2 a 3.3 m   **b** 100 Hz
4 0.56 m
5 336 m s⁻¹

Spread 6.18

1 b 2620 m s⁻¹
2 c 1330 Hz

Spread 6.19

1 a $0.87A_0$, $0.75I_0$
b $0.71A_0$, $0.50I_0$

Spread 6.20

4 a 60 cm, inverted, $m = 3$
b −100 cm, erect, $m = -5$
c −6.7 cm, erect, $m = -0.33$

5 a 1.85 cm, 54 dioptres; 2.0 cm, 50 dioptres (assuming eye length of 2.0 cm)

Spread 6.21

1 a 10°   **b** 0.7°
c b only
2 a $v = 60$ mm, $m = 5$
b 42 mm
c 30 ×
4 a 7.5°

Exam questions

1 b i 10 mm   **ii** 2.0 m
4 c 600 nm
7 b i 80.5°   **ii** 14°
8 a i 1.4   **ii** 30.3°
11 a 7.2×10^{-7} m
b 39°   **c** 0° (no first order)
12 a 2.4×10^{-4} m
13 b i 4   **ii** 9
iii 14.5°
14 a 261 Hz   **b** 780 m s⁻¹
16 c 75 m   **d** 200
17 b i 2×10^{-3} s   **ii** 30 mm
iii 15 m s⁻¹
18 b ii 970 m s⁻¹
20 a ii 4.0×10^{-5} m
c 1.6×10^{-12} m
21 a $L = mvr$
b $F = \dfrac{e^2}{4\pi\varepsilon_0 r^2}$
c $v = \dfrac{e}{\sqrt{4\pi m \varepsilon_0 r}}$
d i 0.53×10^{-10} m
ii 2.16×10^{-18} J
iii -2.16×10^{-18} J

CHAPTER 7

Practice questions

Spread 7.2

2 +5 J
3 0 J, +30 J

Spread 7.3

1 13.8 cm, 44.4 °C
3 $R_0 = 10\,\Omega$, $a = 4.209 \times 10^{-3}$ K⁻¹, $b = -2.083 \times 10^{-6}$ K⁻¹

Spread 7.4

2 a 1.95×10^5 Pa   **b** 653 Pa °C⁻¹
3 i 85 kPa   **ii** 160 kPa
iii 70 kPa

Spread 7.5

1 2.27×10^4 cm³
2 a 3.84×10^{-2} mol
b 225 °C
3 3.5×10^{-13} mol = 2.1×10^{11} molecules
4 a 9.67×10^5 Pa   **b** 255 °C

Spread 7.6

1 500 m s⁻¹
2 2×10^{23} molecules

Spread 7.7

1 a 490 m s⁻¹   **b** 532 m s⁻¹
c 500 m s⁻¹

Spread 7.8

1 3.65 kJ, 6.07×10^{-21} J
2 125 J
3 127 °C
4 748 J
5 a 7.6 kJ   **b** 25 J

Spread 7.9

1 0.4 °C s⁻¹
2 300 J min⁻¹

Spread 7.10

1 a 6.8 kJ
b 2.3×10^3 J (for 1 g of steam)
d 123 g

Spread 7.11

1 21%, 4.7 kW
3 a 11 J   **b** 11 J
4 a 1.25×10^{-3} m³
b 187 J   **c** +31 J
d 21 J   **e** 52 J

Spread 7.14

1 a 8.30×10^{-20} J
c 20.1, 19.4
d doubles
2 of order 10^4 K

Spread 7.15

1 650 J m⁻² s⁻¹
2 a 2.3 W m⁻² K⁻¹   **b** 540 W

Spread 7.16

1 a 4.0×10^{26} W
b 1.5×10^{11} m
2 18.4 W
3 b 150 °C

Exam questions

2 b i 12.5 J mol⁻¹ K⁻¹
ii 20.8 J mol⁻¹ K⁻¹
iv 5×10^{-4} m³
3 b i 26.3 °C   **ii** 42.6 °C
c i 14.4 Ω   **ii** 33 A
iii 2.26 m   **iv** 2000 W
4 a 940 W   **b** 12.6 °C
c 490 W
5 b i 1110 J   **ii** 1660 J
7 b i 1.7 mol   **ii** 27
iv 5.0×10^2 m s⁻¹
12 a i 6.2 mol   **ii** 1670 K
iii 82%
b i 60 kJ   **ii** 174 kW
iii 29%
15 a 0.1 Ω m
c $A = 2.7 \times 10^6$ mA, $\Delta E_g = 0.33$ eV (5.3×10^{-20} J)

CHAPTER 8

Practice questions

Spread 8.2

3 a 5.9×10^6 m s⁻¹, yes, $0.02c$
b 1.9×10^7 m s⁻¹, yes, $0.06c$
c 5.9×10^7 m s⁻¹, no, $0.2c$

Spread 8.3

1 a 9.4×10^6 m s⁻¹, 4.0×10^{-17} J
b 2.7×10^{15} m s⁻²
c 1.9×10^5 V m⁻¹
2 8.4×10^{-4} T

Spread 8.4

2 5×10^{18} s⁻¹ (assuming $\lambda_{av} = 500$ nm and efficiency = 10%)
5 microwaves: 1×10^{10} Hz, 4.1×10^{-5} eV, 6.6×10^{-24};
light: 6×10^{14} Hz, 2.5 eV, 4.0×10^{-19} J;
X-rays: 3×10^{18} Hz, 12 keV, 2.0×10^{-15} J

Spread 8.5

1 d none are emitted; photon energy is 4.1 eV, which is less than the work function for zinc
e 0.19 V

2 1.4 V

3 $8.1 \times 10^5 \, \text{m s}^{-1}$

Spread 8.6

1 a $1.2 \times 10^{-10} \, \text{m}$
 b $3.9 \times 10^{-11} \, \text{m}$
 c $3.9 \times 10^{-12} \, \text{m}$
 d $3.9 \times 10^{-13} \, \text{m}$

2 17 V

4 a 0.213 nm, 0.123 nm
 c for d_1: $\theta_1 = 16.8°$, $\theta_2 = 35.0°$;
 for d_2: $\theta_1 = 30.0°$, $\theta_2 = 87.3°$
 e 3.0 cm, 5.8 cm, 7.0 cm, 17.3 cm

Spread 8.9

1 d $2.1 \times 10^{14} \, \text{Hz}$

Spread 8.10

2 120 nm, 100 nm, 97 nm, 95 nm;
 all ultraviolet

3 a $9.2 \times 10^9 \, \text{Hz}$
 b 9.2×10^9

6 91 nm

Spread 8.12

1 a 0.5

2 a about 1.3 y

3 c U-238: 76%, U-235: 24%

Spread 8.13

1 a 10 000 y
 b i X: 1.3 g, Y: 2.5 g
 ii X: 14 g, Y: 8.4 g
 c 20 000 y

2 a $1.0 \times 10^{-6} \, \text{s}^{-1}$
 b $4.6 \times 10^{12} \, \text{Bq}$
 c 27 d

3 c i 80 000 c.p.s.
 ii 20 000 c.p.s.
 iii 13 000 c.p.s.

4 $1.4 \times 10^5 \, \text{Bq}$ or $3.9 \, \mu\text{Ci}$

5 a 15 mins $\quad 7.6 \times 10^{-4} \, \text{s}^{-1}$
 b 6.6×10^{10}

Spread 8.16

1 a $1/2^4 = 1/16$
 b 3/4

Spread 8.17

1 $7.8 \times 10^6 \, \text{y}$

3 10 000 y

Spread 8.18

4 b 180 000

Spread 8.19

2 false

4 a 1.46×10^5, 5.1 MeV

Spread 8.21

5 b 17 mm

Spread 8.24

2 a 230 y $\quad$ **c** 0.995c
 d 22.9 ly

Spread 8.25

1 a 15%

2 a 500 keV, 0.50 MeV, $8.0 \times 10^{-14} \, \text{J}$
 c 100%
 d 1.0 MeV/c^2
 e 100%
 g $4.2 \times 10^8 \, \text{m s}^{-1}$; no! – this exceeds c:
 relativistic equations are needed.

Spread 8.26

1 a $7.5 \times 10^{-15} \, \text{m}$

5 a $3.6 \times 10^{-15} \, \text{m}$

Spread 8.27

1 a $4.2 \times 10^5 \, \text{J}$
 b $4.7 \times 10^{-12} \, \text{kg}$

3 d 5.8 MeV

Spread 8.28

1 a 101 MeV $\quad$ 7.2 MeV/nucl.
 124 MeV $\quad$ 7.8 MeV/nucl.
 479 MeV $\quad$ 8.6 MeV/nucl.
 1580 MeV $\quad$ 7.7 MeV/nucl.
 1759 MeV $\quad$ 7.4 MeV/nucl.

Spread 8.29

3 a 2.6×10^{24}
 b $3.2 \times 10^{-11} \, \text{J}$
 c $8.3 \times 10^{13} \, \text{J}$
 d 540 kg
 e 14 tonnes

Spread 8.32

2 g 27 MeV

Spread 8.33

1 $1.9 \times 10^{-11} \, \text{m}$

2 5 kV

5 b $6.7 \times 10^{-34} \, \text{Js}$

Spread 8.34

1 a $9.4 \times 10^{15} \, \text{s}^{-1}$
 b 74.3 W
 c $2.5 \times 10^{-11} \, \text{m}$

3 a 2.3°

4 $1.2 \times 10^{-12} \, \text{m}$

Exam questions

3 b i 0.0091 m_u
 ii 4.24 MeV

4 c i 2900

5 c ii 866 m

8 a i 95; 241 $\quad$ **iv** $2.7 \times 10^5 \, \text{m s}^{-1}$
 b iii $6.6 \times 10^{-15} \, \text{m}$

12 b 35; 96
 e $3.5 \times 10^{-28} \, \text{kg}$

13 b i $3.36 \times 10^{-19} \, \text{J}$
 ii 2.1 eV
 c $21 \, \text{kV m}^{-1}$

14 a $5.1 \times 10^{-8} \, \text{A}$
 b $1.02 \times 10^{-7} \, \text{A}$

15 a $3.0 \times 10^{19} \, \text{J}$

17 b i $2.2 \times 10^{18} \, \text{J}$
 ii $6.6 \times 10^{-7} \, \text{m}$
 iii 12.08 eV, 10.19 eV, 1.89 eV

18 c i $6.2 \times 10^{-21} \, \text{J}$
 ii $8 \times 10^{-22} \, \text{J}$

19 b i 27.5 ns $\quad$ **ii** 15.6 m

21 a 3.1 MeV
 b 0.12 m

22 c i $3.8 \times 10^{-12} \, \text{J}$
 ii $1.19 \times 10^8 \, \text{s}$

23 b i 4 μs $\quad$ **ii** 2.2 μs
 iii 1100 m
 c i $1.9 \times 10^{-28} \, \text{kg}$
 ii 90 MV

24 b ii 17.6 MeV
 c i 1.87 fm $\quad$ **ii** 0.410 MeV
 d i $3.2 \times 10^9 \, \text{K}$

25 c i $3.4 \times 10^{-9} \, \text{A}$
 ii 2.1×10^{10}
 e 1.1×10^{16} atoms

26 b i 1.8420 u

27 b i $2.65 \times 10^7 \, \text{m s}^{-1}$

29 iv $1.1 \times 10^{-9} \, \text{m}$

CHAPTER 9

Practice questions

Spread 9.1

1 uud: $Q = (+2/3) + (+2/3) + (-1/3)$
 $= +1$, M $= 1/3 + 1/3 + 1/3 = 1$
 ddu: $Q = (-1/3) + (-1/3) + (+2/3)$
 $= 0$, M $= 1/3 + 1/3 + 1/3 = 1$

2 uuu $\qquad Q = 2$, M $= 1$
 uud $\qquad Q = 1$, M $= 1$
 udd $\qquad Q = 0$, M $= 1$
 ddd $\qquad Q = -1$, M $= 1$
 uus $\qquad Q = 1$, M $= 1.17$
 uss $\qquad Q = 0$, M $= 1.3$
 sss $\qquad Q = -1$, M $= 1.5$
 uds $\qquad Q = 0$, M $= 1.17$
 dds $\qquad Q = -1$, M $= 1.17$

3 9.1×10^{-37}

Spread 9.2

1 a newtonian: $2.46 \times 10^{-26} \, \text{kg ms}^{-1}$,
 relativistic: $2.46 \times 10^{-26} \, \text{kg ms}^{-1}$
 b newtonian: $246 \times 10^{-22} \, \text{kg ms}^{-1}$,
 relativistic: $5.64 \times 10^{-22} \, \text{kg ms}^{-1}$

2 $3.5 \times 10^{-10} \, \text{J}$, $1.0 \times 10^{-18} \, \text{kg m s}^{-1}$

4 45 GeV/c

5 0.14 GeV/c^2

Spread 9.3

1 543.8 GeV

2 67.5 GeV

3 67.5 MeV/c, 67.5 MeV

4 $1.63 \times 10^{22} \, \text{Hz}$, $1.84 \times 10^{-14} \, \text{m}$

Spread 9.4

1 b i $5.1 \times 10^{-3} \, \text{rad}$
 ii 2800 m $\quad$ **iii** $1.24 \times 10^{-23} \, \text{kg}$
 iv 7100% $\quad$ **v** $7 \times 10^{12} \, \text{eV}$

Spread 9.8

1 50–133 ms

2 a yes $\qquad$ **b** no
 c yes $\qquad$ **d** yes
 e no

Spread 9.9

2 $\bar{\text{u}}\bar{\text{u}}\bar{\text{d}}$: $Q = (-2/3) + (-2/3) + (-1/3) = -1$

3 $\bar{\text{u}}\bar{\text{u}}\text{d}$

4 $\nu_\mu + \text{n} \rightarrow \text{p} + \mu^-$, L_μ: $1 + 0 = 0 + 1$;
 $\bar{\nu}_\mu + \text{p} \rightarrow \text{n} + \mu^+$,
 L_μ: $(-1) + 0 = 0 + (-1)$

5 $\mu^- \rightarrow \text{e}^- + \text{X} + \text{Y}$
 a L_μ: $1 = 0 + 1 + 0$
 L_e: $0 = 1 + 0 + (-1)$
 X $= \nu_\mu$, Y $= \bar{\nu}_\text{e}$
 b $\mu^+ \rightarrow \text{e}^+ + \nu_\text{e} + \nu_\mu$

Spread 9.10

1 $4.5 \times 10^{52} \, \text{kg} = 2.3 \times 10^{22} \, \text{Suns}$

2 1.29 MeV

Spread 9.11

1

	Q	B	S
u u s	+1	1	−1
d $\bar{\text{s}}$	0	0	+1
u u u	+2	1	0
d s s	−1	1	−2
s s	−1	1	−3

2 a u $\bar{\text{s}}$
 b $\bar{\text{u}}$ s
 c u d s
 d $\bar{\text{d}}\bar{\text{s}}\bar{\text{s}}$
 e u u d

Spread 9.12

1 c strangeness not conserved,
 B not conserved
 d strangeness not conserved,
 B not conserved

f baryon number
g electron number (NB weak interaction, so strangeness does not need to be conserved)

2 a Δ^0: uds, P: uud; π^-: $\bar{u}d$
  i remains the same
  ii remains the same

Spread 9.13

1 X cannot be a lepton: it does not conserve lepton number; X cannot be a baryon, $\therefore$ X must be a meson; not strong decay: too slow, not electromagnetic, $\therefore$ weak decay; actually X is a K^+

Spread 9.14

1 $1.6 \times 10^{-17}\,\text{kg m s}^{-1}$, (hint: calculate γ from $\gamma m_0 c^2$ then assume the electron is moving at the speed of light); $4.1 \times 10^{-17}\,\text{m}$

2 non-relativistic at this energy, so use $KE = p^2/2m$: $p = 1.46 \times 10^{-19}\,\text{kg m s}^{-1}$; $4.55 \times 10^{-15}\,\text{m}$

3 a yes
  b no: does not conserve charge
  c no: does not conserve charge
  d yes

Exam questions

3 a i uud
  ii $Q = 1, B = 1, S = 0$
  iii $u\bar{d}$
 b i u   **ii** d
  iii W^+ boson, β^+, neutrino ν,
  iv D, E

4 b i B, D

7 a i 2.3 N   **iii** gluon
  iv rest mass = 0, charge = 0

8 b $K^- \rightarrow s\bar{u}$
  $\pi^+ \rightarrow u\bar{d}$
  $p \rightarrow uud$

10 iv 205 MeV/c

13 a $1.3 \times 10^5\,\text{J}$
  b i 90 μs   **ii** 1.3 μs
  c i $40\,\text{m s}^{-1}$ **ii** $1.05 \times 10^{-8}\,\text{Pa}$

14 a i $2.5\,\text{MV m}^{-1}$
  ii $8 \times 10^{-15}\,\text{J}$
  iv $2.2 \times 10^6\,\text{m s}^{-1}$
 b ii 0.16 m

CHAPTER 10

Practice questions

Spread 10.1

1 $1.6 \times 10^8\,\text{Pa}$
2 a $7.1 \times 10^{10}\,\text{Pa}$ **b** 880 μm
  c 30 μm
3 significantly > 1.9 mm

Spread 10.4

1 a 6   **b** 0.52

Spread 10.12

1 a $1.0 \times 10^{-3}\,\text{m}^3$
  b $-2.0 \times 10^{-7}\,\text{m}^3$
2 a decreases by 0.6 μm
3 b it increases

Spread 10.16

1 $2.3 \times 10^{-5}\,\text{m s}^{-1}$, $4.1 \times 10^{-9}\,\text{V}$
2 0.85 A

Spread 10.19

4 a 10^{10} bits
  b 8 hours (in practice less than 2 hours)

Exam questions
1 b $4.0 \times 10^{10}\,\text{Pa}$
8 b $6 \times 10^{23}\,\text{mol}^{-1}$
11 a i 110 kg

CHAPTER 11

Practice questions

Spread 11.1

1 7.5 N, 92.5 N, 0.23
2 1.65 kN, 6.8°

Spread 11.4

1 a 0.14 m   **b** 0.17 m
2 a converging lens, 3D
  b converging lens, 3.8D; diverging lens, –2D

Exam questions

4 a 6.55, 6.43, 6.32, 6.21
  c i 780 counts minute^{-1} ml^{-1}
    ii 92.5–94 minutes
6 b i $10^{-2}\,\text{W m}^{-2}$ **ii** 40 dB
  c $1.5 \times 10^{-15}\,\text{W}$
7 b 24%
8 $2.3 \times 10^8\,\text{N m}^{-2}$
9 a ii compressive: 160 N; shear: 190 N
  b i $1.6 \times 10^6\,\text{N m}^{-2}$
10 a i 144
  b ii $2.9 \times 10^9\,\text{s}$
  c i 4.5×10^{20} **iii** 0.088–0.097 W
11 c i $0.21\,\text{m s}^{-1}$
    ii $3.2 \times 10^{-7}\,\text{m}^3$
12 a ii 400–700 nm
  b i 54

CHAPTER 12

Practice questions

Spread 12.2
2 a 48 mins

Spread 12.3
1 $2.0 \times 10^{30}\,\text{kg}$

Spread 12.7
2 22
4 a $3.4 \times 10^9\,\text{m}$
  b $3.8 \times 10^5\,\text{m s}^{-1}$
5 $4.6 \times 10^{16}\,\text{W}$

Spread 12.9
1 $6.8 \times 10^8\,\text{m}$, $1.7 \times 10^6\,\text{m}$
2 a $6.18 \times 10^{18}\,\text{m}$ **b** 660 ly
  c $4.2 \times 10^7\,\text{AU}$
4 $3.1 \times 10^{15}\,\text{m}$
5 $9.5 \times 10^{17}\,\text{m}$

Spread 12.10
1 a $1.7 \times 10^{-7}\,\text{m}$
  b blue
  c $3.2 \times 10^9\,\text{m}$
2 a 1.1 mm
3 around 5800 K
4 316 times further away

Spread 12.11
1 a magnitude 3
  c 16
3 6.7
4 $3.2 \times 10^{34}\,\text{W}$
5 a 4.28   **b** 35.0 pc
  c 13.8 pc

Spread 12.13
3 a yes (18× from non-luminosity law)

b 1.4
c 11 000 K (assume its density is similar to that of the Sun)

Spread 12.16

1

Sirius	2.65 pc	8.64 ly
Deneb	490 pc	1600 ly
Altair	4.94 pc	16.1 ly
Arcturus	10.3 pc	33.5 ly

3 6
5 8.4 d

Spread 12.17

1 a –3.2
  b 5.2
  c 110 pc
2 a $6.3 \times 10^5\,\text{pc}$

Spread 12.18

2 a redshift
  c 0.158
  d $0.158c$ (away)
  e 0.86, 14 billion years old

Spread 12.19

1 b 160 Mpc, 510 Mpc
2 a 360 Mpc

Spread 12.22

3 b $1.68 \times 10^{-27}\,\text{kg m}^{-3}$
  $1.86 \times 10^{-26}\,\text{kg m}^{-3}$

Exam questions

5 c i 1730 s
7 c $6.09 \times 10^7\,\text{m s}^{-1}$
  d $3.97 \times 10^9\,\text{ly}$
8 a ii $3.8 \times 10^{17}\,\text{s}$
10 a ii $0.196 \times 10^{-7}\,\text{m}$
  iii $0.15 \times 10^{-8}\,\text{m s}^{-1}$
16 b ii 2.96 km
18 a iii $f_o = 24\,\text{cm}$; $f_e = 2\,\text{cm}$
  b i 2.2 cm from lens
  iii 3.6 cm
20 a iii $8.18 \times 10^{16}\,\text{m}$
21 a iii $7.3 \times 10^9\,\text{m}$
23 b ii 302 pc

CHAPTER 13

Practice questions

Spread 13.9

1 18 μV

Exam questions

6 d 8 kHz
7 d 0.0199 W
10 a ii power = $7.56 \times 10^{-17}\,\text{W}$
    voltage = 0.65 mV
18 a iii 630 kHz – 2.3 MHz

Index

electron–hole pairs 153
electronic balance 161
electronic circuits 570
electronic noise source 570
electron-neutrino 428
electrostatic focusing 336, 337
electrostatic force 128, 430
electrostatic loudspeakers 204
electrostatic plates 586
electroweak force 429
eluate 497
eluent 497
email 573
emission spectra 258
encoder 569
encoding 568
   16-bit 580
end correction 267
endoscope 589
endurance limit 459
energy 66, 400, 427
   band 464
   band structure of a semiconductor
      467
   conservation 66, 308
   diagram for the decay of cobalt-60
      362
   equivalent of one atomic mass unit
      374
   flow 312
   gap 464, 465
   in the gravitational field 176
   level 306, 314, 338
   and momentum in fixed targets and
      colliders 406
   renewable 600, 601
   sources 598, 600, 601
   stored in a capacitor 162
   supply (primary sources) 598
   transfers in a nuclear reactor 382
   transformations 142
enrichment 596
entropy 285, 289, 308
   and energy 306
   and living systems 308
   and thermal contact 307
   of the Universe 309
enzymes 494
epicentre 238
epicycles 514
equant 514
equation(s) 612
   continuity 76
   of motion 48
equilibrium 51, 52, 284, 285
   of coplanar forces 54
equipartition of energy 297
equipotential 158, 190, 312
   curve 159
   lines 186, 187
   surface 171, 188
   volume 188
Eris 525
error bars 38
error-correction codes 581
escape velocity 179, 543
estimating uncertainties 35
ether 366
Eudoxus 514
European Laboratory for Particle
   Research (CERN) 370, 398, 404,
   407, 414, 429, 573, 606
     budget 398

European Southern Observatory
   (ESO) 521
European Space Agency Satellite
   (ERS-1) 242
European Very Large Telescope 521
evaporation 310, 311
   cooling by 300
event horizon 543
evidence for Big Bang theory 559
exchange particles 399, 431
excited state 338
exhaust 304
   stroke 304
exothermic reactions 374
exotic matter 557
exotic particles 561
expansion 553, 559
   accelerating 557
   of spacetime 558
   of the universe 548, 556
explosions 92, 304
exponential(s) 620
   curve 159
   decay curve 344
   decay equation 346
exposure 495
extension 60
external combustion engine 304
extrapolation 38
extrasolar planets 603
extrinsic semiconductors 467
F
face-centred cubic (f.c.c.) structure
   444
far point 276, 488
farad (F) 156
Faraday, Michael 14, 156, 261
Faraday cages 188, 189
Faraday's field lines 14
Faraday's law 213, 214, 216
fate of the universe 556
fatigue 459
   strength 459
FDM see frequency-division
   multiplexing (FDM)
feedback 583
femur 482
Fermi, Enrico 416
fermi (fm) 28
Fermilab 30, 422
fermion 606
ferrite aerial 577
ferromagnetism 463
ferrous core 200
fertile material 596
Feynman, Richard 16, 66, 202, 420,
   424, 426, 452
Feynman diagrams 426, 428, 429
fibre optic 246, 572
fibre-optic communications 588
   advantages 588
fibrillation 154, 493
field(s) 185, 200
   gravitational 174
   lines 171, 175, 186, 188
   of a point charge 182
   and potential 177
   and potential gradient 186
   due to spherical objects 195
   strength 171, 174, 197
   theory 14, 15
   between two charged plates 190
   of two charges 183
   winding 203

field-line diagram 182
filaments 18
file transfer protocol (FTP) 573
film badge 495
fine-beam tube 325, 327
   method for measuring the charge-
     to-mass ratio of an electron 327
finite but unbounded universe 552,
   556
Finnegan's Wake 418
first harmonic 265
first law of thermodynamics 287, 302
fish-eye lens 246
fissile material 596
fixed-target experiments 406
Fizeau, Armand 260
flattening filter 501
Fleming's left-hand rule 197, 201, 326
Flettner ship 78
flight 79, 80
flotation 74
fluid friction 91, 94
fluorescence 325
flux 212, 214
   cut 213
   density 533
   linkage 213, 215, 216, 219
FM see frequency modulation (FM)
focal point 274
focusing magnets 407
force(s) 185, 291
   on a current-carrying wire 197
   in buildings 58
   and collisions 402
   in equilibrium 52
   of gravity 172, 484
   on the hip joint 482
   on insulators 193
   laws and fields 195
   between two current-carrying wires
     199
   as a vector 52
forced (driven) oscillator 115
forced oscillations 115
force–time graphs 89
formation of Solar System 526
forming hypotheses 40
forward-biased diode 152
fossil fuels 598, 599
Foucault, Jean 260
four-stroke cycle 304
four-stroke internal combustion
   engine 304
fovea 487
fracking 598
frame of reference 212
Franklin, Benjamin 130
Fraunhofer, Joseph von 258, 536
free electrons 130, 131, 152, 197
free fall 62, 484
free holes 152
free oscillator 115
free-body diagram 52, 53, 91
frequency 100, 150, 236, 240
frequency modulation (FM) 568, 574
frequency-division multiplexing
   (FDM) 571
friction 91, 94, 458
   between surfaces 94
   coefficient of 94
   and internal energy 94
frictional drag 27
frictional forces 90
Friedman, Jerome 424